P9-DTT-511

Marriages, Families, and Relationships

Marriages, Families, and Relationships

MAKING CHOICES IN A DIVERSE SOCIETY

Twelfth Edition

Mary Ann Lamanna
University of Nebraska, Omaha

Agnes Riedmann
California State University, Stanislaus

Susan Stewart
Iowa State University

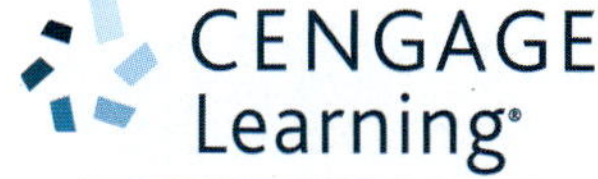

Australia • Brazil • Mexico • Singapore • United Kingdom • United States

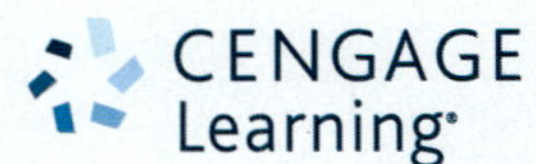

Marriages, Families, and Relationships: Making Choices in a Diverse Society, Twelfth Edition
Mary Ann Lamanna, Agnes Riedmann, and Susan Stewart

Product Manager: Seth Dobrin

Product Development Manager: Erik Fortier

Content Coordinator: Jessica Alderman

Product Assistant: Coco Bator

Media Developer: John Chell

Marketing Manager: Kara Kindstrom

Content Project Manager: Cheri Palmer

Art Director: Caryl Gorska

Manufacturing Planner: Judy Inouye

Rights Acquisitions Specialist: Thomas McDonough

Production Service: Jill Traut, MPS Limited

Photo Researcher: Padma Soundararajan, PreMediaGlobal

Text Researcher: Punitha Rajamohan, PreMediaGlobal

Copy Editor: S.M. Summerlight

Illustrator and Composition: MPS Limited

Text Designer: Diane Beasley

Cover Designer: Caryl Gorska

Cover Image: Gerard Fritz/Getty Images

Design Images: ARENA Creative/Shutterstock.com; aPERFECT/Shutterstock.com; Sunny studio-Igor Yaruta/Shutterstock.com; VLADGRIN/Shutterstock.com

Library of Congress Control Number: 2013950500

ISBN-13: 978-1-285-73697-6
ISBN-10: 1-285-73697-4

Cengage Learning
200 First Stamford Place, 4th Floor
Stamford, CT 06902
USA

Cengage Learning is a leading provider of customized learning solutions with office locations around the globe, including Singapore, the United Kingdom, Australia, Mexico, Brazil, and Japan. Locate your local office at **www.cengage.com/global**

Cengage Learning products are represented in Canada by Nelson Education, Ltd.

To learn more about Cengage Learning Solutions, visit **www.cengage.com**

Purchase any of our products at your local college store or at our preferred online store **www.cengagebrain.com**

Printed in the United States of America
1 2 3 4 5 6 7 18 17 16 15 14

dedication

to our families, especially

Larry, Valerie, Sam, Janice, Simon, and Christie

Bill, Beth, Angel, Chris, Natalie, Alex, and Livia

Gwendolyn, Gene, Lee, Christine, Mom and Dad

About the Authors

Mary Ann Lamanna is Professor Emerita of Sociology at the University of Nebraska at Omaha. She received her bachelor's degree in political science Phi Beta Kappa from Washington University (St. Louis); her master's degree in sociology (minor in psychology) from the University of North Carolina, Chapel Hill; and her doctorate in sociology from the University of Notre Dame.

Research and teaching interests include family, reproduction, and gender and law. She is the author of *Emile Durkheim on the Family* (Sage Publications, 2002) and co-author of a book on Vietnamese refugees. She has articles in law, sociology, and medical humanities journals. Current research concerns the sociology of literature, specifically "novels of terrorism" and a sociological analysis of Marcel Proust's novel *In Search of Lost Time.* Professor Lamanna has two adult children, Larry and Valerie.

Agnes Riedmann is Professor of Sociology at California State University, Stanislaus. She attended Clarke College in Dubuque, Iowa. She received her bachelor's degree from Creighton University and her doctorate from the University of Nebraska. Her professional areas of interest are theory, family, and the sociology of body image. She is author of *Science That Colonizes: A Critique of Fertility Studies in Africa* (Temple University Press, 1993). Dr. Riedmann spent the academic year 2008–09 as a Fulbright Professor at the Graduate School for Social Research, affiliated with the Polish Academy of Sciences, Warsaw, where she taught courses in family, as well as in social policy and in globalization. She has two children, Beth and Bill; two granddaughters, Natalie and Livia; and a grandson, Alex.

Susan Stewart is Associate Professor of Sociology at Iowa State University. She received her bachelor's degree from The State University of New York at Fredonia and her doctorate from Bowling Green State University. Her professional areas of interest are family, gender, demography, and research methods. She has published articles on nonresident parenting, stepfamilies, adoption, and childhood obesity. She is author of *Brave New Stepfamilies* (Sage Publications Inc., 2007). She teaches courses in family and intimate relationships, gender, demography, and research methods. Dr. Stewart is thrilled to have been added as an author of the twelfth edition of this book. She used the third edition of the Lamanna and Riedmann textbook in the sociology of family class she took as an undergraduate. She lives in Ames, Iowa, with her 10-year-old daughter.

Brief Contents

Contents

Chapter 1 Making Family Choices in a Changing Society 3

Chapter 2 Exploring Relationships and Families 29

Chapter 3 Our Gendered Identities 53

Chapter 4 Our Sexual Selves 79

Chapter 5 Love and Choosing a Life Partner 107

Chapter 6 Living Alone, Cohabiting, Same-Sex Unions, and Other Intimate Relationships 135

Chapter 7 Marriage: From Social Institution to Private Relationship 165

Chapter 8 Deciding about Parenthood 193

Chapter 9 Raising Children in a Diverse Society 221

Chapter 10 Work and Family 249

Chapter 11 Communication in Relationships, Marriages, and Families 277

Chapter 12 Power and Violence in Families 301

Chapter 13 Family Stress, Crisis, and Resilience 331

Chapter 14 Divorce and Relationship Dissolution 357

Chapter 15 Remarriages and Stepfamilies 387

Chapter 16 Aging and Multigenerational Families 417

Boxes

A Closer Look at Family Diversity

As We Make Choices

Facts about Families

Issues for Thought

Preface

As we complete our work on the twelfth edition of this text, we look back over eleven earlier editions. Together, these represent thirty-five years spent observing and rethinking American families. Not only have families changed since we began our first edition but so has social science's interpretation of family life. It is gratifying to be a part of the enterprise dedicated to studying families and to sharing this knowledge with students.

Our own perspective on families has developed and changed as well. Indeed, as marriages and families have evolved over the last three decades, so has this text. In the beginning, this text was titled *Marriages and Families*—a title that was the first to purposefully use plurals to recognize the diversity of family forms—a diversity that we noted as early as 1980. Now the text is titled *Marriages, Families, and Relationships*. We added the term *relationships* to recognize the increasing incidence of individuals forming commitments outside of legal marriage. At the same time, we continue to recognize and appreciate the fact that a large majority of Americans are married or will marry. Hence, we consciously persist in giving due attention to the values and issues of married couples. Of course, the concept of marriage itself has changed appreciably. No longer necessarily heterosexual, marriage is now an institution to which same-sex couples in a growing number of states have legal access.

Meanwhile, the book's subtitle, *Making Choices in a Diverse Society*, continues to speak to the significant changes that have taken place since our first edition. To help accomplish our goal of encouraging students to better appreciate the diversity of today's families, we present the latest research and statistical information on varied family forms, lesbian and gay male families, and families of diverse race and ethnicity, socioeconomic, and immigration status, among other variables.

We continue to take account not only of increasing race/ethnic diversity but also of the fluidity of the concepts *race* and *ethnicity* themselves. In this edition, we give greater direct attention to the socially constructed nature of these concepts. We integrate these materials on family diversity throughout the textbook, always with an eye toward avoiding stereotypical and simplistic generalizations and, instead, to explaining data in sociological and sociohistorical contexts.

Interested from the beginning in the various ways that gender plays out in families, we have persistently focused on areas in which gender relations have changed and continue to change, as well as on areas in which there has been relatively little change. In keeping with our practice of reviewing and reevaluating every single word for a new edition, we have in this revision given concerted attention to discussions that may now be better presented in gender-neutral context and language. However, we hasten to add that assuredly not all topics lend themselves to gender-neutral language. For example, research indicates that intimate violence perpetrated by heterosexual men is qualitatively different from that perpetrated by heterosexual women.

In addition to our attention to gender, we have studied demography and history, and we have paid increasing attention to the impact of social structure on family life. We have highlighted the family ecology perspective in keeping with the importance of social context and public policy. We cannot help but be aware of the cultural and political tensions surrounding families today. At the same time, in recent editions and in response to our reviewers, we have given more attention to the contributions of psychology and to a social psychological understanding of family interaction and its consequences.

We continue to affirm the power of families as they influence the courses of individual lives. Meanwhile, we give considerable attention to policies needed to provide support for today's families: working parents, families in financial stress, single-parent families, families of varied racial/ethnic backgrounds, stepfamilies, same-sex couples, and other nontraditional families—as well as the classic nuclear family.

We note that, despite changes, marriage and family values continue to be salient in contemporary American life. Our students come to a marriage and family course because family life is important to them. Our aim now, as it has been from the first edition, is to help students question assumptions and reconcile conflicting ideas and values as they make choices throughout their lives. We enjoy and benefit from the contact we've had with faculty and students who have used this book. Their enthusiasm and criticism have stimulated many changes in the book's content. To know that a supportive audience is interested in our approach to the study of families has enabled us to continue our work over a long period.

The Book's Themes

Several themes are interwoven throughout this text: People are influenced by the society around them as they make choices, social conditions change in ways that may impede or support family life, there is an interplay between individual families and the larger society, and individuals make family-related choices throughout adulthood.

Making Choices throughout Life

The process of creating and maintaining marriages, families, and relationships requires many personal choices; people continue to make family-related decisions, even "big" ones, throughout their lives.

Personal Choice and Social Life

Tension frequently exists between individuals and their social environment. Many personal troubles result from societal influences, values, or assumptions; inadequate societal support for family goals; and conflict between family values and individual values. By understanding some of these possible sources of tension and conflict, individuals can perceive their personal troubles more clearly and work constructively toward solutions. They may choose to form or join groups to achieve family goals. They may become involved in the political process to develop state or federal social policy that is supportive of families. The accumulated decisions of individuals and families also shape the social environment.

A Changing Society

In the past, people tended to emphasize the dutiful performance of social roles in marriage and family structure. Today, people are more apt to view committed relationships as those in which they expect to find companionship, intimacy, and emotional support. From its first edition, this book has examined the implications of this shift and placed these implications within social scientific perspective. Individualism, economic pressure, time pressures, social diversity, and an awareness of committed relationships' potential impermanence are features of the social context in which personal decision making takes place today. With each edition, we recognize again that, as fewer social guidelines remain fixed, personal decision making becomes even more challenging.

Then too, new technologies continue to create changes in family members' lives. Discussions about technological developments in communication appear throughout the book—for example, a lengthy discussion of how technology and social media impact families in Chapter 1, maintaining ties between college students and their parents (Chapter 9), sexting and cyber adultery (Chapter 4), Internet matchmaking (Chapter 5), reproductive technology (Chapter 8), parental surveillance of children (Chapter 9), working at home versus the office (Chapter 10), and how noncustodial parents keep in touch with children through technology (Chapter 14).

The Themes throughout the Life Course

The book's themes are introduced in Chapter 1, and they reappear throughout the text. We developed these themes by looking at the interplay between findings in the social sciences and the experiences of the people around us. Ideas for topics continue to emerge, not only from current research and reliable journalism, but also from the needs and concerns that we perceive among our own family members and friends. The attitudes, behaviors, and relationships of real people have a complexity that we have tried to portray. Interwoven with these themes is the concept of the life course—the idea that adults may change by means of reevaluating and restructuring throughout their lives. This emphasis on the life course creates a comprehensive picture of marriages, families, and relationships and encourages us to continue to add topics that are new to family texts. Meanwhile, this book makes these points:

- People's personal problems and their interaction with the social environment change as they and their relationships and families grow older.
- People reexamine their relationships and their expectations for relationships as they and their marriages, relationships, and families mature.
- Because family forms are more flexible today, people may change the type or style of their relationships and families throughout their lives.

Marriages and Families—Making Choices

Making decisions about one's family life begins in early adulthood and lasts into old age. People choose whether they will adhere to traditional beliefs, values, and attitudes about gender roles or negotiate more flexible roles and relationships. They may rethink their values about sex and become more informed and comfortable with their sexual choices.

Women and men may choose to remain single, to form heterosexual or same-sex relationships outside of marriage, or to marry. They have the option today of staying single longer before marrying. Single people make choices about their lives, ranging from decisions about living arrangements to those about whether to engage in sex only in marriage or committed relationships, to engage in sex for recreation, or to abstain from sex altogether. Many unmarried individuals live as cohabiting couples (often with children), an increasingly common family form.

Once individuals form couple relationships, they have to decide how they are going to structure their lives as committed partners. Will the partners be legally married? Will they become domestic partners? Will theirs be a dual-career union? Will they plan periods in which one partner is employed, interspersed with times in which both are wage earners? Will they have children? Will they use new reproductive technology to become parents? Will other family members live with them—siblings or parents, for example, or, later, adult children?

Couples will make these decisions not once, but over and over during their lifetimes. Within a committed relationship, partners also choose how they will deal with conflict. Will they try to ignore conflicts? Will they vent their anger in hostile, alienating, or physically violent ways? Or will they practice supportive ways of communicating, disagreeing, and negotiating—ways that emphasize sharing and can deepen intimacy?

How will the partners distribute power in the marriage? Will they work toward relationships in which each family member is more concerned with helping and supporting others than with gaining a power advantage? How will the partners allocate work responsibilities in the home? What value will they place on their sexual lives together? Throughout their experience, family members continually face decisions about how to balance each one's need for individuality with the need for togetherness.

Parents also have choices. In raising their children, they can choose the authoritative parenting style, for example, in which parents take an active role in responsibly guiding and monitoring their children, while simultaneously striving to develop supportive, mutually cooperative family relationships.

Many partners face decisions about whether to separate or divorce. They weigh the pros and cons, asking themselves which is the better alternative: living together as they are or separating? Even when a couple decides to separate or divorce, there are further decisions to make: Will they cooperate as much as possible or insist on blame and revenge? What living and economic support arrangements will work best for themselves and their children? How will they handle the legal process? The majority of divorced individuals eventually face decisions about recoupling. In the absence of firm cultural models, they choose how they will define stepfamily relationships.

When families encounter crises—and every family will face *some* crises—members must make additional decisions. Will they view each crisis as a challenge to be met, or will they blame one another? What resources can they use to handle the crisis? Then, too, as more and more Americans live longer, families will "age." As a result, more and more Americans will have not only living grandparents but also great grandparents. And increasingly, we will face issues concerning giving—and receiving—family elder care.

An emphasis on knowledgeable decision making does not mean that individuals can completely control their lives. People can influence but never directly determine how those around them behave or feel about them. Partners cannot control one another's changes over time, and they cannot avoid all accidents, illnesses, unemployment, separations, or deaths. Society-wide conditions may create unavoidable crises for individual families. However, families *can* control how they respond to such crises. Their responses will meet their own needs better when they refuse to react automatically and choose instead to act as a consequence of knowledgeable decision making.

Key Features

With its ongoing thorough updating and inclusion of current research and its emphasis on students' being able to make choices in an increasingly diverse society, this book has become a principal resource for gaining insights into today's marriages, relationships, and families. Over the past eleven editions, we have had four goals in mind for student readers: first, to help them better understand themselves and their family situations; second, to make students more conscious of the personal decisions that they will make throughout their lives and of the societal influences that affect those decisions; third, to help students better appreciate the variety and diversity among families today; and fourth, to encourage them to recognize the need for structural, social policy support for families. To these ends, this text has become recognized for its accessible writing style, up-to-date research, well-written features, and useful chapter learning aids.

Up-to-Date Research and Statistics

As users have come to expect, we have thoroughly updated the text's research base and statistics, emphasizing cutting-edge research that addresses the diversity of marriages and families, as well as all other topics. In accordance with this approach, users will notice several new tables and figures. Revised tables and figures have been updated with the latest available statistics—data from the U.S. Census Bureau and other governmental agencies, as well as survey and other research data.

Features

The several themes described earlier are reflected in the special features.

Former users will recognize our box features. The following sections describe our four feature box categories:

As We Make Choices. We highlight the theme of making choices with a group of boxes throughout the text, for example, "Ten Rules for a Successful Relationship," "Looking for Love on the Internet" "Disengaging from Power Struggles," "Selecting a Child Care Facility," "Ten Keys to Successful Co-Parenting," and "Tips for Step-Grandparents." These feature boxes emphasize human agency and are designed to help students through crucial decisions.

A Closer Look at Diversity. In addition to integrating information on cultural and ethnic diversity throughout the text proper, we have a series of features that give focused attention to instances of family diversity—for example, "African Americans and 'Jumping the Broom'," "Diversity and Child Care," "Family Ties and Immigration," "Parenting LGBT Children," and " Do You Speak Stepfamily," among others.

Issues for Thought. These features are designed to spark students' critical thinking and discussion. As an example, the Issues for Thought box in Chapter 16 explores "Filial Responsibility Laws" and encourages students to consider what might be the benefits and drawbacks of legally mandating filial responsibility. Similarly, in the Issues for Thought box in Chapter 4, "Bisexual or Just "Bi-Curious"? in Chapter 4, students are asked to think about whether there are different standards of same-sex attraction and behavior for women versus men.

Facts about Families. This feature presents demographic and other factual information on focused topics such as "How Family Researchers Study Religion from Various Theoretical Perspectives" (Chapter 2), on "Six Love Styles" (Chapter 5), on transracial adoption (Chapter 8), and on "Foster Parenting" (Chapter 9), among others.

Chapter Learning Aids

A series of chapter learning aids help students comprehend and retain the material.

- **Chapter Summaries** are presented in bulleted, point-by-point lists of the key material in the chapter.
- **Key Terms** alert students to the key concepts presented in the chapter. A full glossary is provided at the end of the text.
- **Questions for Review and Reflection** help students review the material. Thought questions encourage students to think critically and to integrate material from other chapters with that presented in the current one. In every chapter, one of these questions is a policy question. This practice is in line with our goal of moving students toward structural analyses regarding marriages, families, and relationships.

Key Changes in This Edition

In addition to incorporating the latest available research and statistics—and in addition to carefully reviewing every word in the book—we note that this edition includes many key changes, some of which are outlined here. We have shortened several chapters in order to make chapter length more uniform throughout the text. In shortening long chapters, we have not omitted topics; rather we have consolidated discussions and eliminated wordiness. Because the pertinent information changes often and can now be readily found online, all the appendices have been deleted.

In this twelfth edition, we have again revisited and somewhat restructured the chapter outline and order. We have dropped the former Chapter 3, "American Families in Social Context," and integrated material from former Chapter 3 into relevant other chapters. Moreover, in response to reviewers, we have returned the chapter on marriage to its earlier placement after the chapter on living alone, cohabiting, same-sex unions, and other intimate relationships.

As with previous revisions, we have given considerable attention not only to chapter-by-chapter organization, but also to *within*-chapter organization. Our ongoing intents are to streamline the material presented whenever possible and to ensure a good flow of ideas. In this edition, we have also continued to consolidate similar material that had previously been addressed in separate chapters.

Meanwhile, we have substantially revised each and every chapter. Every chapter is updated with the latest research throughout. We mention some (but not all!) specific and important changes here.

Chapter 1, Marriage, Relationships, and Family Commitments: Making Choices in a Changing Society, continues to present the choices and life course themes of the book, as well as points to the significance for the family of larger social forces. A new section, "A Sociological Imagination: Personal Troubles and Some Social Conditions That Impact Families," introduces a discussion of the sociological imagination as it relates to issues of marriage and family, incorporating historical information and demographic characteristics from Chapter 3 in the previous edition.

Chapter 2, Exploring Relationships and Families, continues to portray the integral relationship between family theories and methods for researching families, with new examples to drive home the theoretical perspectives.

Chapter 3, Our Gendered Identities, has been significantly updated with information on gender identities and expectations, including a new section on "Race/Ethnic Diversity and Gendered Expectations." New sections on "Biology-Based Arguments" and "Society-Based Arguments" explore the emergence of gender roles.

Chapter 4, Our Sexual Selves, presents a new feature, "Issues for Thought: Bisexual or Just 'Bicurious'? The Emergence of Pansexuality." This box invites students to consider how our understandings of gender and sexual identity have become increasingly fluid in society and are different for women and men. The statistics on sexuality have been substantially revised to reflect new surveys on sexual behavior, infidelity, HIV/AIDS, and pornography use, with special focus on gender differences in each. Given the proliferation of pornography, now easily accessible on the Internet, there is a new section on how pornography affects one's own sexuality as well as intimacy between couples. With the Obama administration's withdrawal of federal funding to abstinence only sex education, we have deleted the section on the politics of sex and sex education.

Chapter 5, Love and Choosing a Life Partner, contains a new section on definitions, perceptions, and experiences of love and how they differ for women and men. There is a new section on dating and relationship development and the emergence of various patterns of "nondating" among adolescents and young adults. This discussion reflects the ever increasing age at first marriage, the lengthening time young adults remain unmarried, and how they navigate sexual and intimate relationships during this period. We have added a new box, "Looking for Love on the Internet," that incorporates technological changes in how people search for a mate. We have also revised our box on "Acquaintance Rape" to include "Sexual Assault," to highlight increased recognition of the various forms that sexual assault and abuse can take, as well as programs and campaigns geared toward educating men.

Chapter 6, Living Alone, Cohabiting, Same-Sex Unions, and Other Intimate Relationships, discusses demographic, economic, technological, and cultural reasons for the increasing proportion of unmarrieds, with updated statistics on unmarrieds in America. New sections on the numbers, age, and characteristics of cohabitors look further into the growing trend. This chapter also includes extensive, expanded, and thoroughly updated sections on trends in legal marriage for same-sex couples, including discussion of the Supreme Court decision on the Defense of Marriage Act (DOMA).

Chapter 7, Marriage: From Social Institution to Private Relationship, has been thoroughly updated with new statistics and research findings. This chapter explores the changing picture regarding marriage, noting the social science debate regarding whether this changing picture represents family change or decline. As part of our updated exploration of this question, we thoroughly explore the selection hypothesis versus the experience hypothesis with regard to the benefits of marriage known from research.

Chapter 8, Deciding about Parenthood, now includes data analysis on international and transracial adoptions. A new boxed feature, "Conception, Pregnancy, and Childbirth—the Basics" integrates key information previously presented in an appendix.

Chapter 9, Raising Children in a Diverse Society, like all the chapters in this edition, has been thoroughly updated with the most current research. As in recent prior editions, after describing the authoritative parenting style, we note its acceptance by mainstream experts in the parenting field. We then present a critique that questions whether this parenting style is universally appropriate or simply a white, middle-class pattern that may not be so suitable to other social contexts. We also discuss challenges faced by parents who are raising religious- or ethnic-minority children in potentially discriminatory environments.

We continue to emphasize the challenges that all parents face in contemporary America, especially given our economic downturn. We have expanded sections on single mothers, single fathers, and nonresident fathers. We have given more attention to relations with young-adult children as more and more of them have "boomeranged" home in this difficult economy. As with all other chapters in this text, we keep in mind the linkage between structural conditions and personal decisions. Hence, there is added discussion of the parenting beliefs and practices in working-class families. We have also added a new table in this twelfth edition, "The American Academy of Pediatrics Position Against Spanking, Versus the American College of Pediatricians' Distinction between Disciplinary Spanking and Corporal Punishment." We have added a new section on parents in transnational families.

Chapter 10, Work and Family, has been significantly reorganized and considerably shortened. All research and statistics are updated. We continue to follow the National Institute of Child Health and Human Development study of child care. There are new boxes, including "Issues for Thought: When One Woman's Workplace Is Another's Family." Another new box looks at things for parents to think about if leaving kids home alone: "As We Make Choices: Self-Care (Home Alone) Kids."

Chapter 11, Communication in Relationships, Marriages, and Families has been reorganized with shortened, better-clarified sections. Additional information on relationship counseling previously included in an appendix has been included in this chapter.

Chapter 12, Power and Violence in Families, has been reorganized and shortened considerably, but without deleting any topics explored in previous editions. There is more and ongoing emphasis on power relations within the context of growing family race/ethnic diversity. This chapter consolidates the classic research on family power, while current research on marital and partner power has been expanded to include issues of household work and money management, as well as decision making per se. A discussion of equality and equity concludes the part of the chapter on marital and partner power. Additionally, analysis of power differential between citizens and their immigrant spouses is introduced.

A clearer distinction has been made between intimate terrorism and common couple violence. The section on abuse among same-gender, bisexual, and transgender couples has been updated and expanded, as has the section on violence among immigrant couples. Finally, several new boxes have been added to this chapter: "A Closer Look at Diversity: Mobile Phones, Migrant Mothers, and Conjugal Power," "As We Make Choices: Domination and Submission in Couple Communication Patterns," "Facts about Families: Major Sources of Data on Intimate Partner Violence (IPV) and Child Maltreatment," and "Facts about Families: Signs of Intimate Terrorism."

Chapter 13, Family Stress, Crisis, and Resilience, continues to emphasize and expand discussion of the growing body of research on resilience in relation to family stress and crises and has been updated with many new examples. Figure 13.2 on boundary ambiguity has been simplified to better clarify the concept.

Chapter 14, Divorce and Relationship Dissolution, has a new title to reflect the increase in committed nonmarital relationships and an awareness that these, too, often dissolve. There is new section on starter marriages and silver divorces that reflects variations in the divorce rate by age. There is also a new section that highlights the lack of information available to divorcing couples regarding the legal process of divorce ("The Black Box of Divorce"), and a new section on the consequences of divorce for divorcing couples, extended families and circles of friends (divorce fall-out). There are now separate sections for economic consequences of divorce for women, men, and children with a heavily revised section covering the socioemotional consequences of divorce for each. All reflect the very latest theories and empirical findings on these topics. New information has been added on the experience of joint and father custody, reflecting growing incidence of these practices. Finally, the section on coparenting has been revised to reflect growing acceptance of this term to describe parental relationships after divorce.

Chapter 15, Remarriages and Stepfamilies, continues to stress diversity within stepfamilies, reflecting continued growth of nonmarital childbearing, cohabitation, father custody, racial/ethnic diversity, and same sex couples with stepchildren. There is a new box, "Issues for Thought: What Makes a Stepfamily?," that focuses on how these shifts are changing societal definitions of stepfamilies. We also consider diversity in how members of stepfamilies define themselves with a new box, "A Closer Look at Diversity: Do You Speak Stepfamily?" We continue to pay attention to micro-level stepfamily dynamics with new sections on dating with children, the process through which people become stepparents, and the challenges of day-to-day living in stepfamilies, including the complex legal and financial issues they face. We continue to add new research findings to our discussion of the short- and long-term financial, social, and emotional well-being of stepfamily members, especially children. Finally, we have additional suggestions for how society can better meet the needs of stepfamilies in our section on creating supportive stepfamilies.

Chapter 16, Aging Families, has a new thematic emphasis on multigenerational families, ties, and obligations in a cultural content of individualism. Additionally, this chapter includes a new discussion of caregiver ambivalence coupled with multigenerational families as safety nets for all generations.

MindTap™: The Personal Learning Experience

MindTap for Lamanna/Riedmann/Stuart's *Marriages, Families, & Relationships: Making Choices in a Diverse Society*, twelfth edition from Cengage Learning represents a new approach to a highly personalized, online learning platform. A fully online learning solution, MindTap combines all of a student's learning tools—readings, multimedia, activities, and assessments into a singular Learning Path that guides the student through the introduction to sociology course. Instructors personalize the experience by customizing the presentation of these learning tools to their students, even seamlessly introducing their own content into the Learning Path via "apps" that integrate into the MindTap platform. Learn more at www.cengage.com/mindtap.

MindTap for Lamanna/Riedmann/Stuart's *Marriages, Families, & Relationships: Making Choices in a Diverse Society*, twelfth edition, features Aplia assignments, which help students learn to use their sociological imagination through compelling content and thought-provoking questions. Students complete interactive activities that encourage them to think critically in order to practice and apply course concepts. These valuable critical thinking skills help students become thoughtful and engaged members of society. Aplia for *Marriages, Families, & Relationships: Making Choices in a Diverse Society* is also available as a standalone product. Login to CengageBrain.com to access Aplia for *Marriages, Families, & Relationships: Making Choices in a Diverse Society*.

MindTap for Lamanna/Riedmann/Stuart's *Marriages, Families, & Relationships: Making Choices in a Diverse Society*, twelfth edition is easy to use and saves instructors time by allowing them to:

- Seamlessly deliver appropriate content and technology assets from a number of providers to students, as they need them.

- Break course content down into movable objects to promote personalization, encourage interactivity and ensure student engagement.
- Customize the course—from tools to text—and make adjustments "on the fly," making it possible to intertwine breaking news into their lessons and incorporate today's teachable moments.
- Bring interactivity into learning through the integration of multimedia assets (apps from Cengage Learning and other providers), numerous in-context exercises and supplements; student engagement will increase leading to better student outcomes.
- Track students' use, activities, and comprehension in real-time, which provides opportunities for early intervention to influence progress and outcomes. Grades are visible and archived so students and instructors always have access to current standings in the class.
- Assess knowledge throughout each section: after readings, in activities, homework, and quizzes.
- Automatically grade all homework and quizzes.

Instructor Resources

Online Instructor's Resource Manual with Test Bank. This thoroughly revised and updated Instructor's Resource Manual contains detailed lecture outlines; chapter summaries; and lecture, activity, and discussion suggestions; as well as film and video resources. It also includes student learning objectives, chapter review sheets, and Internet exercises. The test bank consists of a variety of questions, including multiple-choice, true/false, completion, short answer, and essay questions for each chapter of the text, with answer explanations and references to the text.

Cengage Learning Testing Powered by Cognero®. Cengage Learning Testing Powered by Cognero is a flexible, online system that allows you to author, edit, and manage test bank content from multiple Cengage Learning solutions, and create multiple test versions in an instant. You can deliver tests from your LMS, your classroom or wherever you want—no special installs or downloads needed.

Online PowerPoints. These vibrant, Microsoft PowerPoint lecture slides for each chapter assist you with your lecture, by providing concept coverage using images, figures, and tables directly from the textbook.

The Sociology Video Library Vol. I – IV. These DVDs drive home the relevance of course topics through short, provocative clips of current and historical events. Perfect for enriching lectures and engaging students in discussion, many of the segments on this volume have been gathered from BBC Motion Gallery. Ask your Cengage Learning representative for a list of contents.

CourseReader for Sociology. Easy-to-use and affordable access to primary and secondary sources, readings, and audio and video selections for your courses with this customized online reader. CourseReader for Sociology helps you to stay organized and facilitates convenient access to course material, no matter where you are.

Acknowledgments

This book is a result of a joint effort on our part; we could not have conceptualized or written it alone. We want to thank some of the many people who helped us. Looking back on the long life of this book, we acknowledge Steve Rutter for his original vision of the project and his faith in us. We also want to thank Sheryl Fullerton and Serina Beauparlant, who saw us through early editions as editors and friends and who had significant importance in shaping the text that you see today.

As has been true of our past editions, the people at Cengage Learning have been professionally competent and a pleasure to work with. We are especially grateful to Seth Dobrin, Product Manager, who has guided this edition, and to Erin Mitchell, former Sociology Editor, who oversaw the initiation of this current revision. Huge thanks to Erik Fortier, Product Development Manager, Social Sciences, who provided the constant consultation, encouragement, and feedback to our author team that enabled this edition to come to completion on schedule. Media Developer John Chell, Content Coordinator Jessica Alderman, and Editorial Assistant Nicole Bator have been important to the success of this edition. Tom McDonough, Rights Acquisitions Specialist, made sure we were accountable to other authors and publishers when we used their work.

Jill Traut, Project Manager for MPS Limited, led a production team whose specialized competence and coordinated efforts have made the book a reality. She was excellent to work with, always available and responsive to our questions, flexible, and ever helpful. She managed a complex production process smoothly and effectively to ensure a timely completion of the project and a book whose look and presentation of content are very pleasing to us—and, we hope, to the reader.

The internal production efforts were managed by Cheri Palmer, Content Project Manager. Copy Editor S.M. Summerlight did an outstanding job of bringing our draft manuscript into conformity with style guidelines and was amazing in terms of his ability to notice fine details—inconsistencies or omissions in citations, references, and elements of the manuscript. Padma Soundararajan, Photo Researcher (PreMediaGlobal), worked with us to find photos that captured the ideas we presented in words.

Diane Beasley developed the overall design of the book, one we are very pleased with. Caryl Gorska, Art Director, oversaw the design of new edition. Heather Mann proofread the book pages, and Edwin Durbin compiled the index. Once it is completed, our textbook needs to find the faculty and students who will use it. Kara Kindstrom, Sr. Marketing Manager, captured the essence of our book in the various marketing materials that present our book to its prospective audience.

Closer to home, Agnes Riedmann wishes to acknowledge her late mother, Ann Langley Czerwinski, PhD, who helped her significantly with past editions. Agnes would also like to acknowledge family, friends, and professional colleagues who have supported her throughout the thirty-five years that she has worked on this book.

Sam Walker has contributed to each edition of this book through his enthusiasm and encouragement for Mary Ann Lamanna's work on the project. Larry and Valerie Lamanna and other family members have enlarged their mother's perspective on the family by bringing her into personal contact with other family worlds—those beyond the everyday experience of family life among the social scientists!

Mary Ann Lamanna and Agnes Riedmann continue to acknowledge one another as coauthors for thirty-five years. Each of us has brought somewhat different strengths to this process. We are not alike—a fact that has continuously made for a better book, in our opinion. At times, we have lengthy email conversations back and forth over the inclusion of one phrase. Many times, we have disagreed over the course of the past thirty years—over how long to make a section, how much emphasis to give a particular topic, whether a certain citation is the best one to use, occasionally over the tone of an anxious or frustrated email. But we have always agreed on the basic vision and character of this textbook. And we continue to grow in our mutual respect for one another as scholars, writers, and authors. We have now been joined by Susan Stewart as coauthor. She brings a fresh perspective to the book as well as a comprehensive knowledge of research in the field. Her expertise has especially contributed to revision of Chapters 4, 5, 14, and 15.

As a new author, Susan Stewart would like to acknowledge Agnes Riedmann and Mary Ann Lamanna for their unwavering support, mentoring, and enormous patience as she learned the art and science of textbook writing. She would also like to acknowledge her daughter, Gwendolyn, who provided rich experiences that contributed to her insight about parent-child relationships, relationships with her own parents and sisters, and relationships with her ex-spouse and in-laws.

Reviewers gave us many helpful suggestions for revising the book. Peter Stein's work over the years as a thorough, informed, and supportive reviewer has been an especially important contribution. Although we have not incorporated all suggestions from reviewers, we have considered them all carefully and used many. The review process makes a substantial, and indeed essential, contribution to each revision of the book.

Twelfth Edition Reviewers

Chuck Baker, Delaware County Community College; Adriana Bohm, Delaware County Community College; John Bowman, University of North Carolina at Pembroke; Jennifer Brougham, Arizona State University–Tempe; Shaheen Chowdhury, College of DuPage; Diana Cuchin, Virginia Commonwealth University; James Guinee, University of Central Arkansas; Amy Knudsen, Drake University; Wendy Pank, Bismarck State College; Rita Sakitt, Suffolk County Community College; Tomecia Sobers, Fayetteville Technical Community College; Richard States, Allegany College of Maryland; and Scott Tobias, Kent State University at Stark.

Eleventh Edition Reviewers

Rachel Hagewen, University of Nebraska, Lincoln; Marija Jurcevic, Triton College; Sheila Mehta-Green, Middlesex Community College; Margaret E. Preble, Thomas Nelson Community College; Teresa Rhodes, Walden University.

Tenth Edition Reviewers

Terry Humphrey, Palomar College; Sampson Lee Blair, State University of New York, Buffalo; Lue Turner, University of Kentucky; Stacy Ruth, Jones County Junior College; Shirley Keeton, Fayetteville State University; Robert Bausch, Cameron University; Paula Tripp, Sam Houston State University; Kevin Bush, Miami University; Jane Smith, Concordia University; Peter Stein, William Paterson University.

Of Special Importance

Students and faculty members who tell us of their interest in the book are a special inspiration. To all of the people who gave their time and gave of themselves—interviewees, students, our families and friends—many thanks. We see the fact that this book is going into a twelfth edition as a result of a truly interactive process between ourselves and students who share their experiences and insights in our classrooms; reviewers who consistently give us good advice; editors and production experts whose input is invaluable; and our family, friends, and colleagues whose support is invaluable.

Marriages, Families, and Relationships

MAKING CHOICES IN A DIVERSE SOCIETY

1

Making Family Choices in a Changing Society

Janine Wiedel Photolibrary/Alamy

Learning Objectives

1. Understand why researchers and policy makers need to define family, even though definitions are not always agreed upon and can be controversial.
2. Relate ways that family structure, or form, is increasingly diverse.
3. Explain why there is no typical American family.
4. Describe and give examples of various society-wide, structural conditions that impact families.
5. Discuss why the best life course decisions are informed ones made consciously.
6. Explain and give examples of how families provide individuals with a place to belong.
7. Understand why there is a tension in our culture between familistic values on the one hand and individualistic values on the other hand.

This text is different from others you may read. It isn't necessarily intended to prepare you for an occupation. Although it could help you in a future career, this text has three other goals as well: to help you (1) appreciate the variety and diversity among families today, (2) understand your past and present family situations and anticipate future possibilities, and (3) be more conscious of the personal decisions you make throughout your life and of the societal influences that affect those decisions.

Families are central to society and to our everyday lives. Families undertake the pivotal tasks of raising children and providing family members with support, companionship, affection, and intimacy. Meanwhile, what we think of as family has changed dramatically in recent decades. This chapter explores *family* definitions while noting the many and varied structures or forms that families take today. This chapter also describes some society-wide conditions that impact families: ever-new biological and communication technologies, economic conditions, historical periods of events, and demographic characteristics such as age, religion, race, and ethnicity.

Later in this chapter, we'll note that when maintaining committed relationships and families, people need to make informed decisions. Chapter 1 introduces concepts to be explored much more fully throughout this textbook. The theme of knowledge plus commitment is integral to this book. Finally, we end this chapter with a discussion of four themes that characterize this text. You'll see that these four themes comprise the text's four learning goals, which are listed in the Preface. We begin with a working definition of family—one that we can keep in mind throughout the course.

Defining Family

As shown in Figure 1.1, people make a variety of assumptions about what families are and are not. We've noticed when teaching this course that some students, when asked to list their family members, include their pets. Are dogs, cats, or hamsters family members?

Some individuals who were conceived by artificial insemination with donor sperm are tracking down their "donor siblings"—half brothers and sisters who were conceived using the same man's sperm. They may define their "donor relatives" as family members (Shapiro 2009). Indeed, there are many definitions given for the family, not only among laypeople but also among family scientists themselves (Weigel 2008). We, your authors, have chosen to define **family** as follows: A family is any sexually expressive, parent-child, or other kin relationship in which people—usually related by ancestry, marriage, or adoption—(1) form an economic or otherwise practical unit and care for any children or other dependents, (2) consider their identity to be significantly attached to the group, and (3) commit to maintaining that group over time.

How did we come to this definition? First, caring for children or other dependents suggests a function that the family is expected to perform. Definitions of many things have both functional and structural components. Functional definitions point to the purpose(s) for which a thing exists—that is, what it does. For example, a functional definition of an iPhone would emphasize that it allows you to make and receive calls, take pictures, connect to the Internet, and access media. Structural

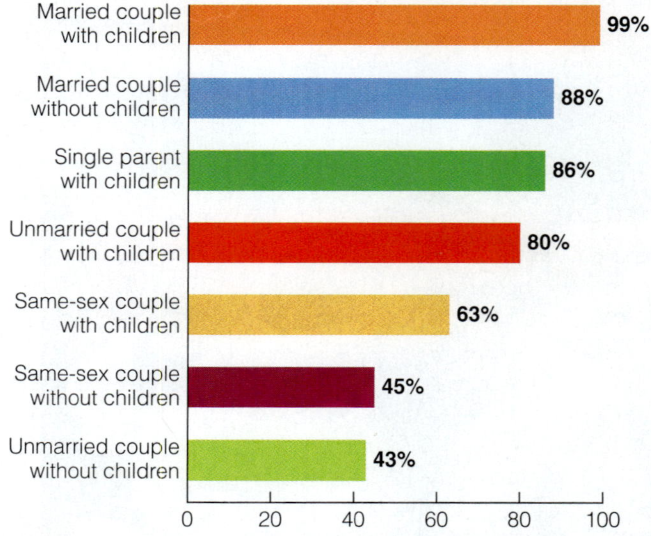

FIGURE 1.1 What is a family? Percent saying each of these is a family, 2010.

Source: Pew Research Center 2010a, p. 40.

definitions emphasize the *form* that a thing takes—what it actually is. To define an iPhone structurally, we might say that it is an electronic device, small enough to be handheld, with a multimedia screen, and with components that allow sophisticated satellite communication. Concepts of the family comprise both functional and structural aspects. We'll look now at how the family can be recognized by its functions, and then we'll discuss structural definitions of the family.

Kevin Dodge/Corbis

We can define families by their functions—raising children, providing economic support for dependents, and offering emotional support for all family members. These three look as if they're doing all of that. But functional definitions of *family* aren't enough. We also need to consider the group's structure. This family consists of a heterosexual couple and their child. They may be married or cohabiting.

Family Functions

Social scientists usually list three major functions filled by today's families: raising children responsibly, providing members with economic and other practical support, and offering emotional security.

Family Function 1: Raising Children Responsibly If a society is to persist beyond one generation, adults have to not only bear children but also feed, clothe, and shelter them during their long years of dependency. Furthermore, a society needs new members who are properly trained in the ways of the economy and culture and who will be dependable members of the group. These goals require children to be responsibly raised. Virtually every society assigns this essential task to families.

A related family function has traditionally been to control its members' sexual activity. Although there are several reasons for the social control of sexual activity, the most important one is to ensure that reproduction takes place under circumstances that help to guarantee the responsible care and socialization of children. The universally approved locus of reproduction remains the married-couple family (Cherlin 2005). "Throughout history, marriage has first and foremost been an institution for procreation and raising children. It has provided the cultural tie that seeks to connect the father to his children by binding him to the mother of his children" (Wilcox et al. 2011a, p. 82). Nevertheless, in the United States and other industrialized countries the child-raising function is often performed by divorced, separated, never-married, or cohabiting parents, and sometimes by grandparents or other relatives.

Family Function 2: Providing Economic and Other Practical Support A second family function involves providing economic support. Throughout much of our history, the family was primarily a practical, economic unit rather than an emotional one (Shorter 1975; Stone 1980). Although the modern family is no longer a self-sufficient economic unit, virtually every family engages in activities aimed at providing for such practical needs as food, clothing, and shelter.

Family economic functions now consist of earning a living outside the home, pooling resources, and making consumption decisions together. In assisting one another economically, family members create some sense of material security. For example, family members offer one another a kind of unemployment insurance. If one family member is laid off or can't find work, others may be counted on for help. Family members care for each other in additional practical ways too, such as nursing and transportation during an illness.

Family Function 3: Offering Emotional Security Although historically the family was a pragmatic institution involving material maintenance, in today's world the family has grown increasingly important as a source of emotional security (Cherlin 2008; Coontz 2005b). Not just partners or parents but children, siblings, and extended kin can be important sources of emotional support (Waite et al. 2011). This is not to say that families can solve all our longings for affection, companionship, and intimacy. Sometimes, in fact, the family situation itself is a source of stress as in the case of parental conflict, alcoholism, drug abuse, or domestic violence. But families and committed relationships are meant to offer important emotional support to adults and children.

Family may mean having a place where you can be yourself, even sometimes your worst self, and still belong.

Defining a family by its functions is informative and can be insightful. For example, Laura Dawn, in her book of stories about people who took in survivors of Hurricane Katrina, describes "how strangers became family" (Dawn 2006). But defining a family only by its functions would be too vague and misleading. For instance, neighbors or roommates might help with child care, provide for economic and other practical needs, or offer emotional support. But we still might not think of them as family. An effective definition of family needs to incorporate structural elements as well.

Structural Family Definitions

Traditionally, both legal and social sciences have specified that the family consists of people related by blood, marriage, or adoption. In their classic work *The Family: From Institution to Companionship*, Ernest Burgess and Harvey Locke (1953 [1945]) specified that family members must "constitute a household," or reside together. Some definitions of the family have gone even further to include economic interdependency and sexual–reproductive relations (Murdock 1949).

The U.S. Census Bureau defines a family as "a group of two or more persons related by blood, marriage, or adoption and residing together in a household" (U.S. Census Bureau 2012c, p. 6). It is important to note here that the Census Bureau uses the term **household** for any group of people residing together. Not all households are families by the Census Bureau definition—that is, persons sharing a household must also be related by blood, marriage, or adoption to be considered a family.

Family structure, or the form a family takes, varies according to the social environment in which it is embedded. In preindustrial or traditional societies, the family structure involved whole kinship groups. The **extended family** of parents, children, grandparents, and other relatives performed most societal functions, including economic production (e.g., the family farm), protection of family members, vocational training, and maintaining social order. In industrial or modern societies, the typical family structure often became the **nuclear family** (husband, wife, children), which was better suited to city life. Until about fifty years ago, social attitudes, religious beliefs, and law converged into a fairly common expectation about what form the American family should take: breadwinner husband, homemaker wife, and children living together in an independent household—the nuclear-family model.

Nevertheless, the extended family continues to play an important role in many cases, especially among recent immigrants and race/ethnic minorities. Furthermore, to cope with hardships associated with the current economic recession, more families of all races/ethnicities are doubling up—that is, relatives are moving in together to create more multigenerational or otherwise extended-family households. About 15.5 million, or 13 percent of American households are occupied by extended or multifamily groups—an increase of 12 percent or more since the onset of the recession in late 2007 (Mykyta and Macartney 2012, p. 2 and Table A-1). "Accordion" family households that expand or contract around more or fewer family members depending on family need perform important economic and often emotional social functions (Newman 2012).

Meanwhile, today's families are not necessarily bound to one another by legal marriage, blood, or adoption. The term *family* can identify relationships in addition to spouses, parents, children, and extended kin. Individuals fashion and experience intimate relationships and families in many forms. As social scientists take into account this structural variability, it is not uncommon to find them referring to the family as *postmodern* (Stacey 1990).

Ariel Skelley/Photographer's Choice/Getty Images

The extended family—grandparents, aunts, and uncles—can provide occasion for good times as well as an important source of security, its members helping each other, especially during crises.

Postmodern: There Is No Typical Family

Barely half of U.S. adults are married—a record low (Taylor et al. 2011). Think of television shows in which single parents, interracial couples, lesbian or gay male couples, and still other family variations increasingly appear. Just 6 percent of families now fit the 1950s nuclear-family ideal of married couple and children, with a husband-breadwinner and wife-homemaker (U.S. Census Bureau 2012b, Tables F1, FG8). The past several decades have witnessed a proliferation of relationship and family forms: single-parent families, stepfamilies, families with children of more than one father, two-earner couples, stay-at-home fathers, cohabiting heterosexual couples, gay and lesbian marriages and families, three-generation families, and communal households, among others (Cherlin 2010; Dorius 2012). It appears that individuals can construct a myriad of social forms in order to address family functions. The term **postmodern family** came into use to acknowledge the fact that families today exhibit a multiplicity of forms and that new or altered family forms continue to emerge and develop.

Figure 1.2 displays the types of households in which Americans live. Just 20.9 percent of households are nuclear families of husband, wife, and children, as compared with 31 percent in 1980 and with 44 percent in 1960 (Casper and Bianchi 2002, p. 8; U.S. Census Bureau 2012c, Table 59). The most common household type today is that of married couples without children: The children have grown up and left or the couple has not yet had children or doesn't plan to.

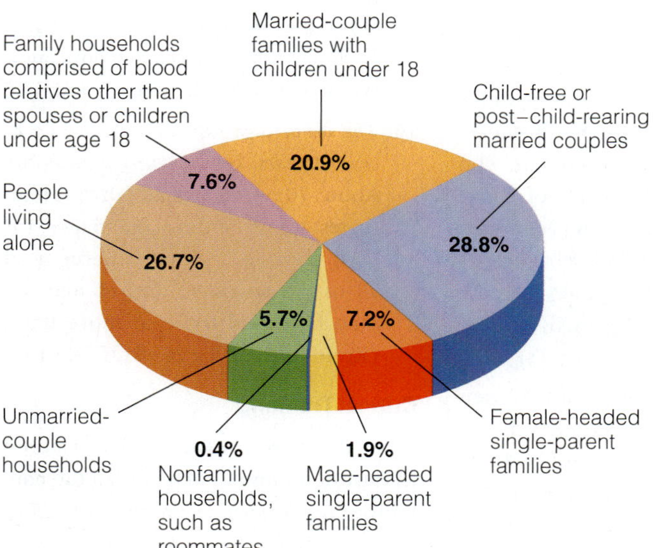

FIGURE 1.2 The many kinds of American households, 2010.* A household is one or more persons who occupy a dwelling unit. This figure displays both family and nonfamily households.**

*This is the most recent year for which all of the data for this figure are available.

**Unmarried-couple households may be composed of same-sex (10.7 percent) or heterosexual couples (89.3 percent) (calculated from Lofquist et al. 2012, Table 3); Census Bureau classifies unmarried-couple households as nonfamily households.

Source: U.S. Census Bureau 2012c, Tables 59, 63.

More households today (26.7 percent) are maintained by individuals living alone than by married couples with children. There are also female-headed (40 percent) and male-headed (1.9 percent) single-parent households, unmarried-couple households (5.3 percent), and family households containing relatives other than spouses or children (7.6 percent). "Facts about Families: American Families Today" presents additional information about families. Today we see historically unprecedented diversity in family composition or form.

As one result of this diversity, law, government agencies, and private corporations such as insurance companies must now make decisions about what they once could take for granted—that is, what a family is. If rent policies, employee-benefit packages, and insurance policies cover families, decisions need to be made about what relationships or groups of people are to be defined as a family. The September 11th Victim Compensation Fund of 2001 struggled with this issue in allocating compensation to victims' survivors. New York State law was amended to allow awards to unmarried gay and heterosexual partners (Gross 2002). President George W. Bush subsequently signed a federal bill extending benefits to domestic partners of firefighters and police officers who lose their lives in the line of duty (Allen 2002).

Adapting Family Definitions to the Postmodern Family

As family forms have grown increasingly variable, social scientists have proposed—and often struggled with—new, more flexible definitions for the family. Sociologist David Popenoe (1993) defined today's family as "a group of people in which people typically live together in a household and function as a cooperative unit, particularly through the sharing of economic resources, in the pursuit of domestic activities" (1993, p. 528). Sociologist Frank Furstenberg writes as follows: "My definition of 'family' includes membership related by blood, legal ties, adoption, and informal ties including *fictive* or socially agreed upon kinship" (2005, p. 810, italics in original).

Legal definitions of family have become more flexible as well. In the past few decades, judges, when defining the family in cases that come before them, have used the more intangible qualities of stability and commitment along with the more traditional criteria of common residence and economic interdependency (*Dunphy v. Gregor 1994*). From this point of view, the

Facts about Families

American Families Today*

What do U.S. families look like today? Statistics can't tell the whole story, but they are an important beginning. As you read these ten facts, remember that the data presented here are generalizations and do not consider differences among various sectors of society. We explore social diversity throughout this textbook, but for now let's look at some overall statistics.

1. *Marriage is important to Americans—but not to the extent that it was fifty years ago.* Sixty-one percent of never-married adults say they want to marry; another 27 percent is not sure. But 39 percent of us (44 percent of 18- to 29-year-olds and 32 percent of Americans age 65 and older) see marriage as becoming obsolete (Taylor et al. 2011).
2. *A smaller proportion of people is married today.* Between 51 and 56 percent of Americans age 18 and older were married in 2010, compared to 59 percent in 2000, 62 percent in 1990, and 72 percent in 1960. Twenty-seven percent have never married; 10 percent are divorced, and 6 percent widowed (Cohn et al. 2011; Taylor et al. 2011; U.S. Census Bureau 2012c, Table 56).
3. *Young people are postponing marriage.* In 2011, the median age at first marriage was 26.5 for women and 28.7 for men, as compared with 20.8 for women and 23.5 for men in 1970 (U.S. Census Bureau 2011b, Table MS-2). Today's average age at marriage is the highest recorded since the 1890 census.
4. *Cohabitation has become a fairly acceptable family form as well as a transitional lifestyle choice.* The number of cohabitating adults has increased more than tenfold since 1970—and by 40 percent just since 2000 (U.S. Census Bureau 2012b, Table UC-1; Lofquist et al., Table 2). Nearly 40 percent of cohabiting couples lived with children under 18 in 2011—either their own or those from a previous relationship or marriage. Unmarried-couple families are only 5 percent of households at any one time, but more than 50 percent of first marriages are preceded by cohabitation (Lofquist et al. 2012).
5. *Fertility has declined.* At 1.9 in 2011, the total fertility rate (TFR)—the average number of births that a woman will have during her lifetime—had dropped by almost 4 percent from 2009 (Martin et al. 2012, p. 9; Mather 2012). After a high of 3.6 in 1957, the total TFR has been at about 2 over the past twenty years (U.S. Census Bureau 2012c, Table 83). A society requires a TFR of at least 2.1 in order for the population numerically to replace itself, so the current TFR is below replacement level.
6. *Particularly among college-educated women, parenthood is often postponed.* The average age for a woman's first birth increased by about 2 years between 1970 and 2010—from age 21 to about 23. But the statistics differ according to education. For instance, nearly 60 percent of women who had not finished high school had a first birth by age 20, compared with 4 percent of women with a bachelor's degree or higher (Martinez, Daniels, and Chandra 2012, Figure 3).
7. *Compared to 4 percent in 1950, the nonmarital birthrate is high* with 40 percent of all U.S. births today being to unmarried mothers. Unlike 1950, however, between one-quarter and one-half of nonmarital births today occur to cohabitating couples (Carter 2009). Along with overall fertility, the unmarried-mother birthrate fell slightly between 2009 and 2010. We don't know whether this situation marks the beginning of a new trend or simply reflects a temporary response to the current recession (Martin et al. 2012, p. 10).
8. *Same-sex-couple households increased* by 80 percent between 2000 and 2010 (Homan and Bass 2012). It is difficult to precisely quantify the number of same-sex-couple households in the United States (Lofquist 2012). However, the 2010 U.S. Census counted about 646,464 same-sex couple households. Of these, 131,729—a little more than 25 percent—were married-couple households (U.S. Census Bureau 2011a). It is estimated that between 16 percent and 19 percent of same-sex households (about 14 percent of male and 27 percent of female) include children (Gates 2011; Krivickas and Lofquist 2011; Lofquist 2011).
9. *The divorce rate is high.* The divorce rate doubled from 1965 to 1980. Then it dropped and has fallen more than 30 percent since 1980 (U.S. Census Bureau 2012c, Table 78). Still, it is estimated that only about half of recent first marriages will last twenty years, although the likelihood of divorce declines with more years of education (Copen et al. 2012, Tables 5 and 6).
10. *The remarriage rate has declined in recent decades but remains significant.* About 70 percent of all current marriages are first-time marriages for both spouses. About 4 percent of all married women and of married men have wed three or more times (Kreider and Ellis 2011b, Table 10).

Critical Thinking

What do these statistics tell you about the strengths and weaknesses of the contemporary American family and about family change?

*The Census Bureau defines households as people living together in the same domicile; family households are domiciles housing persons related by blood, marriage, or adoption. Figures differ depending on whether household or family is the unit of analysis. For example, married-couple family households are 50 percent of all households, but married-couple families are 74 percent of all families (U.S. Census Bureau 2012c, Table 59).

definition of family "is the totality of the relationship as evidenced by the dedication, caring and self-sacrifice of the parties" (*Judge Vito Titone in Braschi v. Stahl Associates Company 1989,* quoted in Gutis 1989). More recently, a few state legislatures have provided that legal status and rights can be enjoyed by more than two—that is, by three or four—parents in one family. What would be an example of a family like this? Here's one: Two children spend three nights a week with their partnered gay fathers. The other nights they stay with their lesbian mothers, who live nearby (Lovett 2012).

Many employers have redefined family with respect to employee-benefit packages. Just more than half of the Fortune 500 companies, as well as many state and local governments, offer domestic partner benefits to persons in an unmarried couple who have registered their relationship with a civil authority (Appleby 2012). President Barack Obama signed an executive order granting federal employees and their domestic partners some of the rights (but, importantly, neither health insurance nor retirement benefits) enjoyed by married couples (Miles 2010). As we write, legislation that would extend domestic partner benefits to all federal civilian employees is moving through Congress (Broverman 2012). Meanwhile, federal practices permit low-income unmarried couples to qualify as families and live in public housing. Several states allow same-sex marriage, and several others provide some spousal rights to same-sex couples. Same-sex marriage is discussed in particular in Chapter 6 and elsewhere in this text as well.

We, your authors, began this section with our definition of family. Our definition recognizes the diversity of postmodern families while paying heed to the essential functions that families are expected to fill. Our definition combines some structural criteria with a more social–psychological sense of family identity. We include the commitment to maintaining a relationship or group over time as a component of our definition because we believe that such a commitment is necessary in fulfilling basic family functions. It also helps to differentiate the family from casual relationships, such as roommates, or groups that easily come and go.

We have worked to balance an appreciation for flexibility and diversity in family structure and relations with the concern that many policy makers and social scientists express about how well today's families perform their functional obligations. Ultimately there is no one correct answer to the question, "What is a family?"

Relaxed Institutional Control over Relationship Choices: "Family Decline" or "Family Change"?

Public opinion polls show that overall about 30 percent of Americans reject today's trend toward the postmodern family while about the same proportion accept new family forms. Another 37 percent accept some aspects of family change but are concerned about others (Morin 2011). Similarly, a 2010 Pew Research Center survey found that 34 percent of Americans saw the growing variety of family types as a good thing, while 29 percent thought it was a bad thing. The remainder either didn't answer or saw the changes as creating no difference (Pew Research Center 2010a, p. 3).

In 2012, 59 percent of Americans saw unmarried (heterosexual) sex as morally acceptable, but 38 percent saw it as morally wrong. Those numbers had changed from 53 percent and 42 percent in 2001. Sixty-seven percent of Americans today see divorce as morally acceptable, whereas in 2001 that figure was 59 percent. Nevertheless, a significant minority (25 percent) continue to see divorce as morally wrong. Americans are somewhat more evenly split regarding having a baby outside marriage: 54 percent say doing so is morally acceptable today, compared with 45 percent in 2002. Forty-two percent of us think that having a baby outside marriage is morally wrong; 50 percent thought so in 2002 ("Marriage" 2012). Today Americans are also fairly evenly split regarding whether same-sex marriage should be legally valid, whereas just before the turn of the twenty-first century only about one-third felt that same-sex unions should be legally valid (Gallup Poll 2012b). Americans are strongly opinionated about family change; we can better understand why if we understand that the family has historically been understood as a **social institution**.

Social institutions are patterned and largely predictable ways of thinking and behaving—beliefs, values, attitudes, and norms that are organized around vital aspects of group life and serve essential social functions. Social institutions are meant to meet people's basic needs and enable the society to survive. Earlier in this chapter, we described three basic family functions. Because social institutions prescribe socially accepted beliefs, values, attitudes, and behaviors, they exert considerable social control over individuals.

During the 1960s, however, family formation became increasingly less predictable; demographers noted dramatic social transformations:

> Since the end of the postwar baby boom in 1964, age at first marriage has increased, marital childbearing has decreased, nonmarital childbearing has increased, divorce rates have risen, and cohabitation has become common among young adults. The most dramatic shifts in families and households occurred in the 1970s and 1980s, and the magnitude of most changes since then has been smaller and more gradual. (Jacobsen and Mather 2010, p. 9)

Furthermore, same-sex marriage has become legally available in 10 percent or more of states. Combined with increased longevity and lower fertility rates, these changes have meant that a smaller portion of adulthood is spent in traditionally institutionalized marriages and families (Cherlin 2004, 2008).

Critics have described the relaxation of institutional control over relationships and families as "family decline" or "breakdown." Those with a **family-decline perspective** claim that a cultural change toward excessive individualism and self-indulgence has hurt relationships, led to high divorce rates, and undermines responsible parenting (Whitehead and Popenoe 2006):

> According to a marital decline perspective ... because people no longer wish to be hampered with obligations to others, commitment to traditional institutions that require these obligations, such as marriage, has eroded. As a result, people no longer are willing to remain married through the difficult times, for better or for worse. Instead, marital [or other relationship] commitment lasts only as long as people are happy and feel that their own needs are being met. (Amato 2004, p. 960)

© Rob Marmion/Shutterstock.com

In a world of demographic, cultural, and political changes, there is no typical family structure. Today's postmodern family includes cohabiting families, single-parent families, lesbian and gay partners and parents, and remarried families. Interracial families are more evident, too, and their increasing social acceptance may result in their experiencing greater community support.

Moreover, fewer family households contain children. According to the family-decline perspective, this situation "has reduced the child centeredness of our nation and contributed to the weakening of the institution of marriage" (Popenoe and Whitehead 2005, p. 23; Wilcox et al. 2011a,b). "Facts about Families: Focus on Children" provides some statistical indicators about the families of contemporary children.

Not everyone concurs that the family is in decline: family change, yes, but not decline (Coontz 2005a). Scholars and policy makers with a **family-change perspective** sometimes point out that some family changes can be for the better. Longer life expectancy can mean more positive years with parents, grandparents, and great-grandparents. Easier access to divorce than was the case fifty years ago means that family members have alternatives to living with domestic violence. With 86 percent of Americans approving black–white marriages (Jones 2011), increasing tolerance for interracial unions in general can translate to greater acceptance for particular mixed-race families so that they experience more supportive or less hostile communities. Family flexibility can be functional in times of economic crisis as extended families expand to take in needy relatives.

Family-change scholars argue that we need to view the family from a historical standpoint. In the nineteenth and early twentieth centuries, American families were often broken up by illness and death, and children were sent to orphanages, foster homes, or already burdened relatives. Single mothers, as well as wives in lower-class, working-class, and immigrant families, did not stay home with children but went out to labor in factories, workshops, or domestic service. The proportion of children living only with their fathers in 1990 wasn't much different from that of a century ago (Kreider and Fields 2005, p. 12).

Family-change scholars posit that today's family forms need to be seen as historically expected adjustments to changing conditions in the wider society, including the decline in well-paid working-, middle-, and even upper-middle-class jobs that used to provide solid economic family support. Family-change sociologists do not ignore the difficulties that separation, divorce, and nonmarital parenthood present to families, children, and the broader society. However, these social scientists view the family as "an adaptable institution" (Amato et al. 2003, p. 21) and argue that it makes more sense to provide support to families as they exist today rather than to attempt to turn back the clock to an idealized past (Cherlin 2009a; McHale, Waller, and Pearson 2012).

Today's American families struggle with new economic and time pressures that affect their ability to realize their

Facts about Families

Focus on Children

In many places throughout this text, we focus particularly on children in families. As with our population as a whole, the number of children in the United States is growing. Today approximately 74 million children under age 18 are living in the United States (U.S. Federal Interagency Forum on Child and Family Statistics 2012, Table Pop1). However, the proportion of today's population that is under age 18—about 24 percent—represents a substantial drop from the 1960s, when more than one-third of Americans were children (U.S. Census Bureau 2012c, Table 7). Here we look at five statistical indicators regarding U.S. children's living arrangements and well-being.

Charles Thatcher/Getty Images

The faces of America's children provide evidence of increasing ethnic diversity. The child population of the United States is more racially and ethnically diverse than the adult population. Making up about one-third of the U.S. population today, racial/ethnic minorities are projected to reach 50 percent of the total population by about 2042. Mostly due to rapid growth in Latino families, the population under age 18 is projected to reach this point by 2023 (Mather 2009).

1. At any given time, a majority of children live in two-parent households. In 2010, 70 percent of children under 18 lived with two parents—and 68 percent with two married parents. Twenty-six percent of children lived with only one parent (23 percent with mother; 4 percent with father), and another 4 percent did not live with either parent (U.S. Census Bureau 2011b, Table SO901).
2. Many children experience a variety of living arrangements while they're young. A child may progress through living in an intact two-parent family, a single-parent household, with a cohabitating parent, and finally in a remarried family. About half of all American children are expected to live in a single-parent household at some point in their lives, most likely in a single-mother household (Kreider and Ellis 2011a, p. 24; U.S. Federal Interagency Forum 2005, p. 8, Figure POP6-A).
3. Children are more likely to live with a grandparent today than in the past. In 1970, 3 percent of children lived in a household containing a grandparent. By 2011 that rate had more than tripled to 10 percent (U.S. Census Bureau 2011c, Table C4). In about a quarter of the cases, grandparents had sole responsibility for raising the child, but many households containing grandparents are extended-family households that include other relatives as well (Edwards 2009).
4. Although most parents are employed, children are more likely than the general population to be living in poverty. The poverty rate of children under age 18 stood at about 18 percent over the ten years prior to the 2008 onset of the recent recession. By 2010 that percentage had risen to 22, while that of the general adult population is about 14 percent and that of those older than age 65 is approximately 9 percent. About 16.1 million American children under age 18 live in poverty—an increase of approximately 1 million from 2009 to 2011 (DeNavas-Walt, Proctor, and Smith 2012, Table 3).
5. A growing number of U.S. children have a foreign-born parent. The percentage of children under age 18 living with at least one foreign-born parent rose from 14 percent in 1994 to 23 percent in 2011—almost a quarter of all U.S. children. Twenty-one percent of children were native-born children with at least one foreign-born parent, and 3 percent were foreign-born children with at least one foreign-born parent. Having parents who were born outside the United States can affect the language spoken at home. In 2010, 22 percent of children ages 5 to 17 spoke a language other than English at home, up from 18 percent in 2000 (Wallman 2012).

Critical Thinking

Perhaps the greatest concern Americans have about family change today is its impact on children. What do these family data tell us about the family lives of children?

family values. Even amid global recession, many European countries remain committed to paid family-leave policies that enable parents to take time off from work to be with young children and that provide relatively generous economic support for families in general (Human Rights Watch 2011). Family-change scholars "believe that at least part of the increase in divorce, living together, and single parenting has less to do with changing values than with inadequate support for families in the U.S., especially compared to other advanced industrial countries" (Yorburg 2002, p. 33). Placing an individual's or family's private troubles within a society-wide context is the crux of what sociologists call a **sociological imagination**.

A Sociological Imagination: Personal Troubles and Some Social Conditions That Impact Families

People's private lives are affected by what is happening in the society around them. In his classic book, *The Sociological Imagination* (1959), sociologist C. Wright Mills developed the principle that private, or personal, troubles are connected to events and patterns in society. Many times what seem to be personal troubles are shared by others, and these troubles often reflect societal influences. For example, when a family breadwinner is laid off or quits looking for work after many months of searching, the cause does not likely lie in his or her lack of ambition but in the economy's inability to provide employment. As another example, the difficulty of juggling work and family is not usually just a personal question of individual time-management skills but of society-wide influences—the totality of time required for employment, commuting, and family care in a society that provides limited support for working families. As a final example, many families were separated by the destruction that resulted when Hurricane Katrina hit New Orleans in 2005; a good number of them remain divided and probably will never be reunited (Rendall 2011).

In this section we'll look at five social factors that affect families:

1. ever new biological and communication technologies;
2. economic conditions;
3. historical periods or events;
4. demographic characteristics (statistical facts about the make-up of a population), such as age, religion, and race or ethnicity; and
5. family policy.

Ever-New Biological and Communication Technologies

The pace of technological change has never been faster; new technologies will continue to alter not only family relationships but also how we define families. Here we'll look at two types of technological change that impact family life—biological and communication technologies.

Biological Technologies Since the 1960s invention of the birth-control pill and the 1978 arrival of the first "test-tube baby," modern science has expanded our options regarding both preventing pregnancy and enhancing fertility. These developments are further addressed in other chapters of this text, particularly in Chapter 8. Here we introduce the point that ever-new biological technologies dramatically impact families' daily lives (Farrell, VandeVusse, and Ocobock 2012).

"Mommy, Mommy, when I grow up, I want to be a mommy just like you. I want to go to the sperm bank just like you and get some sperm and have a baby just like me" (6-year-old quoted in Ehrensaft 2005, p. 1). Science continues to develop new techniques that enable individuals or couples to have biological children. The more common infertility interventions involve prescription drugs and microscopic surgical procedures to repair a female's fallopian tubes or a male's sperm ducts (Ehrenfeld 2002). More widely publicized *assisted reproductive technology* (ART) offers increasingly successful reproductive options (U.S. Centers for Disease Control and Prevention 2008).

In general, ART involves the manipulation of sperm or egg or both in the absence of sexual intercourse, often in a laboratory. ART procedures include:

- *artificial insemination* (male sperm introduced to a female egg without sexual intercourse),
- *donor insemination* (artificial insemination with sperm from a donor rather than from the man who will be involved in raising the child),
- *in vitro fertilization* (sperm fertilizes egg in a laboratory rather than in the woman's body),
- *surrogacy* (one woman gestates and delivers a baby for another individual who intends to raise the child),
- *egg sale or donation* (by means of a surgical procedure a woman relinquishes some of her eggs for use by others), and
- *embryo transfers* (a laboratory-fertilized embryo is placed into a woman's womb for gestation and delivery).

ART allows otherwise infertile heterosexual couples to have biological children. ART also allows singles or lesbian and gay male couples to become biological parents. Furthermore, the ability to freeze eggs, sperm, or fertilized embryos enables individuals to become biological parents later in life, after careers are launched, after undergoing medical treatments that will leave them infertile, or even after death. Anticipating either contact with hazardous materials or death, men deployed to Iraq have banked sperm before their departures. At least one baby has been conceived by a father who was killed in Iraq before his child's conception (Lehmann-Haupt 2009; Oppenheim 2007). A few grandparents, eager for grandchildren, have offered to finance egg freezing for their grown children (Gootman 2012).

On a somewhat different note, by testing a male's blood, it is now possible to confirm the paternity of a likely biological father as early as the eighth or ninth week of pregnancy. "Besides relieving anxiety, the test results might allow women to terminate a pregnancy if the preferred man is not the father—or to continue it if he is." Then, too, "men who clearly know they are the father might be more willing to support the woman financially and emotionally during the pregnancy which some studies suggest might lead to healthier babies" (Pollack 2012a).

Moreover, in 2012, for the first time researchers determined virtually the entire genome of a fetus using only a blood sample from the pregnant woman and a saliva specimen from the biological father. Now it's fairly easy to know the complete DNA blueprint of a fetus months before the baby is born. Thousands of genetic diseases can now be detected prenatally, a situation allowing parents to address these conditions while pregnant—either by fixing problems, accepting that the child will have a genetic disease, or aborting the fetus (Pollack 2012b).

Although biological technologies expand options, they also raise new possibilities for thorny relationship or ethnical issues. As a statistically rare, but real example, a spouse may choose to change his or her sexual anatomy with the help of drugs or surgery (Daniel 2011). As a somewhat more common example, many states have laws by which sperm donors, with the exception of the husband, have no parental rights, but this barrier between sperm donors and their biological children is gradually being broken. Some sperm donors are sought out by their "children" as they enter adolescence or young adulthood (Harmon 2007b). Moreover, some fertility-enhancing procedures and extensive DNA fetal mapping raise issues surrounding abortion. Policy and ethical issues associated with biological technologies are more fully addressed in Chapter 8.

Communication Technologies Communication technologies have dramatically changed the way that family members interact. Today we can video record family events such as a birthday party or a bris (the ritual circumcision of a Jewish son) on our cell phones and then send the images to family members around the world. Developments such as texting, e-mail, websites, webcams, blogs, Facebook, Skype, and Twitter facilitate communication in ways that we would never have dreamed possible thirty years ago. Many relationships now begin in cyberspace, minimizing the need for geographical proximity at first meeting. Family members, including grandparents, stay in contact on Facebook. With cell phone calls and text messaging, parents can monitor teens when they aren't home. Technologies installed in family automobiles allow parents to monitor their children's driving speeds, and Global Positioning Systems (GPS) can tell parents where their children have driven. Some young adults away at college or elsewhere text their parents once or more daily. Meanwhile, Internet access is changing power relations in some families as tech-savvy youth become information experts for their families, a skill that can enhance their power relative to other family members (Belch, Krentler, and Willis-Flurry 2005).

Social support for challenges from infertility to living in stepfamilies to caring for someone with a chronic illness can be found on the Internet. Using cell phones and social media or playing video games together can enhance family connection (Padilla-Walker, Coyne, and Fraser 2012). However, the Internet can also be a source of frustration and conflict for partners or parents who experienced another family member's emotional absence because of social networking or online game playing. Some families have dealt with easier access to pornography or cyberinfidelity, for instance.

Florian Franke/Alamy

Much is said about how social media separate family members—particularly teenagers who text or Facebook during what a parent hoped would be a family-togetherness event. But research shows that communication and Internet technologies can also bring families together. Using cell phones and social media or playing video games together can enhance family connection.

Furthermore, social networking sites such as Facebook have made breaking up and divorce potentially more hurtful as partners publish details on their pages (Luscombe 2009).

Moreover, communication technology results in a digital divide between those who have access to computers and the 20 percent of American households that don't and hence cannot access the benefits of computer use, such as filling out online job applications (Crawford 2011). Although 97 percent of households with annual incomes of at least $15,000 have Internet broadband access, just 37 percent of households with annual incomes of $15,000 or less do (U.S. Census Bureau 2012c, Table 1155). Social scientists also have noted a "new digital divide" among children with computer access. Virtually all youngsters use computer technology mainly for social networking and to play games, but children with college-educated parents are more likely than others to use the Internet for educational activities (Richtel 2012). This new digital divide is one of countless examples of how socioeconomic conditions—both those in the larger society and those of individual families—impact family life.

Economic Conditions

Families are facing very stressful economic times. As you will see throughout this text—and probably already know from your own experience—the economy has important consequences for family relationships. Regardless of our current economic situation, the overall long-term trend in U.S. household income has been upward (see Figure 1.3). That overall pattern masks a situation of growing inequality, however. During the post–World War II decades of the 1950s and 1960s, incomes grew rapidly and at about the same rate—almost 3 percent annually—for families at all income levels. From 1970 to 2000, however, the pattern changed sharply. Incomes of the top 1 percent grew more than threefold (300 percent), while median household income grew less than 15 percent.

Although the U.S. economy was good for some Americans during the 1990s, others experienced lost benefits, longer workdays, and more part-time and temporary work. Over recent decades, job restructuring (with the goal of employing fewer workers to accomplish a task) and outsourcing, or sending jobs to other countries where labor is cheaper, have resulted in diminished job security and lower wages for many Americans. A concern is that multinational corporations "are the new countries" inasmuch as they exist beyond any one nation's borders and detach themselves from any one country's national Interest. As one Apple executive interviewed about outsourcing put it, "We don't have an obligation to solve America's problems" (Foroohar 2012).

Over the past thirty-five years, the inequality gap has increased. In 2006, Princeton economist Paul Krugman stated, "The income gap is now as extreme as it was in the 1920s, wiping out decades of rising equality" (p. 46). The gap has continued to grow, with the poorest 20 percent of the population earning $20,000 or less annually, and the 20 percent of the population with the highest incomes earning $100,000 or more. In fact, since 1970 the annual income of top corporate

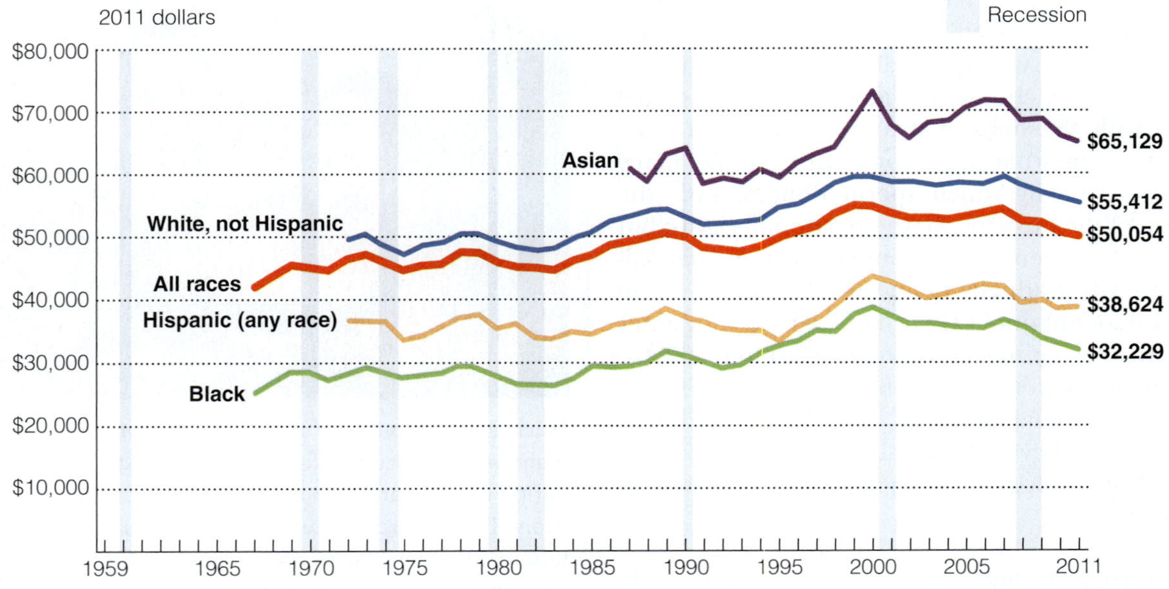

FIGURE 1.3 Real median household income by race and Hispanic origin: 1967 to 2011.

Source: DeNavas-Walt et al. 2011, Figure 1, p. 8.

CEOs has increased from an average of $1 million to $13 million dollars—an increase of 1,200 percent (DeNavas-Walt et al. 2011, p. 10). In 2011, the top one-fifth of U.S. households received slightly more than half (51.5 percent) of the nation's total income, whereas the poorest one-fifth received just 3.4 percent (DeNavas-Walt, Proctor, and Smith 2012, Table 2).

Moreover, in percentage terms, the bursting of the housing market bubble and the recession that followed took a far greater toll on the middle and working classes than on the wealthy. Home ownership is the principal source of wealth for the working class and many in the middle class, whereas for wealthier Americans investment in stocks is also significant. Since the official end of the recession in mid 2009, the housing market has remained in a slump while the stock market has recaptured much of what it lost between 2007 and 2009 (Pew Research Center 2011b).

Household *wealth* differs from income. Income is the annual inflow of wages, interest, profits, or other sources of earning. Wealth is the accumulated sum of assets (houses, cars, savings and checking accounts, stocks and mutual funds, retirement accounts, etc.) minus the sum of debt (mortgages, auto loans, credit card debt, etc.). Wealth gaps between the richest few and the rest of us have always been greater than income gaps. However, wealth gaps have grown to higher and higher levels, resulting in what many economists describe as the shrinking of the middle class.

Furthermore, a substantial proportion of American families live in poverty. As a result of President Lyndon Johnson's War on Poverty measures in the 1960s, poverty rates fell dramatically. The current poverty rate is considerably higher, partly—but assuredly not entirely—due to the recent recession. Having risen gradually from 13.2 percent in 2000, the poverty rate reached 15.0 percent in 2011 (DeNavas-Walt, Proctor, and Smith 2012, Figure 4). That poverty rate—fifteen of every 100 Americans—"matches brief peaks after recessions in the early 1980s and 1990s but otherwise hasn't occurred since 1965" (Kiviat 2011). The *child* poverty rate is 21.9 percent—more than one child in five—much higher than child poverty rates in other industrialized nations (DeNavas-Walt, Proctor, and Smith 2012, p.13). Moreover, approximately one-third of all U.S. families (10.2 million) can be classified as "working poor": at least one wage earner is employed full-time, but the family still lives with very low annual income (Roberts, Povich, and Mather 2011–2012).

Income, wealth, and poverty rates diverge by race/ethnicity, education, and parents' education. NonHispanic whites had the lowest poverty rate in 2011 (9.8 percent), followed by Asian Americans (12.3 percent). Hispanics (25.3 percent) and African Americans (27.6 percent) have higher rates of poverty. Although the poverty rate of nonHispanic whites is low, they compose 41.4 percent of the total number of persons in poverty because they are such a large part of the population (DeNavas-Walt, Proctor, and Smith 2012, Table 3).

Income varies by gender as well. Women have gained more than men since about 1980, while men's wages have been largely stagnant (DeNavas-Walt, Proctor, and Smith 2012, Figure 2). Still, access to a male wage remains an advantage, a situation explored further in Chapters 3 and 10. Incomes also vary by family type. Married-couple households had the highest incomes in 2011—$74,130 compared to $49,567 for unmarried male-headed households and $33,637 for unmarried female-headed households (DeNavas-Walt, Proctor, and Smith 2012, Table 1). Experts debate the extent to

Dog housing inequality? Yes, indeed. Whether or not an effective definition of *family* can include pets, the lifestyle of the family pooch pretty much matches that of its owner. Economic inequality is rising in the United States. Both lower-income sectors and the middle class are losing ground.

which more female-headed, single-parent households contribute to poverty. Chapter 6 examines this question in some detail.

Meanwhile, the recession that began in late 2007 caused uncertainty and change in virtually all families. The overall unemployment rate climbed through the first decade of this century, from 4.0 percent in 2000 to 5.8 in 2008, then shot up to 9.3 in 2009, 9.6 in 2010 and to 9.8 percent in 2011 (U.S. Census Bureau 2012c, Table 622). By August 2012, the overall rate had declined to 8.1, but analysis showed that most of that decline resulted from discouraged individuals ending their searches for work (U.S. Bureau of Labor Statistics 2012a; National Employment Law Project 2012). In January 2013, the unemployment rate stood at 7.9 (U.S. Bureau of Labor Statistics 2013).

Unfortunately, *long-term* unemployment also increased. The proportion of the unemployed who were without work for at least six months climbed from 11 percent in 2000 to 43 percent in 2010; it stood at 38 percent in January 2013 (U.S. Bureau of Labor Statistics 2013; U.S. Census Bureau 2012c, Table 622). Overall rates that take everyone into consideration mask the situations of specific age or ethnic groups. For instance, among those under age 25, the unemployment rate was 16.2 percent in 2011, whereas for those 25 to 34 that rate—more than 10 percent in 2009—was 9.3 in 2011 (Jacobsen and Mather 2011, Figure 8).

From 1950 to 1970, a middle-class person had to work 42 hours a month to meet the monthly rent on a median-priced dwelling. In 2000, the average employee had to work 67 hours a month to put his or her family into mid-range housing (Frank 2011). "In no state can an individual working full-time at the minimum wage afford ... a two-bedroom apartment for his or her family" (Children's Defense Fund 2012, p. 18). In fact, in many states it takes more than two full-time jobs at minimum wage to afford that apartment—and in California, the District of Columbia, Maryland, New York, and New Jersey, it would take three jobs (Children's Defense Fund 2012, Table 9). It would take almost four and one-half full-time jobs at minimum wage to rent that two-bedroom apartment in Hawaii—probably a good start to explaining why Hawaii has the highest proportion of multifamily households (Lofquist et al. 2012, Table 6).

Recession has made things worse for many of us. "Recession means worry—all too tangible worry" (Bazelton 2009). With the Great Recession that began in late 2007, housing prices dropped and many Americans lost their jobs and homes. With fewer tax dollars available, state governments cut services, many of them important to poor, working-, or middle-class families. Although policy makers declared the recession over in 2009 and defined what followed as a period of recovery, lost jobs and lowered median family income persisted (Pew Research Center 2012a). Indeed, this has been a "low-wage recovery" as relatively high-wage jobs that were lost, often to outsourcing, have been "replaced" with jobs paying much less (National Employment Law Project 2012).

Because many people put off marriage until they can earn enough to support a family, more marriages were delayed or foregone during the recession, and the birthrate declined as well between 2009 and 2010 (Haub 2011; Mather 2012). Husband unemployment can mean power shifts in families as wives become sole or primary breadwinners (Rosin 2012b). Meanwhile, the divorce rate may drop, at least temporarily: "[F]ewer unhappy couples will risk starting separate households. Furthermore, the housing market meltdown will make it more difficult for them to finance their separations by selling their homes" (Cherlin 2009b).

Young adults' difficulties in finding jobs mean that more of them are cohabiting rather than marrying (Kreider 2010) or are living in their parents' homes. Job losses and housing evictions have meant not only more homeless families but also more extended-family and intergenerational households as older parents and their adult children move in together. Between 2000 and 2012, the proportion of adults ages 25 to 34 living with their parents rose from about one in ten to about one in three (Jacobsen and Mather 2011, Figure 10; Parker 2012). Figure 1.4 gives young adults' answers to survey questions asking what they've done in response to the current recession. On a positive note, many family members—including young adults who've moved back home—may find new ways to interact together with activities that don't cost money. As one middle-class mother said, "We have more time now. We talk. We may not go anywhere but at least we're all home together" (in Stetler 2009).

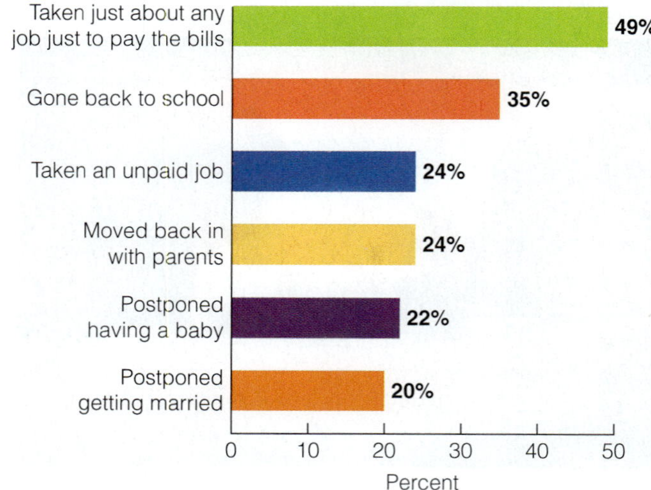

FIGURE 1.4 How economic conditions have affected young adults' lives.

Source: Pew Research Center 2012b, p. 5, survey question 29.

Life chances—the opportunities one has for education and work, whether one can afford to marry, the schools that children attend, and a family's health care—all depend on family economic resources. Money may not buy happiness, but it does afford a myriad of options: sufficient and nutritious food, comfortable residences, better health care, keeping in touch with family and friends through the Internet, education at good universities, vacations, household help, and family counseling. As this discussion on the economy implies, historical events (periods of recession, for instance) impact families.

Historical Periods and Events

In early twentieth-century United States, the shift from an agricultural to an industrial economy brought people from farms to cities and thereby helped to change family household composition as well as attitudes and behaviors. Later, family life was experienced differently by people living through the Great Depression of the 1930s, World War II in the 1940s, the optimistic fifties, the tumultuous sixties, the economically constricted seventies and eighties, the time-crunched nineties, or war and the threat of terrorism throughout the 2000s (Carlson 2009). For example, during the Great Depression, couples delayed marriage and parenthood and had fewer children than they wanted to (Elder 1974). Similar trends are in evidence today. During World War II, married women were encouraged to get defense jobs and place their children in day care. Families in certain nationality groups—Japanese and some Italians—were sent to internment camps and had their property seized even though most were U.S. citizens or long-term residents (Taylor 2002b; Tonelli 2004).

The end of World War II was followed by a spurt in the divorce rate, when hastily contracted wartime marriages proved to be mistakes. After the war, the 1950s saw an expanding economy and a postwar prosperity based on the production of consumer goods. Marriage and childbearing rates rose (Kirmeyer and Hamilton 2011). The GI bill enabled returning soldiers to get a college education, and the less educated could get good jobs in automobile and other factories. In those prosperous times, people could afford to get married young and have larger families. Most white men earned a "family wage" (enough to support a family), and most white children were cared for by stay-at-home mothers. Divorce rates slowed their long-term increase. The expanding economy and government subsidies for housing and education provided a strong foundation for white, middle-class family life (Coontz 1992).

The large baby boom cohort, children born after World War II (1946–1964), has had a powerful impact on American society, giving us the cultural and sexual revolutions of "the sixties" as they moved from adolescence to young adulthood in the Vietnam War era. Baby boomers are now reshaping aging as they enter their senior years. The (white) baby boomers had relatively secure childhoods in both family and economic terms. As one indicator, 86 percent of baby boomers grew up with both parents. The generations that followed have encountered more challenging economic and family environments. Of those born between 1965 and 1982, 79 percent grew up with both parents; 69 percent of those born between 1983 and 2001 did (Carlson 2009, Table 1).

Today an employee is much less likely to earn a family wage. Partly for that reason, more married families comprise two earners. Then, too, the feminist movement opened opportunities for women and changed ideas about women's and men's roles in the family and workplace. As young people prepare for a competitive economic environment, both sexes are delaying marriage and going further in school. In the late 1960s and through the 1970s, marriage rates declined and divorce rates increased dramatically—perhaps in response to a declining job market for working-class men, the increased economic independence of women, and the cultural revolution of the 1960s, which encouraged more individualistic perspectives. These trends, as well as the sexual revolution, contributed to a dramatic rise in nonmarital births. Today many families cope with the effects of the U.S. war against terrorism and deployment in militarized zones abroad (Wadsworth 2010).

It is no surprise that the historical period in which an individual is raised impacts that person's attitudes about family-related issues. For example, 33 percent of Americans who grew up in the 1950s favor allowing same-sex partners to marry legally, compared with 59 percent of Americans who became adults in the twenty-first century (Crowley 2011). Among other demographic characteristics, age affects family behaviors and attitudes.

Demographic Characteristics: Age Structure

A dramatic demographic development has been the increased longevity of our population. Life expectancy in 1900 was forty-seven years, but an American child born in 2008 is expected to live to seventy-eight (U.S. Census Bureau 2012c, Table 107). Furthermore, in 2011 the oldest baby boomers—that noticeably large number and proportion of Americans born between 1946 and 1964—began turning 65 (Jacobsen, Kent, Lee, and Mather 2011).

Among the positive consequences of increased longevity are more years invested in education, longer marriages for those who do not divorce, a longer period during which parents and children interact as adults, and a long retirement during which family activities and other interests may be pursued or second careers launched. More of us will have longer relationships with grandparents or grandchildren; some of us will know our great-grandparents or grandchildren.

At the same time, the increasing numbers of elderly must be cared for by a smaller group of middle-aged and young adults. Furthermore, divorce and remarriage may change family relationships in ways that affect the willingness of adult children to care for their parents (Bergman 2006). Then, too, as the ratio of retired elderly to working-age people grows, so will the problem of funding Social Security and Medicare.

At the other end of the age structure, a declining proportion of children is likely to affect social policy support for families raising children. Fewer children may mean less attention and fewer resources devoted to their needs in a society under pressure to provide care for the elderly. Economic opportunities, resources, and obligations are an important aspect of the American society in which families are embedded.

Demographic Characteristics: Religion

Religious affiliation and practice is a significant influence on family life, ranging from which holidays are celebrated to the placement of family relations into a moral framework. For example, research shows that when children and adolescents have deeper religious connections, they tend to have less premarital sex and to be older when they have their first sexual experience and, as adults, more willing to care for their aging parents (Eggebeen and Dew 2009; Gans, Silverstein, and Lowenstein 2009, but see Stark 2009; Wildeman and Percheski 2009).

The historically dominant religion in the United States has been Protestantism, especially "mainstream" denominations such as Presbyterianism and Methodism. Catholics, Latter-day Saints, and Jews have been traditionally present and visible as well. With relatively recent heightened immigration from the Middle East and Asia, the numbers of Muslims, Hindus, and Buddhists have increased in the United States. The resulting extent of religious diversity is illustrated by the fact that a Midwestern city such as Omaha, Nebraska, has a Buddhist center, a Hindu temple, and a mosque.

Religion offers rituals to mark important family milestones such as birth, coming of age, marriage, and death. Religious affiliation provides families with a sense of community, support in times of crisis, and a set of values that give meaning to life. Membership in religious congregations is associated with age and life cycle; young people who have not been actively religious tend to become so as they marry and have children. Research suggests that "religious couples are less prone to divorce because, on average, they enjoy higher marital satisfaction, face a lower likelihood of domestic violence, and perceive fewer attractive options outside the marriage than their less religious counterparts" (Vaaler, Ellison, and Powers 2009, p. 930). Some studies show that prayer in relationships, especially praying together or for the partner's well-being, is related to greater couple happiness and commitment (Fincham and Beach 2010). What seems to be important overall is not which religion family members belong to, but the fact that family members hold religious beliefs and attend services together (Miller 2000; Vaaler, Ellison, and Powers 2009). Religious beliefs do vary and do affect attitudes, marriages, and families. For instance, Latter-day Saints, evangelical Christians, Jehovah's Witnesses, and Muslims reject homosexuality perhaps more strongly than some others. Conservative Protestant Christians and Latter-day Saints are strongly

AP Photo/Susan Walsh

At Arlington National Cemetery, Buddhist monks—their lives dramatically impacted by the historical period in which they live—escort the coffin of an American soldier killed in Iraq. There has been a Buddhist presence in the United States since at least the nineteenth century, and Buddhist practices have been followed by many Americans of non-Asian backgrounds. But the number of Buddhists more than doubled from 1990 to 2001 as the Asian American population increased through immigration.

opposed to abortion, whereas Muslims and Catholics remain almost evenly split (Pew Forum on Religion & Public Life 2008, p. 135). We'll examine many other examples throughout this text.

U.S. families of religions out of the mainstream face the challenge of maintaining a religiously proper family life in the context of a culture that not only does not share their beliefs but also may be inclined to stereotype them (Hirji 2012). Dating, marital choice, child raising, dress, and marital decision making can be religious issues, according to which the morally correct way diverges from mainstream American culture. Muslim (and occasionally other immigrant) families have the added burden of facing suspicion and hostility in the wake of 9/11. For many Americans, finding a balance between participating in the larger society and preserving religious values is a challenge in a society characterized by religious freedom rather than religious establishment.

Phil Schermeister/Corbis

Many social factors condition people's options and choices. One such factor is an individual's place within our culturally diverse society. These rural Navajo reservation children are learning to weave baskets to sell to tourists. Even within a race/ethnic group, however, families and individuals may differ in the degree to which they retain their original culture. Many Navajo live in urban settings off the reservation or go back and forth between the reservation and towns or cities. Less than 1 percent of the population is American Indian or Alaska Native.

Demographic Characteristics: Race and Ethnicity

Race is a social construction reflecting how Americans view different social groups. "Race is a real cultural, political, and economic concept, but it's not biological," says biology professor Alan Templeton ("Genetically, Race Doesn't Exist" 2003, p. 4). The term *race* implies a biologically distinct group, but scientific thinking rejects the idea that there are separate races clearly distinguished by biological markers. Features such as skin color that Americans and many others use to place someone in a racial group are superficial, genetically speaking.

In this text, we use the race/ethnic categories formally adopted by the U.S. government because we draw on statistics collected by the U.S. Census Bureau and other agencies. In the census, racial identity is based on self-reporting. Beginning in 2000, individuals were permitted to indicate more than one race. Many Hispanics see themselves as "not white" or "not black." However, the Census Bureau defines *Hispanic* or *Latino* as an ethnic identity, not a race. Hispanics may be of any race. **Ethnicity** has no biological connotations; instead, it refers to cultural distinctions often based in language, religion, foodways, and history. Because race and ethnicity exist simultaneously in each individual and are generally difficult to separate in practice, we, your authors, often use the term *race/ethnicity*. For census purposes, there are two major categories of ethnicity: Hispanic and nonHispanic. This situation means that data on ethnicities other than Hispanic—Arabs or Portuguese, for example—come from surveys other than those done by the Census Bureau.

Social scientists and policy makers also may group African Americans, Hispanics, American Indians, Asians, and other non-whites into a category termed **minority group** or **minority**. This term conveys the idea that persons in non-white race/ethnic categories experience some disadvantage, exclusion, or discrimination in American society as compared to the politically and culturally dominant nonHispanic white group. *Minority* in a sociological context does not have its everyday meaning of less than 50 percent. Regardless of size, if a group is distinguishable and in some way disadvantaged within a society, sociologists consider it a minority group. The term can be controversial, viewed by some as demeaning and as ignoring differences among groups and variation in the self-identities of individuals (Gonzalez 2006a; Wilkinson 2000). We, your authors, will avoid using it other than when speaking of numerical differences or in reporting Census Bureau data that is so labeled.

Race/Ethnic Diversity Of particular interest is the increasing race/ethnic diversity of U.S. families. About one-fifth of U.S. families speak a language other than or in addition to English at home. Approximately 62 percent of them speak Spanish, with the remaining 48 percent speaking any one of forty or more other languages (U.S. Census Bureau 2012c, Table 53). Four Hispanic surnames—Garcia, Rodriguez, Martinez, and

A Closer Look at Diversity

Family Ties and Immigration

There is more racial and ethnic diversity among American families than ever before, and much of this diversity results from immigration. Depending on calculations, the U.S. foreign-born population grew by between 600,000 and 1.5 million between 2009 and 2010 (Cohn 2012b) and now constitutes 13 percent of the U.S. population (Patten 2012).

The United States admits approximately 1 million legal immigrants each year. Asia, Latin America, and the Caribbean—not Europe—are now the major sending regions, with the highest percentage of new immigrants now arriving from Asia (Pew Research Center 2012c). In addition to legal immigrants, approximately 11.9 million undocumented immigrants (not legal residents) reside in the United States, with approximately 8.3 million in the U.S. labor force (Passel and Cohn 2009, p. 2). Substantial numbers are from countries such as Canada, Poland, and Ireland, but the majority are from Mexico, Central America, and the Caribbean (Martin and Midgley 2006). A recent Gallup poll found that 66 percent of respondents in a nationwide random sample said they thought immigration is a good thing for the United States today. At the same time, 53 million felt that controlling U.S. borders to halt the flow of illegal immigrants into the country was extremely important (Gallup Poll 2012a).

Some immigrants, particularly recent Asian immigrants, are highly educated professionals (Pew Research Center 2012c). However, many immigrants leave a poorer country for a richer one in hopes of bettering their families' economic situations. As immigrants establish themselves, they send for relatives—in fact, the majority of legal immigrants enter the United States through family sponsorship (Martin and Midgley 2006). As a result, more Americans maintain **transnational families** whose members bridge and maintain relationships across national borders. They may experience back-and-forth changes of residence, family visits, money transfers, the placement of children with relatives in the other country, or the search for a marriage partner in the home country.

Moreover, many immigrant families are **binational**, with nuclear-family members having different legal statuses. One partner or spouse may be a legal resident, the other not. Children born in the United States are automatically citizens, even though one or both parents may be undocumented (illegal) residents. In fact, almost one-third of all immigrant children come from such mixed-status families (Fortuny et al. 2009). Problematically, the undocumented or unauthorized immigrant parents of many native-born American children in binational families increasingly face deportation (Golash-Boza 2012; Yoshikawa and Suarez-Orozco 2012). Of serious concern are the estimated 3.1 million children who, legal citizens themselves, have seen their undocumented parents deported (Preston 2007).

Immigrant families pay payroll, Social Security, property, and sales taxes even though some have only limited access to government benefits. Both costs and benefits are not evenly distributed. Most immigrant family tax dollars go to the federal government, whereas the costs of immigrants' schooling or emergency health care are largely paid by local governments (Martin and Midgley 2006). Transnational and binational families are explored in several places throughout this text.

Critical Thinking

What are some strengths exhibited by immigrant families? What are some challenges they face? At the society-wide level, what benefits does recent increased immigration offer the United States? What challenges does it bring?

Hernandez—rank among the fifteen most common in the United States (U.S. Census Bureau 2012d). The most recent national population statistics show that in 2009, the nation was 65 percent nonHispanic white, 12 percent black, and 4.5 percent Asian. In 2012, for the first time in our nation's history, nonHispanic white births accounted for 49.6 percent of all births and hence were no longer the majority (Tavernise 2012). Over the past fifty years, relatively low fertility rates among nonHispanic whites (compared to higher rates among racial and ethnic minorities) and immigration combined to "put the United States on a new demographic path" (Mather 2009; U.S. Census Bureau 2011c, Table 7). "A Closer Look at Diversity: Family Ties and Immigration" discusses immigration further. Hispanics are now 15.8 percent of the population, surpassing blacks as the largest race/ethnic group after nonHispanic whites. Hispanics and Asians are the fastest-growing segment of the population (U.S. Census Bureau 2012c, Table 6).

The 2011 child population estimate is more diverse than our adult population: 54 percent are nonHispanic white, 23 percent Hispanic, 15 percent black, and 4 percent Asian. Four percent of children are American Indian, Alaska Native, Native Hawaiian, or of more than one race (U.S. Census Bureau 2012b, Table C3). Race/ethnic minorities comprise more than one-third of the U.S. population and 46 percent of the child population. By 2042 they are expected to make up half of the population (Mather and Pollard 2009).

Note that no category system can truly capture cultural identity. As race/ethnic categories become more fluid and as the identity choices of individuals with a mixed heritage vary, race/ethnic identity may come to be seen as voluntary—"optional" rather than automatic, especially for young adults (Saulny 2011a). A further point is that considerable diversity exists within major race/ethnic groupings. There are Caribbean and African blacks, for example, as well as those descended from U.S. slave populations. There are Chinese, Japanese, Korean, Indian, and other Asians. There are Salvadoran, Nicaraguan, Costa Rican, Chilean, and other Hispanics. Within-group diversity makes generalizations about race/ethnic groups somewhat questionable. For instance, "Hispanic" or "Latino" categories are "useful for charting broad demographic changes in the United States... [but they] conceal variation in the family characteristics of Latino groups [Cubans and Mexicans, for example] whose differences are often greater than the overall differences between Latinos and non-Latinos" (Baca Zinn and Wells 2007, pp. 422, 424). Moreover, there are areas of social life in which race/ethnic differences seem minor—if they exist at all. Little difference in family patterns is apparent between blacks and whites serving in the military, for example (Lundquist 2004).

Race/Ethnic Stratification Meanwhile, race/ethnic stratification persists. For one thing, a history of racial discrimination affects wealth stratification today. As one example, the GI bill, mentioned earlier, was available to returning black soldiers as well as to whites, but many colleges did not accept African Americans, and one had to be accepted into a college program in order to qualify for the GI bill's college assistance. Likewise, the GI bill did not officially discriminate against African Americans' desire for home ownership, but the bill was of little use to them because of the many restrictive covenants against black residents and because real estate agents often did not show listed properties to black customers (Reed and Strum 2008). On average, the income and wealth of Asian and of nonHispanic white households are much higher and poverty rates significantly lower than those of African American, Hispanic, or Native American households.

The experiences we have are shaped by the **social class** in which we reside, as well as our race and gender. Social theorist Pierre Bourdieu (1977[1972]) refers to *habitus* as "one's experience and perception of the social world." The perceptions we form via those experiences impact the ways in which we interact with the world, including our families. "A child develops a set of bodily and mental procedures that frames perceptions, appreciations, and actions vis-à-vis familial and intimate external environments" (Gerbrandt 2007, p. 57). In other words, the class position and racial characteristics of our family impact our childhood experiences, which will impact the decisions we make and how we experience the world as we mature into adulthood, as well as the advantages or disadvantages that we encounter.

Children born to interracial and inter-ethnic unions further add to America's diversity. Although the growth in race/ethnic intermarriage rates for Asians and Hispanics has declined somewhat since the 1990s, their numbers continued to rise. Interracial and inter-ethnic marriage and cohabitation rates involving African Americans have continued to increase significantly (Qian and Lichter 2011). As a result, the proportion of interracial children is significant (U.S. Census Bureau 2012c, Tables 10 and 13).

We'll see throughout this text that social class is often more important than race/ethnicity in shaping people's families. Yet, race/ethnic heritage—the family's place within our culturally diverse society—affects preferences, options, and decisions, not to mention opportunities. For instance, ethnicity can influence options and decisions about whether or when to marry, where the family will live, employment, wives' work preferences, preferred parenting practices, caring for aging parents, and so on. As the U.S. population changes, policy makers need to recognize the complexity and diversity of the growing minority population. We return to issues of racial and ethnic diversity throughout this textbook.

This text assumes that people need to understand themselves and their problems in the context of the larger society. Individuals' choices depend largely on the alternatives that exist in their social environment and on cultural values and attitudes toward those alternatives. Moreover, if people are to shape the kinds of families they want, they must not limit their attention just to their own relationships and families. This is a principal reason why we explore social policy issues throughout this text.

Family Policy: A Family Impact Lens

Family policy involves all the procedures, regulations, attitudes, and goals of programs and agencies, workplace, educational institutions, and government that affect families. *Family policy* encompasses policies that directly address the main functions of families—family formation, partner relationships, economic support, childrearing, adoption, child care, family violence, juvenile crime, and long-term care. Issues regarding same-sex couples' separation, divorce, and child custody, as well as determining the legal status for lesbian parents who used ART, are all social policy matters (Hare and Skinner 2008; Oswald and Kuvalanka 2008). Whether the federal government should prohibit farm children under age 16 from driving tractors or working other dangerous agricultural equipment is a matter of family policy—and hotly debated in some states ("Parents Defend...." 2012). The federal government and states have developed programs to encourage and support marriage, to encourage father involvement in fragile families, to discourage teen sexual activity, and to move single mothers from welfare to work.

Family policy expert Karen Bogenschneider urges that political decisions regarding families be scrutinized through a family policy or a **family impact lens** (Bogenschneider et al. 2012), by which we ask how the policy in question impacts families. As one example, workplace and government maternity leave policies influence new mothers' employment patterns (Laughlin 2011). Another example: Of concern have been the many young adults whose undocumented parents brought them to the United States when they were children and who are therefore not legal residents but have no connections in their country of origin (Gonzalez 2006b). In June 2012, President Obama issued an executive order allowing these individuals to stay in the United States without fear of deportation (but also without legal citizenship status) and to be able to work. The order is estimated to affect some 800,000 youth and their families who presumably will experience diminished stress associated with the fear of being abruptly separated (Preston and Cushman Jr. 2012).

Looking through the family impact lens reveals that "laws place some families in the margins of society while privileging others" (Henderson 2008, p. 983). Federal family policy has privileged heterosexual marriages by defining same-sex unions as "not-marriage," a situation that negatively affects many children in LGBT families who may not have access to a non–adoptive parent's employer-provided health care benefits (Movement Advancement Project 2011). As another example, "racial profiling, mandatory minimum sentences, and especially the disparities in drug laws [which more heavily penalize crimes involving drugs typically used by blacks] have had a dramatic effect on the incarceration rates of young male [family members], especially in urban inner-city neighborhoods" (Clayton and Moore 2003, p. 86).

AP Photo/East Valley Tribune, Heidi Huber

Two volunteers at the American Muslim Women's Association work on a craft project to benefit poorer immigrants and refugees. The Arab American population is slightly more than 1.5 million. Contrary to what many think, 65 percent of Arab Americans are Christian, and most are second- or third-generation American citizens. Arabs who have immigrated since the 1950s are likely to be Muslim. Employing a *family impact policy lens,* media scholar Professor Jack Shaheen examined American movies depicting Arabs or Arab Americans and found that generally they presented negative stereotypes of "barbarism" and "buffoonery" (Beitin, Allen, and Bekheet 2010). Some modern young Muslim women have recently adopted the head scarf to express an intensified identification with Islam in the context of experiences of discrimination or challenges to their religious community.

Given the social and political diversity of American society, all parents or political actors are unlikely to agree on the best courses of action. Americans are not only not in agreement on the role government should play vis-à-vis families but also divided on what "family" means. Indeed, the diversity of family lifestyles in the United States makes it extremely difficult to develop family policies that would satisfy all, or even most, of us. Making well-informed family decisions can mean getting involved in national and local political debates and campaigns. One's role as family member, as much as one's role as citizen, has come to require participation in society-wide decisions to create a desirable context for family life and family choices.

The Freedom and Pressures of Choosing

Social factors influence people's personal choices in three ways. First, it is usually easier to make the common choice. In the 1950s and early 1960s, when people tended to marry earlier than they do now, it felt awkward to remain unmarried past one's mid-twenties. Now, staying single longer is a more comfortable choice. Similarly, when divorce and nonmarital parenthood were highly stigmatized, it was less common to make these decisions than it is today. As another example, contemporary families usually include fewer children than historical families did, making the choice to raise a large family more difficult than in the past (Zernike 2009).

A second way that social factors can influence personal choices is by expanding people's options. For example, the availability of effective contraceptives makes limiting one's family size easier than in the past, and it enables deferral of marriage with less risk that a sexual relationship will lead to pregnancy. Then, too, as we have seen, new forms of reproductive technology provide unprecedented options for becoming a parent.

However, social factors can also limit people's options. For example, American society has never allowed polygamy (more than one spouse) as a legal option. Those who would like to form plural marriages risk prosecution (Janofsky 2001). Until the 1967 *Loving v. Virginia* U.S. Supreme Court decision, a number of states prohibited racial intermarriage. As we will discuss in Chapter 7, the possibility of same-sex marriage is currently being contested in various courts throughout the

United States, and outcomes will either expand or limit couples' options. More broadly, economic changes of the last thirty-five years, which make well-paid employment more problematic, have limited some individuals' marital options (Sassler and Goldscheider 2004).

As families have become less rigidly structured, people have made fewer choices "once and for all." Of course, previous decisions do have consequences, and they represent commitments that limit later choices. Nevertheless, many people reexamine their decisions about family—and face new choices—throughout the course of their lives. Thus, choice is an important emphasis of this book.

The best decisions are informed ones. It helps to know something about all the alternatives; it also helps to know what kinds of social pressures affect our decisions. As we'll see, people are influenced by the beliefs and values of their society. There are **structural constraints**, economic and social forces, that limit personal choices. In a very real way, we and our personal decisions and attitudes are products of our environment.

But in just as real a way, people can influence society. Individuals create social change by continually offering new insights to their groups. Sometimes social change occurs because of conversation with others. Sometimes it requires becoming active in organizations that address issues such as abortion, racial equality, immigrant rights, gay rights, or stepfamily supports, for example. Sometimes influencing society involves many people's living their lives according to their values, even when these differ from more generally accepted group or cultural norms.

We can apply this view to the phenomenon of "living together." Fifty years ago, it was widely believed that cohabiting couples were immoral. But in the 1970s, some college students openly challenged university restrictions on cohabitation, and subsequently many more people than before—students and nonstudents, young and old—chose to live together. As cohabitation rates increased, societal attitudes became more favorable. Over time, cohabitation became "mainstream" (Smock and Gupta 2002). Although some religions and individuals continue to object to living together outside marriage, a majority of Americans today agree that a cohabiting couple who have lived together for five years or more is just as committed as a married couple (Gallup Poll 2012b). It is now significantly easier for people to choose this option. We are influenced by the society around us, but we are also free to influence it, which we do every time we make a choice.

Making Informed Decisions

By taking a course in marriage and the family, you may become more aware of your alternatives and how a decision may be related to subsequent options and choices. All people make choices, even when they are not conscious of it. Sometimes we "slide" into a situation rather than make a conscious decision. We can think of these two ways of dealing with choices as **deciding versus sliding** (Stanley 2009) (see Figure 1.5). A good way to make choices is to be well informed—that is, to do so knowledgeably.

FIGURES 1.5 The process of informed decision making: deciding versus sliding.

An important component of informed decision making involves recognizing as many options as possible (Meyer 2007). In part, this text is designed to help you do that. A second component in making well-informed decisions involves recognizing the social pressures that can influence our choices. Some of these pressures are economic, whereas others relate to cultural norms. Sometimes people decide that they agree with socially accepted or prescribed behavior. They concur in the teachings of their religion, for example. Other times, people decide that they strongly disagree with socially prescribed beliefs, values, and standards. Whether they agree with such standards or not, once people recognize the force of social pressures, they can choose whether to act in accordance with them.

A third aspect of deciding about, rather than sliding into, a situation involves considering the consequences of each alternative rather than just gravitating toward the one that initially seems easier or most attractive. For example, we've seen that as a result of the recession, a growing number of young adults live with their parents. Someone deciding whether to move back into his or her parents' home may want to list the consequences. In the positive column, moving home might mean being able to help with family finances as well as save money that would have otherwise gone toward separate rent. In the negative column, returning to one's parental home could result in more cramped family space and increased family conflict. Listing positive and negative consequences of alternatives helps one see the larger picture and thus make a more informed decision.

Part of this process might also require being aware of research findings concerning your options. It might help to know, for instance, that the well-respected Pew Research Center surveyed young adults who'd moved back home and found that one quarter said the situation was bad for their relationship with their parents. Another quarter said moving home was good for their relationship, and about half said moving home made no difference (Parker 2012).

If we're going to decide, not slide, we also need to be aware of our values and understand how they relate to each of our options (Meyer 2007). We'll note here that contradictory sets of values exist in American society. For instance, standards regarding nonmarital sex range from abstinence to recreational sex. Contradictory values can cause people to feel ambivalent about what they want for themselves. Clarifying one's values involves cutting through this ambivalence in order to decide which of several standards are more strongly valued.

It is important to respect the so-called gut factor—the emotional dimension of decision making. Besides rationally considering alternatives, people have subjective (often almost visceral) feelings about what for them is right or wrong, good or bad. Respecting one's feelings is an important part of making the right decision. Following one's feelings can mean grounding one's decisions in a religious or spiritual tradition or in one's cultural heritage, for these have a great deal of emotional power and often represent deep commitments.

Two other important components of decision making are considering how the decision will affect your future and thinking about how it is likely to affect other people. Underlying this discussion is the assumption that individuals cannot have everything. People cannot simultaneously have the relative freedom of a childfree union and the gratification that often accompanies parenthood, for instance. Every time people make an important decision or commitment, they rule out alternatives—for the time being and perhaps permanently.

It is true, however, that people can focus on some goals and values during one part of their lives, then turn their attention to different ones at other times. Adulthood is a time with potential for continued personal development, growth, and change. In a family setting, development and change involve more than one individual. Multiple life courses must be coordinated, and the values and choices of other members of the family will be affected if one member changes. Moreover, life in American families reflects a cultural tension between family solidarity and individual freedom (Amato 2004; Cherlin 2009a).

Families of Individuals

Americans place a high value on family. It is hardly surprising that a vast majority of Americans report family is extremely important to them (Carroll 2007b, "Marriage" 2008). Why?

Families As a Place to Belong

Families create a place to belong, serving as a repository or archive of family memories and traditions (Cieraad 2006). **Family identity**—ideas and feelings about the uniqueness and value of one's family unit—emerge via traditions and rituals: family dinnertime, birthday and holiday celebrations, vacation trips, and perhaps family hobbies like working together in the garden. Family identities typically include members' cultural heritage. For example, all the children in one family may be given Irish, Hispanic, Asian Indian, or Russian names.

Families provide a setting for the development of an individual's **self-concept**—basic feelings people have about themselves, their abilities, characteristics, and worth. Arising initially in a family setting, self-concept and identity are influenced by significant figures in a young child's life, particularly those in the parent role, together with siblings and other relatives.

How family members and others interact with and respond to us continues to impact self-concept and identity throughout life (Cooley 1902, 1909; Mead 1934; Yeung and Martin 2003). A child who is loved comes to think he or she is a valuable and loving person. A child who is given some tasks and encouraged to do things comes to think of him- or herself as competent.

Hill Street Studios/Blend Images/Getty Images

Families are comprised of individuals, each seeking self-fulfillment and a unique identity, but individuals can find a place to learn and express togetherness, stability, and loyalty within the family. Families also perform the important function associated with providing emotional support—they give us a place to belong. Events, rituals, and histories become intrinsic parts of each individual.

Early childhood also marks the onset of learning social roles. Children connect certain behaviors to the different roles of mother, father, grandmother, grandfather, sister, brother, and so on. Much of young children's play consists of imitating these roles. Role-taking, or playing out the expected behavior associated with a social position, is how children begin to learn behavior appropriate to the roles they may play in adult life. Behavior and attitudes associated with roles become internalized, or incorporated into the self. Meanwhile, expressing our individuality within the context of a family requires us to negotiate innumerable day-to-day issues. How much privacy can each person have at home? What family activities should be scheduled, how often, and when? What outside friendships and activities can a family member sustain?

Familistic (Communal) Values and Individualistic (Self-Fulfillment) Values

Familistic values such as family togetherness, stability, and loyalty focus on the family as a whole. They are *communal* or *collective* values; that is, they emphasize the needs, goals, and identity of the group. Many of us have an image of the ideal family in which members spend considerable time together enjoying one another's company. Furthermore, the family is a major source of stability. We believe that the family is the group most deserving of our loyalty (Connor 2007). Those of us who marry vow publicly to stay with our partners as long as we live. We expect our partners, parents, children, and even our more distant relatives to remain loyal to the family unit.

But just as family values permeate American society, so do **individualistic** (*self-fulfillment*) **values**. These values encourage people to think in terms of personal happiness and goals and the development of a distinct individual identity. An individualistic orientation gives more weight to the expression of individual preferences and the maximization of individual talents and options.

The contradictory pull of both familistic and individualistic values creates tension in society (Amato 2004, Cherlin 2009a)—and tension within ourselves that we must resolve. "It is within the family... that the paradox of continuity and change, the problem of balancing individuality and allegiance, is most immediate" (Bengston, Biblarz, and Roberts 2007, p. 323).

American society has never had a remarkably strong tradition of familism, the virtual sacrifice of individual family members' needs and goals for the sake of the larger kin group (Sirjamaki 1948; Lugo Steidel and Contreras 2003). Our national cultural heritage prizes individuality, individual rights, and personal freedom. But, on the other hand, an overly individualistic orientation puts stress on relationships when there is little emphasis on contributing to other family members' happiness or postponing personal satisfactions in order to attain family goals.

People As Individuals and Family Members

The changing shape of the family has meant that family lives have become less predictable than they were in the mid-twentieth century. The course of family living results in large part from the decisions two adults make, moving in their own ways and at their own paces through their lives. A consequence of ongoing developmental change in individuals is that the union or family may be put at risk. If one or more individuals change considerably over time, they may grow apart instead of together. A challenge for contemporary relationships is to integrate divergent personal change into the relationship while nurturing any children involved.

How can people make it through their own and each other's changes and stay connected as a family? Two guidelines may be helpful. The first is for family members to take responsibility for their own past choices and decisions rather than blaming previous "mistakes" on others in the family. In addition, it helps to recognize that a changing family situation—for example, a college graduate's returning home to live with parents, a partner's deciding to quit his or her job and attend graduate school, a preteen's getting used to a new stepparent—may mean that family living will be difficult for awhile. Family relationships need to be flexible enough to allow for each person's individual changes—to allow family members some degree of freedom. At the same time, it's good to remember the benefits of family living and the commitment necessary to sustain it. Individual happiness and family commitment are not inevitably in conflict; research shows that committed family bonds have significant positive impacts on individual well-being (Waite and Gallagher 2000; Wilcox et al. 2011a).

> On the one hand, people value the freedom to leave unhappy unions, correct earlier mistakes, and find greater happiness with new partners. On the other hand, people are concerned about social stability, tradition, and the overall impact of high levels of marital instability on the wellbeing of children. The clash between these two concerns reflects a fundamental contradiction within marriage itself; that is, marriage is designed to promote both institutional and personal goals.... To make marriages with children work effectively, it is necessary for spouses to find the right balance between institutional and individual elements, between obligations to others and obligations to the self. (Amato 2004, p. 962)

Throughout this text we will continue to explore the tension between individualistic and familistic values and discuss creative ways that partners and families can alter committed, ongoing relationships in order to meet their changing needs.

Marriages and Families: Four Themes

In this chapter we have defined the term *family* and discussed diversity and decision making in the context of family living. We can now state explicitly the four themes of this text.

1. Personal decisions must be made throughout the life course. Decision making is a trade-off; once we choose an option, we discard alternatives. No one can have everything. Thus, the best way to make choices is knowledgeably.
2. People are influenced by the society around them. Cultural beliefs and values influence our attitudes and decisions. Societal or structural conditions can limit or expand our options.
3. We live in a society characterized by considerable change, including increased ethnic, economic, and family diversity; by tension between familistic and individualistic values; by decreased marital and family permanence; and by increased political and policy concern about the needs of children and families. This dynamic situation can make personal decision making more challenging than in the past—and more important.
4. Personal decision making feeds into society and changes it. We affect our social environment every time we make a choice. Making family decisions can also mean choosing to become politically involved in order to effect family-related social change. Making family choices consciously, according to our values, gives our family lives greater integrity.

We will revisit these topics throughout this text, and we, your authors, believe that they provide a strong foundation for the subject of marriages and families.

Summary

- We, your authors, define family as any sexually expressive, parent-child, or other kin relationship in which people—usually related by ancestry, marriage, or adoption—(1) form an economic or otherwise practical unit and care for any children or other dependents, (2) consider their identity to be significantly attached to the group, and (3) commit to maintaining that group over time.
- Social scientists usually list three major functions filled by today's families: raising children responsibly, providing members with economic and other practical support, and offering emotional security.
- With relaxed institutional control, family diversity has progressed to the point that there is no typical family form today.
- Whether we are in an era of "family decline" or "family change" is a matter of debate.
- Families exist in a social context that affects many aspects of family life. Families are affected by ever-new biological and communication technologies, economic conditions, historical periods, and demographic characteristics such as age, religion, race, and ethnicity.
- Marriages and families are comprised of individuals. Our culture values both families and individuals.

Families provide members a place to belong and help ground identity development. Meanwhile, finding personal freedom within families is an ongoing, negotiated process.

- People make choices, either by consciously deciding or by sliding into situations; the best decisions are informed ones consciously made. Our decisions are limited by social structure, and at the same time they are causes for change in that structure.
- Change and development continue throughout adult life. Because adults change, relationships, marriages, and families are far from static.

Questions for Review and Reflection

1. Without looking at ours, write your definition of family and then compare it to ours. How are the two similar? How are they different? Does your definition have some advantages over ours?
2. Why is the family a major social institution? Does your family fulfill each of the family functions identified in the text? How?
3. What important changes in family patterns do you see today? Do you see positive changes, negative changes, or both? What do they mean for families, in your opinion?
4. What are some examples of a personal or family problem that is at least partly a result of problems in the society? Describe one specific social context of family life as presented in the text. Does what you read match what you see in everyday life?
5. **Policy Question**. What, if any, are some changes in law and social policy that you would like to see put in place to enhance family life?

Key Terms

binational 20
deciding versus sliding 23
ethnicity 19
extended family 6
familistic (communal) values 25
family 4
family-change perspective 10
family-decline perspective 10
family identity 24
family impact lens 22
family policy 21
family structure 6
household 6
individualistic (self-fulfillment) values 25
life chances 17
minority 19
minority group 19
nuclear family 6
postmodern family 7
race 19
self-concept 24
social class 21
social institution 9
sociological imagination 12
structural constraints 23
transnational families 20

2

Exploring Relationships and Families

Richard G. Bingham II/Alamy

Learning Objectives

1. Understand how scientific knowledge differs from that gained through personal experience.
2. Discuss various theoretical perspectives on families, noting their main contributions and critiques.
3. Describe why rules for research are essential to science.
4. Discuss several data gathering techniques.
5. Describe some ethical principles associated with scientific research.

- "What's happening to the family today?"
- "What's a good family?"
- "How do I make that happen?"

In Chapter 1 we said that the best decisions are informed ones made consciously. Throughout this textbook we, your authors, point out many facts that are supported by evidence about relationships and families. We base what we write on published information that we trust is accurate. We provide citations to the sources of our information, then give the complete reference that goes with each citation in the reference section at the back of this book. If you wish, you can find the article or book that we've cited, then read it for yourself and see whether you agree with our interpretation.

Where does the information in the article or book come from? Mainly, it results from social scientists' use of theoretical perspectives and research methods designed to explore family life. This chapter invites you into the world of social science so that you can understand and share this way of examining family life.

First we'll discuss how science differs from simply having an opinion or a strongly held belief. Next we will examine various theoretical perspectives used by social scientists. After that we'll explore some important things to know about scientific research, then discuss various ways that family scientists gather data. Throughout, we need to keep in mind that studying a phenomenon as close to our hearts as family life can be a knotty challenge.

Science: Transcending Personal Experience

The great variation in family forms and the variety of social settings for family life mean that few of us can rely only on firsthand experience when studying families. Although we "know" about the family because we have lived in one, the beliefs we have about the family based on personal experience may not tell the whole story. We may also be misled by media images and common sense—what "everybody knows." What "everybody knows" can misrepresent the facts.

The Blinders of Personal Experience

Although personal experience provides us with information, it may also act as blinders. We may assume that our own family is normal or typical. If you grew up in a large family, for example, in which a grandparent or an aunt or uncle shared your home, you probably assumed (for a short time at least) that everyone had a big family. Perceptions like this are usually outgrown at an early age. However, some family styles may be taken for granted or assumed to be universal when they are not.

In looking at family customs around the world, we can easily see the error of assuming that all marriage and family practices are like our own. Common American assumptions about family life not only fail to hold true in other places but also frequently don't even describe our own society well. Lesbian or gay male families; black, Latino, and Asian families; Jewish, Protestant, Catholic, Latter-day Saints (Mormon), Islamic, Buddhist, and nonreligious families; upper-class, middle-class, and lower-class families; urban and rural families—all represent differences in family lifestyle.

Nevertheless, the tendency to use only our experiential knowledge as a yardstick for measuring things is strong. Therefore, science has developed norms for transcending the blinders of personal experience. The central aim of scientific investigation is to find out what

Does this family look like yours? If "yes" or "somewhat yes," in what ways do these folks look like your family? If not, how does your family look? Researchers work to get actual facts about families, not stereotyped images. Some American families do look like this one, but—as discussed in Chapter 1—they are not the numerical majority.

A Closer Look at Diversity

Studying Families and Ethnicity

As men and women from diverse race/ethnic backgrounds came into the field of family studies, they pointed out how limited and biased our theoretical and research perspectives had been. For many years, research on African Americans focused almost exclusively on poor, single-parent households in the inner city and ignored middle-class blacks (Hymowitz 2006). Overlooking many other topics, research on Latinos often investigated Mexican immigrants' assumed "patriarchal" culture (Baca Zinn and Wells 2007; Taylor 2007).

Following the negative reaction to the earlier, limited portrayal of race/ethnic family differences, researchers began to report on the strengths of families of color, multiracial families, and multi-ethnic families, pointing to strong extended-family support, more egalitarian spousal relationships, and class, regional, and rural/urban diversity. For example, a substantial proportion of African American single-mother households contain other adults who take part in raising the children (Taylor 2007).

As another example, Annette Lareau (2003a) points to the rich family life of working- and lower-class children whose parents are less focused on educational and achievement goals and activities and hence have considerable time to spend with relatives. Research on extended–family ties illuminates the great amount of instrumental help that Hispanic extended families provide to their members. This means that workplace policies that presume only nuclear–family members need the flexibility to provide family care does not take into account the real lives of Hispanic families (Sarkisian, Gerena, and Gerstel 2006).

As yet another example, social theorist Professor Edward W. Said, in his book *Orientalism* (1979), noted that European, then American, scholars have long presented people from the Middle East in ways which stereotyped Arabs as exotic, mysterious, and dangerous. Following in Said's footsteps, media scholar Professor Jack Shaheen examined more than 1,000 American studio films depicting Arabs or Arab Americans. He found an unchanging and rigid stereotype that presents an image of "barbarism" and "buffoonery." In addition, it may be that much of the scholarly marriage and family literature in the United States focusing on Arab families tends to view this ethnically and religiously diverse group as monolithic and through the lens of Euro-American superiority (Beitin, Allen, and Bekheet 2010).

Furthermore, research now using a comparative approach has shown us that the same family phenomenon may have different outcomes in different racial/ethnic settings. For example, communication processes vary by family types, with multiracial and multi-ethnic families developing unique forms of communication that assist in maintaining solidarity among the family members (Soliz, Thorson, and Rittenour 2009, p. 829).

Today's research on family and ethnicity tends to be more complex and sophisticated than in the past. Concern about family fragility and individual disorganization is balanced by recognition of diversity and of community and family strengths. Multiple influences on race/ethnic families are acknowledged: (1) mainstream culture, (2) ethnic settings, and (3) the negative impact of disadvantaged neighborhoods or family circumstances that can produce behaviors that are inappropriately viewed as a "minority culture" (S. Hill 2004). Structural influences—that is, economic opportunity—are seen as a powerful influence on family relations and behavior. The role of "agency," or the initiative of families, is recognized: "What happens on a daily basis in family relations and domestic settings also constructs families.... Families should be seen as settings in which people are agents and actors, coping with, adapting to, and changing social structures to meet their needs" (Baca Zinn and Wells 2007, p. 426; see also S. Hill 2004).

Critical Thinking

Does your family heritage or your observation of families make you think of family patterns that seem different from people's assumptions about families? How might your insights or observations help researchers learn more about families in a variety of family settings?

is actually going on, as opposed to what we assume is happening. **Science** can be defined as "a logical system that bases knowledge on... systematic observation" and on *empirical evidence*—facts we verify with our senses (Macionis 2006, p. 15). The central purpose of the *scientific method* is to overcome researchers' blinders, or biases. ("A Closer Look at Diversity: Studying Families and Ethnicity," discusses race/ethnic bias in research.) Scientific researchers are ever cognizant of the need to gather data that accurately correspond with reality. "We must be dedicated to finding the truth *as it is* rather than as we think it *should be*" (Macionis 2006, p. 18, italics in original).

Scientific Norms To transcend personal biases, scientists follow certain norms (Babbie 2007; Merton 1973 [1942]). Of course, researchers are expected to be honest, never fabricating results. Scientists are expected to publish their research. Publishers are required to evaluate submissions only on merit, never taking into account the researcher's social characteristics, such as race/ethnicity, gender, socioeconomic class, religion, or institutional affiliation. To accomplish this, publishers have reviewers, or "referees," who evaluate submissions "blind" (without knowing the name or anything else about the researcher submitting the article for publication).

Publishing allows research results to be reviewed and critiqued by others. In this way science becomes *cumulative*: Findings from various research projects build on one another. Over time, a particular conclusion will be seen to have more evidence behind it than others (e.g., Amato 2012; Carey 2012; Marks 2012). It is well established, for example, that marriage carries many benefits for the individual, the couple, and their children (Waite and Gallagher 2000; Wilcox et al. 2011a, b). It is also well established that the arrival of children is associated with at least an initial decline in marital happiness, probably from less leisure time as well as the challenges of child raising and concomitant modifications to the couple's relationship (Clayton and Perry-Jenkins 2008).

This last is a conclusion that is not so pleasing to hear, but an important scientific norm involves having *objectivity*: "The ideal of objective inquiry is to let the facts speak for themselves and not be colored by the personal values and biases of the researcher" (Macionis 2006, p. 18). To do this, scientists use rigorous methods that follow a carefully designed research plan. We return to a discussion of scientific methods later in this chapter.

"In reality, of course, total neutrality is impossible for anyone" (Macionis 2006, p. 18). However, following standard research practices and submitting the results to review by other scientists is likely in the long run to correct the biases of individual researchers. At the same time, there are many visions of the family and relationships; what an observer reads into the data depends partly on his or her theoretical perspective.

Theoretical Perspectives on the Family

Theoretical perspectives are ways of viewing reality. As a tool of analysis they are equivalent to lenses through which observers view, organize, and then interpret what they see. A theoretical perspective leads family researchers to identify those aspects of families and relationships that interest them and suggests possible explanations for why patterns and behaviors are the way they are.

There are several different theoretical perspectives on the family. It is useful to think of each as a point of view. As with a physical object such as a building, when we see a family from different angles, we have a better grasp of what it is than if we look at it only from one, fixed position. Often theoretical perspectives on relationships and families complement one another and may appear together in a single piece of research. In other instances, the perspectives appear contradictory, leading scholars and policy makers into heated debate.

In this section, we describe nine theoretical perspectives related to families:

1. family ecology perspective
2. the family life course development framework
3. the structure–functional perspective
4. the interaction–constructionist perspective
5. exchange theory
6. family systems theory
7. conflict and feminist theory
8. the biosocial perspective
9. attachment theory

We will see that each perspective illuminates our understanding in its own way. Table 2.1 presents a summary of these theoretical perspectives. (Chapter 10's Table 10.1 applies several of these theoretical perspectives to the topic of unpaid household labor.)

The Family Ecology Perspective

The **family ecology perspective** explores how a family is influenced by the surrounding environment. The relationship of work to family life, discussed in Chapter 10, is one example of an ecological focus. Sociologists might look at how nonstandard work schedules affect family relationships, for example (Davis et al. 2008). We use the family ecology perspective throughout this book when we stress that, although society does not determine family members' behavior, it does present constraints for families as well as opportunities. The concept *sociological imagination*, introduced in Chapter 1, is in line with the family ecology perspective. Families' lives and choices are affected by economic, educational, religious, and cultural institutions, as well as by historical circumstances such as the development of the Internet, war, recession, and immigration patterns.

Every family is embedded in "a set of nested structures, each inside the next, like a set of Russian dolls" (Bronfenbrenner 1979, p. 3). Each "nested structure" includes events, social policies, social characteristics, and culture—structures that exist outside families and influence them. We can think of these various outside influences as radiating outward from the family as follows: (1) the neighborhood; (2) the workplace; (3) the community, town, or city; (4) the state, including state laws and policies; (5) the country, including national laws and policies; (6) the world, especially in an era of globalism; and (7) Earth's physical environment. All

Table 2.1 Theoretical Perspectives on the Family

Theoretical Perspective	Theme	Key Concepts	Current Research
Family Ecology	The ecological context of the family affects family life and children's outcomes.	Natural physical-biological environment Social-cultural environment	Effect on families of economic inequality in the United States Race/ethnic and immigration status variations Effect on families of the changing global economy Family policy Neighborhood effects
Family Life Course Development Framework	Families experience predictable changes over time.	Family life course Developmental tasks "On-time" transitions Role sequencing	Emerging adulthood Timing of employment, marriage, and parenthood Pathways to family formation
Structure–Functional	The family performs essential functions for society.	Social institution Family structure Family functions Functional alternatives	Cross-cultural and historical comparisons Analysis of emerging family structures in regard to their comparative functionality Critique of contemporary family
Interaction–Constructionist	By means of interaction, humans construct sociocultural meanings. The internal dynamics of a group of interacting individuals construct the family.	Interaction Symbol Meaning Role making Social construction of reality Deconstruction Postmodernism	Symbolic meaning assigned to domestic work and other family activities Deconstruction of reified categories
Exchange Theory	The resources that individuals bring to a relationship or family affect the formation, continuation, nature, and power dynamics of a relationship. Social exchanges are compiled to create networks and social capital.	Resources Rewards and costs Family power Social networks Social support	Family power Entry and exit from marriage Family violence Network-derived social support
Systems Theory	The family as a whole is more than the sum of its parts.	System Equilibrium Boundaries Family therapy	Family efficacy and crisis management Family boundaries
Feminist Theory	Gender is central to the analysis of the family. Male dominance in society and in the family is oppressive of women.	Male dominance Power and inequality	Work and family Family power Domestic violence Deconstruction of reified gender categories Deconstruction of definition of marriage as necessarily heterosexual Advocacy of women's issues
Biosocial Perspective	Evolution of the human species has put in place certain biological endowments that shape and limit family choices.	Evolutionary heritage Genes, hormones, and brain processes Inclusive fitness	Connections between biological markers and family behavior Evolutionary heritage explanations for gender differences, sexuality, reproduction, and parenting behaviors Development of research methods that can explore the respective influences of "nature" and "nurture"
Attachment Theory	Early childhood experience with caregiver(s) shape psychological attachment styles.	Secure, insecure/anxious, and avoidant attachment styles	Attachment style and mate choice, jealousy, relationship commitment, separation, or divorce

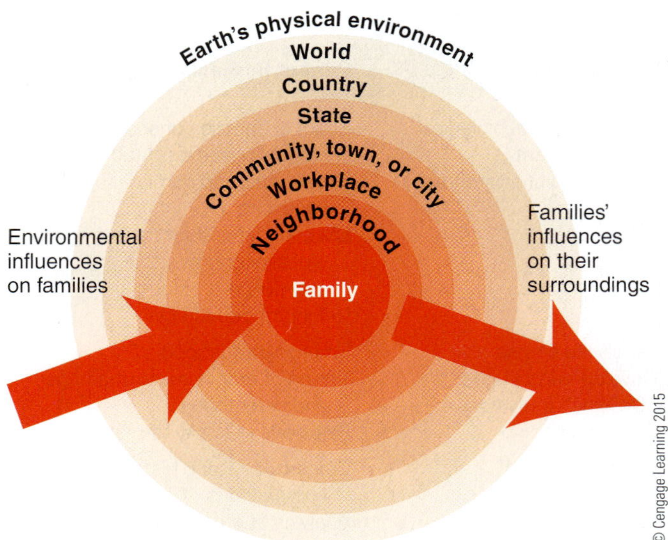

FIGURE 2.1 Various outside influences radiate outward from the family, influencing it and being influenced by it.

parts of the model are interrelated and influence one another (Bubolz and Sontag 1993; and see Figure 2.1).

Earth's physical environment—climate and climate change, soil, plants, animals—provides an essential backdrop against which all family living is played out. Family ecologists stress the interdependence of all the world's families—not only with one another but also with our planet's physical environment. In this vein, in the early 1990s the Family Energy Project at Michigan State University began to study families' energy usage, and subsequent research focuses on how to motivate families to conserve (Bubolz and Sontag 1993). Although it is crucial, the interaction of families with the physical environment is beyond the scope of this text. Our interest centers on families in their sociocultural environments. The social–cultural ecology of families may be examined historically. Ways that historical periods affect individuals, relationships, and families are explored in Chapter 1.

This perspective also analyzes the environments of contemporary families at various levels, from the global to the neighborhood. On the global level, for instance, the September 11, 2001, terrorist attacks on the United States and the subsequent wars in Afghanistan and Iraq have been part of a global conflict affecting American family life in many ways. More recently, the economic recession that began in 2007 ended many jobs filled by immigrants, who consequently wrestled with decisions about returning to their home countries (Schuman 2009). More generally, economic globalization—with the increasing outsourcing of jobs to regions with lower labor costs—has affected breadwinning and consumption in many American families.

On the national level, American policies and culture that emphasize military intervention worldwide impact military families (Wadsworth 2010). Meanwhile, families are impacted by federal Head Start, food stamp, and other programs designed to help those in poverty. The national Social Security program, coupled with Medicare, greatly influences elderly family members' retirement and housing choices. Sometimes researchers compare the relative effects of various countries' family environments. One fairly recent study compared how various countries' national policies about men's and women's paid work affect family households' division of labor (Cooke and Baxter 2010). We return to this discussion in Chapters 3 and 10.

Furthermore, family ecologists often stress the importance of workplace, town or city, state, and national policies on family living. An example is state policy that may or may not recognize same-sex marriages. As another example, Arizona legislation that focuses on deporting undocumented immigrants impacts many families in that state. On the community level, the availability of reasonably priced mass transit affects access to work.

Neighborhoods impact family well-being as well (Bowen et al. 2008). Homeless children in poor

All else being equal, residing in a supportive and helpful neighborhood translates into less family stress than otherwise. Meanwhile, families need to participate in the neighborhood in order to help create a cooperative environment for themselves. This group is organizing a Neighborhood Watch program. People in this neighborhood join together in activities that benefit families who live there.

neighborhoods are at greater risk for negative social, educational, economic, and health outcomes, and early death from gun violence (Conger, Conger, and Martin 2010; Edin and Kissane 2010; Gültekin 2012). Mothers raising children amidst neighborhood poverty express fears about letting them play outdoors (Kimbro and Schachter 2011). Whether in the neighborhood, workplace, community, or broader society, culture can influence families. For example, a review of the literature that examined marital well-being among U.S. couples of Mexican origin concluded that navigating the often contradictory expectations of two cultures—Mexican and Anglo—can produce challenges that may help to explain relatively high divorce rates among Mexican American couples (Helms, Supple, and Proulx 2011).

Ecologists have also examined the sociocultural settings of relatively privileged families (Swartz 2008). Examining the kinds of economic and social advantages enjoyed by the middle and upper levels of society is uncommon but needs to be encouraged. It may provide insight into the conditions that would enable *all* families to succeed. Moreover, elements in the social–cultural environment of upper-socioeconomic-level families—excessive achievement pressure or the isolation of children from busy, accomplishment-oriented parents—can be problematic. For instance, "the silence (in the community and in academia) surrounding domestic violence in affluent communities jeopardizes the health and safety of [affluent, victimized] mothers and their children..." (Haselschwerdt 2012, p. F16). The ecology perspective helps to identify factors that are important to societal and community support for all families. Exploring family life through this perspective leads to interest in family policy (the various laws and other regulations and procedures that impact families), which is discussed in Chapter 1.

Contributions and Critiques of the Family Ecology Perspective This perspective first emerged in the late nineteenth century, a period marked by concern about family welfare. The family ecology model resurfaced in the 1960s with the War on Poverty, a federal program directed toward the elimination of the high levels of poverty that then existed. The family ecology perspective makes an important contribution today by challenging the idea that family satisfaction or success depends solely on individual effort. Furthermore, the perspective turns our attention to family social policy—what may be done about social issues or problems that affect relationships and families.

A possible disadvantage of the family ecology perspective is that it is so broad and inclusive that virtually nothing is left out. One research agenda can hardly account for the family's sociocultural environment on all levels, from the global to the neighborhood. More and more, however, social scientists are exploring family ecology in concrete settings. For example, Canadian researchers Phyllis Johnson and Kathrin Stoll (2008) investigated how Sudanese refugee men continued to enact the breadwinner role for their families in Africa while resettling alone in western Canada. As a second example, U.S. researchers affiliated with the Children's Defense Fund continue to examine the ongoing plight of Louisiana children who suffer the disastrous effects of Hurricane Katrina (Cass 2007).

The Family Life Course Development Framework

Whereas family ecology analyzes relationships, families, and the broader society as interdependent parts of a whole, the **family life course development framework** focuses on the family itself as the unit of analysis (Sassler 2010; White and Klein 2008). The concept of the *family life course* is central here, based on the idea that the family changes in fairly predictable ways over time.

Typical stages in the *family life course* are marked by (1) the addition or subtraction of family members (through birth, death, and leaving home), (2) the various stages that the children go through, and (3) changes in the family's connections with other social institutions (retirement from work, for example, or a child's entry into school). Each stage has requisite *developmental tasks* that must be mastered before family members transition successfully to the next stage. Therefore, this perspective has tended to assume that families perform better when life course stages proceed in orderly fashion.

Traditionally, this perspective assumed that families begin with marriage. The *newly established couple* stage ends when the arrival of the first baby thrusts the couple into the *families of preschoolers* stage, followed later by the *families of primary school children stage,* and still later by the *families with adolescents stage* (Crosnoe and Cavanagh 2010). *Families in the middle years* help their offspring enter the adult worlds of employment and their own family formation. Later, parents return to a couple focus with the time and money to pursue leisure activities (if they are fortunate!). Still later, *aging families* must adjust to retirement and perhaps health crises or debilitating chronic illness. The death of a spouse marks the end of the family life course (Aldous 1996).

Role sequencing, the order in which major life course transitions take place, is important to this perspective. The *normative order hypothesis* proposes that the work–marriage–parenthood sequence is thought to be best for mental health and happiness (Jackson 2004; Wilcox et al. 2011a). Then, too, *"on-time" transitions*—those that occur when they are supposed to, rather than "too early" or "late"—are generally considered most likely to result in successful role performance during subsequent life course stages (Booth, Rustenbach, and

Jeff Greenberg/The Image Works

According to the family life course development framework, this father is in the *families of primary school children stage*. Like other family life course stages, this stage has particular tasks that need to be performed—tasks for which previous life course stages, if completed successfully, have helped to prepare him.

McHale 2008; Hogan and Astone 1986). For example, researchers have found that now having a baby in one's twenties may be seen as "out of step," or as a risky life course move (Jong-Fast 2003).

Using this theoretical framework, sociologists Jeremy Uecker and Charles Stokes (2008) explored the incidence of "early marriage"—that is, marrying before age 23—in the United States today. Uecker and Stokes argue that "[s]cholars and policy makers... should pay adequate attention to understanding and supporting these individuals' marriages" (p. 835). Because marrying in one's early twenties is an exception to the rule today, researchers of the life course perspective have turned their attention to the current, "emerging" transition to adulthood (Bentley 2007).

Emerging adulthood is a stage in individual development that precedes and affects entry into the family life course. The concept conveys a sense of ongoing development, a period "when the scope of independent exploration of life's possibilities is greater for most people than it will be at any other period of the life course" (Arnett 2000, p. 469). Transition to adulthood is now completed more gradually and later than it has been in the past—usually by age 30 (Arnett 2004; Furstenberg 2008). A principal reason for this change: As discussed in Chapter 1, it takes longer today to earn enough to support a family (Gibson-Davis 2009). Emerging adulthood is further explored in Chapters 6 and 7.

In addition to examining the transition to adulthood, researchers using the family life course development framework also extensively study the various transitions, or "pathways," to family formation (Amato et al. 2008; Lichter and Qian 2008). "The scope of research on intimate partnering now includes studies of 'hooking up,' Internet dating, visiting relationships, cohabitation, marriage following childbirth, and serial partnering, as well as more traditional research on transitions into marriage" (Sassler 2010, p. 557). In this vein, researchers note "the continued 'decoupling' of marriage and childbearing" (Smock and Greenland 2010). Other life course researchers have tackled the question how lifelong childlessness affects well-being (Umberson, Pudrovska, and Reczek 2010). Thus, the family life course development framework meets the postmodern family.

Contributions and Critiques of the Family Life Course Development Framework The family life course development framework directs attention to various stages that relationships and families encounter throughout life. Hence this perspective encourages us to investigate various family behaviors over time. For instance, research consistently finds that women are more likely to work on maintaining family relationships. Building on this research, three Belgian sociologists asked, "Is this true for all life stages?" They found the answer to be yes (Bracke, Christiaens, and Wauterickx 2008). As another example, researchers looked at reasons for calling telephone crisis hotlines across the life course. They found that "issues of loneliness increased with age whereas depression-related calls decreased" (Ingram et al. 2008).

Furthermore, this perspective directs our attention to how particular life course transitions affect family interaction. For example, researchers have investigated how transitions to parenthood or from cohabitation to marriage affect the time partners spend on housework (Baxter, Hewitt, and Haynes 2008). Then, too, as we have seen, the perspective brings our attention to how transitioning "early," "late," or "on-time" affects relationships and families (Booth, Rustenbach, and McHale 2008). This perspective also prompts researchers to look at interactions among family members who are in different life course stages, such as a study of ongoing affection between grandparents and young adults (Monserud 2008).

Critics note remnants of the traditional tendency within this perspective toward assumptions of life course standardization, possibly suggesting a white, middle-class bias. Moreover, because of economic, ethnic, and cultural differences, two families in the same life cycle stage may be very different. For these reasons, the family development perspective is perhaps somewhat less popular now than it was half a century or more ago.

The Structure–Functional Perspective

The **structure–functional perspective** investigates how a given social structure functions to fill basic societal needs. As discussed in Chapter 1, families are principally accountable for three vital family functions: to raise children responsibly, to provide economic support, and to give family members emotional security. *Social structure* refers to the ways that families are patterned or organized—that is, the form that a family may take.

As discussed in Chapter 1, there is no typical American family structure today. Instead, families evidence a variety of forms including, among others, same-sex families, cohabitating families, single-parent families, and transnational families whose members bridge and maintain relationships across national borders. The structure–functional perspective encourages researchers to ask how well a particular family structure performs a basic family function. For example, there is considerable research into how well single parents or cohabitating couples perform the function of responsible child rearing (P. Brown 2004; Carlson 2006; Manning and Lamb 2003). Results of this research are explored in Chapters 6, 7, and 14.

The structure–functional perspective may encourage a family researcher to think in terms of *functional alternatives*—alternate structures that might perform a function traditionally assigned to the nuclear family. A study among recent immigrants found that *fictive kin*—relationships "based not on blood or marriage but rather on religious rituals or close friendship ties, that replicates many of the rights and obligations usually associated with family ties"—can serve as a functional alternatives to the nuclear family. Results showed that "functions include assuring the spiritual development of the child and thereby reinforcing cultural continuity, exercising social control, providing material support, and assuring socio-emotional support" (Ebaugh and Curry 2000, pp. 189, 199).

The term *dysfunction* emerged from the structure–functional perspective (Merton 1968 [1949]) as a focus on social patterns or behaviors that fail to fulfill basic family needs. Obviously, domestic violence is dysfunctional in that it opposes the family function of providing emotional security. Although the term *dysfunctional family* is often used by laypeople and in counseling psychology, sociologists seldom use the term, which is considered too vague and imprecisely defined.

The structure–functional perspective might also encourage one to ask, "Functional for whom?" when examining a particular social structure (Merton 1968 [1949]). For instance, traditional male authority and higher prestige may be functional for fathers—and in some cultures for brothers—but not necessarily for mothers or sisters. Separating may seem to be functional for one or both of the adults involved, but it's not necessarily so for the children (Amato 2000, 2004).

Contributions and Critiques of the Structure–Functional Perspective

Virtually all social scientists agree on the one basic premise underlying structure–functionalism: that families are an important social institution performing essential social functions. The structure–functional perspective encourages us to ask how well various family forms do in filling basic family needs. Furthermore, the perspective can be interpreted as encouraging us to examine ways in which functional alternatives to the heterosexual nuclear family may perform basic family functions.

However, as it dominated family sociology in the United States during the 1950s, the structure–functional perspective gave us an unrealistic image of smoothly working families characterized only by shared values. Furthermore, the perspective once argued for the functionality of specialized gender roles: the *instrumental* husband-father who supports the family economically and wields authority inside and outside the family, and the *expressive* wife-mother-homemaker whose main function is to enhance emotional relations at home and socialize young children (Parsons and Bales 1955).

Then, too, the structure–functional perspective has generally been understood to define the heterosexual nuclear family as the "normal" or "functional" family structure. As a result, many social scientists, particularly feminists, rebuke this perspective (Anderson and Sabatelli 2007; Stacey 2006). The vast majority of family sociologists today rarely reference structure–functionalism directly.

The Interaction–Constructionist Perspective

As its name implies, the **interaction–constructionist perspective** focuses on *interaction,* the face-to-face encounters and relationships of individuals who act in awareness of one another. Often this perspective explores the daily conversation, gestures, and other behaviors that go on in families. By means of these interchanges, something called "family" appears (Berger and Kellner 1970). Family identity, traditions, and commitment emerge through interaction, with the development of relationships and the generation of *rituals*—recurring practices defined as special and different from the everyday (Byrd 2009; Oswald and Masciadrelli 2008).

Sometimes this perspective explores family *role-making* as partners adapt culturally understood roles—for example, uncle, mother-in-law, grandmother, or stepfather—to their own situations and preferences. One study looked at how older Chinese and Korean immigrants remade family roles on immigrating to the United States (Wong, Yoo, and Stewart 2006). A Korean grandmother described remaking her mother-in-law role:

> Once I immigrated I realized there are cultural differences between the U.S. and Korea especially when it comes to family dynamics. For example, I can't always say what I would like to say to my daughter-in-law. I follow the American ways and have given up trying to tell her what to do . . . I would like to tell my daughter-in-law to punish the grandchildren when they misbehave. But in America, us elders do not have the right to say this. I just keep these thoughts to myself. (p. S6)

This African American family is celebrating Kwanzaa, created in the 1960s by Ron Karenga and based on African traditions. An estimated 10 million black Americans now celebrate Kwanzaa as a ritual of family, roots, and community. The experience of adopting or creating family rituals fits the interaction–constructionist perspective on the family.

Lawrence Migdale/Photo Researchers, Inc./Science Source

This point of view also examines how family members interact with the outside world in order to manage family identity. An example is a study of interaction strategies used by couples who had chosen to remain childfree. Feeling potentially stigmatized, some claimed that they were biologically unable to have children. Others aggressively asserted the merits of a childfree lifestyle (Park 2002). The couples worked to construct how others would define their not having children.

Reality as Constructed This approach explores ways that people, by interacting with one another, *construct,* or create, meanings, symbols, and definitions of events or situations. A respondent in the study of Chinese and Korean immigrants saw family photographs as symbols of her changing (reconstructed) family role:

> My children got married and started to have a family of their own. . . . We are now no longer the center, but on the peripheries of their families. Even when we take pictures, we don't stand in the center but on the side. It's totally different in China. Even when we took pictures, parents would be pictured in the middle. (Wong, Yoo, and Stewart 2006, p. S6)

As people "put out" or *externalize* meanings, these meanings come to be *reified,* or made to seem real. Once a meaning or definition of a situation is reified, people *internalize* it and take it for granted as "real," rather than viewing it as a human creation (Berger and Luckman 1966). For example, many newlyweds take it for granted that a honeymoon should follow their wedding; they don't think about the fact that the idea of a honeymoon is socially constructed (Bulcroft et al. 1997). Sociologists James Holstein and Jaber Gubrium (2008) combine this perspective with the family life course development framework to investigate how individuals gradually construct their life course.

Unlike structure–functionalism, in which analysis begins with one or more family forms that are understood as given, the interaction–constructionist perspective focuses on the processes through which family forms are constructed and maintained. For instance, we typically think of the "battered woman" as having been abused by a male, thereby maintaining the social construction of domestic life as heterosexual (VanNatta 2005). Our values and beliefs about divorce, childbearing outside marriage, and single-parent families can also be understood as socially constructed (Thornton 2009). Exposing the ways that symbols and definitions are constructed is called *deconstruction*, a process typically identified with postmodern theory.

Postmodern Theory **Postmodern theory** can be understood as a special focus within the broader interaction–constructionist perspective (Kools 2008). Gaining recognition in the social sciences since the 1980s, postmodern theory largely analyzes *social discourse* or *narrative* (public or private, written or verbal statements or stories). The analytic purpose is to demonstrate that a phenomenon is socially constructed (Gubrium and Holstein 2009). A principal goal involves debunking *essentialism*—the idea that categories really do exist in nature and are not simply reifications. Examples include analyses of the concepts of *gender* and *race*. Formerly taken for granted as essentially "real," these categories are now generally recognized—at least within the social sciences—as social constructions. (Chapter 3's "Issues for Thought: Challenges to Gender Boundaries," further explores the social construction of gender.)

When applied to relationships, postmodern theory posits that beliefs about what constitutes a "real" family are nothing more than socially fabricated narratives, having been constructed through public discourse.

Contributions and Critiques of the Interaction–Constructionist Perspective The interaction–constructionist perspective alerts us to the idea that much in our environment is neither "given" nor "natural," but socially constructed by humans—those in the past and those around us now. In this way, the perspective can be liberating. If a social structure, definition, value, or belief is oppressive, it can be challenged: Constructed by human social interaction, phenomena can also be changed by such interaction. Social movements advocating legalization of same-sex marriage proceed from this beginning point. At the family level, this perspective leads researchers to focus on family members' interaction patterns, along with emergent definitions, symbols, rituals, and the consequences thereof.

Critics ask, "Where do we go from here?" (Wasserman 2009). Once the taken-for-granted is deconstructed, then what? "If everything is socially constructed, then we gain nothing by employing the term. It has become a mantra that explains very little" (Stacey 2006, p. 481). Moreover, it is virtually impossible to conduct traditional social science research in the absence of agreed-upon social categories (Cockerham 2007).

Exchange Theory

Exchange theory applies an economic perspective to social relationships. A basic premise is that when individuals are engaged in social exchanges, they prefer to limit their costs and maximize their rewards. Chapter 1 discusses making informed decisions as a process of "deciding" rather than "sliding." According to exchange theory, which also emphasizes decision making, we choose among options after calculating potential rewards against costs and weighing our alternatives. Those of us with more resources, such as education or good incomes, have a wider range of options from which to choose. This orientation examines how individuals' personal resources, including physical attractiveness and personality characteristics, affect the formation and continuation of relationships.

According to this perspective, an individual's dependence on and emotional involvement in a relationship affects her or his relative power in the relationship. When alternatives to a relationship seem slim, one wields less power in the relationship. According to the *principle of least interest*, the partner with less commitment to the relationship is the one who has more power (Waller 1951). Those with more resources and options can use them to bargain and secure advantages in relationships. People without resources or alternatives to a relationship typically defer to the preferences of the other and are less likely to leave (Sprecher, Schmeeckle, and Felmlee 2006). From this point of view, responses to domestic violence and decisions to separate or divorce are affected by partners' relative resources.

The relative resources of participants shapes power and influence in families and impacts household communication patterns, decision making, and division of labor. Relationships based on exchanges that are equal or equitable (fair, if not actually equal) thrive, whereas those in which the *exchange balance* feels consistently one-sided are more likely to dissolve or be unhappy. Dating relationships, marriage and other committed partnerships, divorce, and even parent-child relationships show signs of being influenced by participants' relative resources (Nakonezny and Denton 2008).

Social Networks Exchange theory also focuses on how everyday social exchanges between and among individuals accumulate to create social networks. Elizabeth drives Juan to the airport, Juan babysits for Maria, Maria proofreads an assignment for Elizabeth, and so forth until a network of social exchanges emerges. The Internet offers opportunities for building social networks ranging from the local to the international level, such as those on Facebook.

Among other things, *social network theory*, a middle-range subcategory within the exchange perspective, examines how social networks provide individuals with *social capital*, or resources (friendship, people with whom to exchange favors), that result from their social contacts. Social capital is analogous to financial capital, or money, inasmuch as we can "spend" it to acquire rewards, such as a romantic partner, a job, or emotional support (Benkel, Wijk, and Molander 2009; Wejnert 2008).

Contributions and Critiques of Exchange Theory The exchange perspective provides a framework from which to draw specific hypotheses about weighing alternatives and making decisions regarding relationships. Furthermore, this perspective leads us to recognize that inequality, or an unfavorable balance of rewards and costs, gradually erodes positive feelings in a relationship. The perspective also encourages us to recognize the social capital brought about by membership in social networks. Exchange theory is subject to the criticism that it assumes a human nature that is unrealistically rational and even cynical at heart about the roles of love and responsibility.

Family Systems Theory

Family systems theory views the family as a whole, or *system*, comprised of interrelated parts (the family members) and demarcated by boundaries. Originating in natural science, systems theory was applied to the family first by psychotherapists and was then adopted by family scholars.

A *system* is a combination of elements or components that are interrelated and organized into a whole. Like an organic system (the body, for example), the parts of a family compose a working system that behaves fairly predictably. The ways in which family members respond to one another can show evidence of patterns. For example, whenever Jose sulks, Oscar tries to think of something fun for them to do together.

Furthermore, systems seek *equilibrium*, or stable balance and symmetry. Change in one of the parts sets in motion a process to restore equilibrium. For example, in the body system, if one hand becomes disabled, the other must adjust to do the work of both. In family dynamics, this tendency toward equilibrium puts pressure on each member to retain his or her fairly predictable role. A changing family member is subtly encouraged to revert to her or his original behavior within the family system. For change to occur, the family system as a whole must change. Indeed, that is the goal of family therapy based on systems theory. The family may see one member as the problem, but if the psychologist draws the whole family into therapy, the family *system* should begin to change.

Social scientists have moved systems theory beyond its therapeutic origins to employ it in a more general analysis of families. They are especially interested in how family systems process information, deal with challenges, respond to crises, and regulate contact with the outside world. Researchers have elaborated and explored concepts such as *family boundaries* (ideas about who is in and who is outside the family system). This perspective also prompts researchers to investigate such things as *family boundary ambiguity*, wherein it is unclear who is in the family and who isn't. Stepfamilies have been researched from this point of view: Do children of divorced parents belong to two (or more) families? Are former spouses and their relatives part of the family (Boss 1997; Stewart 2005a)?

David Young-Wolff/PhotoEdit

Even before the recession that began in 2007, delayed marriage, high housing costs, and other serious financial pressures made it more common for young adults to continue living with their parents or to move back home. The recession accelerated this trend. Family systems theory tells us that when an adult child moves back home, the family system changes and the entire system of family roles will need to readjust to maintain balance and restore equilibrium.

Contributions and Critiques of Family Systems Theory When working with families in therapy, this perspective has proven very useful. By understanding how their family system operates, individuals can make desired personal or family changes. Systems theory often gives family members insight into the effects of their behavior. It may make visible the hidden motivations behind certain family patterns. For example, doctors were puzzled by the fact that death rates were higher among kidney dialysis patients with *supportive* families. Family systems theorists attributed the higher rates to the unspoken desire of the patients to lift the burden of care from the close-knit family they loved (Reiss, Gonzalez, and Kramer 1986).

Envisioning the family as a system can be a creative perspective for research. Rather than seeing only the influence of parents on children, for example, system theorists are sensitized to the fact that this is not a one-way relationship and have explored children's influence on family dynamics (Crouter and Booth 2003).

A criticism of systems theory is that it does not sufficiently consider a family's economic opportunities, race/ethnic and gender stratification, and other features of the larger society that influence internal family relations. When used by therapists, systems theory has been criticized as tending to diffuse responsibility for conflict by attributing dysfunction to the entire system, rather than to culpable family members within the system. This situation can lead to "blaming the victim," as well as making it difficult to extend social support to victimized family members while establishing legal accountability for others, as in situations involving incest or domestic violence (Stewart 1984).

Conflict and Feminist Theory

We like to think of families as beneficial for all members. For decades sociologists ignored the politics of gender and differentials of power and privilege within relationships and families. Beginning in the 1960s, conflict and feminist theorizing and activism began to change that oversight, as issues of latent conflict and inequality were brought into the open.

A first way of thinking about the **conflict perspective** is that it is the opposite of structure–functional theory. Not all of a family's practices are good; not all family behaviors contribute to family well-being. Family interaction can include domestic violence as well as holiday rituals—sometimes both on the same day.

Conflict theory calls attention to power—more specifically, unequal power. It explains behavior patterns such as the unequal division of household labor in terms of the distribution of power between husbands and wives. Because power within the family derives from power outside it, conflict theorists are keenly interested in the political and economic organization of the larger society.

The conflict perspective traces its intellectual roots to Karl Marx, who analyzed class conflict. Applied to the family by Marx's colleague, Friedrich Engels (1942 [1884]), the conflict perspective attributed family and marital problems to class inequality in capitalist society.

In the 1960s, a renewed interest in Marxism sparked the application of the conflict perspective to families in a different way. Although Marx and Engels had focused on economic classes, the emerging feminist movement applied conflict theory to the sex/gender system—that is, to relationships and power differentials between men and women in the larger society and in the family.

Although there are many variations, the central focus of the **feminist theory** is on gender issues. A unifying theme is that male dominance in the culture, society, families, and relationships is oppressive to women. *Patriarchy*, the idea that males dominate females in virtually all cultures and societies, is a central concept (Hesse-Biber 2007).

Unlike the perspectives described earlier, which were developed primarily by scholars, feminist theories emerged from political and social movements over the past fifty years. As such, the mission of feminist theory is to use knowledge to actively confront and end the oppression of women and related patterns of subordination based on social class, race/ethnicity, age, or sexual orientation.

The feminist perspective has contributed to political action regarding gender and race discrimination in wages, sexual harassment, divorce laws that disadvantage women, rape and other sexual and physical violence against women and children, and reproductive issues, such as abortion rights and the inclusion of contraception in health insurance. Feminist perspectives promote recognition of women's unpaid work, the greater involvement of men in housework and child care, efforts to fund quality day care and paid parental leaves, and transformations in family therapy so that counselors recognize the reality of gender inequality in family life and treat women's concerns with respect (C. Baker 2008; Friedman and Valenti 2008; Gupta and Ash 2008; Mollen 2008). The feminist perspective has combined with the family life course development framework to analyze aging and gender issues (Ross-Sheriff 2008).

Since the publication of Naomi Wolf's 1991 classic, *The Beauty Myth: How Images of Beauty Are Used Against Women*, feminist theory has given considerable attention to eating disorders and body image issues (Latham 2008). For example, a study that combines the feminist with the interaction–constructionist perspective investigated the process through which a young woman internalizes an identity as a "fat girl" and thereby socially "unfit," or unacceptable for romantic relationships (Rice 2007). Feminist scholars also consider whether a decision to have cosmetic surgery evidences a woman's agency or the unrecognized influence of a patriarchal construct of feminine beauty (Tiefer 2008).

In recent years, feminist theory has embraced postmodern analyses, deconstructing formerly taken-for-granted concepts such as *gender dichotomy* (the idea that there are naturally two very distinct genders) or the idea that marriage must naturally be heterosexual (Dreger and Herndon 2009). Having co-opted a pejorative term from the popular culture, some feminists refer to this kind of analysis as *queer theory* (Eaklor 2008; Stacey 2006). From the feminist perspective, championing the traditional heterosexual nuclear family at the cost of both heterosexual and lesbian women's equality and well-being is unconscionable (Harding 2007).

Contributions and Critiques of Feminist Theoretical Perspectives

By calling attention to women's experiences, feminist theory has encouraged us to see things about relationships and family life that had been overlooked before the 1960s. Women's domestic work was largely invisible in social science until the feminist perspective began to treat household labor as work that has economic value. The feminist perspective brought to light issues of wife abuse, marital rape, child abuse, and other forms of domestic violence.

According to some social scientists, feminist theory is too political, value laden, or adversarial to be considered a valid academic approach (Landau 2008; Lloyd, Few, and Allen 2007). The concept of patriarchy has been criticized as being unscientifically vague and ahistorical. Posited to exist in virtually all societies, patriarchy loses meaning as an analytic category when it minimizes differences between America in the twenty-first century and ancient Rome, where husbands allegedly had life-and-death power over women. Moreover, inasmuch as some feminist theory embraces postmodernism, it is subject to the same criticisms as postmodernism, which were described above. "Feminist engagements with science have never been straightforward, and are less so all the time" (Valentine 2008, p. 355).

The Biosocial Perspective

The **biosocial perspective** is characterized by "concepts linking psychosocial factors to physiology, genetics, and evolution" (Booth, Carver, and Granger 2000, p. 1018). This perspective argues that human physiology, genetics, and hormones predispose individuals to certain behaviors (Bearman 2008). In other words, biology interacts with the social environment to affect much of human behavior and, more specifically, many family-related behaviors (Booth et al. 2006). "[Q]uantitative genetic studies have increasingly... found major interplay between genetic and non-genetic [environmental] factors, such that the outcomes cannot sensibly be attributed just to one or the other, because they depend on both" (Rutter 2002, pp. 1–2).

According to the biosocial (or evolutionary psychology) perspective, much of contemporary human behavior evolved in ways that enable survival and continuation of the human species. Successful behavior patterns are encoded in the genes, and this *evolutionary heritage* is transmitted to succeeding generations. The survival of one's genetic material into future generations is paramount. Hence human behavior has biologically evolved to be oriented to the survival and reproduction of all close kin who carry those genes (D'Onofrio and Lahey 2010).

Evolutionary explanations are offered for many contemporary family patterns. For instance, research suggests that children are more likely to be abused by nonbiologically related parents or caregivers than by biological parents. Nonbiological parent figures are less likely to invest money and time in their children's development and future prospects (Case, Lin, and McLanahan 2000; Wilcox 2011). The biosocial perspective explains this by arguing that parents "naturally" protect the carriers of their genetic material. Accordingly, although he acknowledged that there were many successful stepfamilies and adoptions, sociologist David Popenoe (1994) found that these family forms were not supported by our evolutionary heritage. He concluded that "we as a society should be doing more to halt the growth of stepfamilies" (p. 21).

From its early days, some proponents of the biosocial perspective have held that certain human behaviors, because they evolved for the purpose of human survival, were both "natural" and difficult to change. It has been asserted, for example, that traditional gender roles evolved from patterns shared with our mammalian ancestors that were useful in early hunter–gatherer societies. Gender differences—males allegedly more aggressive than females, and mothers more likely than fathers to be primarily responsible for child care—are seen as anchored in hereditary biology (Rossi 1984; Udry 1994, 2000).

However, biosociologists emphasize that biological predisposition does *not* mean that a person's behavior cannot be influenced or changed by social structure (Bearman 2008). "Nature" (genetics, hormones) and "nurture" (culture and social relations), they argue, interact to produce human attitudes and behavior (Horwitz and Neiderhiser 2011). As an example, research on testosterone levels in married couples found high levels of the husbands' testosterone to be associated with poorer marital quality when their role overload was high but with better marital quality when role overload was low. In other words, "testosterone enables positive behavior in some instances and negative behavior in others" (Booth, Johnson, and Granger 2005, p. 483; see also Booth et al. 2006).

Contributions and Critiques of the Biosocial Perspective

This perspective encourages scientists to investigate research questions regarding relationships and families that would otherwise be overlooked: Is there a genetic basis for human family and relationship behaviors and attitudes? If so, to what extent can those attitudes and behaviors be changed? To what degree do social forces (nurture) and biological predispositions (nature) interact to result in human behavior and attitudes?

Over the past twenty-five years, the biosocial perspective has emerged as a significant theoretical perspective on the family. Researchers have employed this point of view to examine such phenomena as gender differences, sexual bonding, mate selection, jealousy, parenting behaviors, marital stability, and male aggression against women (D'Onofrio and Lahey 2010; Dorius et al. 2011). In the words of two of the

Martin Rodgers/Stock Boston

These individuals are waiting for medical attention in a neighborhood clinic. How might scholars from different theoretical orientations see this photograph? *Family ecologists* might remark on the quality of the facilities—or speculate about the family's home and neighborhood—and how these factors affect family health and relations. They might compare this crowded and understaffed clinic for the poor with the better equipped and staffed doctors' offices that provide health care to wealthier Americans. Scholars from the *family life course development framework* would likely note that this woman is in the child-rearing stage of the family life cycle. *Structure–functionalists* would be quick to note the child-raising (and, perhaps, expressive) function(s) that this woman is performing for society. *Interaction–constructionists* might explore the mother's body language: What is she saying nonverbally to the child on her lap? What might he symbolize to her? *Exchange theorists* might speculate about this woman's personal power and resources relative to others in her family. *Family system theorists* might point out that this mother and child are part of a family system: Should one person leave or become seriously and chronically ill, for example, the roles and relationships among all members of the family would change and adapt as a result. *Feminist theorists* might point out that typically it is mothers, not fathers, who are primarily responsible for their children's health—and ask why. The answer from a *biosocial perspective* might be that women have evolved a stronger nurturing capacity that is partly hormonally based. *Attachment theorists* might surmise whether the mother is interacting with her child in a way that promotes a secure, insecure/anxious, or avoidant attachment style. How would you interpret this photo?

perspective's proponents, "Genetically informed studies that have examined family relations have been critical to advancing our understanding of gene-environment interplay" (Horwitz and Neiderhiser 2011, p. 804). However, this perspective was once used to justify gender inequality as biologically based and hence "natural." More recently, evolutionary perspectives have been the basis for criticism of nonreproductive sexual relationships and the employment of mothers as contrary to nature (Daly and Wilson 2000). It is therefore not surprising that many distrust this perspective or that it has been politically and academically controversial. We explore and appraise the biosocial approach, or evolutionary psychology, when discussing gender (Chapter 3), extramarital sex (Chapter 4), child care (Chapter 10), and children's well-being in stepfamilies (Chapter 15).

Attachment Theory

Counseling psychologists often analyze individuals' relationship choices in terms of attachment style. **Attachment theory** posits that during infancy and childhood a young person develops a general style of attaching to others (Ainsworth et al. 1978; Bowlby 1988; Pittman 2012). Once a youngster's attachment style is established, she or he unconsciously applies that style, or "state of mind," to later adult relationships (Mikulincer and Shaver 2012).

A child's primary caretakers (usually parents and most often the mother) evoke a *style* of attachment in him or her. The three basic attachment styles are *secure*, *insecure/anxious*, and *avoidant*. Children who can trust that a caretaker will be there to attend to their practical and emotional needs develop a secure attachment style. Children who feel uncared for or abandoned develop either an insecure/anxious or an avoidant attachment style (Crespo 2012).

In adulthood, a secure attachment style involves trust that the relationship will provide ongoing emotional and social support. An insecure/anxious attachment style entails concern that the beloved will disappear, a situation often characterized as "fear of abandonment." Someone with an avoidant attachment style dodges emotional closeness either by avoiding relationships altogether or demonstrating ambivalence, seeming preoccupied, or otherwise establishing distance in intimate situations (Knudson-Martin 2012; Rauer and Volling 2007).

Attachment theory has grown in importance and prominence in family studies over the past several decades (Bell 2012; Pittman 2012). Some researchers combine this perspective with the family life course development framework to look at stability or variability of attachment styles throughout an individual's life (Klohnen and Bera 1998). Attachment theory is also used by counseling psychologists. The assumption is that if a client learns to recognize a problematic attachment style, he or she can change that style (Ravitz, Maunder, and McBride 2008; Weissman, Markowitz, and Klerman 2007).

Contributions and Critiques of Attachment Theory

This perspective prompts us to look at how personality impacts relationship choices, from initiating to maintaining them. Attachment theory also encourages us to ask what kind of parenting best encourages a secure attachment style. These are important research questions. Critics argue that an attachment style might depend on the situation in which a person finds him or herself rather than on a consistent personality characteristic (Fleeson and Noftle 2008; and see Knudson-Martin 2012). Of course, when therapists employ this point of view, they recognize that even if it is a relatively stable personality characteristic, one's attachment style can be changed over time.

The Relationship Between Theory and Research

Theory and research are closely integrated, ideally at least. Theory should be used to help direct research questions and to suggest useful concepts. Often when designing their research, scientists employ one or more theoretical perspectives from which to generate a *hypothesis* or "educated guess" about the way things are. Scientists then test these hypotheses by gathering data. At other times, to interpret data that has already been gathered, scientists ask themselves what theoretical perspective best explains the facts (Lareau 2012). Over time, our understanding of family phenomena may change as social scientists undertake new research and modify theoretical perspectives. Even when theory is not directly spelled out in a study, it is likely that the research fits into one or more of the theoretical perspectives described above. "Facts About Families: How Family Researchers Study Religion from Various Theoretical Perspectives" illustrates ways that researchers have used these theoretical perspectives when studying the broad topic of religion and family. We'll turn our attention now to various methods that researchers use to gather information, or data, on family life.

Students take entire courses on research methods, and obviously we can't cover the details of such methods here. However, we do want to explore some major principles so that you can think critically about published research discussed in this text or in the popular press. As the subtitle of one textbook says, research methods provide "a tool for life" (Beins 2008). We invite you to think this way as well.

Designing a Scientific Study: Some Basic Principles

At the onset of a scientific study, researchers carefully design a detailed research plan. Some research is designed to gather *historical data.* Professor Steven Ruggles (2011) at the University of Minnesota analyzed nineteenth-century U.S. population statistics back to 1850 to examine intergenerational households. He was curious to see whether the majority of intergenerational households formed in order to care for elderly family members or whether, on the other hand, the households evidenced a reciprocal relationship—one in which each generation participated to help take care of the other. He found that nineteenth-century U.S. intergenerational households were mainly reciprocal. Historical studies of marriage and divorce in the United States portray a picture of the past that, contrary to common belief, was not necessarily stable or harmonious (Cott 2000; Hartog 2000). Although family history is an important area of research, this textbook does not allow space for us to fully explore it.

Research designs can also be *cross-cultural,* comparing one or more aspects of family life among different societies. A study described in Chapter 5, which asked students in ten different countries whether it's necessary to be in love with the person you marry, is an example of cross-cultural research (Levine et al. 1995).

Scientists consider many questions when designing their research: Will the study be cross-sectional or longitudinal? Deductive, inductive, or a combination of the two? Will the study be mainly quantitative or qualitative? Will the sample be random and the data generalizable? Because the goal of all research is to transcend our personal blinders or biases, as was discussed at the beginning of this chapter, scientists must meticulously define their terms and take care not to overgeneralize. This section looks briefly at these considerations.

Cross-Sectional Versus Longitudinal Data When designing a study, researchers must decide whether to gather cross-sectional or longitudinal data. *Cross-sectional studies* gather data just once, giving us a snapshot-like, one-time view of behaviors or attitudes. *Longitudinal studies* provide long-term information as researchers continue to gather data over an extended period of time.

Facts about Families

How Family Researchers Study Religion from Various Theoretical Perspectives

Research topics can be studied from different points of view. Here we see how the topic religion and family life has been investigated with different theoretical perspectives and by the use of various research methods.

- **The Family Ecology Perspective.** Loser et al. (2008) conducted qualitative interviews with highly religious parents and children in sixty-seven families who belonged to the Church of Jesus Christ of Latter-day Saints. The researchers concluded that religion is often personally internalized and should not be understood only as a component in a family's sociocultural environment.
- **The Family Life Course Development Framework.** Pearce (2002) analyzed longitudinal data from a Detroit survey of white mothers and children to find that early childhood religious exposure later influenced childbearing attitudes during transition to adulthood. Young adults with Catholic mothers or mothers who frequently attended religious services were especially likely to resist the idea of not having children.
- **The Structure–Functional Perspective.** Schottenbauer, Spernak, and Hellstron (2007) found that parents' use and modeling of religiously based coping skills, along with family attendance at religious or spiritual programs, was functional in enhancing children's health, social skills, and overall behavior.
- **The Interaction–Constructionist Perspective.** Hirsch (2008) used naturalistic observation to understand how young, actively Catholic women in Mexico creatively interpret their religion's proscription against birth control while choosing to use it. As one "grassroots theologian" explained, "[E]ven in the bible it says 'help yourself, so I can help you'; even the priests tell you that" (pp. 98–99).
- **Exchange/Network Theory.** Christian Smith (2003) used secondary analysis of the national Survey of Parents and Youth to find that participation in religious congregations increases the likelihood that family members will benefit from sharing a network that includes parents, their children, their children's friends and teachers, and their children's friends' parents.
- **Family Systems Theory.** Lambert and Dollahite (2008) conducted qualitative research with fifty-seven religious couples and concluded that these respondents saw God as a third partner in an otherwise dyadic family system—a third system member whose presence enhanced their marital commitment.
- **Feminist Theory.** Feminist social historians Carr and Van Leuven (1996) edited a cross-cultural anthology whose works examine the implications of religion for female family members of various religious cultures. Overall, the book argues that women's oppression originates not in religion itself but in the exploitation of religion as a subjugation tool by patriarchal religio-cultural systems.
- **The Biosocial Perspective.** Wright (1994) argued that humans have evolved as "the moral animal," a situation that facilitates our species' cooperation toward the goal of survival.
- **Attachment Theory.** Reinert (2005) surveyed seventy-five Catholic seminarians, presenting them with "Awareness of God" and "Attachment to Mother" scales. He found that a seminarian's early childhood attachment to his mother to be a key influence in the degree of his attachment to a personal God.

Critical Thinking

Think of a family-related topic and consider how you might study it. What theoretical perspective would you use to help frame your research questions? What research methods and data-gathering techniques would you use?

For example, to understand how psychotherapy may modify attachment insecurity over the life course, researchers designed longitudinal studies that followed respondents' attachment styles over thirty years from childhood into adulthood (Klohnen and Bera 1998). Other researchers monitored nonresident fathers' involvement with their children for three years and found finances and relations with their children's mothers to be significant causes for changes over time (Ryan, Kalil, and Ziol-Guest 2008). A difficulty encountered in longitudinal studies, besides cost, is the frequent loss of subjects to death, relocation, or loss of interest. Social change occurring over a long period of time may make it difficult to ascertain what, precisely, has influenced family change. Yet cross–sectional data (one-time comparison of different groups) cannot show change in the same individuals over time.

Deductive Versus Inductive Reasoning *Deductive reasoning* in research begins with a hypothesis that has been derived (i.e., deduced) from a theoretical point of view. "Reasoning down" from the abstract to the concrete, a researcher designs a data–gathering strategy in order to test whether the hypothesis can be supported by observed facts. Researchers who use *inductive reasoning* observe detailed facts, then induce, or "reason up," to arrive at generalizations grounded in the observed data. Inductive studies do not begin with a preconceived hypothesis. Instead, researchers begin their observations with open minds about what they'll see and find. Typically, deductive reasoning is associated with quantitative research and inductive reasoning with qualitative research.

Quantitative Versus Qualitative Research In *quantitative research,* the scientist gathers, analyzes, and reports data that can be quantified or understood in numbers. Quantitative research finds numerical incidences in a population—for example, the average size of a family household or what percent of Americans are currently cohabitating. Statistical facts and findings, such as those in Chapter 1's "Facts About Families" boxes, are examples of quantitative data. Quantitative research also uses computer-assisted statistical analysis to test for relationships between phenomena. For instance, quantitative analysis has found a statistically demonstrated correlation between being raised by a single parent and teen pregnancy (Albrecht and Teachman 2003).

When performing *qualitative research,* the scientist gathers, analyzes, and reports data primarily in words or stories (Matthews 2012). For example, social scientists interviewed eight mothers in three rural trailer parks who described their lives in detail (Notter, MacTavish, and Shamah 2008). In their subsequently published article, the researchers quote the women in their own words:

> I pay attention to how my mother raised me and I try to do it different. I try to teach him [her son] how to take care of himself. He knows how to do chores and how to cook. I had to learn all of that on my own. I try to teach him how to state his opinion. My mother never taught me to do that.... (p. 619)

As a second example, sociologist Gina Miranda Samuels (2009) conducted qualitative research with black adults raised by white parents. Samuels's findings are reported in narrative using respondents' own words (p. 89):

> But I remember there was one girl named Ebony and she could not BELIEVE I had been adopted by white people. She was like, "WOW! You were adopted by white people?! Are they nice to you? Do they treat you well?" And that was a shock to me because that was the first time I realized that black people might not get treated well by white people.... (Justine, 28)

> I was in my salon and I didn't even know what a hot comb was. That was my giveaway! And he [the stylist] was like, "Were you raised by white people?" And then, he was like, "OH. I was able to tell that by the way you talked and by the way you carried yourself." (Crystal, 24)

The aim of qualitative research is to gain in-depth understandings of people's experiences, as well as the processes they go through when defining, adapting to, and making decisions about their situations. Qualitative research typically employs the interaction-constructionist theoretical perspective described earlier in this chapter.

When designing studies, researchers must also carefully define their terms: What precisely is being studied and how exactly will it be measured (Bickman and Rog 2009)?

Defining Terms Researchers scrupulously define the concepts that they intend to investigate, then report those definitions in their published studies so that readers know precisely what was investigated and how. For example, researchers once considered all (heterosexual) cohabitators as fitting one general definition. They found cohabitation before marriage to be statistically related to divorce later (Dush, Cohan, and Amato 2003). However, as definitions of cohabitation were further refined to differentiate serial from one-time cohabitators, results began to show that cohabitating only with your future marriage partner was *not* more likely to end in divorce (Lichter and Qian 2008; Teachman 2003).

As another example, looking at how children of same-sex parents fare, researchers have generally found no differences from those raised in other family forms (Amato 2012). However, a recent study by Mark Regnerus at the University of Texas at Austin found otherwise. Regnerus's research found that "adult children of parents who have same-sex relationships" evidenced a list of more negative outcomes, such as lower employment and income levels and poorer mental and physical health (Regnerus 2012a).

Following the norm that research results are open to review by other scientists, scholarly critics pointed out that Regnerus had compared apples to oranges, so to speak. He had considered a parent who had *ever* had a lesbian or gay male relationship as a same-sex parent. Although some of the adult children in the study had grown up in planned same-sex parent families, most of Regnerus's respondents had spent some of their childhood in other family types—for instance, with heterosexual parents who later divorced, in single-parent households, or in stepfamilies (Osborne 2012).

Erroneously, Regnerus compared this category of adult children with those who had grown up in intact heterosexual families with two biological parents who stayed together (Schumm 2012). "What we really need in this field is for strong skeptics to study gay, stable parents and compare them directly to a similar group of heterosexual, stable parents," concluded New York University sociologist Judith Stacey (in Carey 2012). Put another way, the two samples that Regnerus used were not similar enough to be accurately compared.

Samples and Generalization You may have noticed that in the study of women in trailer parks mentioned above, the researchers interviewed only eight mothers. We cannot expect the situations of these few respondents to correspond with all American women living in trailer parks. For one thing, all eight mothers were white (Notter, MacTavish, and Shamah 2008). We cannot possibly conclude from this research that all women who live in trailer parks are white. Rather than a nationwide demographic portrayal, the purpose of interviewing these eight women was to learn about the experiences and processes that mothers can go through when residing in trailer parks.

To gather data that can be *generalized* (applied to a population of people other than those directly questioned), a researcher must draw a sample that accurately reflects, or represents, that population in important characteristics such as age, race, gender, and marital status. Results from a survey in which all respondents are college students, for example, cannot be interpreted as representative of Americans in general.

Gallup polls are examples of research that uses *representative samples* that reflect the national adult population. When a Gallup poll reports that most Americans would be unwilling to forgive an unfaithful spouse, we know that the findings from their sample can be generalized to the whole national population with only a small probability of error (Jones 2008). To draw a representative sample of a population, everyone in that population must have an equal chance to be selected. The best way to accomplish this is to have a list of every individual in the population and then randomly choose from the list (see Babbie 2009). A national *random sample* of approximately 1,500 people may validly represent the U.S. population.

Sometimes there are no complete lists of members of a population. For instance, researchers were interested in the ramifications of living with a compulsive hoarder (one who continuously acquires yet fails to discard large numbers of possessions). They located 665 respondents who reported having a family member or friend with hoarding behaviors (Tolin et al. 2008). How did they accomplish this? The researchers had made several national media appearances about hoarding. As a result, more than 8,000 people had contacted them for guidance or information. Drawing from this group, the researchers e-mailed potential participants, inviting them to take part in the study and asking them to forward the invitation to others in similar situations.

Ultimately, these researchers found that living with the clutter associated with hoarding often causes depression and isolation, partly because one is embarrassed to invite friends home. Although the findings were based on a fairly large sample, however, they cannot be generalized to all people who live with a compulsive hoarder because the sample was not random: Not everyone who lives with a compulsive hoarder had the same chance of being chosen for the study. It is reasonable to argue that those who contacted the researchers for guidance or information were more distraught than those who didn't. As a result, the findings may show greater difficulty in living with a compulsive hoarder than is generally experienced by all those in this population. Nevertheless, this is valuable research inasmuch as it lends insight into what living with a compulsive hoarder entails, at least for many.

There are many occasions when it is impossible to find a random sample for the topic one wants to investigate. The study of blacks raised by whites, discussed above, provides a second case in point. In this instance, the researcher recruited volunteer respondents by using Web-based and print advertisements to African American and multiracial adoption agencies (Samuels 2009). In general, volunteers do not result in a random sample.

In addition to these considerations, designing a study involves decisions about the techniques by which the data will be collected, or gathered.

Data-collection Techniques

We will refer to various **data-collection techniques**—interviews and questionnaires, naturalistic observation, focus groups, experiments and laboratory observation,

Rayes/Riser/Getty Images

A research team plans data collection and analysis for a survey of how families spend their time together.

case studies—throughout this text, so we will briefly describe them here. Each technique has strengths and weaknesses. However, the strengths of one technique can compensate for the weaknesses of another. To get around the drawbacks of a given technique, researchers may combine two or more in one study (Clark et al. 2008).

Interviews, Questionnaires, and Surveys The most common data-gathering technique in family research involves personally interviewing respondents or asking them to complete self-report questionnaires about their attitudes and past or present behaviors. When conducting interviews, researchers ask questions in person or by telephone. Gallup polls use telephone surveys. Alternatively, a researcher may distribute paper-and-pencil or Web-based questionnaires that respondents complete by themselves. Increasingly viewed as comparable in reliability to paper-and-pencil questionnaires, Internet surveys use e-mail or Web-based formats (Coles, Cook, and Blake 2007). Examples of the latter can be found at Surveymonkey.com, which facilitates the design, distribution, and some analysis of online surveys.

Questions can be *structured* (or *closed-ended*). After a statement such as "I like to go places with my partner," the respondent chooses from a set of fixed answers, such as "always," "usually," "sometimes," or "never." Researchers spend much time and energy on the wording of such questions because they want all respondents to interpret them in the same way. Also, word choice can influence responses. For example, respondents tend to be more favorable to the phrase "assistance to the poor" than they are to the term "welfare" (Babbie 2007, p. 251).

A *survey* is a quantitative data-gathering tool that comprises a series of structured, or closed-ended, questions. Once completed, survey responses are tallied and analyzed, usually with computerized coding and statistical programs. Questionnaires are inappropriate for young children, of course. Face-to-face interviews and focus groups offer alternatives in studies involving young children (Freeman and Mathison 2009).

Sociologists often engage in *secondary analysis* of large data sets—the result of fairly comprehensive surveys administered to a national representative sample. Once completed and tallied, the responses are made public, often via the Internet, so that other researchers (who had nothing to do with designing the questions) can analyze the data. A myriad of data sets are available for secondary analysis. The National Survey of Family Growth (NSFG) contains data from a national sample of nearly 11,000 U.S. women between ages 15 and 44. To investigate the breadwinner role cross-culturally, researchers analyzed data previously collected in the International Social Survey, conducted annually in more than twenty countries (Yodanis and Lauer 2007). The longitudinal national survey, "Marital Instability Over the Life Course," whose findings are described periodically throughout this text, is also available for secondary analysis (Amato and Booth 1997; Hawkins and Booth 2005). As part of their study design, researchers who use secondary analysis determine how they might analyze answers to already conceived questions in order to yield new information and insights. A drawback to secondary analysis is that researchers, because they do not design their own questions, can investigate only topics and details about which survey questions have been designed by others. Nevertheless, secondary analysis is popular, and many studies that use secondary analysis on other large data sets are described throughout this textbook.

Surveys are an efficient way to gather data from large numbers of people. Different respondents' standardized answers to structured questions can be readily compared. However, because they allow only predetermined answers to standardized questions, surveys may miss points that respondents would consider important but cannot report. For this reason, some researchers ask unstructured questions.

Unstructured (also called *open-ended*) questions do not offer a limited, preset range of answers. Instead, the purpose is to allow the respondent to talk freely. Interviewers using open-ended questions learn to listen and then probe for more detail. Samuels's study of blacks adopted by whites used unstructured, *open-ended questions*:

> I began the interviews by asking participants to share their adoption stories, including what they knew about their birth families. Participants described their childhood communities, how they were raised to think about their racial heritages and adoptions, and if their insights or identities changed as they became adults. (2009, p. 84)

Questioning respondents—whether quantitatively or qualitatively, whether interviewing or using a self-report questionnaire format—is the most common data-gathering technique used by family researchers. Nevertheless, there are limitations. Valid responses depend on participants' honesty, motivation, and ability to respond. Some individuals—for example, toddlers or people who suffer from the advanced stages of Alzheimer's disease—are not appropriate subjects for questioning.

Naturalistic Observation In **naturalistic observation** (also called "participant observation" or "field research"), the researcher spends extensive time with respondents and carefully records their activities, conversations, gestures, and other aspects of everyday life. This data-gathering technique often accompanies the interaction–constructionist theoretical perspective. The researcher attempts to discern family relationships and communication patterns and to draw implications and conclusions for understanding family behavior in general. The study of women living in trailer parks employed naturalistic observation as well as interviews. Over a period of sixteen months, researchers spent twelve to twenty hours in each woman's home, sometimes taking part in family meals (Notter, MacTavish, and Shamah 2008).

The principal advantage of naturalistic observation is that it allows one to view family behavior as it actually happens in its own natural—as opposed to artificial—setting. The most significant disadvantage is that what is recorded and later analyzed depends on what one or a few observers think is significant. Another drawback is that naturalistic observation requires enormous amounts of time to observe only a few families who cannot be assumed to be representative of family living in general. Moreover, not all research topics lend themselves to naturalistic observation. Explaining her decision to interview rather than directly observe blacks raised by whites, Samuels points out that "[t]here are not 'sides of town' or neighborhoods where multiracials or transracially adopted families and individuals reside, in which a researcher can become immersed, gain access, and conduct naturalistic inquiry" (2009, p. 84).

Focus Groups A focus group is a form of qualitative research in which, in a group setting, a researcher asks a gathering of ten to twenty people about their attitudes or experiences regarding a situation. Researchers have used focus groups to explore how parents feel about their overweight children, for example (Jones et al. 2009). Participants are free to talk with each other as well as to the researcher or group leader. Focus group sessions last between one and two hours and are typically electronically recorded and then transcribed or entered into a computer for later analysis. The study of Chinese and Korean immigrants, discussed earlier in this chapter, is based on data collected in eight San Francisco area focus groups (Wong, Yoo, and Stewart 2006).

Frank Siteman/PhotoEdit

Conducting focus groups is one way to gather data when doing social science research. Here teenagers are being asked their opinions on a topic of interest to researchers—their attitudes about attending college perhaps.

Group discussion produces data and insights that would be less forthcoming otherwise. Also, researchers can capture the participants' everyday speech in order to better understand their life situations—and perhaps to include some of their language in subsequent interviews or questionnaires. Focus groups are useful when researchers do not feel they know enough about a topic to design a set of closed-ended questions. Focus groups can also be successful when working with children (Freeman and Mathison 2009; Gibson 2012).

However, there are disadvantages to the method. Researchers have less control in a group setting than they do in a one-on-one interview, and hence focus groups can be time consuming, given the amount of usable data recorded. Then, too, data can be difficult to analyze because focus group conversation is casual and flows in response to others' comments. Furthermore, the researcher can easily influence responses by inadvertent comments that lead respondents to say things that they may not really mean.

Experiments and Laboratory Observation In an experiment or in laboratory observation, behaviors are carefully monitored or measured under controlled conditions. In an **experiment**, subjects from a pool of similar participants are randomly assigned to groups (*experimental* and *control groups*) that are then subjected to different experiences (*treatments*). For example, families with children who are undergoing bone marrow transplants may be asked to participate in an experiment to determine how they may best be helped to cope with the situation. One group of families may be assigned to a support group in which the expression of feelings, even negative ones, is encouraged (the experimental group). A second group may receive no special intervention (the control group). If at the conclusion of the experiment the groups differ according to measures of coping behavior, the outcome is presumed to be a result of the experimental treatment. Because no other differences are presumed to exist among the randomly assigned groups, the results of the experiment provide evidence of the effects of therapeutic intervention.

The experiment just described takes place in a field (real-life) setting, but experiments are often conducted in a laboratory setting because researchers have more control over what will happen. The laboratory setting allows the researcher to plan activities, measure results, determine who is involved, and eliminate outside influences. A true experiment incorporates the features of random assignment and experimental manipulation of the important variable (the treatment).

Laboratory observation, on the other hand, simply means that behavior is observed in a laboratory setting, but it does not involve random assignment or experimental manipulation of a variable. Family members may be asked to discuss a hypothetical problem or play a game while their behavior is observed and recorded. Later those data can be analyzed to assess the family's interaction style and the nature of their relationships. Laboratory methods are useful in measuring physiological changes associated with anger, fear, sexual response, or behavior that is difficult to report verbally. In the 1970s, social psychologist John Gottman began studying newly married couples in a university lab while they talked casually or wrestled with a problem. Video cameras recorded the spouses' gestures, facial expressions, and verbal pitch and tone. Some couples volunteered to let researchers monitor shifts in their heart rates and chemical stress indicators in their blood or urine as a result of their communicating with each other (Gottman 1996). Gottman's findings are discussed in detail in Chapter 11.

Laboratory observation has advantages and disadvantages. An advantage is that social scientists can watch human behavior directly, rather than depending on what respondents *tell* them. A disadvantage is that the behaviors being observed often take place in an artificial situation, and whether an artificial situation is analogous to real life is debatable. A couple asked to solve a hypothetical problem through group discussion may behave differently in a research laboratory than they would at home.

Clinicians' Case Studies We also obtain information about families from *case studies* compiled by clinicians—psychologists, psychiatrists, marriage counselors, and social workers—who counsel people with marital and family problems. As they see individuals, couples, or whole families over a period of time, these counselors become acquainted with communication patterns and other interactions within families. Clinicians offer us knowledge about family behavior and attitudes by describing cases or reporting conclusions based on a series of cases.

The advantages of case studies are the vivid detail and realistic flavor that enable us to experience vicariously the family life of others. Clinicians' insights can provide hypotheses for further research. However, case studies have important weaknesses. There is always a subjective or personal element in the way the clinician views the family. Inevitably, any one person has a limited viewpoint. Clinicians' professional training may lead them to misinterpret aspects of family life. Psychiatrists, for example, used to assume that the career interests of women were abnormal and caused the development of marital and sexual problems (Chesler 2005 [1972]).

Furthermore, people who present themselves for counseling may differ in important ways from those who do not. Most obviously, they may have more problems. For example, throughout the 1950s psychiatrists reported that gays in therapy had many emotional difficulties. Subsequent studies of gay males not in therapy concluded that gays were no more likely to have mental health problems than were heterosexuals (American Psychological Association 2007).

Ideally, a number of scientists examine one topic by using several different methods. The scientific conclusions in this text are the results of many studies from various and complementary research tools. Despite the drawbacks and occasional blinders, the total body of information available from sociological, psychological, and counseling literature provides a reasonably accurate portrayal of marriage and family life today. Although imperfect, the methods of scientific inquiry bring us better knowledge of the family than does either personal experience or speculation based on media images.

The Ethics of Research on Families

Exploring the lives of families the way social science researchers do carries responsibility. Researchers must do nothing that would negatively impact respondents, a principle summarized as "do no harm." Researchers also must show respect to those being studied and take into consideration the needs of their respondents. Feminist theorists in particular argue that researchers should be attuned to how their findings might help their respondents as well (McGraw, Zvonkovic, and Walker 2000). To help accomplish these standards, most research plans must be reviewed by a board of experts and community representatives called an *institutional review board* (IRB). No federally funded research can proceed without an IRB review, and most institutions require one for all research on human subjects (Cohen 2007a).

The IRB scrutinizes each research proposal for adherence to professional ethical standards for the protection of human subjects. These standards include *informed consent* (the research participants must be apprised of the nature of the research and then give their consent); lack of coercion; protection from harm; confidentiality of data and identities; the possibility of compensation of participants for their time, risk, and expenses; and the possibility of eventually sharing research results with participants and other appropriate audiences. Other than ensuring that the research is scientifically sound enough to merit the participation of human subjects, IRBs do not focus on evaluating the research topic or methodology.

In this chapter we've looked at how social scientists explore—that is, think about, research, and study—families. Throughout the remainder of this text you're encouraged to recall perspectives, facts, and issues raised in this chapter as we examine theory and research on a variety of family topics.

Summary

- Scientific investigation—with its ideals of objectivity, cumulative results, and various methodological techniques for gathering empirical data—is designed to provide an effective and accurate way of gathering knowledge about the family.
- Different theoretical perspectives—family ecology, the family life course development framework, structure–functional, interaction–constructionist, exchange, family systems, feminist, biosocial, and attachment theory—illuminate various features of families and provide a foundation for research.
- Research designs can be historical, cross-cultural, longitudinal, qualitative or quantitative, and inductive or deductive.
- Data-collection techniques include various ways of questioning respondents (interviews, questionnaires, survey instruments), focus groups, laboratory observation and experiments, naturalistic observation, and clinicians' case studies.
- Researchers need to be guided by professional standards and ethical principles of respect for research participants.

Questions for Review and Reflection

1. Choose one of the theoretical perspectives on the family and discuss how you might use it to understand something about life in your family.
2. Choose a magazine photo and analyze its content from one of the perspectives described in this chapter. Analyze the photo from another theoretical perspective. How do your insights differ depending on which theoretical perspective is used?
3. Discuss why science is often considered a better way to gain knowledge than personal experience alone. When might this not be the case?
4. Think of a research topic, then review the data-gathering techniques described in this chapter to decide which of these you might use to investigate your topic.
5. **Policy Question.** What aspect of family life would it be helpful for policy makers to know more about as they make law and design social programs? How might this topic be researched? Is it controversial?

Key Terms

attachment theory 43
biosocial perspective 42
conflict perspective 41
data-collection techniques 47
exchange theory 39
experiment 49
family ecology perspective 32
family life course development framework 35
family systems theory 40
feminist theory 41
interaction–constructionist perspective 38
naturalistic observation 48
postmodern theory 39
science 31
structure–functional perspective 37
theoretical perspective 32

3

Our Gendered Identities

Learning Objectives

1. Distinguish between the concepts *sex* and *gender*.
2. Contrast the ideas of gender as a "natural," individual characteristic versus gender as negotiated, enacted, or displayed.
3. Explain the biosocial perspective regarding cultural gender expectations.
4. Distinguish between biology-based and society-based arguments concerning how gender roles originated.
5. Describe how historical periods influenced gender roles.
6. Explain ways that gendered attitudes and norms are socialized and internalized.
7. Discuss the women's movement and three types of men's movements.
8. Speculate about the future of gendered differentiation.

During the 2012 Obama–Romney presidential campaign, *Family Circle* magazine ran a contest: Michelle Obama and Ann Romney competed for best cookie baker. Critics found it insulting that two university educated women, one with a Harvard law degree, were asked to prove themselves as worthy first ladies by baking. Did the contest endorse "retro" values? Was it sexist? The magazine editors said no. They said that when a woman finally represents her political party to run for president, they'll ask her husband to bake cookies too. They evidently assumed she'd be heterosexually married. (Actually women have represented "third parties"—political parties other than Democrat or Republican—to run for president.) National Public Radio ran the cookie-baking story. Listener Gary Edwards commented online, "Oh come on!" "It is for fun," he wrote. "There is no sex-politics involved." Another listener, D. Skully, got more serious: "A mom who never tried to bake a decent cookie for her kid is kind of pathetic" (Cornish 2012). Or was he kidding?

You may have thought that economically developed societies such as the United States had finished with gender issues. The fact is that we're all experiencing gender in unprecedented ways. Women couldn't vote anywhere in the world until the late nineteenth century—fewer than 150 years ago, a dot in the long line of human history. Fewer than 100 years ago, U.S. women became eligible to vote. Thanks in large part to advances resulting from the women's movement, discussed later in this chapter, both women's and men's roles are more flexible than ever before. Some argue that we've entered a "postfeminist" era in which gender no longer matters, that we're too busy with work and family to care about it anymore (Read 2011; Showden 2009). Despite dramatic and unprecedented change, however, gender issues remain with us.

On the one hand, examples of females in nontraditional roles abound: women as chief executive officers, astronauts, or military officers and enlisted personnel, a quarter of them exposed to combat—even before January 2013 when the Pentagon officially allowed women soldiers to engage in combat (Pew Research Center 2010b; Rosenthal 2013). We take it for granted that most working-age women, including many mothers, are employed, often in demanding careers. Men have begun to move into traditionally female occupations such as nursing or elementary school teaching, and married men are doing more at home than they used to (Bianchi, Robinson, and Milkie 2006; Dewan and Gebeloff 2012; Mundy 2012). While many fathers are examining how they themselves were fathered and committing themselves to be more emotionally present, research uncovers heightened levels of caring among male adolescents (Marsiglio 2012; Schalet 2012).

On the other hand, many of us continue to see women and men as fundamentally different in personality and aptitudes. And we inhabit a world in which talk show personality Rush Limbaugh calls a female law student advocating for government-subsidized birth control a "slut," refers to Secretary of State Hillary Clinton as "sex-retary Clinton," and dubs women's rights activists "feminazis" (Dowd 2012; Limbaugh 1992). More generally, research uncovers continuing disadvantages for women both at home where many of them do more than their fair share of the housework and in the workplace where there are relatively few women in top corporate positions and where, on average, women are paid less than men (Sandberg 2013b; Slaughter 2012). Studies also uncover gender-related disadvantages for men (Coontz 2013). All else being equal, fathers are significantly less likely to be awarded custody in the case of divorce, for instance.

We seem to have taken renewed interest in demanding that males prove themselves to be "real men," while expecting women to show that they're "mom enough" (Pickert 2012). Recall the public radio listener's comment at the beginning of this chapter that a mom who doesn't bake cookies is "kind of pathetic." The fact is that gender expectations and behaviors, as both traditionally practiced and as they are changing, continue to impact virtually every aspect of family life.

This chapter examines various society-wide, or *macro* aspects of gender as they affect us on the *micro* level as individual family members. In doing so, we'll explore personality traits and cultural scripts typically associated with masculinity and femininity and consider that the boundary between what's masculine and what's feminine is both culturally prescribed and ambiguous. We'll analyze gender in the social institutions of religion, politics, education, and economics. We will explore possible influences on gendered behaviors and examine how people are socialized to gender roles. We'll look at the women's movement along with men's movements that have emerged around gender issues. To begin, we'll look at what it means to have a gendered identity.

Gendered Identities

Each of us has a gendered identity that fundamentally influences our sense of who we are. The concept of **gender identity** refers to the degree to which an individual sees her- or himself as feminine or masculine. Most of us probably take for granted that gender is a dichotomy, or dichotomous: You are either male and masculine or female and feminine. Yet social scientists increasingly see gender as a continuum—an imaginary line along which individuals vary between the opposite poles of masculinity and femininity. In this chapter, we will often use the term *gender* rather than *sex* for an important reason. The word **sex** is used in reference to male or female anatomy and physiology. We use the terms **gender** or **gender role** more broadly to describe societal attitudes and behaviors expected of and associated with the two sexes (Duck and Wood 2006; Oakley 1972).

Meanwhile, a complication occurs because a small proportion of people do not feel at ease with their sex as recorded at birth. These persons may have been born with ambiguous genital anatomy (**intersex** individuals) or not (**transsexual** or **transgendered** individuals). In any case, they are uncomfortable with the gender that society has assigned them. "Issues for Thought: Challenges to Gender Boundaries" addresses these variations in gender identity. Then, too, in a process called **gender bending**, some people challenge a gender mandate on purpose. For example, in 2012 San Francisco Giants relief pitcher Brian Wilson was televised wearing red nail polish. On a more serious level, the increasing number of individuals in jobs or professions thought to be only for the "other sex" shows widespread gender bending. Nevertheless, whether we adhere to them strictly or not, cultural expectations about how boys and girls, men and women should behave and relate to one another influence our gendered identities, family roles, and life choices.

Cultural Gender Expectations

In all societies that we know of, humans are to some extent differentiated, or thought of as separate and different, according to gender. This **gender differentiation** is apparent in our cultural expectations about how men and women should behave. You can probably think of "masculine" characteristics associated with being male, along with "feminine" traits associated with being female. Men are assumed to have **agentic** (from the root word *agent*) or **instrumental character traits**—strength, confidence, self-reliance, assertiveness, and ambition—that enable them to accomplish difficult tasks or goals. Heterosexuality is also associated with being masculine (Sallee 2011). A "real man" avoids all things feminine or "sissy" (David and Brannon 1976). Accordingly, some

The Advertising Archives

AP Photo/Marcio Jose Sanchez

Mass media presents us with idealized, one-dimensional (typically airbrushed, unrealistic, and sexualized) stereotypes of ourselves. This has important connotations for our relationships, how we see ourselves, and how we see each other. Although less often, the media also gives us *gender bending* images that challenge culturally accepted gender expectations. San Francisco Giants relief pitcher Brian Wilson attracted attention when he was televised wearing red and black nail polish during one game of the 2012 Major League Baseball playoffs.

Issues for Thought

Challenges to Gender Boundaries

Intersex and transgender individuals, along with transfamilies (families in which one partner is transgendered), "throw existing taxonomic classification systems of identity into perplexing disarray" (Pfeffer 2012, p. 574). Somewhere between 1 and 4 percent of live births are intersex—that is, the infants have some anatomical, chromosomal, or hormonal sexual variation from what is considered "normal" and characteristic of the vast majority of the population. Indeed, "chromosomes, hormones, the internal sex structures, the gonads and the external genitalia all vary more than most people realize" (Fausto-Sterling 2000, p. 20). In the 1950s, intersex babies (then termed *hermaphrodites*) were assigned a gender identity by doctors, and parents were advised to treat them accordingly (Preves 2010; and see Colapinto 2000). The children typically underwent surgery to give them genitals more closely approximating the assigned gender. Following this model, intersex infants and children are often currently treated with "corrective" surgery, hormonal therapy, or both (Spack et al. 2012).

Intersexuality emerged as an area of political activism with the formation of the Intersex Society of North America (ISNA) in 1993. Members demanded societal acceptance of gender ambiguity and demonstrated against what they saw as arbitrary gender assignment and the surgical "correction" of intersexed infants (Preves 2010). Some medical ethicists take the position that "the various forms of intersexuality should be defined as normal" (Lawrence McCullough, quoted in Fausto-Sterling 2000, p. 21) and that surgical "corrections" are unethical and "reinforce the stigma through degradation and shame" (Gough et al. 2008, p. 494; Warne and Bhatia 2006).

Meanwhile, the terms transsexual and transgender describe an identity adopted by individuals whose genitalia suggest a clear sexual identity at birth, yet they are uncomfortable with the sexual identity to which they were assigned at birth. Transgendered adults may want to change their anatomic sex through *sexual reassignment surgery* (SRS) or hormonal treatment in order to conform their bodies to their preferred gender identity. Such transformations can prove difficult for the people undergoing reassignment, not only because of the physical and psychological transformations but also because of the responses to the changes by friends and family—and even the public. In 1998, classical pianist David Buechner became Sara Buechner. Before 1998, Buechner was a famous musician. After her gender reassignment, the public, friends, and family did not support the changes, and her career stalled. Eventually, family, friends, and even the public may become accustomed to the new person. As more people choose to redefine themselves in this way, the rest of society will follow with relative levels of acceptance (Marikar 2009; Winerip 2009).

Other transsexuals may simply adopt the dress and demeanor of the sex with which they identify or one that is not gender identified (Preves 2010). A relatively few parents allow their young children to experiment with transgendered behaviors—a kindergarten boy sports hot pink socks and sparkly sneakers, for instance, or a 4-year-old girl likes to be called "handsome prince" and has persuaded her mother to get her a Mohawk haircut (Hoffman 2011; Padawer 2012). The majority of transgendered individuals are older, however, and some universities have established gender-neutral housing at the request of transgendered students (F. Bernstein 2004; Raymond and Gordon 2008). Suggesting the beginning of societal accommodation of a more complex sex/gender system, some bureaucratic forms now include a "transgender" box as well as those for "male" and "female."

Although we tend to think of transgendered individuals as adults, recent media attention has uncovered the issue of boys as young as toddlers who seem to strongly prefer girls' clothing, toys, and behaviors, and who as children insist that they belong to the "other" gender (Spiegel 2008b). As a pre-adolescent's potential transgendered identity becomes an issue, some parents consider hormonal therapy that delays puberty (Hoffman 2011). Delaying puberty either (1) allows a child more time to accept his or her identity assigned at birth or (2) makes it less difficult later (at about age 16) to medically alter a person's anatomy to match the preferred gender identity (Spiegel 2008a).

Parents, physicians, mental health professionals, and educators must address ethical issues regarding surgical "correction" and possible gender reassignment for intersex or transsexual/transgendered individuals (P. L. Brown 2006; Hoffman 2011; Weil 2006). Meanwhile, the biological, psychological, and social realities presented challenge the notion of clearly demarcated masculine and feminine genders.

Critical Thinking

Have transgendered individuals been politically visible in your school or community? What are your own thoughts about whether gender is a dichotomy or a continuum along which individuals may vary?

---

coaches and military officers motivate males through insults, calling them "women," "ladies," or "girls."

A relative absence of agency characterizes femininity, and women are expected to embody relationship-oriented or **expressive character traits**: warmth, sensitivity, the ability to express tender feelings, and placing concern about others' welfare above self-interest (Parsons and Bales 1955; Sallee 2011). In one interesting study, communications theorist Sut Jhally (2009) examined how people's hands are portrayed in television, film, and magazines. He noted that female "hands are shown not as assertive or controlling of their environment but as letting the environment control them" (p. 6; also see Goffman 1979). We should add that cultural expectations for femininity involve culturally defined beauty standards, whereas a "real man" is expected to be tall and muscular.

Masculinities Three major culturally defined obligations for men involve (1) group leadership, (2) protecting group territory and weaker or dependent others, and (3) providing resources, typically by means of occupational success (Farrell 1974; Kimmel 2000). In the 1980s, a complementary cultural message emerged: The "new man" was both financially successful and emotionally sensitive, valuing tenderness and equal relationships with women (Messner 1997). More recently, still another transformation of the ideal male image appeared, partly in response to first responders' widely televised brave behavior after the 9/11 terrorist attacks: the unafraid "can-do" man who tackles traumatic events head-on but feels free to shed tears after doing so (Adelman 2009, p. 279).

Many expectations for masculinity are positive: bonding with others and managing conflict through shared activities, humor, and fun; caring for others by providing for and protecting them; developing self reliance, inner strength, bravery, courage, and heroism; and banding together toward common goals (Kiselica and Englar-Carlson 2010). A qualitative study titled "Good Guys with Guns" found that some men associated carrying a concealed handgun with the need to protect their family: "[W]hen you get married,... you have roles.... And if we were ever attacked or accosted or something then, then it's up to me to protect her until she can, you know, be safe" (Stroud 2012, p. 225).

A further masculine cultural message glorifies outwitting others (usually, but not necessarily men) in competitions—for example, wartime combat, contact sports, or barroom brawls (Katz 2006; Sullivan and McHugh 2009): "If I've got a stake or a pool cue, I will own your ass" (in Stroud 2012, p. 225). Furthermore, if a male finds legitimate avenues to occupational success blocked to him due, for example, to race/ethnic discrimination or lack of education, he might compensate, or "make it," through alternative routes such as body building, acts of aggression, intense sports or other highly risky behaviors, subordinating women and nonmasculine men, or striking a "cool pose" (Crook, Thomas, and Cobia 2009; Ezzell 2012; Smith and Beal 2007). The latter involves dress and postures manifesting fearlessness and detachment and has been adopted by some race/ethnic minority males for emotional survival in a discriminatory and hostile society. White supremacy movements can be seen as a similar response to status disadvantage for less-advantaged white males (Swain 2002).

Femininities As with masculinities, a variety of cultural messages depict **femininities**. Traditionally, the pivotal expectation for a woman has required her to offer emotional support. The ideal woman has been expected to be physically attractive, not too competitive, a good listener, adaptable, and a man's always supportive helpmate. She was further expected to be a "good mother," putting her family's and children's needs before her own. As an example, a study titled "No Vacation for Mother" examined gender depictions in 1950s travel literature and focused on cultural expectations surrounding women's domestic labor and the care of husbands and children during family vacations. She found that the literature "reinforced rigid, traditional gender roles" (Morin 2012, p. 436). Generally, "the qualitative research in the mid-twentieth century shows quite clearly that women were expected by others, including their husbands,... and expected themselves... to take care of their men, their families, and their homes. That's what a good wife did, whether or not she earned a living as well" (Risman 2011, p. 20). This cultural message persists although to lesser degrees.

In the 1980s, as more women entered the workforce and came to value career success, the *professional woman* image emerged: independent, ambitious, self-confident. This cultural expectation combined with the older caregiving and sacrificing one to form the *superwoman* message. In this case, a good wife and mother—"hair flying as she rushed around, attaché case in one arm, a baby in the other"—efficiently attained career success *and* supported her children, perhaps by herself. "The Superwoman could have it all, but only if she did it all" (Gottesdiener 2012). In the words of a contemporary women's magazine writer, career women

> were being challenged to go big or go home, as the saying goes, and this didn't even take into account the ongoing messaging in both ads and magazine content that we had to be beautiful and fit as well as hyperachieving.... But while today's ads have backed off this messaging significantly, we're still faced with the ever-present pressure to do whatever it is we do—career woman, stay-at-home mom, or independent spirit—looking beautiful, ageless and thin, much like the women who grace the covers of these magazines. (Nelson 2012)

More recently, another feminine image emerged to depict the *satisfied single* woman—either heterosexual or lesbian, usually employed, and perhaps a parent—who is happy not to be in a serious relationship.

The Relative Values of Masculinity versus Femininity Feminist theorist Dorothy Smith (1987) has argued that mainstream culture values masculinity more highly than femininity. For instance, the ideal career path follows a masculine model, according to which occupational dedication is paramount and family caregiving is secondary, rather than being defined as a form of success (Slaughter 2012). In a qualitative study of Native American business managers, a respondent explained that she had not been raised to be assertive or competitive but to behave in ways very different from those required at work: "How we were brought up is very different and continues to be a struggle on a daily basis . . . as a young girl, out of respect for adults, you listen and not interrupt and not have eye-to-eye contact. . . . I certainly have had to work hard in being able to do the opposite of those things" (Muller 1998, p. 20). Smith argues that a socially structured "line of fault" separates many females' lived experience from cultural definitions of success. A woman lives with a **bifurcated consciousness**—a divided perception according to which she is aware of and often troubled by two conflicting messages: first, that caregiving is most important for her; and second, that caregiving is not as highly valued across society as is career success. How women navigate this bifurcated consciousness is a question for ongoing research.

Kumar Sriskandan/Alamy

Pink is for girls, and blue is for boys. This image simultaneously shows social change and resistance to change. Ken is cooking: social change. But he's using pink kitchen utensils. Are they Barbie's? Should kitchen utensils be Barbie's and not Ken's? What do you think images like these convey to children?

Cultural Expectations and Role Performance The interactionist-constructionist perspective (see Chapter 2) prompts us to see expectations about what's gender appropriate as socially fashioned. **Dramaturgy**, a theoretical subcategory within the interactionist-constructionist perspective, sees individuals as enacting culturally constituted scripts and socially prescribed roles in front of others (everyday-life audiences). However, persons do not blindly follow a given cultural script. Instead, they ad lib, modifying or renegotiating prescribed roles (Goffman 1959). We "perform" or "do" gender (Butler 1988; Sallee 2011). When Michelle Obama and Ann Romney publicly baked cookies, they were "doing gender" in that they performed according to a traditionally feminine script. On the other hand, when San Francisco Giants pitcher Brian Wilson painted his fingernails just for fun, he renegotiated the masculine script. The idea of "doing gender" raises the question: To what extent do people behave in accordance with cultural expectations?

To What Extent Do Women and Men Follow Cultural Expectations?

Do individuals exhibit gender-differentiated behaviors according to cultural script expectations? Research on actual behavioral gender differences suggests that there are fewer than we might think. In fact, "men and women, as well as boys and girls, are more alike than they are different" (Hyde 2005, p. 581). Analysts have found evidence of gender differences in (1) motor performance, especially in boys' greater throwing distance and speed; (2) sexuality, especially in male's greater incidence of masturbation and acceptance of casual sex; and (3) physical aggressiveness, with males generally more violent than females (Else-Quest, Hyde, and Linn 2010; Hyde 2007; Maccoby and Jacklin 1974). Research also suggests that women have greater feelings of connectedness in interpersonal relations. Women do occupy the caregiving professions in greater numbers than men, whereas men tend to be in more competitive occupations (Webster and Rashotte 2009).

Psychologists studying gender-differentiated emotions find girls and women more likely to exhibit sadness and anxiety—"submissive emotions ... that do not threaten interpersonal interaction in most cases." Boys and men show anger more often than girls do, and in school or business settings, "even laughter at the expense of others." This type of "masculine" behavior involves emotions that potentially threaten relationships (Chaplin, Cole, and Zahn-Waxler 2005, p. 80; see also Cassano and Zeman 2010). However, there is considerable individual variation in emotional displays. The situational context accounts for much of the difference (Gormley and Lopez 2010; Meier, Hull, and Ortyl 2010). A man may reveal deep sadness at a funeral, for instance, or over a divorce; a woman may get very angry should her car be hit by a careless driver.

Behavior vis-à-vis cultural gender expectations generally fits a pattern that can be illustrated as two overlapping distribution curves (see Figure 3.1). For instance, although the majority of men are taller than the majority of women, the area of overlap in men's and women's heights is considerable, and some men are shorter than some women. It is also true that differences *among* women or *among* men (*within-group variation*) are usually greater than the average difference *between* men and women (*between-group variation*). We return to this concept of within-group versus between-group variation later in this chapter and throughout this text.

Race/Ethnic Diversity and Gendered Expectations

Images of men as instrumental and women as expressive are based primarily on white, middle-class heterosexuals. Especially in past decades, researchers' habit of using middle-class whites as the norm caused ignorance regarding gender and race/ethnic diversity. Researchers in the 1970s tended to see women as a homogeneous category disadvantaged compared with men, also a homogenous category. Now scholars recognize that men are not equally privileged and examine gender in relation to its linkages to race/ethnicity and class (Ferree 2010; and see McIntosh 1988). For example, a black woman immigrant from Haiti who is a single mother and domestic worker has a very different life from that of her employer, an Asian Indian woman lawyer who is an Ivy League graduate with a professional husband. Furthermore, the professional husband's life is much different from that of a white male high school dropout who remains single into his forties because he finds himself financially unable to marry (Porter and O'Donnell 2006). "When race, social class, sexual orientation, physical abilities, and immigrant or national status are taken into account, we can see that in some circumstances 'male privilege' is partly—sometimes substantially—muted" (Baca Zinn et al. 2004, p. 170, citing Kimmel and Messner 1998).

Moreover, studies show that race/ethnic differences in role expectations and behaviors are actually not as strong as stereotypes suggest. The importance of the male provider role is a powerful theme in *all* race/ethnic groups (Cunningham 2008; Taylor, Tucker, and Mitchell-Kernan 1999). Chapter 1 discusses how historical periods and events influence family behavior. In this section, we'll see that a good deal of the race/ethnic diversity in gendered behavior has its roots in groups' histories. In the United States, this historical impact is evident regarding immigration.

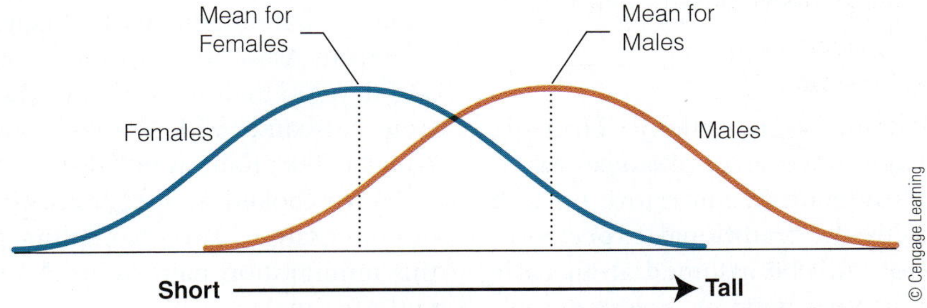

FIGURE 3.1 How females and males differ on one trait, height, conceptualized as overlapping normal distribution curves. Means (averages) may differ by sex, but trait distributions of men and women occupy much common ground.

Immigration and Gender Marriages of more recent immigrants appear to be less egalitarian, or equal, than those of couples of similar background whose families have been in the United States longer. This situation arises because migration tends to change masculine–feminine roles among those from cultures where family life involves females' dependence on and acceptance of decisions of male family heads. Male heads of household typically lose status when masculine privilege and authority in the United States is not what it was in the home country, and they may have to take jobs of much lower status than they had at home. Women who enter the labor force after coming to the United States begin to experience an independence and autonomy that carry over into the negotiation of new gender roles and decision-making patterns. This situation has been found among Hispanics, Asians, and Arabs, particularly Muslims (Hondagneu-Sotelo and Messner 1994; Jo 2002; Kibria 2007).

Regarding the latter, Shireen Zaman of the Muslim Research Institute for Social Policy and Understanding, explains that, "What we're seeing now in America is ... a quiet or informal empowerment of women. In many of our home countries, socially or politically it would've been harder for Muslim women to take a leadership role. It's actually quite empowering to be Muslim in America" (Knowlton 2010). Feminist activist and scholar Naomi Wolf, author of the 1991 classic *The Beauty Myth: How Images of Beauty Are Used Against Women,* recently commented on the Muslim practice of women wearing a full covering called a *hijab* or a head scarf. Her comments are worth considering here:

> Westerners should be aware of being presumptuous in assuming they know that a hijab means oppression to a woman wearing it.... [M]any feminists I've heard from including young women in Western Europe ... said very intriguingly ... that when they wore the head scarf or modest clothing like Hijab, they chose to do so ... [because] it made [them] feel freer in the Western context than wearing Western clothing, freer of objectification, freer of sexual harassment, freer of having to worry all the time about how [they] looked compared to fashion models. (Wolf 2012)

Hispanics and Gender In high school, Hispanic women, or Latinas, experience cross-pressures when the desire to succeed academically and move on to a career is in tension with the traditional expectation that the wife-mother role will be assumed at an early age. Young Latinas seem torn between newer models for women as mothers *and* career women versus the traditional model of marriage and homemaking exclusively (Canedy 2001). On average among Mexican immigrants, women do more of the housework and child care and have less decision-making power than men—but they have more power than their counterparts in Mexico (Pinto and Coltrane 2009). Salvadoran immigrant women, who typically were employed in their home country, nevertheless remark on their greater autonomy in the United States. They feel freer to come and go without a husband's close monitoring and also feel more likely to get help dealing with an abusive spouse (Baca Zinn and Wells 2007; Oropesa and Landale 2004; Zentgraf 2002).

Nevertheless, cultural images of Latinas depict them as submissive to their men (Andersen and Collins 2007a; Covert and Dixon 2008; Gewertz 2009). Meanwhile, Hispanic men, or Latinos, are stereotyped as following the *machismo* cultural ideal of extreme masculinity and male dominance (Ramos-Sánchez and Atkinson 2009). The relatively few Latinos whose behavior corresponds with this stereotype may be reacting to U.S. Latino history. According to William Carrigan and Clive Webb (2009), racial prejudice against Mexicans in the late 1800s through the early 1900s emphasized Mexican men as having more feminine attributes such as cowardice and a preference for wearing "fancy" clothing. Overcoming this demeaning legacy may be an important part of the Latino psychological and cultural heritage that has contemporary ramifications.

Asian Americans and Gender Male dominance may continue to be characteristic of some recent Asian immigrants (Ishii-Kuntz 2000). However, scholars note increased independence of Asian women in the United States. Ironically, they attribute this development to discrimination. To begin with, the image of Asian women as subordinate to men in patriarchal households was not always the reality. Historically, in the United States, Asian women entered the labor force because of the low wages of Asian men. Furthermore, President Franklin D. Roosevelt's authorized internment, or movement, to War Relocation Camps of Japanese Americans during World War II eroded Japanese husbands' provider role and undercut their authority over women and children. Although there is still evidence of gendered division of labor, contemporary Japanese couples evidence greater equality than in the past (Takagi 2002). Similarly, as American Asian Indian women obtain more education and develop their own careers, they demand more help from husbands with domestic chores and child care (Bhalla 2008; Kallivayalil 2004).

We've looked at race/ethnicity within the context of immigration. Two race/ethnic groups that do not fit this immigration pattern are Native American Indians and African Americans.

Native American Indians and Gender Native American Indians have a complex heritage that varies by tribe and may include an organizational structure in which

Erma J. Vizenor is chairwoman of the White Earth Reservation, the largest tribe in Minnesota. Vizenor, who holds a bachelor's degree in elementary education and a Harvard doctorate in administration, planning, and social policy, is one of the approximately 20 percent of Native American tribal leaders who are female. She works closely with the national organization, American Indian Women of Proud Nations, whose mission is to "support American Indian women's efforts to build healthier lives for themselves, their families, and their communities in a spirit of holistic inquiry and empowerment" (Conference for American Indian Women of Proud Nations 2012).

women own the family's house, tools, and land. Historically, Native American women's political power declined with European invasion and the subsequent spread of patriarchally organized social institutions in what is now the United States. Colonial Europeans in influential positions refused to recognize female Native American leaders, and Native Americans' forced movement to reservations further undermined gender equity (Young n. d.). Recently, Native American women have begun to regain their power (Davey 2006). Meanwhile, the increase in female-headed families has been "the most significant role change in recent times," one that has enhanced female authority and status in the family and the tribe (Yellowbird and Snipp 2002, p. 243). The number of women tribal leaders has doubled over the last thirty years, and it seems that one result is greater attention to child welfare, other social services, and education (Davey 2006).

African Americans and Gender At least partly because of a history of slavery in which both genders labored and partly because of postslavery discrimination against black men in the labor force, African American women have had higher employment rates than their white counterparts (Corra et al. 2009; Durr and Hill 2006). Although black wives do more housework and child care than their husbands, on average African American husbands do more than other U.S. husbands (McLoyd et al. 2000; Taylor 2002a). Research finds that black couples experience and prefer more role flexibility and power sharing than do whites (Corra et al. 2009; Cowdery et al. 2009). Studies also show that African American boys and girls are fairly equally socialized in both employment and child care (Brown et al. 2009; Staples and Boulin Johnson 1993; Theran 2009).

However, greater role flexibility in black families does not lessen the priority of provider-role expectations for men, despite the difficulties encountered by black males in fulfilling this role. Interestingly, this situation may be related to the fact that many African American families see the ability to have traditional gender relationships as evidence of economic success (Cowdery et al. 2009; Furdyna, Tucker, and James 2008). Having examined cultural messages regarding gender and having considered the degrees to which various race/ethnic groups behave according to these cultural expectations, we turn to an investigation of how gendered expectations and roles generally emerged.

How Did Gender Roles Emerge?

Scholars, policy makers, and many of the rest of us have been asking—since the beginning of time?—how gender differentiation and inequities came to be. In this section, we'll look at two broad types of arguments designed to determine whether differentiation is biology based or society based. Table 3.1 (on page 62) summarizes the arguments.

Biology-Based Arguments

As discussed in Chapter 2, the biosocial or evolutionary psychology perspective sees basic aspects of human behavior as partly biological, having evolved to facilitate human survival (Berenbaum, Blakemore, and Belz 2011). Gender differences—males on average larger, more assertive, and more physically aggressive than females and females more likely responsible for child care—are seen as anchored in hereditary biology (D' Onofrio and Lahey 2010; Hasan and Fauzi 2012; Udry 1994, 2000). To the extent that biosocial explanations for gender differentiation are correct, gendered differences will be difficult to change. However, biological *predisposition* does *not* mean that a person's behavior cannot be influenced or changed by social structure (Bearman 2008). Biology ("nature") and society ("nurture") interact to produce human behavior (McIntyre and Edwards 2009; Rossi 1984).

Biological theories of gender difference were initially offered by ethologists, who study humans as an evolved

Table 3.1 Focus on Gendered Behavior as Primarily Biological versus Socially Structured

Gendered Behavior as Biological and "Natural"	Gendered Behavior as Socially Structured
1. Gendered behaviors developed to meet family system needs and are natural and functional.	1. Gender-role differences depend on individuals' "doing" or performing gender as they interact with others to negotiate their own individuality amid structured opportunities and constraints along with cultural expectations.
2. Functional needs of (heterosexual, nuclear) family result in differentiation into an instrumental leader (father) and an expressive leader (mother). The two roles—instrumental and expressive—describe division of labor and other aspects of interaction within families.	2. Although they do not have to, our socially structured institutions value—and reward with money and power—instrumental, agentic behaviors over expressive ones.
3. Childhood gender socialization entails learning either the instrumental (male) or expressive (female) role; by late adolescence, gender-appropriate attitudes and behaviors usually have been thoroughly internalized.	3. Socialization includes learning to instrumental and expressive behaviors and attitudes, as well as learning to ascertain when best to display either and to what degree.

animal species (e.g., Tiger 1969). Lionel Tiger, who primarily studied baboons, found male baboons to be dominant and argued that *Homo sapiens* inherited this condition through natural selection. Newer data on non-human primates shows that species vary in their behavior, with males not consistently dominant (Bartlett 2009; Haraway 1989; Vogel et al. 2009). Subsequent biological theory has focused on genes. Most often found in the field of evolutionary psychology, this theoretical model sees humans' behavior as primarily associated with the (often unconscious) need to pass on one's genetic material to future generations. To pass on their genes, persons act to maximize their reproduction or that of close kin, a goal requiring divergent behaviors on the part of the sexes. Because of anatomical differences, males seek to impregnate as many females as possible, while females prioritize guaranteeing the best conditions for nurturing their relatively small number of offspring (Buss 2009; Buss and Shackelford 2008; Dawkins 1976).

Furthermore, because of their greater physical strength and freedom from reproductive responsibility, men were more likely to hunt, especially large game. Women, who might be pregnant or breast-feeding, gathered food that was available close to home while they also cared for children. Divergent circumstances elicited different adaptive strategies and skills from men and women, and these became encoded in the genes—greater aggression and spatial skills for men, nurturance and domesticity for women. According to this perspective, these traits remain part of our genetic heritage and are today the foundation of gender differences in personality, abilities, and behavior (Maccoby 1998).

Moreover, genetic heritage is expressed through **hormonal processes**. **Hormones** are chemical substances secreted into the bloodstream; they influence the activities of cells and body organs. The primary female hormone is *estrogen*, secreted by the female ovaries primarily. Although present in both women and men, estrogen is typically evident at higher levels in women of reproductive age. Among other functions, estrogen is instrumental in breast development and other secondary sex characteristics (anatomical characteristics other than genitalia associated with of the male or the female sex). After childbirth and while breast-feeding, women also secret *oxytocin*, a hormone associated with bonding and nurturance. Both men and women secret oxytocin at tender or loving times, such as when cuddling or during intercourse (Ditzen et al. 2009). Oxytocin is further discussed in Chapter 11.

The primary male sex hormone is *testosterone*, produced in the male testes. Females also secrete testosterone but in smaller amounts. As with estrogen in women, testosterone levels in men peak in adolescence and early adulthood and then slowly decline over the remainder of the life cycle. Research shows that testosterone level has significant impact on aggression levels in both sexes (Carré and McCormick 2008; Mehta, Jones, and Josephs 2008). Sex hormones also cause *sexual dimorphism*—that is, physical differences between the sexes in body structure and size, muscle development, fat distribution, hair growth, voice quality, and the like. Sociologist Randal Collins has proposed that males' greater average physical strength, resulting partly from higher testosterone levels, resulted in violence or threats of violence that helped create male dominance in past (and in some current) societies (Coltrane and Collins 2001).

A related area of biology-based scholarship focuses on brain organization and functioning. The argument is that male and female brains differ at birth due to greater amounts of testosterone secreted by a male fetus (Blum 1997; Hines 2011; Springer and Deutsch 1994). Studies in this area have found not only biological but also social causes for brain differences (Jordan-Young 2012; Liu et al. 2009). Neuroscientist Lise Eliot's *Pink Brain, Blue Brain* (2009) reviews the research and concludes that, "infant brains are so malleable that small differences at birth become amplified over time, as parents, teachers, peers—and the culture at large—unwittingly reinforce gender stereotypes" (book jacket; see also Jordan-Young 2012).

Despite documented hormonal influences on human behavior, the relationship between hormones and behavior goes in both directions: What's happening in one's environment can influence hormone secretion levels. Several studies have found, for example, that the hormonal levels of males in romantic relationships and those of new fathers undergo changes parallel to those associated with maternal behavior (e.g., lower testosterone and cortisol and detectable levels of estradiol, an estrogen-like hormone) (Gray, Ellison, and Campbell 2007; McIntyre et al. 2006; van Anders and Watson 2006, 2007). On the contrary, testosterone rises in men in response to insults or to athletic and other forms of competition (Carré and McCormick 2008; Mehta, Jones, and Josephs 2008). *Sociologists* who work from a biosocial perspective—for instance, Alan Booth and colleagues (Booth, Carver, and Granger 2000; Booth, Granger, Mazur, and Kivlighan 2006; Booth, Johnson, and Granger 2005)—are finding complex interactions among gender, social roles, and biological indicators rather than categorical gender differences. Biology-based arguments leave plenty of room for culture and the influence of social factors (Eliot 2009; Lewontin and Levins 2007; McIntyre and Edwards 2009; Nielsen 2009).

Society-Based Arguments

Society-based arguments examine how conditions in the broader society affect gender. We can look at the emergence of gender roles throughout four broad economic stages of human history: (1) foraging and hoe, (2) agricultural, (3) industrial, and (4) postindustrial societies.

Foraging and Hoe Societies In both foraging (hunting and gathering) and hoe societies (where digging sticks were used to break ground for minimal levels of cultivation), food production was relatively compatible with pregnancy, childbirth, and breast-feeding; thus, women played an important part in the economy. Among foragers, 60 percent to 80 percent of the food comes from gathering activities performed predominantly by women, and both sexes may hunt small game (Linton 1971). With women fully participating, males are less dominant than in agricultural or industrial societies (Chafetz 1989).

Agricultural Societies Agricultural societies—based on plow cultivation, the domestication of animals, and land ownership (hoe cultivators were nomadic)—developed about 5,000 years ago. Plow agriculture requires greater physical strength, full-time labor, and work farther away from the family dwelling, conditions less compatible with pregnancy and nursing (Basow 1992). Although women continued to contribute significantly to the family economy, men did the plowing and other heavy work, thereby making women's productive labor less visible. About this time, **patriarchy**—a form of social organization based on the supremacy of fathers and inheritance through the male line—became firmly established. As it became possible to accumulate wealth through large landholdings, concern with property inheritance—and hence with the legitimacy of offspring—increased the social control exerted over women. Chapter 4 explores patriarchy as it applies to sexual attitudes and behaviors.

Industrial Societies With industrialization, beginning about 200 years ago in Europe, economic production gradually shifted from agriculture to mechanized production of manufactured goods. The status of women declined further as industrialization separated work from home and family life, transferring work traditionally done by women (such as clothing production) from homes to factories. Because factory-based industrial work could not easily be combined with domestic tasks and the supervision of children, women at home no longer contributed directly to economic production; their indirect contribution to the economy through domestic support and reproduction of the labor force (that is, by bearing and raising children) became virtually invisible.

In the middle and upper classes, an ideology of *separate spheres* arose to support this separation of men's and women's roles. Men came to be seen as possessing the instrumental traits referred to earlier and as being more comfortable than women with the competition and harshness of the **public sphere**, or world outside the home. Middle- and upper-class (white) women were "angels of the house," whose delicacy and passivity were appropriate to their sheltered lives in the complementary **private sphere**. [We should note that some women were able to use the moral authority given to them by a separate-spheres ideology to lead antislavery campaigns, temperance crusades, and suffrage movements (Rossi 1973).]

Women's submissiveness and emotional sensitivity, the expressive qualities described earlier, supposedly led them to provide a haven for their wage-earner husbands to bring up innocent children (Cancian 1987). These breadwinner and housewife roles applied mainly to middle- and upper-class white men's and women's work—and for only a brief period, from the late nineteenth century to the mid-twentieth century. Most immigrant, black, and working- and lower-class women did not have the housewife option, but instead needed employment in domestic service, factories, or home-based activities such as piecework performed for money. Nevertheless, although the traditionally idealized portrait of men, women, work, and personality is changing, we have retained remnants of it to the present.

Postindustrial Societies A postindustrial society is dominated by information-based and service work, not

BobPeterson/UpperCut Images/Getty Images

Today's postindustrial labor force is more inclusive of women, minimizing gender differentiation and ushering in further gender-role changes. Illustrating this, more women are entering nontraditional occupations, such as the military. Women comprise about 15 percent of today's military (Patten and Parker 2011). Some counselors have wondered about the difficulty soldier-mothers may have in returning from the battlefield to the "normal" life of parent and child (L. Alvarez 2009; St. George 2006).

the manufacture of things. Processing ideas and information in complex areas (for example, media, law, financial services, computer technology) and providing services that range from medical care to telemarketing now dominate the U.S. economy. The idea that economic arrangements underlie gender norms still holds, however. Because a postindustrial labor force is more inclusive of women, it minimizes gender differentiation and spurs further changes in gender roles. Having examined how historical periods have impacted gender roles, we turn to another conceptualization of how the broader society affects individuals' gendered options—gender structures.

Gender Structures

A theme of this text is that opportunities available in the social structure (that is, *socially structured opportunities*) have an effect on men's and women's options and behaviors. The "deceptive differences" (Epstein 1988) we observe or think we observe typically involve men and women assigned to different social roles. A woman secretary, for example, is expected to be compliant and supportive of her male boss's decisions. To observers, she seems to have a gentle and submissive personality, while he is seen to have leadership qualities (Meier, Hull, and Ortyl 2010;Webster and Rashotte 2009).

Moreover, institutional structures—such as family, church, state, the economy, and education—play critical roles in influencing the ways that people enact gender (Carrigan, Connell, and Lee 1985).

Gender, like other personal attributes, is embedded in social structure, in the broader society. Sociologist Barbara Risman (2011) has suggested that gender roles are influenced by a society's sociocultural environment. Just as every society has a political structure, such as democracy or monarchy, so too every society has a **gender structure** "from patriarchal to at least hypothetically egalitarian."

Gender structures have implications for socialization and for the development of identities and selves (e.g., the individual level). But gender structures also shape the social roles women and men are expected to follow, what "doing gender" means in any given interactional encounter, and how marriage is understood and defined. Gender structures also are formalized into institutional laws, rules, and organizational norms (Risman 2011, pp. 19–20).

Institutions in virtually every society have been characterized by patriarchy and **male dominance**, by which males assume authority over the females. On the personal level, male dominance involves his wielding greater power in a heterosexual relationship. On the societal level, male dominance is the assignment to men of greater control and influence over society's institutions and benefits. Only very recently in human history—within the last one hundred years—has patriarchal dominance begun to break down. Today we see it more clearly in some social institutions than in others.

Gender expectations, behaviors, and options are embedded in our five basic social structures, or institutions: (1) family, (2) religion, (3) government or politics, (4) education, and (5) the economy. Moreover, expectations and practices in one institution affect those in the others. Ways that gender is imagined and scripted in education, for instance, can impact how gender is enacted in family kitchens. Throughout this textbook, we examine ways that gender is performed within families. In this section, we address gender in the four major social institutions other than family. Of course, family is the focus of this text and therefore discussed throughout.

Religion

Taken as a whole, the institution of religion evidences male dominance. Although most U.S. congregations have more female than male participants, men more often hold positions of authority. Nevertheless, there is strong evidence of change. Women have been elected as bishops and denomination leaders in the African Methodist Episcopal, Anglican, United Methodist, and Presbyterian churches (Chaves, Anderson, and Byassee 2009; Gott 2010).

The effects of personal religious involvement on daily family lives are complex. On the one hand, growth in evangelical Protestantism, Islamic fundamentalism, and the Latter-day Saints religion, along with the charismatic renewal in the Catholic Church, has fostered a traditional family ideal of *male headship*—the husband as provider and decision-making "head" of the family—and a corresponding "emphasis on domesticity for women—the belief that a woman's role is to care for children, her husband, and the home—and the ascribing of men's roles to the public sphere and women's roles to the private" (Bulanda 2011, p. 180). Research shows that in households where the husband is more religious than the wife, particularly in Islam and evangelical Protestantism, the more strictly gendered divisions of labor can lead to family tension and decreased marital satisfaction, particularly for the wife, and increase the risk of divorce (Duba and Watts 2009; Vaaler, Ellison, and Powers 2009; Zink 2008).

On the other hand, actual practice among conservative Christian couples appears more egalitarian than formal doctrine would suggest. As feminist sociologist Myra Marx Ferree put it, "conservative Christians . . . are more or less consciously remaking patriarchy for themselves" (Ferree 2010, p. 429, citing Wilcox 2004). For one thing, there is a diversity of viewpoints on gender roles *within* conservative Christian religions (Bartkowski 2001; Beaman 2001; Gallagher 2004). For some, "headship has been reorganized along expressive lines, emptying the concept of virtually all of its authoritativeness" (Wilcox 2004, p. 173). Put another way, the ideas of a *servant* head and *mutual* submission (rather than simply a wife's submission to her husband) lead to more egalitarian decision making in day-to-day family life.

Evangelical women believe that male headship provides them with love, respect, and security (Bulanda 2011). Although they may expect to do less housework than other husbands, conservative Christian men, more than mainstream Protestant men, do seem more emotionally expressive with their wives and children and more committed to their marriages. Conservative Christian husbands also have lower rates of domestic violence than average (Gallagher 2004; Wilcox 2004). Economic need shapes a pragmatic approach to the employment of women, but conservative Christian women are more likely to scale back their work hours and job level upon marriage or childbirth or both (Glass and Nath 2006).

There is also a growing feminist movement among American Muslim women who seek to combine their religiocultural heritage with equal rights for females (Knowlton 2010). According to Muslim family life educator Aliya Hirji (2012), for instance,

> Marriage is a partnership between two people. It is written [in the Qur' an 30:21] that among God's signs is that God "created you for your mates from among yourselves that ye may dwell in tranquility with them, and He has put love and mercy between your (hearts)." (p. F12)

Hirji emphasizes the ability to honor one's conviction to Islam while working toward increased gender equality (Hirji 2012).

Government and Politics

As of March 2013, women held ninety-seven of the 535 seats in the 113th session of the U.S. Congress—twenty women or 20 percent of the 100 seats in the Senate and seventy-seven women or 18 percent of the 435 seats in the House of Representatives (Center for American Women and Politics 2013). In addition, three women serve as delegates from Guam, the Virgin Islands, and Washington, D.C. Twenty-nine, or 30 percent, of the ninety-seven females in the 113th Congress are women of color, all except one serving in the house: thirteen African Americans, seven Latinas, and seven Asian or Pacific Islanders. In the 112th Congress, five female senators and seven female representatives served in leadership positions, including as committee chairs (Center for American Women and Politics 2013).

Of our nine U.S. Supreme Court justices, three are women. Women have become more visible in the executive branch of government as well. Three women have served as secretary of state: Madeleine Albright, appointed in 1997; Condoleezza Rice, appointed in 2005; and Hillary Rodham Clinton, appointed in 2009 and a presidential contender herself. We see change, although not equality, when women—51 percent of the total population—comprise just 17 percent of Congress and a minority of the cabinet and Supreme Court.

In 2008 when Hillary Clinton was competing for the Democratic party nomination to run for president, surveys showed that 71 percent of the public said it was willing to vote for a woman president—but only 56 percent thought that family members, friends, or coworkers would be willing to do so (Rasmussen Reports 2008). According to political strategists, voters "have grown more accustomed to women in powerful positions" (Toner 2007). Nevertheless, statistics point to the fact that women, although slightly more than 50 percent of the population, are significantly

Michael Temchine/The Washington Post/Getty Images

Over the past century and a half, 36 women received their political party's nomination and ran for U.S. president. The first female candidate was Victoria Woodhull of the Equal Rights Party. She ran in 1872. In 2012, Chicago-born physician and environmental author-activist Jill Stein (on the right in this photo) was the Green Party's candidate for U.S. president. Her running mate (left) was Minneapolis-born antipoverty and social justice advocate Cheri Honkala. Comedian and author Roseanne Barr, born in Utah Salt Lake City of Jewish and Mormon background/heritage, ran for the Peace and Freedom Party and focused on economic issues. Her running mate was Cindy Sheehan, American antiwar activist whose son was killed in Iraq. Sheehan's memoir, *Peace Mom: A Mother's Journey Through Heartache to Activism*, was published in 2006. Peta Lindsay, Virginia-born antiwar activist, ran for the Party of Socialism and Liberation, with Yari Osorio as her running mate.

underrepresented in high government positions. Males of color are underrepresented as well. Although African Americans are 13 percent of the U.S. population, the forty-four blacks in the House of Representatives—and none in the Senate!—comprised just 10 percent of the 112th Congress. More dramatically, the twenty-five Hispanics in the House of Representatives and the two in the Senate comprised 5 percent of Congress, although Hispanics made up about 10 percent of the population old enough to be elected to these positions ("Men and Women in the U.S. Congress" 2012; U.S. Census Bureau 2012c, Table 10).

Race and gender inequities in government representation impact our daily lives because individuals of different races, ethnicities, or gender have, on average, different opinions about important issues. A recent survey by the Pew Research Center found that 14 percent of blacks, compared with 38 percent of whites, saw the police as likely to treat blacks fairly. Forty-three percent of blacks, compared with 13 percent of whites, said there is a lot of discrimination against blacks, and 58 percent of African Americans saw news coverage of blacks as too negative, compared with nearly half that proportion—31 percent—for whites (Pew Research Center 2012e). In another recent survey, U.S. respondents were asked what should be the priority for dealing with illegal immigration. Compared with 29 percent of the general population, 10 percent of Hispanics said the answer lay in better border security. Compared with 24 percent of the general population, 42 percent of Hispanics thought the government should create a path to citizenship for undocumented workers (Lopez, Gonzalez-Barrera, and Motel 2011, Figure 4).

Gender differences are also apparent in opinion polls, with women more often wary of nuclear power; seeing government as doing "not enough" for the elderly, children, and the poor; supporting government subsidized birth control, same-sex marriage, and protection of the natural environment; and more likely than men to say that peace is best achieved through diplomacy as opposed to armed combat (Pew Research Center 2012d). Inasmuch as women and men, as well as race/ethnic minorities, tend to have differing positions on critically important topics, the relative proportions of men and women in political positions impacts social policy.

Education

This section explores college and university education as a gender structure. Education before college is addressed later in this chapter. Women have been the majority of college students since 1979 and now surpass men in the proportion of the total population who are college graduates (Fry and Cohn 2010, p. 21). In 2009, women earned 57 percent of bachelor's and 60 percent of master's degrees, 49 percent of first professional degrees, and 48 percent of doctorates (U.S. Census Bureau 2012c, Table 299). Men's college enrollments have *not* declined, but they have not increased as rapidly as women's have (Corbett, Hill, and St. Rose 2008; Fry and Cohn 2010). Nevertheless, some colleges now actively recruit men (Jaschik 2009; Meyer 2009).

The changing gender balance in higher education—indeed, in high school completion, with higher proportions of males than females dropping out (U.S. Census Bureau 2012c, Table 273)—has led to interest in men's and women's changing economic and power balances both in the broader society and within families. Recently, the Pew Research Center found that in 40 percent of families with children (including single-parent families, which are mostly female-headed), women are either the sole or principal breadwinners (Wang, Parker, and Taylor 2013).

What about the college climate itself? In 2007, 45 percent of college faculty members were white males, 32 percent white females, 3 percent black males, 3 percent black females, 5 percent Asian males, 3 percent Asian females, 2 percent Hispanic males, 2 percent Hispanic females, and less than 1 percent each Native American males and females (Snyder, Dillow, and Hoffman 2009). Women were 43 percent of faculty in 2009 compared with 32 percent in 1995 (U.S. Census Bureau 2012c, Table 296). Relative to men, women have gained faculty positions and are expected to continue doing so as more females earn academic credentials. Nevertheless, the number of *full-time* female faculty members continues to lag, especially at elite institutions such as Harvard and Yale (E. Kramer 2011; Lloyd-Thomas and Wang 2012).

Although women students outnumber men in colleges and universities today, there is still gender differentiation in their choice of majors, as many men and women continue follow "traditional" expectations. For instance, in 2009, 14 percent of associate's degrees in engineering went to women, compared with 87 percent of those in nursing (U.S. Census Bureau 2012c, Table 301). Women's master's degrees are primarily in education, nursing, and social work (traditional female areas), while men still dominate in business, computer programming, management, math, and science—fields that tend to pay more (Coontz 2013). Faculty in "traditionally" female or male occupations become role models, nonverbally communicating by their relative numbers what is the appropriate major for students. With fewer female faculty in math, sciences, and engineering, women are less likely to feel encouraged to pursue those topics. With fewer male faculty in early childhood education, nursing, dental hygiene, or social work departments, men may feel less comfortable pursuing these options. Then, too, sexual harassment can be a problem on college campuses—for men as well as for women, and for gays and lesbians (AAUW Educational Foundation 2006; Dziech 2003).

An interesting qualitative study (Sallee 2011) used participant observation in one university aerospace and engineering department to uncover situations that discourage women from entering that field. The researchers interviewed the two female and twelve male doctoral students, as well as the one female and four male faculty in the department. The study concluded that, both as a discipline and as a specific department, aerospace and engineering is characterized by a culture of "invisible masculinity"—that is, by values of hierarchy (winning, being the best) and competition. Tim, a graduate student in engineering, told an interviewer that he'd rather talk with his female than male classmates. "With women, Tim felt that he was free to discuss whatever topics were of interest to either of them. In contrast, with men, 'conversation is more of a contest to win'" (p. 205).

Moreover, women faculty and students suspected as not being truly deserving to be in the department were noticed and remarked on for their appearance. After a new female faculty member's seminar talk, her male colleagues talked of "drooling" over her good looks, an exception in engineering, according to stereotype. As a male student told his interviewer,

> I think my [undergraduate] graduating class [in engineering] had three [girls], but two of them could play linebacker for the Bears.... No, engineering is not the place to go if you want to pick up on women. Every once in a while you'll see a cute girl walking down the E-Quad [engineering section of campus]. She's lost. (pp. 206–207)

Uncomfortable in this climate of "invisible masculinity," female faculty and students reported feeling marginalized. Interestingly, departmental efforts to be fairer to women were perceived by the female students as mostly insincere attempts to be politically correct or "careful." This situation pointed to, rather than normalized, the students' gendered presence.

We know of no comparable study in subjects such as elementary education, nursing, or dental hygiene—fields numerically dominated by women (Dewan and Gebeloff 2012). But we might speculate that an analogous "invisible femininity" characterizes these disciplines and departments. For instance, social work curricula emphasize collaboration, encouraging, and building on other people's strengths. Perhaps males, more comfortable with a masculine than with a feminine cultural script, are inclined to feel marginalized in departments like these. In sum, the gender structure in education has definitely changed with regard to males' and females' relative participation. However, evidence of gender differentiation remains, with males and females more often found in "masculine" or "feminine" majors, respectively.

Economics

According to the Pew Research Center and the Bureau of Labor Statistics, only 4 percent of husbands had wives who brought home more income than they did in 1970, a share that rose to 29 percent in 2010 (Fry and Cohn

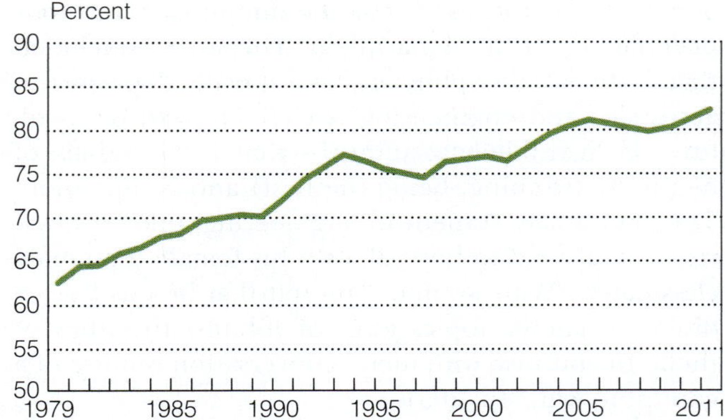

FIGURE 3.2 Women's earnings as a percent of men's, full-time wage and salary workers, 1979–2011 annual averages.

U.S. Bureau of Labor Statistics 2012b, Chart 1.

2010; U.S. Bureau of Labor Statistics 2013, Table 25). Although the situation is gradually changing, men on average have been and continue to be dominant economically. As depicted in Figure 3.2, among full-time employees women earn about 82 percent of what men do. As shown in Figure 3.3, nonHispanic white women earn about 82 percent of what white men do. Asian women earn 77 percent of what Asian males earn. Black and Hispanic female–male comparisons are more favorable to women, at 91 percent, but this is primarily because black and Hispanic men have much lower earnings than do Asian or white men (U.S. Bureau of Labor Statistics 2012b). Moreover, although women now comprise 49 percent of the paid workforce, the percentage of female chief executive officers at Fortune 500 companies was 4.2 in 2013—up from 1.4 percent in 2003 but still very low (Luscombe 2013, p. 36).

Some people have argued that men's earnings are higher because jobs more often held by males are more difficult, require more training, or have less-favorable working conditions. Assumptions that jobs typically filled by women are less challenging can affect pay scales. Furthermore, as women enter the job pool for certain professional jobs, the field becomes more crowded and wages and salaries are less competitive (Blau, Brinton, and Grusky 2006). A woman also contends with employers' assumptions—and perhaps her own assumptions—that she will opt out of the labor force to take care of her children or other family members (Sandberg 2013a). As a result, an employer may be less likely to select even highly ambitious and fully committed women employees for further training or positions with advancement potential (Laff 2007; Pinto 2009). It's also possible that even ambitious women, valuing family and motherhood roles, may pull back a bit from pursuing the high leadership positions. Facebook executive Sheryl Sandberg, who caused quite a stir in 2013 with her controversial book *Lean In: Women, Work, and the Will to Lead* (2013a), explains this very point:

> From an early age, girls get the message that they will likely have to choose between succeeding at work and being a good wife and mother. By the time they are in

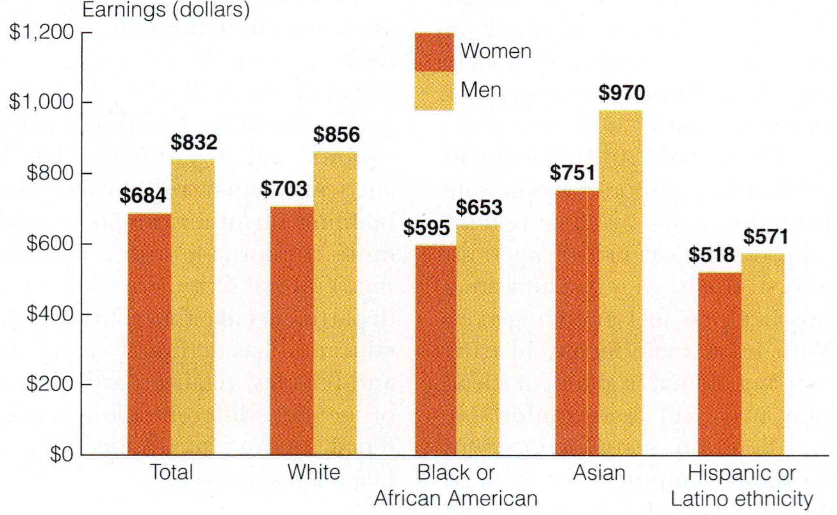

FIGURE 3.3 Median usual weekly earnings of full-time wage and salary workers by sex, race, and Hispanic or Latino ethnicity, 2011 annual averages.

U.S. Bureau of Labor Statistics 2012b, Chart 2.

Note: Persons identified as Hispanic may be of any race.

> college, women are already thinking about the trade-offs.... [Consequently, once employed in a career, they] make a lot of small decisions [that impact their career trajectory]. A law associate might decide not to shoot for partner because someday she hopes to have a family. A sales rep might take a smaller territory or not apply for a management role. A teacher might pass on leading curriculum development for her school. Often without even realizing it, women stop reaching for new opportunities. (Sandberg 2013b, p. 44)

Some analysts suggest that the male–female earnings difference is related to choice of occupational specialty and practice setting (Gauchat, Kelly, and Wallace 2012). However, as evidenced by the differences in pay for the same profession at both ends of occupational earnings, this explanation does not tell the whole story. The numbers show that in the same occupational categories women continue to earn less on average than do men. For instance, in the highest-paying occupation, that of chief executive, women's average salary was $1,464 weekly in 2011 whereas men averaged $2,122. Among elementary and middle school teachers, women and men averaged $933 and $1,022 per week, respectively. Among food-preparation workers, women earned an average $375 per week while men made $395 (U.S. Bureau of Labor Statistics 2012b, Table 2).

Not all of the wage gap is explainable by factors such as men's and women's employment histories and skills or women's own choices. After various factors associated with earnings are taken into account, research has found a varying differential in earnings (by perhaps as much as 22.9 percent) that is usually attributed to discrimination, albeit of a nonobvious sort (Bergmann 2011; Institute for Women's Policy Research 2009). We return to explanations for gendered earning disparities in Chapter 10.

Note, however, that despite the persistence of gendered economic differentiation, there is also considerable evidence of convergence in men's and women's earnings. As shown in Figure 3.2, women's earnings as a percent of men's rose from 62 percent in 1979 to 82 percent in 2011. Furthermore, younger employed women today, those between ages 25 and 34, earn a larger proportion—92 percent—of what men do (U.S. Bureau of Labor Statistics 2012b). Unfortunately, some degree of this convergence results from falling wages for men. At the same time, the narrowing gap also comes from rising wages for women as they have increased their education and skills (England 2006).

Both the gendered divergence in earnings that persists to a degree for most men and women, as well as the trend toward earnings convergence that we have seen over the past several decades, have dramatic and unprecedented implications for family interaction, such as division of household labor and family power dynamics. Having examined four gender structures—religion, politics, education, and economics—we turn to a discussion of how individuals learn the roles they are expected to play in these institutions and their families.

Gender and Socialization

As people in a given society learn to talk, think, and feel, they *internalize* cultural expectations about how to behave (Mead 1934); that is, they come to accept these expectations as their own. The process by which society influences members to internalize attitudes and expectations is called **socialization**. Interactionists point out that individuals do not automatically absorb; rather, they negotiate cultural attitudes and roles. Furthermore, they do so throughout life. Therefore, gendered behavior varies from individual to individual. Nevertheless, in various ways society encourages people to adhere, often unconsciously, to culturally acceptable gender roles.

Cultural images in language and in the media convey the gendered expectations described earlier in this chapter. Soon after birth, for instance, most infants receive either a masculine or a feminine name—not to mention a pink or a blue blanket. Besides first names, titles, adjectives, nouns, pronouns, and verbs remind people that males and females differ—and in stereotypic ways (Twenge, Campbell, and Gentile 2012). A new mother may be told that she has either a "lovely girl" or a "strong boy." In her book about the differences in masculine and feminine communication styles, linguist Deborah Tannen noted the following: "If I wrote, 'After delivering the acceptance speech, the candidate fainted,' you would know I was talking about a woman. Men do not [more delicately] faint; they pass out" (Tannen 1990, pp. 241–242).

A study of 6,000 children's books published between 1900 and 2000 found that—even in recent years—males were central characters in 57 percent of the stories; females were central in 31 percent. When the characters were animals, the proportion that were male was twice that for female animals. When a child's name was in the book title, the name was most often a boy's (McCabe et al. 2011). In her cleverly titled book, *Cinderella Ate My Daughter*, journalist and gender scholar Peggy Orenstein (2011) shows how intensive twenty-first-century marketing of a "new girlie-girl culture" encourages girls at "younger and younger ages... to define themselves through appearance—from the outside in rather than from the inside out" (Orenstein 2012). In the adult media, music videos typically portray females as trying to get a man's attention. Some videos broadcast shockingly violent misogynist (hateful of women) messages (Conrad, Dixon, and Zhang 2009). Social scientists have offered various theories on how we absorb gendered messages in language and the media.

Gender Socialization Theories

In the classic **interaction-constructionist perspective** (see Chapter 2) children develop self-concepts based on feedback from those around them. Play is not idle time; it is a significant vehicle through which children develop ideas about appropriate gender roles for themselves both as children and as adults (Mead 1934). In play, preschool children imitate adult roles, such as parent, teacher, or health care provider. Little girls are more likely to play "mommy" with their dolls and (pink!) kitchen sets, whereas little boys more often play with cars or action figures. It is now likely that little girls as well as little boys play "going to work."

Among school-age children, studies show that boys and girls tend to play separately and differently (Aydt and Corsaro 2003; Maccoby 2002; McIntyre and Edwards 2009). Girls play in one-to-one relationships or in small groups of two and three; their play is relatively cooperative, emphasizes turn taking, requires little competition, and has relatively few rules. In "feminine" games such as jump rope or hopscotch, the goal is skill rather than winning (Munroe and Romney 2006). Boys more often play in fairly large groups, characterized by more fighting and attempts to effect a hierarchical pecking order. Boys also seem to exhibit high spirits and having fun (Maccoby 1998; Munroe and Romney 2006). Meanwhile, organized youth sports play a role in socialization (Hardin and Greer 2009). Now girls have more organized sports available to them, as well as more media models of women athletes. Girls who take part in sports have greater self-esteem and self-confidence (Andersen and Taylor 2002; Dworkin and Messner 1999).

According to **social learning theory** (Bandura and Walters 1963), a theory within the interaction-constructionist perspective, children (and adults) learn aspects of gender roles as they are taught by parents, teachers, peers, friends, partners, and the media. Children imitate models for behavior and are rewarded by parents and others for whatever is perceived as sex-appropriate behavior. As children grow older, toys, talk of future careers or marriages and parenthood, and admonitions about "sissies," "tomboys," "ladies," or "dykes" communicate parents' ideas about appropriate behavior for boys and girls. However subtle, the rewards and punishments that parents and others assign to gender expectations are assumed to be key to behavior patterns.

Some psychologists think that what comes first is not rules about what boys and girls should do but rather the child's awareness of being a boy or a girl. In this **self-identification theory** (also termed *cognitive-developmental theory*), children categorize themselves as male or female, typically by age 3. They then identify behaviors in their families, in the media, or elsewhere that are appropriate to their sex and adopt these behaviors. In effect, children socialize themselves from available cultural materials (Kohlberg 1966; Signorella and Frieze 2008).

Somewhat similar to self-identification theory, gender schema theory posits that children develop a framework of knowledge (a **gender schema**) about what girls and boys typically do (Bem 1981). Children then use this framework to organize how they interpret new information and think about gender. Once a child has developed a gender schema, the schema influences how she or he processes new information, with gender-consistent information remembered better than gender-inconsistent information. For example, a child with a traditional gender schema might generalize that physicians are men even though the child has sometimes had appointments with female physicians. Overall, gender schema theorists see gender schema as maintaining traditional stereotypes. This theoretical framework continues to be tested and still shows that younger children maintain more rigid gender stereotypes than do adolescents (Crouter et al. 2007; Signorella and Frieze 2008). In addition to offering theories on the processes by which gender expectations become internalized, social scientists have examined various settings for socialization. The next sections examine two of these: families and schools.

Gender Socialization in Families

Liza Mundy, a former *Washington Post* reporter and author of the 2012 book *The Richer Sex,* suggests that changing family structure is important in socializing girls and young women away from economic dependency. With 40 percent of babies born to unmarried women, the number and proportion of single-parent and cohabiting families is significant. Hence "young women are acutely aware that they may be the sole earner in their household" (Mundy 2012, p. 34). Regardless of family structure, certainly parents influence their children's gender attitudes and behavior (Epstein 2011). An interesting study of Mexican American families found that adolescents in families that espoused traditional division of labor and attitudes were more likely themselves to espouse traditional gender attitudes (Lam, McHale, and Updergraff 2012). Despite some recent studies, however, the majority of our findings on family gender socialization have been based on research with white, middle-class, two-parent families. From the 1970s on, these parents have reported treating their sons and daughters similarly (Wood and Eagly 2002).

However, differential socialization by parents and other family members does continue to exist, although it is typically not conscious (Kane 2012a). Instead, family gender socialization reflects the extent to which parents unconsciously "accept the general societal roles for men and women" (Kimmel 2000, p. 123). A study of 120 babies' and toddlers' rooms found that girls had more dolls, fictional characters, children's furniture, and the color pink; boys had more sports equipment, tools, toy vehicles, and the colors blue and red (Pomerleau et al. 1990). Parents encourage exploratory behavior more in boys than in girls. Toys considered appropriate for boys encourage physical activity and independent play, whereas "girl toys" elicit closer physical proximity and more talk between child and caregiver (Athenstaedt, Mikula, and Bredt 2009; Orr 2011).

The same parents who support nonsexist child raising for their daughters are often concerned if their sons are "not competitive enough" or are "too sensitive." While family members increasingly encourage girls to develop instrumental skills, boys are still discouraged from or encounter family ambivalence about cultivating tenderness (Blakemore and Hill 2008). Sociology professor Emily Kane interviewed preschoolers' parents from a wide variety of backgrounds and asked them to talk about gender and their children. In a public radio interview, she noted

> the ways in which most people were, at least to some extent, reinforcing traditional patterns like preparing their girls more for the sort of nurturance we associate with parenting, and preparing their boys more for the kind of competition we might associate with the workplace.... I think in some ways even those early decisions on whether you're going to encourage your daughter to hold a doll and nurture it, whether you're going to encourage your son to be competitive in sports, to be more independent, to be outside with a wider range of activity can end up reinforcing paths that lead women to have more obligations for family care, for example, and I should say in some ways really constrain boys and men too. (Cohen and Martinez 2012)

Courtesy Deb Glover and Celeste Wheeler

Children learn a lot about gender roles from their parents, whether they are taught consciously or unconsciously. Children also internalize messages from media and other cultural influences. Toys send messages about gender roles. "One consequence of children developing gender-stereotyped toy preferences is that it may constrain their experiences, since different types of toys facilitate different types of skill-building" (Goldberg, Kasby, and Smith 2012, p. 512). What does this toy say? What skills does it facilitate?

Furthermore, we saw earlier in this chapter, parents more often allow girls to express feelings of anxiety or sadness; boys, on the other hand, are more commonly allowed to express anger (Chaplin, Cole, and Zahn-Waxler 2005). Fathers in particular more easily accept a school-age daughter being a tomboy than a son displaying behaviors thought to be feminine (Marks, Lam, and McHale 2009). Many parents consciously try to avoid gender stereotyping as they raise their children. However, Kane found that even among those who "considered gender expectations problematically limiting to their children, I heard reports of an everyday world teeming with social pressures, judgments from friends, relatives, and even strangers if their kids didn't stick to a narrowly gendered path" (Kane 2012b).

There are exceptions, however. Some recent research and media attention has turned to parents faced with dilemmas about socializing children as young as 4 years old who display **gender variance**—that is, they are determined to dress and behave more like the "other" gender. (*Gender variance* differs from *gender bending*, mentioned earlier in this chapter; *gender variance* is a more expansive concept, encompassing gender-related attitudes and behaviors much more broadly.) Parents, teachers, and pediatricians wrestle with how early individuals should be allowed to define themselves (Hoffman 2011; Padawer 2012). The fact that this dilemma is considered an ethical question underscores our society's mostly invisible commitment to the idea that gender is dichotomous. Parental solutions? Often, let the child dress as he or she likes at home but limit what's presented in public. As one father told his kindergarten son, "We'll paint your nails [at home today] and wash them off before you go to school the next day" (Hoffman 2011). This tactic grows more difficult, of course, as the youngster gets older (Padawer 2012).

Gender Socialization in Schools

Although relations in the family provide early feedback and help shape a child's developing identity, peer groups and teachers become important as children grow to school age (Rose and Rudolph 2006). A practice evident among schoolmates involves **borderwork**—the process, or work, of monitoring and maintaining the conceptualized border between appropriately masculine boys or men and acceptably feminine girls or women (Aydt and Corsaro 2003; Thorne 1993). Educators have observed and documented borderwork in fifth- and eleventh-grade classrooms, among others. Students patrolled each other's behavior according to accepted gender expectations and scripts. The more common punishments (sanction) designed to keep classmates within appropriate gender boundaries were laughter, homophobic name-calling, and social exclusion, especially directed at males who were thought to be acting feminine (Godley 2006; Lee and Troop-Gordon 2011; Poteat, O'Dwyer, and Mereish 2012).

Moreover, teachers may reinforce the idea that males and females are more different than similar. At times boys and girls interact comfortably—in the school band, for example. But teachers often pit girls and boys against each other in spelling bees or math contests, for instance. A recent study by Pennsylvania State University psychologists found that by the ways that they structure their classrooms—for example, lining up children by gender or having them put their completed work on different bulletin boards—many (although not all) preschool teachers emphasize dichotomous gender differentiation. The researchers further found that when teachers differentiate between boys and girls, young children are more likely to negatively stereotype and prefer not to play with the other sex (Hilliard and Liben 2010). On the other hand, young children engage in fewer stereotypical behaviors when they play in gender-mixed groups (Goble et al. 2012).

Furthermore, studies show that teachers pay more attention to males than to females, and males tend to dominate learning environments from nursery school through college (Lips 2004; Zaman 2008). Compared to girls, boys have been more likely to receive a teacher's attention, to call out in class, to demand help or attention from the teacher, to be seen as model students, or to be praised by teachers. Boys are also more likely to be disciplined harshly by teachers (Epstein 2007). Peggy Orenstein, also previously discussed, spent one year observing pupils and teachers in two California middle schools, one mostly white and middle class and the other predominantly African American and Hispanic and of lower socioeconomic status. Orenstein (1994) found that in both schools, girls were subtly encouraged to be quiet and nonassertive whereas boys were rewarded for boisterous and even aggressive behaviors.

Interestingly, African American school girls were louder and less unassuming than nonHispanic white girls. For example, they called out in class as often as the boys did. But Orenstein noted that teachers' reactions differed. The participation and even antics of white boys in the classroom were considered inevitable and rewarded with extra teacher attention, whereas the assertiveness of African American girls was defined as "menacing, something that, for the sake of order in the classroom, must be squelched" (Orenstein 1994, p. 181; see also Theran 2009). Orenstein further found that Latinas, along with Asian American girls, had special difficulty being heard or even noticed. Probably socialized into quiet demeanor at home, these girls' scholastic or leadership abilities largely went unseen. In some cases, their classroom teachers did not even know who the girls were when Orenstein mentioned their names.

Treating girls and boys differently in school is detrimental to both genders (Corbett, Hill, and St. Rose 2008; Sullivan, Riccio, and Reynolds 2009). For one thing, we're beginning to see a difference between males and females in goals and attitudes toward schooling. Boys have poorer study habits and less concern about doing well in their studies (Tyre 2006). A 2003 *USA Today* poll found that 84 percent of girls but only 67 percent of boys found it important to continue beyond high school ("Boys' Academic Slide" 2003). More boys fall behind grade level, are suspended, and are placed in special education classes. Boys have a greater incidence of diagnosis of emotional disorders, learning disorders, attention-deficit disorders, and teen deaths (Conlin 2003; Goldberg 1998; Sommers 2000a, 2000b).

Concerns about Boys in School Largely thanks to the women's movement, discussed below, over the past several decades girls have been the primary focus of attention in examining possible bias in educational institutions. However, girls are now doing relatively well, even in math and science courses where traditionally many teachers stereotyped them as less competent and they therefore underperformed (Eccles 2011; Riegle-Crumb 2012). Analysts concerned about boys note a mismatch between their higher levels of physical activity and school expectations about sitting still and following detailed rules (Orr 2011; Poe 2004). Accordingly, they propose accepting a certain level of boys' rowdy play as *not* deviant, allowing more physical movement in classrooms, and encouraging activities shared by boys and girls (Kindlon and Thompson 1999).

Other analysis points to the relative lack of male role models in elementary and secondary education. Approximately 85 percent of elementary teachers and 59 percent of secondary teachers are women (Coopersmith 2009, Table 3). Still other analysts, such as psychiatrist William Pollack, focus on dysfunctional aspects of gender differentiation: Boys act tough to protect themselves, but such a stance interferes with scholastic performance

(Goldberg 1998; also see Kimmel 2001; Orr 2011). Proposed solutions involve redefining the male role to include emotional expression, encouraging boys to more directly express emotions other than anger, and cracking down on bullying (Totura et al. 2009; Zaman 2008).

Gender and Social Change

The increasing convergence of men's and women's social roles, although incomplete, reflects a dramatic change from the far more strictly gender-differentiated world of the 1950s and before. Fewer people would agree today that "sons in a family should be given more encouragement to go to college than daughters," an expression of **traditional sexism**—the belief that women's roles should be confined to the family and that women are not as qualified as men for leadership positions (Sherman and Spence 1997). Particularly in mainstream American culture, beliefs such as these have declined since the 1970s (Twenge 1997).

For many people today, however, a more subtle **contemporary sexism** has replaced traditional sexism (Sherman and Spence 1997). Contemporary sexists agree with statements such as, "Discrimination in the labor force is no longer a problem" and, "In order not to appear sexist, many men are inclined to overcompensate women" (Campbell, Schellenberg, and Senn 1997). The listener's comment at this chapter's beginning that a cookie-baking contest for presidential candidates' wives was only for fun with no sex politics involved might be interpreted as an example of contemporary sexism. Contemporary sexism denies that gender discrimination persists and may include the idea that women are asking for too much, a situation that results in resistance to their demands (Douglas 2009). Nonetheless, the apparent advances for today's women result not only from structural forces (especially educational and economic) that have meant women's increased labor force participation but also from active change efforts by women and their allies in the women's movement.

The Women's Movement

The "first wave" of feminism began with an 1848 convention on women's rights at Seneca Falls, New York, and came to an end when a major convention goal, voting rights for women (that is, *women's suffrage*), was achieved in 1920 (Rossi 1973). From about 1920 until the mid-1960s, there was virtually no activism regarding women's rights or roles. Women did make some gains in education, and they were encouraged to take employment in factories during the early 1940s when men were fighting in World War II. However, after the war ended in 1945, media glorification of the housewife role for women and the breadwinner role for men helped to make these seem natural and generally desirable. However, by the 1960s, higher levels of education for (white, upper-middle and middle-class) females left college-educated women with a significant gap between their abilities and the housewife role assigned to them. Betty Friedan's book *The Feminine Mystique* (1963) captured this dissatisfaction and made it a topic of public discussion or discourse.

Meanwhile, employed women chafed at the unequal pay and often sexist conditions in which they worked and began demanding equal opportunity. The civil rights movement of the 1960s provided a model of activism. In this climate, dissatisfaction with traditional roles precipitated a social movement—the second wave of the women's movement, which challenged traditional gender roles and strove to increase gender equality. Among other changes, federal legislation and executive orders declared discrimination against women in federal government contracts illegal and mandated that educational institutions finance sports for females just as they did for males (Title IX). Grassroots feminist groups with a variety of agendas developed across the country.

The National Organization for Women (NOW) was founded in 1966 and continues its activism today. Early on, NOW had multiple goals: opening educational and occupational opportunities to women and establishing support services such as child care. Many people see the movement as advancing the interests and status of women as mothers and caregivers, as well as workers (Showden 2009). The organization also came to support reproductive rights, including abortion, and same-sex unions.

Today we are in the third wave of feminism that "tends to be much more pluralistic" and "much less dogmatic" about issues surrounding sexuality (for example, whether pornography is necessarily demeaning to women), personal expression (for example, whether breast-enhancement surgery is demeaning to women), and fashion choices (for example, whether wearing makeup is demeaning to women) (Heilmann 2011; Wolf 2012). The idea that feminism should stress *intersectionality*—structural connections among race, class, and gender is characteristic of third-wave feminism (Robinson 2007; Yuval-Davis 2006). If *Ms. Magazine*, which celebrated its fortieth birthday in 2012, represents second-wave feminism, *Bust Magazine*, established in 1993 "for women with something to get off their chests," may be more characteristic of third-wave feminism.

Women's attitudes vary regarding feminism. A small proportion disagree with most feminist principles and encourage traditional marriage and motherhood as the best path to female fulfillment (Bulanda 2011; Passno 2000). Some white, working-class women and women of color feel that the women's movement has focused too much on emotional oppression or abuse and on professional women's thwarted career opportunities rather than on "the daily struggle to make ends meet that is faced by working class women" (Aronson 2003, p. 907). Many black women have seen the movement as irrelevant to them, because—always having been employed, first under slavery and later because of financial necessity—they were never housewives in large numbers (Collins 2000; Lessane 2007).

Meanwhile, Chicano/Chicana (Mexican American) activism gave *la familia* a central place as a distinctive cultural value. Hispanics of both genders placed a high value on family solidarity, with individual family members' needs and desires subsumed to the collective good; as a result, Chicana feminists' critiques of unequal gender relations often met with Hispanic hostility (Manago, Brown, and Leaper 2009). Nevertheless, a Chicana feminism slowly emerged, and the Mexican American Women's National Association (now MANA) was established in 1974. Generally, Chicanas support women's economic issues, such as equal employment and day care, while showing less support for abortion rights than do Anglo women (Segura and Pesquera 1995).

In some ways, such as their relatively low wages and experience with race/ethnic discrimination, black women and Chicanas are more like black and Mexican American men, who are also subordinated, than they are like nonHispanic white women. African American and Hispanic women consider race/ethnic as well as gender discrimination in setting their priorities (Arnott and Matthaei 2007). In fact, it is more precise to say their feminist views are characterized by intersectionality (Baca Zinn et al. 2007).

The concept of *intersectionality* also applies to Muslim women and Arabic women, who view feminism through the lens of culture, religion, and community. Strict adherence to religio-cultural traditions can be at odds with Western feminist views. However, many Muslim women are using the Qur'an to, challenge sharia law, question patriarchy, and demand women's rights (Karim 2009; Sayeed 2007). "Muslims coming to North America are often seeking an egalitarian version of Islam," according to Ebrahim Moosa, Duke University Islamic Studies professor. "That forces women onto the [women's rights] agenda ..." (Knowlton 2010).

Media sometimes assert that a younger "postfeminist" generation does not support a women's movement. The assumption is that younger women may have a negative image of feminism or be latently feminist but believe that women's rights' goals have already been achieved. Sometimes such articles assert that younger women are simply too busy with work and family to have the time to be active (Showden 2009). Nevertheless, research suggests that "postfeminism" is a myth and that many young women support feminist goals (Bolzendahl and Myers 2004; Center for the Advancement of Women 2003; Hall and Rodriguez 2003). "One might note that many of the ideologies associated with feminism have become relatively common place and speak to the success of feminism in attaining much broader acceptance of gender equality" (Schnittker, Freese, and Powell 2003, pp. 619–20).

Men's Movements

Some men responded to feminism by initiating men's movements. Sociologist Michael Kimmel (1995) had divided men's movements into three fairly distinct camps: profeminists, antifeminists, and masculinists. Profeminists support feminism in its opposition to patriarchy. An example is the National Organization for Men Against Sexism (NOMAS). In 2012, in partnership with the National Conference on Domestic Violence, NOMAS cosponsored the Thirty-Seventh Annual Conference on Men and Masculinity (NOMAS 2012a).

Antifeminists work to counter—or at least to complement—feminist interests and resultant social trends by attending to men's rights. Declining economic opportunities for men, coupled with accusations of male privilege, lead some men to feel unfairly picked on (Hebert 2007). According to its website, the current goal of the National Organization for Men (NOM) is "to help ensure fair media coverage to men's perspectives" (National Organization for Men). A representative article titled "The New Double Standard" by copresident Marty Nemko reads as follows:

> [W]hen few women are in a lucrative profession, for instance computer science or engineering, most universities and large employers install reverse discrimination policies to avoid organizations such as NOW tarring them with the dreaded epithet, "sexist!" Yet, where is NOW's and the media's outrage when women are overrepresented in a desirable career such as a pharmacist? Where's the outrage over the fact that only men must register for the draft, the obligation to risk getting one's head blown off? (Nemko n. d.)

Not so outraged, *masculinists* work to develop a positive image of masculinity, one combining strength with sensitivity. Masculinists have influenced university curricula to add courses in men's, masculinity, or gender studies, rather than only in women's studies. An article cited earlier in this chapter that emphasizes positive aspects of the male gender role is written from a masculinist point of view (Kiselica and Englar-Carlson 2010). Women's and men's movements are society-wide, or macro, indicators of social change. As individuals, we experience these changes at the personal and family, or micro, level.

Gender and Family in the Future

Change is difficult. Despite dramatic and unprecedented change over the past fifty years, society persists in emphasizing the public sphere as more important to masculinity and the private sphere to femininity. Both genders find it difficult to juggle job or career success with parenthood and otherwise supportive relationships (Jarrell 2007). Men often encounter more resistance than women do when they try to exercise "family-friendly" options in the workplace (Neal and Hammer 2007; Peterson 2004). They also may face prejudice when they take jobs traditionally considered women's (Snyder and Green 2008)—although many men do confront early socialization and macho messages to make nonstereotyped career decisions (Hancock 2012).

Expansion in opportunities for both men and women can lead to ambivalence or to conflicts, not only between partners and other family members but also within ourselves as we confront the "lived messiness" of gender in contemporary life (Heywood and Drake 1997, p. 8). In fact, many Americans support separate gender roles; although gender expectations have changed and continue to change, they have not done so completely (Pew Research Center 2010a). Although about 70 percent of Americans believe making money and cooking family meals should be equally shared by women and men, wives are still far more likely to do the laundry and husbands the yard work (Newport 2008).

In one scholar's opinion, "Despite the extraordinary improvements in women's status over the past two centuries, ... people still think about women and men differently and ... men still occupy most of the highest positions of political and economic power" (Jackson 2006, p. 229). Inasmuch as Americans continue to see women and men through a lens of differentiated competence and persist in staking a major component of their identity in appropriate gender performances, gender differences in tasks and relative power will remain (Ridgeway 2006). But following traditional gender roles can be costly.

The Costs of Following Traditional Gender Expectations

Both men and women pay a price for gender as traditionally structured. Biases and gender stereotypes thwart both males' and females' career opportunities, men's confidence in nontraditional family roles, and both sexes' ability to communicate supportively with one another (Bobbitt-Zeher 2011; Hancock 2012).

As one indicator of costs to men, we can examine life-expectancy statistics (Sorenson 2011). Overall life expectancy for U.S. men is about five years shorter than it is for women (about 76 versus about 81 years); from birth onward, males have higher death rates (U.S. Census Bureau 2012c, Tables 104, 110). Among other causes, difference in longevity has been attributed to males' (culturally encouraged) riskier behavior, including more smoking and alcohol use ("Percent of Adults Who Smoke by Gender" 2010; Nolem-Hoeksema 2004). Also, perhaps tapping into the culturally expected male trait of self-reliance, men have far fewer physical checkups than women, and middle-aged men who are more traditional regarding gender visit the doctor still less (Dotinga 2009; Painter 2006; Rabin 2006).

Table 3.2 compares men's and women's death rates for 15- to 24-year-olds and for 25- to 34-year-olds. In reviewing this table, keep in mind that for these age groups accidents are the leading cause of death—37 percent of all deaths—followed by homicides (12–13 percent) and suicides (10–12 percent) (U.S. Census Bureau 2012c, Table 122).

Table 3.2 Death Rates[a] of 15- to 24-Year-Olds and 25- to 34-Year-Olds by Gender and Race/Ethnicity, 2008

15- to 24-Year-Olds		25- to 34-Year-Olds	
Male		**Male**	
Total	110	Total	142
NonHispanic White	103	NonHispanic White	133
Black	159	Black	225
Asian/Pacific Islander	50	Asian/Pacific Islander	55
American Indian	151	American Indian	199
Hispanic[b]	104	Hispanic[b]	107
Female		**Female**	
Total	40	Total	63
NonHispanic White	38	NonHispanic White	58
Black	50	Black	98
Asian/Pacific Islander	23	Asian/Pacific Islander	35
American Indian	59	American Indian	99
Hispanic	31	Hispanic	43

[a]Rates are per 100,000 individuals.
[b]Hispanics can be of any race.
Source: U.S. Census Bureau 2012c, Tables 110, 112.

For both age categories, the male death rate is more than twice that for females. Regarding race/ethnicity, in both age categories black males have the highest death rate, followed by American Indian males. Among females, American Indians have the highest death rates, followed closely by blacks. Among all race/ethnicities and both genders in these age groups, Asian/Pacific Islanders have the lowest rates. We can note that for males in each age category, within-group differences are significant: Among 15- to 24-year-olds, for instance, American Indian and black males have death rates three times higher than Asian/Pacific Islanders. Among men ages 25–34, American Indian and black males have death rates nearly four times higher than among Asian/Pacific Islanders. The *within-group* death rates for males are higher than the overall *among-group* differences between males and females. For all that has been said about gender differences, we also need to be concerned about race/ethnic disparities and inequalities *within* genders (Coontz 2013).

Women also suffer costs because of traditional gender structures. Poverty levels are higher for women than for men—16.3 percent for women compared with 13.6 percent for men in 2011 (DeNavas-Walt, Proctor, and Smith 2012, p. 15). This situation of higher poverty among women holds true for blacks and Hispanics as well as for nonHispanic whites and, to a lesser extent, for Asians—that is, for all race/ethnic groups counted (U.S. Census Bureau 2012c, Table 713). The poverty rate for nonmarried, female-headed family households was more than three in 10 (31.2 percent) in 2011, compared with 16.1 percent for analogous male-headed households. In male nonfamily households, the average annual income in 2011 was about $35,000 compared to about $25,000 for those headed by females. Family households headed by nonmarried men (single-parent or cohabiting households or both) earned an annual average of $49,567, compared to $33,637 for those headed by women (DeNavas-Walt, Proctor, and Smith 2012, Tables 1 and 4).

Individuals pay a gender price in countless other ways too, of course. Men's work, child care, and other family-focused opportunities are limited. Men's performance in the realms of child care and housework are often trivialized or the focus of humor in the media. Meanwhile, women still do the majority of household labor and child care even when employed. They assume a disproportionate share of tasks concerning keeping up with extended kin—holiday shopping and remembering birthdays, for example. Then, too, gender-differentiated communication patterns can frustrate both sexes. Family power arrangements can feel problematic. In contrast, "If people think differently about money, power and gender roles, everyone may come out ahead" (Mundy 2012, p. 30). Where will we go from here?

Where to From Here? Virtually all scholars assume that, overall, women will continue to press for greater role flexibility and opportunities in the workplace, in their pursuit of education, in politics, in how they define and redefine their religious beliefs and practices, and in how they fashion family obligations and relations. Sociologist Robert Jackson is convinced that the forces of history will sweep gender inequality aside. In his words, gender differentiation "is fated to end because essential organizational characteristics and consequences of a modern, industrial, market-oriented, electorally governed society are inherently inconsistent with the conditions needed to sustain gender inequality" (Jackson 2006, p. 241). Meanwhile,

> [Another] way to think about the change in the gender structure is to conceptualize gender-conscious actors actively making that change, or "undoing" gender. When young wives remain ambitious workers, committed to their own independent economic success, and expect their husbands to be equal partners at home, these women are undoing gender and changing the gender structure. When men use paternity leaves, when they take responsibility for their share of household labor, they, too, are undoing gender, including the male privilege that has defined gender as a stratification system and a structural aspect of our society. (Risman 2011, p. 21)

Many couples are redefining the masculine provider role to include provision of a man's guidance. With flexible gender roles, says one parent, "your priority is to provide for your family—the love, the affection, the nurturing. For us, it's about what's best for the family" (in Mundy 2012, p. 30). Men partnered with women in high-paying positions gain the option of relinquishing their jobs or becoming secondary earners in order to devote themselves to family and child care (Mundy 2012). Other men make more subtle changes, such as breaking through previously learned isolating habits to form more intimate friendships and deeper family relationships.

However, among others, author Liza Mundy (2012), quoted earlier suggests that,

> In the face of women's rising power and changing expectations, many men may experience an existential crisis. When the woman takes on the role of primary breadwinner, it takes away an essential part of many men's identity: that of the provider, the role he was trained, tailored and told to do since he could walk and talk. His heroes are likely all successful in this area.... So when you take that away, men have nowhere to turn for guidance. There's no map through that wilderness. (p. 32)

Mundy suggests that men can respond to women's changing roles in several ways. Many will "rise to the challenge, trying harder in the classroom, competing with women but in a good way. It means adapting but also broadening the definition of masculinity to include new skills and pleasures. Hunting but also cooking. Gold but also child care. We are always too quick to think of masculinity as finished" (2012, p. 32).

Gendered expectations and behaviors—both as they have and have not changed—underpin virtually all the topics that you will explore in this course. Gender is important to sexuality, to communication, to parental and other family roles, and to family power, as well as to income and poverty issues. Future chapters explore these topics.

Summary

- Roles of men and women have changed, but living in our society remains a somewhat different experience for women and for men. Gendered cultural messages and social structure influence people's behavior, attitudes, and options. In many respects, however, men and women are more alike than different.
- Traditional masculine expectations require that men be confident, self-reliant, and occupationally successful. During the 1980s, the "new man" image emerged, according to which men are expected to value tenderness and equal relationships with women.

- Individuals vary in the degree to which they follow cultural models for gendered behavior. The extent to which men and women differ from each other and follow these cultural messages can be visualized as two overlapping normal distribution curves.
- There are race/ethnic and class differences in both gendered cultural expectations and behaviors. This and other diversity is at times expressed in responses to the women's and men's movements.
- A society's gender structures impact ways that individuals play gender roles—that is, how they "display" or "perform" gender. Gender structures include all five basic social institutions: family, religion, politics or government, education, and the economy.
- Biology interacts with culture to produce human behavior, and the two influences are not really separable. Sociologists give considerable attention to the socialization process. Advocates have expressed concern about barriers to the opportunities and achievements of both boys and girls.
- Turning our attention to the actual lives of adults in contemporary society, we find women and men negotiating gendered expectations and making choices in a context of change at work and in the family. New cultural ideals are far from realization, and efforts to create lives balancing love and work involve conflict and struggle but promise fulfillment as well.
- Whether gender will continue to be a moderator of economic opportunities and life choices in the future is uncertain as men's and women's roles and activities converge, but men remain more advantaged. It is likely that gender identity will continue to be important to both men and women.

Questions for Review and Reflection

1. What are some of the characteristics generally associated with males in our society? What traits are associated with females? How might these affect our expectations about the ways that men and women should behave?
2. Do you think men are dominant in major social institutions such as politics, religion, education, and the economy? Or are they no longer dominant? Give evidence to support your opinion.
3. Women and men may renegotiate and change their gendered attitudes and behaviors as they progress through life. What evidence do you see of this in your own life or in others' lives?
4. Choose two gender structures discussed in this chapter and discuss how each works to influence ways that an individual might perform gendered expectations.
5. **Policy Question.** What family law and policy changes of recent years do you think are related to the women's and men's movements? What policies do you think would be needed to promote greater gender equality or more satisfying lives for men and women?

Key Terms

4

Our Sexual Selves

Learning Objectives

1. Describe how one's sexual identity develops and the different ways in which sexual identity may be expressed.
2. Explain the interpersonal exchange model of sexual satisfaction and the interactionist perspective on human sexuality.
3. Describe how historical periods influenced our understanding of sexuality.
4. Compare sexual values for people in committed and noncommitted relationships.
5. Discuss trends in infidelity and how it affects intimate relationships.
6. Understand how politics affects sexuality and sexual expression.

From childhood to old age, people are sexual beings. Sexuality has a lot to do with the way we think about ourselves and how we relate to others. It goes without saying that sex plays a vital role in marriages and other intimate partner relationships. Despite the pleasure it may give, sexuality may be one of the most baffling aspects of our selves. When we're with others, finding mutually satisfying ways of expressing sexual feelings can be a challenge.

In this chapter, we define sexual identity and then examine the diversity in sexual identities that exists today. We will review the historically changing cultural meanings of sexuality. We will discuss sex as a pleasure bond that requires open, honest, and supportive communication and look at the role sex plays in intimate relationships over the life course.

We will look at some challenges that are associated with sexual expression. What happens when one or both partners has an affair? How have changes in technology and the media (such as increasing accessibility of pornography) altered sexual norms, attitudes, and behavior? Where can people find accurate information about sex?

Before we discuss sexuality in detail, we want to point out our society's tendency to reinforce the differences between women and men and to ignore the common feelings, problems, and joys that make us all human. The truth is, men and women aren't really so different (Peterson and Hyde 2011). Many physiological parts of the male and female genital systems are either alike or analogous, and sexual response patterns are similar in men and women (Masters and Johnson 1966).

Sexual Development and Identity

Knowledge about children's sexual development and the emergence of sexual identity is not as extensive as we would like, but we do know some things.

Children's Sexual Development

"Human beings are sexual beings throughout their entire lives" (DeLamater and Friedrich 2002, p. 10). As early as twenty-four hours after birth, male newborns get erections, and infants may touch their genitals. Research indicates that between ages 2 and 5, a substantial number of children engage in "rhythmic manipulation" of their genitals, which researchers consider to be a "natural form of sexual expression" (DeLamater and Friedrich 2002, p. 10). Children may also "play doctor," examining one another's genitals (Friedrich et al. 1998).

Early sexual behavior peaks at age 5 and then declines until sexual attraction first manifests itself around age 11 or 12. Children now mature about two years earlier than 100 years ago (Brink 2008). According to data from the National Health and Nutrition Examination Study, the average age at **menarche**, a girl's first menstrual cycle, dropped during the twentieth century for all races and ethnicities and is now about 12, with African American girls experiencing the greatest decline (McDowell, Brody, and Hughes 2007; Walvoord 2010). Breast development is also occurring at younger ages, especially for minority girls (Biro et al. 2010). Puberty before age 10 is has become increasingly common, and the effects on children's health and psychosocial development are unclear, leaving many parents wringing their hands and doctors shaking their heads (Weil 2012). Reasons for the decline in age at puberty may include better nutrition as well as negative changes such as obesity, decline in physical activity, and pollution (Anderson and Must 2005; Herman-Giddens 2007; Walvoord 2010).

Unfortunately, earlier sexual maturity is associated with earlier onset of romantic involvement, sexual intercourse, depression and anxiety, behavior problems, smoking, and alcohol use in adolescence as well as accelerated entrance into cohabitation and marriage (Cavanagh 2011; DeRose et al. 2011; Richards and Oinonen 2011). Moreover, as the age of puberty has declined, the age at marriage has risen, leaving a more extended period during which sexual activity may occur among adolescents and unmarried adults, requiring them to make decisions about sex such as whether to remain abstinent or to have sex and with whom (see later discussion of sexual values).

Sexual Identity

As we develop into sexually expressive individuals, we may be drawn to a partner of the same sex, the opposite sex, or both sexes. **Sexual identity** (or *sexual orientation* or *sexual orientation identity*) refers to whether one is attracted to one's own gender or a different gender (American Psychological Association Task Force on Appropriate Therapeutic Responses to Sexual Orientation 2009; Sexual Identity and Gender Identity Glossary 2012). **Affectional orientation** is an even newer

Issues for Thought

Bisexual or Just "Bi-Curious"? The Emergence of Pansexuality

In 2008, Katy Perry's hit single "I Kissed a Girl" shot straight to number 1 on the Billboard Hot 100 list and is largely responsible for her meteoric rise to fame and her current place as one of the world's leading female pop stars. What was it about that song that struck such a chord with her fans, mostly young adolescent girls? The popularity of the song reflects our society's growing movement toward pansexuality.

We tend to think of sexual identity as a dichotomy: One is either "gay" or "straight." Actually, sexual identity is a continuum. Freud, Kinsey, and many contemporary psychologists and biologists maintain that humans are inherently bisexual; that is, we all have the latent physiological and emotional structures necessary for responding sexually to either sex.

From the interactionist point of view (see Chapter 2), the very concepts "bisexual," "heterosexual," and "homosexual" are social inventions. They emerged in scientific and medical literature in the late nineteenth century (Katz 2007, pp. 10–12; Seidman 2003, pp. 46–49, 56–58). Although same-sex sexual relations existed all along, the conceptual categories and the notion of sexual identity itself were cultural creations. Developing a sexual orientation today may be influenced by the resultant tendency to think in dichotomous terms. In fact, the social pressure to view oneself as either straight or gay may inhibit latent bisexuality or inconsistencies (Gagnon and Simon 2005; Katz 2007, pp. 25–27; Rosario et al. 2006; Schwartz 2007). In recognition of this problem, a new concept has emerged that encompasses a wide range of sexual identities. Pansexual is an identity of someone who has the potential to be sexually attracted to various gender expressions, including those outside the gender-conforming binary (LGBTSS at Iowa State University 2012). *Pan* means all, and *pansexual* captures both the fluidity and diversity of one's sexual identity. It also emphasizes the importance of emotional as opposed to purely physical attraction and one's desire to be with a particular person. One student put it best: "It's about hearts, not parts."

Back to Katy Perry and the "I Kissed a Girl" phenomenon. College students report that it is not uncommon to see two women kissing and "making out" at bars and parties. Are they lesbians? Bisexual? Or perhaps just "bi-curious." In other words, are they simply trying out or experimenting with sexual behaviors? Some have referred to this phenomenon as LUG (lesbian until graduation). However, it is a stereotype that same-sex experimentation occurs mostly on college campuses. A study of 13,500 women conducted by the U.S. Centers for Disease Control and Prevention found that women with a college diploma were less likely than women with less education to have engaged in a sexual experience with another woman (Lewin 2011).

Their behavior might also be a "performance." There is often alcohol involved, which lowers inhibitions, and oftentimes the behavior takes place in public in response to the whoops and cheers of men who reward them with free drinks. Women have much greater latitude in their sexual behavior than do men, and many men find it a "turn-on" to watch two women engaging in sexual behavior with one another. Regardless of why they do it, many adolescents and young adults (both men and women) experiment sexually with members of the same sex. Pansexuality provides a useful alternative to bisexuality for describing the complexity and range of sexual identity.

Critical Thinking

Have you ever seen two women kissing at a bar or party? Did you consider them lesbian, bisexual, or something else? In your opinion, does our society view sexual behavior between women differently than that between men? Should pansexuality become recognized as a legitimate sexual orientation?

term also used in conjunction with sexual identity and sexual orientation. It's preferred by some because the term is intended to include emotional and physical attractions beyond sexual attraction (LGBTSS, Iowa State University 2012).

Heterosexual has been the traditional way to describe people attracted to opposite-sex partners whereas **homosexual** was used to describe people attracted to same-sex partners. Because they are associated with negative stereotypes, use of each of these terms has declined, having been replaced by *straight*, *gay*, and *lesbian*. Whereas *gay* can be used in reference to both men and women, *lesbian* generally refers only to women (Sexual Identity and Gender Identity Glossary 2012). **Bisexuals** are attracted to people of both sexes. **Pansexual** is an identity of someone who has the potential to be sexually attracted to various gender expressions, including those outside the gender-conforming binary. *Pansexuality* is described in greater detail in the "Issues for Thought: Bisexual or Just 'Bi-Curious'?" box. Keep in mind that language reflects the norms and values of the prevailing culture, and terminology with respect to sexual identity is continually evolving.

It is also important to remember that sexual identity should not be confused with *gender identity*, which refers to the degree to which an individual sees herself or himself as feminine or masculine. The pervasive stereotype of gay men is that they are feminine or "flamboyant" and of lesbians that they are masculine or "butch." Yet most gay men feel like men and have a gender identity that is "male," and most lesbians feel like women and have a gender identity that is "female." Moreover, a person's sexual identity does not necessarily predict his or her sexual behavior; abstinence is a behavioral choice, as is sexual expression with partners of the nonpreferred sex.

The reason some individuals develop a gay sexual identity has not been definitively established—nor do we yet understand the development of heterosexuality. The American Psychological Association (APA) takes the position that a variety of factors impact a person's sexuality. The most recent literature from the APA says that sexual orientation is not a choice that can be changed at will, and that

> sexual orientation is most likely the result of a complex interaction of environmental, cognitive and biological factors ... is shaped at an early age ... [and evidence suggests] biology, including genetic or inborn hormonal factors, play a significant role in a person's sexuality. (American Psychological Association 2010)

A sense of one's sexual identity often begins in childhood and progresses through adolescence (Schwartz 2012). Of 2,560 California high school students surveyed over a three-year period, 11 percent reported a gay or lesbian identification, approximately 12 percent identified as bisexual, and nearly 5 percent noted that they were questioning their sexuality (Russell, Clarke, and Clarey 2009). How sexual identity unfolds depends on various social factors, an important one being whether one experiences family support (Rosario, Scrimshaw, and Hunter 2008).

The American Psychological Association does not support *sexual orientation change efforts* (SOCE) that try to convert gay men and lesbians back to heterosexuals through psychotherapy, support groups, or religious programs and retreats, noting studies finding side effects such as loss of sexual feeling, depression, anxiety, and suicidality (American Psychological Association Task Force on Appropriate Therapeutic Responses to Sexual Orientation 2009). California recently passed a law prohibiting licensed mental health professionals from practicing therapies aimed at making gay and lesbian teenagers straight (Associated Press 2012). However, some "ex-gay" men argue that so-called conversion therapies work and have successfully ridded them of their homosexual desires. One man said, "my homosexual feelings have nearly vanished. In my 50s, for the first time, I can look at a woman and say 'she's really hot'" (Eckholm 2012).

Deciding who is to be categorized as **gay**, **lesbian**, or bisexual for research purposes is not easy: How much experience? How exclusively homosexual? Noting that estimating the percentage of the population that is gay is difficult, a University of California demographer estimates that 3.5 percent of adults identify as lesbian, gay, or bisexual and estimates that 0.3 percent are transgender. This represents 9 million **lesbian, gay, bisexual, and transgender (LGBT)** adults (Gates 2011; see also Gates and Newport 2012). Nonwhites were more likely than whites to identify as LGBT, as are women compared to men, and younger as opposed to older Americans (Gates and Newport 2012). Finally, it is important to note that more people have reported having taken part in sexual behavior with a partner of the same-sex than self-identify as gay or lesbian. Dr. Gates estimated that 19 million or 8.2 percent of Americans have engaged in sexual behavior with someone of the same sex and 25.6 million or 11 percent have acknowledged at least some sexual attraction (Gates 2011).

What Does It Mean to be Transgendered? Gay, lesbian, and bisexual individuals are considered part of the LGBT community. The "T" in LGBT stands for **transgendered**. *Transgendered* refers to people who switch gender roles, just once or more than once, and includes both **transsexuals** and **transvestites**. *Transsexuals* switch physical sexes through surgery, hormone therapy, electrolysis (hair removal), and other treatments, as discussed in Chapter 3, "Issues for Thought: Challenges to Gender Boundaries." *Transvestites* are people who dress as the opposite gender because it feels erotic, empowering, or rebellious or for other reasons (Sexual Identity and Gender Identity Glossary 2012). In an MSNBC documentary, transgender youth described feeling like they were "born in the wrong body" (MSNBC 2012). These youth may initially think they may be gay or lesbian, because (according to their physical anatomy) they are attracted to the incorrect sex. Later, they may realize that it is their genital anatomy and not whom they desire that is "incorrect."

Grossman and D'Augelli (2006) assembled a focus group of twenty-four trangendered adolescents and young adults ages 15 to 21. The participants were, on average, about 10 years old when they became aware that their gender identity did not match their biological sex. Their statements reveal an early awareness that something was not "right":

> "I used to play baseball and hangout with the boys, but I always felt like a girl."
>
> "I know that I was biologically a girl, but ever since I was little, I always wanted to be a man so bad. Other people said I want to be a lawyer, a doctor, and I said I want to be a man."
>
> "Since I was young, and I would see people getting married, I always pictured myself in the groom's place instead of the bride's." (Grossman and D'Augelli 2006, pp. 121–122)

Few trangendered men and women make the full physical transformation to their preferred sex. **Sexual reassignment surgery** (SRS) is very expensive. It is performed by a very small number of doctors and is generally only available in large U.S. cities and typically not covered by insurance. Many transsexuals go abroad for their surgery (to Thailand, for example) because it is cheaper (Conway 2012). SRS is also major surgery with risks, one of which may be loss of sensitivity to the genitals. The transition from male to female is much more common than female to male. Since SRS began in the 1960s, it has been estimated that there are now between 30,000 and 40,000 transsexual women in the United States (Conway 2012). Transgender men and women still have choices though. Some opt to have "top," but not "bottom" surgery, or choose limit their treatment to hormones and cosmetic procedures, which can be administered by local physicians. A problem sometimes faced by transgender Americans is what to do when they are required to present a birth certificate or identification that does not match their current gender. Two transgender New Yorkers are suing the city so that they can have their sex "corrected" on their birth certificates (Eligon 2011). Parents of transgender children sometimes find themselves battling schools so that their children may dress how they would like (Frosch 2013). On the other hand, a number of top universities have adopted student health plans that cover the cost of sex-change treatments (Pérez-Peña 2013). "Issues for Thought: Challenges to Gender Boundaries," in Chapter 3 further explores this topic.

Asexuality A small number of Americans referred to as **asexuals** do not experience sexual attraction to others. This situation differs from *celibacy* or *abstinence*, which is a *decision* not to have sexual relations, at least for a time, but not from a lack of desire. Asexual individuals have emotional feelings and may desire intimate relationships with others, just not sexual ones. Some asexuals may even experience physical arousal or even masturbate but feel no desire for partnered sexuality (Asexual Visibility and Education Network 2012). The Asexual Visibility and Education Network (AVEN) was founded in 2001 as a networking and information resource (www.asexuality.org). This group would like to see *asexuality* become a recognized sexual identity so that absence of sexual desire is not treated as dysfunctional but as a "normal" alternative (AVEN 2012). Clinicians vary in whether they agree that asexuality is a clear sexual identity, but as one example, Dr. Irwin Goldstein, director of the Boston University Center for Sexual Medicine, considers that "[l]ack of interest in sex is not necessarily a disorder or even a problem... unless it causes distress" (Duenwald 2005, p. 2).

Lack of sexual desire *can* be a problem for many men and women, as a result of aging, obesity, smoking, drug use, diabetes, cancer treatments, and certain prescription medications (Schick et al. 2010). Age is associated with less sexual desire, greater difficulty with lubrication, lower levels of arousal, and lower overall sexual functioning (Tracy and Junginger 2007). Many antidepressants and blood pressure medications inhibit sexual desire and orgasm (Cohen et al. 2007).

Not receiving treatment for sexual problems affects individuals and couples. Some men are so embarrassed and distressed that they were unable to discuss the issue openly with their partner let alone a doctor (Lodge and Umberson 2012). Among the female partners of the men who had erectile dysfunction (ED), a study titled "The Female Experience of Men's Attitudes to Life Events and Sexuality" (MALES) found that those whose partners did not receive treatment reported less sexual satisfaction and fewer orgasms than the partners of the men who did (Fisher et al. 2005). Interest in and treatment of sexual dysfunction should increase as the U.S. population continues to age and as older people remain healthier and desire to be sexually active longer (see Chapter 16).

Theoretical Perspectives on Human Sexuality

There are various theoretical perspectives concerning marriage and families, as we saw in Chapter 2. Many of these have been applied to human sexuality. For example, we can look at sexuality using a structure–functional perspective. In this case, we see sex as a focus of norms designed to regulate sexuality so that it serves the societal function of responsible reproduction. Those looking at sexuality from a biosocial perspective consider that humans—like the species from which they evolved—are designed so that they can efficiently transmit their genes to the next generation. According to this biosocial perspective, men are naturally promiscuous, seeking multiple partners so as to distribute their genes widely, whereas women, who can generally have only one offspring a year, are inclined to be selective and monogamous (Dawkins 1976; 2006).

Both these perspectives have their limits. The structure–functional perspective tells us little about the emotions and pleasures of sexual relationships, whereas the biosocial perspective argues a genetic determinism that is contradicted by historical and cross-cultural variation in sexual behavior and relationships. Two more useful ways of looking at sexual relationships in a sociological perspective are exchange theory and interaction theory, both introduced in Chapter 2.

The Exchange Perspective: Rewards, Costs, and Equality in Sexual Relationships

From a general *exchange theory* perspective, women's sexuality and associated fertility are resources that can be exchanged for economic support, protection, and status in society. However, an exchange theory perspective that brings sex closer to our human experience is the **interpersonal exchange model of sexual satisfaction** (Lawrence and Byers 1995).

Figure 4.1 shows us that in the interpersonal exchange model of sexual satisfaction, satisfaction depends on the *costs* and *rewards* of a sexual relationship, as well as the participant's *comparison level*—what the person expects out of the relationship. Also important is the *comparison level for alternatives*—what other options are available,

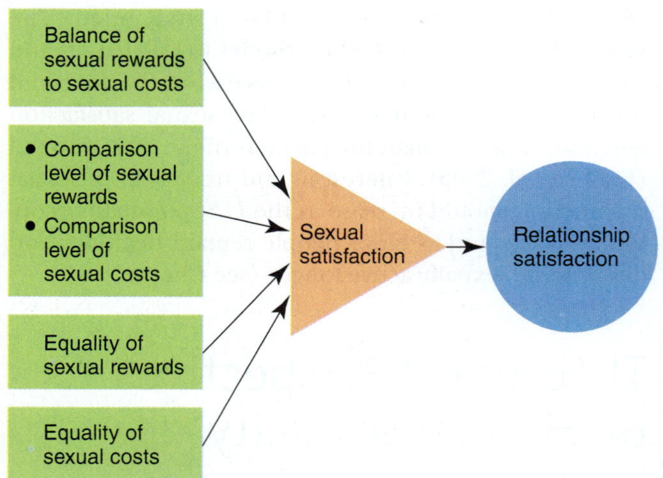

FIGURE 4.1 Models of factors associated with sexual and relationship satisfaction.

and how good are they compared to the current relationship? Finally, in this day and age, expectations are likely to include some degree of *equality*. Research to test this model found that these elements of the relationship did indeed predict sexual satisfaction in married, cohabiting, and dating couples (Byers 2005; Mark and Murray 2012; Kisler and Christopher 2008); however, social class (Neff and Harter 2003), race (Stanik and Bryant 2012), and gender (Mark and Murray 2012) are important variables to take into consideration within this exchange model of sexual satisfaction.

The Interactionist Perspective: Negotiating Cultural Messages

The *interactionist* perspective emphasizes the interpersonal negotiation of relationships in the context of sexual scripts: "*That* we are sexual is determined by a biological imperative toward reproduction, but *how* we are sexual—where, when, how often, with whom, and why—has to do with cultural learning, with meaning transmitted in a cultural setting" (Fracher and Kimmel 1992). Cultural messages give us legitimate reasons for having sex, as well as who should take the sexual initiative, how long a sexual encounter should last, how important experiencing orgasm is, what positions are acceptable, and whether masturbating is appropriate, among other things. Recently, cultural messages have concerned what sexual interaction or relationships are appropriately conducted over the Internet, as well as with the newer phenomenon of "sexting" via cell phones with picture and video capabilities (see Figure 4.2).

An **interactionist perspective on human sexuality** holds that women and men are influenced by the **sexual scripts** that they learn from their culture (Gagnon and Simon 2005; Lyons et al. 2011; Masters et al. 2013). They then negotiate the particulars of their sexual encounter and developing relationship (Backstrom, Armstrong, and Puentes 2012; MacNeil and Byers 2009, VanderLaan and Vasey 2009).

Sex partners assign meaning to their sexual activity—that is, sex is symbolic of something, which might be affection, communication, recreation, or play, for example. Whether each gives their sexual relationship the same meaning has a lot to do with satisfaction and outcomes. For example, if one is only playing while the other is expressing deep affection, trouble is likely. A relationship goal for couples becoming committed is to establish a joint meaning for their sexual relationship.

Sex has different cultural meanings in different social settings. In the United States (and elsewhere), messages about sex have changed over time.

Changing Cultural Scripts

From colonial times until the nineteenth century, the purpose of sex in America was defined as reproductive. A new definition of sexuality emerged in the nineteenth century and flourished in the twentieth. Sex became significant for many people as a means of communication and intimacy (D'Emilio and Freedman 1998).

Early America: Patriarchal Sex

In a patriarchal society, descent, succession, and inheritance are traced through the male genetic line, and the socioeconomic system is male-dominated. Sex is defined as a physiological activity, valued for its procreative potential. **Patriarchal sexuality** is characterized by many beliefs, values, attitudes, and behaviors developed to protect the male line of descent. Men are to control women's sexuality. Exclusive sexual possession of a woman by a man in monogamous marriage ensures that her children will be legitimately his. Men are thought to be born with an urgent sex drive, whereas women are seen as naturally sexually passive; orgasm is expected for men but not for women. Unmarried men and husbands whose wives do not meet their sexual needs may gratify those needs outside marriage. Sex outside marriage is considered wrong for women, however.

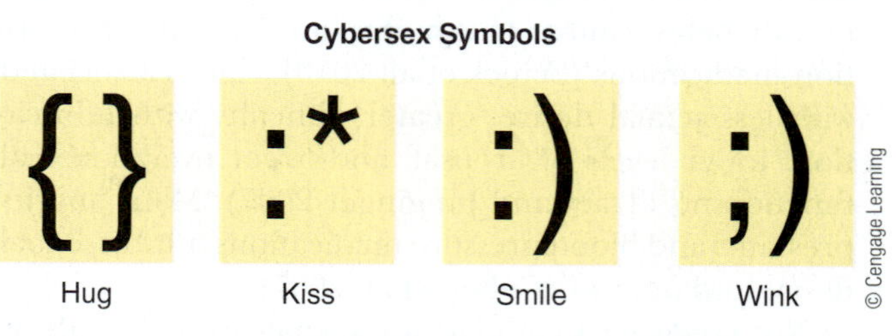

FIGURE 4.2 Cybersex. Is it sex—cyberstyle—or is it abstinence? From an interactionist perspective, we might say that society is still constructing the answer.

Although the patriarchal sexual script has been significantly challenged, it persists to some extent and corresponds to traditional gender expectations. If masculinity is a quality that must be achieved or proven, one arena for doing so is sexual accomplishment or conquest. The National Survey of Sexual Health and Behavior (NSSHB) conducted in 2009 by the Center for Sexual Health Promotion at Indiana University provides the most recent and comprehensive snapshot of the sexual activities of Americans. Based on 5,865 adolescents and adults between ages 14 and 94, this study found that men were considerably more likely than women to perform or "do" sex. For example, the percentage reporting having had masturbated alone at least once in the last month was roughly 50 percent higher for men than women in all age groups. Whereas 91.3 percent of men reported having had an orgasm at their most recent partnered sexual encounter, only 64.4 percent of women reported having done so. One might return to a biosocial perspective to explain these differences except that they are less pronounced among the youngest cohorts. Nevertheless, many other remnants of patriarchal sexuality continue to exist in society, such as pornography, sexual harassment, and sexual violence, which will be discussed later in this and subsequent chapters.

The Twentieth Century: The Emergence of Expressive Sexuality

A different sexual message has emerged as the result of societal changes that include the decreasing economic dependence of women and the availability of new methods of birth control. Because of the increasing emphasis on couple intimacy, women's sexual expression is more encouraged than it had been earlier (D'Emilio and Freedman 1998). With **expressive sexuality**, sexuality is seen as basic to the humanness of both women and men; there is no one-sided sense of ownership. Orgasm is important for women as well as for men. Sex is not only, or even primarily, for reproduction but also an important means of enhancing human intimacy. Hence, all forms of sexual activity between consenting adults are acceptable.

The 1960s Sexual Revolution: Sex for Pleasure

Although the view of sex as intimacy continues to predominate, in the 1920s, an alternative message began to emerge wherein sex was seen as a legitimate means to individual pleasure, whether or not it was embedded in a serious couple relationship. Probably as a result, the generation of women born in the first decade of the twentieth century showed twice the incidence of nonmarital intercourse (sex outside marriage) as those born earlier (D'Emilio and Freedman 1998). This probably occurred mostly in relationships that anticipated marriage (Zeitz 2003). Further liberalization of attitudes and behavior characterized the sexual revolution of the 1960s.

What was so revolutionary about the sixties? For one thing, the birth-control pill became widely available; as a result, people were freer to have intercourse with more certainty of avoiding pregnancy. Laws regarding sexuality became more liberal in the 1960s (the exception being laws regulating sex between same-sex partners). Until the U.S. Supreme Court decision in *Griswold v. Connecticut* (1965) recognized a right of "marital privacy," the sale or provision of contraception was illegal in some states. The idea that sexual and reproductive decision making belonged to the couple, not the state, was extended to single individuals and minors by subsequent decisions (*Eisenstadt v. Baird* 1972; *Carey v. Population Services International* 1977).

This period initiated a long term shift in Americans' attitudes and behavior regarding sex. In 1959, about four-fifths of Americans surveyed said they disapproved of sex outside marriage (Smith 1999). By 2006, only 25 percent said it was "always wrong" (Schott 2007). Not only did attitudes become more liberal, but also behaviors (particularly women's behaviors) changed. Between 1943 and 1999, the rate of nonmarital sex and the number of partners rose while age at first intercourse dropped (Wells and Twenge 2005). In 2011, 6.2 percent of students in grades nine through twelve from the National Youth Risk Behavior Survey had had sexual intercourse before age 13 and 15.3 percent had experienced sexual intercourse with four or more persons (U.S. Centers for Disease Control and Prevention 2011). The trend toward higher rates of nonmarital sex has continued. Now "almost all Americans have sex before marrying" (Finer 2007, p. 73).

Today, sexual activity often begins in the teen years. Table 4.1 shows the percentages of sexually experienced teens in each of the major racial/ethnic groups according to the Youth Risk Behavior Surveillance System, a national high school–based survey conducted by the U.S. Centers for Disease Control and Prevention (Child Trends 2012a, b). Not surprisingly, sexual experience increases with age and grade level. In 2011, almost half (47 percent) of high school students had had sexual experience (49 percent of males and 47 percent of females), and 41 percent were currently sexually active. Data from the National Survey of Family Growth also provides information on teens' prior sexual activity. For the majority (70 percent) of teens who had experienced sex, their first experience was with a romantic partner, although some 10.8 percent of teens said they "really didn't want it to happen at the time" (Martinez, Copen, and Abma 2011). Perhaps the most significant change the sexual revolution ushered in, among heterosexuals at least, has been in marital sex. "Today's married couples have sexual intercourse more often, experience more sexual pleasure, and engage in a greater variety of sexual

Table 4.1 Sexual Experience of High School Students by Race/Ethnicity and Gender, 2011

	Percentage Who ...			
	"Ever Had Sexual Intercourse"		"Are Currently Sexually Active"	
Ethnicity	Males	Females	Males	Females
White	44	45	30	35
Black	67	54	46	37
Hispanic	53	44	35	32
Total	49	47	33	34

Source: Child Trends 2012a, Figure 2; Child Trends 2012b, Appendix 1.

activities and techniques than people surveyed in the 1950s" (Greenberg, Bruess, and Haffner 2002, p. 437). In the NSSHB study, 83 percent of men said they experienced "quite a bit" to "extreme" pleasure during their most recent sexual event, which was further enhanced when the sex occurred within a married, cohabiting, or committed relationship (Herbenick et al. 2010b).

The 1980s and 1990s: Challenges to Heterosexism

If the sexual revolution of the 1960s focused on freer attitudes and behaviors among heterosexuals, recent decades have expanded that liberalism to include gay men and lesbians. Until several decades ago, most people thought about sexuality almost exclusively as between men and women. In other words, our thinking was characterized by **heterosexism**—the taken-for-granted system of beliefs, values, and customs that places superior value on heterosexual behavior and denies or stigmatizes nonheterosexual relations. However, once the Stonewall riots—events sparked by a 1969 police raid on a New York gay bar—galvanized the gay community into advocacy, gay males and lesbians have not only become increasingly visible but also have challenged the notion that heterosexuality is the one proper form of sexual expression. A number of high-profile people in government and the media have publicly come out as gay, as have countless actors, actresses, and celebrities. For example, in 2012, Representative Tammy Baldwin, a Wisconsin Democrat in the House of Representatives, was elected as the first openly gay U.S. senator (O'Brien 2012).

Gay men and lesbians have won legal victories, new tolerance by some religious denominations, greater understanding on the part of some heterosexuals, and sometimes positive action by government. Although states, communities, and corporations have passed sexual orientation antidiscrimination laws, you can still get fired from your job for being gay in most states. In 2011, President Barack Obama repealed "Don't ask, don't tell," the 17-year-old law that banned openly gay men, lesbians, and bisexuals from military service (Bumiller 2011). In 2012, President Obama publicly supported gay rights, including gay marriage. See Chapter 6 for more information on gay marriage in the United States.

The public's attitudes toward homosexuality, though never as favorable as toward nonmarital sex generally, became more favorable in the 1990s, after earlier high rates of disapproval. In the early 1970s, about 70 percent thought homosexual relations were "always wrong." In 1986, the Supreme Court decision in *Bowers v. Hardwick* declined to extend privacy protection to gay male or lesbian relationships, and homosexual conduct remained criminalized in some states. Then in a landmark 2003 case (*Lawrence et al. v. Texas*), the Supreme Court reversed its earlier decision, striking down a Texas law criminalizing homosexual acts, thus legalizing same-sex sexual relations. Figure 4.3 shows that

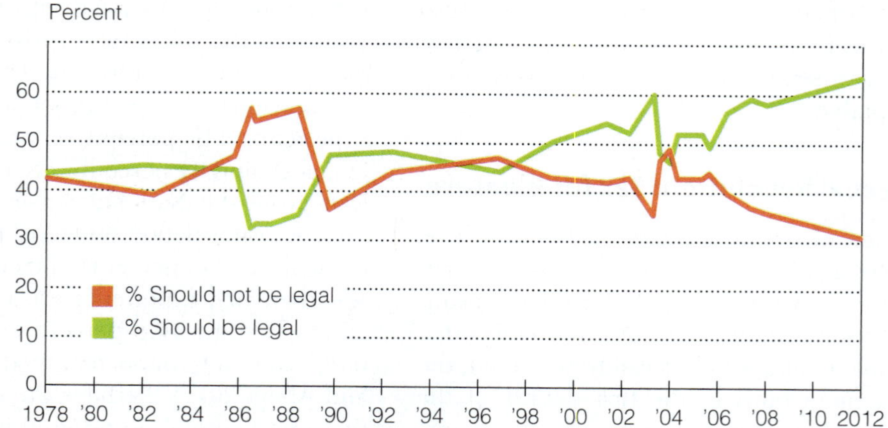

FIGURE 4.3 Do you think gay or lesbian relations between consenting adults should or should not be legal?

Source: Saad 2012c.

in 2012, 63 percent of those surveyed in a Gallup poll agreed that "gay/lesbian relations should be legal" (Saad 2012c).

Americans are divided over whether gays and lesbians choose their sexual orientation, a split that shapes attitudes. In 2012, 40 percent of Americans surveyed thought that being gay or lesbian is something a person is born with whereas 35 percent thought that it was a result of a person's upbringing or environment (Saad 2012c). People who see being gay as a choice are less sympathetic to lesbians or gay men regarding jobs and other rights (Loftus 2001). Americans are more likely to approve of civil rights protections for gays and lesbians than of a gay or lesbian lifestyle, and that approval has continued to strengthen since the 1970s (Saad 2006a). The workplace is "becoming friendlier" to gay, lesbian, and bisexual employees, and the vast majority of Forbes 500 companies include sexual orientation in their antidiscrimination policies (Fidas and Luther 2010).

Homophobia—viewing homosexuals with fear, dread, aversion, or hatred—is still present in American society. Although support for gays and lesbians has never been higher, 50 percent of Americans oppose gay marriage and 42 percent consider gay and lesbian relations morally wrong (Saad 2012c). Employers still discriminate against job applicants who appear to be gay (Tilcsik 2011). When people know someone who is gay, lesbian, or bisexual, however, their support for gay and lesbian rights increases ("Gay and Lesbian Rights" 2009; Herek 2009a, 2009b).

Semantics are also important when Americans are contemplating gay rights. For example, a 2009 CBS News–*New York Times* poll found that a majority (59 percent) of Americans favored allowing "homosexuals" to serve in the U.S. military, while the same poll showed that an even stronger majority (71 percent) favored allowing "gay men and lesbians" to serve in the U.S. military.

Comparing the Sexual Behaviors of Gays and Lesbians and Heterosexuals Sexual desires, behaviors, and satisfaction have much to do with cultural trends. Leonore Tiefer notes that how we develop our desires and our expectations have very much to do with our "socially produced expectations [which] affect meaning and satisfaction" (Tiefer 2007, p. 246).

In their landmark study, Philip Blumstein and Pepper Schwartz (1983), who studied a large national sample of 12,000 volunteers from the Seattle, San Francisco, and Washington, D.C., areas before the HIV/AIDS epidemic, described lesbian relationships as the "least sexualized" of four kinds of couples: heterosexual cohabiting and married couples, lesbians, and gay men.

Lesbians have sex less frequently than gay men, although it may be difficult to make comparisons because lesbians' physical relationship can take the form of hugging, cuddling, and kissing, not only genital contact (Peplau, Fingerhut, and Beals 2004; Tracy and Junginger 2007). Nevertheless, lesbians report slightly greater sexual satisfaction than do heterosexual

© sepavo/Shutterstock.com

Lesbian and gay male couples and families have become increasingly visible over the past decade. Meanwhile, discrimination and controversy persist.

women (Holmberg, Blair, and Phillips 2010). Gay men appear to be more accepting of nonmonagamous relationships than are lesbians—or heterosexuals (Adam 2007). For example, in a national study of 782 same-sex and heterosexual couples, 43.7 percent of gay men who had discussed sex outside the relationship with their partner reported that they had agreed that "under some circumstances it is all right," compared to 5.1 percent of lesbians, 6.0 percent of heterosexual males, and 3.3 percent of heterosexual females (Gotta et al. 2011). However, a comparison of two national studies indicates that monogamy among gay males has substantially increased—from 17.4 of gay male couples reporting they were monogamous in 1975 to 40.6 percent in 2000 (Gotta et al. 2011). The authors speculate this may be a result of increased awareness of HIV/AIDS and other sexually transmitted infections. Others argue that longer-term, monogamous, and committed relationships have become the norm among men and lesbians, especially among younger cohorts (Green 2009). Casual sex among lesbians appears to be relatively rare.

We have discussed differences, but gay and lesbian relationships resemble those of heterosexual marriage and cohabitation in many ways. For example, for all couple types—gay, lesbian, heterosexual cohabitants, and heterosexual married couples—sexual satisfaction is associated with general relationship satisfaction and sexual frequency (Holmberg and Blair 2009; Tracy and Junginger 2007). "Despite variability in structure, close dyadic relationships work in similar ways" (Kurdek 2006, p. 509).

The Twenty-First Century: Risk, Caution—and Intimacy

Although pleasure seeking was the icon of sixties sexuality, warnings to be cautious in the face of risk characterize contemporary times. Many heterosexual young adults acknowledge the risk of AIDS, but decide to take their chances. "Most emerging adults ... say that fear of AIDS has become the framework for their sexual consciousness, deeply affecting their attitudes toward sex and the way they approach sex with potential partners," perhaps asking for a test result or insisting on condom use (Arnett 2004, p. 91). In addition, teenagers from the National Survey of Family Growth (boys more so than girls) said the fear of avoiding pregnancy and disease were the main reason they had not yet had sex (Schalet 2012). Meanwhile, a number of singles have multiple partners over time; males tend to report higher rates of multiple partners, with 23.2 percent reporting "fifteen or more female sexual partners in lifetime" compared to 9.2 percent of females (Chandra et al. 2011). Among teenagers aged 15 to 19, males are also more likely than females to have had sex with more than one partner. According to the 2006–2010 National Survey of Family Growth, 35 percent of sexually experienced women had just one partner compared to 30 percent of men. Similarly, 49 percent of women had three or more partners compared to 55 percent of men (Martinez , Copen, and Abma 2011). See "Facts about Families: How Do We Know What We Do? A Look at Sex Surveys."

Recent data from the 2009 NSSHB indicates that men and women of all ages engage in a wide variety of sexual behaviors: masturbation (alone and with a partner), oral sex, and anal intercourse (Herbenick et al. 2010a). Oral sex is the most frequently reported behavior, with 57 to 91 percent of men and 52 percent to 72 percent of women age 25 to 49 having engaged in it in the past year. Although a large percentage of men and women engaged only in one sexual behavior during their last sexual encounter (penile–vaginal intercourse being most common), the majority engaged in multiple sexual activities (Herbenick et al. 2010b). A greater number of sexual activities was associated with increased likelihood of orgasm for both men and women.

Sexting may be a new word to many older generations, but is a well-known and growing phenomenon among young people of all genders, sexualities, classes, and belief systems in our modern technological era. *Sexting* is defined as sending or posting sexually suggestive text messages and images, including nude or seminude photographs, via cell phones or smartphones or over the Internet [Federal Bureau of Investigation (FBI) 2012]. According to a 2008 survey, many teens and young adults have sent sexually provocative photographs and text messages over their cell phones to people they don't know or people they want to date or hook up with. In addition, 71 percent of teen girls, 67 percent of teen boys, 83 percent of young adult women, and 75 percent of young adult men admit to sending sexually suggestive photos and text messages of themselves to people with whom they are in a romantic relationship.

Sexting can have important legal consequences. Teenagers under age 18 can be sentenced to juvenile court for possessing and sending nude photos. There are social and emotional consequences as well. In one notable case, an 18-year-old girl committed suicide after nude photos intended for her boyfriend were distributed to hundreds of her classmates (FBI 2012). Sexing is thought to be part of a broader social phenomenon referred to as **pornification**, in which sending, receiving, and viewing sexually explicit material is seen as a normal thing to do among young people (Paul 2005). This trend is discussed

Facts about Families

How Do We Know What We Do? A Look at Sex Surveys

The pioneer surveys on sex in the United States were conducted by Alfred Kinsey and his associates (Kinsey, Pomeroy, and Martin 1948, 1953). Kinsey's two-volume study *Sexual Behavior in the Human Male* and *Sexual Behavior in the Human Female* provided one of the first looks into the sexual lives of American men and women. It was groundbreaking for its time because it was published during a period of political and cultural conservatism. Kinsey used volunteers, including people from his own social circle. He believed that a statistically representative survey of sexual behavior would be impossible because many of the randomly selected respondents would refuse to answer or would lie. On the other hand, because he relied on volunteers, including inmates and prostitutes, his research is criticized for not being representative of mainstream sexuality. His work is most useful in terms of representing the broad range of sexual behaviors in American society.

More recent scientific studies on sexual behavior have used random samples that are more representative of sexual behaviors of the U.S. population. These include the General Social Survey (GSS) and other studies conducted by the National Opinion Research Center (NORC) at the University of Chicago National Health, the National Health and Social Life Survey (NHSLS), and the National Social Life, Health, and Aging Project. These data sources are of high quality. For example, 80 percent of people contacted to participate in the NHSLS agreed to be interviewed—an impressively high response rate considering the topic (Laumann et al. 1994). To provide some anonymity for the more sensitive part of the interview—questions about oral and anal sex, for example—specific sexual behavior questions were asked by means of a written questionnaire. Findings of the National Health and Social Life Survey were then welcomed as the first-ever truly scientific nationwide survey of sex in the United States. Other national studies that have provided information on the sexual lives of Americans include the National Survey of Families and Households (NSFH), the National Survey of Family Growth (NSFG), and the National Study of Adolescent Health (Add Health). The newest data on sexuality comes from the National Study of Sexual Health and Behavior (NSSHB), conducted by researchers at the Center for Sexual Health Promotion at Indiana University. This is a national sample of Americans aged 14 to 94. Other data on sexuality come from surveys conducted by government agencies such as the U.S. Centers from Disease Control and Prevention.

Conclusions based on survey research on sensitive matters such as sexuality must always be qualified by an awareness of their limitations—the possibility that respondents have minimized or exaggerated their sexual activity or that people willing to answer a survey on sex are not representative of the public. Nevertheless, with data from these national sample surveys, we have far more reliable information than ever before.

Critical Thinking

Have you ever been curious about the sexual behavior of those around you? To what extent should we trust data on sexuality? If you were given a survey asking personal questions about your sexual behaviors, would you be truthful?

in more detail later in the chapter. See also "As We Make Choices: Sexting—Five Things To Think About Before Pressing 'Send.'"

Today, there is risk in a sexual encounter. Today's sexually liberated environment is especially disadvantageous for women in terms of experiencing negative emotions, a negative reputation, and unwanted sex (Katz, Tirone, and van der Kloet 2012). At the same time, a more liberal sexual environment offers the potential for expressive sexuality for both women and men and true sexual intimacy. People now have more knowledge of the principles of building good relationships (even if they do not always put them into practice). Now that the possibility of satisfying sexual relationships seems more attainable, how do men and women negotiate those sexual relationships inside and outside of marriage?

Although relationships between the sexes are more equal today than in the past, many—though assuredly not all—women and men today may have internalized divergent sexual scripts or messages. Today's heterosexuals negotiate sexual relationships in a context in which new expectations of equality and similarity coexist within a heritage of gender-related difference. Men may be somewhat more accepting of recreational sex than women are, and studies continue to show that women are more interested than men in romantic preliminaries (Purnine and Carey 1998).

At the same time, men are becoming more sensitive, less boastful about sexual conquests, and more romantic (Schalet 2012; Seal and Ehrhardt 2003). After the sexual revolution of the 1960s, men became more interested in communicating intimately through sex, whereas women showed more interest than before in physical pleasure (Pietropinto and Simenauer 1977; Seal, O'Sullivan, and Ehrhardt 2007). Nearly forty years ago, pioneering sex researchers Masters and Johnson (1976) argued that more equal gender expectations led to better sex.

As We Make Choices

Sexting—Five Things To Think about Before Pressing "Send"

1. Don't assume anything you send or post is going to remain private.
2. There is no changing your mind in cyberspace—anything you send or post will never truly go away.
3. Don't give into the pressure to do something that makes you uncomfortable, even in cyberspace.
4. Just because a message is meant to be fun doesn't mean the person who gets it will see it that way... many teen boys (29 percent) agree that girls who send such content "are expected to date or hook up in real life."
5. Nothing is truly anonymous.

Source: Adapted from *Sex and Tech: Results from a Survey of Teens and Young Adults* 2009, p. 2.

This discussion points again to the fact that cultural messages vary and that sexual relationships are negotiated in a social context. "Since early in the twentieth century the bonds between marriage and sexual activity have been unraveling" (T. Smith 2006, p. 26). In the next sections, we discuss sexual activity outside of marriage and committed relationships. After that, we will examine sexuality within marriage and committed relationships—where most sexual activity takes place—and then look at what is known about racial/ethnic diversity in sexual expression. Keep in mind that sexual expression may include more than intercourse—activities ranging from genital intercourse, to oral and anal sex, to kissing and cuddling.

Race/Ethnicity and Sexual Activity

As discussed previously in this chapter, there is variation between racial and gender groups when it comes to sexual activities (refer to Table 4.1). A study in the *American Journal of Public Health* found that the vast majority of men, 89 percent, from all racial groups had only one sex partner at a time. The prevalence, however, of maintaining multiple sexual partnerships varied substantially by race/ethnicity. According to the 2006–2008 NSFG, black and Hispanic men (32 percent and 22 percent, respectively) were more likely to have had more than two sexual partners in the last year than were nonHispanic white men (16 percent) (Chandra et al. 2011; see also Adimora, Schoenbach, and Doherty 2005). Interviews with heterosexual African American men suggest that having concurrent sex with multiple women provides proof of manhood (Bowleg et al. 2011). Among both males and females, whites are more likely to have experienced oral and anal sex than have African Americans or Hispanics (Mosher, Chandra, and Jones 2005, Tables 1, 2, p. 12). This is also the case among adolescents (Lindberg, Jones, and Santelli 2008).

However, African Americans, Hispanics, and non-Hispanic whites are more similar than dissimilar in most aspects of their sexual behavior, and any dissimilarities are most likely related to socioeconomic status rather than race (Chandra et al. 2011; Dodge et al. 2010; Knox and Zusman 2009). Asians have more conservative sexual attitudes than whites and Hispanics, report fewer sexual partners over a lifetime, and tend to have their first sexual experiences later (Ahrold and Meston 2010). Gay and lesbian sexuality, like heterosexual behavior, has been explored among African Americans mostly in the context of problems (e.g., AIDS) and at the lower end of the social scale. An exception is found in the analysis of the 2000 census data on black same-sex households, which make up 14 percent of all such households. Black same-sex households tend to be less well off economically than other same-sex households. They are also more likely to be raising children (Dang and Frazer 2004).

Social scientists writing about gay black male sexuality believe it is not as visible as among whites because blacks may find white gay subcultures alien, and they may tend to be integrated into heterosexual communities and extended families that strongly disapprove of homosexuality (Bowleg et al. 2003; Bowleg et al. 2011). Similar issues are found in the more traditional Latino and Asian cultures where heterosexuality is emphasized (Balsam et al. 2011; Calzo and Ward 2010). As part of that social integration, as well as the racism that men of color perceive in the gay community (Han 2008), black and Latino gay men may be more likely to identify as bisexual than exclusively homosexual even when engaged primarily in same-sex relations (Sandfort and Dodge 2008). This pattern of engaging in sex with other men while maintaining a straight masculine identity has been labeled "the down low" (Denizet-Lewis 2003; Malebranche 2007). However, a recent study of college students indicates that disapproval of black gays and lesbians may be lessening (Whitley, Childs, and Collins 2011). Still, African-American gay and bisexual men may feel uncomfortable seeking potentially supportive social networks, such as LGBT groups on college campuses (Goode-Cross and Tager 2011).

Black lesbians are relatively invisible because of their smaller numbers and integration into extended-family relationships. One early study of black lesbians was a qualitative study of 530 lesbians and 66 bisexual women who were well-educated, middle-class women in their thirties, who first became conscious of their attraction to women at around age 14, with first same-sex sexual experiences at age 19. Their adult relationships were generally satisfying and close (Mays and Cochran 1999). Other black women adopt a hypermasculine "stud" role as a defense mechanism against sexism, racism, and homophobia (Lane-Steele 2011). Latina lesbians also face disapproval from their communities, especially if they are parenting children (Ricon and Lam 2011). Many Latina lesbians are embedded in traditional communities of recent immigrants and conservative Christians (Asencio 2009; Barbosa, Torres, Silva, and Khan 2010).

Sexual Values Outside Committed Relationships

These days young people generally do not marry until their late twenties. What this means is that they have many years, perhaps as long as a decade if not longer, in which to experiment sexually. Below we discuss four sexual values or standards of behavior for singles not in committed relationships (either marriage or cohabitation). Later we'll discuss extramarital sex—that is, sexual relations of married or cohabiting people outside of their committed relationship.

Abstinence

Many contemporary groups, especially the conservative Christian and Islamic communities, encourage **abstinence** (refraining from sex) as a moral imperative. Agnostics and atheists were found to have significantly less conservative sexual attitudes than those identifying with a Christian, Jewish, or Hindu religious affiliation (Ahrold et al. 2011; Ellison, Wolfinger, and Ramos-Wada 2012). Among different racial and ethnic groups, Asian Americans exhibit the most conservative sexual attitudes (Ahrold and Meston 2010).

People may be surprised to learn that teens today are less likely to be having sex than in the past. In the last 20 years, the percentage of sexually experienced teenagers has significantly declined. Among females, in 1988 51 percent reported ever having sex compared to 43 percent in 2006–2010. Among males, the percentage having had sex declined from 60 percent to 42 percent during that same period (Martinez, Copen, and Abma 2011). The most frequent reason for not having sex was "against religion or morals" (41 percent for females and 31 percent of males); this has remained unchanged since 2002 (Martinez, Copen, and Abma 2011).

Similarly, aside from a small, unexplained increase in 2006, rates of teen pregnancy and teen births have shown a steady, downward trend for decades. Teen birthrates had declined 34 percent from 1991 to 2005 (Hamilton, Martin, and Ventura 2010), with accompanying declines in teen pregnancy rates (Dyess 2009; Gavin et al. 2009; Schlesinger 2010). Today, teen pregnancy and teen birthrates in the United States are at an all-time low for all races and ethnicities, though they are higher than in other industrialized countries (U.S. Centers for Disease Control and Prevention 2011).

Generally, teens who do not engage in sexual activity give conservative values or fear of pregnancy, disease, or parents as their reasons (Martinez, Copen, and Abma 2011; Rasberry and Goodson 2009). A study of college students suggests that women who refrain from sexual activity do so because of an absence of love, a fear of pregnancy or sexually transmitted infections (STIs), a belief that use of contraception will cause infections or cancer, or a belief system that endorses nonmarital virginity (Kaye, Sullentrop, and Sloup 2009). Men's hesitancy to pursue sexual involvement comes more often from feeling "inadequate or insecure" (Christopher and Sprecher 2000, p. 1,009). Individuals choose abstinence for other reasons as well. Some women and men have withdrawn from nonmarital sexual relationships entirely to avoid bad experiences. Some withdraw from sexual risk, at least for a time, rather than feel vulnerable in the open sexual climate of the sexual revolution (Rasberry and Goodson 2009). The feminist movement is cited as empowering women to be abstinent if they wish (Ali and Scelfo 2002). Finally, abstinence not only occurs outside of committed relationships, but many committed couples choose to remain abstinent until marriage for religious or cultural reasons.

Sex with Affection

Sexual intercourse between unmarried men and women is now widely accepted, provided they have a fairly stable, affectionate relationship. About 60 percent of adults approve of sexual intercourse outside of marriage (Gallup Poll 2012b). We mainly have sex with people we know and care about. Data from the 2009 NHHSB indicate that among both men and women the majority of sex occurs within committed relationships (Herbenick et al. 2010b). Most unmarried teens and adults engage in a pattern known as *serial monogamy*, with most partners demonstrating sexual exclusivity in dating relationships (Furstenberg et al. 1983). Most Americans have only one sexual partner in the course of a year (Chandra et al. 2011). Among college-age women, sexual enjoyment and orgasm is significantly higher in relationships than in causal encounters (Armstrong, England, and Fogarty 2012).

Sex Without Affection and Recreational Sex

About fifteen years ago, a *New York Times* article surprised many by describing sexual behavior patterns that adolescents termed **friends with benefits** and **hooking up** (Denizet-Lewis 2004). The basic idea in hooking up and friends with benefits and other forms of casual sex is that a sexual encounter means nothing more than just that—sexual activity. *Hooking up* can occur with no prior acquaintance between the parties and no further contact afterward. *Friends with benefits* is described as sex that takes place with friends but without expectations for romantic love or commitment (Furman and Shaffer 2011). Psychologists Jocelyn J. Wentland and Elke D. Reissing (2011) conducted focus groups with male and female college students from a large, urban university in eastern Canada and discovered that what we routinely call *hooking up* incorporates a range of relationship types. Based on these discussions, Wentland and Reissing placed various casual sex relationships on a continuum according to different dimensions. For example, *booty calls* are similar to friends with benefits in that they involve repeated sexual encounters but require less communication, emotional investment, and responsibility. One important commonality, however, was the presence of alcohol. These sexual scripts may have emerged because many of today's young adults—who are delaying marriage, going to college, and developing careers—still want to have sex. They want some intimate connection without risking romantic disappointment and emotional loss.

Today, sociologists see this as a growing trend among adolescents and young adults. Sixteen percent of female teens and 28 percent of male teens had sex for the first time with someone they had just met or someone they considered "just a friend." Another 9 percent of teen women and 12 percent of teen men first had sex with someone they were "going out with once in a while" (Martinez, Copen, and Abma 2011). New technologies such as the cell phone have bolstered this trend by making it easy to get in touch with potential hookups on short notice.

According to Dr. Kathleen Bogle, a sociologist who studies this phenomenon, dating used to be something that led up to sex. In the hookup era, the sex (or sexual activity) happens first, which may or may not ever lead to dating or romantic involvement (B. Wilson 2009; see also Bogle 2008). *Hooking up* does not necessarily mean intercourse; oral sex and *making out* (kissing and hugging) may be just as common (Bogle 2008). The idea of the hookup has also moved beyond the college to people of all ages, as well as those who have diseases, disfigurements, and even psychological issues and who have difficulty finding others who can understand their needs (Alexander 2010; Durham 2010; Herbenick et al. 2010b).

What does this development mean for young lives and future marriage prospects? Reactions of social scientists range from alarm at the disappearance of a courtship path to marriage (Glenn and Marquardt 2001; Stepp 2007) to acceptance of the thought that hooking up is simply the new way to begin a relationship that may become serious despite its initial intent (England and Thomas 2007; Epstein et al. 2009). By carefully outlining the positives and negatives of both dating and hooking up, one observer reminds us that dating also has drawbacks as a relationship development process (Bogle 2004/2008).

Meanwhile, some observers note that men, as they did in the dating era, continue to have greater relationship power in hookup settings (discussed further in the next section; Bogle 2008). Moreover, now that there are more women than men on most college campuses, male power is reinforced as heterosexual women are forced to compete for the much smaller group of available men (Williams 2010). More seriously, though, researchers' interest in time should lead to learning more about how the gap between hooking up and married with children is to be bridged.

The Double Standard

According to the **double standard**, women's sexual behavior must be more conservative than men's. Despite the sexual revolution of the 1960s and 1970s, numerous studies conducted since then indicate the double standard continues to exist (Crawford and Popp 2003). In its original form, the double standard meant that women should not have sex before or outside marriage, whereas men could. Within the context of marriage and committed relationships, femininity "is typically framed in terms of being sexually desirable rather than sexually desiring whereas masculinity connotes sexual aggression and prowess" (Elliott and Umberson 2008, p. 392). The continued existence of the double standard alongside the prevailing pattern of hooking up makes navigating sexual relationships challenging for young women who do not want to be perceived as promiscuous (Bogle 2004; Stepp 2007). In a study of female and male college students, women who had hooked up were more than twice as likely as men to report feeling less respected by the person afterward (54 percent compared to 24 percent; Armstrong, England, and Fogarty 2012). In another study of college students, women had significantly more worries or romantic thoughts about their hookups than did men (Townsend and Wasserman 2011). Other research notes, however, that this is not necessarily the case; in fact, many men find these ever-transforming sexual scripts difficult to negotiate. Young men seem to be as emotionally vulnerable as women and are frequently using hookups to find lasting and meaningful relationships (Epstein et al. 2009; Manning, Giordano, and Longmore 2006; Pollack 2006; Smiler 2008). Moreover, a recent study of adolescents found that girls with a higher number of sexual partners were not less popular, did not lack friends, or did not suffer from lower self-esteem (Lyons et al. 2011).

The power imbalance in hooking up can be seen from the perspective of sexual satisfaction. Both female and male college students report that is typical for men to completely disregard the woman's pleasure in a hookup and that equality is not expected (Armstrong, England, and Fogarty 2012). Men are much more attentive to the sexual needs of women in relationships than with casual sex partners; they often feel responsible for their girlfriends' orgasms and express pride in their ability to bring them to orgasm (Armstrong, England, and Fogarty 2012). We return to the topic of gender and dating relationships in Chapter 5.

Sexual Values for Committed Relationships

Up to this point, we've been examining scripts for sex in noncommitted relationships. It might surprise you that various aspects of nonmarital sex are more likely to be studied than are sex within marriage and that sexual activities of teens receive more research attention than those of adults (Fisher et al. 2010). "More is known about sexuality in marriage at this time than has ever been true in the past. But we still have only a limited view of how sexuality is integrated into the normal flow of married life" (Christopher and Sprecher 2000, p. 1013). Now we will look at a different form of sex outside marriage and committed relationships—sexual infidelity or the "affair."

Monogamy and Sexual Infidelity

As explored in Chapter 7, marriage typically involves promises of sexual exclusivity—that spouses and committed partners will have sexual relations only with each other. Cohabiting and other committed relationships also involve expectations of fidelity, although to a somewhat lesser degree (Treas and Giesen 2000). Americans believe in fidelity and sexual exclusivity regardless of a legally binding commitment to one another, but it seems that beliefs and actions do not always quite match, and some people find themselves unable to completely adhere to their own expectations.

Although infidelity is found in virtually any society and throughout history, the proscription against extramarital sex is stronger in the United States than in many other parts of the world. Ninety-two percent of Americans consider extramarital affairs "morally wrong" (Newport 2009). Cohabiting couples also generally expect each other to be sexually faithful (94 percent, compared to 99 percent of married couples). However, the rate of sexual infidelity is higher among cohabiting couples than among married couples (Treas and Giesen 2000).

Some researchers distinguish among emotional infidelity, sexual infidelity, and combined emotional and sexual infidelity. The latter is most disapproved of. Emotional (without sexual) infidelity is least disapproved of (Blow and Hartnett 2005). Sexual infidelity is engaging in sexual relations with someone who is not one's own marriage or committed partner. Some researchers define emotional infidelity as an "intense, primarily emotional, nonsexual relationship" with someone who is not one's own marriage or committed partner (Potter-Efron and Potter-Efron 2008, p. 2). Both of these forms of infidelity have long-term, sustained negative impacts on a marriage or committed relationship, including increased risk of divorce (Meier, Hull, and Ortyl 2009; Previti and Amato 2004). Data from the General Social Survey collected between 1991 and 2008 indicate that half of men and and half of women who have had extramarital sex eventually separate or divorce from their spouses (Allen and Atkins 2012).

Statistics on sexual infidelity are based on what people report: Some spouses or committed partners hesitate to admit an affair; others boast about affairs that didn't really happen. In surveys, men and women tend to underreport infidelity. In one study, married women were six times more likely to admit to infidelity in a computer-assisted self-interview than in a face-to-face one with an interviewer (Whisman and Snyder 2007). Therefore, estimates of prevalence of extramarital sex vary and are not very precise.

Contrary to media and Internet hype—which tends to focus on a small number of notable cases of infidelity among famous politicians, actors, and athletes—there is no evidence of an "infidelity epidemic" (Watson 2011). Looking at the research as a whole, lifetime prevalence rates of sexual infidelity ranges somewhere between 15 percent and 25 percent (Blow and Hartnett 2005; Drexler 2012). Infidelity has increased in the last two decades. Data from General Social Surveys (GSS) showed an increase in infidelity between 1991 and 2006 across all age cohorts, especially among older men and women (Atkins and Furrow 2008). Men are more likely to engage in extramarital sex than are women (Drexler 2012; Watson 2011). Of married men, 3.4 percent reported a lifetime experience of affairs with other men; 5.3 percent of cohabiting men reported such experiences. Of married women, 7.2 percent reported some sexual experience with other women, and 10.8 percent of cohabiting women did so (Mosher, Chandra, and Jones 2005, Tables 1, 2, 8). Men are more forgiving of partners who have had affairs with other women than other men, whereas women are more forgiving of men who have had heterosexual affairs than homosexual ones (Freeman 2011). Some husbands may even encourage or pressure their wives' sexual relationships with other women—for example, in couples engaging in "swinging" (Praver 2009).

Risk Factors Sociologists Judith Treas and Deirdre Giesen (2000) developed a conceptual model of risk factors for extramarital sex. They found that entering an extramarital affair is a rational decision—that is, affairs are generally *not* spontaneous (the result of too much alcohol, for example) nor are they the consequence of overwhelming romantic passion. Rather, "[p]eople contemplating sexual infidelity described considered decisions" (Treas and Giesen 2000, p. 49). In a recent study, researchers also found that loneliness is an important factor in one's decision to be unfaithful. This study found that the conditions of many undocumented workers in the United States is such that loneliness, as well as fear of deportation, led to unsafe sexual practices and an increased risk of sexually transmitted disease (Hirsch et al. 2009). Similarly, military veterans have higher levels of infidelity. In a study also based on the 1992 National Health and Social Life Survey, 32 percent of veterans reported having had an extramarital sexual relationship compared to 17 percent of nonveterans (London, Allen, and Wilmoth 2012). Infidelity was one factor underlying veterans' higher divorce rates. Sexual infidelity is more common with couples in which the wife is pregnant, and when the husband scored higher on neuroticism (Whisman, Gordon, and Chatav 2007).

Previous researchers have found gender differences evident in the analysis of patterns of extramarital sex (Harris 2003). If a wife has an affair, she is more likely to do so because she feels emotionally distanced by her husband. Men who have affairs are far more likely to do so for the sexual excitement and variety they hope to find. Moreover, "men feel more betrayed by their wives having sex with someone else; women feel more betrayed by their husbands being emotionally involved with someone else" (Glass 1998, p. 35; Begley 2007; Blow and Hartnett 2005).

Relationship dissatisfaction is a motive more important to women. Sexual dissatisfaction and declines in frequency are also associated with affairs, especially for men (Blow and Hartnett 2005). Among women, childhood sexual abuse is associated with sexual infidelity (Whisman and Snyder 2007). In fact, research shows that the risk of extramarital affairs for both genders is "significantly higher among marriages characterized by spousal violence, divorce proneness, a past experience of marital separation, or the practice of spending relatively little time together," as opposed to marital and sexual satisfaction (DeMaris 2009, p. 605).

Historically, societies have depended on community pressures to control any sort of disapproved of sexual activity. Shared social networks of family and friends, as well as church attendance, seemed to operate as social controls in discouraging infidelity (Hutson 2009). Couples who lead separate lives and who have jobs requiring travel are more likely to have extramarital affairs (Treas and Giesen 2000). Women are working longer hours, traveling more, and have the same access to cell phones, text messaging, and so on that men have to create and nurture intimate connections outside of marriage or committed relationship (Parker-Pope 2008).

The 1990s saw the emergence of a new brand of marital infidelity—adultery on the net, or **cyberadultery**. The Internet and cell phone technology that allows texting and sending and receiving pictures and video has created new opportunities for individuals to develop secret relationships (Hertlein 2012). The emotional connection may lead to a meeting—and then perhaps to a sexual relationship. Because they are often conducted in secret, reliable statistics on cyberadultery are hard to find (Smith 2011). According to one study, 30 percent of men and 34 percent of women had engaged in cybersex (Daneback, Cooper, and Mansson 2005). Another study indicates that 65 percent of people who look for sex on the Internet had intercourse with their online partner and fewer than half used a condom (Reitmeijer, Bull, and McFarlane 2001). Even if the couple does not meet, men and women still tend to think of online relationships as a form of betrayal that is associated with relationship problems (Hertlein and Piercy 2006; Schneider, Weiss, and Samenow 2012; Whitty 2005). Addiction to cybersex, including among those in committed relationships, is increasing (Ross, Månsson, and Daneback 2012).

Union duration of marriage or cohabitation, which can be a measure of both *investment* in the relationship and *habituation*, showed a positive relationship with likelihood of extramarital sex during the union. This provides some support for the **habituation hypothesis**—that is, that familiarity reduces the reward power of a sexual encounter with a spouse or partner compared to a new relationship (Liu 2000). At the same time, union duration is also a simple measure of exposure to the risk of an extramarital affair (Treas and Giesen 2000).

Effects of Sexual Infidelity The secrecy required by an affair erodes the connection between partners. When discovered, the betrayal may spark jealousy—or it may create a crisis that motivates a search for the resolution

Cusp/SuperStock

It is important for married and committed to make time for each other. Regular "date nights" have been shown to enhance relationship and sexual satisfaction.

children stay up later and, by the time they are teenagers, parents no longer have any private evening time together. One woman's solution to this problem:

> Our house shuts down at 9:30 now. That doesn't mean we say "It's your bedtime, kids. You're tired and you need your sleep." It means we say, "Your dad (or your mom) and I need some time alone." The children go to their rooms at 9:30. Help with homework, lunch money, decisions about what they'll wear tomorrow—all those things get taken care of by 9:30 or they don't get taken care of. (Monestero 1990)

The important thing, these therapists stress, is that partners don't lose touch with either their sexuality or their ability to share it with each other. In other words, as Patti Newbold, who lost her husband, sadly notes, "Marriage isn't about my needs or his needs or about how well we communicate about our needs. It's about loving and being loved" (Marano 2010, p. 71).

Sexual Relationships and Pornography

Pornography—sexually explicit images and descriptions in books, magazines, film, and cyberspace—has seeped into American life to the extent that many of us are no longer disturbed by it (Paul 2005). The average Internet porn user is a middle-aged, married man of median income (Frontline 2012). Among adolescents and young adults, men are more likely to view sexually explicit material than women (Morgan 2011; Shaughnessy, Beyers, and Walsh 2011). But women read and view porn, too, as sales of the 2011 best-selling erotic novel *50 Shades of Grey* (James 2011) attest.

Some women who view pornography report several benefits: It helps them explore their sexuality, gives them ideas to use in real-life sex, and helps them learn how to look and act sexy (Paul 2005). Furthermore, pornography, along with other sexual games such as role-playing, sometimes enhance and "liven up" sexual encounters and may increase sexual satisfaction and overall intimacy (Hertlein 2012; Pratt, Brody, and Gu 2011).

On the other hand, pornography may contribute to a heterosexual couple's gender inequality and the objectification of the female partner. Most mainstream pornography can be considered **misogynistic**, exhibiting hatred, dislike, mistrust, mistreatment, or general disregard for women. The sex portrayed is generally that of a dominant male in charge of a sexual encounter with a submissive female doing as she is told. Furthermore, pornographic sex typically reflects the manner in which males are more likely to come to orgasm (relatively fast straightforward penile thrusting into vagina or anus or by oral sex) as opposed to the way that most females experience orgasm (resulting from a combination of fast and slow touching and various forms of sexual stimulation over a longer period of time). Viewing pornography can alter one's sexual preferences: More-frequent viewers are more likely to want to engage in sex acts portrayed online (Morgan 2011). Some women report that having pornography around the house makes them uncomfortable with their own bodies and body image, but they feel unable to complain. Moreover, some men admit that their regular, perhaps addictive, use of porn has hurt their ability to enjoy sex in real life (Paul 2005). Among college students, more-frequent viewing of pornography was associated with lower age at first intercourse, more frequent casual sex, a greater number of partners, and lower sexual and relationship satisfaction (Morgan 2011).

We have been talking about human sexual expression as a pleasure bond. It is terribly unfortunate that sexuality can also be associated with exploitation, violence, disease, and even death. Indeed, the fact that it is so difficult to make a transition to the next topic points to the multifaceted, even contradictory, nature of contemporary human sexual expression.

The Politics of Sex

One of the most striking changes over the past several decades has been the emergence of sexual and reproductive issues as political controversies. Religious and political conservatives and secular and religious individuals and organizations with a more liberal set of values—more open to nonmarital sexuality, for example—regularly confront one another. Adding to the political mix, public health professionals approach policy from a research-based and pragmatic perspective, seeking the most effective means to achieve sexual and reproductive health goals.

Controversies over AIDS—and over how to educate youth about AIDS and about sex in general—illustrate the conflict between a morally neutral and pragmatic public health policy and a religious fundamentalist moral approach. Other sexuality issues engendering political conflict include abortion and contraception.

Political controversy has influenced both research and education about sexuality in the United States over the last few decades.

Adolescent Sexuality and Sex Education

Many adolescents are sexually active; 47 percent of high school students responding to the 2011 Youth Risk Behavior Surveillance have had sexual intercourse (Child Trends 2012b). What may surprise you is that teen sexual intercourse has declined since 1991; in that year, 54 percent of high school youth had had intercourse. The percent of high school males and females who were virgins is very similar, at 51 percent and 53 percent, respectively, in 2011 (Child Trends 2012b). The downward trend in adolescent sexual intercourse (and births) predates the emphasis on "abstinence-only" sex-education programs (Dailard 2003; "Improvements in Teen Sexual Risk Behavior Flatline" 2006).

Experts attribute the decline in sexual intercourse to comprehensive sex education and to fear of sexual disease. Data from the National Survey of Family Growth (1995 and 2002 waves) indicate that 86 percent of the decline in pregnancy risk is the result of improved contraception (regular use, better methods), whereas only 14 percent results from delayed sex (Santelli et al. 2007). These declines in pregnancy risk reversed in 2005, but the downward trend in teen births resumed in 2008 (Hamilton, Martin, and Ventura 2010). Other factors associated with a delay in the onset of sexual activity among teens, including minorities, are growing up in an intact family, family income, religiosity, communicating with parents about sex, having a good relationship with their parents, and feeling a greater connection with one's school (Van Campen and Romero 2012). Studies show that early and current sexual activity among teens is associated with a greater risk of engaging in delinquent and externalizing behaviors or "acting out," substance abuse, depression, having multiple partners, having nonconsensual and unwanted sex, having lower educational attainment, and using less contraception (Child Trends 2012a,b). Sexual activity in adolescence can have consequences in relationships in adulthood. Early sexual activity has been linked to lower marital quality and an increased risk of divorce (Paik 2011; Sassler, Addo, and Lichter 2012).

In assessing the decline of teen sexual activity, it is important to note that a lot depends on one's definition of sexual activity. Yes, sexual intercourse is less frequent among adolescents than in the past. It is also true that teens exhibit surprisingly high rates of oral sex: 55 percent of males ages 15 through 19 and 54 percent of females in the same age group engage in oral sex (Lindberg, Jones, and Santelli 2008). Many adolescents do not consider oral sex to be "sex." In one study, only 20 percent of college students at one university considered oral sex to be actual sex (Hans, Gillen, and Akande 2010). Indeed, a "virgin" is still widely considered someone who has not had vaginal intercourse—other sexual behaviors are allowed (Higgins et al. 2010). They also seem to be attracted to oral sex because it does not present a risk of pregnancy (true) or of sexually transmitted disease (not true). However, it does not appear that adolescents are substituting oral sex for vaginal sex. Most oral sex takes place among couples who have already had intercourse (Lindberg, Jones, and Santelli 2008). Only 17 percent of teen men and 13 percent of teen women reported having had oral sex but not intercourse (Child Trends 2011). Ten percent of adolescents have engaged in anal sex, the most important predictor of which was vaginal sex (Lindberg, Jones, and Santelli 2008).

Sex Education Deciding how much and what kind of information schools should provide about sex has always been controversial. In 1996, the federal government took the official position that abstention from sexual relations unless in a monogamous marriage is the only protection against sexually transmitted disease and pregnancy—and that abstinence is the only morally and rationally appropriate principle of sexual conduct. These so-called abstinence-only programs, which emphasize no sex until marriage, received nearly $1.3 billion from the federal government between 2001 and 2009 (Jayson 2009). Abstaining from sex can have positive effects. For example, delaying first intercourse increases the likelihood that adolescent girls will graduate high school (Sabia and Rees 2009).

However, there is no evidence that abstinence-only *programs* are effective in delaying sex or preventing pregnancy, nor are they scientifically accurate. The federal government's own study of the effectiveness and accuracy of its federally funded abstinence-only education programs found them wanting ("The Abstinence-Only Delusion" 2007; Begley 2007; Brody 2004; "Conclusions Are Reported" 2007; Kirby, Laris, and Rolleri 2006; Crosse 2008, p. 5). In 2010, President Obama eliminated funding for community-based abstinence only education. In its place, he allocated $200 million to a new teen pregnancy prevention initiative (Crile 2011).

Although abstinence-only education is ineffective in delaying sex, some research found that virginity pledges taken in certain circumstances seemed to delay adolescent sexual activity (Bearman and Brückner 2001), but later research has found no difference in delay of sexual activity between those who take the pledge and those who do not (Rosenbaum 2009; Tanne 2009; Thomas 2009). Research also found that once these teens become sexually active, they do so without precautions and have higher rates of pregnancy and sexually transmitted diseases (STDs) than other teens (Altman 2004; Rosenbaum 2009, p. e114). Among young adults, contraceptive use and rates of intercourse did not vary between virginity pledgers and nonpledgers. "By the time they become young adults, some 81 percent of pledgers have engaged in some type of sexual activity" (Rector and Johnson 2005, p. 13).

Sex education needs to take into account the teen propensity to engage in oral sex and to consider it risk-free despite high rates of STDs among youth (Halpern-Felsher et al. 2005). Moreover, the first experience for a small proportion of teens—7.8 percent—is forced sex (Eaton et al. 2008). Some others' experiences are ambivalent as to how much they wanted sex (Houts 2005). For example, more than half of teen females (59 percent) said they "didn't really want it to happen at the time" or "had mixed feelings" the first time they had sex. A substantial percentage (38 percent) of teen males had similar feelings about their first sexual encounter (Martinez, Copen, and Abma 2011).

Sex-education programs may emphasize peer pressure, troubled families and neighborhoods, or hormonal processes and rarely consider the broad array of teen motivations for sexual activity. It seems that both adolescent males and females seek sex because they expect it to meet needs for intimacy, sexual pleasure, and social status (Ott et al. 2006). These issues need to be taken into account in sex-education programs. Long-time sex-education researcher Douglas Kirby and his colleagues have identified some programs that seem effective in discouraging early sexual activity and sexual risks (Kirby, Laris, and Rolleri 2006).

Girls receive more sexual education than boys, both formally (in school) and informally (from parents), especially concerning menstruation. Allen, Kaestle, and Goldberg (2011) collected twenty-three written narratives from male college students on the topic of menstruation. The following comes from Kyle, who first heard about menstruation when he was about 10 years old:

> The elementary school I was attending decided to separate the guys and girls into two separate classes for the afternoon. The boys' class was supervised while we played board games and we were told that the girls were doing the same thing. This was, of course, the school's weak attempt at covering up sex education for the girls, where they were being taught about menstruation. After the class, school was out and while on the bus many of my girl friends that were in the class were talking about it. Looking back it was pretty funny because they were trying to tell me that girls had periods. But they were too shy to actually describe it; they just kept saying, "you know; a period, girls have periods." All I could think about was the punctuation mark because up till then that's all I thought about when I heard the word period. (Allen, Kaestle, and Goldberg 2011, p. 141)

They found that men who had one-to-one conversations with sisters, girlfriends, and parents were more empathetic and were less likely to have disdainful attitudes, crack jokes about menstruation, and state that periods were "gross."

Most adults feel that sex education should be shared by parents and schools. As Figure 4.4 indicates, 80 percent of adults said that sex education was primarily the responsibility of parents (Rasmussen Reports 2009). Children receive most of their sexual knowledge at home from their mothers and female family members (Grange, Brubaker, and Corneille 2011; Hutchinson and Cedarbaum 2011). Yet one study revealed that a sizable percentage of mothers of young female adolescents—especially those with less sexual knowledge, comfort, and self-efficacy—had no intention of discussing sexual health with their children in the next six months (Beyers and Sears 2012). Children, including girls, can benefit from discussions with their fathers, even if those discussions are incomplete (Hutchinson and Cedarbaum 2011). One woman explains:

> He gave me a philosophy about sex that has served me very well. His philosophy was that sex is not dirty but rather a serious issue with lots of emotional and physical ramifications. He didn't tell me not to have sex but rather to know what I was getting myself into and to consider heavily the consequences of my decision. (Hutchinson and Cedarbaum 2011, p. 559)

Percentage of adults who say that sex education is primarily the responsibility of parents (not schools)**	80%
Percentage of married adults who say that sex education is primarily the responsibility of parents (not schools)**	83
Percentage of unmarried adults who say that sex education is primarily the responsibility of parents (not schools)**	74
Percentage of adults who also approve of school health classes that include sex education**	74

Percentage of parents of high school students who say sex education should cover . . .

HIV/AIDS	99
How to talk with parents about sex and relationship issues*	98
The basics of how babies are made, pregnancy, and birth	97
Waiting to have sexual intercourse until older	96
How to get tested for HIV and other STDs	96
How to deal with the emotional issues and consequences of being sexually active*	96
Waiting to have sexual intercourse until married*	94
How to talk with a girlfriend or boyfriend about "how far to go sexually"*	94
Birth control and methods of preventing pregnancy	93
How to use and where to get contraceptives	85
Abortion*	83
How to put on a condom*	79
That teens can obtain birth control pills . . . without permission from a parent*	73
Homosexuality and sexual orientation*	73

* Questions marked with an asterisk were asked of only half the sample.

** Questions marked with a double asterisk were asked in a 2009 telephone survey by Rasmussen Reports.

FIGURE 4.4 Who should teach children about sex, and what parents want sex education to teach their children.

Source: Survey of 1,001 parents of children in seventh to twelfth grade sponsored by National Public Radio (NPR)/Kaiser Family Foundation/Kennedy School of Government (2004). The survey was conducted in September/October 2003.

Notes: The high school parent subsample = 450. Rasmussen Reports National Survey of 1,000 Adults, conducted January 12–13, 2009.

Reports National Survey of 1,000 Adults, conducted January 12–13, 2009.

Parents also want schools to play a role in educating children about sex. Nearly three-quarters (74 percent) of adults approved of health classes that include sex education, with very high percentages approving of coverage of HIV/AIDS, birth control, and abortion, as well as how to communicate with parents and boyfriends and girlfriends about sex and waiting to have sex (Rasmussen Reports 2009).

Sexual Responsibility

People today are making decisions about sex in a climate characterized by political conflict over sexual issues. Premarital and other nonmarital sex, same-sex relationships, abortion, sex education, and contraception represent political issues as well as personal choices. Public and private communication must rise to new levels. The AIDS epidemic has brought the importance of sexual responsibility to our attention in a dramatic way.

Making knowledgeable choices is a must. Because there are various standards today concerning sex, all individuals must determine what sexual standard they value, which is not always easy. Today's adults may be exposed to several different standards throughout the course of their lives. Even when individuals feel that they have clear values, applying those values in particular situations can be difficult. A person who believes in the standard of sexual permissiveness with affection, for example, must determine when a particular relationship is affectionate enough.

Making these choices and feeling comfortable with them requires recognizing and respecting one's own values instead of just being influenced by others in a sexual situation. Anxiety may accompany the choice to develop a sexual relationship, and there is considerable potential for misunderstanding between partners. This section addresses some of the principles of sexual responsibility that may serve as guidelines for sexual decision making.

Risk of Pregnancy

One obvious responsibility concerns the possibility of pregnancy. Partners should plan responsibly whether, when, and how they will conceive children and then use effective birth-control methods accordingly. Contraceptive use has slightly increased among teens. Among teens aged 15 to 19, 78 percent used a method of contraception the first time they had sex in 2006–2010 compared to 75 percent in 2002. By far, the most common method was the condom (68 percent in 2006–2010), followed by the pill (16 percent) and other hormonal methods such as a contraceptive patch, Lunelle injectable, Nuva-Ring, or implanon (6 percent). Overall, since 1995 more than 96 percent of sexually active teens reported having used a contraceptive method (Martinez, Copen, and Abma 2011). The Alan Guttmacher Institute and Planned Parenthood are excellent sources of information on all matters related to sexual health, including contraception, pregnancy, abortion, sexually transmitted infections, and relationships, and both have information specifically geared toward men, adolescents, and racial and ethnic minorities.

Sexually Transmitted Infections

A second responsibility concerns the possibility of contracting STIs or transmitting them to someone else. Sexually transmitted infections are infections that can be transmitted through sexual contact. Space in this text does not allow a full description of types, symptoms, diagnoses, treatment, and prevention. For the most comprehensive and up-to-date information on STIs and HIV/AIDS and sexual health, students should refer to the National Center for HIV/AIDS, Viral Hepatitis, STD, and TB Prevention at the U.S. Centers for Disease Control and Prevention, as well as the Alan Guttmacher Institute and Planned Parenthood.

A theme of this text is that social, political, economic, and cultural conditions affect people's choices. The consequences of decisions about sexual activity intersects with other social characteristics. In Chapter 1, we noted that poor people (including a substantial portion of people of color) have lower rates of education; when they do have access to education, it is often of poorer quality. Poverty denies access to education about sexually transmitted diseases such as HIV/AIDS as well as how to prevent them. Furthermore, poverty denies access to medical care and resources, which would help prevent these and other transmittable (but preventable) diseases. Individuals should be aware of the facts concerning HIV/AIDS and other STIs. Sexually active adolescents are at higher risk of contracting sexually transmitted infections than adults are (U.S. Centers for Disease Control and Prevention 2010).

The Special Case of HIV/AIDS **HIV/AIDS** has now been known for more than thirty years. The *human immunodeficiency virus* (HIV) that produces an *acquired immunodeficiency syndrome* (AIDS) has existed longer than that, but it was only in 1981 that the virus was recognized as the cause of a rapidly increasing number of deaths. If untreated, an HIV infection eventually progresses to full-blown AIDS, a viral disease that destroys the immune system. With a lowered resistance to disease, a person with AIDS becomes vulnerable to infections and other diseases that other people easily fight off.

Following a period of increase from 1987 through 1994, deaths from HIV/AIDS reached a plateau in 1995. Subsequently, the rate for this disease decreased 33 percent per year from 1995 through 1998 and 6.5 percent per year from 1999 through 2010 (Hoyert and Xu 2012). As of 2011, HIV/AIDS was no longer among the fifteen leading causes of death in the United States. Nevertheless, it remains an important public health concern, especially among younger adults and racial and ethnic minorities. African Americans are the most severely affected. Although they represent only 14 percent of the U.S. population, in 2009 they accounted for 44 percent of all new HIV infections (U.S. Centers for Disease Control and Prevention 2011).

An estimated one in sixteen black men and one in thirty-two black women will be diagnosed with HIV/AIDS in their lifetime. AIDS is now the ninth leading cause of death among blacks and the third leading cause of death among black men and women ages 35 to 44 (U.S. Centers for Disease Control and Prevention 2011).

Recent incidence estimates from 2006 through 2009 suggested overall stability in new HIV infections, with the exception of persons ages 13 to 29, who saw a 21–percent increase (U.S. Centers for Disease Control and Prevention 2012b). Roughly 50,000 people in the United States are newly infected with HIV each year. In 2010 (the most recent year that data are available), there were an estimated 47,500 new HIV infections. Most (63 percent) of these new infections occurred in gay and bisexual men (U.S. Centers for Disease Control and Prevention 2012a).

Figure 4.5 shows the estimated percentages of diagnoses of HIV infection among adults and adolescents in 2010, by gender and transmission category. An estimated 77 percent of diagnosed HIV infections among males were attributed to male-to-male sexual contact, and 86 percent of diagnosed infections among females were attributed to heterosexual contact (U.S. Centers for Disease Control and Prevention 2012c).

Both the survival rate and average life expectancy of those with HIV/AIDS are increasing (U.S. Centers for Disease Control and Prevention 2012c). However, people diagnosed with HIV have an average life expectancy that is 21.1 years shorter than the general population (Harrison, Song, and Zhang 2010). Moreover, there are large disparities in the life expectancy by gender (shorter for men) and race/ethnicity (shorter for blacks and Hispanics).

As the mood in the gay community has lightened since the earliest days of HIV/AIDS in the 1980s, some gay men have continued to have unprotected sex (so-called barebacking) with many and anonymous partners,

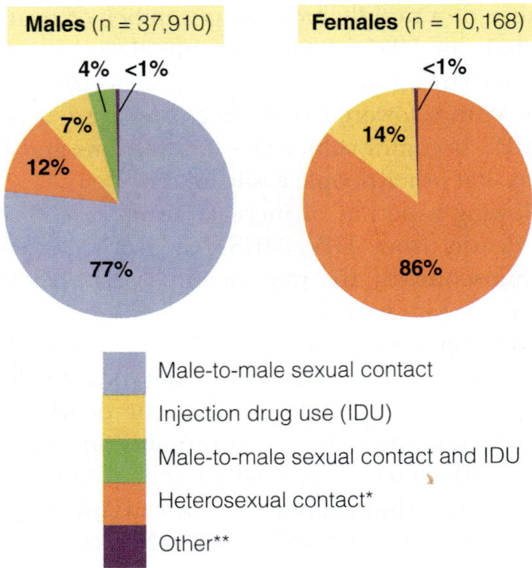

Note. All displayed data have been statistically adjusted to account for reporting delays and missing risk-factor information, but not for incomplete reporting.

* Heterosexual contact with a person known to have, or to be at high risk for, HIV infection.

** Includes hemophilia, blood transfusion, perinatal exposure, and risk factor not reported or not identified.

FIGURE 4.5 Diagnoses of HIV infection among adults and adolescents, by sex and transmission category, 2010: 46 states and 5 U.S. dependent areas.

Source: U.S. Centers for Disease Control and Prevention 2012c. *HIV Surveillance Report,* 2010; vol. 22. http://www.cdc.gov/hiv/topics/surveillance/resources/reports/. Published March 2012. Retrieved January 25, 2013.

and we have seen a surge of HIV infections among younger gay men ("HIV/AIDS Epidemic in the United States" 2013). Gay activists and public health professionals have expressed concern that drug ads with pictures of relatively healthy gay men convey misleading messages about the difficulties of living with AIDS, messages that may reduce caution and prevention (Cooter and Stein 2010, pp. 14–15; M. Gross 2006). In response, in 2012 the federal government allocated about $28 billion to combating HIV, including funds for care, research, cash and housing assistance, prevention, and the international AIDS epidemic ("HIV/AIDS Epidemic in the United States" 2013). Still, 21 percent of an estimated 1.1 million Americans living with HIV don't know they are infected and therefore may be having unprotected sex (U.S. Centers for Disease Control and Prevention 2013a, b, c).

Responsibility to Sexual Partners

A third responsibility concerns one's sexual partners. In sharing sexual pleasure, partners realize that sex is something partners do with each other, not to or for each other. Each partner participates actively as an equal in the sexual union. This includes communicating with partners or potential sexual partners. As we've seen in this chapter, sex may mean many different things to different people. A sexual encounter may mean love and intimacy to one partner and be a source of achievement or relaxation to the other. Honesty lessens the potential for misunderstanding and hurt feelings between partners.

Further, each partner is responsible for his or her own sexual response. When this happens, the stage is set for conscious, mutual cooperation. Partners feel freer to express themselves sexually. Women, for example, often cannot have an orgasm through intercourse alone (Herbenick et al. 2010b). Rather than "faking it," it is a woman's responsibility to communicate to her partner what "turns her on." Two different studies of female college students found sexual assertiveness to be positively associated with receiving oral sex (Backstrom, Armstrong, and Puentes 2012; Bay-Cheng and Fava 2011). Others suggest a sexual script in which women should orgasm before men and that men should be more attuned to pleasing their partner (Muehlenhard and Shippee 2010).

Responsibility to Oneself

A fourth responsibility is to oneself. In expressing sexuality today, each of us must make decisions according to our own values. A person may choose to follow values held on the basis of religious commitment or as put forth by ethicists, psychologists, or counselors. People's values change over the course of their lives, and what's right at one time may not appear to be later. Despite the confusion caused both by internal changes as our personalities develop and by the social changes going on around us, it is important for individuals to make thoughtful decisions about sexual relationships.

Summary

- Social attitudes and values play an important role in the forms of sexual expression that people find appropriate and enjoyable.
- Despite decades of conjecture and research, it is still unclear just how sexual identity develops and whether it is genetic or socially shaped. Recent decades have witnessed increased acceptance of gay, lesbian, bisexual, and transgender (LGBT) individuals, though some disapproval, discrimination, and hostility remain.
- Whatever one's sexual identity, sexual expression is negotiated amid cultural messages about what is sexually permissible or desirable. In the United States, these cultural messages have moved from patriarchal

sex, based on male dominance and on reproduction as its principal purpose, to a message that encourages sexual expressiveness in myriad ways for both genders equally.

- Values outside of committed relationships include abstinence, permissiveness with affection, recreational sex, and the double standard—the latter diminished since the 1960s but still alive.
- Sex in marriage and between committed partners changes throughout the life course. Young spouses have sex more often than do older mates. Although the frequency of sexual intercourse declines over time and through the length of a marriage, many older adults have active and satisfying sex lives.
- Fidelity remains a strong value for committed partners and spouses, but extramarital sex does occur and represents a serious challenge to relationship trust.
- Making sex a pleasure bond, whether a couple is married or not, involves cooperation in a nurturing, caring relationship. To fully cooperate sexually, partners need to develop high self-esteem, break free from restrictive gendered stereotypes, communicate openly, and make time for each other.
- Pornography, the risk of pregnancy, sexually transmitted infections, and HIV/AIDS has had an impact on relationships, marriages, and families.
- Sexuality, sexual expression, and sex education are current public issues, and different segments of American society have divergent views.
- Whatever the philosophical or religious grounding of one's perspective on sexuality, there are certain guidelines for personal sexual responsibility that we should all heed.

Questions for Review and Reflection

1. Give some examples to illustrate changes in sexual behavior and social attitudes about sex. What might change in the future?
2. Do you think that sex is changing from "his and hers" to "theirs"? What do you see as some difficulties in making this transition?
3. Discuss what you've learned about the nature of sexual relationships. What did you find useful and relevant to everyday life? What seems remote from real-world experience?
4. How do you think HIV/AIDS affects sex and sex relationships—or does it?
5. **Policy Question**. What role, if any, has government policy played in sex education, research and information on sex, and sexual regulation?

Key Terms

abstinence 91
affectional orientation 80
asexual 83
bisexual 81
cyberadultery 94
double standard 92
expressive sexuality 85
friends with benefits 92
gay 82
habituation 98
habituation hypothesis 94
heterosexism 86
heterosexual 81
HIV/AIDS 103
homophobia 87
homosexual 81
hooking up 92
interactionist perspective on human sexuality 84
interpersonal exchange model of sexual satisfaction 83
lesbian 82
LGBT (lesbian, gay, bisexual, and transgender) 82
menarche 80
misogynistic 99
pansexual 81
patriarchal sexuality 84
pornification 88
pornography 99
sexting 88
sexual identity 80
sexual reassignment surgery 82
sexual scripts 84
transgendered 82
transsexuals 82
transvestites 82

5

Love and Choosing a Life Partner

RubberBall/Alamy

Learning Objectives

1. Discuss the various ways of defining love.
2. Describe different models for understanding love.
3. Compare and contrast arranged and free-choice marriages.
4. Define assortative mating and the process of finding a committed partner.
5. Describe dating from early adolescence to adulthood.
6. Discuss ways of nurturing committed relationships.

Almost three-quarters of young adults believe in "one true love," and more than 90 percent would like a "soul mate" (Robison 2003; Whitehead and Popenoe 2001). We all want to be loved, and most of us expect to be in a committed relationship—if not now, then in the future. Being in a committed relationship involves selecting someone with whom to become emotionally and sexually intimate and, often, with whom to raise children. Accordingly, the choice of a partner is a major life course decision. Figure 1.5 in Chapter 1 illustrates that making intentional, knowledgeable decisions requires an awareness of one's personal beliefs and values, as well as conscious consideration of alternatives and serious thought about the probable consequences. You might want to refer to Figure 1.5 as you study this chapter.

Research suggests that the best way to choose a life partner is to look for someone to love who is socially responsible, respectful, and emotionally supportive. It is also important that the person is committed both to the relationship and to the value of staying together. Lastly, it helps if that person also demonstrates good communication and problem-solving skills (Bradbury and Karney 2004; Hetherington 2003).

Equally important is looking for a mate with values that resemble one's own, because similar values and attitudes are strong predictors of ongoing happiness and relationship stability. Although romantic love is usually a very important ingredient (Amato 2007), successful life partnerships are also based on partners' maturity, common goals, qualities of friendship, and the soundness of their reasons for getting together (Gaunt 2006; Lacey et al. 2004).

Nearly three-quarters of women and almost two-thirds of men marry by age 30; by age 40, more than 80 percent of Americans have married (Kreider and Ellis 2011b). By 24, more Americans are married than cohabiting (Saad 2008b). Therefore, the topic of choosing a marriage partner is critically important. Meanwhile, we need to note here that research and published counseling advice on choosing a committed life partner have focused almost solely on heterosexual, *marital* mate selection. One reason is that marriage is more easily identifiable for researchers—one is either married or not—whereas the existence and nature of committed, lifelong relationships outside marriage are harder to identify and research. However, long-term cohabiting relationships have increased in frequency, and this trend is particularly true after divorce. The process of choosing a second, or third, marital or cohabiting partner is covered in Chapter 15.

In this chapter, we'll look at some things that influence the choice of a life partner and subsequent relationship satisfaction. We will examine new research on how the process of selecting a potential marriage partner unfolds. Within potential partnerships, we will examine how a relationship develops and proceeds from first meeting to commitment. We will also discuss interreligious and interracial unions. To begin, we explore some things that we know about love.

Love and Commitment

When most people think of love, they think of "romantic love" as opposed to love between parents and children, family members, and friends. Choosing a partner for a lifelong commitment usually starts with romantic love (although some relationships start with platonic friendship that becomes romantic later). Yet, "[w]ith respect to love, the gap between everyday people and family scholars is surprisingly wide.... [M]ost researchers have avoided the topic.... Yet attitude surveys reveal that the great majority of Americans view love as the primary reason for getting and staying married" (Amato 2007, p. 306). Why the lack of research on love?

Defining Love

Love is exceedingly difficult to define, although many have tried. Researchers tend to feel more comfortable measuring the concepts of attachment, intimacy, compassion, infatuation, which are *related to* and often treated as *measures* of love, as opposed to love itself (Langeslag, Muris, and Franken 2012; Neff and Beretvas 2013). There is an emerging area of "love" research from the physical sciences, such as studies of levels of oxytocin and serotonin in the blood of romantic couples (Langeslag, van der Veen, and Fekkes 2012; Schneiderman et al. 2012). The definition of love many of us are familiar with is commonly heard at Christian wedding ceremonies and comes from the Bible

> Love is patient, love is kind. It does not envy, it does not boast, it is not proud. It does not dishonor others, it is not self-seeking, it is not easily angered, it keeps no record of wrongs. Love does not delight in evil but

> rejoices with the truth. It always protects, always trusts, always hopes, always perseveres (Corinthians 13:4, New International Edition).

One family researcher defined love as "a strong emotional bond with another person that involves sexual desire, a longing to be with the person, a preference to put the other person's interests ahead of one's own, and a willingness to forgive the other person's transgressions" (Amato 2007, p. 206). As you can see, the definitions are very similar.

With the obvious exceptions of physical and emotional abuse, loving involves the acceptance of partners for themselves and "not for their ability to change themselves or to meet another's requirements to play a role" (Dahms 1976, p. 100). People are free to be themselves in a loving relationship and to expose their feelings, frailties, and strengths (Armstrong 2003). Related to this acceptance is caring or *empathy*—the concern a person has for the partner's growth and the willingness to "affirm [the partner's] potentialities" (May 1975, p. 116; Jaksch 2002).

Psychologist Erich Fromm (1956) chastises Americans for their emphasis on wanting to *be loved* rather than on learning to *love*. Many of the ways we make ourselves lovable, Fromm writes, "are the same as those used to make oneself successful, 'to win friends and influence people.' As a matter of fact, what most people in our culture mean by being lovable is essentially a mixture between being popular and having sex appeal" (p. 2). However, loving is not about being focused on oneself but about showing empathy and compassion for others (Ciaramigoli and Ketcham 2000; Neff and Beretavas 2013). Maintaining a loving relationship also requires commitment of both partners (Dixon 2007). Committing oneself to another person involves working to develop a relationship "where experiences cover many areas of personality; where problems are worked through; where conflict is expected and seen as a normal part of the growth process; and where there is an expectation that the relationship is basically viable and worthwhile" (Altman and Taylor 1973, pp. 184–87; Amato 2007).

Committed lovers have fun together; they also share more tedious times. They express themselves freely and authentically (Smalley 2000). Committed partners view their relationship as worth keeping, and they work to maintain it despite difficulties or disagreements (Amato 2007; Love 2001). **Commitment** is characterized by this willingness to work through problems and conflicts as opposed to calling it quits when problems arise. In this view, commitment involves consciously investing in the relationship (Etcheverry and Le 2005). Then too, committed partners "regularly, routinely, and predictably attend to each other and their relationship no matter how they feel" (Peck 1978, p. 118).

David Freund/E+/Getty Images

Marriages between individuals with a relatively secure attachment style that take place after about age 25 and are between partners who grew up in intact (nondivorced) families are the most likely to be satisfying and stable. But having grown up as a child of divorce does *not* mean that an individual will *necessarily* have an unhappy or unstable marriage.

Gender Differences in Love

Research on love is experiencing a resurgence, especially regarding gender differences in the meaning and experience of love. Love, and the need for love, is thought to be the domain of women. However, an emerging field of research examining love from the perspective of men is calling this assumption into question. Amy Schalet, the author of *Not Under My Roof: Parents, Teens, and the Culture of Sex* (2011), examined sex and relationships among teens from the United States and the Netherlands. Her findings suggest that boys, at least a subset of them, are becoming less focused on casual sex and more focused on romance. As one boy from her study says, "[My girlfriend] is the only one I ever want to have from now on .... We're just so happy together and I couldn't imagine being with anyone else. My first priority is being in love with my girlfriend, and giving her everything I can" (Schalet 2011, p. 158). Other research based on a random sample of 1,316 adolescent boys and girls from Toledo, Ohio, by Giordano, Longmore, and Manning (2006) similarly challenges the notion that girls want romance and boys want sex. The boys and girls in their study scored equally a scale of "passionate love," developed by psychologist Elaine Hatfield and sociologist Susan Sprecher more than 25 years ago (Hatfield and Sprecher 2010; Hatfield 2013). They also conducted in-depth interviews with some of the boys in the study. A 17-year-old boy responded to the question "How important is your relationship to [your girlfriend] in your life?" by saying

> As important as you get. You know, well, you think of it as this way, you give up your whole life, you know, know, to save Jenny's life, right? That's how I feel. I'd give up my whole life, to save any of my friends' life too. But it's a different way. Like, if I could save Jon's life, and give up my own, I would, because that is something you should, have in a friend, but I wouldn't want to live without Jenny, does that make some sense? (Giordano, Longmore, and Manning 2006, p. 277)

Men also appear to fall in love more quickly than women. Whereas women take an average of 134 days to say "I love you," men take about half that time, only 88 days. Moreover, 39 percent of men say "I love you" within the first month of seeing someone, compared to 23 percent of women (DeLacey 2013). Moreover, a survey of 1,611 men and women ages 18 to 23 by Simon and Barrett (2010) indicates that the breakup of romantic relationships take a much greater toll on men than women. When it comes to love, women appear to be more resilient (Paul 2010). As the authors of the study wrote, "It appears that young men benefit more than women from support, and that they are more harmed than women by strain in ongoing romantic relationships" (Paul 2010). These studies indicate that, despite gendered expectations of a tough exterior, love is important and meaningful in the lives of men. In the typical engagement proposal, sociologists David Schweingruber, Sine Ahahita, and Nancy Burns (2004) found that men are responsible for, and successfully perform, various "romantic rituals" (asking the father's permission, getting down on one knee), as well as executing elaborate demonstrations of love, often in a public setting. Nevertheless, some commentators suggest that romance has gone the wayside among today's pragmatic, risk-averse, and work-focused couples (Douthat 2009).

Sternberg's Triangular Theory of Love

In research on relationships varying in length from one month to thirty-six years, psychologist Robert Sternberg (1988a, 1988b; Sternberg and Sternberg 2008) found three components necessary to authentic love: intimacy, passion, and commitment. According to **Sternberg's triangular theory of love**, *intimacy* "refers to close, connected, and bonded feelings in a loving relationship. It includes feelings that create the experience of warmth in a loving relationship ... [such as] experiencing happiness with the loved one; ... sharing one's self and one's possessions with the loved one; receiving ... and giving emotional support to the loved one; [and] having intimate communication with the loved one" (Sternberg 1988a, pp. 120–21).

Passion "refers to the drives that lead to romance, physical attraction, sexual consummation, and the like in a loving relationship" (Sternberg 1988a, pp. 120–21). *Commitment*—the "decision/commitment component of love"—consists of not only deciding to love someone but also deciding to maintain that love. **Consummate love** (see Figure 5.1), composed of all

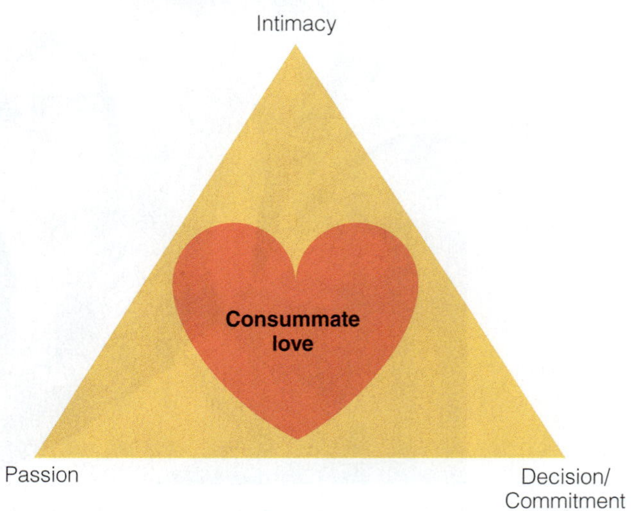

FIGURE 5.1 The three components of love: triangular theory.

Source: Adapted from Sternberg 1988a, Figure 6.1, p. 121. In Robert J. Steinberg and Michael L. Barnes (eds.), *The Psychology of Love.*

three components, is "complete love, ... a kind of love toward which many of us strive, especially in romantic relationships" (Sternberg 1988a, pp. 120–21).

The three components of consummate love develop at different times as love grows and changes. "Passion is the quickest to develop, and the quickest to fade.... Intimacy develops more slowly, and commitment more gradually still" (Sternberg, quoted in Goleman 1985). Passion, or "chemistry," peaks early in the relationship but generally continues at a stable, although fluctuating, level and remains important both to our good health (Kluger 2004) and to the long-term maintenance of the relationship (Love 2001). Intimacy, which includes conveying and understanding each other's needs, listening to and supporting each other, and sharing common values, becomes increasingly important as time goes on. In fact, psychologist and marriage counselor Gary Smalley (2000) argues that a couple is typically together for about six years before the two feel safe enough to share their deepest relational needs with one another. Commitment is essential; however, commitment without intimacy and some level of passion is hollow. In other words, all these elements of love are important. Because these components not only develop at different rates but also exist in various combinations of intensity, a relationship is always changing, if only subtly (Sternberg 1988b). In addition to Sternberg's model, attachment-theory scholars (see Chapter 2) offer insight into loving relationships.

Attachment Theory and Loving Relationships

Recall from Chapter 2 that attachment theory posits that during infancy and childhood a young person develops a general style of attaching to others (Ainsworth et al. 1978; Bowlby 1988; Pittman et al. 2011). Applying attachment theory to loving relationships, we can presume that people with more secure attachment styles would have less ambivalence about emotional closeness and commitment. "Secure attachment, in part, depends on regarding the relationship partner as being available in times of need and as trustworthy" (Kurdek 2006, p. 510).

We might therefore conclude that those with a *secure* attachment style have stronger interpersonal skills and are better prospects for a committed relationship (Jenkins-Guarnien, Wright, and Hudiburgh 2012; Rauer and Volling 2007). An *insecure/anxious* attachment style entails "fear of abandonment" with consequent possible negative behaviors such as unwarranted jealousy or attempts to control one's partner. An *avoidant* attachment style leads one to pass up or shun closeness and intimacy either by evading relationships, demonstrating ambivalence, seeming preoccupied, and, among men, by rejecting romance and expressing hostile attitudes toward women (Fletcher 2002; Hart, Hung, Glick, and 2012; Hazen and Shaver 1994). Attachment style is associated with young adults' use of communication technology in romantic relationships (Marshall et al. 2013; Morey et al. 2013). Those with the avoidant style were more likely to use e-mail as opposed to phone calls and texting (Morey et al. 2013). Those with an anxious attachment style were more likely look at their partners' Facebook pages (Marshall et al. 2013)—perhaps to check up on their partners?

Indeed, romantic couples' use of social media and communication technology (e.g., Facebook, e-mail, texting, cell phones) is high and varied. These technologies can be used both positively and negatively in relationships. For example, romantic couples use cell phones and texting most often to "express affection." However, texting was also used to avoid confrontation and to hurt one another (Coyne et al. 2011). However, whereas e-mail was associated with more conflict, texting was associated with more positive relationships. For those with insecure/anxious attachment, the use of social networks sites has been found to be associated with greater intimacy and support but also greater jealousy (Marshall et al. 2013; Morey et al. 2013). Couples who viewed sexual information together expressed greater intimacy, increased sexual behavior and openness to trying new things, and were less judgmental of one another's bodies. However, "cybersex" (engaging in online sexual behavior with a person other than one's romantic partner) compromises relationships and is associated with sex addiction (Hertlein 2012). The attachment style of one's partner can either magnify or lessen the effects of one's own attachment style. For example, if both individuals are insecure and anxious, the relationship will be characterized that way as well. On the other hand, a person with an insecure attachment style who is in a committed relationship with someone having a secure attachment style may gradually learn to feel more secure (Banse 2004). Individual or relationship therapy may help people change their attachment styles.

Attachment theory and Sternberg's triangular theory of love are not the only ways of looking at love, of course. "Facts about Families: Six Love Styles" analyzes love in yet a third way. A fourth way to better understand love is to think about what love is *not*. We turn now to an examination of three things love isn't: martyring, manipulating, and limerence.

Three Things Love Is Not

Love is not inordinate self-sacrifice. And loving is not the continual attempt to get others to feel or do what we want them to—although each of these ideas is frequently mistaken for love. Nor is love all the crazy feelings you get when you can't get someone out of your mind. We'll examine these misconceptions in some detail.

Facts about Families

Six Love Styles

Relationships evidence different characteristics or personalities. John Alan Lee (1973) classified six love styles, initially based on interviews with 120 white, heterosexual respondents of both genders. Lee subsequently applied his typology to same-sex relationships (Lee 1981). Researchers then developed a Love Attitudes Scale (LAS): eighteen to twenty-four questions that measure Lee's typology (Hendrick, Hendrick, and Dicke 1998). Although not all subsequent research has found all six dimensions, this typology of love styles has withstood the test of time, has proven to be more than hypothetical, and may even have cross-cultural relevance (Lacey et al. 2004; Le 2005; Masanori, Daibo, and Kanemasa 2004).

Love styles are sets of distinctive characteristics that loving or lovelike relationships take. The word *lovelike* is included in this definition because not all love styles amount to genuine loving as defined in this chapter. People may incorporate different aspects of several styles into their relationships. What are Lee's six love styles?

1. *Eros* (AIR-ohs) is a Greek word meaning "love"; it forms the root of our word *erotic*. This love style is characterized by intense emotional attachment and powerful sexual feelings or desires. Sustained relationships established by erotic couples are characterized by continued and emotionally intense sexual interest. A sample question on the LAS designed to measure eros asks respondents to agree or disagree with the following statement: "My partner and I have the right chemistry between us" (Hendrick, Hendrick, and Dicke 1998).
2. *Storge* (STOR-gay) is an affectionate, companionate style of loving. This love style focuses on deepening mutual commitment, respect, friendship over time, and common goals. Storgic lovers' basic attitudes to their partners are one of familiarity: "I've known you a long time, seen you in many moods" (Lee 1973, p. 87). Storgic lovers are likely to agree that "I always expect to be friends with the one I love" (Hendrick, Hendrick, and Dicke 1998).
3. *Pragma* (PRAG-mah) is the root word for *pragmatic*. Pragmatic love emphasizes the practical element in human relationships and rational assessment of a potential partner's assets and liabilities. Arranged marriages are often examples of pragma. So is a person who decides very rationally to get married to a suitable partner. The following is one LAS statement that measures pragma: "A main consideration in choosing a partner is/was how he/she would reflect on my family" (Hendrick, Hendrick, and Dicke 1998).
4. *Agape* (ah-GAH-pay) is a Greek word meaning "love feast." Agape emphasizes unselfish concern for a beloved's needs even when that requires personal sacrifice. Often called *altruistic love*, agape emphasizes nurturing others with little conscious desire for a return other than the intrinsic satisfaction of having loved and cared for someone else. Agapic lovers would likely agree that, "I try to always help my partner through difficult times" (Hendrick, Hendrick, and Dicke 1998).
5. *Ludus* (LEWD-us) focuses on love as play or fun. Ludus emphasizes the recreational aspects of sexuality and enjoying many sexual partners rather than searching for one serious relationship. Of course, ludic flirtation and playful sexuality may be part of a more committed relationship based on one of the other love styles. LAS statements designed to measure ludus include: "I enjoy playing the game of love with a number of different partners" (Hendrick, Hendrick, and Dicke 1998).
6. *Mania*, a Greek word, designates a wild or violent mental disorder, an obsession, or a craze. Mania involves strong sexual attraction and emotional intensity, as does eros. However, mania differs from eros in that manic partners are extremely jealous and moody, and their need for attention and affection is insatiable. Manic lovers alternate between euphoria and depression. The slightest lack of response from a love partner causes anxiety and resentment. Manic lovers would be likely to say, "When my partner doesn't pay attention to me, I feel sick all over" or "I cannot relax if I feel my partner is with someone else" (Hendrick, Hendrick, and Dicke 1998). Because one of its principal characteristics is extreme jealousy, we may learn of manic love in the news when a relationship ends violently. Of Lee's six love styles, mania least fits our definition of love as described earlier.

How do these love styles influence relationship satisfaction and continuity? Psychologists Marilyn Montgomery and Gwendolyn Sorell (1997) administered the LAS to 250 single college students and married adults of all ages. They found that eros can last throughout marriage and is related to high satisfaction. Agape is also positively associated with relationship satisfaction (Neimark 2003). Interestingly, Montgomery and Sorell found storge to be important only in marriages with children. Ludus did not necessarily diminish relationship satisfaction among those who are mutually uncommitted. However, ludic attitudes have been empirically associated with diminished long-term relationship and marital satisfaction (Le 2005; Montgomery and Sorell 1997).

Critical Thinking

Thinking about relationships (if you are in one), what is your and your partner's "love style"? Is this a different love style than you have had in other relationships?

Martyring **Martyring** involves maintaining relationships by consistently minimizing one's own needs while trying to satisfy those of one's partner. Periods of self-sacrifice are necessary through difficult times. However, as a premise of a relationship, excessive self-sacrifice or martyring is unworkable. Martyrs may have good intentions, believing that love involves doing unselfishly for others without voicing their own needs in return. Consequently, however, martyrs seldom feel that they receive genuine affection. A martyr's reluctance to express his or her needs is damaging to a relationship because it prevents openness and intimacy.

Manipulating Manipulators follow this maxim: If I can get her [or him] to do what I want done, then I'll be sure she [or he] loves me. **Manipulating** means working to control the feelings, attitudes, and behavior of your partner or partners in underhanded ways rather than by directly (not abusively!) stating your case. When not getting their way, manipulators are likely to find fault with a partner, sometimes with verbal abuse. "You don't really love me," they may accuse. Manipulating, like martyring, can destroy a relationship.

Limerence Have you ever been so taken with someone that you couldn't get him or her out of your mind? Although the object of your attention may be unaware of your feelings, you review every detail of the last time you saw him or her and fantasize about how you might actually develop a relationship. Psychologist Dorothy Tennov (1999 [1979]) named this situation *limerence* (LIM-er-ence). She makes the following points: First, limerence is not just "lust" or sexual attraction. People in limerence fantasize about being with the limerent object in all kinds of situations—not just sexual ones. Second, many of us have experienced limerence. Third, limerence can possibly turn into genuine love, but more often than not, it doesn't.

People discover love; they don't simply find it. The term *discovering* implies a process—developing and maintaining a loving relationship require seeing the relationship as valuable; committing to mutual needs, satisfaction, and self-disclosure; engaging in supportive communication; and spending time together. We now turn to factors that affect how that love plays into the selection of a life partner.

Mate Selection: The Process of Selecting a Committed Partner

The popular notion of finding one's lifelong partner is that it happens as if being struck by Cupid's arrow—suddenly, unexpectedly, and by chance. Realistically, settling on a mate is a much more complex process. One such model of mate selection is discussed below.

A Sequential Model of Mate Selection

Designed to apply to same-sex unions as well as to heterosexual marriages, Figure 5.2 depicts a model of factors that affect relationship stability—whether partners remain together over time (Kurdek 2006). Relationship

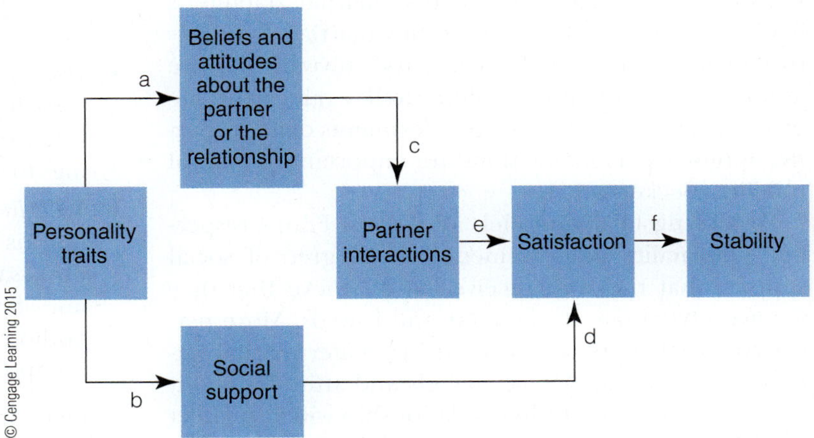

FIGURE 5.2 A time-ordered sequential model of relationship outcomes.

"The model has six components that form a time-ordered sequence of six linkages (letters a through f). The [personality traits] component refers to personality traits partners bring to their relationships that affect both the manner in which the relationship events are appraised (... [beliefs and attitudes about the relationship]; Link a) and the quality of perceived social support that is received (Link b) .... The [beliefs and attitudes about the partner or the relationship] component refers to beliefs and attitudes about the partner or the relationship that affect how partners interact with each other (Link c) .... The component social support underscores the view that intimate relationships coexist with other personal relationships, particularly those involving friends and family members .... The partner interactions component represents how partners behave toward one another and [along with social support, Link d] forms another basis for overall satisfaction with the relationship (Link e) .... Relationship satisfaction refers to the overall level of positive affect experienced in the relationship and the extent to which important personal needs are being met in the relationship and is one determinant of relationship stability (Link f)."

Source: Adapted from Kurdek 2006, pp. 510–511.

stability, happiness, and satisfaction depend on how the partners interact with each other and on the perceived social support the couple receives from family members, friends, and the community in general. How partners interact with each other, in turn, depends on a person's ideas about the partner and the relationship. These beliefs and attitudes depend, at least partly, on the personality traits that each partner brings to the union (Kurdek 2006, p. 510).

As an example, let's say that Fran and Maria are considering marriage. Each wonders about the odds of staying married. First, Maria and Fran need to take their personality traits into account. Is Fran thoughtful, dependable, reliable, and honest? Is Maria? Is one or both prepared to support a family? What beliefs and attitudes do they have about each other and about their relationship? Does Fran believe that marriage to Maria is likely to result in marital satisfaction? Or is Fran marrying Maria despite misgivings about her or the relationship? Does Maria believe that marital stability is likely for her? Or does she see this marriage as likely to end in divorce but "worth a try" anyway? Positive attitudes about the relationship, similar values and life goals, and realistically positive assessments of a prospective spouse's personality traits are important to marital stability (Becker 2012).

According to this model, Maria and Fran's respective personality traits influence the degree of social support that they will receive—and believe that they receive—from family members and friends. More perceived social support will result in greater marital satisfaction. Fran and Maria's beliefs and attitudes about each other and about their relationship will also affect how they interact with each other. Will they—do they now—interact primarily in supportive ways? Do they handle conflict well? Supportive interaction results in greater marital satisfaction. Greater marital satisfaction, in turn, results in the greater likelihood of marital stability.

Minimizing Mate Selection Risk

Other chapters in this text focus on various aspects of partner interactions and social support. Here we focus on choosing a partner who is best predisposed psychologically to maintain a stable and committed relationship.

Psychologists and counselors advise choosing a partner who is integrated into society by means of school, employment, a network of friends, and who fairly consistently demonstrates supportive communication and problem-solving skills (Cotton, Burton, and Rushing 2003). We're reminded that "[h]eavy or risky drinking is associated with a host of marital difficulties including infidelity, divorce, violence and conflict" (L. Roberts 2005, p. F13). The same can be said for other forms of substance abuse (Kaye 2005, F15). Furthermore, research shows that relationships are more likely to be stable when partners' parents have not been divorced. This particular issue is discussed further in Chapter 14.

Another step in minimizing mate selection risk is to let go of misconceptions we might have about love and choosing a partner. According to psychologist Beverly Smallwood (2013), some of the most common love myths are:

- *The right person will meet all my needs.* In reality, it is too much to expect one person to meet your every need. One needs "God, friends, a strong sense of purpose, healthy self-esteem, and a willingness to take responsibility for your own happiness."
- *I can change my partner.* There is only one person you can change: yourself.
- *Love will conquer all.* Face differences in values, behavioral choices, backgrounds, and personal habits *before* making a lifelong commitment.
- *Love is a feeling.* According to Smallwood, "love is a verb. It's about doing—even in those temporary times when you inconveniently don't have wonderful feelings to stimulate the positive action."
- *We'll live happily ever after.* This is the idea that "real love" is something you won't have to work at. Smallwood says, "A marriage certificate is really a work permit." She says that "even the best relationships have potholes, tragedies, and disappointments" and that it is important to keep your eyes wide open.

Later in this chapter, the section "Some Things to Talk About" gives other ideas on assessing your own and a prospective partner's values and attitudes. Of course, it's important to be truthful when relating with a potential partner, as well as to ascertain how truthful a partner is being (S. Campbell 2004). Some couples go to counseling to assess their future compatibility and commitment (Carlson et al. 2012). Others may access marital compatibility tests on the Internet. Although we, your authors, are unable to attest to the efficacy of these, they do stimulate couple discussions about important topics. At this point, we turn to the social science analogy of choosing a mate in a marketplace.

The Marriage Market

Imagine a large marketplace in which people come with goods to exchange for other items. In nonindustrialized societies, a person may go to market with a few chickens to trade for some vegetables. In more industrialized societies, people attend hockey-equipment swaps, for example, trading outgrown skates for larger ones. People choose partners in much the same way: They enter the *market* (traditionally called the **marriage market**) armed with resources—personal and social characteristics—and then they bargain for the best "buy" that they can get.

Arranged and Free-Choice Marriages

In much of the world, particularly in parts of Asia and Africa that are less Westernized, parents have traditionally arranged their children's marriages. In **arranged marriage**, future spouses can be brought together in various ways. For example, in India, parents typically check prospective partners' astrological charts to ensure future compatibility. Traditionally, the parents of both prospective partners (often with other relatives' or a paid matchmaker's help) worked out the details and then announced the upcoming marriage to their children. The children may have had little or no say in the matter, and they may not have met their future spouse until the wedding. Today it is more common for the children to marry only when they themselves accept their parents' choice. Unions like these, sometimes called *assisted marriages*, can be found among some Muslim groups and other recent immigrants to the United States (Ingoldsby 2006b; MacFarquhar 2006). Research shows that—at least at first—couples who have had more input report greater marital satisfaction (Madathil and Benshoff 2008).

Arranged marriage was observed throughout most of the world into the twentieth century—and well into the eighteenth century in Western Europe. We can still find arranged marriages in many parts of the world today. The majority of young couples in cultures that have traditionally practiced arranged marriage continue to heed extended family members' opinions about a prospective mate (Zhang and Kline 2009). One modern twist on arranged marriage is "speed dating" among Muslims families residing in the United States (Ellick 2011).

The fact that marriages are arranged doesn't mean that love is ignored by parents. Indeed, marital love may be highly valued. However, couples in arranged marriages are expected to develop a loving relationship *after* the marriage, not before (Tepperman and Wilson 1993). A study by Myers, Madathil, and Tingle (2005) compared marital satisfaction among arranged marriages in India to those more freely chosen in the United States and found no differences in marital satisfaction between the two groups. According to the authors, "Although this is not a case in favor of arranged marriages, it provides no support for a position opposing this tradition" (p. 189). Meanwhile, with global Westernization, arranged marriages are less and less common, especially among those with higher education (D. Jones 2006; Hoelter, Axinn, and Ghimire 2004; Zang 2008).

The United States is an example of what cross-cultural researchers call a **free-choice culture**: People choose their own mates, although often they seek parents' and other family members' support for their decision. Immigrants who come to the United States from more collectivist cultures, in which arranged marriages have been the tradition, face the situation of living with a divergent set of expectations for selecting a mate. Some immigrant parents from India, Pakistan, and other countries arrange for spouses from their home country to marry their offspring. Either the future spouse comes to the United States to marry the young person, or the young person travels to the home country for a wedding ceremony, after which the newlyweds usually live in the United States (Dugger 1998). In this case, the marriage is typically characterized by the greater Westernization of one partner (the young person who has lived in the United States) and the spouse's simultaneous need to adjust not only to marriage but also to an entirely new culture.

The incidence of **cross-national marriages** like these may decline because the required visa for an immigrating spouse has been far harder to obtain since September 11, 2001. Because marriage is now the most common path to U.S. citizenship, authorities are increasingly aware of, and prosecuting, marriage fraud (Seminara 2008). Furthermore, American-born children of immigrants may see themselves as too Americanized for this approach to finding a partner (MacFarquhar 2006). Whether unions are arranged or not, we can think of choosing a marital partner as taking place in a market. On the other hand, international matchmaking websites, such as Shaadi.com, might increase cross-national marriages.

Regarding arranged marriages, parents go through a bargaining process not unlike what takes place at a traditional village market. They make rationally calculated choices after determining the social status or position, health, temperament, and, sometimes, physical attractiveness of their prospective son- or daughter-in-law. Sometimes, as in the Hmong culture, the exchange involves a *bride price*, money or property that the future groom pays the future bride's family so that he can marry her. More often, the exchange is accompanied by a *dowry*, a sum of money or property the female brings to the marriage. Wealthier grooms demand a higher price, and older brides must pay more. Even in countries where dowries are illegal, the practice remains widespread (Self and Grabowski 2009; Srinivasan and Lee 2004).

With arranged marriage, the bargaining is obvious. The difference between arranged marriages and marriages in free-choice cultures may seem so great that we are inclined to overlook an important similarity: *Both* involve bargaining. What has changed in free-choice societies is that individuals, not family members, do the bargaining.

Fabio Cardoso/Corbis

JAMES P. BLAIR/National Geographic Stock

Although the arranged marriage of this couple in northern India (right) may seem to be a world apart from the more freely chosen marriage of this couple in the United States (left), bargaining has occurred in both unions. In arranged marriages, families and community do the bargaining, based on assets such as status, possessions, and dowry. In freely chosen marriages, the individuals perform a more subtle form of bargaining, weighing the costs and benefits of personal characteristics, economic status, and education.

Social Exchange

The ideas of bargaining, market, and resources used to describe relationships come to us from exchange theory, which is discussed in Chapter 2. A basic idea of exchange theory is that whether relationships form or continue depends on the rewards and costs they provide to the partners. Individuals, it is presumed, want to maximize their rewards and avoid costs, so when they have choices they will pick the relationship that is most rewarding or least costly. This analogy is to economics, but in relationships, individuals are thought to have other sorts of resources to bargain besides money: physical attractiveness, intelligence, educational attainment, earning potential, personality characteristics, family status, the ability to be emotionally supportive, and so on. Individuals may also have costly attributes, such as belonging to the "wrong" social class, religion, or racial/ethnic group, being irritable or demanding, and being geographically inaccessible (a major consideration in modern society). The increasing number of single people with children in today's market may find parenthood to be a costly attribute (Goldscheider, Kaufman, and Sassler 2009). An unbalanced sex ratio, an imbalance in number of males and females, can also affect the marriage market. For example, China's one-child policy combined with a strong preference for sons has made it difficult for men of marriage age to find a wife (Frank 2011). A lack of men both numerically and in terms of "marriagability" is a problem among African Americans. Higher rates of homicide and drug use among black men and higher rates of college completion among black women diminish the dating pool available to black women (Stanley 2011). Nevertheless, the tendency toward homogamy is strong. Even in the context of unbalanced sex ratios that would favor marriage outside one's racial, cultural, or ethnic group, men and women still tend to marry within their own grouping (Kalmijn and Van Turbergen 2010).

The Traditional Exchange Historically, women have traded their ability to bear and raise children, coupled with domestic duties, sexual accessibility, and physical attractiveness, for a man's protection, status, and economic support. Evidence from dating websites shows that the traditional exchange still influences heterosexual relationships. England and McClintock (2009) note "the double of standard of aging," in which age more negatively affects marriage prospects for women than men, results in the ability of men to marry younger women. Men are more likely to advertise for a physically attractive woman; women, for an economically stable man. Although women increasingly have their own employment and income, women continue to expect greater financial success as a category from prospective husbands than vice versa (Buss et al. 2001; Fitzpatrick, Sharp, and Reifman 2009). National data that looked at black, Hispanic, and white males found that the probability of a man getting married largely depends on his earning power (Edin and Reed 2005; Lichter, Qian, and Mellott 2006; Schoen and Cheng 2006).

Bargaining in a Changing Society "As gender differences in work and family roles blur, individuals' criteria for an acceptable mate are likely to change" (Raley and Bratter 2004, p. 179). For instance, research that looked at heterosexual mate preferences in the United States over the past sixty years showed that men and women—but especially men—have increased the importance that they put on potential financial success in a mate, while a woman's domestic skills have declined in importance (Siegel 2004; Sweeney and Cancian 2004). One study indicates that, for a young man today, a woman's high socioeconomic status increases her sexiness (Martin 2005).

As gender roles become more alike, exchange between partners may increasingly include "expressive, affective, sexual, and companionship resources" for both partners. In fact, a high-earning woman might bargain for a nurturing, housework-sharing husband, even if his earning potential appears to be lower than hers (Press 2004; Sprecher and Toro-Morn 2002). As college-educated young women approach occupational and economic equality with potential mates, the exchange becomes more symmetrical than in the past, with both genders increasingly looking for physical attractiveness, emotional sensitivity, and earning potential in one another (Buss et al. 2001; Montoya 2008).

Desiring wives who can make good money, college-educated men are now much more likely to marry college-educated women than a few decades ago. In fact, today *both* men and women are likely to want a spouse with more education or who earns more than they do (King and Allen 2009; Raley and Bratter 2004).

The fact that college-educated women and men tend to marry one another points to another concept associated with mate selection—assortative mating.

Assortative Mating: A Filtering Out Process

Individuals gradually filter, or sort out, those who they think would not make the best life partner or spouse. Research has consistently shown that people are willing to date a wider range of individuals than they would live with or become engaged to, and they are willing to live with a wider range of people than they would marry (Jepsen and Jepsen 2002). For instance, one study has found that women are less likely to consider the economic prospects of their male partners when deciding whether to cohabitate than when deciding about marriage (Manning and Smock 2002). Social psychologists call this process **assortative mating** (or, sometimes, *assortive mating*). Assortative mating raises another factor shaping partner choice—the tendency of people to form committed, and especially marital, relationships with others with whom they share certain social characteristics. Social scientists term this phenomenon **homogamy**.

Homogamy: Narrowing the Pool of Eligibles

Individuals tend to make relationship choices in socially patterned ways, viewing only certain others as potentially suitable. The market analogy would be to choose only certain stores or websites at which to shop. Each shopper has a socially defined **pool of eligibles**: a group of individuals who are the all-unmarried or all-unpartnered individuals (see Figure 5.3).

Geographic Availability **Geographic availability** (traditionally known in the marriage and family literature as *propinquity* or *proximity*) has historically been a reason that people meet others who are like themselves (Harmanci 2006; Travis 2006). For instance, as the size of various immigrant communities in the United States grows, the geographic availability of eligibles in the same ethnicity increases, resulting in ethnic homogamy (Gowan 2009; Qian and Lichter 2007). Geographic segregation, which can result from either discrimination or strong community ties, contributes to homogamous marriages (C. Gallagher 2006; Iceland and Nelson 2008; Lichter et al. 2007). Intermarriage patterns within the American Jewish community are an example. Only about 6 percent of Jews

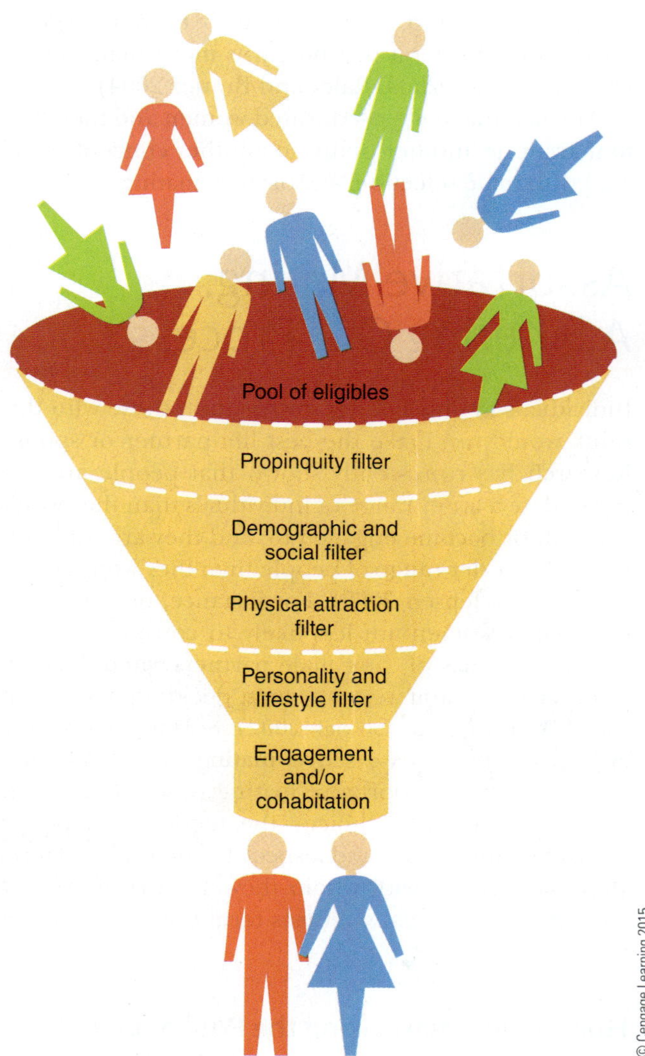

FIGURE 5.3 Choosing a spouse.
How does one go about the complex process of choosing a spouse? The assortative mating theory posits that mate selection involves narrowing down the pool of eligibles (currently unmarried men and women) until a suitable partner is found. The process operates not unlike a funnel. The pool of prospective mates starts out large. For example, the large pool of perspective mates is made smaller by propinquity (geographical area), desired demographic and social characteristics, physical attraction, and personal and lifestyle factors. Finally, the couple announces their intentions by getting engaged, moving in together, or both. This is the final step before a marital commitment is made. Whereas some couples move through these stages quickly, other couples take years to complete this process. Some people, of course, do not make it through these stages and therefore start the process again until the desirable mate is found.

married non-Jews in the late 1950s. Now that the barriers that once excluded Jews from certain residential areas and colleges are gone, about half marry gentiles (Sussman 2006).

Geographic availability also helps to account for educational and social class homogamy. Middle-class people tend to socialize together and send their children to the same schools; upper- and lower-class people do the same. Recent Princeton alumna Susan Patton sparked controversy by recommending that women find their husband in college, which was widely misinterpreted as recommending that women marry young (Lara 2013). In reality, she was simply describing the current marriage market for women. She said "For most of you, the cornerstone of your future and happiness will be inextricably linked to the man you marry," Patton warns. "You will never again have this concentration of men who are worthy of you. Here's what nobody is telling you: Find a husband on campus before you graduate" (Lucas 2013).

Today, first encounters may occur in cyberspace, and people meet others as far away as other continents. Some websites facilitate matches between men and women of different races and ethnicities. However, the Internet may actually encourage endogamy among religious or racial/ethnic groups, who can advertise online for homogamous dating partners (see also Desmond-Harris 2010; Wimmer and Lewis 2010). Illustrating these points, Russel K. Robinson, a black gay man writing in the *Fordham Law Review*, describes his experience as follows:

> Although I lived on the wealthy, predominantly white west side of [Los Angeles], the Internet created opportunities for me to interact with men in [less wealthy areas of the city]—men I almost certainly would not meet randomly while going through my daily routine.... Even as the Internet increases romantic opportunity, it also channels interactions.... Like many dating websites, Match.com prompts the user to indicate which races he will and will not date.... [I]f a white user is interested only in white romantic partners, he can easily structure his screen so that he never even has to view nonwhite profiles. (Robinson 2008, 2791–92)

Demographic and Social Filter People tend to form committed relationships with people of similar race, age, education, religious background, and social class (Kossinets and Watts 2009). As an example, the Protestant, Catholic, and Jewish religions, as well as the Muslim and Hindu religions, have all traditionally encouraged **endogamy**: marrying within one's own social group.

Why is it that men and women with similar characteristics tend to partner with each other? For one thing, people often find it easier to communicate and feel more at home with others from similar education, social class, and racial or ethnic backgrounds (Lewin 2005). They are likely to have attitudes, mannerisms, and vocabulary similar to one another. They feel comfortable in one another's surroundings. There is also

As We Make Choices

Looking for Love on the Internet

Before the Internet, e-mail, and social media, the process of finding a future mate proceeded in a relatively straightforward fashion: A couple would meet through friends, family, classes, or "by chance" and then might chat about their backgrounds and common interests, at which time they might set up a date to get to know each other better. These days, ways of finding romantic partners has been greatly expanded by technology, through dating websites, social media such as Facebook, various apps that arrange meetings between individuals and groups of friends, online video games, and the chat rooms of various interest and support groups that allow real-time interaction between strangers in different communities, states, and countries.

Such dating mechanisms have many advantages. Young people who have relocated to a new community for employment are often cut off from friends and relatives who would help facilitate meetings with suitable partners. Technology improves dating efficiency (Slater 2013). Dating sites allow participants to select mates with desirable characteristics before meeting so that participants don't "waste time" on unsuitable partners. There are special websites that cater to men and women wanting a partner who shares their religion, age, or educational level. Single men and women who are parents usually lack both time and the child care necessary to get out and meet people in traditional ways; instead, they can meet online. Indeed, populations in the "thin" marriage market (middle-aged heterosexuals, gay men and lesbians, physically disabled) are especially likely to find partners online (Barrow 2010; Rosenfeld and Thomas 2012). Typical settings for finding mates remain geared toward young adults and revolve around alcohol, which not everyone, especially older singles, is comfortable with. Likewise, some people are less outgoing and at ease in public settings, which hampers their success on the dating scene.

There are also disadvantages to dating online. Online profiles are incomplete—participants play up positive qualities and downplay or omit negative ones (Hertlein 2012). "Chemistry" and physical attraction can be difficult to assess. Scientific studies have shown that many elements of attraction are biological and unconscious in their origins and may have developed as human beings evolved. Online dating requires good written communication skills, which individuals possess in varying degrees (Bucior 2012). Prescreening potential mates on demographic, physical (e.g., blonde hair, blue eyes), and social factors may cause participants to miss out on partners who they might mesh with in terms of personality, life goals, and dreams for the future. On other hand, because profiles are public, members can be inundated with "winks" and e-mails from people they are not interested in. As opposed to being known by people in one's social network, online dates are "virtual strangers" who may introduce risks of physical, emotional, or financial harm (Rosenfeld and Thomas 2012). Therefore, dating websites advise members to meet potential dates in a public place, get their full name, and let a family member or friend know who they are going out with and when and where. It is recommended that members of dating websites use state sex-offender registries and online public court records to check out dates before a first meeting. Most singles don't rely exclusively on technology, however. Online dating may be best thought of as part of a single person's "tool kit" in the quest for love.

Critical Thinking

Have you, or someone you know, ever tried online dating? Would you recommend it to a single friend or family member? Do you think dating websites take the romance out of dating?

Men and women of all ages are increasingly "looking for love" online. There are many different Internet dating sites available catering to the needs of different races and ethnicities, religions, and ages. Dating sites encourage homogamy by first matching partners on a range of characteristics, including education, income, desire for children, hobbies and interests, and even preferences for a particular height and body type. Only then are partners allowed to get to know each other through e-mails, instant messaging, or face-to-face meetings.

social pressure from friends and family members. Compared to men, women are more sensitive to their parents' feelings about a potential mate (Dubbs, Buunk, and Li 2012). Friendship networks tend to be highly homogeneous, encouraging similar people to meet and interact (Wimmer and Lewis 2010). Interethnic relationships are more likely to develop when young adults are relatively independent of parental influence or when one's parents have an ethnically diverse network of friends (Rosenfeld and Byung-Soo 2005).

Sometimes, social pressure results from a group's concern for preserving its ethnic or cultural identity. Arab, Asian, or Hispanic immigrants may pressure their children to marry within their own ethnic group to preserve the culture (Kitano and Daniels 1995; Lee 1998). Whether blatant or subtle, social pressure toward homogamy can be forceful. Making knowledgeable choices involves recognizing the strength of social pressure and deciding whether to act in accordance with others' expectations.

Physical Attraction Filter Physical attraction is a step in assortative mating that is difficult to measure because it involves both social and biological factors. Physical attractiveness is a powerful resource in the marriage market. Physically attractive men, for instance, use this as a mating strategy—they are found to have a greater number of sex partners than less-attractive men (McClintock 2011). On the other hand, more physically attractive women are more successful than less-attractive women in using this quality as a strategy to delay sexual intercourse and secure committed relationships.

Nevertheless, a look at engagement photos, as well as a large body of empirical research, indicates that most individuals marry a partner of similar physical attractiveness as their own. Based on the National Longitudinal Study of Adolescent Health's Romantic Pair data, Carmalt, Cawley, Joyner, and Sobal (2008) found physical attraction to be the dominant mate-selection criterion. This extended not only to facial appearance but also to body weight. Being obese was found to be more detrimental to women than to men, and it was more of a barrier to white women than to black women in obtaining an attractive partner. However, this can be offset by greater education, a good personality, and attention to grooming.

Personality and Lifestyle Filter Personality traits can set the tone of the "emotional climate" of marriages, and negative patterns of interaction between partners during courtship have less-satisfying marriages (Markman et al. 2010; Wilson and Huston 2013). In order for relationships to flourish in the long term, couples need not only be physically and emotionally compatible but also intellectually, ideologically, and spiritually compatible (De La Lama, De La Lama, and Wittgenstein 2012).

The Final Filter: Cohabitation and Engagement Engagement is the way most couples indicate to others that they are "serious" in their intentions toward one another (Schweingruber et al. 2004). As such, engagement activities often take place in public settings, from picking out rings to the proposal itself. The couples' engagement provides family and friends a final opportunity to approve the relationship before going forward with the marriage. The need for others' approval is less true of cohabitation, which is discussed in Chapter 6. Nevertheless, cohabitation has become increasingly important "filter factor" in the decision to marry. Roughly half of married couples live together first (Saad 2008b). Many cohabitors mention the importance of "testing out" the relationship before making the final step toward marriage (although some cohabitors have no intention of getting married).

It is somewhat unclear whether and to what extent cohabitation is related to marital success. In the decades since cohabitation has become widespread, numerous studies have found that cohabitation before marriage increases the likelihood of divorce (Jones 2010). However, more recent studies are finding that the positive relationship between cohabitation and divorce may be waning and may increasingly depend on the couples' education, income, and number of previous cohabitations (Miller, Sassler, and Kusi-Appouh 2011; Teachman, Tedrow, and Hall 2006).

Cohabitation serves different purposes for different couples: Living together "may be a precursor to marriage, a trial marriage, a substitute for marriage, or simply a serious boyfriend–girlfriend relationship" (Bianchi and Casper 2000, p. 17). Since the 1970s, the proportion of marriages preceded by cohabitation has grown steadily; by 1995, a majority of marriages followed this pattern (Cherlin 2010). Cohabitation as a "substitute for marriage" is discussed at length in Chapter 6. Here we address cohabitation as a stage in choosing a spouse. Specifically, we will explore this question: How does cohabiting affect subsequent marital quality and stability?

Since about 1990, the proportion of cohabitors who eventually married their partners has declined (Seltzer 2000). This situation is largely because cohabiting has become more socially acceptable, a cultural change that "contributes to a decline in cohabiting partners' expectations about whether marriage is the 'next step' in their own relationship" (Seltzer 2000, p. 1,249). Another reason that fewer cohabitors are marrying has to do with economics: Lower-income cohabiting couples are less likely to marry (Gibson-Davis 2009; Lichter, Qian, and Mellott 2006).

Nevertheless, at least half of today's married couples between ages 18 and 49 report having lived together before their wedding (Saad 2008b). Many began cohabiting with definite plans to marry their partner. "Thus, first-time cohabitors often believe their union is part of the marriage process" (Guzzo 2009a, p. 198). On the other hand, cohabitors may gradually come to believe that they'll marry eventually. One study found that cohabitors who had talked about future marriage had "generally been living with their partners for about two years, indicating that the issue of greater permanence in their relationships surfaces over time" (Sassler 2004, p. 501). All else being equal, cohabiting couples who identify as conservative Protestants are more likely than other cohabitors to marry (Eggebeen and Dew 2009).

Many young people today follow the intuitive belief that "cohabitation is a worthwhile experiment for evaluating the compatibility of a potential spouse, [and therefore] one would expect those who cohabit first to have even more stable marriages than those who marry without cohabiting" (Seltzer 2000, p. 1,252; see also Manning 2009; "Marriage" 2008). Among high school seniors, about two-thirds agreed that "[i]t is usually a good idea for a couple to live together before getting married in order to find out whether they really get along" (National Marriage Project 2009, Figure 18). At this point, research is inconclusive on whether doing so is really a good idea. Interestingly, however, a study of 120 heterosexuals cohabiting for about one year found that those who lived together to test their compatibility had more negative couple communication and generally poorer quality relationships than those who reported cohabiting for other reasons (Rhoades, Stanley, and Markman 2009).

So far, there has not been much research on whether cohabitation that is limited only to one's future spouse increases or decreases the odds of marital success. We can report that findings from a national representative sample show the divorce rate for serial cohabitors to be twice that for women who cohabited only with their eventual husbands (Lichter and Qian 2008). A second study that also used a national representative sample (of 6,577 women) found that premarital cohabitation limited to the woman's future husband did not increase the couple's likelihood of divorce (Teachman 2003). This situation also appears to be true for remarriages: Cohabiting with only one's future second spouse did not increase divorce likelihood (Teachman 2008a). Cohabiting couples with similar understandings of the nature and goals of the relationship (such as expectations for future marriage) is associated with a lower likelihood of divorce (Wilson and Huston 2013).

Meanwhile, research over the past twenty years has consistently shown that marriages preceded by more than one instance of cohabitation are *more* likely to end in separation or divorce than are marriages in which the spouses had not previously cohabited at all (Xu, Hudspeth, and Bartkowski 2006). However, one study shows that these findings apply to nonHispanic whites but not to African Americans or Mexican Americans, for whom cohabiting may be a more normative life course event, as discussed in Chapter 8 (Phillips and Sweeney 2005). Why might serial cohabitation before marriage be related to lower marital stability? Hypotheses to answer this question can be divided into two categories—*experience* and *selection*—both of which are supported to some degree by research.

First, the **experience hypothesis** posits that cohabiting experiences themselves affect individuals so that, once married, they are more likely to divorce (Amato 2010). For example, serial cohabitation may adversely affect subsequent marital quality and stability inasmuch as the experiences actually weaken commitment because "'successful' cohabitation demonstrates that reasonable alternatives to marriage exist" (Thomson and Colella 1992, p. 377). There is also evidence that "young adults become more tolerant of divorce as a result of cohabiting, whatever their initial views were," possibly because "cohabiting exposes people to a wider range of attitudes about family arrangements than those who marry without first living together" (Seltzer 2000, p. 1,253; see also Dush, Cohan, and Amato 2003; Popenoe and Whitehead 2000).

A related hypothesis suggests that some cohabiting couples who would not have married if they had been simply dating but not living together do end up marrying just because getting married seems to be the expected next thing to do. We can assume that it is less difficult to end an unsatisfactory dating relationship than a cohabiting one. Furthermore, research has found that cohabitors who marry after having a nonmarital birth experience lower marital relationship quality than do nonparent cohabitors who eventually marry (Tach and Halpern-Meekin 2009). This finding may result from the fact that cohabiting parents are more likely to marry mainly because they feel that they should. Choosing by default, couples may "slide" from cohabiting into marrying, rather than making more deliberative decisions (Stanley, Rhoades, and Markman 2006; Stanley 2009).

Second, the **selection hypothesis** assumes that individuals who choose serial cohabitation (or who "select" themselves into cohabitating situations) are different from those who do not; these differences translate into higher divorce rates. Serial cohabitors are more likely to have low relative education and income as well as less effective problem-solving and communication skills—factors related to divorce (Amato et al. 2008; Lichter and Qian 2008; Thornton, Axinn, and Xie 2007). Furthermore, those who choose serial cohabitation may have more-negative attitudes about marriage in general and more-accepting attitudes toward divorce.

An important international study lends considerable support to the selection hypothesis (Liefbroer and Dourleijn 2006). This study looked at the effects of cohabitation on marital stability in several countries and found that cohabiting had no negative effect on marital stability in countries such as Norway, where cohabiting is more common than in the United States. The researchers reasoned that in societies where cohabitation is about as common as marriage, those who live together before marrying would not be significantly different from those who do not. Therefore, no selection effect would be operating. The fact that this research found no negative effect of cohabitation on marital stability in societies where there would be little selection effect supports the selection hypothesis. Other support for the selection hypothesis is the finding that negative effects of cohabitation apply more strongly for nonHispanic whites than for blacks or Mexican Americans, the latter two racial/ethnic groups having a higher percentage of cohabitors (Phillips and Sweeney 2005).

Heterogamy in Relationships

The opposite of endogamy is **exogamy**, marrying outside one's group, or **heterogamy**—that is, choosing someone dissimilar in race, age, education, religion, or social class. With "the loosening of relationship conventions," more older women, for example, are dating or marrying men at least five years younger—the media-hyped "cougar" phenomenon (Kershaw 2009b). With regard to socioeconomic class and education, although people today are marrying across small class distinctions, they still are not doing so across large ones. For instance, individuals of established wealth or high education levels seldom marry those who are poor or who have low educational achievement (Fu and Heaton 2008). This "loosening of relationship conventions" evidences itself in other types of heterogamy as well—namely, relationships that cross racial, ethnic, and religious lines.

Interracial and Interethnic Heterogamy

As young adults experience increased independence from family influence, we can expect a rise in interracial and interethnic unions (Rosenfeld 2008). **Interracial marriages** include unions between partners of the white, black, Asian, or Native American races with a spouse outside their own race. Interracial unions have existed in the United States throughout our history (Maillard 2008). However, not until 1967 (*Loving v. Virginia*) did the U.S. Supreme Court declare that interracial marriages must be considered legally valid in all states.

As defined by the U.S. Census Bureau, Hispanics are not a separate race but, rather, an ethnic group. Unions between Hispanics and others, as well as between different Asian/Pacific Islander, Hispanic, or black ethnic groups (such as Thai–Chinese, Puerto Rican–Cuban, or African American–black Caribbean) are considered **interethnic marriages**.

Although interracial marriage has grown, available statistics show that the proportion of marriages that are interracial or interethnic remains fairly small (between 7 and 8 percent of the adult U.S. population, or 4.5 million couples) (U.S. Census Bureau 2010b, Table 60). Perhaps due to an expanding marriage pool, Asian American couples appear to be bucking this trend. Their rates of marrying outside their race declined 10 percent between 2008 and 2010 (Swarns 2012). Asians who do this often do so in the face of strong disapproval from parents (Farr 2011).

Note that there can be cultural diversity within racial and ethnic categories. For example, Asians include individuals from a variety of nations and cultures. Immigration has an important part to play in the marriage market and the path to citizenship for this racial group. Most form racially homogenous unions, but there is a high rate of marriage, and even higher rate of cohabitation, between recent arrivals and native-born (Qian, Glick, and Batson 2012). Whereas for Asians intermarriage has cultural benefits, the opposite is true for Native Americans who are the most likely of any racial or ethnic group to out-marry. Because they must be at least one-quarter Indian, many children resulting from these unions are vulnerable to losing their federal tribal benefits (Ahtone 2011).

Indeed, rates of interracial and interethnic marriage vary by race and ethnicity. As shown in Figure 5.4 of all interracial marriages in 2010 (this does not count Hispanic–nonHispanic unions), only about 23 percent (558,000 couples) were black–white (U.S. Census Bureau 2012b, Table 60). The vast majority of the remainder were combinations of whites with Asians, Native Americans, and others. There are also gender

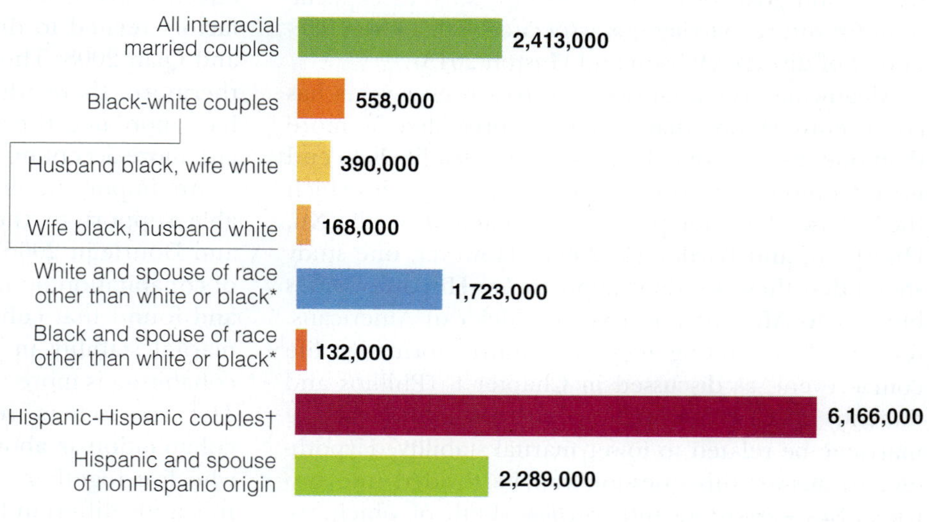

FIGURE 5.4 Number of interracial and Hispanic–nonHispanic married couples, 2010.

Source: U.S. Census Bureau 2012b, Table 60.

differences in interracial marriages. About 70 percent of black–white marriages involved black men married to white women. Most Asian-white marriages involved white men and Asian women. Among Hispanic marrieds, 27 percent have a spouse of nonHispanic origin (U.S. Census Bureau 2010b, Table 60). If we count cohabiting couples, the percentage today of racially or ethnically heterogeneous couples would be somewhat higher than the statistics for married couples because (due to the assortative mating process) cohabiting couples are less homogamous than married couples (Batson, Qian, and Lichter 2006; Joyner and Kao 2005).

Reasons for Interracial and Interethnic Relationships

Much attention has been devoted to why people marry or form other romantic unions interracially. One apparent reason among racial/ethnic groups that are relatively small in number is simply that they have a smaller pool of eligibles in their own race/ethnicity and are also more likely than larger racial/ethnic groups to interact with others of different races (Qian and Lichter 2007; Robinson 2008).

Another explanation is the **status exchange hypothesis**—the argument that an individual might trade his or her socially defined superior racial/ethnic status for the economically or educationally superior status of a partner in a less-privileged racial/ethnic group (Kalmijn 1998). In this regard, racial stereotypes may play a part:

> [A] society dominated by Euro-Americans will unsurprisingly privilege a standard of beauty and cultural styles that is a mirror image of itself, even if that image is a media distortion.... [I]nterviews with Asian men and women [found] that a sizable minority of respondents preferred whites as potential or current mates because of their preference for "European" traits including tallness, round eyes, buffness for men and more ample breasts for women. (C. Gallagher 2006, p. 150)

Vstock LLC/Getty Images

Some ethnic groups, particularly those consisting of a large proportion of recent immigrants, strongly value homogamy. Nevertheless, an increasing number of Americans enter into ethnically heterogamous unions. Although "feeling at home"—a factor that encourages homogamy—may be difficult at first, some individuals thrive on the cultural variety characteristic of interracial or interethnic relationships.

Applying the status exchange hypothesis to black–white intermarriage would suggest marrying "up" socioeconomically on the part of a white person who, in effect, trades socially defined superior racial status for the economically superior status of a middle- or upper-middle-class black partner. Little research has been done to test this hypothesis, but a recent study of intermarriage among native Hawaiians, Japanese, Filipinos, and Caucasians in Hawaii supports the hypothesis (Fu and Heaton 2000). That those with higher education levels are more likely to out-marry has also been found among immigrants to the United States, especially those with smaller populations (Kalmijn 2012).

Some African American sociologists have expressed concern about black men—especially educated black men—choosing spouses from other races (Crowder and Tolnay 2000; Staples 1994, 1999). Robert Davis, a past president of the Association of Black Sociologists, believes that black men are inclined to see white women as "the prize" (Davis, quoted in "Why Interracial Marriages" 1996). Some African American women view black males' interracial relationships as "selling out"—sacrificing allegiance to one's racial heritage to date someone of higher racial status (Paset and Taylor 1991). This issue can be a problem on college campuses, where black men, especially athletes, are perceived as only pursuing relationships with white women (Wilkins 2012). According to one black student on a football scholarship:

> I'm not gonna lie, bein' young, I thought that havin' sex with white women was also a sense of accomplishment... because my whole life I've been told that is off-limits. (Wilkins 2012, p. 175)

As with homogamous couples, those in heterogamous relationships find their partners by means of social exchange in a marriage market (Fryer 2007). Because of the shortage of black men, the Council on Contemporary Families (2011) suggests that intermarriage may be a good thing for black women as well, and possibly a better route to marriage.

It is often assumed that interracial couples marry *in spite of* their partner being of a different race or ethnicity. In fact, some men and women purposefully marry someone of a different race, religion, or culture. One woman explains:

> Part of the fact that he speaks a different language and part of the fact that, like, a whole culture and that difference is what I like about him so much... the whole

> element of a different culture and different languages is a huge part of what attracts me to him. (Yodanis, Lauer, and Ota 2012, p. 1029–1030)

One recent study of unmarried interracial couples in college found *higher* relationship satisfaction compared to same-race couples (Troy, Lewis-Smith, and Laurenceau 2006). A comparison of Mexican American–nonHispanic white marriages with those of homogamous white and homogamous Mexican American couples found little difference in marital satisfaction among the three groups (Negy and Snyder 2000).

Interfaith Relationships

In 2010, Riley (2013) commissioned a poll of a nationally representative sample of 2,450 Americans that included an oversample of **interfaith marriages**. She found that approximately 45 percent of married couples married outside their religion (Riley 2013). Being highly educated seems to lessen individuals' commitment to religious homogamy (Petersen 1994). Religions that see themselves as the one true faith and people who adhere to a religion as an integral component of their ethnic/cultural identity (for example, some Catholics, Jews, and Muslims) are more likely to encourage homogamy, sometimes by pressing a prospective spouse to convert (Bukhari 2004). Often, religious bodies are concerned that children born into the marriage will not be raised in their religion (Sussman 2006). Because Americans are becoming less religious, perhaps religious homogamy can be expected to decline further.

Interfaith and intercultural wedding ceremonies are also on the rise. For example, 23 percent of Catholics, 33 percent of Jews, and 21 percent of Muslims are married to people of different faiths (Duba and Watts 2009; Lara and Duba Onedera 2008). Such ceremonies often blend the traditions of two different cultures. These couples often include the dress, music, and readings of both cultures in the wedding ceremony (Trickey 2011). In spite of a trend toward less religious homogamy and a lessened tendency of Asian, Hispanic, and European Americans—such as Irish, Italians, or Poles—to marry within their own ethnic groups, homogamy is still a strong force (Fu and Heaton 2008). In the United States, this is especially so among ethnicities and cultures with a history of early marriage (Kalmijn and Van Turbergen 2010).

Heterogamy and Relationship Quality and Stability

How does marrying someone from a different religion, social class, or race/ethnicity affect a person's chances for a happy union? In general, research suggests that marriages that are homogamous in age, education, religion, and race are the happiest and most stable (Bratter and King 2008; Jones 2010; Larson and Hickman 2004). This is also true for cohabiting couples (Hohmann-Marriott and Amato 2008).

Marriages that are homogamous are more likely to be stable because partners are more likely to share the same values and attitudes when they come from similar backgrounds (Durodoye and Coker 2008; Lincoln, Taylor, and Jackson 2008). Heterogamous marriages may create conflict between the partners and other groups, such as parents, relatives, and friends. Continual discriminatory pressure from the broader society may create undue psychological and marital distress (Bratter and Eschbach 2006; Childs 2008; Yancey 2007). Although opinion polls and other research show that Americans are becoming less disapproving of interracial dating and marriage (Carroll 2007a; J. Jones 2005), among both blacks and whites, a minority of individuals strongly disapprove of interracial marriage (Jacobson and Johnson 2006). The risk of divorce to interracial couples varies by race and ethnicity. Black–white couples face the greatest level of disapproval, especially from the white community, which consequently negatively affect the survival of these marriages (Zhang and Van Hook 2009, p. 105). Also, a higher divorce rate among heterogamous marriages may reflect the fact that these partners are likely to be less conventional in their values and behavior, and unconventional people may divorce more readily than others (see Hohmann-Marriott and Amato 2008).

As an example, Islamic marriage counselor Aneesah Nadir suggests that, in a homogamous Muslim marriage, both spouses—not just one—are likely to find value in referring to the Quran for answers to disagreements (Nadir 2009).

However, a recent study based on analysis of data from 23,139 married couples "failed to provide evidence that interracial marriage per se is associated with an elevated risk of marital dissolution." Instead, the risk of divorce or separation among the interracial couples sampled was similar to that of the race of the spouse from the more divorce-prone race. Accordingly, "Mixed marriages involving blacks were the least stable followed by Hispanics, whereas mixed marriages involving Asians were even more stable than endogamous white marriages" (Zhang and Van Hook 2009, p. 104).

Interreligious marriages also tend to be more stressful and less stable than homogamous ones (Mahoney 2005; Riley 2013). One probable reason that religious homogamy improves chances for marital success involves value consensus. Religion-based values and attitudes may come into play when negotiating leisure activities, child-raising methods, investments and expenditures of money, and appropriate spousal roles (Curtis and Ellison 2002; Lambert and Dollahite 2006). Meanwhile, analysis of data from a national random telephone survey of Protestant and Catholic households concluded that although marital satisfaction was less for interdenominational couples, the difference disappeared when the interdenominational couple had generally similar religious orientations, good communication skills, and similar beliefs about child raising (Hughes and Dickson 2005; Williams and Lawler 2003). Some recent research shows a declining effect of religious differences on marital satisfaction over the past several decades due to the greater effect of couples' gender, work, and co-parenting concerns (Myers 2006; Williams and Lawler 2003).

Again, we see that private troubles—or choices—are intertwined with public issues. Some ethnic and religious groups strongly value homogamy. Meanwhile, it is also true that if people are able to cross racial, class, or religious boundaries and at the same time share important values, they may open doors to a varied and exciting relationship. Chapter 9 explores raising children in interracial families. In our society, choice of life partners—whether homogamous or heterogamous—typically involves developing an intimate relationship and establishing mutual commitment. The next section examines these processes.

Meandering Toward Marriage: Developing the Relationship and Moving Toward Commitment

Social scientists have been interested in the process through which a couple develops their relationship and mutual commitment. What first brings people together? What keeps them together?

By 2011, the median age at first marriage was 26.5 for women and 28.7 for men—the highest recorded averages since the 1890 census. These figures compare with 20.8 for women and 23.2 for men in 1970 (U.S. Census Bureau 2011b, Table MS-2).

Young people today "meander toward marriage," feeling that they'll be ready to marry when they reach their late twenties or so (Arnett 2004, p. 197). Assuming that adolescents begin exploring and pursuing romantic relationships in their teen years, the period in which premarital relationships can occur may last five, ten, or even twenty years and sometimes even longer (Wilson 2009; see also Meier and Allen 2009; Wallace 2007). Experiencing unprecedented freedom, today's young adults often express the need to explore as many options as possible before "settling down." As one young woman explained:

> I think everyone should experience everything they want to experience before they get tied down, because if you wanted to date a black person, a white person, an Asian person, a tall person, short, fat, whatever, as long as you know you've accomplished all that, and you are happy with who you are with, then I think everything would be OK. I want to experience life and know that when that right person comes, I won't have any regrets. (Quoted in Arnett 2004, p. 113)

In Chapter 4, you read about *sexual scripts* or norms governing sexual behavior. There are also **dating scripts** that govern behavior in the getting-to-know-you stage of dating relationships. These are highly gendered, with men and women having far different expectations about what happens during and after a date (Littleton et al. 2009). Whereas men are more likely to desire and pursue sexual activity, women are more likely to look at dating in terms of the possibility of a committed relationship. Moreover, dating scripts can vary by race and ethnicity and other factors. For example, deaf university students' dating scripts were mostly similar to hearing university students, except deaf students mentioned more group activities and lower sexual expectations (Gilbert, Clark, and Anderson 2012). Compared to whites, African Americans were more likely to include "meeting the family" and exhibited larger gender differences in dating expectations, especially regarding sexual behavior (Jackson et al. 2011). African American males' internalization of the rigid "street code"—intense concern about toughness, respect, reputation, revenge (Anderson 1999)—is associated with less commitment, satisfaction, and greater hostility and conflict in romantic relationships (Barr, Simons, and Stewart 2013).

Contemporary Dating

Now that the period of dating has been greatly extended due to late marriage, research on dating has flourished. But what is "dating"? These days there is considerable variation in premarital romantic relationships, making dating very difficult to define. Therefore, dating tends to be defined broadly by researchers. For example, after extensive pretesting and taking into account gender, class, and race differences adolescents' understandings of premarital romantic relationships, Giordano, Longmore, and Manning (2006), in their study of adolescents from Toledo, Ohio (the Toledo Adolescent Relationships Study known as TARS), defined it this way, "Now we are interested in your own experiences with dating and the opposite sex. When we ask about 'dating' we mean when you like a guy, and he likes you back. This does not have to mean going on a formal date" (p. 268).

The TARS data revealed something interesting. This definition not only captured "dating" relationships but also captured what the researchers refer to as "nondating" relationships between adolescents that include some elements of dating (namely, sexual behavior) but lack commitment associated with being "boyfriend" and "girlfriend." These situations are discussed below.

Dating versus "Nondating"

The traditional dating script that most of us think of as "dating" emerged in the middle of the twentieth century. This form of dating was facilitated by widespread access to automobiles, which removed interaction between young men and women from the front parlor of the young woman's home, under the watchful eye of parents, to places free from parental control such as movies or the local malt shop. Therefore, dating replaced old-fashioned "courtship" as the process of finding a mate. These days the key difference between courtship and dating is that finding a marriage partner may or may not be the end goal. Note, too, that this form of dating was typical of white, middle class adolescents, and may not have been representative of the dating patterns of different race, class, or religious groups (Jackson et al. 2011).

With all the attention to the current "hooking up culture" and casual sex discussed in Chapter 4, it is easy to forget that young couples still engage in traditional dating. Bartoli and Clark (2006) asked college students to describe a "typical" date. Both men and women in the study said that a typical date involved (1) initiation—meeting in a public place (class, a party, or bar); casual talking, finding common interests, and calling for date; (2) the date itself—movies, dinner, talking (women only), or shared interest; and (3) an outcome—talking, going back to the house, kissing goodnight, going home, and developing a relationship. Research by Giordano, Longmore, and Manning (2006) suggests that some adolescent boys find this script as desirable as do girls. There are differences between men and women, though, that reflect traditional gender roles. For example, when it comes to eating out, Amiraian and Sobal (2009) found that women were significantly more likely than men to mention nonfattening and nonmessy foods (e.g., salad). Men have higher sexual expectations for the first date, especially if they pay for the date (Emmers-Sommer et al. 2010).

It is also easy to forget that parents are involved in overseeing their children's behavior, including dating relationships. Schalet (2011), in her study of U.S. and Dutch adolescents, found that American parents tended to view their children as unable to control their impulses and therefore requiring a great deal of protection and control, which often led to the children keeping their dating lives secret. One girl said this:

> They don't want to know that I'm doing it. It's kind of like, "Oh, my god, my little girl is having sex kind of thing." ... it's just really overwhelming to them to know that their little girl is in their house having sex with a guy. That's just scary to them .... [They] won't even let me have a guy in my room without the door open. (Schalet 2011, p. 113)

Dutch parents, on the other hand, had higher expectations for their children's ability to control themselves and make good decisions on their own. One Dutch teenager explained:

> [My parents] do not say, "You are not allowed to [smoke, drink, do drugs]." They know: if kids want to do it, they will do it anyway. I am free to do it. The give me a lot of information about what drugs, alcohol, and smoking do to you. Not to disapprove, but... to reach your own conclusions that it is bad for you. (Schalet 2011, pp. 148–149).

Another girl explains her mother's attitude about birth control. She said, "if you want to go on the pill, I will allow you to. Because I'd rather you go on the pill than come home pregnant really young" (Schalet 2011, p. 141).

On the other hand, many adolescents and young adults engage in nondating as well. Nondating is generally sexual in nature and takes various forms such as "hooking up" or being "friends with benefits" (Claxton and van Dulmen 2013; discussed in detail in Chapter 4). Among adolescents from the TARS study (Manning, Giordano, and Longmore 2006), there were some key differences between daters and nondaters. Daters referred to themselves as boyfriend and girlfriend, were close in age, reported that sex brought them closer, had relationships that tended to last a few months, and told their friends about the relationship. Whereas daters followed the traditional pattern, nondaters did not. Nondaters reported not wanting a boyfriend or girlfriend; their partners included friends and exes; their relationships lasted days or years; there was a greater gap in ages; sex did not bring the couple closer; the couple didn't always tell their friends; half the sexual unions were for one time only; and the relationships were less exclusive. As discussed in Chapter 4, hooking up is associated with many disadvantages for women; women are more likely than men to prefer dating to hooking up (Bogle 2008; Bradshaw, Kahn, and Saville 2010).

Despite its growing appeal among college students, hooking up has unfortunately been empirically linked to known risky behaviors such as alcohol abuse and engaging in sexual intercourse without using a condom

Issues for Thought

Sexual Assault and Acquaintance Rape

Unfortunately, an awareness of sexual assault and rape is crucial for young people navigating a large, complicated, and highly diverse marriage market. **Sexual assault** is any type of sexual contact or behavior that occurs without the explicit consent of the recipient and includes fondling, groping, digital penetration (with fingers), forced oral and anal sex (sodomy), forced sexual intercourse (rape), and attempted rape (U.S. Department of Justice 2013). Adolescents and young adults between the ages of 12 and 24 are disproportionately the victims of sexual assault and rape—they are two to three times more likely to be sexually assaulted than adults age 25 and older (Child Trends 2012). Rape and sexual assault also disproportionately affect women, although men can be victims of sexual assault as well (the perpetrators of sexual assault are almost exclusively men however). Among high school students, females are more than twice as likely as males to report being raped, at 12 percent and 5 percent. However, both male and female victims experience similar consequences of rape and sexual assault—from physical injuries and symptoms (fatigue, chronic headaches) to emotional problems, including depression and suicide.

Contrary to the impression that we are likely to get from news media, most rape victims, especially younger ones, know their rapists (Child Trends 2013). In **acquaintance rape**, also known as *date rape*, someone becomes the victim of an unwanted sexual encounter with a date or other acquaintance. **Sexual coercion** is very common among perpetrators of acquaintance rape. Such coercion uses verbal or emotional pressure (threatening violence, tricking, lying, or using guilt), one's position of power (being a boss, teacher, coach, or other adult), or other means to manipulate the victim into sexual activity. Sexual coercion is often hard to detect and can include comments such as "We've had sex before, so you can't say no now"; "If you love me, you would have sex with me"; "If you don't have sex with me, I will find someone who will"; or "I'm not sure I can be with someone who doesn't want to have sex with me" (McCoy and Oelschlager 2013). Excessive use of alcohol is involved in sexual victimization (Foran and O'Leary 2008; Peralta and Cruz 2006). Findings from various research studies over the past decade show that sexually coercive men tend to dismiss women's rejection messages regarding unwanted sex and differ from noncoercive men in their approach to relationships and sexuality: They date more frequently; have higher numbers of sexual partners, especially uncommitted dating relationships; prefer casual encounters; and may "take a predatory approach to their sexual interactions with women."

Many female victims blamed themselves, at least partially—a situation that can result in still greater psychological distress (Breitenbecher 2006). One reason victims blame themselves has to do with **rape myths**: beliefs about rape that function to blame the victim and exonerate the rapist (Cowan 2000). Rape myths include the ideas that (1) rapists are violent strangers lurking in the shadows looking for victims; (2) the rape was somehow provoked by the victim (for example, she "led him on" or wore provocative clothes); (3) men cannot control their sexual urges, a belief that consequently holds women responsible for preventing rape; and (4) rapists are mentally ill. These beliefs encourage potential victims to feel safe with someone they know, no matter what (Cowan 2000; Littleton et al. 2009). The problem is that women who believe in such myths are less likely to recognize and leave potentially risky situation and may even interpret a sexual assault as just a "bad hookup" (Franklin 2012; Littleton et al. 2009).

Increasingly, college men report that they know men are responsible for rape. This could be the result of rape-prevention information and workshops that reach into male dorms as well as campaigns geared toward men (Cardiff 2013; Domitrz 2003).

Critical Thinking

What can you do to help prevent sexual assault and acquaintance rape? What should you do if you or a friend is raped or assaulted? What would or should you do if a friend or acquaintance of yours was known to be the perpetrator of a sexual assault?

Courtesy of SAVE (Sexual Assault Voices of Edmonton)

Rather than targeting the victims of rape, this date-rape campaign targets *potential rapists*. It also presents various previously ambiguous situations, such as a man having sex with a woman who is too intoxicated to consent, in "no uncertain terms" (Cardiff 2013).

(Downing-Matibag and Geisinger 2009; Norris et al. 2013). Casual dating such as this can be associated with date or acquaintance rape, as discussed in "Issues for Thought: Sexual Assault and Acquaintance Rape."

The majority of couples meet for the first time in face-to-face encounters, such as at school, work, a game, or a party. However, more couples today, especially those who are older, meet through online singles' ads and dating websites (Ellin 2009). Development of a face-to-face romantic relationship moves from initial encounter to discovery of similarities and self-disclosure (Knobloch, Solomon, and Theiss 2006). However, meeting for the first time online is a bit different.

Dating websites that match couples on demographic and social traits take credit in their advertising for creating thousands if not millions of relationships and marriages. Does meeting through a dating website affect marital quality and stability? Empirical research on this issue is in its early stages. A study based on data from the new "How Couples Meet and Stay Together" survey of 4,002 couples found no difference in relationship quality between couples who met online as opposed to in person (Rosenfeld and Thomas 2012). Other research indicates that online communication enhances commitment and trust because there is greater self-disclosure than what can be said in person (Hertlein 2012).

From Dating to Commitment

However individuals meet, what is it that draws and keeps them together? One way to explore this process involves the idea of a developing relationship as moving around a wheel.

From an interaction-constructionist perspective (see Chapter 2), qualitative research with serious dating couples shows that they pass through a series of fairly predictable stages (Sniezek 2002, 2007) by means of which they further define their relationship. Hinting, testing, negotiating, joking, and scrutinizing the partner's words and behavior characterize this process. As one example, a thirty-two-year-old emergency room worker described reading her partner's joking about marriage—and his use of the word *yet*—as a possible sign that he had considered marrying her:

> It was right here in our kitchen I put it (the food) down on his plate and I'm like "prison food." And he said "Um gee and we're not even married yet." And that was like the first joke. It stuck in my mind. (Sniezek 2007)

As the relationship progresses toward an eventual wedding, the "marriage conversation" is introduced. In the research being described here, women were more likely to initiate marriage talk—cautiously and indirectly: "Yeah it's like so what are you thinking? Where is this relationship heading?" In other cases, the marriage conversation began more directly. One woman raised the question of marriage when her partner suggested that they live together:

> When he asked me to move in with him, I told him I felt uncomfortable living with someone and not being married—a moral issue for me. So we talked about [the probability of getting married] at that time. (Sniezek 2007)

Once marriage talk is initiated, the couple faces negotiating a joint definition of the relationship as premarital. If one partner rejects the idea that the relationship should lead to marriage, several responses can occur. In some cases, one partner's marriage hopes may be relinquished, although the relationship continues. In other cases, a partner may deliver an ultimatum. Sometimes an ultimatum results in marriage; in other cases, stating an ultimatum causes an irreparable rift in the relationship.

Finally, most couples do not define themselves as "really" engaged until one or more ritualized practices take place—buying rings, setting a wedding date, public announcements to family and friends, holding engagement parties, and so on. These practices make the redefinition of the relationship increasingly public and "hardened" (Sniezek 2007).

In another qualitative study on this subject, two social scientists conducted lengthy interviews with 116 individuals in premarital relationships. They examined the process by which these partners gradually committed to marriage (Surra and Hughes 1997). From the interviews, the researchers classified the respondents' relationships in two categories: *relationship-driven* and *event-driven.*

In event-driven relationships, partners vacillated between commitment and ambivalence. Often they disagreed on how committed they were as well as why they had become committed in the first place. The researchers called this relationship type *event-driven* because events—fighting, discussing the relationship with one's own friends, and making up—punctuated each partner's account.

It's probably no surprise that event-driven couples' satisfaction with the relationship fluctuated over time. Although often recognizing their relationships as rocky, they do not necessarily break up because positive events (for example, a discussion about getting married or an expression of approval of the relationship from others) typically follow negative ones. At least some event-driven couples would probably be better off not getting married. Relationship-driven couples follow an evolving, wheel-like pattern that is described by the wheel of love theory.

The Wheel of Love According to this theory, the development of love has four stages in a circular process—a **wheel of love**—that can continue indefinitely. The four stages—rapport, self-revelation, mutual dependency, and personality need fulfillment—are shown in Figure 5.5, and they describe the span from attraction to love.

Rapport Feelings of **rapport** rest on mutual trust and respect. A principal factor that makes people more likely to establish rapport is similarity of values, interests, and

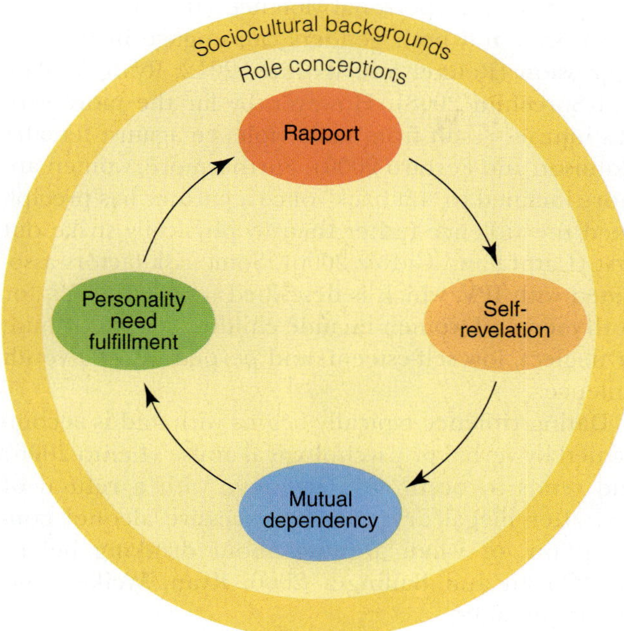

FIGURE 5.5 Reiss's wheel theory of the development of love.

Source: From *Family Systems in America*, 3rd ed., by I. Reiss © 1980 Wadsworth, a division of Cengage Learning.

background (Gottlieb 2006)—social class, religion, and so forth. The outside circle in Figure 5.5 conveys this point. However, rapport can also be established between people of different backgrounds, who may perceive one another as an interesting contrast to themselves or see qualities in one another that they admire.

Self-revelation, or *self-disclosure*, involves gradually sharing intimate information about oneself. People have internalized different views about how much self-revelation is proper. The middle circle of Figure 5.5, role conceptions, signifies that ideas about social class-, ethnic-, or gender-appropriate behaviors influence how partners self-disclose and respond to each other's self-revelations and other activities.

For most of us, love's early stages produce anxiety. We may fear that our love won't be returned. Maybe we worry about being exploited or are afraid of becoming too dependent. One way of dealing with these anxieties is, ironically, to let others see us as we really are and to share our motives, beliefs, and feelings (Peck and Peck 2006). As reciprocal self-revelation continues, an intimate relationship may develop while a couple progresses to the third stage in the wheel of love: developing interdependence or mutual dependency.

In the **mutual dependency** stage of a relationship, the two people desire to spend more time together and thereby develop interdependence or, in Reiss's terminology, *mutual dependency*. Partners develop habits that require the presence of both partners. Consequently, they begin to depend on or need each other. For example, watching a good movie may now seem lonely without the other person because enjoyment has come to depend not only on the movie but also on sharing it with the other. Interdependency leads to the fourth stage: a degree of mutual personality need fulfillment.

In the **need fulfillment** stage of relationship development, two people find that they satisfy a majority of each other's emotional needs. As a result, rapport increases, leading to deeper self-revelation, more mutually dependent habits, and still greater need satisfaction. The relationship is one of ongoing emotional exchange and mutual support. Along this line, social scientist Robert Winch (1958) once proposed a theory of complementary needs: We are attracted to partners whose needs complement our own (Malakh-Pines 2005). Sometimes people take this idea to mean that "opposites attract." This may make intuitive sense to some of us, but needs theorists more often argue that we are attracted to others whose strengths are *harmonious* with our own (Klohnen and Mendelsohn 1998; Schwartz 2006). As partners develop mutual need satisfaction and interdependence, they gradually define their relationship. Part of defining a relationship should involve, according to counselors, talking about serious questions (Pendley 2006). What are some things that deserve discussing?

Some Things to Talk About Talking about the relationship that you and your partner want may bring up differences, many of which can be worked out. If value differences are uncovered and cannot be worked out—for example, about whether or not to have children—it might be better to end the relationship before committing to a union that cannot satisfy either partner. Openly and honestly discussing matters like the following is important to successful mate selection (Pendley 2006):

- When should a relationship be dissolved and under what circumstances? How long and in what ways would you work on an unsatisfactory relationship before dissolving it?
- What are your expectations, attitudes, and preferences regarding sex?
- Do you want children? If so, how many? Whose responsibility is birth control?
- If you have children, how will you allocate child-rearing responsibilities? Do the two of you agree on child-raising practices, such as whether to spank a child?
- What is the financial situation of each partner as he or she comes into the union? How will the couple manage any previous debt, credit problems, and their existing financial situations in general?

- Will partners equally share breadwinning and homemaking responsibilities or not? How will money be allocated? Who will be the owner of family property, such as family businesses, farms, or other partnerships?
- Do you expect your partner to share your religion? Will you attend religious services together? If you are of a religion different from that of your mate, where will you worship? What about the children's religion?
- What are your educational goals? How about your prospective partner's?
- How will each of you relate to your own and to your partner's relatives?
- What is your attitude toward friendships with people of the opposite sex? How about cyberfriends? Would you ever consider having sex with someone other than your mate? How would you react if your partner were to have sex with another person?
- How much time alone do you need? How much are you willing to allow your partner?
- Will you purposely set aside time for each other? If communication becomes difficult, will you go to a marriage counselor?
- What are your own and your partner's personal definitions of *intimacy*, *commitment*, and *responsibility*?

Discussing topics such as these is an important part of defining a couple's relationships.

We turn now to an even more serious issue—that of dating violence.

Dating Violence: A Serious Sign of Trouble

Sometimes we need to make decisions about continuing or ending a relationship that is characterized by physical violence or verbal abuse. Physical violence occurs in 20 percent to 40 percent of dating relationships (Luthra and Gidycz 2006)—most incidents of aggression involve pushing, grabbing, or slapping. Between 1 percent and 3 percent of college students have reported experiencing severe violence, such as beatings or assault with an object. Researchers are concerned that dating violence among teens and young adults is widespread and that many teens—as well as others—apparently minimize violence or expect it in certain situations (Hoffman 2009; Prospero 2006; Sears et al. 2006). Data from Add Health indicates that overall, nearly half (47 percent) of young adults ages 18 to 27 who were currently in relationships had experienced either unidirectional (21 percent as perpetrator and 25 percent as victim) or reciprocal (54 percent) intimate partner violence (Renner and Whitney 2012). Most interpersonal violence (IPV) is therefore reciprocal, and both genders can engage in physical aggression (Renner and Whitney 2012; Ryan, Weikel, and Sprechini 2008). However, by far the more serious injuries result from male violence against females (Johnson and Ferraro 2000). Furthermore, women are more inclined to "hit back" once a partner has precipitated the violence rather than to physically strike out first (Luthra and Gidycz 2006). Some risk factors associated with IPV, which is described in Chapter 12, for both men and women include childhood sexual abuse or neglect, low self-esteem, and perpetration of youth violence.

Dating violence typically begins with and is accompanied by verbal or psychological abuse (Lento 2006) and tends to occur over jealousy, with a refusal of sex, after illegal drug use or excessive alcohol consumption, or when arguing about drinking behavior (Cogan and Ballinger 2006; Ryan, Weikel, and Sprechini 2008).

Researchers have found it discouraging that about half of abusive dating relationships continue rather than being broken off (Few and Rosen 2005). Given that the economic and social constraints of marriage are not usually applicable to dating, researchers have wondered why violent dating relationships persist. Evidence suggests that having experienced domestic violence in one's family of origin—even chronic verbal abuse in the absence of physical violence—is significantly related to both being abusive and accepting abuse as normal (Cyr, McDuff, and Wright 2006; Tshann et al. 2009). A recent qualitative study of twenty-eight female undergraduates in abusive dating relationships found that some of these women felt "stuck" with their partner (Few and Rosen 2005). A majority had assumed a "caretaker identity" similar to martyring. As one explained: "I always was a rescuer in my family. I felt that I was rescuing him [boyfriend] and taking care of him. He never knew what it was like to have a good, positive home environment, so I was working hard to create that for him" (p. 272).

Others felt stuck because they wanted to be married, and their dating partner appeared to be their only prospect: "I think near the end, one of the reasons I was scared to let go was: 'Oh, my God, I'm twenty-seven.' I was worried that I was going to be like some lonely old maid" (p. 274).

What are some early indicators that a dating partner is likely to become violent? A date who is likely to become physically violent often exhibits one or more of the following characteristics:

- handles ordinary disagreements or disappointments with inappropriate anger or rage;
- has to struggle to retain self-control when some little thing triggers anger;

- goes into tirades;
- is quick to criticize or to be verbally mean;
- appears unduly jealous, restricting, and controlling; and
- has been violent in previous relationships.

Dating violence is never acceptable. Making conscious decisions about whether to marry a certain person raises the possibility of not marrying him or her. Letting go of a relationship can be painful. Next, we'll look at the possibility of breaking up.

The Possibility of Breaking Up

Returning to Reiss's wheel theory of love, we note that once people fall in love, they may not necessarily stay in love. Relationships may "keep turning," or they may slow down or reverse themselves. Sometimes love's reversal, and eventual breakup, is a good thing: "Perhaps the hardest part of a relationship is knowing when to salvage things and when not to" (Sternberg 1988a, p. 242). Being committed is not always noble, as in cases of relationships characterized by violence or consistent verbal abuse, for example (partner abuse is discussed in Chapter 13). Committed love will require some sacrifices over the course of time. However, as one therapist put it, "Love should not hurt" (Doble 2006). Breaking up is difficult, and men and women may remain emotionally invested in their partners for some time (Spielmann, MacDonald, and Tackett 2012). Partners must regain their sense of self after a breakup; failure to do so contributes to prolonged emotional distress (Mason et al. 2012).

According to the exchange perspective, dating couples choose either to stay committed or to break up by weighing the rewards of their relationship against its costs (Spielmann et al. 2012). As partners go through this process, they also consider how well their relationship matches an imagined ideal one. Partners also contemplate alternatives to the relationship, the investments they've made in it, and barriers to breaking up. (This perspective is also used when examining people's decisions about divorce, as discussed in Chapter 14.)

An individual is likely to remain committed when a partner's rewards are higher than the costs, when there are few desirable alternatives to the relationship, when the relationship comes close to one's ideal, when one has invested a great deal in the relationship, and when the barriers to breaking up are perceived as high. However, couples are more likely to break up when costs outweigh rewards, when there are desirable alternatives to the relationship, when one's relationship does not match one's ideal, when little has been invested in the relationship in comparison to rewards, and when there are fewer barriers to breaking up.

Even when a couple does not break up, recent research on dating couples has found support for the *principle of least interest* whereby the less-involved partner wields more power in and control over the continuation or ending of the relationship (Waller 1951). Some related research has found that the lesser-valued partner in a relationship is more inclined toward jealousy and yet also more willing to forgive the more highly valued partner for relationship indiscretions (Sidelinger and Booth-Butterfield 2007). However, high relationship satisfaction and stability are associated with equal emotional involvement (Crawford, Feng, and Fischer 2003; Sprecher, Schmeeckle, and Felmlee 2006).

Love is a process of discovery that involves continual exploration and sharing. Choosing a supportive partner is an important factor in developing a satisfying long-term relationship. Creating and maintaining that union involves recognizing that challenges will arise and committing to face and overcome them.

Nurturing Loving and Committed Relationships

Maintaining a satisfying long-term relationship is challenging, if only because two people, two imaginations, and two sets of needs are involved. Differences *will* arise because no two individuals have exactly the same points of view. Relationships can more often be permanently satisfying, counselors advise, when spouses learn to care for the "unvarnished" other, not a "splendid image" (Van den Haag 1974, p. 142). Gary Chapman (2007) the author of *The Five Love Languages*, maintains that individuals each have their own "love language" or styles of love that make them feel loved and secure in their relationships: (1) words of affirmation, (2) quality time, (3) receiving gifts, (4) acts of service, and (5) physical touch. He says that it less important that couples have similar love styles than for each to *understand* each other's particular love language and therefore strive to meet their partner's needs, as opposed to their own.

Choosing a supportive partner is an important factor in developing this kind of long-term love and relationship satisfaction.

Summary

- Loving is a caring, responsible, and sharing relationship involving deep feelings, and it is a commitment to intimacy. Genuine loving in our competitive society is possible and can be learned.
- Love should not be confused with martyring, manipulating, or limerence.
- People discover love; they don't simply find it. The term *discovering* implies a process—developing and maintaining a loving relationship require seeing the relationship as valuable, committing to mutual needs satisfaction and self-disclosure, engaging in supportive communication, and spending time together.
- Historically in Western cultures, marriages were often arranged in the marriage market as business deals. In some of the world that is less Westernized, some marriages are still arranged. Some immigrant groups in the United States (and other Westernized societies) today practice arranged or "assisted" marriage.
- Whether marriage partners are arranged, "assisted," or more freely chosen, social scientists typically view people as choosing marriage partners in a marriage market; armed with resources (personal and social characteristics), they bargain for the best deal they can get.
- Although gender roles and expectations are certainly changing, some aspects of the traditional marriage exchange remain such as a man's providing financial support in exchange for the woman bearing and raising children, her domestic services, and her sexual availability. Nevertheless, couples today are increasingly likely to value both partners' potential for financial contribution to the union.
- An important factor shaping partner choice is homogamy, the tendency of people to select others with whom they share certain social characteristics. Despite the trend toward declining homogamy, it is still a strong force, encouraged by geographical availability, social pressure, and feeling at home with people like ourselves.
- Committed relationships develop through building rapport and gradually negotiating the relationship as premarital, leading to marriage.
- Serial cohabiting (although not necessarily cohabiting before marriage only with one's future spouse) has been shown to increase the likelihood of divorce. The suggested reasons for this involve the selection hypothesis and the experience hypothesis.
- Couples today "meander toward marriage" and usually date a number of potential partners before settling on a permanent mate.
- Committed relationships require steady nurturing and expressions of love toward one's partner.

Questions for Review and Reflection

1. Sternberg offers the triangular theory of love (Figure 5.1). What are its components? Are they useful concepts in analyzing any love experience(s) you have had?
2. Discuss the reasons why marriages are likely to be homogamous. Why do you think homogamous unions are more stable than heterogamous ones? How might the stability of interracial or interethnic

relationships change as society becomes more tolerant of these?

3. If possible, talk to a few married couples you know who lived together before marrying and ask them how their cohabiting experience influenced their transition to marriage. How do their answers compare with the research findings presented in this chapter?
4. Think about your own romantic relationships. Were you engaged in dating or "nondating"?
5. This chapter lists topics that are important to discuss before and throughout one's marriage. Which do you think are the most important? Which do you think are the least important? Why?
6. **Policy Question**. What social policies, if any, currently exist to discourage couples who are experiencing dating violence from getting married? What new policies might be enacted to further discourage dating violence?

Key Terms

acquaintance rape (date rape) 127
arranged marriage 115
assortative mating 117
commitment 109
consummate love 110
cross-national marriages 115
dating scripts 125
endogamy 118
exogamy 122
experience hypothesis 121
free-choice culture 115
geographic availability 117
heterogamy 122
homogamy 117
interfaith marriages 124
interethnic marriages 122
interracial marriages 122
manipulating 113
marriage market 115
martyring 113
mutual dependency 129
need fulfillment 129
pool of eligibles 117
rape myths 127
rapport 128
selection hypothesis 121
self-revelation 129
sexual assault 127
sexual coercion 127
status exchange hypothesis 123
Sternberg's triangular theory of love 110
wheel of love 128

6

Living Alone, Cohabiting, Same-Sex Unions, and Other Intimate Relationships

Mode Ian O'Leary/Mode Images/Alamy

Learning Objectives

1. Describe current statistical trends regarding the proportion of singles in the United States.
2. Discuss reasons for the increasing proportion of singles in the United States today.
3. Recognize various living arrangements for singles.
4. Compare and contrast what cohabitation means in different cultures.
5. Understand how financial concerns affect decisions about cohabitation.
6. Analyze the general findings and concerns regarding outcomes for children in cohabiting families.
7. Describe the public debate between proponents and opponents of legal same-sex marriage.
8. Discuss the general findings regarding outcomes for children raised in same-sex couple households.

Danaher, Luke, Rick, and Shyaporn are four heterosexual, 30-something New York City guys who have lived together for eighteen years and plan to go on doing so. When they decide as a group that they need to move, the four look for housing together. Explains social scientist Bella DePaulo, author of the book *Singled Out* (2006), "there are so many variations on how to live" these days, and many adults now follow this "friendship model," according to which committed roommate arrangements become family, or family-like (Howard 2012). In the words of family sociologist Judith Stacey, these men are "unhitched," because, for one thing, living among friends means that "the vagaries of sexual attraction don't disrupt your security and stability" (Stacey, in Howard 2012; Stacey 2011). Would you call these four men a family?

When asked, "How important is family in your life?" 98 percent of Americans say it's "the most important" (76 percent) or "one of the most important" (22 percent) elements (Pew Research Center 2010a, p. 41). Yet we saw in Chapter 1 that today's *postmodern* family is characterized by a diversity of family forms. We will examine some of these family forms in this chapter as we look at various living arrangements of singles: living alone or with one's parents, living communally or in groups, cohabiting, and forming same-sex unions.

Many college students think of "being single" as not being in a romantic relationship. By this way of thinking, a person in a dating relationship or cohabiting would not be single. To the U.S. Census Bureau, however, *single* means *unmarried*. In this chapter, we will examine what social scientists know about the large and growing number of singles. In the process, we'll find that the distinction between being single and being married is not as sharp today as it once was.

For instance, a significant minority of same-sex couples is now legally married. We include same-sex unions in this chapter, however, because at this point in time the majority of same-sex couples are *not* legally married. According to the U.S. Census Bureau, both members of a same-sex couple are single and living in a same-sex cohabiting relationship. Then, too, the distinction between *family* and *household* has become increasingly blurred. Most people who are legally single are embedded in families, even when not sharing the same household with them. To begin exploring these ideas, we'll examine some reasons for the increasing proportion of unmarrieds in our society.

Reasons for More Unmarrieds

Figure 6.1 shows the proportions of never-married, divorced, and widowed individuals by various race/ethnic categories. These percentages are considerably higher than in past decades. In 1970, fewer than 28 percent of U.S. adults were single. Today that number is about 44 percent (Saluter and Lugaila 1998, Figure 1; U.S. Census Bureau 2012c, Table 56). Much of this change is the result of a growing proportion of widowed elderly. However, the high proportion of singles also results from divorce rates that are falling but are still high. Divorce rates generally rose beginning in the early twentieth century and, especially since the 1960s, leveled off and began to fall by about 1980; they have remained relatively high (Pew Research Center 2010a). The high proportion of singles today also results from people postponing marriage until older and older and from a dramatically escalating cohabitation rate (Cherlin 2010). Cohabiters can be never married, divorced, or widowed.

By 2011, the median age at first marriage was 26.5 for women and 28.7 for men—the highest recorded averages since the 1890 census. These figures compare with 20.8 for women and 23.2 for men in 1970 (U.S. Census Bureau 2011b, Table MS-2). As a consequence of postponing marriage, the proportion of singles in their twenties has risen dramatically. In 1970, 36 percent of women age 20 through 24 were never married; by 2010, that figure had more than doubled to 79 percent. The ranks of never-married men age 20 through 24 have increased from 55 percent in 1970 to 89 percent in 2010 (Saluter and Lugaila 1998; U.S. Census Bureau 2012c, Table 57). Several social factors—demographic, economic, technological, and cultural—encourage Americans to postpone marriage, not marry at all, or divorce rather than stay married.

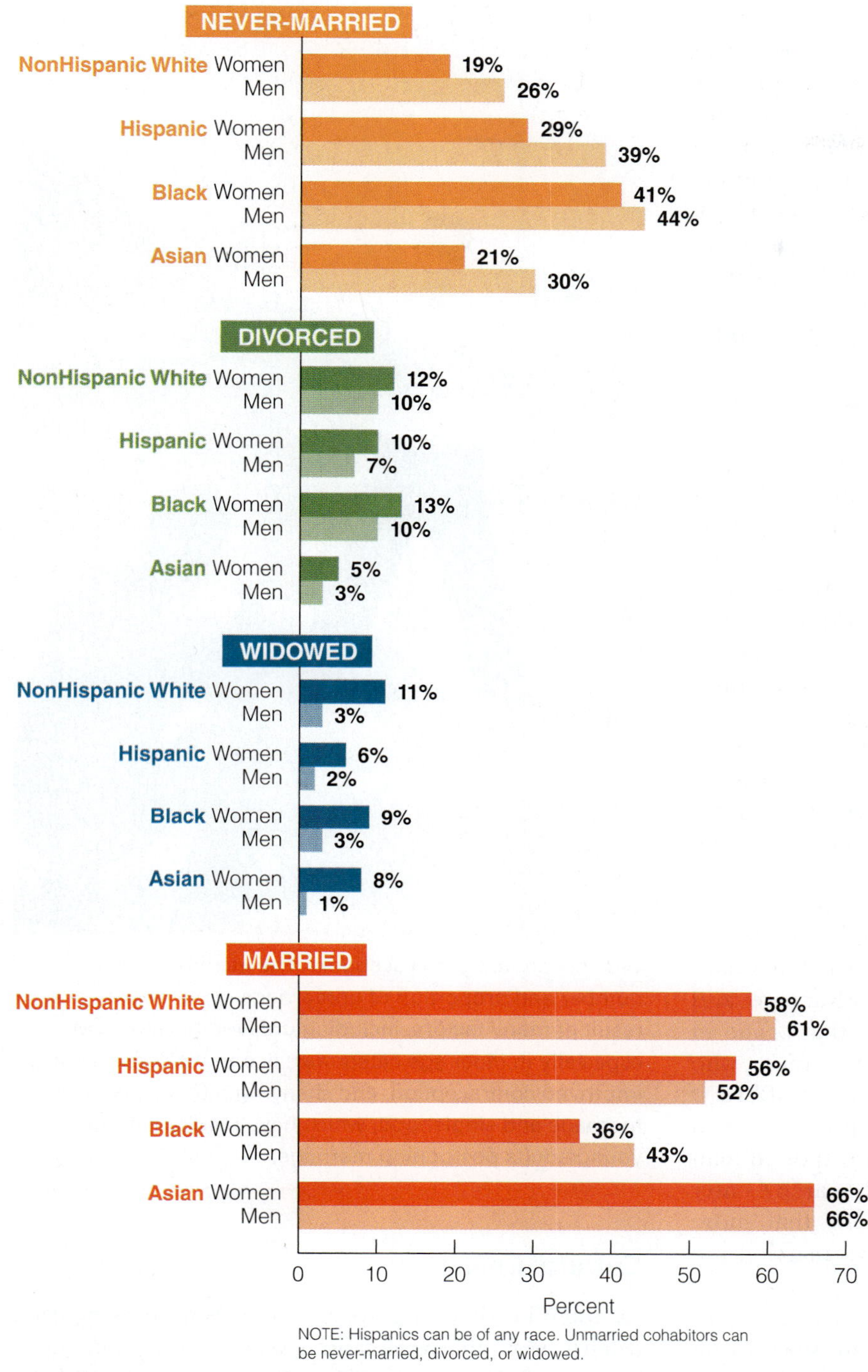

FIGURE 6.1 Marital status of the U.S. population, age 18 and over, 2010, by race/ethnicity.

Source: U.S. Census Bureau 2012c, Table 56.

Demographic, Economic, and Technological Changes

One reason for the growing proportion of singles is *demographic,* or related to population numbers. A high rate of heterosexual marriage presumes that there are matching numbers of marriage-age males and females in the population. Therefore, the **sex ratio**—the number of men to women in a given society or subgroup—influences marital options and singlehood. The sex ratio is expressed in one number: the number of males for every 100 females. Thus, a sex ratio of 105 means that there are 105 men for every 100 women in a given population.

Throughout the nineteenth and early twentieth centuries, the United States had more men than women, mainly because more men than women migrated to this country and, to a lesser extent, because a considerable number of young women died in childbirth. Today, this situation is reversed due to changes in immigration patterns and advances in women's health. In 1910, there were nearly 106 men for every 100 women. The sex ratio was 100, or "even," shortly after World War II ended, in about 1948. In 2010, there were about 97 men for every 100 women (U.S. Census Bureau 2012c, Table 7).

Specialized sex ratios may be calculated—for example, the sex ratio for specific race/ethnic categories at various ages. Beginning with middle age, there are increasingly fewer men than women in every race/ethnic category. Sex ratios differ somewhat for various race/ethnic categories, however. Due to higher death rates at younger ages, the sex ratio is lower for blacks and Native Americans than for Hispanics or nonHispanic whites.

Furthermore, social scientists consider how desired characteristics in a potential mate impact *effective* sex ratios. For instance, relatively high incarceration rates and poor employment prospects particularly affect low-income African American males' odds of marrying (Banks 2011; Chambers and Kravitz 2011; Roberts 2012). When more women than men in a group or category are highly educated or have steady jobs, the effective ratio of (desirable) males to females declines (Bolick 2011b; Dixon 2009; King and Allen 2009).

In addition to demographics, *economic* factors have increased the proportion of nonmarrieds. For one thing, expanded educational and career options for college-educated women over the past several decades have encouraged many of them to postpone marriage (Arnett 2004; Bolick 2011b). Then, too, many middle-aged and older career or divorced women tend to look on marriage skeptically, viewing it as a bad bargain once they have gained financial and sexual independence (Levaro 2009).

Moreover, in contrast to their more casual attitudes about cohabiting, people view marriage as a status that needs to be financially affordable (Mather 2011; Trail and Karney 2012).

Although it might appear at first glance that getting married should not be more expensive than living together, couples tend to have a number of specific concerns. They feel they should be able to afford the trappings of a middle-class lifestyle (e.g., the ability to purchase a home) and a reasonable wedding (rather than go downtown to the courthouse), achieve financial stability, be debt free, and demonstrate fiscal responsibility (Smock and Greenland 2010, p. 582).

Although young adults' debt burden has declined since the onset of the recession in 2007, individual Americans between ages 18 and 35 owe a median average of $15,473; 15 percent of the total debt in that age group is the result of student loans (Fry 2013).

The fact that many men's earning potential has declined relative to women's may make marriage less attractive to both genders (Bolick 2011b; Raley and Bratter 2004). Growing economic disadvantage and uncertainty accompanying the recession that began in 2007 have made marriage less available to many who might want to marry but believe they can't afford it (Gibson-Davis 2009; Mather 2011). As many as one-fifth of 18- to 34-year-olds say that they have postponed marriage due to the recent recession (Pew Research Center 2012b). Difficulty in finding jobs means that more young adults who would otherwise have married are currently cohabiting (Kreider 2010).

In addition to economics, *technological* changes over the past sixty-five years have affected the proportion of singles. Beginning with the introduction of the birth-control pill in the 1960s, improved contraception has contributed to the decision to delay or forego marriage. With effective contraception, sexual relationships outside marriage and without great risk of unwanted pregnancy became possible (Gaughan 2002; Coontz 2005b). Moreover, as described in Chapter 1, reproductive technologies such as artificial insemination offer the possibility for planned pregnancy to unpartnered women as well as to same-sex couples. In addition to these structural reasons for the increasing proportion of nonmarrieds, cultural changes have played a part.

Marcus Mok/Asia Images

Will this photo be on Facebook? The increase in the number and proportion of unmarrieds in our society is a result of many factors, including unfavorable sex ratios, especially in older age categories; economic constrictions; improved contraception; and changing attitudes toward marriage and singlehood, which have resulted in more young adults postponing marriage.

Cultural Changes

As noted in Chapter 2, social scientists have recognized a fairly new life cycle stage called *emerging adulthood*: Young people today spend more time in higher education or exploring options regarding work, career, and family making than in the past (Arnett 2000; Furstenberg 2008). Although all age groups have helped to increase the proportion of singles, emerging adulthood accounts for many unmarrieds today.

Several other cultural changes over the past few decades also account for the growing proportion of nonmarrieds. First, attitudes toward nonmarital sex have changed dramatically over past decades. With about 60 percent of adults of all ages approving, sexual intercourse outside marriage has become fairly widely accepted (Gallup Poll 2012b). Currently, "hooking up"—discussed in Chapter 4 and called "recreational sex" in the 1960s—has gained attention (Rosin 2012a; Wilson 2009). More recently, social scientists have recognized what they call the "stayover," according to which college-educated young adults spend two or three nights a week at their partner's home on a long-term basis but do not actually move in together (Jamison and Ganong 2011).

With proscriptions against nonmarital sex relaxed, and as American culture gives greater weight to personal autonomy, many find that—at least "for now"—singlehood is more desirable than marriage (Furstenberg 2008; Meier and Allen 2009). As one young man explained:

> One day I was at work and my friend called me up from Florida and said, "What are you doing?" I'm like, "Just working," and he said, "Can you come down?"... So I took a week off all of a sudden and went down to Florida. And I know I'd never be able to do that if I was married. (in Arnett 2004, p. 101)

A young woman evidences a similar attitude:

> I hope to be married by the time I'm 30. I mean, I don't see it being any time before that.... Just go out and enjoy life and then settle down, and you'll know you've done everything possible that you wanted to do, and you won't regret getting married. (in Arnett 2004, p. 103)

And from a 25-year-old single mother:

> I think it's just that I'm not out there really looking for someone right now. I'm self-centered... right now. I want to get through school, and I want to get employment.... And I want to get things going for myself before I do anything.... It's like just for me and my son. I got a son to raise. (in Manning, Trella, Lyons, and DuToit 2010, p. 94)

If you're younger than 50, it may be hard to believe, but being single used to be considered deviant, not the acceptable option that it is today. During the 1950s, people, including social scientists, tended to characterize the never-married as selfish, neurotic, or unattractive. The divorced were also stigmatized. Probably it goes without saying that among the vast majority of Americans these views have changed. In 2012, 67 percent of Americans found divorce to be morally acceptable (Gallup Poll 2012b).

Then, too, getting married is no longer just about the only way to gain adult status. Before about 1940, the most legitimate reason for leaving home, at least for women, was to get married. Today, 45 percent of young men and 39 percent of young women say that they first left home for other reasons, often to attend college or "to gain independence." The nationwide General Social Survey (GSS), conducted by the National Opinion Research Center (NORC), found that 72 percent of people see completing school as extremely important to becoming an adult. Having full-time employment and being able to support a family were mentioned as extremely important by 60 percent of those surveyed. No longer living with one's parents was mentioned by 29 percent. Getting married was mentioned as extremely important to becoming an adult by just 19 percent of those surveyed (General Social Survey 2012). Marriage has lost its monopoly as the way to claim adulthood (McLanahan 2010).

Moreover, cohabitation has emerged as a socially accepted alternative to marriage. Fewer than half (42 percent) of U.S. adults today think that it is "morally wrong" to have a baby outside of marriage; 42 percent believe that an unmarried couple that has lived together for one year is just as committed as a couple that has been married for one year (Gallup Poll 2012b). Beliefs such as these are considerably less likely among recent immigrants and members of more conservative religions. Nonetheless, young adults generally are influenced by their parents' attitudes toward marriage—and young adults experience considerably greater independence and less parental pressure to marry than in the past (Arnett 2004; McLanahan 2010; Willoughby et al. 2012).

Finally, the diminished permanence of marriage may render it less desirable now (Cherlin 2009a). Marriage has become less strongly defined as permanent, and some singles fear a possible future divorce of their own (Miller, Sassler, and Kusi-Appouh 2011). In one recent qualitative study, some women said they were reluctant to marry, not because of any stigma associated with divorce but because they feared the divorce process itself: "If I get married [and then divorce], I don't want to have to go through all that stuff" (Manning, Trella, Lyons, and DuToit 2010, p. 94). Fear such as this may reduce the likelihood of marriage (Waller and Peters 2008). If marriage is losing its permanent status, then

> [i]ndividuals, as a result, have less faith that a successful marriage is possible, and they transfer support for marriage into support for other coupling arrangements, such as cohabitation—arrangements that are easier to dissolve if (and when) problems arise. (Willetts 2006, p. 125)

To summarize, it appears that much of the increase in singlehood results from the following:

1. low sex ratios, particularly in certain regions and among specific age and race/ethnic groups;
2. increasing educational and economic options for some, coupled with growing financial disadvantage for others;
3. technological changes regarding pregnancy;
4. changing cultural attitudes toward marriage and singlehood—heightened acceptance of premarital sex, personal autonomy, singlehood, and cohabitation;

5. marriage having lost its monopoly as a way to claim adulthood; and
6. permanence no longer an anticipated benefit of marriage.

We can apply the exchange theoretical perspective (see Chapter 2) to this issue of fewer compelling reasons to marry. Overall, as people weigh the costs against the benefits of being married, marriage now seems to offer fewer benefits relative to being single—although as Chapter 7 points out, there *are* significant benefits to marriage. If fewer Americans are married today, what are singles' various living arrangements?

Singles: Their Various Living Arrangements

Singles make a variety of choices about how to live. Some live alone, others with parents, and still others in groups or communally. Some unmarrieds cohabit with partners of the same or the opposite sex. This section explores these living arrangements. Single parents are addressed at length in Chapter 9 as well as elsewhere throughout this text.

Living Alone

The number of one-person households has increased dramatically over past decades. Individuals living alone now make up 27 percent of U.S. households—up from just 8 percent in 1940 (U.S. Census Bureau 1989, Table 61; U.S. Census Bureau 2012c, Table 59). The likelihood of living alone increases with age for all race/ethnic groups and is markedly higher for older women than for older men (U.S. Census Bureau 2012c, Table 58).

Asians and Hispanics of all ages are less likely to live alone than are blacks or nonHispanic whites. As you can see in Figure 6.1, Asian Americans are most likely to be married and least likely to be divorced. Significantly less likely to be married than other race/ethnic groups, blacks of all social classes are more likely than others to be living by themselves, particularly in older age groups. African Americans are most likely to be never married, followed by Hispanics (see Figure 6.1). More collectivist Asian and Hispanic cultures discourage living alone. Of course, some who live alone are actually involved in committed relationships.

Living Apart Together

An emerging lifestyle choice is **living apart together (LAT)**. Here a couple is committed to a long-term relationship, but each partner also maintains a separate dwelling (DePaulo 2012). The number of these relationships is difficult to ascertain because they are hard to define (Cherlin 2010) and because the U.S. Census Bureau does not measure them. However, according to David Popenoe, codirector of the National Marriage Project at Rutgers University, LAT is clearly an emerging trend in the United States. LAT is at least partly motivated by a desire to retain autonomy. As one LAT woman said, "I like my own life, my own identity and want to keep it. I like having the things I love around me." As one man put it, "I am as devoted as any husband to her, ... but I like my alone time and being around my stuff, not [hers]" (in Brooke 2006).

For an older adult, living apart together "allows for unencumbered contact with adult children from previous relationships while protecting their inheritance and offering freedom from caregiving as a prescribed duty... Separate homes also allow a tangible line of demarcation in terms of gender equity and the distribution of household labor" (Levaro 2009, p. F10). Then, too, some people in LAT relationships are already burdened with enough actual or potential family caregiving—for children still at home, parents, or a disabled sibling, for instance—that they hesitate to take on more caregiving demands that could compromise the ones they have already (Duncan and Phillips 2011). Some young adults in LAT relationships reside with their parents (S. Smith 2006).

Living with Parents

Between 2000 and 2012, the proportion of adults ages 25 to 34 living with their parents rose dramatically—from about one in ten to about one in three (Jacobsen and Mather 2011, Figure 10; Parker 2012). Table 6.1 shows that the percentage of young adults living at home has increased since 1960. In 2011, 59 percent of men and 50 percent of women ages 18 through 24 lived

Table 6.1 Percent Living with their Parents, by Age, Sex, and Year

	Ages 18–24	Ages 25–34
Men		
1960	52	11
1995	58	15
2002	55	14
2008	56	15
2011	59	19
Women		
1960	35	7
1995	47	8
2002	46	8
2008	48	10
2011	50	10

Source: U.S. Census Bureau 2012b Table AD-1.

with their parents. For men and women ages 25 to 34, the proportions are 19 percent and 10 percent, respectively (U.S. Census Bureau 2012b, Table AD-1). Some adults who live with their parents have never moved out, but others—called *boomerangers*—have left home and then returned.

Back in 1940, the proportion of adults under age 30 living with their parents was quite high. Sociologists Paul Glick and Sung-Ling Lin suggest why:

> The economic depression of the 1930s had made it difficult for young men and women to obtain employment on a regular basis, and this must have discouraged many of them from establishing new homes. Also, the birthrate had been low for several years; this means that fewer homes were crowded with numerous young children, and that left more space for young adult sons and daughters to occupy. (Glick and Lin 1986, p. 108)

These same reasons apply to many young people today. Looking at Table 6.1, you will note that the proportion of young adults living with parents declined slightly between 1995 and 2002. Because median household income rose during that time period, we can argue that a better economy helped to reduce the percentage of adults living with their parents. As discussed in Chapter 1, the economic recession that began in 2007 has led to an increase in the number of boomerangers (Jacobsen and Mather 2011). Pew Research results, shown in Chapter 1's Figure 1.4, indicate that 24 percent of the 18- to 34-year-olds polled said they had moved back in with their parents in response to the recent recession (Pew Research Center 2012b). Moreover, even before that, housing in urban areas was too expensive for many singles to maintain their own apartments.

Several recent books argue that today's young adults have been coddled and spoiled since they were toddlers and have not developed the necessary skills to transition to responsible adulthood. Typical of these books is Sally Koslow's *Slouching Toward Adulthood* (2012). Koslow is the mother of two boomerang college-graduate sons, whom—along with the rest of their generation—she dubs "adultescents." On the other hand, other social scientists have argued that young people choosing a slower path to adulthood is generally a functional choice because doing so means more time to prepare for career and family responsibilities (Settersten and Ray 2010).

We need to note here that living with parents can occur in a variety of circumstances. Overall our society highly vales individuality and independence, but some ethnic groups, such as the Hmong, *expect* single women to reside with their parents until marriage. Unmarried women who have babies, especially those who became mothers in their teens, may be living with parents. Formerly married young men and women sometimes return to their parental home after divorce.

Chapter 2 introduces the idea that family members create *boundaries* defining who is in and who is not in the family. Accordingly, an interesting study found that parents are less likely to welcome into their home an adult child who is accompanied by a cohabiting partner (Seltzer, Lau, and Bianchi 2012). Meanwhile, many adult children who live with their parents contribute what they can, financially and otherwise, to the family household (McLanahan 2010). Parents often appreciate the help and companionship of their live-in adult children (Parker 2012; Straus 2009).

According to sociologist Barbara Risman, parents and emerging adults today have more in common in attitudes and values than did baby boomers and their parents (in Jayson 2006a). In the words of two other experts on this issue, "Our research shows that the closer bonds between young adults and their parents should be celebrated, and do not necessarily compromise the independence of the next generation" (Fingerman and Furstenberg 2012). Today's parents may expect to serve as "collaborators in their children's transition to adulthood, acting as scaffolding systems to help young people reach their goals and as safety nets to catch them before they fall too far" (Swartz et al. 2011, p. 427). On the other hand, conflict with parents can precipitate the decision to move out and perhaps to take up residence with a romantic partner (Sassler 2004). Just as economic considerations, desire for emotional support, or the need for help with child raising may lead young singles to choose to live with parents, similar pressures may encourage singles to fashion group or communal living arrangements.

Group or Communal Living

Groups of single adults and perhaps children may live together. Often these are simple roommate arrangements. But some group houses purposefully share aspects of life in common. **Communes**—that is, situations or places characterized by group living—have existed in American society throughout its history and have widely varied in their structure and family arrangements.

In some communes, such as the nineteenth-century American Shakers and the Oneida colony and the traditional twentieth-century Israeli kibbutzim, economic resources have been shared (Kephart 1971; Kern 1981; Spiro 1956). Work is organized by the commune, and commune members are fed, housed, and clothed by the community. Other communes may have some private property. Sexual arrangements also vary among communes, ranging from celibacy to monogamous couples (as in the kibbutzim and some communes in the United States) to the open sexual sharing found in both the Oneida colony and some modern American groups. Children may be under the control and supervision of a parent, or they may be raised more communally, with biological relationships deemphasized and responsibility for discipline and care vested in the entire community.

Living communally has declined in the United States since its highly visible status in the 1960s, when many communes were established as ideological retreats from what their founders saw as the misguided American life characteristic of the 1950s. However, some communes that were established then still exist. Furthermore, small-scale and nonideological versions of group living have more recently surfaced (Jacobs 2006). The recession that began in 2007 prompted 12 percent of single adults between ages 18 and 34 to acquire a roommate, and another 2 percent took in a boarder (Wang and Morin 2009).

Matt Kramer

Cohousing started in Denmark and spread to the United States in the early 1980s. Residents own their own homes, with residences clustered closely to leave open space, which is community owned. Cohousing complexes, which typically combine private areas with communal kitchens—and, often with community gardens—offer alternative living arrangements and can be a way to cope with some of the problems of aging, unattached singlehood, or single parenthood.

Communal living, either in single houses or in cohousing complexes that combine private areas with communal kitchens and "family rooms," may be one way to cope with some of the problems of aging, unattached singlehood, or single parenthood ("Cohousing in Today's Real Estate Market" 2006; Wildman 2011). In a small but growing number of cohousing complexes, people of diverse races, ethnicities, and ages choose to reside together, sharing some meals and recreational activities. Communal living is designed to provide enhanced opportunities for social support and companionship. More commonly, financial considerations and the desire for companionship encourage romantically involved singles to share households. We turn now to a discussion of "living together," or cohabitation.

Cohabitation and Family Life

Cohabitation, or nonmarrieds living together and sexually available to each other, is "widely viewed as one of the most important changes in family life in the past forty years, dramatically altering the marital life course by offering a prelude to or a replacement for marriage" (S. Smith 2006, p. 7). Not only in the United States but also in other industrialized nations, "living together" has dramatically increased. In this country, the cohabitation trend spread widely in the 1960s, took off sharply in the 1970s, and has risen steadily ever since, as Figure 6.2 illustrates. Today, about 7.6 million U.S. heterosexual couples cohabit (U.S. Census Bureau 2012e, Table UC3). This number may be an undercount, because cohabiters do not necessarily move into separate housing. Instead, they may live together in a parental home or reside with roommates and therefore would not be included in a census count (Manning and Smock 2005; Pilkauskas 2012).

Cohabitation: The Numbers

In the early twentieth century, living together outside marriage was illegal in every state and relatively uncommon. Due to laws enacted at the turn of the twentieth century, unmarried cohabitation has remained illegal in a handful of states, although the laws are seldom enforced. The American Civil Liberties Union has sued to overturn anticohabitation laws in states where they still exist (Alternatives to Marriage Project 2012; Jones 2006). In Florida, one of the states where cohabitation remains illegal, a 2011 push to abandon the old law met with opposition from social conservatives, who declared that they were "not ready to give up on monogamy and a cultural statement that marriage still matters" (M. Hartman 2011; see also Haughney 2011).

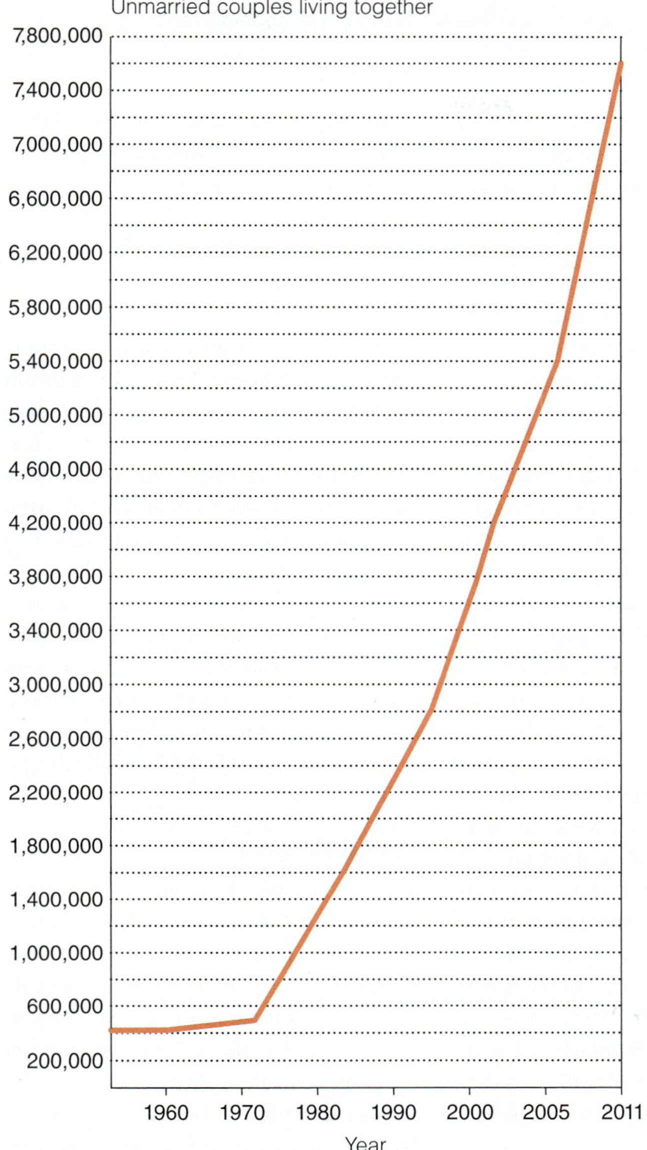

FIGURE 6.2 Unmarried heterosexual couples living together in the United States, 1960–2011.

Source: Edwards 2009; Glick and Norton 1979; Schneider 2003; Simmons and O'Connell 2003; U.S. Census Bureau 2012e, Table UC3.

Nevertheless, the number of cohabitating adults has increased more than ten times since 1970—and by 40 percent just since 2000 (U.S. Census Bureau 2012b, Table UC-1; Lofquist et al. 2012, Table 2). By 2008, an estimated 58 percent of 30- to 44-year-olds had lived with an opposite sex partner at some time in their lives—up from 33 percent in 1987 (Fry and Cohn 2011). Approximately 7 percent of adults in this age group are cohabiting at any one time, and opposite-sex cohabiting households comprise between 5 percent and 6 percent of U.S. households (U.S. Census Bureau 2012e, Table S1101). Associated with the recession that began in 2007, cohabitation rates increased more than usual between 2009 and 2010 as people began to postpone marriage until they could better afford it (Kreider 2010).

Six percent of households with children are headed by cohabiting partners (Kreider and Elliott 2009b, pp. 6, 8). In 2011, nearly 40 percent of cohabiting couples lived with children under 18—either their own or those from a previous relationship or marriage (U.S. Census Bureau 2011c, Table UC3). Unmarried-couple families are only 5 percent of households at any one time, but more than 50 percent of first marriages are preceded by cohabitation (Lofquist et al. 2012).

The incidence of cohabitation is expected to escalate further as new generations grow up in cohabiting families or in families with permissive attitudes about cohabitation; hence these new generations may be socialized to take cohabiting for granted (Seltzer 2004; Willoughby and Carroll 2012). As respondent Alan, age 27, told an interviewer, "My great grandma said you got to test drive the car before you buy it. So cohabitation is a good way to really get to know someone" (Manning, Cohen, and Smock 2011, p. 138).

Cohabitation and Age

Approximately three-quarters of cohabiters are younger than 45 (U.S. Census Bureau 2008, Table UC4). Nevertheless, the proportion of middle-aged cohabiters has increased over the past two decades. Cohabitation among individuals age 50 and older more than doubled between 2000 and 2010—from 1.2 million to 2.75 million. Middle-age and older cohabiters are generally in relationships of longer duration and—as you might expect—are more likely to have been divorced (Brown, Bulanda, and Lee 2012). Older singles express less desire to marry (or remarry) than do younger singles (Mahay and Lewin 2007; Swarns 2012a). Older couples have found that living together absent legal marriage may be economically advantageous, because they can retain some financial benefits that are contingent on not being married (Ebeling 2007). "When it comes to marriage and money, widows often say straight out that they are unwilling to risk the financial security of a departed husband's pension." Then, too, many cohabiting older couples "are aware of their children's fear for their inheritance and for that reason may... refrain from marriage" (Levaro 2009, p. F10). "Even if your will leaves everything to your kids, a second spouse can claim what's known as an 'elective share' of your estate—typically a third.... After you're married, it takes more lawyering to disinherit a spouse" (Ebeling 2007).

Characteristics of Cohabiters

Comparing marrieds to cohabiters, analyses of data from several sources show that, as a category, cohabiters are less educated, earn less income, are less likely

A Closer Look at Diversity

The Different Meanings of Cohabitation for Various Race/Ethnic Groups

You probably have an idea of what cohabitation means to you, and you may assume that living together signifies the same thing to all of us. But researchers who have analyzed national survey data to study cohabitation among various race/ethnic groups have uncovered interesting differences (Manning 2004; Ojeda 2011). Cohabiting means different things in different cultures and ethnic groups (Fomby and Estacion 2011; Landale, Schoen, and Daniels 2009; Liu and Reczek 2012).

For instance, Puerto Rican women have a long history of **consensual marriages** (heterosexual, conjugal unions that have not gone through a legal marriage ceremony). The tradition of consensual marriages probably began among Puerto Ricans because of a lack of economic resources necessary for marriage licenses and weddings: "Although nonmarital unions were never considered the cultural ideal, they were recognized as a form of marriage and they typically produced children" (Manning and Landale 1996, p. 65). Therefore, for Puerto Ricans in the continental United States, cohabitation symbolizes a committed union much like marriage, and they don't necessarily feel the need to marry legally should the woman become pregnant because they have already defined themselves as (consensually) married. Compared to Mexican Americans, Puerto Ricans are more likely to agree that "[i]t's all right for an unmarried couple to live together if they have no plans to marry" (Oropesa 1996).

Meanwhile, compared to Puerto Ricans (and to nonHispanic whites), Mexican Americans were more likely to agree that "[i]t is better to get married than go through life being single." Mexican Americans weigh marriage very positively, and a couple that has plans to marry significantly increases the likelihood that Mexican Americans will approve of cohabitation. These findings are especially strong for foreign-born Mexican Americans. However, economic barriers to marriage apparently induce many low-income Mexican Americans to continue to cohabit, rather than to marry, and to raise several children in cohabiting families (Lloyd 2006; Wildsmith and Raley 2006). Furthermore, as a result of their exposure to general cultural values and attitudes in the United States, we can expect second- and third-generation Mexican Americans to embrace cohabitation in increasing proportions (Fomby and Estacion 2011; Oropesa and Landale 2004). Similarly, African Americans consistently tell pollsters that they value marriage highly. However,

> cohabitation and marriage have very different meanings for Whites and Blacks.... Whites are more likely to see cohabitation as a trial marriage.... In contrast, cohabitation is more prevalent and is perceived as an alternative to marriage for Blacks.... (Liu and Reczek 2012)

For Puerto Ricans, "living together" is likely to symbolize a committed union virtually equal to marriage. Among Mexican Americans, cohabitation is less valued than marriage but allowable if the couple plans to marry—although economic barriers to marriage can thwart those plans (Fomby and Estacion 2011). Our diverse American society encompasses many ethnic groups with family norms that sometimes differ from one another. Together with structural and economic factors, cultural meaning systems play a part in how people define cohabitation.

Critical Thinking

The U.S. Census Bureau groups Puerto Ricans and Mexican Americans, along with other Hispanics, into one ethnic category. What does the previously presented information tell you about diversity *within* the ethnic category of Hispanic? Would you suppose that the various groups of Asians, such as the Hmong or Asian Indians, who are also categorized together, differ as well? What about African Americans, or whites?

to own their own homes, less likely to be nonHispanic white, likely to have experienced more transitions in living arrangements as children, and likely to have relatively permissive attitudes toward sex (Fry and Cohn 2011; S. Ryan et al. 2009; Schoen et al. 2009). Research shows that some of these trends begin in adolescence: Cohabiting women were found to have had lower academic achievement and fewer parental resources in adolescence (Amato et al. 2008; Willoughby and Carroll 2012). Especially in Southern states, relatively conservative religious affiliation is negatively related to cohabitation, and the highest cohabitation rate is among those without any religious affiliation (Eggebeen and Dew 2009; Gault-Sherman 2012). Nevertheless, people from all social classes, educational categories, and religious persuasions have cohabited.

Cohabitation As an Acceptable Living Arrangement

As "A Closer Look at Diversity: The Different Meanings of Cohabitation for Various Race/Ethnic Groups" suggests, cohabiting means different things to different people. Generally, however, cohabitation "is very much a family status, but one in which the levels of certainty about commitment are less than in marriage" (Bumpass, Sweet, and Cherlin 1991, p. 913). Put more strongly, cohabitation "does not institutionalize commitment in a way that is easily understood and honored by romantic partners and their friends and family" (Wilcox et al. 2011b). British demographer Kathleen Kiernan (2002) has described a society-wide, four-stage process through which cohabitation becomes a socially acceptable living arrangement that is equal in status to marriage.

In the first stage, the vast majority of heterosexuals marry without living together first. We saw this stage in the United States until the late 1970s. In the second stage, more people live together but mainly as a form of courtship before marriage, and almost all of them marry with pregnancy. Today, some cohabitants consider their lifestyle a means of courtship, or *premarital cohabitation* (Gold 2012; Willoughby, Carroll, and Busby 2011). As one young woman explained, "We wanted to try it out and see how we got along, because I've had so many long-term relationships. I just wanted to make sure we were compatible. And he's been married before, and he felt the same way" (in Arnett 2004, p. 108). Living together as a means of selecting a committed life partner is explored in Chapter 5.

In the third stage, cohabiting becomes a socially acceptable alternative to marriage. A couple no longer feels it necessary to marry with pregnancy or childbirth, and people routinely take an unmarried partner to work or family get-togethers. Nevertheless, in stage three, legal and social differences remain between marriage and "just" living together. In the fourth stage, cohabitation and marriage become virtually indistinguishable, both socially and legally. In this stage, the numbers of married and cohabiting couples are about equal, and a cohabiting couple may have several children (Kiernan 2002).

Social historian Stephanie Coontz (2005b) characterized the United States as "transitioning from stage two to stage three at the end of the twentieth century" (p. 272). Perhaps in some large metropolitan areas of this country—where "cohabitation has replaced marriage as a first union experience for a growing majority of young adults" (Lichter and Carmalt 2009, p. F12)—cohabitation has fully reached stage three. The Netherlands, Sweden, and Norway are societies in stage four (Perelli-Harris and Gassen 2012). Interestingly, in societies where cohabitation is common and virtually institutionalized, cohabiters' relationships are more like those of marrieds than they are in the United States (Hansen, Moum, and Shapiro 2007). However,

On average, cohabiting relationships are relatively short-term. Half last less than one year, because the couple either breaks up or marries. Cohabiting men with intentions to marry their partner are likely to do more housework than other cohabiting males. Cohabiting women in a relationship where both partners definitely intend to marry do less housework than do other cohabiting women (Kuperberg 2012). Why do you think this would be?

many European countries remain in stage three, as evidenced by the fact that a significant majority of cohabiting women marry after their first birth, "suggesting that marriage remains the predominant institution for raising children" (Perelli-Harris et al. 2012). Will the United States ever get to stage four? Coontz is skeptical, because "people [in the United States] still place much more importance on getting married than [many Europeans] do" (2005b, p. 272).

Cohabitation As an Alternative to Both Unattached Singlehood and Marriage

In a qualitative study of 120 heterosexual cohabiters, nearly two-thirds ranked the reason "I wanted to spend more time with my partner" as first for moving in together (Rhoades, Stanley, and Markman 2009). Some couples begin to live together shortly after their first date; others wait for months or longer (Sassler 2004). Accounts of how cohabitation begins suggest that cohabiting does not always result from a well-considered decision. As one 23-year-old woman who had been living with her parents explained,

> I was looking for my own apartment at this time.... He was like, "Why don't you just move in with me?" I was like, "Let's give it some time," or whatever. So I dated him for like a month and then finally all my stuff ended up in his house. (in Sassler 2004, p. 496)

Some cohabitants view living together as an alternative to dating or unattached singlehood. As one respondent told her interviewer:

> Um, he had came over, and we had talked and... he had spent the night and then from then on he had stayed the night, so basically... he just honestly never went home. I guess he had just got out of a relationship, the person he was living with before, he was staying with an uncle and then once we met, it was like love at first sight or whatever and um, he never went home, he stayed with me. (in Manning and Smock 2005, p. 995)

Psychologist Jeffrey Arnett (2004) has dubbed those who live together as an alternative to being single *uncommitted cohabiters*.

Other cohabiters view living together as an alternative to marriage (Cherlin 2010). As they construct their own definitions of commitment, we can think of these couples as *committed cohabiters* (Arnett 2004; Byrd 2009). As one cohabiter explained:

> We've been together 10 years. We met at college.... We graduated and started living together.... We never say never, but we certainly don't have any plans to [marry]. We're very happy being unmarried to each other. (in Sachs, Solot, and Miller 2003)

And as a 32-year old woman who has been in a monogamous relationship for ten years explained:

> I have friends who have been married and divorced already in the time that we have been together.... And I think I like the luxury of the fact that every day that we are together I know we are together because we both choose to be and not because we feel some artificial obligation to be together. (in Byrd 2009)

People's reasons for living together as an alternative to marrying may include the belief that marriage is too confining, signifies loss of identity, or stifles partners' equality and communication (Gold 2012; Moore, McCabe, and Brink 2001; Willetts 2003). However, today couples with attitudes and beliefs like these tend to be older and are not the statistical majority (Arnett 2004). More commonly, cohabiters who do not plan to marry tend to see marriage as beyond them financially (Fry and Cohn 2011).

The Cohabiting Relationship

As a category, heterosexual cohabiting couples differ from married couples in several ways. First, cohabiters are less homogamous, or alike in social characteristics, than are marrieds. At 12 percent, cohabiting couples are about twice as likely as marrieds to be interracial (Gates 2009, p. ii). Compared with married women, cohabiting women are more likely to earn more and be several years older than their partners (Fields 2004). Cohabiting men spend fewer hours in the labor force than do married men (Kuperberg 2012). Cohabiters have been more likely than marrieds to be nontraditional in many ways, including attitudes about gender roles, and to have parents with nontraditional attitudes or who have divorced (Davis, Greenstein, and Gertelsen Marks 2007; Kuperberg 2012).

On average, cohabiting relationships are relatively short-term. It's estimated that half last fewer than two years because the couple either breaks up or marries (Cherlin 2010). "Compared with married couples, cohabiters are much more likely to break up" (Seltzer 2000, p. 1,252). Reasons include the fact that, for the most part, cohabiting partners are not committed to their relationship in the same way that married partners are. Then, too, cohabitation may not include widely held norms to guide behavior to the degree that marriage does. As a result, the relationship may suffer as partners struggle to define their situation. Finally, lack of social support may negatively impact the stability of cohabitation "as members of the [cohabiter's acquaintance] network... provide the partners possibilities for other intimate relationships" (Willetts 2006, p. 114).

Uncertainty about commitment, together with less-well-defined norms for the relationship, may be reasons that, compared to marrieds, cohabitants pool their finances to a lesser extent (Hamplova and LeBourdais 2009), are less likely to say that they are happy with their relationships and find them less fair with regard to sharing finances (Dew 2011; Rhoades, Stanley, and Markman 2012), place greater importance on sexual frequency (Yabiku and Gager 2009), and have more sex outside the relationship than marrieds do (Treas and Giesen 2000).

However, research that analyzed data from the National Survey of Families and Households has found that the relationship quality of "long-term" cohabiting couples (who were together for at least four years) differed little from marrieds in conflict levels, amount of interaction together, or relationship satisfaction. One thing did differ, however: For both marrieds and long-term cohabiters, relationship satisfaction declined with the addition of children to the household, but this decline was more pronounced for cohabiters (Willetts 2006). Other research has found that, compared with younger cohabiters, older cohabiting couples generally report higher relationship quality. Among younger cohabiters, lack of plans for marriage is associated with lower relationship satisfaction (King and Scott 2005). Research shows that engaged cohabiting couples who have agreed-upon marriage plans have less conflict than do couples experiencing ambiguity about future marriage (Willoughby, Carroll, and Busby 2011). Then, too, cohabiting men with intentions to marry their partner do more housework than do other cohabiting males (Ciabattari 2004; Kuperberg 2012).

Cohabitation and Intimate Partner Violence Evidence also exists of considerable domestic violence, also called *intimate partner violence* (IPV), in cohabiting relationships—more than among marrieds (Anderson 2010). This situation may be partly the result of relatively low commitment (Johnson and Ferraro 2000; Dush 2011) combined with conflict over "rights, duties, and obligations" (Magdol et al. 1998, p. 52). Perhaps more important, **selection effects**, the situation in which individuals "select" themselves into a category being investigated—in this case, cohabitation, largely account for these findings. As we have seen, individuals who live together without marrying tend to be less well educated and poorer than marrieds, and—although domestic violence occurs at all economic and education levels—low income and education are statistically associated with higher levels of couple conflict and domestic violence (Anderson 2010; Dush 2011). Furthermore, differences in selection *out* of cohabitation and into marriage may help account for the statistical difference, because the better cohabiting relationships can be expected to transition into marriage, a situation that leaves a disproportionate share of violent couples among cohabiters (Kenney and McLanahan 2006).

Although not as often as one might expect, IPV can result in the dissolution of the relationship (DeMaris 2001). Research on the economic consequences of cohabiter breakups finds similarities to getting divorced. On average, men experience moderate financial decline, whereas women's economic decline is more pronounced (Avellar and Smock 2005). Counselors stress the importance of being fairly independent before deciding to cohabit, understanding one's motives, having clear goals and expectations, and being honest with and sensitive to the needs of both oneself and one's partner. This is especially necessary when children are involved. "As We Make Choices: Some Things to Know about the Legal Side of Living Together" discusses the legal implications of cohabiting in the United States today.

Cohabiting Parents and Outcomes for Children

Today, between 10 percent and 20 percent of all births—up to half of all nonmarital births—occur to a cohabiting mother (Cherlin 2010; Rackin and Gibson-Davis 2012). Perhaps half or more of births to cohabiters are planned (Lichter and Carmalt 2009). As shown in Figure 6.3, 40 percent of cohabiting heterosexual

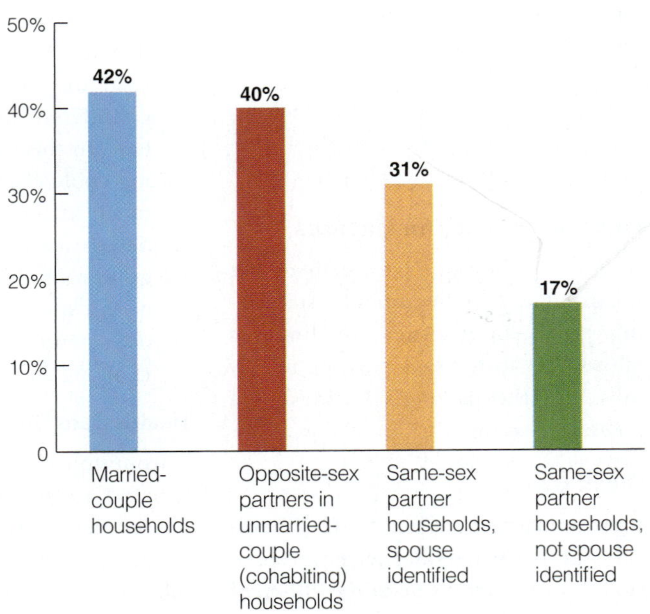

FIGURE 6.3 Percentage of children under age 18 in each of four U.S. household types, 2010.

Source: Gates 2009, Figure 13; U.S. Census Bureau 2012e, Table UC3.; U.S. Census Bureau 2012c, Table 65.

Note: Unmarried partners' children refers to at least one biological child under age 18 of either parent.

As We Make Choices

Some Things to Know about the Legal Side of Living Together

When unmarried partners move in together, they may encounter regulations, customs, and laws that cause them problems; being prepared for these situations can help (Clark 2010). Consulting a lawyer is strongly advised. Some potential trouble spots:

Domestic Partners

- In many areas and employment sites, opposite- and same-sex unmarried couples may register their partnership and then enjoy some rights, benefits, and entitlements that have traditionally been reserved for marrieds, such as access to joint health insurance.
- Registering as domestic partners usually requires joint residence and finances, plus a statement of loyalty and commitment.

Residence

When two unmarried people are renting, landlords may ask each to sign the lease—a legally binding contract—so that each is held responsible for all rent and associated costs.

Bank Accounts

Any couple can open a joint bank account, but one partner can then withdraw money without the other's approval.

Power of Attorney for Finances

Important when one partner becomes incapacitated, a document establishing power of attorney for finances allows the authorized partner to pay bills, run the partner's business, file taxes, and so on.

Credit Cards

If an unmarried couple shares a credit card, both partners are legally responsible for all charges made by either of them, even after the relationship ends. Creditors generally will not remove one person's name from an account until it is paid in full.

Property

A couple can have a written agreement about what happens to property that was purchased together should the relationship end.

Insurance

- The routine extension of auto and home insurance policies to "residents of the household" cannot be presumed to include nonrelatives.
- Anyone may name anyone else as the beneficiary on a life insurance policy. However, insurance companies sometimes require an "insurable interest," generally interpreted to mean a conventional family tie.

Wills and Living Trusts

- If you have no will or living trust when you die, your property will pass to individuals designated by state law—usually, legal spouses and blood relatives. A surviving partner could inherit nothing, not even property that he or she paid for.
- In some states, a cohabiting couple that has filed a joint tax return or used the same last name may be considered married in "common law." In this case, a cohabiting partner could claim a share—perhaps as much as one-third—of your estate, despite a will saying you wanted it to go to your children. It's best to sign a joint statement saying that you do not intend a common law marriage (Ebeling 2007).

Health Care Decision Making

Anyone too ill to be legally competent should have an agent to act for him or her in medical decision making. Many cohabitants want their partners to play this role. To be sure that medical personnel honor this desire, designate your partner as decision maker through a durable power of attorney for health care document.

Children

With the 1972 case of *Stanley v. Illinois* (405 U.S. 645), an unmarried mother is no longer entitled to sole disposition of the child in many states. Although courts have placement discretion, unmarried couples should stipulate in writing that custody is to go to the partner if the other parent dies. Note also that financial obligations for child support do not depend on marital status.

Some—but not all—courts grant visitation rights to a nonbiological, same-sex co-parent should the relationship end. Unmarried parents to a partner's child should consider three documents:

- a co-parenting agreement that spells out the rights and responsibilities of each partner;
- a nomination of guardianship that adds language to a will or living trust; and
- a consent to medical treatment that allows the co-parent to authorize the child's medical procedures.

Breaking Up

Ending a cohabiting relationship is not to be taken lightly:

- Couples who do not stipulate in writing—and preferably with an attorney's assistance—paternity, property, and other agreements can expect legal hassles. See an attorney about the laws in your state.
- Ending a registered domestic partner agreement in California and some other states requires a formal property settlement agreement and a dissolution proceeding in court (California Secretary of State's Business Programs Division 2011).

Critical Thinking

Does having to worry about the legal aspects of cohabitation lessen what appear to be some of the advantages of living together? Why or why not?

households contain children under age 18—a proportion only two percentage points lower than that of married-couple households with children (U.S. Census Bureau 2012e, Table UC3; U.S. Census Bureau 2012c, Table 65). Seven percent of children under age 18 (about 5.1 million) live with a cohabiting parent or parents; of children who live with cohabiting couples, half live with both of their unmarried biological or adoptive parents (U.S. Federal Interagency Forum on Child and Family Statistics 2012).

These statistics describe a situation at one point in time. However, "it is important to keep in mind that as children age, they may spend time in several [living] arrangements" (Kreider and Elliott 2009a, p. 16). More than ten years ago, demographers estimated that at some point during childhood about one in four U.S. children would live in a family headed by a cohabiting couple (Graefe and Lichter 1999, p. 215). Today we estimate this proportion to be considerably higher due to the increasing incidence of cohabiting and of childbearing among unmarried couples (Lichter and Carmalt 2009).

Although a large majority of cohabiting couples with children have one child, a significant number (about 1.5 million cohabiting couples) are raising two or more children (Chandra et al. 2005, Table 9). Research from at least three national samples has found this situation to be more characteristic of black and Hispanic cohabiters than of nonHispanic whites, who are more likely to marry upon becoming pregnant (Chandra et al. 2005, Table 18; Rackin and Gibson-Davis 2012).

A qualitative study with thirty cohabiting working-class childfree couples found that most intended to defer having children—many until after they marry—if they do (Sassler, Miller, and Favinger 2009). In another qualitative study, in-depth interviews with twenty-four childless cohabiters who had had some college found that they associated their desire to be in the middle class with marrying in the event that they had children. For these couples, cohabitation "appears to serve as a staging area, a time when couples can be together, complete schooling, and get fiscally established prior to marrying and beginning families" (Sassler and Cunningham 2008, p. 22).

One study showed that among urban couples who had a child while cohabiting, 28 percent married within five years, another 28 percent separated, and 45 percent remained in the cohabiting relationship (McClain 2011). Although having a child while cohabiting does not necessarily increase a couple's odds of staying together, conceiving a child during cohabitation and then marrying before the baby is born apparently does increase union stability (Rackin and Gibson-Davis 2012). Why would this be? "Although birth in cohabitation indicates a decision to remain together during pregnancy, it also represents a decision not to commit to marriage" (Manning 2004, p. 677). Cohabiting parents who together see a father's involvement in parenting as very important are more likely to stay together (Hohmann-Marriott 2009; McClain 2011). *Unintended* fertility has been found to be generally disruptive for couples, whether cohabiting or married, and unintended fertility is statistically associated with relationship dissolution (Guzzo and Hayford 2012). Perhaps ironically, the fear of divorce among unmarried parents may reduce their likelihood of marrying (Waller and Peters 2008).

Children's Outcomes As pointed out in Chapter 2, with regard to studying families, the *structure–functional theoretical perspective* sometimes asks us to think in terms of *functional alternatives*—alternate structures that might (or might not) well perform a function traditionally assigned to the nuclear family. With regard to living together without marriage (and also with regard to same-sex parents, addressed later in this chapter), many researchers have asked how well cohabiting (or same-sex) adults fill the parenting role when compared to heterosexual marrieds. In this regard, a recent study among Latina mothers whose ethnic origin is in Mexico, Puerto Rico, or the Dominican Republic concluded that, among many Latinos, cohabiting may be similar enough to marriage that children's outcomes do not significantly differ according to whether their parents are married or cohabiting. The researchers suggest that in ethnic groups where cohabitation is more likely to be considered very similar to marriage, cohabiting couples enjoy the benefits of less ambiguity regarding their relationship and more support from extended kin (Fomby and Estacion 2011).

For many families, nevertheless, "[c]ohabitation may not be an ideal childrearing context precisely due to the stress associated with the uncertainty of the future of the union" (S. Brown 2004, p. 353; see also Wilcox et al. 2011a). Because of cohabiting parents' fairly common breakups, children in cohabiting families are likely to have a biological parent living elsewhere, whereas cohabiting parents may have biological children living elsewhere as well (Cherlin 2010). Because cohabiting couples are significantly less likely to stay together than marrieds, many children in cohabiting-couple families will experience a series of changes, or transitions in their family's living arrangements (Osborne, Berger, and Magnuson 2012; Pilkauskas 2012). Then, too, a noteworthy number of parenting couples exhibit **intermittent cohabitation**: Over time, they move in together, then out, then back in (Cross-Barnet, Cherlin, and Burton 2011). In the strong language of one large group of established scholars, "Today, the rise of cohabiting households with children is the largest unrecognized threat to the quality and stability of children's family lives" (Wilcox et al. 2011a, p. 1).

The cumulative instability of cohabitating families as a category is related to problematic outcomes, from lower scholastic performance to greater incidence of marijuana use (Cavanagh 2008; Mandara, Rogers, and Zinbarg 2011; Sun and Li 2011). "Residential and other household changes associated with the formation of new partnerships may disrupt well-established patterns of [parental] supervision" (Thomson et al. 2001, p. 378; see also Magnuson and Berger 2009). Other research has found a relationship between a mother's overall stress (which negatively affects parenting) and her forming a co-residential relationship with a *non*biological father to her child (Cooper, McLanahan et al. 2009; Osborne, Berger, and Magnuson 2012).

Accordingly, many family scholars have expressed concern regarding outcomes for adolescents and younger children living in opposite-sex, cohabiting families (Booth and Crouter 2002; S. Brown 2004; see also Crosnoe and Cavanagh 2010). For example, research that compared economically disadvantaged children from families of various forms found more problem behaviors among children in various types of unmarried families, including cohabiting unions (Carlson 2006). Other research has found that, among couples with comparable incomes, cohabiting parents spend less on their children's education than do marrieds (DeLeire and Kalil 2005).

When compared to those living with two married biological parents, adolescents who have lived with a cohabiting parent are more likely to experience earlier premarital intercourse, higher rates of school suspension, and antisocial and delinquent behaviors coupled with lower academic achievement and expectations for college (Albrecht and Teachman 2003; S. Brown 2004; Carlson 2006). Having a cohabiting male in the household who is not the biological father appears not to enhance adolescents' outcomes when compared with living in a single-mother household (Manning and Lamb 2003). In fact, children living with a single parent who has a cohabiting partner in the household have significantly higher rates of abuse and neglect than do children growing up in other family forms (Sedlak et al. 2010).

Nevertheless, research also shows that, compared to growing up in a single-parent home, children do benefit economically from living with a cohabiting partner, provided that the partner's financial resources are shared with family members (Manning and Brown 2006). As we saw with regard to our earlier discussion about cohabitation and intimate partner violence, a good portion of the negative findings concerning cohabitation and children's outcomes can be attributed to *selection effects.*

> [W]ell-functioning cohabiting couples usually marry when the woman gets pregnant or they consciously decide to start a family.... So most couples who have the social and personal characteristics that foster both stable relationships and good parenting move on to marriage.
>
> This leaves a larger pool of cohabiting couples with economic, psychological, and relationship issues that simultaneously inhibit them from marrying and make them less effective parents—issues like financial instability, low education, infidelity, conflict, violence and lack of trust. (Coontz 2012b)

The scholastic and policy debate regarding the degrees to which selection effects or characteristics of cohabitation itself account for differences continues. Meanwhile, a relatively new area for court determination of child custody concerns children of cohabiting parents who separate. Some courts treat nonmarital relationships as "sufficiently marriage-like" for marital law to apply (Judge Heather Van Nuys in "Court Treats" 2002). Some proponents of the idea that only legally married couples should be treated as married oppose court involvement in custody issues of unmarried parents. However, "courts aren't trying to contribute to the demise of traditional families. But they recognize the reality of families today and functional parents" (Duke University Law School Dean, Katherine Bartlett, quoted in Biskupic 2003, p. 2A). The high and rising incidence of cohabitation reminds us that today's postmodern family encompasses many forms. We turn now to another relationship type that illustrates this point: same-sex couples.

Same-Sex Couples and Family Life

Google "LGBT" (gay, lesbian, bisexual, and transgendered) organizations, and you'll find websites for blacks, Latinos and Latinas, Jews, and Muslims, among others. Contrary to what we're likely to see on television, lesbian and gay singles make up a diverse category of all ages and race/ethnic groups (Gates 2011). Moreover, same-sex couples living together in long-term, committed relationships are not a recent development. According to social historian Samuel Kader (1999), same-sex committed couples date back to the Old Testament, and commitment ceremonies between same-sex partners were not unknown in early Christianity. More recently, scholars

> have uncovered a long and complicated history of gay relationships in nineteenth-century America. Sometimes women passed as men to form straight-seeming relationships; sometimes men or women lived together as housemates but were really lovers; sometimes individuals would marry [heterosexually] but still carry on romantic, sometimes lifelong same-sex intimate relationships. (Seidman 2003, p. 124)

Today there are approximately 500,000 to 600,000 same-sex couple households in the United States, about evenly divided between gay male and lesbian couples (Gates 2009). Arriving at truly accurate numbers of married same-sex couples or households involves complex statistical analysis of census data. After Massachusetts legalized same-sex marriage in 2004, there was hope that the 2010 census would calculate same-sex married, as well as unmarried, couples. However, the 2010 census did not count legally married same-sex couples as married; instead, it counted them as unmarried. However, the Census Bureau has announced a large-scale project to improve future data collection on same-sex couples, and it is anticipated that questions on their marital status will appear in the future (O'Connell and Lofquist 2009; Quinn 2009; "Same-Sex Couples..." 2009).

Families of same-sex couples include lesbian co-parents, as well as an array of combinations of lesbian mothers and biological fathers, surrogate mothers, and gay biological fathers (less frequent) (Goldberg 2010; C. Patterson 2000). One family studies professor describes the diversity apparent in her own lesbian family as follows:

> My partner and I live with our two sons. Our older son was conceived in my former heterosexual marriage. At first, our blended family consisted of a lesbian couple and a child from one partner's previous marriage. After several years, our circumstances changed. My brother's life partner became the donor and father to our second son, who is my partner's biological child. My partner and I draw a boundary around our lesbian-headed family in which we share a household consisting of two moms and two sons, but our extended family consists of additional kin groups. For example, my former husband and his wife have an infant son, who is my biological son's second brother.

M.Sharkey/Contour/Getty Images

In a 2011 *Time* magazine article, author Andrew Solomon introduced us to his "real modern family:" Tammy and Laura, who have two children by in vitro fertilization (IVF) and using their friend John's, sperm; Richard, Andrew's partner, who with IVF has a child with his (Richard's) friend Blaine. Meanwhile, Andrew and Richard wanted a child. Laura offered to serve as surrogate mother; using IVF and Andrew's sperm, little George was born to Andrew and Richard. As Solomon says it in the article's subtitle, "science, friendship, and love created an unconventional clan" (Solomon 2011, p. 32).

Facts about Families

Same-Sex Couples and Legal Marriage in the United States

More than thirty-five states have laws or state constitutional provisions that define marriage as between one man and one woman. The federal government has traditionally recognized the right of individual states to create, interpret, and enforce laws regarding marriage and families. Consequently, the battle over legal marriage for same-sex couples has largely been fought within individual state courts and legislatures.

Lawsuits Claiming Discrimination and Varied State Responses

Beginning in the late 1990s, same-sex couples in several states filed lawsuits claiming that barring lesbians and gays from legal marriage is unconstitutional because it discriminates against same-sex couples. Some (although not all) courts have agreed and ordered their state legislatures to address this problem by passing new, nondiscriminatory laws. Results have been varied. In 2004, Massachusetts became the first state to allow gay and lesbian couples to marry legally.

Many states now allow same-sex marriage. Before June, 2013, same-sex marriage in even these states was "an incomplete legal status, because same-sex marriages were denied federal recognition of their relationship by virtue of the 1996 federal [Defense of Marriage Act, or] DOMA" (Oswald and Kuvalanka 2008, p. 1060). However, a landmark 2013 U.S. Supreme Court ruled that legally married same-sex couples were entitled to federal benefits. (DOMA is described later in this box.) Other states have passed *civil union* laws.

Civil unions allow same-sex partners access to virtually all marriage rights and benefits on the state level but none on the federal level. For instance, a couple would have state-regulated rights to joint property and tenancy, inheritance without a will, and health care decisions for their partners. However, they cannot collect federal Social Security benefits upon a partner's death, nor can a non-U.S. partner become a U.S. citizen upon joining a civil union (Oswald and Kuvalanka 2008).

Meanwhile, some states have amended their state constitutions to stipulate that marriage in that state is to be defined only as heterosexual. As an example, in 2008, California voters (by a margin of 52 percent to 48 percent) passed the California Marriage Protection Act (Proposition 8), a state constitutional amendment declaring that "only marriage between a man and a woman is valid or recognized in California." Proposition 8 overturned a prior California court ruling that said that same-sex couples have a constitutional right to marry.

However, a state constitution does not outrank a federal court ruling, and federal appellate courts have ruled that California's Proposition 8 violates the U.S. Constitution because it discriminates against a category of U.S. citizens without a rational basis for such discrimination (Dolan 2012; McKinley and Schwartz 2010). In 2012, opponents of same-sex marriage appealed the California rulings to the United States Supreme Court (Dolan 2012), the last word in federal law. The Supreme Court refused to hear the case, however—a situation that paved the way for legal same-sex marriage in California (Egelko 2013; Liptak 2013).

> All four sets of grandparents and extended kin related to our sons' biological parents are involved in all our lives to varying degrees. These kin comprise a diversity of heterosexual and gay identities as well as long-term married, ever-single, and divorced individuals. (K. R. Allen 1997, p. 213)

As a second example of same-sex family diversity, Carol conceived Griffin through in vitro fertilization (IVF) with sperm from her friend, George. Monday, Tuesday, Thursday, and Sunday nights George stays in the spare room at Carol's apartment. The other three nights he spends with his domestic partner, David. "It's not like Heather has two mommies," George quips. "It's George has two families" (Kleinfield 2011). Interestingly, as mentioned in Chapter 1, a handful of states have legislated that parental legal status and rights can be enjoyed by more than two parents in one family (Lovett 2012).

In 2004, Massachusetts became the first state to legalize same-sex marriage. (See "Facts about Families: Same-Sex Couples and Legal Marriage in the United States.") Since then, many same-sex couples have married legally in several states. By 2009, approximately 32,000 U.S. same-sex couples had been legally married. Scholars characterize this point in history as "a revolutionary era for same-sex marriage in the United States. Same-sex couples have moved from 'outlaw' status (wherein homosexuality constituted illegal 'sodomy' in some states . . .) to 'in-law' status (with the advent of [legal] same-sex marriage) [in more and more states]" (Shulman, Gotta, and Green 2012, p. 160).

In addition to marrieds, approximately another 80,000 same-sex couples are in **civil unions** or registered as domestic partners. A **domestic partnership** is a legally or policy-defined relationship between two individuals who live together and share a domestic life but are not married.

The terms *domestic partnership* and *civil union* both refer to officially recognized unions in which unmarried couples enjoy some (although not all) rights and

States usually recognize one another's legal decisions. This *principle of reciprocity* would encourage a state to recognize a legal marriage performed in another state. However, states do not have to follow the principle of reciprocity regarding same-sex marriages (Liptak 2013).

The Federal Defense of Marriage Act (DOMA)

Passed in 1996, then reviewed by the United States Supreme Court in 2013, the Defense of Marriage Act (DOMA) was a federal statute declaring that legal marriage must be between "one man and one woman." It thereby denied any federal benefits (such as a surviving spouse's Social Security benefits) to same-sex partners who had married legally in states that allow it. However, in 2013 the U.S. Supreme ruled that regardless of gender, all legally married spouses must have equal access to federal benefits.

Meanwhile, DOMA also relieved states of the obligation to grant reciprocity to marriages performed in another state, and the U.S. Supreme Court left this aspect of the law unchanged. A majority of states have passed laws or state constitutional amendments that refuse to allow legal marriage in that state or to recognize a marriage obtained by same-sex couples in another state (Oswald and Kuvalanka 2008).

A Proposed Federal Amendment to the U.S. Constitution

Meanwhile, a proposed federal amendment to the U.S. Constitution would define marriage as between one man and one woman and ban same-sex marriage in the United States while allowing states to create civil unions or domestic partnerships (Allen and Cooperman 2004; Page and Benedetto 2004). Because the originators of the U.S. Constitution intended for amendments not to be undertaken lightly, they made them very difficult to pass. Passing a U.S. constitutional amendment requires a two-thirds majority in both the U.S. House of Representatives and Senate and then approval by three-fourths of the states. A constitutional amendment that would define marriage as between one man and one woman has not passed in Congress.

This range—from legal same-sex marriage in several states to a possible federal constitutional amendment that would ban same-sex marriages across the country—points up the public divide in the United States regarding same-sex marriage. For further details on legal marriage for same-sex couples in the United States and throughout the world, visit the following websites: American Civil Liberties Union; Partners Task Force for Gay and Lesbian Couples; and DOMA Watch.

Critical Thinking

Today it would be difficult to escape the public debate over, on the one hand, whether the family as institution is threatened or, on the other hand, whether tolerance for diversity is fitting when the issue is legal marriage for same-sex couples. What do you think? Is the institution of marriage and family threatened by same-sex marriage? Why or why not? Can you back up your opinion with facts?

benefits ordinarily reserved for marrieds. However, the term *civil union* is ordinarily used to refer to same-sex couples with a legal status somewhat like marriage. The term *domestic partnership* is typically used to refer to a formal status according to which same- or opposite-sex unmarried partners enjoy benefits (such as health insurance, for instance) offered by some employers, cities, counties, and states. Generally, domestic partnerships grant couples lesser status and fewer benefits than do civil unions. However, to further complicate matters, California, Oregon, and Washington use the term *domestic partnership* for legislation similar or equivalent to civil union laws in other states ("Marriage, Domestic Partnerships, and Civil Unions..." 2009).

Meanwhile, many same-sex couples identify as spouses even though they may not be legally married or in an otherwise legally recognized relationship (Gates 2009). Other than being legally married, same-sex partners may publicly declare their commitment in ceremonies among friends or in some congregations and churches such as the Unitarian Universalist Association or the Metropolitan Community Church, the latter expressly dedicated to serving the gay community. Catholics have access to a union ceremony designed by Dignity, a Catholic support association, although the Catholic Church does not recognize these unions ("Registration of Holy Union..." 2008).

Secular commitment ceremonies for gay men and lesbians are common enough to have sparked a number of wedding-planning businesses for same-sex couples. Same-sex couples use other commitment markers, too, such as joint estate planning, buying a house together, wearing rings, or hyphenating their last names (Porche and Purvin 2008; Reczek, Elliott, and Umberson 2009; Suter, Daas, and Bergen 2008). Registering as domestic partners (described in "As We Make Choices: Some Things to Know about the Legal Side of Living Together") can have emotional significance for same-sex couples who may do so partly as a way of publicly expressing their commitment.

Demographically, same-sex couples who identify as spouses (even if not legally married) are much like heterosexual marrieds in several respects. Same-sex couples who identify as spouses have an average age of 52 and above-average household incomes; 31 percent are raising children. In both couple types, a little more than 20 percent have college degrees, about 80 percent own their own homes, and 6 to 7 percent are interracial (Gates 2009). Interestingly, same-sex unmarried partners and different-sex cohabiting partners have similar rates of being interracial—13 percent of same-sex unmarried partners and 12 percent of different-sex cohabiting partners (Gates 2009, p. 11).

Same-sex couples live in virtually every county of every state (Gates 2009). Often they find community in urban areas that have high concentrations of gays and lesbians, along with strong activist organizations. But lesbians and gays also live in the suburbs and smaller towns (Oswald and Lazarevic 2011). The Internet has changed life for many gay men and lesbians, especially those living in rural areas. Some websites enable homosexuals from all over the world to meet and interact online regardless of geographical boundaries (Gudelunas 2006).

Although discrimination assuredly persists, attitudes toward LGBT rights generally have become more accepting over the past thirty years. Gallup polls have shown a gradual increase—from 34 percent in 1983 to 57 percent in 2008—in agreement with the idea that being gay or lesbian is an "acceptable alternative lifestyle" (Saad 2008a; see also Newman 2007). In 1996, a little more than one-quarter (27 percent) of Americans felt that same-sex marriages should be legally recognized as valid. In 2012, half of Americans said they feel that way (Gallup Poll 2012b; Newport 2012). In addition, more employers, cities, and states have extended various benefits to same-sex couples—health insurance for an employee's partner, for example (Surdin 2009).

As a further example of changing attitudes, in 2012 the Obama administration ordered immigration officials to recognize same-sex partners as family members when one partner is facing deportation. Put another way, the term *family relationships* in immigration policy is now defined to include long-term, same-sex partners (R. Cohen 2012). Later, the Obama administration continued to expanded benefits for same-sex partners (for example, medical and dental benefits for partners of same-sex military personnel) and came out publicly in favor of legalizing same-sex marriage (Shanker 2013). Ultimately, in June 2013 the U.S. Supreme Court ruled that all legally married same-sex couples must be eligible for any federal benefits that are available to marrieds.

The Same-Sex Couple's Relationship

In many respects, same-sex relationships are similar to heterosexual ones. Like heterosexuals, same-sex partners highly value love, faithfulness, and commitment (Meier, Hull, and Ortyl 2009). Research indicates that the need to resolve issues of sexual exclusivity, power, and decision making is not much different in same-sex pairings than among heterosexual partners. However, same-sex partners of both genders are likely to evidence more equality and role sharing than couples in heterosexual marriages (Gotta et al. 2011; Kurdek 2006, 2007; Parker-Pope 2008).

Nevertheless, a recent study of thirty-two New York City black lesbians in stepfamilies where one partner brought her child or children into a same-sex cohabiting relationship found that for these women constructing "family" meant following traditional heterosexual gender roles to some degree. Specifically, biological mothers in these stepfamilies were more involved not only in child care but also in housework. The author argues that following heterosexually "appropriate" marital roles may be especially important for lesbian mothers who previously bore their children in heterosexual contexts, because "they willingly moved from validated relationships with men to ... often stigmatized, same-sex unions. Given this status change, these mothers may find 'appropriate' gender construction that much more important" (Moore 2008, p. 353).

Same-sex couples must daily negotiate their private relationship within a heterosexual—often heterosexist—world (Oswald and Masciadrelli 2008). A comparison of non–legally married same-sex couples with married heterosexual couples found that the former spent more time discussing the state of their relationship. The researchers suggest that this difference may reflect the absence of a legal bond: "It appears that to some degree heterosexual married couples may take for granted that they are bound together through legal marriage, whereas [non–legally married] gays and lesbians must frequently 'take the pulse' of the relationship to assess its status" (Haas and Stafford 2005, p. 56).

A traumatic example of living out a private relationship amid heterosexism occurred in 2009 when a Florida hospital prevented a distraught lesbian from being at her dying partner's bedside ("Federal Court Dismisses Lawsuit ..." 2009). "This is an antigay city and state," the hospital receptionist reportedly explained (Parker-Pope 2009). One year later, President Obama ordered that any hospital that received federal funds through Medicare or Medicaid must grant hospital visiting rights to same-sex partners (Stolberg 2010). Nevertheless, discrimination against same-sex couples persists in many areas of their lives.

Discrimination adds stress for same-sex couples and may result in lowered mental health and relationship quality (Otis et al. 2006). In addition, the potential for discrimination gives partners unique avenues for dealing negatively with couple conflict. "For example, 'outing' one's partner is not an issue for heterosexuals but is a surprisingly common weapon for gay people in an abusive relationship" (Burke and Owen 2006, p. 6; Sorenson and Thomas 2009). Courts have begun to deal with the dissolution of same-sex couple households and possibly accompanying custody battles (Van Eeden-Moorefield et al. 2011). Studies are beginning to examine children's psychological well-being after their same-sex parents' breakup (Gartrell et al. 2011). Chapter 14 addresses issues regarding same-sex couple dissolution.

Same-Sex Intimate Partner Violence Current research suggests that the rates of intimate partner violence (IPV) are about the same for same-sex as for heterosexual couples (Sorenson and Thomas 2009). "Available qualitative data suggests that LGBT victimization is underreported for numerous reasons, including fear that the police are homophobic, and the fear of being 'outed' to family, friends, and coworkers, among other reasons" (Office on Violence Against Women 2010). Meanwhile, society-wide responses to same-sex violence "can be described as neglectful at best" (Sorenson and Thomas 2009, p. 349; Kulkin et al. 2008:

> Restraining orders, a legal remedy that is widely used when a victim is trying to end the relationship that requires the abuser to have either no or only peaceful contact with the victim ... are not available to gay men and lesbians in [some] states. In addition, social services are not always welcoming; lesbian victims of IPV report not feeling comfortable seeking domestic violence services that generally are geared to heterosexual females. And most shelters do not accept male clients, making such services off limits to gay male IPV victims. (Sorenson and Thomas 2009, p. 349)

Furthermore, police can be reluctant to get involved when summoned to a same-sex IPV call (Taranto n. d.).

Gay activists argue that domestic violence laws need to specifically include lesbian and gay partners, and police must be trained to more effectively address intimate partner violence among same-sex couples (Kulkin et al. 2008). In sum, the problem of same-sex IPV "is exacerbated" by a political climate that often "treats gays and lesbians as a marginalized population" (Burke and Owen 2006, p. 7). Increased public education and continued activism are strongly recommended (Taranto n. d.).

Same-Sex Parents and Outcomes for Children

Back in the early 2000s, observers began to note a "gay baby boom" (Johnson and O'Connor 2002). Same-sex couples increasingly became parents through adoption, foster care, or artificial insemination (Biblarz and Savci 2010). A 2008 Census Bureau survey found that 31 percent of same-sex couples who identified themselves as married, and 17 percent of other same-sex households, include children under age 18 (Gates 2009, Figure 13). (See Figure 6.3.) These children were adopted, born to the union, or born in prior heterosexual relationships. Interestingly, now that LGBT identity is less stigmatized than it was twenty or thirty years ago, fewer gays and lesbians are marrying heterosexually, having children, then divorcing later after coming to terms with their sexual identity. Today lesbians and gay males are more likely to form same-sex unions initially. Consequently, fewer same-sex couples now bring children born in prior heterosexual relationships to their union (Gates 2011). Some courts, but certainly not all, permit a lesbian partner to adopt a biological child born by her partner into the same-sex union. Only a minority of states allow to same-sex couples joint adoption of a child born outside the union, and several states restrict or ban such adoptions (Bernard 2012a).

A 2009 poll found 43 percent of respondents saying that gay male and lesbian couples raising children was "bad for society." In contrast, 41 percent found this

Catchlight Visual Services/Alamy

Lesbian couples may take advantage of AID (artificial insemination by donor) technology so that one partner gives birth to a baby they both want. The vast majority of research indicates that children of lesbian or gay male parents are generally well adjusted.

situation "good for society," and another 12 percent said it made no difference (Pew Research Center 2010a, p. 65). Religions vary in their policies regarding same-sex couples raising children. For example, the Catholic Church and the Church of Jesus Christ of Latter-day Saints (Mormons) oppose both legal marriage and adoption by same-sex couples (Egelko 2008). Courts also vary in their receptiveness to same-sex families. Many courts permit a lesbian co-parent to adopt a biological child born to her partner, ensuring legal parenthood to both members of the couple raising a child, and many courts grant joint adoption to gay male couples (Human Rights Campaign 2007). However, adoption by same-sex couples can be difficult due to a potential caseworker's discomfort, among other reasons (Kinkler and Goldberg 2011). Same-sex couple adoption is explicitly prohibited in one state: Mississippi (Garcia 2012).

Among lesbian couples, one partner may give birth to a child that both partners parent. When a couple decides to follow this course, the women face a series of decisions: Who will be the biological mother? How will a sperm donor be chosen? What will they call themselves as parents? How will they negotiate parenthood within a heterosexual society? Where and from whom will they find support? (Chabot and Ames 2004; Goldberg and Smith 2008).

Compared to that regarding heterosexuals, the amount of research on same-sex parents is small but growing. There is less research on gay-male couples who are parenting than on lesbian parents (Biblarz and Savci 2010). However, a recent qualitative study looked at gay-male couples' motivations for becoming parents. One respondent said, "We see so many kids that… haven't gotten a break…. If we could make a difference in just one kid's life, you know, wouldn't it be sad if we didn't?" (Goldberg, Downing, and Moyer 2012, p. 165). Other studies have found that—like lesbian parents—the gay male couples sampled (disproportionately white and middle class) co-parented more compatibly while sharing housework and child care more equally than do heterosexual couples on average (Biblarz and Savci 2010).

One study has found that lesbian partners who become parents are more likely than heterosexual wives to remain committed to full-time work as well as to motherhood (Peplau and Fingerhut 2004). Another study focused on the relationship quality of twenty-nine lesbian couples who gave birth to their first child by means of artificial insemination. These researchers found that, similar to heterosexual couples, lesbian partners' relationship satisfaction declined with the transition to parenthood. This situation largely resulted from having less time to be alone as a couple after the baby was born (Goldberg and Sayer 2006).

"Living in a society fixated on labels and family terminology," lesbian couples may be asked, "Who is the real mom?" (Chabot and Ames 2004, p. 354). In addition, issues of support from the couple's extended families may cause tension:

> Biological mothers' families may undermine the non-biological mother's relationship to the child, seeing her as "less of a mother." … Another possibility is that biological mothers' families … meet or even surpass non-biological mothers' expectations for support, but their frequent presence or greater involvement ultimately causes conflict between the partners. (Goldberg and Sayer 2006, p. 97)

Meanwhile, same-sex parents emphasize their similarity to heterosexual parents: "We go to story time at the library and worry about all the same food groups" (in Bell 2003). "Contrary to stereotypes of these families as isolated from families of origin, most report that children had regular (i.e., at least monthly) contact with one or more grandparents, as well as with other adult friends and relatives of both genders" (C. Patterson 2000, p. 1062). A wide network of adult friends and relatives may even include ex-husbands. A lesbian mother who, like her partner, brought a daughter into the relationship from a heterosexual previous marriage, explains:

> [B]oth of [the girls'] fathers live very close. We stayed right within the same school district that I was in with the little one. Little Monica's father being just in the next school district over. So, the fathers were always there visiting and taking care of [the girls], especially my ex-husband when … I was back in school, so … he always had the responsibility of being there when they got home [from school]. (in Hequembourg 2007, p. 169)

Children's Outcomes The majority of research on the topic finds children of gay male and lesbian parents to be generally well adjusted with few, if any, noticeable differences from children of heterosexual parents in cognitive abilities, school performance, behavior, emotional development, gender identity, or sexual orientation (Amato 2012; Biblarz and Savci 2010; Goldberg 2010; but see Regnerus 2012a, 2012b).

Although not necessarily refuting these findings, sociologists note that the research methodologies of many of these studies are not rigorous, largely because it is very difficult to locate representative samples of gay male and lesbian parents (Eggebeen 2012; Osborne 2012; Regnerus 2012b; Schumm 2012). The research that we do have—which is largely on lesbian, white, and middle- or upper-middle-class parents—concludes that same-sex parents, especially those who identify as spouses, are much like married heterosexuals in their parenting practices (Goldberg 2010). Its members having themselves reviewed the literature, the American Academy of Pediatrics supports gay male and lesbian couples' adopting, bearing, and raising children. "Children thrive in families that are stable and that provide permanent security, and the way we do that is through marriage, [including same-sex marriage]" according to a 2013 American Academy of Pediatrics policy statement (American Academy of Pediatrics 2013). For discussion of a 2012 methodological controversy regarding this body of research, see Chapter 2. General parenting issues are explored in Chapter 9.

Meanwhile, like children of other minority groups, those in same-sex families may experience prejudice from friends, classmates, or teachers. In our heterosexist society, adult children of same-sex parents face dilemmas about coming out regarding their parents (Goldberg 2007, 2010). Children of same-sex parents have formed a support group called COLAGE (Children of Lesbians and Gays Everywhere) and maintain a website. Their purpose is to "engage, connect, and empower people to make the world a better place for children of lesbian, gay, bisexual, and/or transgender parents and families." Allowing same-sex parents to marry legally might be a benefit to their children because marriage is associated with increased "durability and stability of the parental relationship" as well as enhanced in-law, grandparent, and other extended-family investment (Meezan and Rauch 2005, p. 108; Goldberg 2010; Wildman 2010). We turn now to the debate over legal marriage for same-sex couples.

The Debate over Legal Marriage for Same-Sex Couples

In 2000, the Netherlands became the first country to allow same-sex partners to marry. As of this writing, several other countries including Argentina, Belgium, Canada, Iceland, Norway, Portugal, South Africa, Spain, and Sweden—as well as Mexico City—allow legal same-sex marriage (Partners Task Force for Gay and Lesbian Couples 2011).

In the United States, the November 2012 election saw legal marriage endorsed by popular vote for the first time (Bruni 2012). Prior to that, marriage had been legalized in some states by court order. Today in the United States, approximately thirteen states—including California, Connecticut, Iowa, Maine, Maryland, Massachusetts, Minnesota, New Hampshire, New York, Vermont, and Washington—together with the District of Columbia and two Native American tribes, have legalized same-sex marriage. Subsequent to the November 2012 elections (which legalized same-sex marriage in Maine, Maryland, and Washington) and subsequent to a June 2013 Supreme Court Decision that refused to knock down legal same-sex marriage in California, 30 percent of the U.S. population lives in a state that has made same-sex marriage legal or recognizes out-of-state marriages of same-sex couples (Freedom to Marry 2012; Liptak 2013). Scholars predict that same-sex marriage will be legal across the United States at some point in the not too distant future (Powell et al. 2010).

The conservative Family Research Council website, one of several that speak out against same-sex marriage, urges people to "Take a Stand for Marriage!" Other websites, such as Gay and Lesbian Advocates and Defenders (GLAD), the Partners Task Force for Gay and Lesbian Couples, and the National Black Justice Coalition advocate for the other side. Having first emerged as a remote possibility in the 1970s, legal marriage for gay and lesbian couples "became a front-line issue" after 1991 when gay activists formed the Equal Rights Marriage Fund (Seidman 2003; Taylor et al. 2009). "Facts about Families: Same-Sex Couples and Legal Marriage in the United States" outlines political developments regarding legal marriage for same-sex couples.

As shown in Figure 6.4, 52 percent of Americans favor, 42 percent oppose, and 7 percent aren't sure about allowing same-sex couples to adopt children (Pew Research Center 2012f). Just as attitudes have become more accepting about LGBT rights generally, public support for legal same-sex marriage has grown considerably since the mid-1990s, when about only 25 percent of Americans believed that same-sex marriage should be legal (Vestal 2009). Half of Americans (49 or 50 percent, depending on which survey you read) favor legal marriage for same-sex couples, while 44 percent oppose it and another 9 percent are not sure (Newport 2012; Pew Research Center 2012f). Scholars see the national divide over legal same-sex marriage as a mark of cultural ambivalence resulting from conflicting core values: the sanctity of heterosexual marriage versus personal freedom and civil rights (Brumbaugh et al. 2008). What are the arguments against and for legal marriage for same-sex partners?

Arguments for Legal Marriage as Heterosexual Only

Not all religions oppose legal same-sex marriage. The National Association of Evangelicals, Islamic and Orthodox Jewish congregations, and Mormon (LDS) and Catholic churches oppose legal same-sex marriage (Grossman 2010). "On the other side are the Unitarians, the United Church of Christ, the Union for Reform Judaism, the Soka Gakkai branch of Buddhism,

"Do you favor or oppose allowing gays and lesbians to marry legally?"

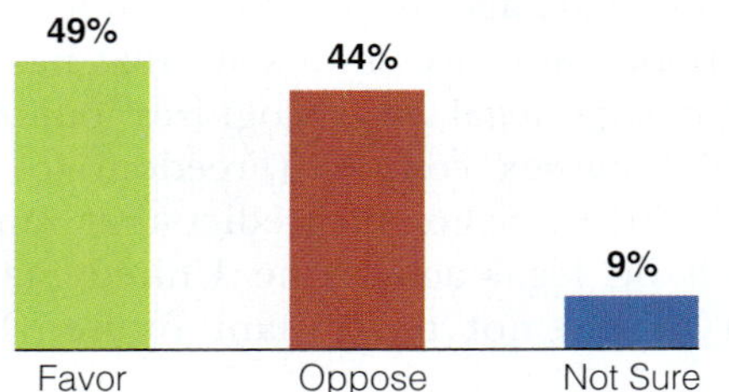

"Would you favor or oppose allowing gays and lesbians to adopt children?"

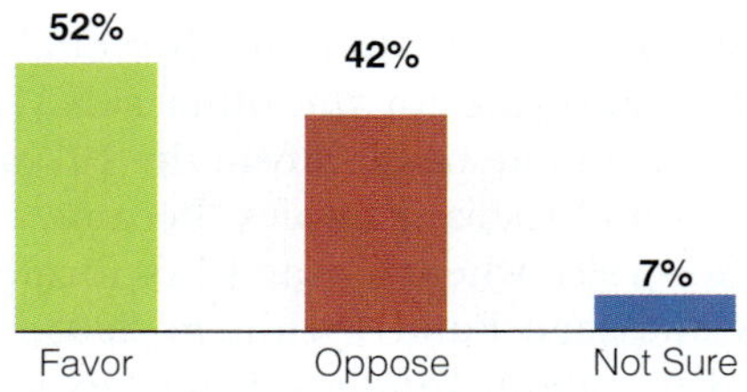

FIGURE 6.4 American public opinion regarding legal marriage and adoption for same-sex couples, 2012. (Percentages do not equal 100 due to rounding.)

Source: Newport 2012; Pew Research Center, 2012f.

and dissident groups of Mormons, Catholics, and Muslims" (Egelko 2008, p. A10). Christian social psychologist David Myers, making a "Christian Case for Gay Marriage," argues that if marriage is good for people and society, as discussed in Chapter 7, then marriage should be an option for everyone, including lesbians and gays (Myers and Scanzoni 2006). However, for religious fundamentalists and other conservative groups, the move to legalize same-sex marriage is an "attempt to deconstruct traditional morality" (Smolowe 1996; J. Wilson 2001).

Those who favor defining legal marriage as only heterosexual are more likely to value traditional gender roles (as well as traditional family structure) and to live in communities in which those values are reinforced through daily communication with like-minded neighbors (McVeigh and Diaz 2009; Powell et al. 2010). Proponents of marriage as only heterosexual argue that heterosexual marriage alone has deep roots in history, as well as in the Judeo-Christian and other religious traditions (Hartocollis 2006; McKinley and Schwartz 2010). They further claim that only heterosexual married parents can provide the optimum family environment for raising children and that legalizing same-sex marriage would weaken an institution already threatened by single-parent families, cohabitation, and divorce (Blankenhorn 2007). Some contend that "[e]ven more ominously, permitting gays to marry would open the door for all sorts of people to demand the right to marry—polygamists, children, friends, kin—even more than two partners" (Seidman 2003, p. 128).

This fear is not entirely unfounded. In 2003, the U.S. Supreme Court struck down laws criminalizing sodomy (interpersonal sexual acts that do not allow for procreation such as oral sex or anal intercourse). At that time, the Supreme Court ruled that individuals have "the full right to engage in private conduct without government intervention." Today some advocates seek the same considerations for proponents of polygamy, or having more than one spouse (Baltzly 2012; Schwartz 2011a). Moreover, changing the legal definition of marriage as necessarily "between one man and one woman" could affect not only the gender stipulation but also the requirement that marriage take place between only two individuals (Kurtz 2003; Stacey and Meadow 2009). Finally, opponents argue that legal marriage for same-sex couples is unnecessary given legislation such as civil unions in some states that gives virtually all the rights of marriage to same-sex couples without the title "married" or "spouse" (Seidman 2003).

Arguments for Legal Same-Sex Marriage In her book, *Beyond (Straight and Gay) Marriage* (2008), American University law professor Nancy Polikoff argues that all family forms need to be valued under the law. Same-sex families comprise a family form that is not going to disappear; if marriage is thought to be good for spouses and children emotionally, financially, and healthwise (see Chapter 7), then to deny these benefits to a significant number of individuals is unethical and socially costly (Movement Advancement Project 2011).

Those who favor legalized same-sex marriage further argue that denying lesbians and gays the right to marry legally violates the U.S. Constitution because it discriminates against a category of citizens (Schwartz 2010, 2011b). Legal marriage yields economic and other advantages. For instance,

> [the] right to divorce is a benefit associated with marriage that often goes unmentioned. Relationships end, and marriage provides the opportunity for a legal chaperon when partners are unable or unwilling to manage their separation, dissolution, and post-divorce parenting in a productive manner. Legal marriage for LGBT partners would protect the interest of all members of the family upon dissolution, just as it does for heterosexual partners. (Allen 2007, p. 181)

Couples who marry in states where same-sex marriage is legal who then return or move to states where it is not legal may encounter difficulties in the unforeseen event of a later desire to divorce (Van Eeden-Moorefield et al. 2011). Same-sex divorce is impossible in a state that does not recognize the marriage, and getting divorced in the state where a couple was married could be challenging. Most states have a residency requirement for divorce, some as long as one year (Clifford, Hertz, and Doskow 2007; K. Hartman 2011). Allowing same-sex legal marriage in all states would facilitate child custody cases in the event of dissolution (Polikoff 2008; Hare and Skinner 2008). Chapter 14 further explores dissolution of same-sex unions and of unmarried heterosexual relationships.

Because the federal **Defense of Marriage Act (DOMA)**, which was in effect between 1996 and 2013, defined marriage as only between "one man and one woman," even same-sex couples who were legally married in states that allowed it could not receive federal benefits designated for married couples (Clifford, Hertz, and Doskow 2007; Oswald and Kuvalanka 2008). More than 1,000 federal laws include marital status as a factor, including the rights to veterans' benefits, for example ("Federal Marriage Benefits Denied..." n. d.; Michon 2012). Other federal laws that apply only to legally married partners include those granting Social Security benefits to a widowed or disabled spouse, the right of legally married partners to inherit from one another without a will, or laws making it possible for the immigrant spouse of a U.S. citizen to also become a citizen. Under federal law, legally married spouses can petition for immigration and citizenship status for their foreign-born husbands or wives. President Obama came out against DOMA, which was in effect when he first became president. In 2011, he announced that under his administration the U.S. Justice Department would no longer defend DOMA in federal courts against charges that it is unconstitutional (U.S. Department of Justice 2011). Then, in June 2013, the U.S. Supreme Court ruled that legally married same-sex couples are entitled to all federal benefits (Liptak 2013).

LGBT activists had argued that creating domestic partnerships or civil unions, instead of allowing marriage for same-sex partners, created second-class citizens (Allen 2007; Leff 2006). As one lesbian spouse, legally married in Massachusetts, said:

> Same-sex marriage not only makes me take my relationship more seriously, it's making me take my country more seriously. I always felt oppressed and not a part of America, not really. But this seems like finally there is a light in the dark, like finally... the government is saying that my relationship counts and I count, too. (in Lannutti 2007, p. 141)

And a lesbian mother explained:

> [Legal same-sex marriage would] help the children accept my partner more.... [T]hey want to call her "stepmom" but because they are being told we are not legally married and she is *not* their step-mom, they don't know what to call her. Hopefully in time they will figure it out for themselves, but if we were legally married they wouldn't need to think about it. (in Shulman, Gotta, and Green 2012, p. 170)

Many children of gay male or lesbian couples also view the legalization of same-sex marriage as giving them security and comfort (P. L. Brown 2004; Goldberg 2007). Relatedly, one mother described her daughter's school experience:

> She would be quite open about [our family] in school and answer, "I have two moms." And kids would say to her, "Well, they can't be married." And she would say we're married because that's how we always represented ourselves. So she had a boy in her class that would say, "They're not married, they can't be married!" So I figured it was such a thrill for her to be able to say, with confidence, "They're married, and yes it is legal," and "I'm just like you." (in Porche and Purvin 2008, p. 153)

Interestingly, however, although many lesbians and gays support the claim for same-sex marriage in principle, in Massachusetts where same-sex marriage has been legal since 2004, a minority of same-sex couples has chosen to marry. A qualitative study of Massachusetts same-sex couples who had been together for twenty years or more sheds light on this development. One reason that committed same-sex couples may not marry when given the option is that they have already "spent thousands of dollars instituting all the legal protections they felt they needed and did not see a need to complicate those arrangements by changing their status through a marriage recognized only by Massachusetts." Attitudes of other participants in this study resembled those of some heterosexual cohabitating couples. They said that "marriage is made 'not by some sort of legal sanctions' but by the commitment of the people in the relationship to each other" (Porche and Purvis 2008, p. 155). Gay men and lesbians themselves have been divided somewhat on the desirability of legalized same-sex marriage, at least for themselves ("Gays Want the Right" 2004).

Dissenting Arguments among Lesbians and Gay Men

With the majority of states offering *no* legal recognition at all for same-sex unions, some LGBT spokespersons argue that too much emphasis is placed on advocating for same-sex marriage when activist resources would be better spent working for wider enactment of civil unions (Leff 2009). Furthermore, some gays and lesbians have opposed legal same-sex marriage in principle. Generally, they have objected to mimicking a traditionally patriarchal institution based on property rights and characterized by a high divorce rate. With irony, San Francisco columnist Mark Morford questioned why same-sex couples would *want* to marry: "Show me a single scientific experiment where fully 50 percent of the results turn out negative and induce collapse and emotional breakdown and childhood therapy and Xanax and alcoholism and screaming, and I'll show you a scientist who will quickly scrap the whole thing and start over"(Morford 2006). Some opponents have further objected to giving the state power to regulate primary adult relationships (Peele 2006), a topic addressed in Chapter 7.

Moreover, opponents have stressed that legalizing same-sex unions would further stigmatize any sex outside marriage, with unmarried lesbians and gay men facing heightened discrimination ("Monogamy: Is It for Us?" 1998; Seidman 2003). We can't say how representative the following statement is, but it does illustrate the viewpoint of at least one lesbian when contemplating the possibility of legalized same-sex marriage in her state:

> I don't want to get married, so this marriage thing is going to make it harder for me to find a person to be in a relationship with. I know that because I don't want to get married, women will think I'm not a good potential partner, and move on.... It sucks, because ... now I have to limit myself to the other non-marrying kinds out there—like finding a great girl wasn't hard enough already! (in Lannutti 2007, p. 145)

Whether and to what extent allowing same-sex couples to legally marry increases their personal life and couple satisfaction is a matter for future research. In the following section, we turn to a discussion of life satisfaction among the unmarried.

Maintaining Supportive Social Networks and Life Satisfaction

Perhaps not surprisingly, life satisfaction is associated with income as well as marital status (Pelham 2008). As illustrated in Chapter 7's Table 7.2, many singles and single parents, particularly women, just do not make enough money. Many work more than one low-paying job and then take care of their homes and children (Huston and Melz 2004). For them, "career advancement" means hoping for a small raise or just hanging on to a job in the face of growing economic insecurity. Pursuing higher educational opportunities means rushing to one or more classes and working full-time. Moreover, research shows that poor women have less-effective private safety nets than do others because their families and friends are also very likely to be poor and overburdened financially and emotionally (Harknett 2006). These women are dealing with work, parenting, and low-income issues.

Vicky Kasala/The Image Bank/Getty Images

Besides a variety of living arrangements, factors such as age, sex, residence, religion, and economic status contribute to the diversity and complexity of single life. An elderly man or woman existing on Social Security payments and meager savings has a vastly different lifestyle from two unmarried professionals living together in an urban area. Also, the experience of being unmarried differs according to whether one is single by choice or involuntarily. Happiness measures consistently show that, among singles, the married but separated are *least* happy, with divorced individuals faring somewhat better and the widowed faring still better.

Research and polls consistently find that, as a group, marrieds are happier than singles (Brown and Jones 2012; Lee and Ono 2012; Wienke and Hill 2009). Research also shows that, regardless of whether they are legally married, people in secure interpersonal heterosexual or same-sex relationships—and those who socialize often with friends and family—are happier than are those who spend considerable time alone (Lee and Ono 2012; Wienke and Hill 2009).

If we think of the various living arrangements of unmarrieds as forming a *continuum of social attachment*, we realize that not all singles are socially unattached, disconnected, or isolated (Ross 1995). Research consistently shows that people in close, supportive relationships—whether heterosexually married, same-sex married, heterosexually cohabiting, same-sex cohabiting, or living alone—tend to enjoy better health and be happier and less depressed than those with no intimate partner at all (Liu and Reczek 2012; Musick and Bumpass 2012; Oswald and Lazarevic 2011; Pelham 2008). Note that the relationship between being involved and not being depressed may hold *only* for those in happy, or supportive, arrangements (Ross 1995).

Meanwhile, for unattached singles, living alone can be lonesome. However, living alone does not necessarily imply a lack of social integration or meaningful connections with others (Trimberger 2005). Nevertheless, unattached singles have tended to report feeling lonely more often than have marrieds (Harter and Arora 2008; Pelham 2008). Poor and older singles are especially likely to be lonely, perhaps because the low incomes and ill health that tend to accompany old age make socializing very difficult (Kim and McKenry 2002).

Sociologist E. Kay Trimberger (2005) argues that the "heaviest thing" for unattached, middle-aged women is the "idea of the couple, and that's so internalized." Trimberger identifies the following "pillars of support" for unattached single women: a nurturing home, satisfying work, satisfaction with their sexuality, connections to the next generation, a network of friends and possibly family members, and a feeling of community.

Some research has found cohabitants to be midway between unattached singles and marrieds in mental and physical well-being (Kurdek 1991), while other studies have shown no difference between cohabitants and other singles, "suggesting that the protection effects of marriage are not as applicable to cohabitation" (Kim and McKenry 2002, p. 905). However, marriage also involves a set of obligations and the responsibility of coping with both the burdens of other family members and the disappointments that come with family life. Valuing personal autonomy, Americans may find these obligations emotionally stressful (Gove, Style, and Hughes 1990). There are some areas in which nonmarrieds may feel better off than the married. Less irritation and a greater sense of control over one's life can be among the advantages of being single (Hughes and Gove 1989).

All of us need support from people we are close to and who care about us. Isolation increases feelings of unhappiness, depression, and anxiety (Umberson et al. 1996), whereas being socially connected "seems to keep stress responses... from running amok," according to UCLA psychologist Shelley Taylor (quoted in "Save the Date" 2004). Among unmarried mothers, a social network that can function as "a parent safety net" is important to positive parenting practices (Ryan, Kalil, and Leininger 2009).

Maintaining close relationships with parents, siblings, and friends is associated with positive adjustment and life satisfaction (Kurdek 2006; Soons and Liefbroer 2009; Spitze and Trent 2006). A recent study of family and community support experienced by same-sex couples concludes with the following policy advice:

> Community initiatives to strengthen families could emphasize the importance of staying in touch and getting along despite disagreements among family members; these campaigns could identify sexual orientation as a topic where adult family members can learn to disagree with each other while agreeing to provide a loving environment.... (Oswald and Lazarevic 2011, p. 383)

A crucial part of one's support network involves valued friendships. Despite changing gender roles, men remain less likely than women to cultivate psychologically intimate relationships with siblings or same-sex friends (Levy 2005; Weaver, Coleman, and Ganong 2003). Indeed, a man may be more open and disclosing with a woman friend (Wagner-Raphael, Seal, and Ehrhardt 2001). One study of men in the construction industry found that many of them, rather than building truly supportive friendships, talked instead about horseplay, alcohol consumption, risk taking, and physical prowess, and generally engaged in one-upmanship (Iacuone 2005). Men (as well as women) who do not establish friendships based on emotional honesty run the risk of feeling socially isolated. In addition to friendships, other sources of support for singles include group living situations, religious fellowships, and volunteer work (Mustillo, Wilson, and Lynch 2004).

A survey conducted by the Pew Research Center during the recent recession found that for the most part young adult single men and women are optimistic about their futures despite hard economic times (Pew Research Center 2012b). Singles' optimism can

Michael Ventura/Alamy

When we think of singlehood as a continuum, we realize that not all singles—even those who live alone—are socially unattached, disconnected, or isolated. It's important to develop and maintain supportive social networks of friends and family. Single people place high value on friendships, and they are also major contributors to community services and volunteer work.

no doubt be attributed partly to their reaching out to their families and friends. "There are so many ways to live and love. The sentimentalized image of Mom, Dad, and the kids gathered around the hearth has had its day. A new American experiment has begun. We're not all going nuclear anymore" (DePaulo 2012). However a person chooses to live the single life, establishing a sense of belonging by maintaining supportive social networks is crucial.

Summary

- Since the 1960s, the number of unmarrieds has risen dramatically. Much of this increase has resulted from young adults postponing marriage coupled with a marked rise in cohabitation.
- One reason people are postponing marriage today is that increasing alternatives may make marriage less attractive. The recession that began in 2007 has led many to postpone marriage.
- The low sex ratio—fewer men for women of marriageable age—has also caused some women to postpone marriage or put it off entirely.
- Attitudes toward marriage and singlehood have changed, so that being unmarried is more often viewed as preferable, at least "for now."
- More and more young unmarrieds are living in their parents' homes, usually at least partly as a result of economic constraints.
- Some unmarrieds have chosen to live in communal or group homes.
- A substantial number and growing percentage of heterosexual unmarrieds are cohabiting.
- As heterosexual cohabitation becomes more acceptable, increasing numbers of cohabiting households include children either born to the union or from a previous relationship.
- The relative instability of heterosexual cohabiting unions has led to concern for the outcomes of children living in cohabiting families.
- Whether legally married or not, same-sex couples must daily negotiate their private relationship within a heterosexual—and often heterosexist—world.
- Research finds that children raised by same-sex couples are not significantly different from those raised by heterosexual parents.
- Congruent with the emergence of the postmodern family, we are witnessing a national (and global) debate over whether legal marriage should be extended to include lesbians and gay men.
- However one chooses to live the single life, it is important to maintain supportive social networks.

Questions for Review and Reflection

1. Individual choices take place within a broader social spectrum—that is, within society. How do social factors influence an unmarried individual's decision regarding his or her living arrangements?
2. What do you see as the advantages and disadvantages of cohabitation compared to marriage?
3. What does current research tell us about the outcomes generally of children raised in homes with two married biological parents compared to those raised in cohabiting families?
4. On average, do the outcomes of children raised by heterosexual parents differ from the outcomes of those raised by same-sex couples?
5. **Policy Question.** Do you think that legalizing same-sex marriage is a good idea? Give arguments based on facts to support your opinion.

Key Terms

civil union 152
cohabitation 142
commune 141
consensual marriages 144
Defense of Marriage Act (DOMA) 159
domestic partnership 152
intermittent cohabitation 149
LAT—living apart together 140
selection effects 147
sex ratio 137

7

Marriage: From Social Institution to Private Relationship

Learning Objectives

1. Analyze the American value of marriage in the midst of changing attitudes about it.
2. Describe the marriage premise and its components, including permanence and expectations of fidelity.
3. Explain why marriage has historically been a public, rather than simply a private, ceremony.
4. Describe the debate between policy makers and scholars who see marriage as "in decline" versus those who see marriage as "changing" but not necessarily in decline.
5. Describe some research-established benefits of marriage.
6. Contrast the selection hypothesis with the experience hypothesis regarding the benefits of marriage.
7. Discuss the origin, goals, and issues surrounding the National Marriage Initiative.

Why a party for a wedding? A wedding marks a couple's public announcement that—as opposed to many other options—the two have chosen marital commitment to define themselves, their relationship, and their lives. Although 12 percent of the never-married say they don't want to marry, and another 27 percent say they aren't sure, 61 percent of the never-married tell pollsters that they do want to marry (Cohn et al. 2011). Americans say that the top three "very important" reasons to marry are for love, to make a lifelong commitment, and companionship. A large majority of Americans (77 percent) say that, compared with other family forms, it's easier for marrieds to raise a family (Pew Research Center 2010a, p. 22). All else being equal, they're right (Wilcox et al. 2011a,b).

Academic research and opinion polls consistently show marrieds to be happier and healthier than unmarrieds and more satisfied with family life than singles, including cohabiters (Pew Research Center 2010a, p. 18; Wilcox et al. 2011a). Yet, ironically, 39 percent of Americans believe that "marriage is becoming obsolete." Among adults between ages 18 and 29, 44 percent say marriage is becoming obsolete. Indeed, more than one-third of those 50 to 64 years old see marriage as becoming obsolete (Pew Research Center 2010a, pp. 3, 12).

Why would people who want to get married say that marriage is becoming obsolete? Some must feel that they don't *really* need to be married. Others believe that they can't afford it. Nevertheless, although the situation is much less clear-cut now than several decades ago, in the United States marriage remains the most socially acceptable and stable gateway to family life (Wilcox et al. 2011a,b).

In this chapter, we explore marriage as a changing institution. We'll examine what distinguishes marriage from other couple relationships and then look at the changing nature of marriage. We'll further look at research on the benefits of marriage for adults and children and examine government initiatives to strengthen marriage. This chapter ends with an exploration of recently married couples' relationships. We will see that getting married announces a personal life course decision to one's relatives, to the community, and, yes, to the state. Despite wide variations, marriages today have an important element in common: the commitment that partners make publicly—to each other and to the institution of marriage itself (Cherlin 2004; Goode 2007 [1982]). Put another way, getting married—as opposed to cohabiting, for instance—is not only a private relationship but also a publicly proclaimed commitment. We begin by looking at marital status in the United States today.

Marital Status: The Changing Picture

Because of a high divorce rate, postponing marriage until older ages, and increased cohabitation, the proportion of Americans age 18 and older who are married has declined significantly over the past fifty years—from 72 percent in 1960 to between 51 and 56 percent in 2010 (Cohn et al. 2011; Taylor et al. 2011; U.S. Census Bureau 2012c, Table 56). Some of this decline can be attributed to our aging population and the fact that older Americans are more likely to be widowed. However, the proportion married has declined for Americans in all age groups. Eighty-two percent of those ages 25 through 34 were married in 1960, compared with 44 percent in 2010. Similarly, 86 percent of those between age 35 and 44 were married in 1960, compared with 62 percent fifty years later (Cohn et al. 2011).

Throughout the first half of the twentieth century, the trend was for more people to marry and at increasingly younger ages. For men, median age at first marriage in 1890—the year when the government first began to calculate and report this statistic—was 26.1. For both women and men, median ages at first marriage fell from 1890 until 1960, when they began to rise again. As pointed out in Chapter 6, the median age at first marriage had increased to 26.5 for women and 28.7 for men by 2011—higher than any ages ever recorded (U.S. Census Bureau 2011b, Table MS-2).

Around 1950, family sociologists described a standard pattern of marriage at about age 20 for women and 22 for men (Aldous 1978). Moreover, about 80 percent of early twentieth-century unions lasted until the children left home (Scanzoni 1972). In the 1960s, that trend reversed; since then, the tendency has been for smaller and smaller proportions of Americans to be married.

One reason for this change in marriage rates is economic: Americans want to be sure that they can afford to be married before they "tie the knot," and poorer individuals are forgoing marriage until they feel that they can afford it (Smock and Greenland 2010). As one indicator of this point, among Americans age 18 and older, 64 percent of college grads are married. This figure compares with fewer than half (48 percent) of those with a high school education or less (Pew Research Center 2010a, p. 11).

A second reason may seem ironic at first glance—the idea that we increasingly expect to find love in marriage. How would expecting to find love in marriage be associated with fewer of us being married? The following sections answer this question. To begin, we examine the time-honored marriage premise with its expectations for permanence and sexual exclusivity.

Zigy Kaluzny-Charles Thatcher/The Image Bank/Getty Images

Marking a couple's commitment, weddings are public events because the community has a stake in marriage as a social institution. Publicly proclaiming commitment to the marriage premise can help to enforce a couple's mutual trust in the permanence of their union. More and more, however, marriage seems to be reserved for the middle and upper classes—those who feel that they can afford this component of the American dream.

The Time-Honored Marriage Premise: Permanence and Sexual Exclusivity

Why does a marriage today require a wedding, witnesses, and a license from the state? Around 400 years ago in Western Europe, the government, representing the community, officially became involved in marriage (House 2002; Thornton 2009). For about one century before that, Roman Catholic Canon Law included rules, or canons, that regulated European marriage—although the canons, difficult to enforce in widely separated rural villages, were often ignored (Halsall 2001; House 2002; Therborn 2004).

The Netherlands first enacted a civil marriage law in 1590 (Gomes 2004). England passed its first marriage act in 1653 but did not require a legal marriage license until 1754 (House 2002). Shortly after Europeans established colonies in the United States, they enacted rules for marriage similar to those that they had known in Europe (Cott 2000). In the 400 years since then, our federal and state governments have generated a massive number of marriage-related laws and court decisions. For instance, polygamy has been illegal in the United States since 1878; and due to laws enacted at the turn of the twentieth century, unmarried cohabitation is still illegal in some states, although the laws are seldom enforced (Hartsoe 2005). Also, before issuing a marriage license, some states require blood tests for various communicable diseases. Many states have waiting periods, ranging from seventy-two hours to six days, between the license application date and the wedding ("Chart: State Marriage License" 2006).

Even in the absence of Canon Law, communities throughout the world, represented by kinship groups or extended families, had always claimed a stake in two important marriage and family functions: (1) guaranteeing property rights and otherwise providing economically for family members, and (2) ensuring the responsible upbringing of children (Ingoldsby 2006a).

Partly because of these essential social functions, also discussed in Chapter 1, social scientists have defined the family as a **social institution**—a fundamental component of social organization in which individuals, occupying defined statuses, are "regulated by social norms, public opinion, law and religion" (Amato 2004, p. 961). Social scientists typically point to five major social institutions: family, religion, government or politics, the economy, and education. In the vast majority of cultures around the world, a wedding marked a couple's passage into institutionalized family roles, usually well monitored by in-laws and extended kin.

Marriage marked the joining, not just of two individuals but of two kinship groups (Sherif-Trask 2003; Thornton 2009). From the couple's perspective, marriage had much to do with "getting good in-laws and increasing one's family labor force" (Coontz 2005b, p. 6). Family as a social institution has historically rested on the time-honored **marriage premise** of permanence, coupled in our society with expectations for monogamous sexual exclusivity.

The Expectation of Permanence

With few cross-cultural or historical exceptions, marriages have been expected to be lifelong undertakings—"until death do us part." **Expectations of permanence** derive from the fact that marriage was historically a practical institution (Coontz 2005b). Economic agreements between partners' extended families, as well as society's need for responsible child raising, required marriages to be "so long as we both shall live."

In the United States today, marriage seldom involves merging two families' properties. In other ways, too, marriage is less critically important for economic security. Furthermore, marriage today is less decisively associated with raising children, although marriage remains significantly related to better outcomes for children (Amato 2005; Furstenberg 2003; Popenoe 2008; Whitehead and Popenoe 2006)—a point that we will return to later in this chapter.

Meanwhile, another function of marriage—providing love and ongoing emotional support—has become key for most people (Cherlin 2004; Coontz 2005b). We explore how expectations for love in marriage affect those for permanence later in this chapter. Here we note that marriage is considerably less permanent now than in the past (Cherlin 2009a; Miller, Sassler, and Kusi-Appouh 2011). However, more than any other nonblood relationship, marriage holds the hope for permanence. At this point, we'll turn to the second component of the time-honored marriage premise: sexual exclusivity.

The Expectation of Sexual Exclusivity

Every society and culture that we know of has exercised control over sexual behavior. Put another way, sexual activity has virtually never been allowed simply on impulse or at random. Meanwhile, anthropologists have found an amazing array of permissible sexual arrangements. **Polygamy** (having more than one spouse) is culturally accepted in many parts of the world. Polygamy can be divided into two types. *Polygyny*, a form of polygamy whereby a man can have multiple wives, "is a marriage form found in more places and at more times than any other" (Coontz 2005b, p. 10). However, polygyny is not always that frequent, because many men cannot afford multiple wives. *Polyandry*—the second type of polygamy, wherein a woman has multiple husbands—is still less frequent (Stephens 1963).

Marriage in the United States legally disallows both forms of polygamy and requires monogamy, along with **expectations of sexual exclusivity**, in which spouses promise to have sexual relations only with each other. (There are exceptions to our cultural expectation for monogamy, however, and three of these exceptions are touched on in "Issues for Thought: Three Very Different Subcultures with Norms Contrary to Sexual Exclusivity.")

In Europe, requirements for women's sexual exclusivity emerged to maintain the patriarchal line of descent; the bride's wedding ring symbolized this expectation. The Judeo-Christian tradition eventually extended expectations of sexual exclusivity to include not only wives but also husbands. Over the last century, as "the self-disclosure involved in sexuality [came to] symbolize the love relationship," couples began to see sexual exclusivity as a mark of romantic commitment (Reiss 1986, p. 56). A 2007 poll asked Americans what makes a marriage work. Ninety-three percent of respondents listed "faithfulness" as very important. Seventy percent said a "happy sexual relationship" was very important, and 62 percent said sharing household chores was very important (Pew Research Center 2007).

A recent analysis compared findings from two important prior studies—one conducted in 1975 and one in 2000 (Gotta et al. 2011). Both studies compared aspects of heterosexually married, lesbian, and gay male relationships (Blumstein and Schwartz 1983; Solomon, Rothblum, and Balsam 2005). Results concerning sexual exclusivity are shown in Table 7.1. As you can see in that table, according to these data, gay men are least likely of the four categories to value sexual exclusivity, and lesbians are the most likely, with heterosexually marrieds falling in between. Meanwhile, every category examined—gay men, lesbians, married heterosexual males, and married heterosexual females—were more likely in 2000 than in 1975 to value sexual exclusivity. Among reasons for this change are the appearance in the 1980s of HIV/AIDS and the fact that the 1970s were generally a less conservative historical period (Gotta et al. 2011).

Moreover, expectations of sexual exclusivity have broadened from the purely physical to include expectations of emotional centrality, or putting one's partner first. Indeed, some marriage counselors now speak of "emotional affairs" (Herring 2005; Meier, Hull, and Ortyl 2009). With nearly 90 percent of us believing an affair is morally wrong (Gallup Poll 2012b), Americans are less accepting of extramarital sex than are people in many other monogamous societies. We'll note here that although the vast majority of Americans say that they disapprove of extramarital sex, the picture is

Issues for Thought

Three Very Different Subcultures with Norms Contrary to Sexual Exclusivity

Although a very substantial majority of Americans value monogamy as a cultural standard, there are subcultural exceptions. This box looks at three of these subcultural exceptions, each one very different from the others: polygamy, polyamory, and swinging.

Polygamy

Polygamy has been illegal in the United States since 1878, when the U.S. Supreme Court ruled that freedom to practice the Mormon religion did not extend to having multiple wives (*Reynolds v. United States* 1878). Today, some activists are pursuing U.S. legalization of polygamy (Stacey and Meadow 2009; Whitehurst 2012).

Although a 2006 Gallup poll found that one-quarter of Americans think that most Mormons endorse polygamy (Carroll 2006), this is not the case. The Church of Jesus Christ of Latter-day Saints (LDS) officially outlawed polygamy in 1890. Nevertheless, there are dissident Mormons (not recognized as LDS by the mainstream church) who follow the traditional teachings and take multiple wives (Woodward 2001; Whitehurst 2012). Some multiple wives have argued that polygyny is a feminist arrangement because the sharing of domestic responsibilities benefits working women (D. Johnson 1991; Joseph 1991).

Federal law prohibits prospective immigrants who practice polygamy from entering the United States. However, polygamy has been found in New York among some immigrants from countries in which it is practiced (Bernstein 2007). Civil libertarians argue that the Supreme Court should rescind its *Reynolds* decision on the grounds that the right to privacy permits this choice of domestic lifestyle as much as any other (Slark 2004).

Polyamory

Polyamory means "many loves" and refers to marriages in which one or both spouses retain the option to sexually love others in addition to their spouse (Polyamory Society n. d.).

Deriving their philosophy from the sexually open marriage movement, which received considerable publicity in the late 1960s and 1970s, polyamorous spouses agree that each may have openly acknowledged sexual relationships with others while keeping the marriage relationship primary. Unlike in swinging, outside relationships can be emotional as well as sexual. Couples usually establish limits on the degree of sexual or emotional involvement of the outside relationship, along with ground rules concerning honesty and what details to tell each other (Macklin 1987, p. 335; Rubin 2001). "Polyamorists are more committed to emotional fulfillment and family building than recreational swingers" (Rubin 2001, p. 721).

Some polyamorous couples are raising children. The Polyamory Society's Children Educational Branch offers advice for polyamorous parents and maintains a PolyFamily scholarship fund, as well as the Internet-based "PolyKids Zine" and "PolyTeens Zine," both designed to present "uplifting PolyFamily stories and lessons about PolyFamily ethical living" (Polyamory Society n. d.). Like polygamy, polyamory has received some media attention in the last several years, and polyamorists are working toward greater social acceptance. Some polyamorists want to establish legally sanctioned group marriages and have begun to organize in that direction (Anderlini-D'Onofrio 2004). Conservative groups, such as the Institute for American Values, see such moves as evidence of an emergent "radical sensibility" that threatens American values and harms children (Marquardt n. d., p. 30; Kurtz 2006).

Swinging

Swinging is a marriage arrangement in which couples exchange partners to engage in purely recreational sex. Swinging gained media and research attention as one of several "alternative lifestyles" in the late 1960s and early 1970s (Rubin 2001). At that time, it was estimated that about 2 percent of adults in the United States had participated in swinging at least once (Gilmartin 1977).

Although they receive relatively little research attention now, compared to their heyday in the 1970s, swingers continue to exist as a minority subculture. It has been estimated that there are now about 3 million married swingers in the United States, an increase of about 1 million since 1990. Some of this growth probably results from the Internet, which helps to link potential swingers (Rubin 2001). There is some evidence that the practice is gaining momentum among committed twenty-something couples.

Michael, a 28-year-old construction worker, and Sara, 24, who works in a doctor's office, have been in a committed relationship for more than a year but they do "full swaps," complete with intercourse, but they refuse to kiss strangers. "Sex is more of a primal, more of an urge-based," Michael said. "The kissing is more intimate so we like to keep that for us" (Chang and Lieberman 2012).

Swingers often face the challenge of managing jealousy, but they emphasize what they see as the positive effects—variety, for example (DeVisser and McDonald 2007; Chang and Lieberman 2012). A couple considering a sexually nonexclusive marriage must take into account not only personal values and relationship management challenges but also the increased risk of being infected with HIV. However, condoms are typically available at private parties and swing clubs (Rubin 2001).

Critical Thinking

What do you think about these exceptions to monogamy? Do you see them as threatening to American values? If so, why? If not, why not? Does one or more of them seem reasonable to you while others do not? If so, why? If not, why not?

Table 7.1 Types of Agreements Regarding Sex Outside the Relationship, 1975 and 2000

Percent of research participants who have discussed it and decided that under some circumstances sex outside the relationship is all right.		
Couple Type	**1975**	**2000**
Gay men	66.8	43.7
Lesbians	33.6	5.1
Married heterosexual males	23.0	6.0
Married heterosexual females	20.6	3.3
Percent of research participants who have discussed it and decided that under no circumstances is sex outside the relationship all right.		
Couple Type	**1975**	**2000**
Gay men	13.5	44.2
Lesbians	44.3	85.7
Married heterosexual males	41.9	80.6
Married heterosexual females	43.5	81.8
Percent of research participants who have discussed sex outside the relationship and don't agree with their partner.		
Couple Type	**1975**	**2000**
Gay men	12.4	4.2
Lesbians	15.3	2.7
Married heterosexual males	10.8	1.5
Married heterosexual females	11.4	2.5

Source: Gotta et al. 2011, Table 6, p. 368.

somewhat different in practice (Fincham and Beach 2010; Parker-Pope 2008). Sexual infidelity is explored in Chapter 4.

To summarize, the marriage premise has changed somewhat over the past century. Expectations for permanence have diminished while those for sexual exclusivity have been extended to include not just physical sex but also emotional centrality. The following section explores how these changes came about.

From "Yoke Mates" to "Soul Mates": A Changing Marriage Premise

Chapter 1 points to an individualist orientation in our society. In eighteenth-century Europe, **individualism** emerged as a way to think about ourselves. No longer were we necessarily governed by rules of community. Societies changed from **communal**, or **collectivist**, to **individualistic**. In individualistic societies, one's own self-actualization and interests are a valid concern. In collectivist societies, people identify with and conform to the expectations of their extended kin. Western societies are characterized as individualistic, and individualism is positively associated with valuing romantic love (Dion and Dion 1991; Goode 2007 [1982]). (By *Western*, we mean the culture that developed in Western Europe and now characterizes that region and Canada, the United States, Australia, New Zealand, and some other societies.)

The Industrial Revolution and its opportunities for paid work outside the home, particularly in the growing cities and independent of one's kinship group, gave people opportunities for jobs and lives separate from the family. In Europe and the North American colonies, people increasingly entertained thoughts of equality, independence, and even the radically new idea that individuals had a birthright to "the pursuit of happiness" (Coontz 2005b). These ideas were manifested in dramatically unprecedented political events of the late 1700s, such as the U.S. Declaration of Independence and the French Revolution.

The emergent individualistic orientation meant a generally diminished obedience to group authority, because people increasingly saw themselves as separate individuals, rather than as intrinsic members of a group or collective. Individuals began to expect self-fulfillment and satisfaction, personal achievement, and happiness. With regard to marriage, an emergent individualist orientation resulted in three interrelated developments:

1. The authority of kin and extended family weakened.
2. Individuals began to find their own marriage partners.
3. Romantic love came to be associated with marriage.

Weakened Kinship Authority

Kin, or extended family, includes parents and other relatives, such as in-laws, grandparents, aunts and uncles, and cousins. Some groups, such as Italian Americans, African Americans, Hispanics, and gay male and lesbian families, also have "fictive" or "virtual" kin—friends who are so close that they are hardly distinguished from actual relatives (Furstenberg 2005; Sarkisian, Gerena, and Gerstel 2006). In collectivist, or communal, cultures, kin have exercised considerable authority over a married couple. For instance, in traditional African societies, a mother-in-law may have more to say about how many children her daughter-in-law should bear than does the daughter-in-law herself (Caldwell 1982).

In Westernized societies, however, kinship authority is weaker. By the 1940s in the United States, at least

The New York Public Library/Art Resource, NY

The married couple embedded in this family of Eastern European immigrants who arrived in New York City in 1832 may be in love, but they were not *expected* to find love in marriage. Instead, their union is held together by strong expectations of permanence as bolstered by the social control of the kinship group.

among white, middle-class Americans, the husband–wife dyad was expected to take precedence over other family relationships. Sociologist Talcott Parsons noted that the American kinship system was not based on extended-family ties (1943). Instead, he saw U.S. kinship as comprised of "interlocking conjugal families" in which married people are members of both their **family of orientation** (the family they grew up in) and their **family of procreation** (the one formed by marrying and having children). Parsons viewed the husband–wife bond and the resulting family of procreation as the most meaningful "inner circle" of Americans' kin relations surrounded by decreasingly important outer circles. However, Parsons pointed out that his model mainly characterized the American middle class. Recent immigrants and lower socioeconomic classes, as well as upper-class families, still relied on meaningful ties to their extended kin.

Although the situation is changing, the extended family (as opposed to the married couple or nuclear family) has been the basic family unit in the majority of non-European countries (Ingoldsby and Smith 2005). In the United States, extended families continue to be important for various European ethnic families, such as Italians, and for Native Americans, blacks, Hispanics, and Asian Americans, as well as other immigrant families (Gowan 2009; Kent 2007; Mather 2009; Richardson 2009).

Norms about extended-family ties derive both from cultural influences and from economic or other practical circumstances (Hamon and Ingoldsby 2003; Wong, Yoo, and Stewart 2006). Immigrants from many less-developed nations work in the United States and send money to extended kin in their home countries (Ha 2006). Among Hispanics, *la familia* ("the family") means the extended as well as the nuclear family.

More and more Hispanics today value the primacy of the conjugal bond (Hirsch 2003). Meanwhile, like the Italians that Gans (1982 [1962]) studied in the 1960s, many Hispanics live in comparatively large, reciprocally supportive kinship networks (Sarkisian, Gerena, and Gerstel 2006; Lugo Steidel and Contreras 2003). For example, many Puerto Rican families have lived in "ethnically specific enclaves" and may rely as much on extended kin as on conjugal ties (Wilkinson 1993). Asian immigrants are also likely to emphasize extended-kin ties over the marital relationship (Glick, Bean, and Van Hook 1997).

We can sometimes get a glimpse of mainstream American individualism through the eyes of fairly recent immigrants from more collectivist societies. For example, a Vietnamese refugee describes his reaction to U.S. housing patterns, which reflect nuclear, rather than extended-family, norms:

> Before I left Vietnam, three generations lived together in the same group. My mom, my family including wife and seven children, my elder brother, his wife and three children, my little brother and two sisters—we live in a big house. So when we came here we are thinking of being united in one place. But there is no way. However, we try to live as close as possible. (quoted in Gold 1993, p. 303)

American housing architecture similarly discourages many Muslim families—from India, Pakistan, or Bangladesh, for example—who would prefer to live in extended-family households (Nanji 1993).

All this is not to say that extended-family members are irrelevant to nonHispanic white families in the United States. Nuclear families maintain significant emotional and practical ties with extended kin and parents-in-law (Lee, Spitze, and Logan 2003). A qualitative study with a sample that was 95 percent white showed that uncles often mentor nephews or nieces (Milardo 2005). Extensive data from the Longitudinal Study of Generations show that young adults today highly value their parents and extended families (Bengston, Biblarz, and Roberts 2007). However, as individuals and couples increasingly become more urban—and more geographically mobile—the power of kin to exercise social control over family members declines. If an individualist orientation has weakened kinship authority, it has also led to the desire to find one's own spouse.

Finding One's Own Marriage Partner

Arranged marriage has characterized collectivist societies (Hamon and Ingoldsby 2003; Ingoldsby 2006b; MacFarquhar 2006; Sherif-Trask 2003). Because a marriage joined extended families, selecting a suitable mate was a "huge responsibility" not to be left to the young people themselves (Tepperman and Wilson 1993, p. 73).

Analyzing arranged marriage in contemporary Bangladesh, sociologist Ashraf Uddin Ahmed notes that an individual's finding his or her own spouse "is thought to be disruptive to family ties, and is viewed as a child's transference of the loyalty from a family orientation to a single person, ignoring obligations to the family and kin group for personal goals" (Ahmed, quoted in Tepperman and Wilson 1993, p. 76). Moreover, there is concern that an infatuated young person might choose a partner who would make a poor spouse.

Ahmed argues that the arranged marriage system has functioned not only to consolidate family property but also to keep the family's traditions and values intact. But as urban economies developed in eighteenth-century Europe and more young people worked away from home, arranged marriages gave way to those in which individuals selected their own mates. Love rather than property became the basis for unions (Coontz 2005b).

Marriage and Love

Throughout the first 5,000 years of human history in all the world's cultures that we know of, people probably fell in love, but they weren't *expected* to do so with their spouses. Marriage was thought to be "too vital an economic and political institution to be entered into solely on the basis of something as irrational as love" (Coontz 2005b, p. 7). Love—an intense, often unpredictable, and possibly transitory emotion—was viewed as threatening to the practical institution of marriage. Valuing romance could lead individuals to ignore or challenge their social responsibilities.

An interesting way that Europe's twelfth- and thirteenth-century noblemen and women managed love's threat to marriage as a social institution was the practice of *courtly love.* As we have seen, most marriages in the upper levels of society during this period were based on pragmatic considerations, not love. But, as the saying went then, "marriage is no real excuse for not loving" (quoted in Coontz 2005b, p. 6). Among Europe's noblemen and women, romantic love was expressed in relationships outside marriage in which a knight worshipped his lady, and ladies had their favorites. These relationships involved a great deal of idealization and could be adulterous but were not necessarily sexually consummated (Stone 1980). The distinction between romance and marriage was also evident in the lower classes (Coontz 2005b, p. 7).

With time, the ideology of romantic love came to be expected of middle-class marriages (Meier, Hull, and Ortyl 2009). In family historian Stephanie Coontz's words, basing marriage on love and companionship

> represented a break with thousands of years of tradition.... Critics of the love match argued... that the values of free choice and egalitarianism could easily spin out of control. If the choice of a marriage partner was a personal decision,... what would prevent young people... from choosing unwisely? If people were encouraged to expect marriage to be the best and happiest experience of their lives, what would hold a marriage together if things were "for worse" rather than "for better"? (2005b, pp. 149–50)

To use Coontz's metaphor, couples were no longer yoked together (like field oxen). "Where once marriage had been seen as the fundamental unit of work and politics, it was now viewed as a place of refuge from work, politics, and community obligations—a haven in a heartless world" (Coontz 2005b, p. 146; Lasch 1977). A successful marriage came to be measured by how well the union met its members' emotional needs.

To summarize this section, emergent individualism in eighteenth-century Europe meant that people, increasingly valuing personal satisfaction and happiness, began to associate romantic love with marriage and, hence, to want to find their own marriage partners, a practice that both resulted from and further caused weakened kinship authority. Couples were no longer bound by the yoke of kin control. As you might guess, the nature of marriage changed. We'll explore that change next.

Deinstitutionalized Marriage

Coontz asserts that love and expectations for intimacy have "conquered marriage" (2005b). What does she mean? Coontz is talking about what family sociologist Andrew Cherlin (2004) has called the **deinstitutionalization of marriage**—a situation in which time-honored family definitions and social norms "count for far less" than in the past (p. 853). For instance, childbearing outside of marriage, once severely stigmatized, now "carries little stigma" (Cherlin 2009a, p. 919; but also see Usdansky 2009a, 2009b).

The following sections present and expand upon Cherlin's analysis of the shift from *institutional* to *companionate* to *individualized* marriage. As we discuss these three kinds of marriage, we need to remember that they are abstractions, or *ideal types.* In this context, the word *ideal* indicates that a type exists as an idea, not that it is necessarily good or preferable. Actual marriages approximate these types to varying degrees.

Institutional Marriage

We have witnessed a gradual historical change in Western and Westernized societies away from **institutional marriage**—that is, marriage as a social institution based on dutiful adherence to the time-honored marriage premise, particularly the norm of permanence (Cherlin 2004, 2009a; Coontz 2005b; Thornton 2009).

> Once ensconced in societal mandates for permanence and monogamous sexual exclusivity, the institutionalized marriage in the United States represented the age-old tradition of a family organized around economic production, kinship network, community connections, the father's authority, and marriage as a functional partnership rather than a romantic relationship.... Family tradition, loyalty, and solidarity were more important than individual goals and romantic interest. (Doherty 1992, p. 33)

Institutional marriage generally offered practical and economic security, along with the rewards that we often associate with custom and tradition (knowing what to expect in almost any situation, for example). With few exceptions over the past 5,000 years, institutional marriage was organized according to patriarchal authority, requiring a wife's obedience to her husband and the kinship group. It is also true that, legally, institutional marriage could involve what today we define as wife and child abuse or neglect. Child and wife abuse were not recognized as social problems in this country until the 1960s and 1970s, respectively.

Across cultures, the strength and scope of patriarchal authority varied, however. As an extreme example, in ancient Rome, the *paterfamilias* (family father), having absolute authority over his wife and children, could legally kill them or sell them into slavery. The occasions on which the *paterfamilias* actually exercised his authority to kill family members were rare, however. Nonetheless, no matter how old they were, sons were subject to the authority of the *paterfamilias* until he died. A daughter lived under her father's rule until she married, when her father's authority over her was legally transferred to her husband (Long 1875; Thompson 2006). In the United States, of course, patriarchal authority never approached anything near that of the ancient Roman *paterfamilias.*

Companionate Marriage

By the 1920s in the United States, family sociologists had begun to note a shift away from institutional marriage, and in 1945 the first sociology textbook on the American family (by Ernest Burgess and Harvey Locke) was titled *The Family: From Institution to Companionship.* By **companionate marriage**,

> Burgess was referring to the single-earner, breadwinner-homemaker marriage that flourished in the 1950s. Although husbands and wives in the companionate marriage usually adhered to a sharp division of labor, they were supposed to be each other's companions—friends, lovers—to an extent not imagined by the spouses in the institutional marriages of the previous era.... Much more so than in the 19th century, the emotional satisfaction of the spouses became an important criterion for marital success. However, through the 1950s, wives and husbands tended to derive satisfaction from their participation in a marriage-based nuclear family.... That is to say, they based their gratification on playing marital roles well; being good providers, good homemakers, and responsible parents. (Cherlin 2004, p. 851)

With companionate marriage, middle-class Americans often dreamed of attaining "the white picket fence"—that is, they saw marriage as an opportunity for idealized domesticity within the "haven" of their own single-family home. (This is why we have drawn a picket fence to symbolize the companionate marriage bond in Figure 7.1.)

Companionate marriages of the 1950s

> were exceptional in many ways. Until that decade, relying on a single breadwinner had been rare. For thousands of years, most women and children had shared the tasks of breadwinning with men.... Also new in the 1950s was the cultural consensus that everyone should marry, and that people should do so at a young age. The baby boom of the 1950s was likewise a departure from the past, because birthrates in Western Europe and North America had fallen steadily during the previous 100 years. (Coontz 2005c)

Meanwhile, women's increasing educational and work options, coupled with their expectations for marital love, sowed the seeds for the demise of companionate marriage (Cherlin 2004; Coontz 2005c).

An individualistic orientation views each person (both husband *and* wife) as having talents that deserve to be actualized. In this climate, women in companionate marriages began to pursue opportunities for self-actualization as well as to expect a husband's expressive support for their doing so (Jackson 2007). Furthermore, women challenged centuries of previously ignored domestic violence. Given the tension between gender inequality and expectations for emotionally supported self-actualization, the companionate marriage "lost ground" (Cherlin 2004, p. 852).

By the 1970s, observers noted a movement away from people's finding of personal satisfaction primarily in acceptable role performance—for example, in the role of husband-breadwinner or wife-homemaker. Research on college students showed a shift in self-orientation away from defining themselves according to the roles they played. More and more, they identified themselves

FIGURE 7.1A The institutional marriage bond. Couples are "yoked" together by high expectations for permanence, bolstered by the strong social control of extended kin and community.

FIGURE 7.1B The companionate marriage bond. Couples are bound together by companionship, coupled with a gendered division of labor, pride in performing spousal and parenting roles, and hopes for "the American dream"—a home of their own and a comfortable domestic life together.

FIGURE 7.1C The individualized marriage bond. Spouses in individualized marriages remain together because they find self-actualization, intimacy, and expressively communicated emotional support in their unions.

in terms of their individual personality traits. But individuals' appreciation for the esteem they get from playing their roles well "buttresses the institutional structure" (Turner 1976, p. 1,011; Babbitt and Burbach 1990). As one result, critics began to warn that American culture was becoming "narcissistic": Individuals appeared less focused on commitment or concern for future generations (Bellah et al. 1985; Lasch 1980).

Feminists defined this situation somewhat differently: Attention to domestic abuse, unequal couple decision making, and unfair division of household labor—as well as a wife's ability to more easily leave an intolerable situation through divorce—could be good things (Hackstaff 2007). Some celebrated the fact that American culture would finally begin to make room for "thinking beyond the heteronormative family" (Roseneil and Budgeon 2004, p. 136; Stacey 1996). Coontz summarizes the situation more neutrally: "For better or worse," over the past thirty years, "all the precedents established by the love-based male breadwinner family were ... thrown into question" (2005b, p. 11; 2005c). However one saw it, by the late 1980s, companionate marriage—which had lasted for but a minute in the long hours of human history—had largely given way to its successor, individualized marriage.

Individualized Marriage

Four interrelated characteristics distinguish **individualized marriage**:

1. It is optional.
2. Spouses' roles are flexible—negotiable and renegotiable.
3. Its expected rewards involve love, communication, and emotional intimacy.
4. It exists in conjunction with a vast diversity of family forms.

Partly because marriage is optional today, brides, grooms, and long-married couples have come to expect different rewards from marriage than people did in the past. They continue to value being good partners and, perhaps, parents. However, today's spouses are less likely to find their only, or definitive, rewards in performing these roles well (Byrd 2009). More than in companionate marriages, partners now expect love and emotional intimacy, open communication, role flexibility, gender equality, and personal growth (Cherlin 2004, 2009a; Meier, Hull, and Ortyl 2009). Perhaps they expect more personal autonomy as well: One recent study has found that couples in individualized marriages are less likely to pool finances, although a majority of marrieds still do (Lauer and Yodanis 2011). Over the course of about three centuries, couples have moved "from yoke mates to soul mates" (Coontz 2005b, p. 124).

Jeff Greenberg/PhotoEdit

As an ideal type, the *companionate marriage* that characterized most of the twentieth century emphasized love and compatibility, as well as separate gender roles. However, in reality, couples represent this ideal type to varying degrees. Although these immigrant parents, who own and operate a small candy store in "Little Saigon," Westminster, California, illustrate companionate marriage in *some* ways, they do not fit the definition of companionate marriage in at least one important way: They share the family provider role.

Intense romantic feelings have been associated with greater marital happiness and may serve to get a married couple through bad times (Udry 1974; Wallerstein and Blakeslee 1995). There can be a downside to all this, though. The idealization and unrealistic expectations implicit in individualized marriage can cause problems. Social theorist Anthony Giddens argues that expectations for a relationship based on intimate communication to the extent that "the rewards derived from such communication are the main basis for the relationship to continue" often lead to disappointment: "[M]ost ordinary relationships don't come even close" (Giddens 2007, p. 30). Giddens may be overstating the case. Probably many marriages do come close. However, the fact remains that such high expectations may be associated with the following results:

1. A person decides not to marry because she or he can't find a "soul mate" who can promise this level of togetherness;
2. a high divorce rate (although assuredly there are other reasons for divorce, too, as described in Chapter 14); and
3. a lower birthrate as individuals focus on options in addition to raising children, a topic addressed in Chapter 8.

One theme of this text is that society influences people's options and thereby impacts their decisions. To the extent that they are legally, financially, and otherwise able, people today organize their personal, romantic, and family lives as they see fit (Byrd 2009). Some engage in "dyadic innovation" (Green 2006, p. 182)—that is, they fashion their relationships with little regard to traditional norms. As a 28-year-old woman told an interviewer:

> Marriage, just because it's a piece of paper, doesn't necessarily mean it's a relationship or a long-standing relationship. A long-standing relationship can be a boyfriend. If you're with somebody and you love them, I don't really care about the piece of paper. So marriage really never enters my mind. (in Byrd 2009, p. 324)

To summarize, "How good is your relationship?" is often a question equal in importance to "Are you married?" (Giddens 2007).

In this climate, a wide variety of family forms emerge. What today we call the *postmodern* family (see Chapter 1), characterized by "tolerance and diversity, rather than a single-family ideal," takes many forms (Doherty 1992, p. 35). As noted in Chapter 1, some observers view the deinstitutionalization of marriage as a loss for society, a "decline" that hopefully can be turned around (e.g., Whitehead and Popenoe 2006). Others see the deinstitutionalization of marriage simply as an inevitable historical change (e.g., Coontz 2005b, 2005c).

Individualized Marriage and the Postmodern Family: *Decline* or Inevitable *Change*?

Those who view individualized marriage as a *decline* assert that our culture's unchecked individualism has caused widespread moral weakening and self-indulgence. They say that Americans, more self-centered today, are less likely than in the past to choose marriage, are more likely to divorce, and are less child-centered (Blankenhorn 1995; Popenoe 2007, 2008; Stanton 2004a, 2004b; Whitehead and Popenoe 2008). From this point of view, the American family has broken down.

Others, in contrast, see the deinstitutionalization of marriage as resulting from inevitable social *change*. These thinkers point out that, for one thing, people who look back with nostalgia to the good old days may be imagining incorrectly the situation that characterized marriage throughout most of the nineteenth and twentieth centuries. For instance, higher death rates for parents with young children meant that many children were not raised in two-parent households (Coontz 1992). Moreover,

> [just] as we cannot organize modern political alliances through kinship ties or put the farmers' and skilled craftsmen's households back as the centerpiece of the modern economy, we can never reinstate marriage as the primary source of commitment and caregiving in the modern world. For better or worse, we must adjust our personal expectations and social support systems to this new reality. (Coontz 2005c)

In a climate characterized by debate between spokespersons from these opposing perspectives, researchers and policy makers examine the social consequences of deinstitutionalized marriage.

Deinstitutionalized Marriage: Examining the Consequences

In her seminal 1995 presidential address to the Population Association of America, family demographer Linda Waite (1995) asked rhetorically, "Does Marriage Matter?" She concluded that indeed it does, for both adults and children. After thoroughly reviewing prior research that compared the well-being of family members in married unions with that of those in unmarried households, Waite reported that spouses as a category:

- had greater wealth and assets,
- earned higher wages,
- had more frequent and better sex,
- had overall better health,
- were less likely to engage in dangerous risk taking,
- had lower rates of substance abuse, and
- were more likely to engage in generally healthy behaviors.

Comparing children's well-being in married families with that of those in one-parent families, Waite found that children children in married families as a category:

- were about half as likely to drop out of high school,
- reported more frequent contact and better-quality relationships with their parents, and
- were significantly less likely to live in poverty.

Since her address, many sociologists and policy makers have further researched and debated Waite's findings. In the section following this one, we will examine the responses of policy makers. Here we review a sampling of demographic data and research findings on the question, "Does marriage matter?"

National income and poverty data apparently support Waite's argument that marrieds are financially better off. The median income for married-couple families in 2009 was $71,627, compared with just $41,501 and $29,770 for unmarried and male- and female-headed households, respectively (U.S. Census Bureau 2012c, Table 699). As Table 7.2 indicates, even when a wife is not in the labor force, married-couple households earn about $6,000 more per year than do unmarried male householders. This income gap is dramatically higher when marrieds are compared with unmarried female householders (U.S. Census Bureau 2012c, Table 699). Clearly, these data support the argument that higher income is positively associated with marriage. Furthermore, since Waite's address, studies have continued to find that, compared to unmarrieds, spouses in enduring marriages generally have better physical and mental health (Fincham and Beach 2010; Liu 2009; Popenoe 2008; Williams, Sassler, and Nicholson 2008).

Table 7.2 Median Income of Families by Types of Family in Constant (2009) Dollars: 1990 to 2009

	All Married-Couple Families	Married-Couple Families, Wife in Paid Labor Force	Married-Couple Families, Wife Not in Paid Labor Force	Unmarried Male Family Householder	Unmarried Female Family Householder
1990	$63,469	$74,418	$48,149	$46,210	$26,937
2000	$73,611	$86,236	$49,800	$46,991	$32,031
2009	$71,627	$85,948	$47,649	$41,501	$29,770

Source: U.S. Census Bureau 2012c, Table 699.

However, research also suggests that the association between marriage and positive outcomes is more complex than Waite indicated. For example, marrieds, on average, are less often depressed than the widowed and the divorced. But those in first marriages are not necessarily less depressed than either the remarried or the never-married (Bierman, Fazio, and Milkie 2006; LaPierre 2009). Then, too, in addition to being married, education, a comfortable income, and (among blacks) not having to suffer from society-wide racism improve mental health (Bierman, Fazio, and Milkie 2006; Mandara et al. 2008). Finally, marrieds have more frequent sex than unmarrieds when all unmarrieds are categorized together, but they do not have more frequent sex than cohabiting couples (Waite 1995).

An early criticism of Waite's claims was that much—although not all—of the association between marriage and positive outcomes was due to *selection effects*. In researchers' language, people may "select" themselves into a category being investigated—in this case, marriage—and this self-selection can yield the results for which the researcher was testing. Increasingly, individuals with superior education, incomes, and physical and mental health are more likely to marry (Bierman, Fazio, and Milkie 2006; England and Edin 2007; Goodwin, McGill, and Chandra 2009, Figure 6; Schoen and Cheng 2006; Teitler and Reichman 2008). The **selection hypothesis** posits that many of the benefits associated with marriage—for example, higher income and wealth, along with better health—therefore actually result from the personal characteristics of those who choose to marry (Cherlin 2003). For example, married women are more likely than those who are cohabiting or heading single-family households to inherit wealth (Ozawa and Lee 2006). Being positioned to inherit wealth from one's family of origin is a personal characteristic that *precedes* getting married.

Nevertheless, not all the benefits associated with marriage are accounted for by selection effects. In contrast to the selection hypothesis, the **experience hypothesis** holds that something about the *experience* of being married itself causes these benefits—a point that we will return to at the end of this chapter. Figure 7.2 illustrates the selection and the experience hypotheses. Meanwhile, considerable research has focused on examining the relationships between marriage and the consequences for children.

Child Outcomes and Marital Status: Does Marriage Matter?

The proportion of children under age 18 living with two married parents declined steadily over the past forty years—from 85 percent in 1970 to 77 percent in 1980 and then to 69 percent in 2011 (U.S. Federal Interagency Forum on Child and Family Statistics 2006, 2012). About 20 million children under age 18 (27 percent of all U.S. children) live in single-parent households. Twenty-three percent of all U.S. children reside in single-mother households, with another 4 percent living with single fathers (U.S. Federal Interagency Forum on Child and Family Statistics 2012). In 2010, 30 percent of all U.S. families had one parent: 25 percent maintained by a mother and 5 percent by a father. Forty-six percent of all single-parent families are nonHispanic white. Blacks comprise 28 percent of single-parent families, Hispanics 21 percent, and Asians 2 percent (U.S. Census Bureau 2012c, Table 67).

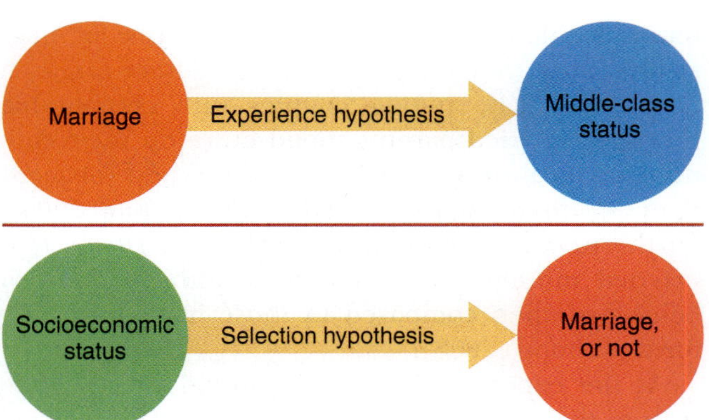

FIGURE 7.2 Causal order: Experience hypothesis, Selection hypothesis.

Source: Adapted from Marsh et al. 2007, p. 739.

Also pointed out in Chapter 6, some "single" parents have cohabiting partners. Of all U.S. children, about 5.1 million (7 percent) live with a parent or parents who are cohabiting. Of children who live with cohabiting couples, about half live with both of their unmarried biological or adoptive parents (U.S. Federal Interagency Forum on Child and Family Statistics 2012).

Considerable research supports Waite's overall conclusion that growing up with married parents is better for children (Wilcox et al. 2011a; Kreider and Elliott 2009a, 2009b; Magnuson and Berger 2009; Sun and Li 2011). For instance, when compared with teens in homes with two married biological parents, those in single-parent and cohabiting families are more likely to experience earlier premarital intercourse, lower academic achievement, and lower expectations for college, together with higher rates of school suspension, delinquency, and marijuana use (Carlson 2006; Mandara, Rogers, and Zinbarg 2011; Manning and Lamb 2003; VanDorn, Bowen, and Blau 2006).

Part of the advantage that marriage gives to children has to do with the better financial circumstances of married families (Fagan 2011). However, studies that compared economically disadvantaged six- and seven-year-olds from families of various types found fewer problem behaviors among children in married families (Ackerman et al. 2001). Other research has found that, among couples with comparable incomes, married parents spend more on their children's education (and less on alcohol and tobacco) than do cohabiting parents (DeLeire and Kalil 2005). Furthermore, analysis of national data shows that married mothers exhibited the healthiest prenatal behaviors when compared to those in other family forms (Kimbro 2008).

As the experience hypothesis would suggest, one reason that, as a category, children with married parents evidence better outcomes may be the *experience* of growing up in a married-couple household. With its presumption of permanent commitment to the family as a whole, marriage "allows caregivers to make relationship-specific investments in the couple's children—investments of time and effort that, unlike strengthening one's job skills, would not be easily portable to another relationship" (Cherlin 2004, p. 855; and see Wilcox et al. 2011a; Popenoe 2008).

However, as with the benefits of marriage for adults, researchers have uncovered complexities when working to unravel the statistical correlation between marriage and positive child outcomes. For instance, findings differ according to how the variable *marriage* is defined. Results differ when the variable *marriage* allows an investigator to compare the effects of having two biological, continuously married parents with those having a remarried stepparent. Using data from the National Longitudinal Survey of Youth, one study (Carlson 2006) compared outcomes for adolescents in several family structures. Similar to prior research, this study found fewer behavior problems among teens who lived with their continuously married biological parents. However, adolescents born outside of marriage but whose biological parents later married or were cohabiting, or whose mother married a stepfather, had more behavior problems than teens whose biological parents were continuously married (Carlson 2006). Furthermore, transitions to and from various family structures have been found to result in poorer outcomes for children (Amato 2012; Bures 2009; Magnuson and Berger 2009).

In addition to refining the marriage variable, researchers have proposed supplementary or alternative causes for children's marriage-associated benefits. For one thing, children raised by married families are less likely to live in poverty—a situation that has serious negative effects on child outcomes (Moore et al. 2009). We know that married parents are less likely to live below poverty level, but factors in addition to marital status are related to poverty as well. For example, in a study focused on Hispanic children, researchers showed that, for Mexican American children, poverty is related to a combination of marital status and the number of children in the household, the latter "an important predictor of poverty regardless of marital status" (Crowley, Lichter, and Qian 2006).

Rhoberazzi/Photodisc/Getty Images

Hispanics and African Americans have higher percentages, or *rates*, of mother-headed, single-parent families. Nevertheless, the majority of mother-headed, single-parent families are nonHispanic white. Also, Hispanics and African American families have higher poverty *rates*, but the majority of families in poverty are nonHispanic white. Although about one-third of mother-headed, single-parent families live below the official poverty level, nearly two-thirds do not.

Besides marital status and poverty, still other factors correlate with children's outcomes—the child's neighborhood and peers, family conflict, parental nurturance and involvement in the child's school activities, parents' participation in religious services, and parents' available social support (Broman, Li, and Reckase 2008; Crawford and Novak 2008; Ryan, Kalil, and Leininger 2009; Wen 2008; Wu and Hou 2008).

Studies have found that *father involvement*—the extent to which a biological father is engaged with his child—is important regardless of whether he is married to his child's biological mother (Bronte-Tinkew et al. 2008; Carlson 2006; Cooper, Crosnoe et al. 2009). Accordingly, sociologist Leslie Gordon Simons and her colleagues distinguished between what they call the *marriage perspective* and the *two-caregivers perspective*: "What we label the marriage perspective rests on the assumption that children are most likely to display healthy growth and development when they are raised by married parents." In contrast, the two-caregivers perspective contends that children do best when raised by two caregivers rather than by a single caregiver (Simons et al. 2006, p. 805).

Analyzing data on 867 African American children from the Family and Community Health Study, Simons and her colleagues found that "child behavior problems were no greater in either mother-grandmother or mother-relative families than in those in intact nuclear families." At least among blacks, these researchers found mother–grandmother families to be "functionally equivalent" (Simons et al. 2006, p. 818). Then, too, other extended kin in black families—uncles, for example—may be involved in child care (Richardson 2009).

Interestingly, research based on a national representative sample of more than 10,000 U.S. teens found that the negative effects of time lived with a single mother were less serious for black and Hispanic adolescents than they were for whites (Mandara, Rogers, and Zinbarg 2011). Why might this be? First, although nonmarried parenthood remains somewhat stigmatized in the United States (Usdansky 2009a, 2009b), it is noteworthy that this stigma may be minimal within the black community where single-parent families are more normative (Heard 2007; Liu and Reczek 2012). Accordingly,

> The common history among blacks allows for the emergence and primacy of social supports, such as women-centered kinship networks, coresidence with extended family, and strong ties to the church, which can buffer the negative effects of stress caused by family instability. (Heard 2007, p. 336)

(For more on blacks and marriage, see "A Closer Look at Diversity: African Americans and 'Jumping the Broom.'")

As with the benefits of marriage for adults, researchers hypothesize that selection effects explain much—although not all—of marriage's advantage for children. We've seen that, on average, individuals who marry are better educated and have higher incomes. As parents, they live in neighborhoods more conducive to successful child raising. Less likely to be stressed because of financial problems, they are more likely to practice effective parenting skills (Manning and Brown 2006; also see Teachman 2008b). In her address to the Population Association of America, Waite acknowledged the contribution of selection effects to child outcomes. She added, however, that "we have been too quick to assign *all* the responsibility to selectivity here, and not quick enough to consider the possibility that marriage *causes* some of the better outcomes we see…" (1995, p. 497, italics in original).

To summarize, a large body of research shows that marriage is associated with benefits for adults and children. However, this relationship is complex, and much of it may be due to variables other than marital status as well as to selection effects (Teachman 2008b). A theme of this text is that research findings expand our knowledge so that we can better make decisions knowledgeably. One thing that the research implies is that, although generally marriage is advantaged, additional factors affect individual outcomes, and the disadvantages for some ethnic groups may not be as severe as for the population as a whole.

Just as researchers have responded in various ways to Waite's address, policy makers have had conflicting reactions. More conservative policy leaders, associated with the *decline* and "family breakdown" perspective, once hoped to effect a "family turnaround" (Whitehead and Popenoe 2003). They noted "a greater emphasis on short-term gratification and on adults' desires rather than on what is good for children" that they attributed to government welfare programs and to decreased attention to religious principles (Giele 2007, p. 76).

Today, these policy makers appear to have given up on a broad "family turnaround." In the words of sociologist David Popenoe (2008):

> [Realistically] there is probably not much that government policies or social action can do to change the situation. If major change is to come about it will have to occur through a broad cultural shift, reflected in the hearts and minds of the citizenry, in the direction of stronger interpersonal commitments and families…. Still, there surely are actions that societies can take to try to improve the situation and not make it worse; actions that discourage cohabitation and encourage marriage, at least when children are involved. (pp. 15–16)

On the other hand, policy makers who see marriage simply as *changing* recognize that many families are struggling but criticize the solutions offered by conservatives and propose their own. The following section explores this policy debate.

A Closer Look at Diversity

African Americans and "Jumping the Broom"

Nationally representative surveys show that, among blacks, husbands and wives, like other Americans, are more likely than unmarrieds to report being "very happy" and satisfied with their finances and family life (Blackman et al. 2006). Although white husbands consistently report the most marital happiness when compared to white wives and to black husbands and wives, there has been a steady decline in this gap (Corra et al. 2009). A significant proportion of African American couples have strong, enduring marriages (Marks et al. 2008). Meanwhile, with 43 percent of black men and 36 percent of black women currently married, African Americans are considerably less likely to be wed than are other U.S. race/ethnic groups (U.S. Census Bureau 2012c, Table 56). A large body of literature, written by both blacks and whites, is accumulating on the structural–cultural reasons for this situation (McAdoo 2007; Banks 2011; Chambers and Kravitz 2011). Contrary to some people's opinions, research also indicates that the availability of welfare is not a significant factor in a black woman's decision not to marry (Berlin 2007; Teachman 2000).

Answering a Gallup poll, 69 percent of African Americans said that it is "very important" for a couple to marry when they plan to spend the rest of their lives together. Asked, "When an unmarried man and woman have a child together, how important is it to you that they legally marry?" college-educated African Americans were *more* inclined than either Hispanics or nonHispanic whites to say that marrying in this situation is "very important." The figures were 55 percent of blacks, 46 percent of Hispanics, and 37 percent of nonHispanic whites (Saad 2006b). Furthermore, attitude surveys consistently show that African Americans value marriage, perhaps more than nonHispanic whites do (Banks 2011; Johnson and Staples 2005; Saad 2006b).

Spencer Grant/Alamy

Although somewhat controversial because it can be a reminder of slavery, jumping the broom at African American weddings is going through some revival as black couples plan wedding celebration rituals designed to incorporate their cultural heritage. Attitude surveys show that African Americans value marriage highly. However, popular media as well as scholarship have focused on the low proportion of married blacks, and we tend to ignore the nearly 40 percent of African American adults who *are* married.

The news media have focused so frequently on poverty-level African Americans and on the relatively low proportion of married blacks of all social classes that we may forget about the 10.6 million (39 percent of) African Americans who *are* married (U.S. Census Bureau 2012c, Table 56). Google "African American marriage" and you'll find several websites marketing wedding products and services to middle-class blacks. One book, *Jumping the Broom* (Cole 1993), is a wedding planner for American blacks. If you're not African American, there's a fairly good chance that you have not heard of jumping the broom. What is it?

For African Americans, the significance of the broom originated among the Asante in what is now the West African country of Ghana. Used to sweep courtyards, the handmade Asante broom was also symbolic of sweeping away past wrongs or warding off evil. Brooms played a part in Asante weddings as well. To culminate their wedding ceremony, a couple might jump over a broom lying on the ground or leaning across a doorway. Jumping the broom symbolized the wife's commitment to her new household, and it was sometimes said that whoever jumped higher over the broom would be the family decision maker (DiStefano 2001; Prahlad 2006).

Among slaves brought to the Americas, jumping the broom continued. Not allowed to marry legally, slaves sometimes jumped a broom as an alternative ceremony to mark their marital commitment. The association of jumping the broom with slavery has

Valuing Marriage: The Policy Debate

Policy advocates from a marital *change* perspective are mainly concerned about the high number and proportions of parents and children living in poverty. They view poverty as causing difficult child-raising environments with resulting negative outcomes for some—though not all—of America's children. From this viewpoint, family struggle results from structural conditions such as recession. Accordingly, these spokespersons argue for structural, or ecological (such as neighborhood-level), solutions.

stigmatized the tradition for some African Americans. However, the ritual is coming back as more middle-class blacks seek culturally relevant wedding celebrations (African Wedding Guide n. d.; Anyiam 2002; DiStefano 2001).

African American Families

African Americans have been increasingly divided between a middle class that has benefited from the opportunities opened by the civil rights movement and a substantial sector that remains disadvantaged.

Childbearing and child rearing are increasingly divorced from marriage. True of all race/ethnic groups, this trend is especially pronounced among blacks, with 72 percent of births in 2011 to unmarried mothers (Child Trends 2012), and—as shown in Chapter 6's Figure 6.1—black couples are far more likely than the national average to have never married (U.S. Census Bureau 2012c, Table 56). As a consequence, only 35 percent of African American children are living with their biological married parents, compared with 75 percent of white (nonHispanic) and 64 percent of Hispanic children (U.S. Federal Interagency Forum 2012).

Differences between African Americans and the national average in the proportion of two-parent families are not new, but as recently as the 1960s, more than 70 percent of black families were headed by married couples, whereas 45 percent were in 2011 (Billingsley 1968; U.S. Census Bureau 2012a, Table F1). Experts do not agree on the cause of the decline in marriage and two-parent families among African Americans, but economic and employment factors are noted in much of the literature as important components of the issue. In fact, research shows that African Americans value marriage. For example, a Gallup poll taken in 2006 finds 69 percent of blacks agreeing that marriage is "very important" "when a man and woman plan to spend the rest of their lives together as a couple"—*higher than the figure for whites* (Saad 2006b).

Given similar values regarding marriage, research consistently suggests that the primary sources of difference in marital patterns are cultural, resulting from enforced family patterns during slavery, and also economic (Banks 2011). Our economy's shift away from manufacturing has meant the elimination of the relatively well-paying entry-level positions that once sustained black working-class families. Low levels of black male employment and income may preclude marriage or doom it from the start (Burton et al. 2009; Holland 2009; Joshi, Quane, and Cherlin 2009; Taylor 2007; W. J. Wilson 2009). Then, too, there is evidence that discrimination plays a part.

African American women have traditionally been employed, and they may be less dependent on the earnings of a spouse for economic survival. Married black women tend to have higher employment rates than their white counterparts (Corra et al. 2009; Durr and Hill 2006). But the *economic independence* explanation of low marriage rates is not supported by research; it appears that the better their earnings, the more likely black women are to marry (Berlin 2007; Tucker 2000). Research also indicates that the availability of welfare is not a significant factor in a black woman's decision to marry (Berlin 2007; Teachman 2000).

Another possible explanation for the lower marriage rates of African Americans is the sex ratio (Taylor 2007). High rates of incarceration, what some black scholars call "the prisonization of black America" (Clayton and Moore 2003, p. 85), as well as poorer health and higher mortality, have taken many African American men out of circulation.

Scholars have noted the strengths of black families (Hill 2003 [1972]; S. A. Hill 2004; Taylor 2007), especially strong kinship bonds. Single-parent families or unmarried individuals are often embedded in extended families and experience family-oriented daily lives. Black extended families are culturally predisposed to accept children regardless of circumstances (Crosbie-Burnett and Lewis 1999). In a child-focused family system, the extended family and community are involved in caring for children; their survival and well-being do not depend on the parents alone (Uttal 1999, p. 855).

Critical Thinking

How do you think that jumping the broom might be used to symbolize the time-honored marriage premise? Why would it be important to incorporate traditions that are relevant to one's own culture into a wedding ceremony? Why do you think we hear relatively little about African Americans' weddings or marriages?

From the *decline* perspective, on the other hand, concerns about "family breakdown" include the high number of federal dollars spent on "welfare" for poverty-level single mothers, coupled with the irresponsible socialization of children (Giele 2007). They define the causes for these concerns as primarily cultural, such as changes in individuals' values and attitudes regarding marriage. Therefore, they offer motivational and educational programs to effect a family "turnaround."

Policies from the Family Decline Perspective

An important goal from the *decline* perspective is to return to a society more in line with the values and norms of companionate marriage. As means to that end, advocates have established programs to encourage marital permanence. Many religions insist on premarital counseling as a way to dissuade couples in inappropriate matches from marrying and to encourage those who do marry to stay together (Nock 2005; Nadir 2009; Ooms 2005).

Covenant Marriage Some conservative Christian organizations and legislators advocate **covenant marriage**. In covenant marriage, partners agree to be bound by a marriage "covenant" (stronger than an ordinary contract) that will not let them get divorced easily ("Covenant Marriages Ministry" 1998). Between ten and twenty years ago, three states—Louisiana, Arizona, and Arkansas—enacted covenant marriage laws. Twenty other states have considered but failed to pass covenant marriage laws (Leon 2009).

How does covenant marriage work? Before their wedding, couples choose between two marital contracts, conventional or covenant. If legally bound by a marriage covenant, couples are required to get premarital counseling and may divorce only after being separated for at least two years or if imprisonment, desertion for one year, adultery, or domestic abuse is proved in court. In addition, a covenant couple must submit to counseling before a divorce (Brown and Waugh 2004).

Typically, fundamentalist Christian religions are enthusiastic about covenant marriage, whereas feminists and other critics are not ("Couple Support" 2006; Leon 2009; Sanchez et al. 2002). For instance, critics point out that proving adultery or domestic abuse in court may be difficult and expensive, whereas living in a violent household can be deadly (Gelles 1996).

Despite promoters' early enthusiasm, covenant marriage has failed to become a serious social movement, never having spread beyond the original three enacting states. Relatively few couples in the states where it is available have opted for covenant marriage ("More Binding Marriage" 2004). The federal Healthy Marriage Initiative (HMI) is another program that encourages marital stability (Chaney 2009; Fincham and Beach 2010).

Government Initiatives With funding from the federal Healthy Marriage Initiative, states have promoted marriage education, some offering money incentives for couples to participate. Other state initiatives include home visitation programs for families that might be targeted for government assistance because of a variety of reasons, such as a birth to a teenager or an unstable marriage; mentoring, marriage counseling, communication skills, and anger management workshops; state-funded resource centers that provide information on marriage; and state websites that include marriage enrichment information and links to service-related sites (Dion 2005; Ooms 2005).

These programs largely began after 2004, when Congress reauthorized the Temporary Assistance for Needy Families (TANF), or "welfare reform" program. The 1996 Personal Responsibility and Work Opportunity Reconciliation Act, or "welfare reform bill," effectively ended the federal government's sixty-year guarantee of assisting low-income mothers and children. The federal Aid to Families with Dependent Children (AFDC) program ended in 1997, and TANF, a different federal program, ensued. TANF limits assistance to five years for most families, with most adult recipients required to find work within two years.

To complement TANF, the Healthy Marriage Initiative emphasized the goal of ending the dependence of single parents on government benefits by promoting not only job preparation but also marriage (Carlton et al. 2009; Healthy Marriage Initiative 2009). Numerous grants have been awarded to researchers interested in looking at ways to encourage marriages that will endure (Fincham and Beach 2010). HMI supporters argue that giving single mothers "accurate information on the value of marriage in the lives of men, women, and children," along with marriage skills education, encourages marriage and reduces divorce (Ooms 2005; Rector and Pardue 2004).

The disparity in marriage rates between the poor and those who are not poor has become significant enough that social scientists have coined a term for this situation—the *marriage gap*. Meanwhile, critics of programs specifically designed to motivate people to marry argue that low-income Americans value marriage and would like to marry, but marriage is difficult to achieve for many of them (Trail and Karney 2012). Low-income single mothers want trustworthy, steadily employed husbands who will help with both finances and child care (Burton et al. 2009; Joshi, Quane, and Cherlin 2009).

As a young, college-educated woman in a qualitative study of African American single mothers explained, "I realized that when I do decide to enter marriage, my partner must have the same ambitions as me or similar to [mine].... Not too many men my age or older have the ambition that I have" (in Holland 2009, p. 173). For rich and poor alike, a wedding symbolizes personal achievement (Cherlin 2004; Mandara et al. 2008). After interviewing women in low-income neighborhoods, two researchers concluded that, for poor women, marriage

> has become an elusive goal—one they feel ought to be reserved for those who can support... a mortgage on a modest row home, a car and some furniture, some savings in the bank, and enough money left over to pay for a "decent" wedding. (Edin and Kefalas 2007, p. 508; also see Gibson-Davis 2009; King and Allen 2009)

Because of declining work opportunities for the less educated and consequent high unemployment rates for men in poor neighborhoods, many potential husbands in these communities cannot promise a steady income (Burton and Tucker 2009; Harris and Parisi 2008; Huston and Melz 2004). Poor women are not necessarily good marriage prospects either. They may have less than desirable economic histories and potential, as well as mental, physical, or substance abuse issues. In addition, they may evidence **multiple partner fertility**—that is, having children by more than one biological father—a situation that renders them less desirable in the eyes of a future male partner (Manning et al. 2010).

Relieving poverty will require solutions other than—or at least in addition to—promoting marriage. Efforts to foster low-income marriages "should directly confront the economic and social realities these couples face" (Trail and Karney 2012, p. 413).

The Relationship between Marriage and Poverty More than 14.7 million U.S. children under age 18 live at or below the poverty line; 22 percent, or more than one in five, of U.S. children live in poverty (U.S. Census Bureau 2012c, Table 712; U.S. Federal Interagency Forum on Child and Family Statistics. 2012). An additional 5 million children live in "near poor" conditions—that is, above the official poverty line but at less than 125 percent of the poverty level (U.S. Census Bureau 2012e, Table S1703). Compared with 1980, poverty rates have declined by about 7 percentage points for African-American children. Poverty declined for Hispanic children as well, but the 2007–2009 recession took back the vast majority of Hispanic gains (U.S. Census Bureau 2012c, Table 712). In 2011, the race/ethnic breakdown of children below 125 percent of poverty level was as follows:

- 14.7 percent of nonHispanic white children,
- 16.8 percent of Asian children,
- 27.5 percent of Native Hawaiian or other Pacific Islander children,
- 34.0 percent of Hispanic children,
- 34.8 percent of African American children, and
- 36.7 percent of American Indian or Alaska Native children (U.S. Census Bureau 2012e, Table S1703).

Figures such as these may lead us to think of poverty in terms of people of color, but the majority of poor children are nonHispanic white. Despite lower *rates* of poverty, nonHispanic whites predominate in sheer numbers, comprising nearly two-thirds (64 percent) of all poor children in the United States (U.S. Census Bureau 2012c, Table 712).

Children growing up in poverty often do not have enough nutritious food; are more likely to live in environmentally unhealthy neighborhoods; have more physical health, socioemotional, and behavioral problems; must travel farther to attain health care; attend poorly financed schools; do less well academically; and are more likely to drop out (Goosby 2007; Moore et al. 2009; Teachman 2008b).

Robert Rector, Senior Research Fellow at the Heritage Foundation, a conservative American think tank, argues that *marriage* is "America's greatest weapon against child poverty" (Rector 2012). From this point of view, the "costs of family instability are not just borne by individuals. They are, to a significant extent, borne by taxpayers who provide income support for many parents and their children, pay substantial administrative costs in ensuring income transfers through the child support system" (Parkinson 2011).

Data that relate child poverty rates to children's living arrangements show that residing with married parents does significantly lessen the likelihood of growing up in poverty (Fagan 2011; Kreider and Fields 2005, Table 2). As you can see from Table 7.3, when all races are taken together, 5.8 percent of married-couple families live below the official poverty line. This figure compares with 16.1 percent of single-male householder families and with 31.4 percent of single-female householder

Table 7.3 Percent of U.S. Families Below Poverty Level, 2011

Family Type	Married Couple	Male Householder, No Spouse Present	Female Householder, No Spouse Present
All races (%)	5.8	16.1	31.4
NonHispanic White (%)	3.8	11.5	24.2
Black (%)	8.6	20.4	38.4
Hispanic (%)	15.9	14.7	40.6
Asian (%)	7.4	11.6	21.5
American Indian/Alaska Native (%)	16.0	**	26.5

**No data available

Sources: DeNavas-Walt, Proctor, and Smith 2012; U.S. Census Bureau 2012e, Table S1702.

families. We might conclude that encouraging people to get married would work *somewhat* to lessen poverty (Amato 2005; Thomas and Sawhill 2005).

However, the association between marriage and poverty is hardly the whole story. Table 7.3 shows that—despite the fact that they are married—3.8 percent of nonHispanic white, 8.6 percent of black, nearly 16 percent of Hispanic, 7.4 percent of Asian, and 16 percent of American Indians or Alaska Natives live in poverty. Obviously, marriage alone is not sufficient to alleviate poverty. For one thing, female householders with no spouse present are almost twice as likely as their male counterparts to live in poverty (31.4 percent, compared with 16.1 percent). In addition to marital status, low wages for women contribute to poverty.

Moreover, a majority of unmarried families is not living in poverty. We must conclude that marriage contributes to a family's economic well-being, but a child does not absolutely need married parents to grow up above the poverty line.

As Figure 7.3 illustrates, the child poverty rate for all races calculated together was about 27 percent in 1959, but beginning with President Lyndon Johnson's **War on Poverty** in the 1960s, it dropped consistently during the 1970s to a low of about 14 percent. You may have heard of War on Poverty programs, such as the Job Corps or the Neighborhood Youth Corps, Head Start, or Adult Basic Education. Although the majority of War on Poverty measures have ended, Head Start and the Job Corps continue to exist.

In the 1970s, a series of economic recessions occurred along with the phasing out of many War on Poverty measures. As a result, the child poverty rate began to rise in the late 1970s and then fell again after about 1993. However, the rate began to rise again in 2000, and the rate rose markedly during the most recent recession. In 2011, the child poverty rate was 22 percent—up from 15.6 percent at the turn of the twenty-first century (Proctor and Dalaker 2003; U.S. Census Bureau 2012c, Table 712; U.S. Federal Interagency Forum on Child and Family Statistics. 2012).

The War on Poverty offered *structural* strategies to decrease poverty, such as community meal programs and health centers, legal services, summer youth programs, senior centers, neighborhood development, adult education, job training, and family planning (Garson n. d.). Commitment to the War on Poverty diminished after the 1970s, with national rhetoric shifting to debates focused on individual responsibility. Today, however, scholars and some policy makers are again insisting that the United States must pay attention to ecological and structural supports for children and families regardless of—or in addition to—concerns about changing family structure (Cherlin 2009a; Moore et al. 2009; Popenoe and Whitehead 2009).

Policies from the Family Change Perspective

Many policy makers maintain that Americans are struggling with economic and time pressures that get in the way of their ability to realize family values (Ozawa and Lee 2006; Teitler et al. 2009). As remedies for poverty, policy leaders in this camp propose structural solutions such as support for education, job training, drug rehabilitation, improved job opportunities, neighborhood improvements, small business development, and parenting skills education (Amato 2005; S. Brown 2004; Ozawa and Lee 2006). Indeed, "[l]ow-income communities have been neglected for so long that the resources needed to rebuild them will require a major shift in

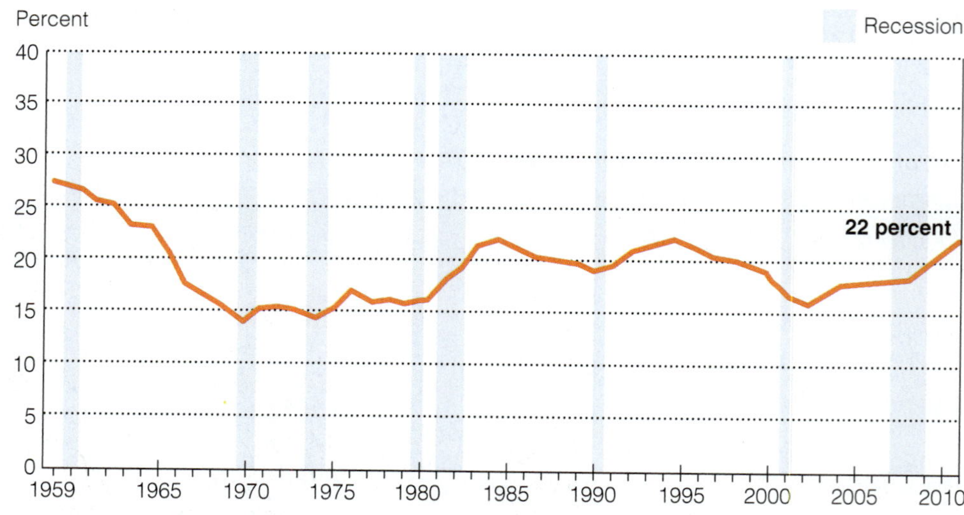

FIGURE 7.3 U.S. poverty rate for children under age 18, 1959 to 2011.

Source: DeNavas-Walt, Proctor, and Smith 2012, Figure 5.

Evelyn Hockstein/MCT/Landov

The disparity in marriage rates between the poor and those who are not poor has become significant enough that social scientists have coined a term for this situation—the *marriage gap*. Some experts, usually from the *family decline* perspective, see marriage as an antidote to poverty. However, low-income Americans generally value marriage and would like to marry, but marriage appears difficult to achieve for many, although certainly not all, of them.

public priorities over an extended period of time, possibly generations" (Huston and Melz 2004, p. 956).

Andrew Cherlin (2009a) argues that in a climate of "contradiction" between the American values of commitment to marital stability and individual freedom and happiness, it is a bit naïve today to think that encouraging people to get or stay married will work to better facilitate raising children responsibly. He finds marital stability virtually impossible to enhance in our American values climate, so he argues that it is not pragmatic to continue to insist on legal marriage as a public policy goal. Instead, he argues for *family stability*—supporting children and therefore their parents in whatever family form they find themselves (Cherlin 2009a; McHale, Waller, and Pearson 2012). An example of policy measures in this vein include community educational interventions designed to promote healthy and supportive coparenting relationships regardless of the parents' marital status (McHale, Waller, and Pearson 2012).

Unfortunately, finding resources for ecological and structural support for families is even more difficult today than before the recession that began in 2007. Not only are resources more scarce, but also politicians and others debate whether (1) "welfare" encourages single parenthood while lessening the motivation to work, or (2) some form of family "safety net" is necessary and should not be stigmatized (DeParle 2009).

Having looked at research and policy on the question of whether marriage matters, we can conclude that marriage does matter, at least for those who can afford to get married.

Happiness, Well-Being, and Life Satisfaction: How Does Marriage Matter?

Although somewhat more true for husbands than for wives, academic research (e.g., Lee and Ono 2012; Wienke and Hill 2009) and opinion polls find marrieds more likely than others to say they are "very happy" and to score higher on well-being indicators such as physical and emotional health (Brown and Jones 2012; Taylor, Funk, and Craighill 2006). Again we face the question whether these findings result from selection or experience effects—that is, are happier people with a higher sense of well-being more likely to get married? (Yes.) Or is there something about the experience of being married that enhances happiness and well-being? (Yes.) Research shows that both are true.

Then, too, researchers have asked whether it is the experience of marital status itself or of couple satisfaction within marriage that makes people happier and adds to well-being. Interestingly, one study found that marital status itself can be significant for husbands whereas marital satisfaction is more necessary for wives (Fincham and Beach 2010). Several chapters throughout the remainder of this text explore possible causes for this finding. Here we look at what it is about the experience of being married that can enhance life satisfaction for both partners.

For one thing, there are some pragmatic reasons that spouses (and cohabitors, but to a lesser degree) benefit from an *economy of scale*. Think of the saying, "Two can live as cheaply as one." Although this principle is not entirely true, some expenses, such as rent, do not necessarily increase when a second adult joins the household (Thomas and Sawhill 2005; Goode 2007 [1982]; Waite 1995). Then, too, the promise of permanence associated with the marriage premise accords spouses the security to develop some skills and to neglect others because they can count on working in complementary ways with their partners (Goode 2007 [1982]; Nock 2005). Furthermore, "[s]pouses act as a sort of small insurance pool against life's uncertainties, reducing their need to protect themselves *by themselves* from unexpected events" (Waite 1995, p. 498).

In addition, marriage offers enhanced social support (Lee and Ono 2012; Manning and Brown 2006). Marriage can connect people to in-laws and a widened extended family, who may be able to help when needed—for instance, with child care, transportation, a down payment on a house, or just an emotionally supportive phone call (Rittenour and Soliz 2009; Wilcox et al. 2011a,b). The enhanced social support that often accompanies marriage works to encourage the union's permanence (Giddens 2007; Lee and Ono 2012). For example, family and friends send anniversary cards, celebrations of the years the couple has spent together and reminders of the couple's vow of commitment. Beginning with a public ceremony, marriage makes for what sociologist Andrew Cherlin (2004) calls *enforced trust*:

> Marriage still requires a public commitment to a long-term, possibly lifelong relationship. This commitment is usually expressed in front of relatives, friends, and religious congregants.... Therefore, marriage ... lowers the risk that one's partner will renege on agreements that have been made.... It allows individuals to invest in the partnership with less fear of abandonment. (p. 854)

Furthermore, marriage offers *continuity*, the experience of building a relationship over time and resulting in a uniquely shared history. And, finally, marriage provides individuals with a sense of obligation to others, not only to their families but also to the broader community (Goode 2007 [1982]; Wolfinger and Wolfinger 2008). This, in and of itself, gives life meaning (Waite 1995, p. 498).

Marital Satisfaction and Choices Throughout Life

Our theme of making choices throughout life surely applies both to couples anticipating marriage as well as to decisions made during the early years of marriage. We'll examine these topics now.

Preparation for Marriage

Given today's high divorce rate, clergy, teachers, parents, policy makers, and others have grown increasingly concerned that individuals be better prepared for marriage. High school and college family life education courses are designed to prepare individuals of various racial/ethnic groups for marriage (Coalition for Marriage, Family and Couples Education 2009; Fincham and Beach 2010). Premarital counseling, which often takes place at churches or with private counselors, is specifically oriented to couples who plan to marry. Many Catholic dioceses require premarital counseling before a couple may be married by a priest. Other Christian denominations, Islam, Judaism, and other religions offer premarital counseling as well. Illustrating the connection between private lives and public interest, some states have established incentives to encourage couples to seek premarital counseling. For example, Minnesota discounts the price of a marriage license when a couple agrees to premarital counseling (Meyer 2011).

Premarital counseling goals involve helping the couple to evaluate whether their relationship should lead to marriage, develop a realistic yet hopeful and positive vision of their future marriage, recognize potential problems, and learn positive problem-solving and other communication skills (Dinkmeyer 2007; Holloway 2008). Some programs have been developed especially for those contemplating postdivorce cohabitation or remarriage, particularly when children are involved (Gonzales 2009). Research designed to assess how effective these programs are shows that they do improve a couple's communication skills and relationship quality, at least in the short term (Blanchard et al. 2009; Fincham and Beach

2010). Unfortunately, however, we have little data on the relationship between premarital counseling and relationship *stability*.

Among other factors, success depends on the personality characteristics of each partner, as well as on couple characteristics, such as the interactional styles with which they begin the program, influences from their families of origin, and their motivation to learn from the program (Murray 2004). One study found that those who become actively involved in premarital counseling are more likely to value marriage and to be kind and considerate to begin with (Duncan, Holman, and Yang 2007). But, overall, family experts see these programs as important, especially for those younger than age 18 and for adult children of troubled or divorced families (Dinkmeyer 2007; Holloway 2008). Psychologist Scott Stanley identifies four benefits of premarital education:

> (a) it can slow couples down to foster deliberation, (b) it sends a message that marriage matters, (c) it can help couples learn of options if they need help later, and (d) there is evidence that providing some couples with some types of premarital training... can lower their risks for subsequent marital distress or termination. (Markman, Stanley, and Blumberg 2001, p. 272)

The idea of "slowing couples down" prompts questions about the relationship between age at marriage and the union's stability.

Age at Marriage, Marital Stability, and Satisfaction

In the 1950s, men tended to marry at about age 24 and women at about 22 (Aldous 1978). We've seen that the median age at first marriage today is about 26 for women and 28 for men—considerably older than at mid-twentieth century. Nevertheless, although much attention has been paid to the increasing age at first marriage in the United States, many Americans continue to marry at young ages. About 2 percent of men and 4 percent of women marry between ages 18 and 20. Another 11 percent of men and 19 percent of women marry between ages 20 and 25 (U.S. Census Bureau 2012c, Table 57). Is there a "right" age to marry?

Over the past several decades, researchers have given considerable attention to the relationship between age at first marriage and marital stability. Findings consistently show that the odds of marital stability increase with age at marriage (Amato et al. 2007). Low socioeconomic origins, poor communication and problem-solving skills, premarital pregnancy, lack of interest in school, and financial struggles are associated with marrying early (Larson and Hickman 2004; Uecker and Stokes 2008). Teen marriages (the majority between 18 and 19 years old) are the least stable (Copen et al. 2012, p. 7).

Typically, policy makers have developed programs encouraging teens to postpone marriage. Meanwhile, when they do occur, early marriages would benefit from recognition and support:

> Given early marriage's known association with marital dissolution, it is important to pay adequate attention to... individuals who marry in early adulthood.... Early marriage comes with its own set of difficulties, however, and if understanding and supporting all marriages—be they early, normative, or late—is a goal of scholarship and policy, this population should garner more attention from both researchers and policy makers. (Uecker and Stokes 2008, p. 844, 845)

Until recently, research on age at first marriage focused solely on marital stability. However, a 2009 analysis of findings from several major national surveys examined the relationship between age at first marriage and marital happiness and satisfaction (Glenn, Uecker, and Love 2009). Findings show that marriages occurring today when spouses are between 22 and 25 are most likely to be not only stable but also happy. Spouses who first married after age 30 reported lower marital satisfaction even as they were likely to stay married (Glenn, Uecker, and Love 2009).

Based on a review of prior research, Glenn, Uecker, and Love (2009) offer possible explanations. For one thing, more "set-in-their-ways" older spouses may find it more difficult to fashion a compatible life together. Also, marrying after about age 30 may mean selecting a spouse from a market in which "lots of the good ones are gone." Or it may suggest that an individual has been searching for the perfect partner—a situation that can only lead to later disappointment. Despite lower satisfaction levels, however, age at first marriage does act as a deterrent to divorce. If they are older, mildly unhappy partners may feel hesitant to reenter singlehood, the dating game, or the marriage market. Some advice from the study:

> The findings of this study *do* indicate that for most persons, little or nothing in the way of marital success is likely to be gained by deliberately delaying marriage beyond the mid twenties. For instance, a 25-year-old person who meets an excellent marriage prospect would be ill-advised to pass up that opportunity only because he/she feels not yet at the ideal age for marriage. Furthermore, delaying marriage beyond the mid twenties will lead to the loss during a portion of young adulthood of any emotional and health benefits that a good marriage would bring.... On the other hand, it is extremely important to stress that the findings of this study should not lead

> anyone of any age to panic and thus make a bad choice of a spouse. (Glenn, Uecker, and Love 2009, pp. 42–43)

The First Years of Marriage

Although marriage is important to happiness and well-being over the life course, the first years of marriage tend to be the happiest, with gradual declines in marital satisfaction afterward (Dush, Taylor, and Kroeger 2008; Tach and Halpern-Meekin 2009). Why this is true is not clear. One explanation points to life cycle stresses as children arrive and economic pressures intensify. Other explanations argue that falling in love and new marriage are periods of emotional intensity from which there is an inevitable decline (Glenn 1998; Whyte 1990). We do know something about the structural advantages of the early years of marriage, and it is likely that these contribute to high levels of satisfaction as well. For one thing, partners' roles are relatively similar or unsegregated in early marriage. Spouses tend to share household tasks and, because of similar experiences, are better able to empathize with each other.

In the 1950s, marriage and family texts characteristically referred to the first months and years of marriage as a period of adjustment, after which spouses had presumably learned to take on or play traditional marital roles. Today we view the first months and years of marriage more as a time of role-*making* than of role-*taking*. From the interaction-constructionist theoretical perspective (see Chapter 2), **role-making** refers to personalizing a role by modifying or adjusting the expectations and obligations traditionally associated it. Role-making involves issues explored more fully in other-chapters of this text. Newlyweds negotiate expectations for sex and intimacy (Chapter 4), establish communication (Chapter 11) and decision-making patterns (Chapter 12), balance expectations about marital and job or school responsibilities (Chapter 10), and come to some agreement about becoming parents (Chapter 8) and how they will handle and budget their money. When children are present, role-making involves negotiation about parenting roles (Chapter 10). Role-making issues peculiar to remarriages are addressed in Chapter 15. Generally, role-making in new marriages involves creating, by means of communication and negotiation, identities as married people (Rotenberg, Schaut, and O'Connor 1993). The time of role-making is not a clearly demarcated period but continues throughout marriage.

Couples must also accomplish certain tasks during this period. In general, "the solidarity of the new couple relation must be established and competing interpersonal ties modified" (Aldous 1978, p. 141; Rotenberg, Schaut, and O'Connor 1993). Getting through this stage requires making requests for change and negotiating resolutions, along with renewed acceptance of each other. Indeed, research by psychoanalyst John Gottman shows that how newlyweds communicate influences their later happiness and the permanence of their marriage (Gottman et al. 1998; Gottman and Levenson 2000, 2002; see also Ledermann et al. 2010).

The couple constructs relationships and interprets events in ways that reinforce their sense of themselves as a couple (Wallerstein and Blakeslee 1995). As they do, relationships with parents change (Sarkisian and Gerstel 2008), and developing supportive relationships with in-laws can be important (Rittenour and Soliz 2009). A qualitative study with lesbian couples found several respondents talking about their committed relationship and parents:

> [My parents] are unsupportive, but that doesn't mean that I don't have a relationship with them. I don't think [my father] particularly wants to hear me talk about [my partner] all the time... but it's something that we kind of deal with because we love each other.... But as far as all of us getting together to have dinner, that doesn't happen. (in Kinkler and Goldberg 2011, p. 397)

As another woman in this study said,

> I don't particularly see our kid having a very close relationship with [my dad]. And certainly not with my mother, but we're so close to [my partner's] family that that kind of makes up for it. I mean I couldn't ask for more fantastic in-laws. They're like my family. (in Kinkler and Goldberg 2011, p. 399)

Perhaps it is no surprise that a study in Switzerland found that newlyweds were happier when relatives were supportive but not interfering (Widmer et al. 2009).

Meanwhile, a national study undertaken by family researchers at Creighton University in Omaha identified three potentially problematic topics for couples in first marriages: (1) money—balancing job and family, dealing with financial debt brought into the marriage by one or both spouses, and what to do with money income; (2) sexual frequency; and (3) agreeing on how much time to spend together—and finding it! Challenges associated with learning to balance work or college courses and a marital relationship are real (Christopherson 2006). Feeling supported by parents and extended kin helps (Kurdek 2005). Other issues the couples in the Creighton study mentioned were expectations about who would do household tasks (and how well), communication challenges, and problems with in-laws (Brennan 2003; Risch, Riley, and Lawler 2004). A more general, necessary goal for couples in early marriage is to create couple connection.

Creating Couple Connection

To have enduring unions, partners need to make their relationship a high priority. Spending time together, communicating supportively, and pursuing leisure activities together are related to couple satisfaction (Gager and Sanchez 2003; Johnson et al. 2005; Kurdek 2005). Couples who make time for shared new experiences are more happily married (Burpee and Langer 2005). As an important psychologist and expert on marital communication, John Gottman offers this advice:

> Happy, solid couples nourish their marriages with plenty of positive moments together.... Too often, families lead complex—even grueling—lives in which they sacrifice the happy times for more materialistic, fleeting goals.... Sundays at the office take the place of Sundays at the park. But if you want to keep your marriage alive, it's essential to rediscover—or perhaps simply make time for—those experiences that make you feel good about your spouse and your marriage. (1994, p. 223)

The above is good advice. Here, though, we want to note also the particular hurdles that some couples face in establishing couple connection. A focus group participant of Mexican American origin explained:

> [T]he husband comes [to the United States] first and later brings his wife. But while he was waiting to earn money to bring his wife over, he brought his cousins and nephews. So this woman is living with ... five of her relatives. So she is playing the role of wife, cousin, friend, and servant in the house, making food for all these people. And then, she has to go to work so that the husband can pay off the money for her trip here.... [I]n Mexico, everyone lives in difficult conditions [too]. But at least they live in their little shacks, ... and just ... the husband, [wife,] and the kids [live there]. And here they have to get used to living with fifteen people. (in Helms, Supple and Proulx 2011, p. 82)

Meanwhile, recent research has focused on how religious conviction can facilitate lasting couple connection (Fincham and Beach 2010). For one thing, a religious belief system called **marital sanctification** encourages spouses to see their marriages as ordained by God and hence having divine significance. Sanctification promotes couple bonding, fosters positive emotions and diminishes negative ones, and facilitates positive attitudes and overall resilience when encountering stress (Ellison et al. 2011).

Comparative analysis of data collected from national samples in 1980 and 2000 revealed that spouses spent less time interacting in 2000 than they did twenty years before that. However, their reported marital satisfaction had not declined significantly, partly because they were more satisfied with the decision-making equality in their marriage (Amato et al. 2003). Nevertheless, increased emphasis on other matters, such as managing debt, job pressures and long work hours, or children's needs, can result in exhaustion, increased conflict, and slow emotional erosion (Dew 2008; Roberts and Levenson 2001). Noting that "[l]ove is not an express lane concept," observers suggest creating daily "connecting moments" when you can be alone together and pay attention to your relationship (Brennan 2003; Brotherson 2003). This idea is further explored in Chapter 11.

Keeping one's marriage vital requires that partners develop identities as married individuals, as well as consciously and continuously building couple commitment—being willing to invest in their union long-term and to persist together through trying times (Byrd 2009). An emotionally meaningful relationship does not develop

Fuse (RF)/Jupiter Images

Because, more than others, a married couple can count on continuity, bolstered by mutual commitment and enforced trust, spouses are freer to plan a future together.

long term "by drift or default" (Cuber and Harroff 1965, p. 145). Satisfaction with the marital relationship has a great deal to do with the choices that partners make. One important set of decisions involves practicing positive communication skills. Research that followed 135 Denver couples over thirteen years, from the time they were engaged and into their marriages, concluded that "the seeds of marital distress and divorce are sown for many couples before they say 'I do.' ... [N]egative premarital and early marital interactions... prime a marriage for the erosion of positivity over time" (Clements, Stanley, and Markman 2004, p. 621). We might add that positive communication and commitment to developing a mutually supportive relationship prime a marriage for long-term success. Building better communication skills is addressed in Chapter 12.

Summary

- The marriage premise involves permanence and—for some societies, particularly Western cultures—monogamous sexual exclusivity.
- New norms for love-based marriage gradually became prevalent in Europe throughout the 1700s and 1800s. Expectations for personal happiness and love in marriage have changed the marriage premise over the past 300 years.
- Marriages and families have become deinstitutionalized. Marriage has changed from institutionalized to companionate and now to individualistic.
- Researchers consistently find a significant correlation between marriage and many positive outcomes for both adults and children, but the relationship is more complicated than it first appears because some of the benefits associated with marriage are the result of selection effects.
- Scholars and policy makers who view individualized marriage and the postmodern family as indications that the institution of marriage and family is in "decline" and "breakdown" have proposed ways to effect a turnaround. Two of these are covenant marriage and the Healthy Marriage Initiative.
- Scholars and policy makers who view individualized marriage and the postmodern family as results of inevitable historical change, resulting in struggle for some families more than for others, have proposed structural solutions to poverty and family struggle—such as higher wages for women, neighborhood development, an adequate minimum wage, and more employment opportunities.
- Optional and less permanent than in the past, marriage continues to offer benefits; married adults are more likely than others to say that they are happy with their lives. Being married continues to help bolster the marriage premise, largely because family and community social support results in enforced trust.
- There is society-wide concern in the United States about preparation for marriage. Premarital counseling and family life education are two approaches that have been developed for preparing people for marriage. They seem to be somewhat successful, at least in improving communication skills.
- Spouses in the first years of marriage engage in role-making, a process that includes—among other things—negotiating issues surrounding money, sexual frequency, and time together.

Questions for Review and Reflection

1. Discuss the marriage premise with its expectations of permanence and sexual exclusivity. Describe how love has changed the marriage premise over the past three centuries. What does family demographer Stephanie Coontz mean when she says that "love conquered marriage"?
2. Pointing out the pros and cons of each, compare and contrast companionate marriage and individualized marriage.
3. Explain the connection between federal "welfare reform" and the National Marriage Initiative.
4. Recognizing the possibility of selection effects, what are some ways that the experience of being married can enhance happiness and life satisfaction?
5. **Policy Question.** How do programs such as the War on Poverty address child poverty? In contrast, how do efforts to promote marriage address child poverty? If you were in a position to devise solutions to child poverty, what would they be? Would they include encouraging more Americans to marry?

Key Terms

8

Deciding about Parenthood

Rana Faure/Fancy/Corbis

Fertility Trends in the United States

History, Fertility Trends, and Family Size

Differential Fertility Rates by Education, Income, and Race/Ethnicity

Things to Consider When Deciding about Parenthood

Rewards and Costs of Parenthood

Facts about Families: Conception, Pregnancy, and Childbirth: The Basics

Issues for Thought: Cesarean Sections: Should a Delivery Be Planned for Convenience?

How Children Affect Couple Happiness

Choosing to Be Childfree

Having Children: Options and Circumstances

Timing Parenthood: Earlier versus Later

Having Only One Child

Nonmarital Births

Multipartnered Fertility

Preventing Pregnancy

Abortion

The Politics of Family Planning, Contraception, and Abortion

Deciding about an Abortion

Involuntary Infertility and Reproductive Technology

Reproductive Technology: Social and Ethical Issues

Reproductive Technology: Making Personal Choices

Adoption

The Adoption Process

Adoption of Race/Ethnic Minority Children

A Closer Look at Diversity: Through the Lens of One Woman, Adopted Transracially in 1962. By Cathy L. Wong, PhD

Adoption of Older Children and Children with Disabilities

International Adoptions

Learning Objectives

1. Summarize U.S. trends in fertility rates, historically and according to race/ethnicity.
2. Describe how educational attainment and income affect fertility decisions.
3. Describe the concept value of children and list some rewards and costs of parenthood.
4. Discuss several options and circumstances regarding deciding about parenthood—for example, having an only child or having children at younger or at older ages.
5. Differentiating abortion from birth control, discuss the debate between pro-life and pro-choice activists.
6. Define involuntary infertility and discuss some of its causes and possible solutions.
7. Explain adoption options in the United States today.

You may be or know someone who has been adopted, perhaps by parents of another race, or who is considering:

- getting pregnant with a first child,
- infertility treatment,
- having an abortion,
- never having children,
- having an only child, or
- adopting a child from overseas.

Each of these decisions is very personal but also influenced by society. Decisions about parenthood reflect people's economic circumstances and cultural beliefs, together with possible external, or social pressures. Decisions about parenthood also proceed from individual needs, values, and attitudes (Lundquist et al. 2009). Throughout this chapter we'll be looking at many aspects of individuals' or couples' decisions about having children.

We address these issues as options related to informed decision making, but we should note that, when asked why they decided to have their first child, nearly half (47 percent) of a Pew Research national representative sample answered, "There wasn't a reason, it just happened" (Livingston and Cohn 2010). Other research suggests that as many as half of U.S. births may be unintended—but also that "it may make more sense to conceptualize pregnancies not as two mutually exclusive categories denoted planned or unplanned, but as points on a continuum of intendedness" (Shreffler et al. 2011, p. F8; see also Mosher, Jones, and Abma. 2012). Put another way, many women are apparently ambivalent about having children: "The strong assumption that all women plan and therefore that unintended pregnancies are unwanted pregnancies obscures variation among women in degrees of planning pregnancy (or not) as well as variation from pregnancy to pregnancy among the same women" (Shreffler et al. 2011, p. F8). Nevertheless, much of this chapter assumes informed decision making, and as you study it, you may want to refer back to Chapter 1's discussion on how we best make informed decisions.

In this chapter we'll look at the rewards and costs of having children, along with how children affect a couple's happiness. We'll examine several options regarding having children—choosing to be childfree, having only one child, and so on. We will also address various circumstances under which Americans have children today. For instance, we'll discuss nonmarital births, consciously choosing to be a single parent, whether to have children at a younger or older age, and multipartnered fertility. We'll also look at issues concerning both preventing pregnancy and involuntary infertility. The chapter closes with a discussion of adoption.

To begin, we'll examine fertility trends in the United States. Then we'll address decision making about whether or not to become a parent.

Fertility Trends in the United States

Beginning early in the nineteenth century, America began to witness significant changes in childbearing or fertility patterns. Demographers use the term **fertility** to refer to live births. The **total fertility rate (TFR)** is the number of live births a typical woman will have during her lifetime. As shown in Figure 8.1, the TFR for the United States dropped sharply from a high of 3.5 in the late 1950s to the lowest level ever recorded—1.7 in 1976, a year marked by dramatically rising fuel and related costs that accompanied the energy crisis of the 1970s. In recent years, the total fertility rate has fluctuated around 2.0. However, people forgo or put off having children in difficult economic times, and in 2011 as a result of the recession that began in 2007 the TFR for the United States fell to 1.9 (Mather 2012).

The TFR is lower in all race/ethnic groups than it was during the baby boom era (1946 to 1964). The overall U.S. fertility today is below **replacement level**, the level of fertility necessary for a society to replace its population. A society needs two children in order to replace every two adults, and demographers peg replacement-level TFR at 2.1 to take into account that not all infants survive and that every woman will reproduce. Therefore, a TFR of 2.0 or below means that a society or race/ethnic category will not replace itself biologically.

History, Fertility Trends, and Family Size

U.S. fertility decline shows a continuous pattern dating back to the early 1800s. As discussed in Chapter 1,

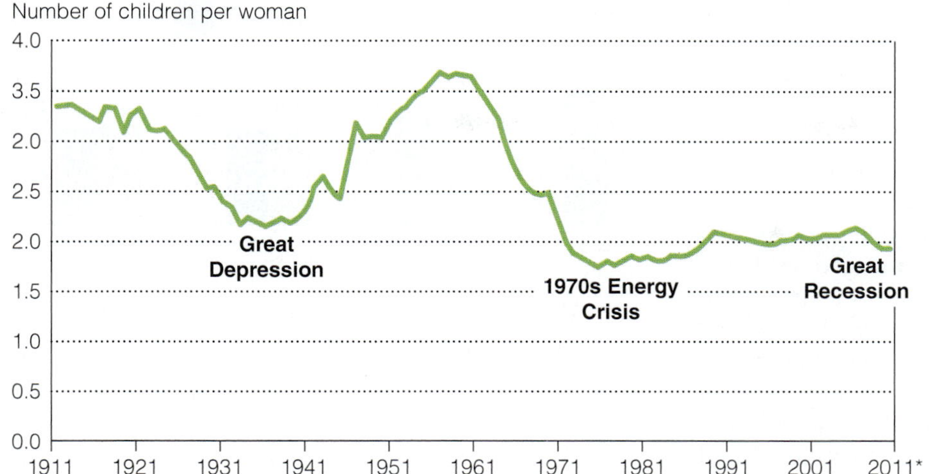

FIGURE 8.1 United States total fertility rate (TFR), 1911 to 2011. U.S. fertility has declined during periods of economic slowdown.

Source: Mather 2012, Figure 1; Population Reference Bureau. World Population Data Sheet 2012.

* Estimated by Population Reference Bureau.

historical periods or events affect individuals' options and decisions. In a preindustrial economy, women could combine productive work on the farm or in a home artisan shop with motherhood. But when work moved from home to factory, the roles of worker and mother were not so compatible. As women's employment increased, fertility declined. Moreover, resulting from improved living conditions, infant mortality declined over time. Gradually, it was no longer necessary to bear many children in order to insure the survival of a few.

In the face of the long-term decline over the past two centuries, the upswing in fertility in the late 1940s and 1950s requires explanation. As adults, those who had grown up during the Great Depression (see Figure 8.1), when family size was limited by economic factors, found themselves in an affluent post–World War II economy. Perhaps compensating for economic deprivations they suffered as children, they later fulfilled dreams of a relatively abundant family life (Easterlin 1987). Marriage and motherhood became dominant cultural goals for

The ideal family size in the United States is now two children. Large families of four or more children are a distinct minority in the United States today—and have been for some time (Kirmeyer and Hamilton 2011). Mothers of large families report feeling stigmatized, seen as uneducated, insufficiently attentive to their children, messy housekeepers, and ignorant of birth control (Hagewen and Morgan 2005). Many larger families are religious and find acceptance in their religious communities (Zernike 2009). A pro–large family movement termed "Quiverfull" is situated in some fundamentalist Christian churches. Thinking of their offspring as "an army they're building for God," Quiverfull families aim for six or more children (Joyce 2006, p. St-1).

American women; men also concentrated their attention on family life. Couples in this generation averaged more than three children, and of course some had more (Kirmeyer and Hamilton 2011).

Today, two children constitute Americans' ideal family size, and this two-child preference has been evident for approximately fifty years (Kirmeyer and Hamilton 2011). Women's extended educational options and demanding work schedules have emerged as barriers to larger family size (Morgan and Rackin 2010). Furthermore, women are waiting longer to have first babies, a situation that ultimately reduces family size because it shortens the years when a female is most biologically able to reproduce. Regarding age and fertility, the birthrate for those ages 15 to 19 has declined to the lowest recorded in sixty-five years of record keeping, a development discussed later in this chapter. Birthrates peak among women in their late twenties, and births to women in their twenties constitute about half of all births.

However, by 2011, a falling birthrate of women in their twenties reached a record low and continued to decline through 2012. The rate for women in their early thirties remained relatively low. Meanwhile, birthrates for women in their late thirties and early forties had risen to unprecedented highs (Hamilton, Martin, and Ventura 2012). As a result, the average age of first birth has increased over the past ten years from an average of 21 to 25 (Martinez, Daniels, and Chandra 2012, p. 6; Martin et al. 2012, p. 8). Since 1980, birthrates for men 45 through 49 have increased by more than 20 percent (Martin et al. 2009). While recognizing overall childbearing trends, we note that fertility rates differ among segments of the population.

Differential Fertility Rates by Education, Income, and Race/Ethnicity

Families with higher education and income tend to have fewer children (Morgan and Rackin 2010). To create a fulfilling life, people with high education and income have alternatives to parenthood—for example, investment in demanding careers, dedication to social activism or service, world travel, or developing latent talents. When people with these options for self-actualization weigh them against having children, the latter may lose out (Weeks 2007). Birthrates also reflect the fact that beliefs and values about having children vary among race/ethnic groups. As Figure 8.2 indicates, U.S. women in each of the four race/ethnic categories shown have lowered their fertility since 1990. However, rates differ according to race/ethnicity.

In 2012, the U.S. Census Bureau "made it official." Accounting for slightly less than 50 percent, nonHispanic white births "are no longer a majority in the United States" (Tavernise 2012). As you can see in Figure 8.2, in 2010 Asian/Pacific Islanders had the lowest TFR at 1.7, followed by nonHispanic whites at 1.8. You'll note that both these figures are below replacement level. In 2010, Hispanics had the highest TFR (2.4) of any U.S. race/ethnic category (Mather 2012). Although the Hispanic birthrate fell sharply—by 5 percent—between 2010 and 2012, Hispanics' relatively high TFR (when compared to other U.S. race/ethnic categories) signifies future growth in this ethnic group's proportional size (Hamilton, Martin, and Ventura 2012, p. 3). Among

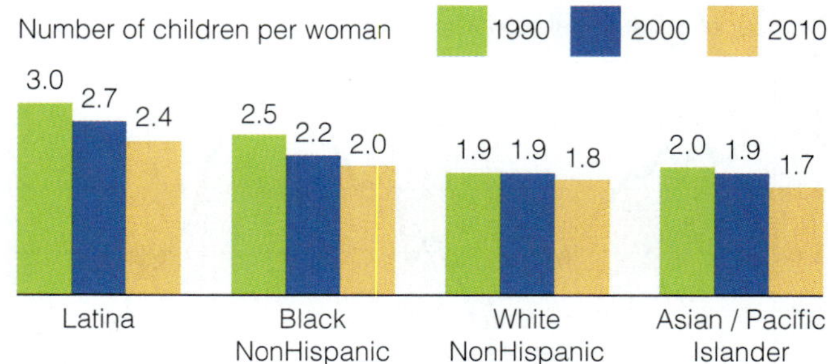

FIGURE 8.2 United States total fertility rates (TFR) by race/ethnicity, 1990, 2000, 2010.

Sources: Mather 2012; Population Reference Bureau. World Population Data Sheet 2012.

Fertility rates of Native Americans/Alaska Natives have declined by more than 20 percent since 1990 to a total fertility rate of 1.4 in 2011 (Hamilton, Martin, and Ventura 2012, Table 1). This rate is considerably below replacement level. With Native Americans/Alaska Natives evidencing lower average educational attainment and income levels, their fertility pattern is an exception to the rule that lower education and income levels are associated with *high* fertility. Native American women who live on reservations have higher fertility than those who do not, possibly because there are relatively limited educational and occupational options available on reservations.

Hispanics, the TFR for Mexican American women has been the highest. Women of Central American and South American background have had fertility rates slightly lower than those of Mexican American women, while women of Puerto Rican and Cuban backgrounds have had still lower rates (Martin et al. 2009, Table 15).

At 2.0 in 2010, the nonHispanic black fertility rate has been higher than either nonHispanic whites or Asian/Pacific Islanders (Mather 2012). One reason offered for Hispanics' and blacks' relatively high birthrates is that Latinas and nonHispanic blacks average lower educational attainment and annual incomes than do nonHispanic white and Asian populations (U.S. Census Bureau 2012c, Tables 224, 695). Moreover, many Hispanics have fairly recently migrated from cultures characterized by high birthrates and by rural and Catholic traditions that encourage large families (Hartnett and Parrado 2012). As immigrants assimilate, their birthrates tend to converge with those of the general population. Variations in birthrates reflect decisions shaped by cultural values and individual attitudes about having children (Hartnett and Parrado 2012). We turn now to a discussion of things to consider when deciding about parenthood.

Things to Consider When Deciding about Parenthood

"Facts about Families: Conception, Pregnancy, and Childbirth" presents some basic facts on the biological processes associated with pregnancy and childbirth. "Issues for Thought: Cesarean Sections: Should a Delivery Be Planned for Convenience?" explores considerations regarding delivery by cesarean section. Understanding facts like these is a first step in making informed decisions about having children. Meanwhile, we note that it is not necessarily easy to choose how many children to have, if any, or when to have them. Furthermore, as we saw in the introduction to this chapter, the degree to which people consciously choose or reject parenthood is uncertain. In a 2002 survey, more than one-third of the women respondents said that their recent births were unintended— 14 percent "unwanted" and 21 percent "mistimed." Additional research suggests that as many as half of all births to American mothers may be unintended (Chandra et al. 2005, p. 1; Trussell and Wynn 2008).

Moreover, social pressure to have children—a cultural phenomenon called **pronatalist bias** (Hagewen and Morgan 2005)—might influence fertility decisions. Interestingly, as same-sex unions become more visible and accepted, some lesbian and gay male couples join straights in feeling pressured to be parents. As one gay man told an interviewer, "Everyone's asking: What's your timetable? What's your plan?" In this case, "everyone" included his parents, his husband's parents, their friends, and work colleagues (Swarns 2012).

Meanwhile, some observers argue that U.S. society is characterized by **structural antinatalism**—that is, not doing what it could to support parents and their children (Huber 1980). Critics of American family policy point out that nutrition, social service, financial aid, and education programs directly affecting the welfare of children are not adequate compared to other nations at our economic level (Children's Defense Fund 2012). Nor do we provide paid parental leave or other support for parents of young children as many other countries do. Children in the United States are more likely to be poor than children in comparable countries, and the United States ranks twentieth out of twenty-one in overall child well-being among advanced industrial nations (Svevo-Cianci and Velazquez 2010; Taubman 2009; UNICEF 2007, p. 1). Given "strong antinatalist forces" that dilute societal support for parents, some scholars have asked, "Why do people choose to have any children?" (Hagewen and Morgan 2005, p. 513). In answer, we look at rewards and costs associated with parenthood.

Rewards and Costs of Parenthood

"From the day children are born they become a source of joy and a source of burdens for their parents" (Nomaguchi and Milkie 2003, p. 372). According to the **value of children perspective**, children historically were economic assets. Even when very young, they provided more working hands in the fields and kitchens. Gradually, the shift from an agricultural to an industrial society and the development of compulsory education transformed children from economic assets to economic liabilities. At school rather than at work in the factory, needing school clothes and fees, children no longer added financially to the family. Instead, they cost their families. But as their economic value declined, children's emotional significance to parents increased, partly because declining infant mortality rates made it safer to become attached to them (Hoffman and Manis 1979; Lamanna 1977). Parents' desire was for "a child to love" and the "joy that comes from watching a child grow" (Morgan and King 2001, p. 11).

Children bring many benefits to parents. Becoming a parent can certify one's attainment of adulthood, although there are other avenues to adult status as well (see Chapter 6). For men as well as women, gay as well as straight, parenthood is an extremely meaningful aspect of personal identity (Games-Evans 2009; Goldberg, Downing, and Moyer 2012). In an important national survey of marrieds, 57 percent of wives and 45 percent of husbands with children at home strongly agreed that "life has an important purpose." These figures compare to just 40 percent of wives and 35 percent of husbands

Facts about Families

Conception, Pregnancy, and Childbirth: The Basics

Conception

Conception begins with *ovulation* when a female's *ovaries* monthly release an egg, or *ovum* after which it travels through her *fallopian tubes* to the uterus (see Figure 8.3a). When a male's *sperm* enter a female's vagina, the sperm move into the fallopian tubes and can live there from two to five days. Conception takes place upon *fertilization*, the joining of sperm and ovum. Subsequently, the fertilized egg, or *zygote*, embeds itself in the lining of the *uterus* and an umbilical cord begins to form.

Pregnancy

The normal length of pregnancy, or gestation, is forty weeks. Early prenatal care is important. In 2006, 69 percent of mothers had begun care in the first trimester of pregnancy. The use of prenatal care beginning in the first trimester increased among all race/ethnic groups throughout the 1990s. However, the percentage receiving no care or late-term care has recently increased, particularly among low-income black and Hispanics (Martin et al. 2009).

By her fourth week, a woman likely notices signs of pregnancy: ceased menstruation, nausea (a physical reaction to the zygote's embedding itself in the uterine wall), increased breast size, and fatigue. From the second until about the eighth week, the embryo's head, skeletal system, heart, and digestive system begin to form. From about the eighth week until birth, the organs and structural system refine themselves and grow (see Figure 8.3b). Toward the end of pregnancy, the fetus typically changes position in the womb so that the head is in the lower part of the uterus. This marks the beginning of preparation for birth.

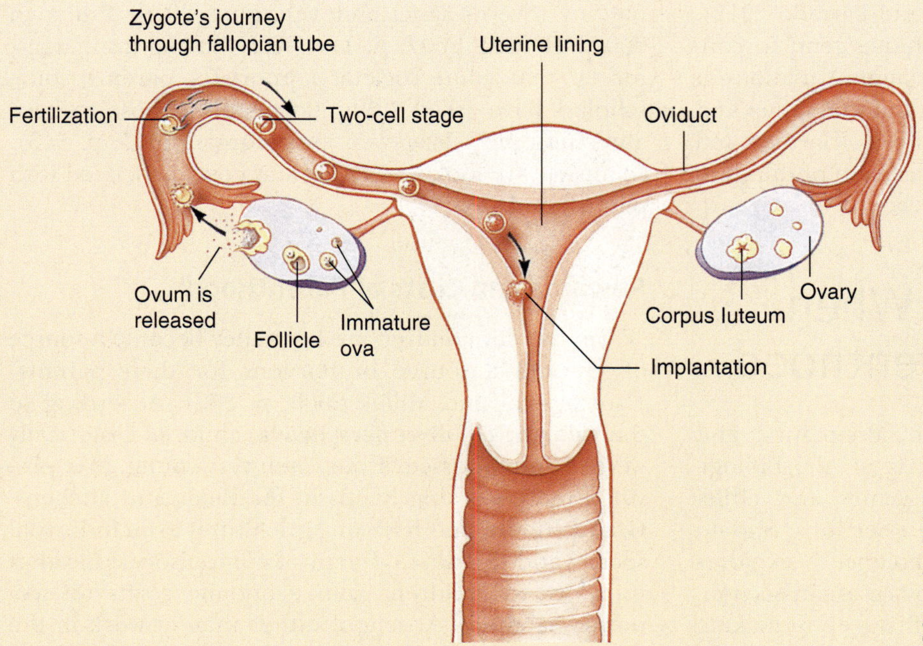

FIGURE 8.3a Ovulation, fertilization, and the germinal period of pregnancy.

Childbirth

Labor is the process by which the baby is propelled from the mother's body through a series of uterine contractions that steadily increase in intensity and duration. In preparation for the

with no children in the home (Wilcox et al. 2011a, p. 16). Indeed, parenthood can provide a sense of commitment and meaning in an uncertain social world (South and Crowder 2010).

Although the rewards of procreation can be immeasurable, raising children is costly. In fact, one recent study concluded that parents psychologically exaggerate to themselves the joys of parenthood in order emotionally to counter the costs (Wilkinson 2011). On a purely financial basis, children decrease a parent's level of living considerably. In husband–wife families with two children, an estimated 42 percent of household expenditures are attributed to children. In fact, the yearly cost of child care for an infant can run upward of $14,480, and costs associated with raising children are rising faster than are incomes. Social researcher Pamela Paul notes that U.S. household incomes rose 24 percent over the last decade, whereas costs associated with child raising rose 66 percent (Paul 2009, p. 5, 7). The average cost of raising a child born in 2011 to age 18 is estimated at $295,560 for middle-income families (Lino 2012, p. 20). For a single mother in poverty, feeding her children well may mean risking her own health as she chooses for herself less-nutritious foods, many of which foster obesity (Martin and Lippert 2012).

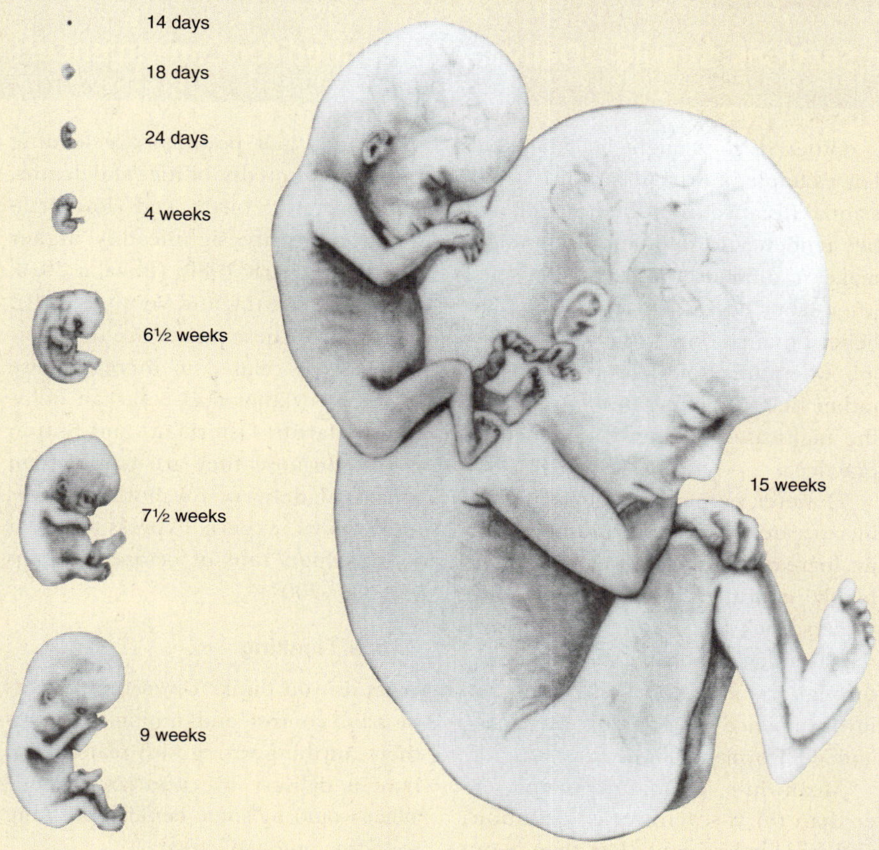

FIGURE 8.3b Prenatal development.

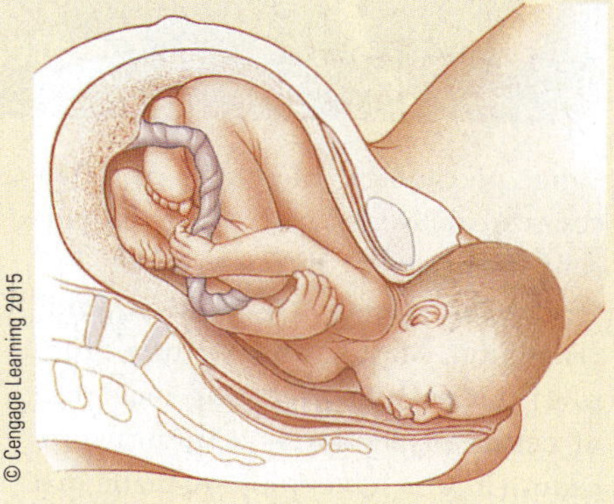

FIGURE 8.3c In the childbirth process, the baby's head begins to emerge during delivery.

baby's passage, the *cervix* dilates to approximately four inches. Full dilation of the cervix marks the beginning of delivery—the expulsion of the fetus, a process that may last from fewer than twenty minutes to (rarely) more than ninety minutes.

When the baby appears at the vaginal opening (*crowning*), its head usually turns so that the back of its skull emerges first, as is shown in Figure 8.3c. If the baby is too large, or if the mother's or baby's physical condition makes the stress of childbirth dangerous, a physician may decide to deliver the child by cesarean section (or C-section), so called after Julius Caesar, who was supposedly born in this way. A *cesarean section* is a surgical operation in which a physician makes an incision in the mother's abdomen and uterine wall in order to remove the infant.

Greater access to prenatal care, hospital delivery, and a doctor's assistance in childbirth contributed greatly to the sharp decrease in infant and maternal deaths during childbirth throughout the twentieth century. The overwhelming majority of births (99 percent) now occur in hospitals and are attended by physicians or—to a much lesser extent—by certified nurse-midwives (Martin et al. 2009).

Critical Thinking

What might be one or more ways, if any, that these facts influence an individual's attitudes or behavior regarding parenthood?

Added to the direct costs of parenting are **opportunity costs**: the economic opportunities for wage earning and investments that parents forgo when rearing children. These costs are more often felt by mothers. Some career women report hiding their pregnancies for as long as possible in order to avoid workplace discrimination, such as failure to be considered for promotion (Quart 2012). A woman's career advancement may suffer as a consequence of becoming a mother, especially in a society that does not provide adequate day care or a flexible workplace (Slaughter 2012). A recent study in Sweden found that men and women were more likely to plan a pregnancy when their workplace policies were more supportive of parent–employee needs (Kaufman and Bernhardt 2012).

A couple in which one partner quits work to stay home with a child or children faces a loss of half or more of its family income. The spouse (more often the woman) who quits work also faces lost pension and Social Security benefits later. All in all, in our society there is "a heavy financial penalty on anyone who chooses to spend any serious amount of time with children" (Crittenden 2001, p. 6). Then, too, with the arrival of children parents experience loss of freedom and flexibility regarding personal activities (Beck et al. 2010).

Issues for Thought

Cesarean Sections: Should a Delivery Be Planned for Convenience?

"The percentage of births through cesarean deliveries rose nearly 60 percent from 1996 through 2009, but this upward trend may be at an end" (Hamilton, Martin, and Ventura 2012, p. 4). Meanwhile, the rising incidence of **cesarean sections**—a surgical procedure in which incisions are made in a woman's abdomen and uterus to deliver her baby—has raised concerns among health experts. Caesarean sections (C-sections) now number nearly one-third of all births, the highest proportion ever reported (Martin et al. 2009; Menacker and Hamilton 2010).

One reason for today's high rates of C-sections involves fetal monitoring, the results of which may trigger necessary intervention. Difficult births may require surgical intervention, and there is no question that cesareans are often lifesaving procedures for both mothers and infants.

Other reasons might be pragmatic. For example, a woman may live a substantial distance from the hospital, and her tendency to deliver quickly might make it difficult to reach a hospital when labor begins. Or a physician may believe that she or he can do the best job when time of delivery is chosen rather than occurring in the middle of the night after a long day of medical practice.

However, perhaps one-quarter of the increase in cesareans occurs unnecessarily. Increasingly, parents, many of them highly educated and in demanding careers, seek to control the birth process as much as possible in order to get through it with minimal discomfort and inconvenience (Declercq, Menacker, and MacDorman 2004).

Meanwhile, premature birth (fewer than thirty-seven weeks' gestation) and low birth weight (less than about five and a half pounds) are leading causes of infant disabilities and deaths. The preterm birth and low-birth-weight rates are significantly higher than in the early 1980s (Bakalar 2010; Hamilton, Martin, and Ventura 2009). Reasons for these trends are not clear but may be related to increased use of induced labor and cesarean deliveries (Martin, Osterman, and Sutton 2010). Because they are riskier than a vaginal delivery for both mothers and infants, experts express alarm at today's high rate of cesarean births (Bakalar 2007).

Critical Thinking

What do you think? Given the benefits of pain control and timing ability, is there anything wrong with planning to have a delivery by caesarian section? What would be some benefits of doing so? Some potential costs?

Studies associate wanting to get pregnant with greater appreciation for parental rewards and decreased emphasis on the costs (East, Chien, and Barber 2012). Still, 68 percent of male and 70 percent of female parents strongly agree that, "The rewards of being a parent are worth it despite the cost and work it takes" (Martinez et al. 2006; Figure 25 see also Schindler 2010).

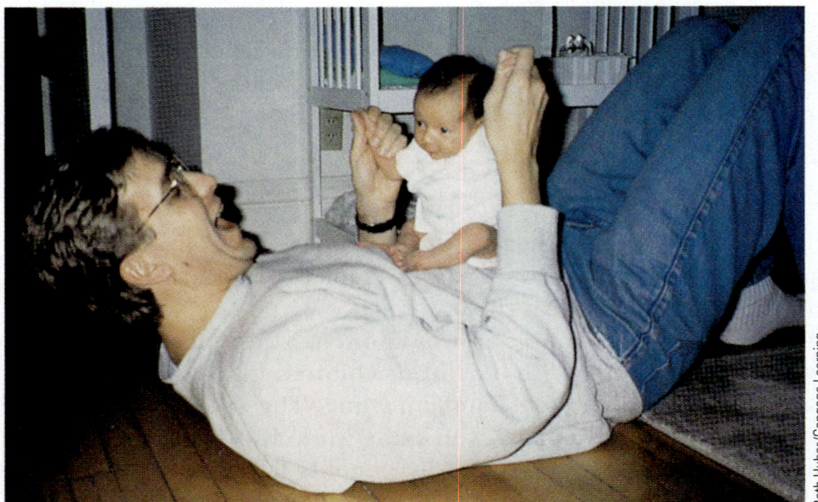

Beth Huber/Cengage Learning

Children can bring vitality and a sense of purpose into a household. Having a child also broadens a parent's role in the world: Mothers and fathers become nurturers, advocates, authority figures, counselors, caregivers, and playmates.

How Children Affect Couple Happiness

Research on how children affect a couple's relationship satisfaction and happiness has focused on marrieds. The extent to which the findings reported in this section apply to cohabitors or other unmarried coparents is unknown and open to speculation—including your own speculation. Among marrieds, evidence shows that children, especially young ones, stabilize a union; that is, parents are less likely to divorce. But a stable marriage is not necessarily a happy one: "[C]hildren have the paradoxical effect of increasing the stability of the marriage while decreasing its quality" (Bradbury, Fincham, and Beach 2000, p. 969). Couple conflict and strain is a common cost of having children.

Survey data, along with a major review of the research in this area, find that not only do parents report lower marital satisfaction than nonparents, but also the lower their marital happiness the more children they have (Doss et al. 2009; Wilcox et al. 2011a). Research involving 1,000 families concluded that even with highly anticipated births,

> most couples [upon becoming parents] become much more traditional in their approach to housework and child care. No matter how much they think the tasks will be shared, most women wind up doing more housework than they did before the birth, and more of the childcare than they expected. The discrepancy between what the couples hoped for and the reality of wives having to take on a "second shift" at home leads to feelings of tension, depression, and sometimes anger in both partners. (Cowan and Cowan 2009, n. p.)

Spouses' reported marital satisfaction tends to decline over time regardless of whether they have children. But conflicts over each partner's employment and home responsibilities may erupt when a child arrives, especially (and perhaps ironically) in "soul-mate" relationships that have to be "nurtured and coddled in order to thrive" (Whitehead and Popenoe 2008, p. 8). In particular, fathers may feel a reduced sense of confidence in their place in the family after the birth of a child, whereas mothers experience higher levels of couple conflict (Doss et al. 2009).

One review of the research through 2000 (Twenge, Campbell, and Foster 2003) concluded that the negative effects of children on marital satisfaction seem to be stronger for today's young adults. Perhaps when children arrive, couples today experience a greater "before–after" contrast than in the past. Couples today often marry and become parents later in their lives than couples did in earlier decades. So for several years they experienced a great deal of personal freedom and a career focus that children interrupt. Moreover, the increased individualism of our culture may make day-to-day responsibility for the care of young children seem less natural than in the 1950s, when social obligations were culturally dominant (Turner 1976). Given these circumstances, some adults decide against parenthood.

Choosing to Be Childfree

Childlessness in general is higher now than a few decades ago. About 21 percent of nonHispanic white, 17 percent of nonHispanic black, 16 percent of Asian, and 12 percent of Hispanic women over age 44 are childless (Jayson 2011). This figure is twice the percentage of childless women in that age group in 1976 (Dye 2008, Table 2; Martinez, Daniels, and Chandra 2012, Table 2).

This section examines **voluntary childlessness**, or being **voluntarily childfree**, the choice of approximately 7 percent of American women (Dye 2008; Kelly 2009). There is little difference in this respect between African Americans and nonHispanic whites (Lundquist et al. 2009, Figure 1). Although some contemporary researchers are uncomfortable with the term *childfree*, it is often used now instead of the more negative-sounding term *childless*. Each term conveys an inherent bias. For that reason and because there are no easy-to-use substitute terms, we use both *childless* and *childfree* in this text.

Different countries and cultures have different norms regarding voluntary childlessness (Merz and Liefbroer 2012). Although not so openly as forty years ago, negative stereotypes of the voluntarily childfree persist (Abma and Martinez 2006; Kelly 2009). At least one contemporary commentator (in Canada) has

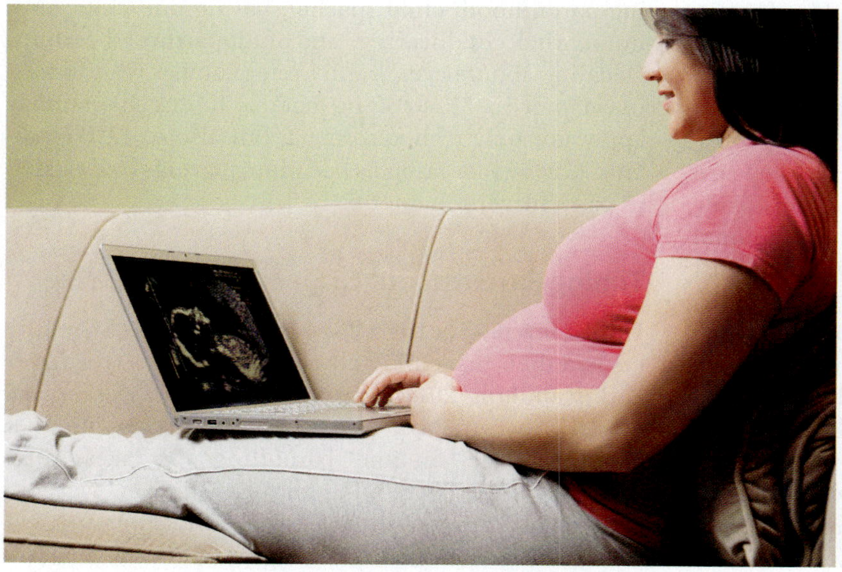

Artiga Photo/Comet/Corbis

In *ultrasound*, sound waves are bounced off the abdomen of a pregnant woman to determine the shape and position of the fetus. The result is a sonogram. Ultrasound is used by doctors to see whether a fetus is developing properly. Sonograms permit prospective parents to observe the fetus. They are often given pictures or videotapes of the fetus to take home. This technology is pushing back parental bonds to before birth. Some commercial firms offer sonograms to provide parents with a very early baby picture. In this photo, a military wife studies the sonogram that she has sent to her husband overseas so he can keep a photo of their future baby with him.

called the "trend of couples not having children just plain selfish" (O'Connor 2012). In 1990, about 70 percent of Americans told pollsters that children were very important to a successful marriage; by 2007 that percentage had dropped to 41 percent (Pew Research Center 2007). The United States appears to have strong, although weakening, fertility norms that continue to encourage two children and discourage childlessness and only-child families.

Partly as a result of these norms, along with social pressure to have children, there is often some ambiguity about the decision to remain childfree. For many, it is a choice that develops gradually over time. For others, remaining childless is a consequence of putting off parenting until it feels "too late now" or until one is less likely to become pregnant because of age (Whitehead and Popenoe 2008). On the other hand, for a number of younger women, not having children represents early commitment to continuing nonreproductivity. "Firm choices to have no children may signal an increasing proportion of women who see the costs of childbearing as too high" (Hagewen and Morgan 2005, p. 522; see also Kelly 2009).

In the 1970s, feminism challenged the inevitability of the mother role. More than 70 percent of women surveyed in 2001 said no to the question of whether "a woman need[s] the experience of motherhood to have a complete life" (Center for the Advancement of Women 2003, p. 8). As we saw earlier in this chapter, research finds childfree couples to be more satisfied with their relationship than parenting couples are (Wilcox et al. 2011a). The voluntarily childless have more education and are more likely to have professional employment and higher incomes. They are often urban, less traditional in gender roles, less likely to have a religious affiliation, and less conventional than their counterparts (Kingston 2009; Lundquist et al. 2009). Childfree women tend to be more highly educated or attached to a satisfying career (Jayson 2011). Couples value their relative freedom to change jobs, pursuing any endeavor they find interesting (Kelly 2009; Park 2005). According to demographer David Foot, a greater ability to control fertility, greater participation of women in paid employment, concern about overpopulation and the environment, or an ideological rejection of the traditional family provide the social context for some people's decisions to remain childless (Kingston 2009).

How happy with being childfree are individuals once they reach older ages? Provided that their being childless is voluntary, childfree elderly are as satisfied with their lives and less stressed than parents—in fact, some studies show them to have lower levels of depression than their counterparts who did have children (Bures, Koropeckyj-Cox, and Loree 2009). They have developed social support networks other than those provided by children (Dykstra and Hagestad 2007; Park 2005). However, perhaps not surprisingly, studies show that childless women who had wanted children are concerned about their identities as women and find some things difficult, such as holidays and family gatherings because, not having children, they feel left out or sad that others do have them (McQuillan et al. 2012). A study of infertile women in the United Kingdom who had wanted children badly enough to have sought infertility treatment found depression levels to be high for those who remained childless. Furthermore, probably due in part to their chronic depression, the women's life expectancies were shorter than women with children (Agerbo, Mortensen, and Munk-Olsen 2012). Meanwhile, individuals who do have children face an array of options and circumstances.

Having Children: Options and Circumstances

Discussions about having children once evoked images of a newly married couple. More and more, however, decisions about becoming parents are being made in a much wider variety of circumstances. In this section, we address childbearing with reference to postponing parenthood, child spacing, having one child only, nonmarital childbearing, and multipartnered fertility. Deciding about parenthood in stepfamilies is addressed in Chapter 15. Much of the material in this chapter may apply not only to heterosexual but also to LGBT parents. Gay men or lesbians becoming parents is discussed in particular in Chapter 6.

Timing Parenthood: Earlier versus Later

Until a few decades ago, the vast majority of U.S. mothers had their first child in their early twenties (Kirmeyer and Hamilton 2011). Applying the *family life course development framework* (see Chapter 2), this situation was considered "on-time" parenthood. Today, because Americans are postponing having children, we consider childbearing in one's early twenties (or younger) as "early." Later age at marriage and many women's desire to complete their education and become established in a career are causes of postponed childbearing. Furthermore, as discussed in Chapter 2, *emerging adulthood* means that childfree individuals postpone family responsibilities (Reifman 2011). With the availability of reliable contraception and the promise of assisted reproduction technology if needed later, many people plan the onset of parenthood for later in their lives (Lehmann-Haupt 2009). We'll look at benefits and drawbacks of having children earlier versus later.

Many couples are postponing parenthood into their thirties, sometimes later. They establish themselves in their careers and get an education and then have time left to bear the preferred two children. The timing of parenthood is a trade-off: Having children when you're younger gives you more potential time with your children, grandchildren, and even great grandchildren. Having children when you're older may allow you to become more psychologically mature and hence a more emotionally steady parent.

Earlier Parenthood Choosing early parenthood means greater certainty of being physically able to have children. Furthermore, parenting early means greater freedom to pursue other activities later, after the children are raised (Christopherson 2006). In addition, earlier parenthood means a greater likelihood of having more time with grandchildren and even great grandchildren.

In one qualitative study, early mothers said that they felt more spontaneity as youthful parents (Walter 1986). "'We wanted to be young parents [said one mother].... We didn't want to be sixty when they got out of high school'" (Poniewozik 2002, pp. 56–57). In a study based on interviews with 114 Canadian expectant mothers, the younger pregnant women (in their twenties) spoke of their physical health as an asset, as well as the health of their parents—they expected to rely on their own parents' for help with their children (Dion 1995).

However, relatively young parents may have to forgo some education and may get a slower start up the career ladder. Early parenthood can create strains on a marriage if the breadwinner's need to support the family means little time to spend at home or if young parents lack the maturity needed to cope with family responsibilities (Gillmore et al. 2008). Moreover, couples who have children early usually start later in their adult lives to save for college or retirement. Often having relatively low incomes, they have to work harder and longer to meet family needs (Jayson 2010; Joshi, Quane, and Cherlin 2009).

Later Parenthood Older mothers tell researchers that they benefited by waiting to have children because they needed a period of time for personal development—not just career development—for themselves and their partners. They felt that delaying parenthood meant greater maturity and preparation for parenthood. Psychiatrists speak of the maturity, patience, and good parenting skills of later-life parents (Tyre 2004). Older fathers see themselves as more patient (Vinciguerra 2007). Some evidence shows that older men and women find more joy in parenthood than do their younger counterparts (Paul 2011). Older mothers felt that they had more confidence in their ability to manage their changed lives because of the organizational skills they had developed in their work. They also had more money with which to arrange support services, and they felt confident about their ability as parents (Jayson 2010).

However, for both men and women, fertility does decline with age, although less dramatically for men. Older age in both women and men also increases the likelihood of having children with genetic abnormalities (Rabin 2007). Older mothers have higher rates of premature or low-weight babies and multiple births—all risks for learning disabilities and health problems. Older mothers also have higher rates of miscarriage, as well as health problems such as diabetes and hypertension (MacDorman and Kirmeyer 2009). Some physicians nevertheless advise that, "while a lot of complications of labor and pregnancy are increased ... the vast majority of [older mothers] do perfectly fine" (Dr. William Gilbert, quoted in "Older Moms" 1999).

Nevertheless, economist Sylvia Ann Hewlett's book *Creating a Life* (2002) based on her survey of 1,168 older women in the top 10 percent of earners reports a high rate of childlessness among successful managerial and professional career women, most of whom had not intended to be childless. Hewlett faults women for focusing on careers based on the assumption that it will be easy enough to have children later in life. An article in the *Journal of Family and Reproductive Health* noted, "As women delay childbearing, there is now an unrealistic expectation that medical science can undo the effects of aging" (Karimzadeh and Ghandi 2008, p. 62). Hewlett's caution about the limits of reproductive technology are valid. Ironically, even women who postponed parenthood found that combining *established* careers with parenting created unforeseen problems (e.g., Slaughter 2012). Career commitments may ripen just at the peak of parental responsibilities.

Family policy scholars (e.g., Pollitt 2002; Mishel, Bernstein, and Shierholz 2009) argue that the real problem is the failure of society-wide or structural support for working families so that younger women could better manage both motherhood and career building.

As they age, older parents often experience a sense of limited time with their children that increases pleasure in parenting: "Everything is more precious." However, it also creates anxiety about the future: That he could die before his daughter reaches adulthood "is a reality that I live with," said one 59-year-old father when his daughter was born (Vinciguerra 2007, p. ST-1). That he may not live to see grandchildren is another reality.

Being born to older parents affects children's lives as well. They usually benefit from the financial and emotional stability that older parents can provide and the attention given by parents who have waited a long time to have children. But children of older parents often experience anxiety about their parents' health and mortality (Vinciguerra 2007). Their parents may become frail before their children have established themselves in their adult lives and before they themselves can serve as active grandparents.

A Word about Child Spacing When parents begin having children later, their options for child spacing may be limited because their desired number of children need to be born before the proverbial biological clock runs out. Parents who have children while younger may want them close together because they believe that having children close in age will encourage their playing together and make parenting easier as the children take part in age-similar activities. However, experts report that for the physical and intellectual health of both mother and child, the optimal years for spacing children is a minimum of three (Mayo Clinic Staff 2011).

"Women with short birth intervals are at higher risk of preterm deliveries, low birth weight, and adverse maternal outcomes" (Martinez, Daniels, and Chandra 2012, p. 7). Moreover, there is evidence that waiting longer to have the next child benefits the older one academically. "This positive effect of a larger gap between kids may be the result of increased time or resources spent on the older child or other factors, but the research, based on a panel of more than 12,000 children, looks pretty solid" (Dell'Antonia 2011). For prospective parents interested in the timing of their parenthood, it's important to have an awareness of the trade-offs. Meanwhile, some prospective parents not only consider the challenges of parenthood to be monumental but also reject the idea of childlessness. For them, the solution is the one-child family.

Having Only One Child

The number of one-child families continues a steady increase, making up 19 percent to 20 percent of American families (U.S. Census Bureau 2012c, Table 64). About 19 percent of women ages 40 through 44 have just one child (U.S. Census Bureau 2010a, Table 1). The proportion of one-child families in the United States appears to be growing because of at least four factors: (1) women's increasing career opportunities and aspirations in a context of inadequate domestic support (child care, etc.), (2) individuals' desire to parent in a context of inadequate workplace support for the parenting role, (3) the high cost of raising a child through college, and (4) peer support—that is, the choice to have just one child becomes easier to make as more couples do so. Divorced, separated, or widowed people who do not form a new reproductive partnership may end up with only one child because the relationship ended before more children were born.

Negative stereotypes present only children as "socially unskilled, dependent, anxious, and generally maladjusted" (Hagewen and Morgan 2005, p. 514). To find out whether there was any basis for this image, psychologists in the 1970s produced a staggering number of studies that generally found no negative effects of being an only child (Pines 1981). More recent studies of kindergartners and adolescents have found that sibling relationships foster the development of interpersonal skills and reduce incidences of depression, particularly in young girls, although the differences between children with siblings and only children is small (Downey and Condron 2004; Kim et al. 2007).

Moreover, research reports only children to be more intelligent and mature, with more leadership skills, and better health and life satisfaction both as children and into adulthood (Hagewen and Morgan 2005; Newman 2011; Watson 2012). One study found that parents of only children had higher educational expectations for their child, were more likely to know their child's friends and the friends' parents, and had more money saved for their child's college education. They reported that they could enjoy parenthood without feeling overwhelmed and tied down, had more free time, and were better off financially than they would have been with more children (Deveny 2008). They were more likely than parents in larger families to share domestic chores and could afford to do more things together (Newman 2011).

Hiroko Masuike/The New York Times/Redux

Sisters who get along well can provide companionship and support for each other as they go through life. More than 700,000 siblings shared a residence in 2000 (Lee 2006).

There are disadvantages, too, in a one-child family. Disadvantages for parents include the fear that the only child might be seriously hurt or might die and the feeling, in some cases, that they have only one chance to prove themselves good parents. For the children, disadvantages include lack of opportunity to experience sibling relationships, not only in childhood but also as adults. The 2000 census indicated that some 700,000 siblings live together. Siblings may provide social support, as well as exchanges of material assistance and someone to rely on in emergencies (Lee 2006; Riedmann and White 1996). Only children may face extra pressure from parents to succeed, and they are sometimes under an uncomfortable amount of parental scrutiny. As adults, they have no help in caring for their aging parents (Watson 2012).

Nonmarital Births

Family sociologists note that increasingly women choose motherhood but not marriage (Gibson-Davis 2011). Approximately 40 percent of all births today are to unmarried women (Hamilton, Martin, and Ventura 2010, p. 5). As with fertility statistics in general, the number of births to unmarried women has declined recently—by 4 percent in 2010, after peaking in 2008—very probably because of the recession that began in late 2007. The birthrate for unmarried women declined by 5 percent between 2009 and 2010. However, childbearing *within* marriage has declined, leaving births outside of marriage a larger proportion of total births (Martin et al. 2012, p. 10). From 1940 to the early 1960s, only 4 percent to 5 percent of births were to unmarried women. By 1980, 18 percent were. Although there was a slight decline in nonmarital births from the mid-1990s to 2002, there has been a marked increase in their proportion of all births (Martin et al. 2009, Table D).

The growing proportion of nonmarital births accompanies changing societal attitudes. In 2002, 45 percent of Americans told pollsters that they felt it was morally acceptable to have a baby outside of marriage. In 2012, more than half—54 percent—found doing so morally acceptable (Gallup Poll 2012b). Individuals are much less likely now to marry with the discovery of a nonmarital pregnancy. Meanwhile, a substantial proportion—on average, those with lower income and education levels—chooses to begin or continue to cohabit (Rackin and Gibson-Davis 2012).

Then, too, "fertility during cohabitation continues to account for almost all of the recent increases in nonmarital childbearing" (Manning 2001, p. 217). Between 40 percent and 50 percent of nonmarital births are to heterosexual cohabiting women, and birthrates for never-married cohabitants are virtually the same as those for married women (Cherlin 2010; Rackin and Gibson-Davis 2012). Cohabiting families (heterosexual and same-sex) are discussed at length in Chapter 6.

As Figure 8.4 shows, in 2011, 72.3 percent of African American births, 66.2 percent of American Indian/Alaska Native births, 53.3 percent of Hispanic births,

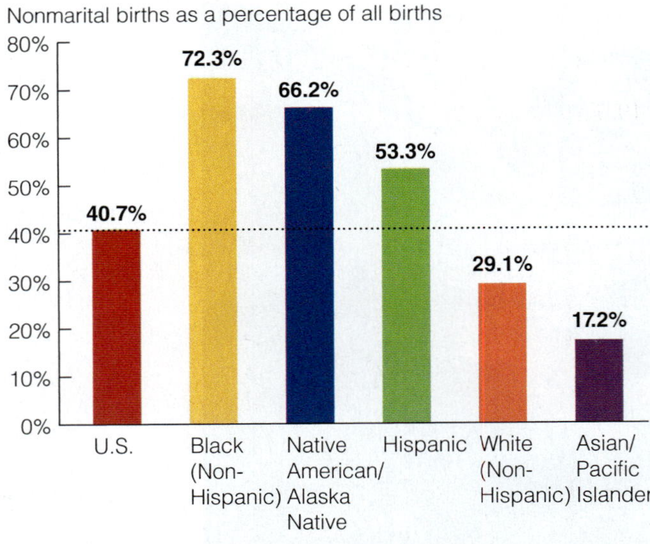

FIGURE 8.4 Births to unmarried women as a percentage of all births, by race and ethnicity, United States, 2011.

Source: Hamilton, Martin, and Ventura, 2012, Table 1.

29.1 percent of nonHispanic white births, and 17.2 percent of Asian/Pacific Islander births occurred outside marriage (Hamilton, Martin, and Ventura 2012, Table 1).

We note here that unmarried parents, including those not living together, may have a more regular relationship than was previously thought. Some are intermittent cohabitors: The children's father moves into and out of the mother's domicile as the couple experiences separations and reunifications (Cross-Barnet, Cherlin, and Burton 2011). The Fragile Families and Child Well-Being Study (McLanahan and Carlson 2002) is a national longitudinal study conducted in twenty large U.S. cities. Initially, it comprised a sample of 3,700 children born between 1998 and 2000 to unmarried parents and 1,200 born to marrieds. One analysis of these data showed that the vast majority of new parents described themselves as "romantically involved on a steady basis." Of these, 50 percent were cohabiting and 33 percent "visiting."

Fathers visited the hospital during delivery or otherwise helped mothers during pregnancy and childbirth. Nearly all fathers said they wanted to be involved in their children's lives, and 93 percent of mothers agreed that they should be. "The myth that unwed fathers are not around at the time of the birth could not be further from the truth" (McLanahan et al. 2001, p. 217). Nevertheless, involvement of these fathers is likely to decline over their child's lifetime, given their limited resources (Mitchell, Booth, and King 2009).

As opportunities grow for women to support themselves and as the permanence of marriage becomes less certain, there is less motivation for a woman to avoid giving birth outside of marriage or a committed relationship because she cannot count on lifetime male support for the child even if she marries. Furthermore, stigma and discrimination against unwed mothers have lessened. Seventy percent of women and 59 percent of men surveyed believe it is "okay for an unmarried woman to have a child" (Martinez et al. 2006, p. 29). Still, because the burden of responsibility for support and care of the child remains on the mother, overall "the economic situation of older, single mothers is closer to that of teen mothers than that of married childbearers the same age" (Foster, Jones, and Hoffman 1998, p. 163).

Single Mothers by Choice Although unwed birthrates are highest among young women in their twenties, they have increased significantly for older women in recent years (Martin et al. 2009; Ventura 2009). Many of these women are **single mothers by choice**. The image is that of an older woman with an education, an established job, and economic resources who has made a choice to become a single mother. Not having found a stable life partner, yet wanting to parent, a woman makes this choice as she sees time running out on her "biological clock" (Lehmann-Haupt 2009).

Whether this is a significant development in terms of numbers is uncertain (Musick 2002), but it has drawn the attention of researchers. Sociologist Rosanna Hertz (2006a) interviewed sixty-five single mothers who had their first child at age 20 or older and who, more significantly, are self-sufficient economically. Not having the "chance" to be in stable, child-rearing marriages, they became mothers through various routes: biological pregnancy, artificial insemination by known or unknown donor, or adoption.

"For the women in this study, single motherhood was never a snap decision" (Hertz 2006a, p. 26):

> I always had in the back of my mind that if I was 30 and not married, then I'd have children on my own. Then it was when I was 32. Then it was when I went back to school at Princeton to get my master's degree. Then it was 36 and I had just broken up with another man. (p. 26)

Women were often surprised to find themselves taking what they saw as an unconventional step:

> Daring to consider getting pregnant on my own just seemed like such an outrageous thing to do. And from that point of thinking about it, to doing it, was the longest stretch because I was kind of shocked that I would think that way, and I wasn't sure of what I really wanted to do. (p. 27)

Once a mother, the parenting practices of single mothers by choice and their sense of family were very traditional. In fact, they saw themselves as exemplifying family values by having chosen parenthood.

Similarly, in two smaller studies of single mothers by choice (Bock 2000; Mannis 1999), researchers interviewed women who adopted children or who purposefully became pregnant. These mothers, usually over age 30, saw themselves as responsible, emotionally mature, and financially capable of raising a child. Rather than viewing themselves as alternative lifestyle pioneers, they saw their choice as conforming to normal family goals. In fact, their decisions to become single mothers were well accepted by their family, friends, employers, clergy, and physicians.

These were white, middle-class, educated women who insisted on the great difference between themselves and "welfare" or teen mothers. What, in fact, are the realities of teen parenthood today?

Births to Adolescents Public concerns about outcomes for the children of unmarried parents intensify when the mother is a teenager. The term *teenage pregnancy* has been associated with the word *problem* for as long as most of us can remember. Adolescent birthrates rose in the late 1960s as sexual behavior liberalized. However, by the time a "teen pregnancy epidemic" was identified, adolescent birthrates had already begun to decline (see Figure 8.5). Declines in the adolescent birthrates have been especially large for young black women. Today's teens are using contraception more regularly. The adolescent abortion rate has dropped also, from a 1990 rate of 40.6 abortions per 1,000 women ages 15 to 19 to a rate of 18.7 in 2007 (U.S. Census Bureau 2012c, Table 102). Although still high when compared to other wealthy industrialized nations, teen pregnancies remain at a historic low for the United States and account for just 23 percent of nonmarital births (Hamilton and Ventura 2012).

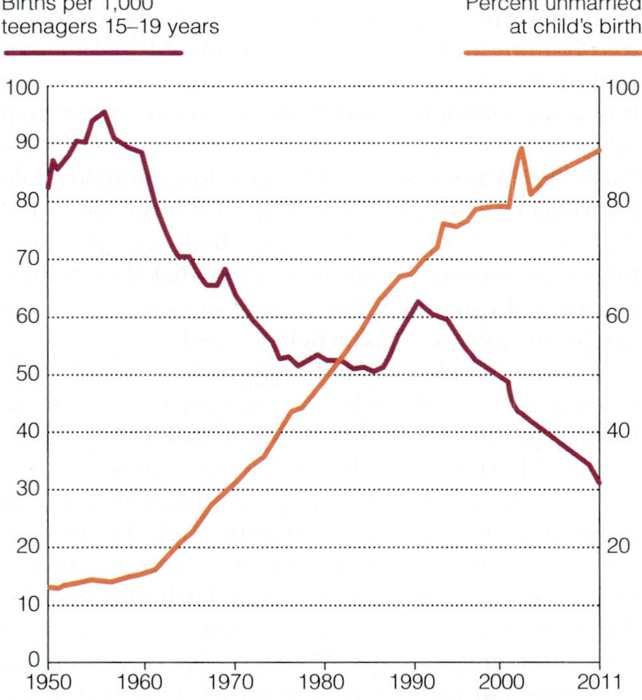

FIGURE 8.5 Birthrate for teen women 15 through 19 years old and percentage of teen births to unmarried women, 1950 to 2011.

Sources: Ventura, Mathews, and Hamilton 2001, Figure 1; Downs 2003, Figure 1; Martin et al. 2009, Tables A, 17; Hamilton, Martin, and Ventura 2012, Tables 1, 5.

"[T]hree-quarters of teen females report that their first sexual experience was with a steady boyfriend, a fiancé, a husband or a cohabiting partner," and a vast majority of them (74 percent) used contraceptives the first time they had sex ("Facts on American Teens" 2010, p. 1). As Figure 8.5 indicates, most teen women giving birth are not married, and so they lack the possible economic and social support of a co-parent. Women as well as men need more education in today's world, and women are expected to seek employment. Teenage parents, especially those with more than one child, face a bleak educational future, limited job prospects, and a very good chance of living in poverty, compared to peers who do not become parents as teenagers (Perper, Peterson, and Manlove 2010). As described earlier in this chapter, the costs associated with raising children can be daunting. Prospects for the children of teen parents have included lower academic achievement and, because of the lack of resources related to poverty, a trend toward a cycle of early unmarried pregnancy themselves (South and Crowder 2010; Bronte-Tinkew, Horowitz, and Scott 2009).

Yet we now recognize that economic or race/ethnic disadvantages may be playing a larger role than age in shaping a teen mother's limited future (South and Crowder 2010). Moreover, outcomes of teen parenthood vary and are not by any means uniformly negative. One longitudinal study of black teen mothers from low-income families in Baltimore concluded that although early childbearing increases the risk of ill effects for mother and child, it is unclear that the risk is so high as to justify the popular image of the adolescent mother as an unemployed woman living on welfare with a number of poorly cared-for children. To be sure, teenage mothers do not manage as well as women who delay childbearing, but most studies have shown that there is great variation in the effects of teenage childbearing (Furstenberg, Brooks-Gunn, and Morgan 1987, p. 142). Similarly, a careful study based on national sample data sets found that "teen childbearing plays no causal role in children's test scores and in some behavioral outcomes of adolescence." Research on other outcomes is inconclusive. "We ... suggest caution in drawing conclusions about early parenthood's overarching effects" (Levine, Emery, and Pollack 2007, p. 105).

Multipartnered Fertility

Multipartnered fertility—having children with more than one biological partner—is a new interest and area of research for family social scientists. Researchers participating in the Fragile Families Study of urban parents at the time of their first birth realized in follow-up that some parents, particularly those unmarried at the birth, went on to have children with new partners.

How frequent is multipartnered fertility? What are the implications for family life and, especially, for the well-being of children? The Fragile Families Study follow-up found that three-fourths of the mothers had children by only one father. Most of the others had children by two fathers, a few by three or four fathers.

Multipartnered fertility is most common in nonmarital families as they have a high rate of breakup; moreover, the participants tend to be younger. NonHispanic black men and women are more likely than those of other race/ethnic groups to have children by more than one partner (Carlson and Furstenberg 2006). Multipartnered fertility seems likely to lead to very complex family systems but weaker ties with extended families. Indeed, multipartnered fertility is associated with less financial, housing, and child care support from kin networks (Carlson and Furstenberg 2006; Harknett and Knab 2007). "[H]aving children by different fathers can present daunting challenges for young mothers. Having to negotiate paternal support and involvement with different men is stressful and may result in different levels of involvement for children who live in the same household but do not share the same father" (p. 37).

A study using data from the National Longitudinal Study of Adolescent Health (Guzzo and Furstenberg 2007) paid particular attention to the policy implications of multipartnered fertility. The lower the levels of education in men, the more likely they are to have fathered multiple children outside of marriage or committed relationships (Bronte-Tinkew, Horowitz, and Scott 2009, Table 1). Much of the foregoing discussion of fertility issues, and especially reports of declining birthrates in some sectors, leads us to the question of preventing unwanted pregnancies.

Preventing Pregnancy

Putting themselves at risk for unintended pregnancy, approximately 4.5 million women who do not want to become pregnant are engaging in sexual intercourse and use no birth control (Gray 2013 Guttmacher Institute 2012). Contraceptive, or birth-control, methods involve (1) physical barriers (for example, latex condoms and diaphragms) that prevent male sperm from reaching and thereby fertilizing a female egg, (2) hormonal substances (for example, "the pill") that prevent ovulation, and (3) physical (for example, IUDs) or pharmaceutical products (for example, "Plan B," or the "morning after" pill) that may prevent fertilization or may in some cases prevent a zygote from implanting itself in the uterine wall (Mayo Clinic Staff 2012).

Only physical barriers help to prevent HIV/AIDS or other sexually transmitted infections or diseases (STIs or STDs). For the most effective protection against both pregnancy and STDs, experts advise combining a barrier with a hormonal contraceptive method. For personally useful information on birth-control methods, see one or more of the following websites: (1) Centers for Disease Control and Prevention, (2) Planned Parenthood, and (3) the Guttmacher Institute.

As early as 1832, a book describing birth-control techniques and devices was published in the United States. The diaphragm and the condom were invented in 1883 (Weeks 2002), but it was not until the contraceptive pill became available in the 1960s that women could be more certain of controlling fertility and did not need male cooperation to do so. Today the pill is the most common birth-control method, followed by surgical sterilization, particularly for women in their thirties and older who feel they have finished with childbearing (U.S. Census Bureau 2012c, Table 98).

Among men, about 50 million have had a vasectomy, a form of birth control that involves surgical sterilization and should be considered permanent (although in some cases it can be surgically reversed); it is more commonly used by nonHispanic whites than other race/ethnic groups (U.S. Census Bureau 2007b, Table 95; "Vasectomy" 2012). The long-awaited male pill remains... long-awaited. Potential formulas would use hormones to "send the [male] body signals to stop producing sperm." Some men attest that they "would definitely do some kind of long-term male contraceptive" if one were available (Belluck 2011).

Because the physical and opportunity costs of childbearing tend to be higher for women than for men, family-planning services have always been oriented to women. However, with the possible exception of surgical sterilization, contraception takes place in a sexual encounter or relational context that affects not only choice of methods but also whether contraception is used at all. The facts that not only birthrates but also abortions have fallen over the past several decades speak to the increasingly effective use of contraception by all age and race/ethnic categories. One reason for this development may be that family-planning organizations have realized that they need to reach out to adolescent young men to provide them with contraceptive and health information and so influence couple decisions as one approach to teen pregnancy prevention (Ball and Moore 2008).

Abortion

Effective contraception prevents unwanted pregnancy. When contraception is not used or fails, a woman who does not want to bear a child may choose to have an induced **abortion**—the surgically or pharmaceutically caused expulsion of a fertilized embryo or fetus from the uterus. Approximately 40 percent of American women have had an induced abortion at some point in their lives (Henshaw and Kost 2008; Kliff 2010). Nevertheless, the rate of abortions has been falling for all age groups since 1980 (U.S. Census Bureau 2007a, Tables 96 and 97; U.S. Census Bureau 2012c, Table 102). Furthermore, more abortions take place early in pregnancy than in the past—61 percent within the first nine weeks (U.S. Census Bureau 2012c, Table 102).

More than four-fifths (85 percent) of abortions are obtained by unmarried women, and this figure includes 29 percent of abortions obtained by cohabiting women. More than half (58 percent) are obtained by women in their twenties, with 17 percent by teens. NonHispanic white women account for the greatest *number* of abortions and for between 36 percent and 38 percent of all abortions (Jones, Finer, and Singh 2010; U.S. Census Bureau 2012c, Table 102).

The percentage of pregnancies that are terminated by abortion (the *abortion ratio*) is highest among black and Asian/Pacific Islander women. Researchers conclude that "women who have abortions are diverse, and unintended pregnancy leading to abortion is common in all population subgroups" (Jones, Darroch, and Henshaw 2002, p. 232). Still, women in poverty account for a disproportionate share of abortions (Boonstra et al. 2006). Slightly less than half (46 percent) have had a previous abortion (U.S. Census Bureau 2012c, Table 102).

Reasons for abortion reported in surveys and interviews at abortion sites include the following:

- having a child would interfere with the woman's education, work, or ability to care for dependents (74 percent);
- not being able to afford a baby at this time (73 percent);
- not wishing to be a single mother or having relationship problems (48 percent);
- the woman or couple had completed childbearing (38 percent); and
- the woman or couple were not ready to have a child (33 percent) (Finer et al. 2005; Boonstra et al. 2006, pp. 8–9).

"Although women who have abortions and women who have children are often perceived as two distinct groups, in reality they are the same women at different points in their lives" (Guttmacher Institute 2006, p. 9; see also Henshaw and Kost 2008). Indeed, 61 percent of women getting an abortion already have at least one child, and this figure includes 34 percent with two or more children (Jones, Finer, and Singh 2010). Rhetorically asking why this "mother majority" among those getting an abortion is seldom discussed, *Slate* blogger Lauren Sandler posits that in a culture seeking to discredit abortion and those getting abortions, "antiabortionists have successfully depicted women who choose to terminate a pregnancy as sexually indiscriminate. It's much easier than to demonize the mother who is struggling to support the kid she already has" (Sandler 2011).

The Politics of Family Planning, Contraception, and Abortion

You may have assumed that contraception had become nonpolitical. At least, you may have assumed that until the 2012 presidential election campaign when several conservative politicians and pundits argued passionately that federal dollars not be spent to provide women with contraceptives (Dowd 2012). Ordinarily, however, the most heated conflict concerns abortion.

Abortion has existed throughout history and was not legally prohibited in the United States until the mid-nineteenth century. Laws prohibiting abortion stood relatively unchallenged until the 1960s when an abortion reform movement resulted in the 1973 U.S. Supreme Court's *Roe v. Wade* decision. *Roe v. Wade* legalized induced abortion without question in the first three months, or first trimester, of pregnancy. But abortion remains subject to regulation in the second trimester and may be outlawed by states after fetal viability (the point at which the fetus is able to live outside the womb), which occurs in the third trimester. Very few abortions—less than half of 1 percent—take place in the third trimester (Henshaw and Kost 2008).

Pro-choice and pro-life activists—those who, respectively, favor or oppose legal abortion—have made abortion a major political issue. Legislation and other public policy responses to abortion have been shaped by this struggle, as has been the availability of abortion services (Pickert 2013). You may recall that, due to abortion politics, funding for Planned Parenthood came into political jeopardy during the 2012 U.S. presidential election campaign (Lepore 2011). Many states have placed various restrictions on access to abortion (Stout 2007). As of 2011, Kansas had only one abortion clinic remaining open in the entire state. The others had been effectively shut down by a plethora of bureaucratic state rules mandating, for instance, the size of recovery, procedure, and janitorial space—requirements that facilities were unable to meet (Levy 2011).

KAREN BLEIER/AFP/Getty Images

Alex Wong/Getty Images

In our society, sexuality and reproduction have become increasingly politicized. Nowhere is this more apparent than in the intensely heated pro-life versus pro-choice debate over abortion, one of the most polarizing issues in America today.

Although the U.S. Supreme Court has upheld many state restrictions on abortion, it has not outlawed the procedure. According to Gallup polls, the majority of Americans (68 percent) believe that *Roe v. Wade* should remain the law of the land. Support for abortion is heavily qualified, however, as you can see in Table 8.1.

The Right to Life movement has claimed that abortion is a threat to both medical and physical health, including threats to future reproductive capacity and an increased risk of breast cancer (National Right to Life 2005, 2006). However, scientific medical evidence strongly indicates otherwise (U.S. Food and Drug Administration 2006). Research indicates that abortion has no impact on a woman's ability to become pregnant later (Boston Women's Health Book Collective 2005). A panel assembled by the National Cancer Institute to review the research concluded that there is no association between induced abortion and breast cancer (Collaborative Group on Hormonal Factors in Breast Cancer 2004; see also Bakalar 2007).

Table 8.1 Percentage of U.S. Adults in May 2012 Saying

ABORTION SHOULD BE ...	
Legal under any circumstances*	25%
Legal only under certain circumstances**	52%
Illegal in all circumstances	20%
No opinion	3%

*Circumstances might involve the woman's physical or mental health or life being endangered, when the pregnancy was caused by rape or incest, or when there is evidence that the baby may be physically impaired (Saad 2003).

**The poll questions did not ask about specific circumstances.

Source: Gallup Poll 2012c.

Deciding about an Abortion

For most women and for many of their male partners abortion is an emotionally charged, often upsetting, experience. Some women report feeling guilty or frightened. Emotional stress is more pronounced for second-trimester than earlier abortions and for women who are uncertain about their decision. Women from religious denominations or ethnic cultures that strongly oppose abortion may have more negative and mixed feelings after an abortion, but outcomes also depend on the woman's emotional well-being before the abortion (Major et al. 2009).

Some women have reported that the decision to abort enhanced their sense of personal empowerment (Kliff 2010). Research has found positive educational, economic, and social outcomes for young women who resolve pregnancies by abortion rather than giving birth (Holmes 1990; Fergusson, Boden, and Horwood 2007). The decision to abort is often very difficult to make and to act on (Major et al. 2009; Steinberg and Russo 2008), but the emotional distress involved in making the decision and having the abortion does not typically lead to severe or long-lasting emotional problems (American Psychological Association 2005a).

Individuals making decisions about abortions are most likely to make them in accordance with their values. A detailed review of religiously or philosophically grounded moral and ethical perspectives on abortion is outside the scope of this text, but as we note in Chapter 1, values provide the context for such decisions.

Abortion decisions are primarily made within the context of unmarried, accidental pregnancy or a failing relationship. However, some happily married couples consider aborting an unwanted pregnancy if, for example, they feel that they have already completed their family or could not manage or afford to raise another child. About 40 percent of unintended pregnancies are aborted (Guttmacher Institute 2006).

The question of abortion can also arise for couples who, through prenatal diagnostic techniques, find out that a fetus has a serious defect. Recent years have seen extraordinary scientific advances in monitoring fetal development, including ultrasound, amniocentesis, and blood screening techniques used to assess a fetus's risk of abnormality. Amniocentesis and other prenatal testing (also discussed in Chapter 1) provide information to parents about their risk of having a child with a birth defect (Pollack 2012b). In amniocentesis, a physician inserts a needle through the abdominal wall into the uterus, withdrawing a small amount of amniotic fluid, and then examining the fluid for evidence of birth defects. As women postpone childbearing to older ages, they are more concerned about the risk of birth defects. Common concerns are Down syndrome, cystic fibrosis, and spina bifida, the latter possibly leading to severe mental and physical disability. In some cases, fetal surgery can correct problems discovered by prenatal testing.

Social scientists working in this field find parents to be troubled when testing reveals the likelihood of a serious problem. Virtually all testing programs include genetic counselors who can advise parents as to the significance of their test results and help them work through their decisions and their emotions. Because detection of an abnormal fetus gives prospective parents the option to choose abortion, antiabortion groups have objected strenuously to prenatal screening and genetic counseling. Some advocates for Down syndrome children "see expanded testing as a step toward a society where children like theirs would be unwelcome" (Harmon 2007a).

But some parents who would reject abortion under any circumstances have undertaken prenatal screening with the thought that, should testing reveal an abnormality, they would have time to prepare to care for their infant. Prenatal testing and the accompanying abortion decision remain ethically and personally difficult choices. Individuals dealing with pregnancy decisions should seek expert medical and ethnical or spiritual advice regarding methods and their risks. For some, concern about fertility means avoiding unwanted births. Other couples and individuals face a different problem. They want to have a child, but either they cannot conceive or they cannot sustain a full-term pregnancy. We turn now to the issue of involuntary infertility.

Involuntary Infertility and Reproductive Technology

Involuntary infertility is the condition of wanting to conceive and bear a child but being physically unable to do so. About 6.7 million, or 11 percent of women—and 6 percent of married women—between ages 15 and 44 are infertile. Male infertility accounts for fertility difficulties in approximately one-third of couples seeking infertility treatment (Centers for Disease Control and Prevention 2012d). Infertility has become more visible because the current tendency to postpone childbearing until one's thirties or forties helps to create a population of infertile potential parents who are intensely hopeful and financially able to seek treatment.

Infertility treatments can involve drug therapy, donor insemination, in vitro fertilization, and related techniques. Louise Brown, the first "test-tube" baby, gave birth to *her* first child in 2006 ("World's 1st" 2007). This look back reminds us of how astonishing the first developments in reproductive technology were. Now, more and more, **assisted reproductive technology (ART)** has become an accepted reproductive option (Chandra and Stephen 2010). ART is discussed at some length in the section, "Ever-New Biological and Communication Techniques," in Chapter 1. You may want to review that section while studying this one.

When faced with involuntary infertility, an individual or a couple experiences a loss of control over life plans and may feel helpless, defective, angry, and often guilty. These emotions also commonly follow miscarriage, the spontaneous *nonvoluntary* abortion or loss of a fetus before its ability to survive outside the womb (Wallach 2010). Although research studies concur that generally infertility is more stressful for women than for men, both are very affected emotionally by the challenge to taken-for-granted life plans and sense of manhood or womanhood (Himanek 2011). Upward of one-half of women undergoing infertility treatments and 15 percent of men say that it "was the most upsetting experience of their lives" ("The Psychological Impact" 2009, p. 1). Going through infertility treatment is costly and not often covered by health insurance.

About one-third of ART procedures result in a live birth (U.S. Census Bureau 2012c, Table 100). When successful, couples face decisions about how much to tell colleagues and friends—and, later, their child—about how they became pregnant (Rauscher and Fine 2012). When infertility treatment is not successful, couples are faced with the decision whether or when to quit trying.

Reproductive Technology: Social and Ethical Issues

Reproductive technologies enhance choices and can reward infertile couples with much-desired parenthood. They enable same-sex couples or uncoupled individuals to become biological parents. But reproductive technologies raise thorny questions as well. For one thing, ART, the fertility-enhancing procedure, raises abortion issues.

ART and Abortion In ART procedures, several eggs typically are fertilized but only one or sometimes two zygotes are allowed to develop. Remaining fertilized eggs may be simply discarded or used in research. Those who view human life as beginning at conception define this practice as abortion. Therefore, some ART parents carry all zygotes to term—hence the increase in multiple births, even multiple births in large numbers of five or more infants. As one mother said, "I, personally, could not bear to think of 'flushing' them [the fertilized eggs]" (in Frith et al. 2011, p. 3333).

Participating in a (Christian-initiated) "embryo adoption program," other couples work to place ART embryos with future parents who will use them in their own "family-building" endeavors: "We felt responsible that we had created these lives and were responsible for finding them good homes if we could not be that home" (Frith et al. 2011, p. 3334). Some of the relinquishing parents hoped to keep communication options open for future contact with a child produced from their relinquished embryo. A relinquishing couple might imagine "for our son to have 'brothers or sisters,' with whom I hoped he could possibly have a special relationship later in life" (Frith et al. 2011, p. 3335). In the words of another relinquishing couple, "We don't want to insert ourselves in [the embryo adopters'] lives but after 18 years old, we want them to have our information so they can get answers to questions. It will also let [our twins] have their questions answered" (Frith et al. 2011, p. 3335).

Inequality Issues Assisted reproductive technology is usually not affordable for those with low incomes. By and large, after initial diagnosis, lower-income couples do not go on to more advanced (and expensive) treatment (Chandra and Stephen 2010). Said one woman, "There need to be some options for people like us who don't have money sitting in the bank" (Becker 2000, p. 20).

Another aspect of inequality issues raised by ART involves the individuals who can afford to buy a woman's eggs, for instance, or hire a surrogate mother to gestate a child for them—compared according to social status, finances, and social power with the woman who sells her eggs or hires herself out as a surrogate. For one thing, a surrogate mother may feel herself to be under the strict surveillance of the woman hiring her—told not to pump her own gas, for instance, because the fumes might damage the fetus (Ali and Kelley 2008). On the other hand, surrogates are also able to exercise power in the relationship. In one fairly recent case, the would-be parents demanded that their surrogate abort the fetus she was carrying after ultrasounds indicated that the child would be born with serious birth defects. The surrogate refused and delivered the baby (who was born with serious birth defects and is now in adoptive care) (Neel 2013).

A surrogate birth can cost future parents between $60,000 and $80,000, with the surrogate mother herself earning between $20,000 and $25,000. Other expenses include medical, legal, and agency costs and fees, as well as maternity clothes for the surrogate or other incidentals (Hinders 2012). Then too, women serving as paid surrogates report feeling disregarded once the baby is born and bequeathed to the social mother (Ali and Kelley 2008). According to one reputable website,

> [A] growing number of American women are choosing to work with [Asian] Indian surrogates [overseas]. Instead of paying up to $70,000 in the United States, a woman can hire an Indian surrogate and pay all associated medical costs for just $12,000.
>
> However, this approach is very controversial. Many people worry that poor women are being exploited by the surrogacy arrangement. There are also concerns that standards of medical care may not be as high as they are in the United States, plus fears of potential legal repercussions if the surrogacy arrangement does not go exactly as planned. If you are thinking of going outside the United States in search of a surrogate, careful research is in order. (Hinders 2012)

Moreover, in a process known as *social surrogacy* (as opposed to biologically necessary surrogacy), some career women, athletes, and models have hired surrogates to gestate a baby fertilized in vitro with the woman's own egg, not because they are unable to carry and deliver the infant themselves but because they would prefer not to (Mayes 2001). Is it appropriate to hire reproductive labor if one has the money to do so?

Who Is a Parent? Then, too, reproductive technology creates "family" relationships that depart considerably from what is possible through unassisted biology (e.g., Hammel 2012; Zernike 2012). Surrogacy, along with embryo transfer, creates the possibility that a child could have three mothers (the genetic mother, the gestational mother, and the social mother), as well as two fathers (genetic and social). In situations like these, how do courts define the "real" parents?

Other concerns regarding one ART procedure, artificial insemination, are related to sperm donors. For one thing, a paid male donor who sells his sperm to one or more fertility clinics or sperm banks in a geographical region may biologically father tens or even hundreds of offspring, all of them genetic half-siblings. What if they should unknowingly meet in adolescence or adulthood and then together have a biologically or genetically compromised child through an incestuous relationship (Mroz 2011)? A second issue involves a donor's financial, social, or emotional responsibility for his offspring. In one reported instance, a male donor, Mr. Russell, produced sperm for a female friend. The resulting child, Griffin, knows his biological father as "Uncle George." But at some point the mother

> intends to tell her son the truth. Mr. Russell worries about that moment. He never wanted to be a parent; he saw the sperm donation as a favor to a friend. He did not attend the birth or Griffin's first birthday party. His four sisters were trying to figure out whether they were aunts. (Kleinfield 2011)

Moreover, officials in one state, Kansas, are seeking child support from the 2009 sperm donor to a lesbian couple, now separated and seeking financial assistance from state welfare agencies (Celock 2013).

An interesting recent development is the emergence of sperm donors as putative fathers, sought out by their biological offspring as they enter adolescence or young adulthood (Harmon 2005b, 2007b). Many states have laws by which sperm donors, with the exception of the husband or partner of the mother, have no parental rights, but this barrier between sperm donors and their biological children is gradually being broken. In fact, the American Society for Reproductive Medicine recommends against secrecy: "It's no longer possible to think of sperm donation without thinking of what the child it produces may someday want" (Talbot 2001, p. 88).

What Kind of Child? As technology advances, the potential to create a child with certain traits expands. Embryo screening—a technology for examining fertilized eggs before implantation to choose or eliminate certain ones—is a boon for prospective parents whose family heritage includes disabling genetic conditions. But embryo screening also raises the possibility of sex selection (Grady 2007; Marchione and Tanner 2006) and perhaps selection for other traits. Those who use sperm donors are already scanning the records to find evidence of traits they would like in their children. Philosophers ponder the implications for parents and children when children can be made to order!

Commercialization of Reproduction A general concern is that the new techniques, when performed for profit, commercialize reproduction. Prospective parents and their bodies are treated as products and thereby dehumanized (Rothman 1999). Examples include the selling of eggs or sperm to for-profit fertility clinics and the marketing of sperm or eggs with certain donor characteristics such as intelligence, physical attractiveness, and athletic ability.

Reports of fraud, overstatement of positive outcomes, failure to warn about the risk of multiple births, and other professional violations (Leigh 2004) make it important to understand that an individual seeking treatment is, in fact, a consumer and should interview the doctor and investigate the facility. Clients may not realize that the average success rate (a baby) is around 27 percent. That varies with the type of procedure and with age; chances are much less for women over 40 and as low as 1 percent for women age 46 or older (Spar 2006, Table 2.2).

We should note that these concerns about reproductive technology are primarily articulated by medical and public health professionals, academics, policy analysts, and ethicists. For the most part, prospective parents are mostly focused on their desire for a child.

Reproductive Technology: Making Personal Choices

Choosing to use reproductive technology depends on one's values and circumstances. Religious beliefs and cultural values may influence decisions. For instance, the official teaching of the Catholic Church prohibits all forms of reproductive technology (Congregation for the Doctrine of the Faith 1988). Furthermore, fertility treatment can be financially, physically, and emotionally draining. The need for frequent physician's visits can interfere with job obligations, and infertility treatment can lead to tensions in a marriage.

On the positive side, the vast majority of children born by means of in vitro fertilization or donor sperm are thoroughly normal. However, there are slightly elevated incidences of birth defects in children born through the use of ART. Although these slight increases intrigue researchers, the chances of bearing children with birth defects, whether using ART or traditional sexual intercourse to get pregnant, remains at approximately 3 percent (Kolata 2009, D1).

Meanwhile, infertility treatments do not work for all who had hoped to become parents. Coming to terms with infertility has been likened to the grief process, in which initial denial is followed by anger, depression, and usually ultimate acceptance. Of course, not everyone has the same experience: "When I finally found out that I absolutely could not have children ... it was a tremendous relief. I could get on with my life" (Bouton 1987, p. 92). Some people gradually choose to define themselves as permanently and comfortably childfree (Koropeckyj-Cox and Pendell 2007). Another way to get on with life yet retain the hope of parenthood is through adoption. Some couples explore their adoption options even as they continue infertility treatments (Ford 2009).

Adoption

Issues specific to same-sex adoption are addressed in Chapter 6. With regard to adoption in general, there are about 1.1 million children with at least one adoptive parent in U.S. households—about 2 percent of all children (Kreider and Ellis 2011a). In terms of *numbers,* there are more adopted children in nonHispanic white families (more than 70 percent of all adopted children). But Asian/Pacific Islander families have the highest *rate* of adoption relative to their population. More girls than boys are adopted. Women, especially single women, prefer to adopt girls, and girls are more likely to be available for adoption. Ninety-five percent of Chinese babies available for adoption, for example, are girls (Kreider 2003; U.S. Census Bureau 2007a, Table 65).

Census data do not distinguish adoptions by biological relatives or stepparents from nonrelative adoptions, but studies show that a majority of adopted children are kinship adoptions—that is, the children are related to their adoptive parents (Child Welfare Information Gateway 2011). Most commonly, those who adopt unrelated children are older, are highly educated, have no other children, and have already used infertility services (Bausch 2006). Some children are adopted informally—that is, the children are taken into a parent's home but the adoption is not legally formalized. **Informal adoption** is most common among Alaska Natives, African Americans, and Hispanics (Kreider 2003).

Adoptions increased through much of the twentieth century, reaching a peak in 1970, but the number has declined since, with between 1 percent and 2 percent of American families now adopting (J. Jones 2008). Reasons for the decline in adoptions include the facts that fewer infants are available due to more effective contraception and to legalized abortion. Moreover, the majority of unmarried mothers who were likely to relinquish their infants in the past are keeping them today (J. Jones 2008, Table 16). Some couples pursue international or transracial adoption, whereas others adopt "special needs" children—those who are older, come with siblings, or are disabled.

The Adoption Process

The experience of legal adoption varies widely across the country, partly because it is subject to differing state laws. Adoptions may be public or private. *Public adoptions* take place through licensed agencies. *Private adoptions* (also called *independent adoptions*) are arranged between the adoptive parent(s) and the birth mother, usually through an attorney. Legal fees and the birth mother's medical costs are usually paid by the adopting couple.

More and more, adoptions are open; that is, the birth and adoptive parents meet or have some knowledge of each other's identities. Even when an adoption is closed, as adoptions used to be, some states now have laws permitting the adoptee access to records at a certain age or under specified conditions. With regard to attitudes about open adoption, an interesting recent study compared straight with same-sex adopting couples. The study found that both couple types favored open adoption but for different reasons. Heterosexuals saw open adoption as virtually their only option because closed adoptions are less and less available. Same-sex couples embraced open adoption because they appreciated "the philosophy of openness whereby they were not encouraged to lie about their sexual orientation in order to adopt" (Goldberg et al. 2011, p. 502).

A concern that arose in recent decades because of some high-profile cases is whether birth parents can claim rights to a biological child after the child has been adopted (Markon 2010). However, of all domestic adoptions in the recent past, less than 1 percent have been contested by biological parents (Stewart 2007).

Another concern of prospective adoptive parents has to do with the adjustment of adopted children—are they likely to have more problems than other children? Considerable research suggests that adopted children, especially males, are at higher risk of problems in school achievement and behavior, psychological well-being, and substance use (Simmel, Barth, and Brooks 2007). As can be the case in social science, a different research review concluded that the overall body of research supports "the view that most adoptive families are resilient" and that positives outweigh negatives (O'Brien and Zamostny 2003, p. 679).

Adoption of Race/Ethnic Minority Children

Today 40 percent of adopted children differ in race/ethnicity from their adoptive parents. (Vandivere,

A Closer Look at Diversity

Through the Lens of One Woman, Adopted Transracially in 1962

By Cathy L.Wong, PhD

In history today we are studying about China. I do not know anyone there, but they might know of me. The teacher seemed to think that I should know about "my people." I tell her I am adopted and she mumbles cautiously that it was "probably for the best." She continues, "at least you will fit in if you were to visit the homeland someday." I'm confused, but wonder if she had stated that because I don't "fit" into America? Or was she implying I did not fit in to this community? I'm confused and saddened by what I think I understand. I look around and see no one who looks like me. I find myself staring intensely at the pictures of all those people climbing the Great Wall and wonder if anyone misses me. I come to realize that I cannot speak Chinese—I feel like a foreigner within both worlds. (Cathy L. Wong, personal journal, 1972, age 11)

I was born in Hong Kong in 1962 and, as an infant, adopted by Americans. Despite the love, attention, and caring of two well-meaning adoptive parents, as long as I can recall, I was left with many unanswered questions as a transracial adoptee. In the midst of all of my unanswered questions and the voids in my own narrative, I have constructed meaning, finding a place for myself within the larger context of the society in which I live.

I am Chinese, my mother is of Swiss and Italian descent, and my father is of Greek origin. In my case, like those of many other transracial adoptees, history played an important role. War and economic strife led to the abandonment of thousands of children in Hong Kong—mostly girls like me—at the end of World War II. This is how I ended up here, in the United States.

In 1972, debates arose in the United States regarding adoptions of children across racial and ethnic lines. Questions were raised such as, "Can white people properly raise children of color?" and "Is this in the best interest of adopted children?" The debate continues today.

At the heart of the debate are white, predominately middle- or upper-middle-class households in first-world countries who are adopting children from second- and third-world countries.

My transracial adoption story is my own and not necessarily reflective of other transracial adoptee experiences. I think that today things are different for many transracial adoptees, because our culture may now be more sensitive to respecting differences.

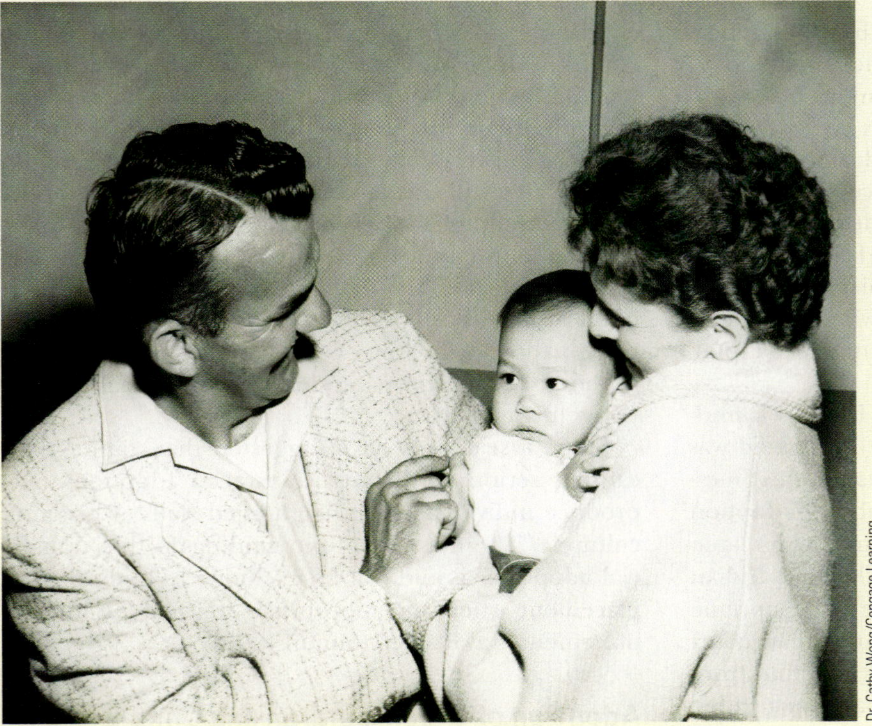

Dr. Cathy Wong/Cengage Learning

1962. Cathy Wong, as an infant, arrives in the United States from Hong Kong after being adopted by her nonHispanic white parents.

Critical Thinking

What are some ways that a society's sensitivity to cultural diversity might influence transracially adoptees' comfort or discomfort in their adoptive family? What would be your answer to the question, "Can white people properly raise children of color?"

Source: Adapted from Wong n. d.

Malm, and Radel 2009, p. 9). The family diversity created by transracial adoption seems in tune with the increasing diversity of American society and now includes four out of ten adoptions in the United States (Vandivere, Malm, and Radel 2009). Yet it has been controversial, as "A Closer Look at Diversity: Through the Lens of One Woman, Adopted Transracially in 1962" suggests.

In 1971, agencies placed more than one-third of their black infants with white parents (Nazario 1990). At that time, the number of black adoptive homes was much smaller than the number of available children, whereas the reverse was true for whites. But interracial adoptions, having increased rapidly in the 1960s and early 1970s, were much curtailed after 1972, when the National Association of Black Social Workers strongly objected. Suggesting that transracial adoption amounted to cultural genocide, race/ethnic minority advocates expressed concern about identity problems and the loss of children from the black community ("Preserving Families of African Ancestry" 2003). Native American activists have often—but not always—successfully asserted tribal rights and collective interest in Indian children (Liptak 2012a; Rayman 2013). In addition to identity concerns, they expressed the fear that coercive pressures might be put on parents to relinquish their children to provide adoptable children to white parents. Indeed, this practice was pervasive through the 1960s (Fanshel 1972).

The Indian Child Welfare Act of 1978 requires that "adoptive placement be made with (1) members of the child's extended family, (2) other members of the same tribe, or (3) other Indian families " so as "to protect the rights of the Indian child as an Indian and the rights of the Indian community and tribe in retaining its children in its society." In practice, outcomes of contested adoption cases have depended on the parents' attachment to the reservation and other circumstances. Tribes have also agreed to placements with white guardians or adoptive parents when they have believed it to be in the child's best interest (Liptak 2012a; Rayman 2013).

In one situation, a Native American father, on finding out that his (non–Native American) girlfriend was pregnant, renounced his parental rights in a text message to her. The girlfriend placed her baby for adoption with non–Native American parents. Three years later, the father wanted his child back under the 1978 Indian Child Welfare Act. The case went to the U.S. Supreme Court, which ruled that—under law—an Indian court could revisit the case, possibly removing the child from its adoptive home. However, the tribal court ruled that the child could remain with her adoptive parents, saying, "It would have been cruel to take [her] from the only mother [she] knew" (Liptak 2012a; and see Rayman 2013 for a recent similar case).

As a result of this controversy, adoption agencies shied away from transracial adoption for many years. In the late 1980s, only about 8 percent of adoptions were interracial, usually adoption of mixed-race, African American, Asian, or Native American children by nonHispanic whites (Bachrach et al. 1990). Congress has had the last word on this matter, however. The Multiethnic Placement Act (1994) and the Adoption and Safe Families Act (1997) prohibit delay or denial of adoption based on race, color, or national origin of the prospective adoptive parents. But some racial issues still arise in adoption decisions. Joseph Crumble typifies the concerns of some black social workers: "For blacks, it's about how confident whites can be with the issues of race when their race is in conflict with the race of the child" (in Clemetson and Nixon 2006, p. A18).

Long-term studies suggest that transracial adoption has proven successful for most parents and children, including with regard to racial issues. Sociologist Rita Simon and social work professor Howard Altstein followed interracial adoptees from their infancy in 1972 to adulthood. They were able to locate eighty-eight of the ninety-six families from the 1984 phase of the study for their latest book (2002). They concluded that, as adolescents and later, transracially adopted children "clearly were aware of and comfortable with their racial identity" (p. 222).

Another longitudinal study of transracial (white parents and African American, Asian, and Latino children) and in-race adoptions (white parents, white children) followed the children from the mid-1970s to 1993, when they were in their early twenties. There were no differences in general adjustment or problem behavior between the two groups. What adjustment difficulties that did exist among the transracially adopted children tended to be connected to racial issues—discrimination and "differentness" of appearance. Not surprisingly, then, researchers found that neighborhood made a difference within the transracial adoptee group; those who were reared in mixed-race neighborhoods were more confident in their racial identity (Feigelman 2000).

Some researchers have suggested that, rather than causing serious problems, transracial adoptions may produce individuals with heightened skills at bridging cultures. "The message of our findings is that transracial adoption should not be excluded as a permanent placement when no appropriate permanent inracial placement is available" (Simon 1990).

Adoption of Older Children and Children with Disabilities

Together with certain race/ethnic minorities, children who are no longer infants and children with disabilities make up the large majority of youngsters now handled by adoption agencies (Finley 2000). Special needs adoptions are pursued not only by couples who are infertile but also for altruistic motives. Gay men have adopted infants with HIV/AIDS, for example (Morrow 1992). In some cases, lesbian and gay male or older couples adopt such hard-to-place children because law or adoption agency policy denies them the ability to adopt other children.

The majority of adoptions of older children and children with disabilities work out well. Disruption and dissolution rates rise with the child's age at adoption. Among adoptions generally, only about 2 percent of agency adoptions end up being *disrupted adoptions* (the child is returned to the agency before the adoption is legally final) or *dissolved adoptions* (the child is returned after the adoption is final). But 4.7 percent of adoptions of children ages 3 to 5 at adoption, 10 percent of those ages 6 to 8, and perhaps as high as 40 percent of children adopted between ages 12 and 17 are disrupted or dissolved (Festinger 2005).

What causes these disrupted and dissolved adoptions? For one thing, some children available for adoption may be emotionally damaged or developmentally impaired due to drug- or alcohol-addicted biological parents, physical abuse from biological or foster parents, or previous broken attachments as they have been moved from one foster home to another. Some develop **attachment disorder**, defensively unwilling or unable to make future attachments (Barth and Berry 1988; Vandivere, Malm, and Radel 2009). Observers have seen attachment disorder among adoptees from Romania and other eastern European orphanages (Mainemer, Gilman, and Ames 1998).

Adoption professionals point out that parents are willing to adopt children with problems as long as they know what they are getting into (Vandivere, Malm, and Radel 2009). Agencies have increasingly tried to gain information about the circumstances of the pregnancy and the child's early life and to match children's backgrounds with couples who know how to help them (Ward 1997).

International Adoptions

International adoptions grew dramatically through 2003 but have slowed down to less than 18,000 adoptions in 2004. In 2007, 60 percent of all children adopted from overseas by American parents were from Asia, especially from China; 15 percent from India, Kazakhstan, Colombia, Ukraine, Philippines, and Ethiopia; 13 percent from Russia; and 11 percent each from Guatemala and Korea (Vandivere, Malm, and Radel 2009, Table 3).

Parents who have adopted internationally have encountered challenges: the expense of travel to a foreign country—and getting time off from work to go to the child's country for an extended stay; difficulty with negotiations and paperwork in a foreign language, and the need to rely on translators and brokers; the uncertainty about being able to choose a child, as opposed to having one thrust upon the parent; the occasional unexpected expansion of adoption fees or expected charitable contributions; the ambivalence and reluctance of a nation to place its children abroad; and the complete failure to bring home a child.

The biological mother's consent is an issue in overseas adoptions because it is more difficult to be sure that the mother has willingly placed her child for adoption rather than being coerced or misled by a baby broker. Romania placed a moratorium on adoptions, fearing corruption of their entire system. Guatemala is revising its process to comply with the Hague Convention on Intercountry Adoption. (The United States ratified this treaty in 2008.) China has moved to impose new rules on foreign adoptions, including not only establishing an age requirement for adoptive parents (under fifty) and stable marriage specifications but also ruling out obese prospective parents (Clemetson 2007; "The Hague Convention" 2010; J. Gross 2007; Yin 2007).

Prospective adoptive U.S. parents who want to adopt internationally face the frustration and emotional pain of getting into the adoption process and even meeting and beginning to bond with a child from overseas, only to have the process aborted by international politics. Situations like this developed in late 2012 with Russian adoptions (Herszenhorn and Kramer 2012). At the time of this writing, Russia's position on adoptions to U.S. parents is uncertain and unpredictable (Herszenhorn 2013).

Meanwhile, international adoptions can pose some of the same problems as the adoption of older children. Conditions in homes and institutions overseas may not be ideal beginnings, and children may have health problems or suffer from attachment disorder (Elias 2005; J. Gross 2006c; Vandivere, Malm, and

In this photo taken in Plainview, Nebraska, 4-year-old adopted daughter Natalie has just been sworn in as a new U.S. citizen. Since 2001, children adopted internationally by U.S. citizens receive their American citizenship automatically.

Radel 2009). But the vast majority of international adoptions are successful (Tanner 2005). A meta-analysis of around 100 studies found that adopted children are referred to mental health services more often than nonadoptees are, perhaps a function of adjustment concerns and high-income parents more than troubling behavior. "Most international adoptees are well adjusted" (Juffer and van Uzendoorn 2005, p. 2,501). They "are underrepresented in juvenile court and adult mental health placements," according to Dr. Laurie C. Miller, editor of the *Handbook of International Adoptive Medicine* (Miller 2005b, p. 2,533; see also Miller 2005a).

Those who adopt internationally say they made this choice for several reasons. They are more apt to be able to adopt a healthy infant, with a shorter wait and often fewer limits in terms of age or marital status. The adoption is perceived to be less risky in that there is little likelihood of a birth mother seeking to reclaim the child (Clemetson 2006; Zuang 2004). To what degree racial preferences enter into the choice of international adoption is difficult to determine—that is, whether wanting a white Eastern European rather than an African American child, for instance, motivates overseas adoptions. On the other hand, at least one nonHispanic white couple in Omaha, Nebraska, that we, your authors, know of have adopted four children—three from South Korea and one child of African American descent.

Today, there are not only more agencies for arranging international adoptions but also more resources for coping with any postadoption difficulties. There are now specialists in "adoption medicine" who can address medical and cognitive problems of children adopted overseas, as well as psychologists who are prepared to address international or transracial adoption issues (J. Gross 2006c; Tuller 2001). There are "culture camps" (Chappell 1996), schools (Zhao 2002), parent groups (Clemetson 2006), and other resources for bridging the cultural gap for a child raised in the United States but conscious of having started life in another country. Most parents try very hard to maintain a bicultural identity for the child (Brooke 2004), and some undertake travel to the child's country of origin.

This chapter has reviewed topics and issues concerning decisions about parenthood. We'll note here that future studies may pay more attention to men's attitudes and feelings about becoming parents, as well as to their responses to infertility, now "largely ignored in the research" (Himanek 2011, p. F10). Moreover,

> Future studies should explore the meanings of parenthood and pregnancies for women and couples. Raising questions about the degree of planning about pregnancy through both fertility and infertility research also opens avenues of research into the importance or lack of importance of the timing of pregnancies, and how women who do not have male partners (e.g., women who are single or lesbian) approach pregnancy compared to women who do have male partners. Questioning the assumption that all women want to be mothers or that pregnancy intentions are generally dichotomous (e.g., planned or unplanned) offers exciting avenues for understanding changing trends in contemporary reproduction. (Shreffler et al. 2011, p. F9)

Summary

- Birthrates have declined for married women, and many women are waiting longer to have their first child. Pregnancy outside of marriage has become increasingly acceptable. Although other nonmarital birthrates have risen in recent decades, teen birthrates have declined.
- Today, individuals have more choices than ever about whether and when to have children and how many to have.
- Although parenthood has become a choice, the majority of Americans continue to value parenthood. Only a small percentage expects to be childless by choice.
- Nevertheless, it is likely that changing values concerning parenthood, a wider range of alternatives for women, the desire to postpone childbearing, and the availability of modern contraceptives and legal abortion will result in a higher proportion of Americans remaining childless or having only one child in the future.
- Some observers believe that societal support for children is so lacking in the United States that it amounts to *structural antinatalism*. They point to the absence of workplace inflexibility, lack of affordable quality day care, and the absence of paid maternal or paternal leave as are provided in Europe and elsewhere.
- Children can add a fulfilling and highly rewarding experience to people's lives, but they also impose complications and stresses, both financial and emotional.
- Deciding about parenthood today can include consideration of postponing parenthood, having a one-child family, engaging in nonmarital births, having new biological children in stepfamilies, adopting, and taking advantage of infertility treatment.

Questions for Review and Reflection

1. What are some reasons that there aren't as many large families now as there used to be?
2. Discuss the advantages and disadvantages of having children. Which do you think are the strongest reasons for having children? Which do you think are the strongest reasons for *not* having children?
3. How would you react to becoming the parent of twins? Triplets? More? If your choice is to take fertility treatments that pose a risk of multiple births or to not have children at all, what would you do—and why?
4. Which reproductive technology would you be willing to use? In what circumstances?
5. **Policy Question.** How is a pronatalist bias shown in our society? Are there antinatalist pressures? What policies might be developed to support parents? Are there any special policy needs of nonparents? Why might a society's social policies favor parents over nonparents?

Key Terms

abortion 209
assisted reproductive technology (ART) 211
attachment disorder 217
cesarean section 200
fertility 194
informal adoption 214
involuntary infertility 211
multipartnered fertility 208
opportunity costs (of children) 199
pronatalist bias 197
replacement level (of fertility) 194
single mothers by choice 206
structural antinatalism 197
total fertility rate (TFR) 194
value of children perspective (on parenthood) 197
voluntarily childfree 201
voluntary childlessness 201

9

Raising Children in a Diverse Society

Learning Objectives

1. Compare and contrast the traditional roles of mother and father and show how these are influenced by gendered expectations.
2. Describe some ways that single motherhood differs from married motherhood.
3. Discuss the diversity among fathers.
4. Distinguish authoritative parenting and authoritarian parenting.
5. Describe two differing positions regarding spanking.
6. Describe the concerted cultivation parenting model typified by middle-and upper-middle-class parents and contrast it with the accomplishment of natural growth parenting model, typified by working-class parents.
7. Discuss nonbiological parent families, such as grandparents raising children and foster families.

"Whoever came up with the Peace Corps motto, 'The toughest job you'll ever love,' probably wasn't a parent" (Picker 2005, p. 46). Actually, *parenting* may be the toughest job you'll ever love. Although raising children may be joyful and fulfilling, parenting takes place in a social context that makes it enormously difficult. For most of human history, adults raised children simply by living with them and thereby modeling adult roles. Children shared the everyday world of adults, working beside and dressing like them. At least in Europe, the concept of childhood as different from adulthood did not emerge until about the seventeenth century (Ariès 1962). Today we regard children as needing special training, guidance, and care.

In this chapter, we will discuss the parenting process in the United States. As you study, we encourage you to think about how parenting is influenced by ways that gender, race/ethnicity, and social class intersect within a family structure. As an example, within the single-parent family structure, parenting is a different experience for a low-income father of color than for a middle-class, nonHispanic white mother who may have deliberately chosen single parenthood, a situation discussed in Chapter 8.

We'll begin by looking at some general characteristics of the parenting process. Next we'll examine how gender affects parenting. We will then describe parenting styles, noting that the authoritative parenting style is advised by child development experts. We'll address ways that parenting differs according to race/ethnicity. We'll describe grandparents who serve as parents.

Other issues related to children appear throughout this text. Child outcomes related to cohabitation and same-sex couples are addressed in Chapter 6. Combining work and parenting is explored in Chapter 10. Suggestions about how best to communicate with children appear in Chapter 11. Violence against children is discussed in Chapter 12. Divorce and children's outcomes are addressed in Chapter 14. Issues unique to stepparents are considered in Chapter 15. Here we address the parenting *process* in a diversity of social circumstances.

Parenting in Twenty-First Century America

As shown in Figure 9.1, married couples comprise fewer than two-thirds (61.4 percent) of families with a joint child under age 18. Single-mother families represent more than one-quarter (25.9 percent) of parenting family groups. The remaining parental family groups include single fathers, unmarried cohabiting couples with at least one joint child under age 18, and grandparent families (U.S. Census Bureau 2012b, Table FG10). According to Census Bureau definitions, a single parent can be either cohabiting or not.

Many parents display a marked fluidity in living arrangements and family structures or forms. Parents move into and out of marriage and cohabitation. A single mother may move from her own apartment to live with her mother and then move one or more times to reside with other relatives or a romantic partner. Then, too, **multipartnered fertility** (a person's having children with more than one partner, see Chapter 8) can mean that a father resides, perhaps temporarily, with one or more of his children but not with others (Carlson 2011; Harknett and Knab 2007).

Hence, the parenting situations discussed in this chapter should be understood as fluid or changeable.

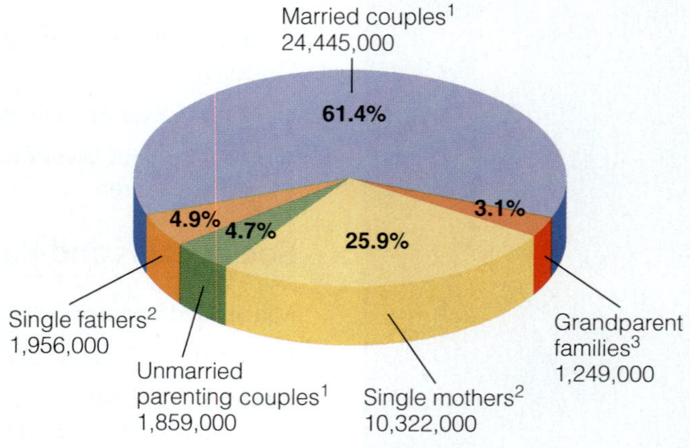

[1]Have at least one joint never-married child under age 18 in the home.
[2]Parent may have a partner, but none of the children is also the child of the cohabiting partner.
[3]Grandparent householder with grandchildren for whom the grandparent is responsible.

FIGURE 9.1 Family groups with children under age 18, 2012.

Source: Calculated from U.S. Census Bureau 2012b, Table FG10.

Regardless of their living arrangements, parents face questions that would not have been imagined several decades ago: Should baby boys be circumcised? How much fast food is too much? Is my child bullying others? Being bullied? Should I believe the teacher who says my child needs medication? Is my child entering puberty too early? Does my youngster spend too much time playing video games? How can I keep my child safe while online? Does my teen text while driving? What should I tell my child about terrorism?

Parenting Challenges and Resilience

We would not want to point out the difficulties facing parents without noting some positives. In general, parents now have more education and are likely to have had some exposure to formal knowledge about child development and child-raising techniques. Many fathers are more emotionally involved than several decades ago (Bianchi, Robinson, and Milkie 2006). Despite media attention to street crime, fewer children are exposed to violent crimes today than twenty years ago (Goode 2012). Cell phones and social media allow parents and children to keep in virtually continual contact and make being in touch with other family members far more likely than a generation ago (Devitt and Roker 2009; Khrais 2012). The Internet offers information for parents dealing with just about any situation.

© iStockphoto.com/ktaylorg

Raising a child with disabilities reminds us that parents and children can evidence resilience, which is enhanced by strong familial bonds.

Nevertheless, parents—even doing their best!—can make mistakes. Sometimes they blame themselves when outcomes for their children are not what they would have wanted (Moses 2010). It helps, though, to know that children can be remarkably **resilient**: Children and adults can demonstrate the capacity to recover from adverse situations and events (Coyle et al. 2009). There is evidence that one caring, conscientious adult can generate a resilient child (Johnson 2000; Soukhanov 1996). Meanwhile, the family ecology perspective (see Chapter 2) leads us to look at ways that society challenges parents today.

Here we list six societal features that make parenting difficult:

1. In our society, the parenting role conflicts with the working role, and employers typically place work demands first (Barnett et al. 2009; Bass et al. 2009; Marshall and Tracy 2009). A majority of parents worry about juggling work and family demands and wish they had more time with their children (Snyder 2007).
2. Today's parents raise their children in a pluralistic society sometimes characterized by conflicting values. Parents are but one of several influences on children. Among others are the children's schools, peers, television, movies, music, and the Internet. Some parents want their children to experience socially diverse school settings. On the other hand, you may know some parents who have responded to this situation by homeschooling their children (Perlstein 2012). Concerns about outside influences may be greater for immigrant parents whose values differ from some that they encounter in the United States (Killoren et al. 2011). As a Muslim mother told a researcher, "As long as ... whatever the children are doing is not in conflict with Islamic values or ways, it is permissible. But when we see it is going to be something against Islamic values, we try to teach our children that this is not correct to our beliefs and practices" (Haddad and Smith 1996, p. 19).
3. Experts have publicized the fact that parents influence their children's health, weight, eating habits, intellectual abilities, behaviors, and self-esteem. Although much parenting advice is useful, the emphasis on how parents influence their children can be overwhelming.
4. Today's parents are often sandwiched between simultaneously caring for children and elderly

parents. Although caregiving can increase life satisfaction, stress builds as family members juggle employment, housework, child care, and parent care (Cullen et al. 2009).

5. Over the past fifty years, parenting has become one lifestyle choice among many, and society-wide support has diminished for the child-raising role, once taken for granted as central. For instance, as a declining proportion of Americans is raising children, some communities show less willingness to support public schools (Whitehead and Popenoe 2008; Wilcox et al. 2011).
6. Today's parents are given full responsibility for successfully raising "good" children, but their authority is often questioned. For example, the state may intervene in parental decisions about schooling, discipline and punishment, medical care, and children's safety as automobile passengers. Immigrant parents from cultures that espouse parenting practices that are illegal or discouraged in the United States face difficulties. For instance, "Parents who try to correct their children's misbehavior through physical punishment [that is defined as extreme in the United States] risk losing their children to the child protection authorities" (Bledsoe and Sow 2011, p. 755).

Due to these factors, being a parent can be far more challenging than many nonparents realize. It's no wonder that parents, especially when employed, are considerably more stressed than nonparents (Carroll 2007c). Then, too, a parent's physical illness, as well as raising a child with special needs, can add stresses peculiar to these situations, and recent government budget cuts have meant diminished resources for children with special needs (Firmin and Phillips 2009).

A Stress Model of Parental Effectiveness

As we saw in Chapters 6 and 7, considerable research shows that growing up with married parents is statistically related to better child outcomes (Osborne, Berger, and Magnuson 2012; Sun and Li 2011; Wilcox et al. 2011a). Researchers attribute much of this finding to stress. Rather than family structure itself, stresses that are peculiar to family forms other than marriage may account for divergent child outcomes. However, being raised in a supportive family is statistically related to more desirable outcomes for children, regardless of family structure (Doohan et al. 2009).

According to a **stress model of parental effectiveness** (see Figure 9.2), parental stress—from job demands, financial worries, concerns about neighborhood safety, feeling stigmatized due to living in a negatively stereotyped family form, or race/ethnic discrimination—causes parental frustration, anger, and depression, increasing the likelihood of household conflict. Parental

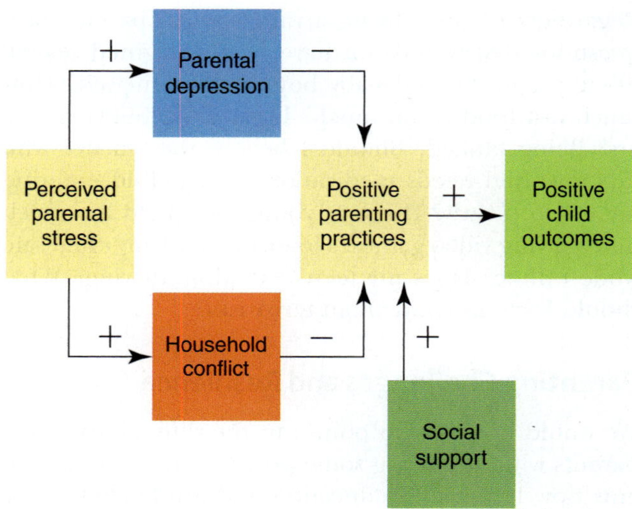

FIGURE 9.2 Stress model of effective parenting. In this figure, plus signs (+) depict positive relationships between variables, and minus signs (–) depict negative ones. Greater use of positive parenting practices results in more positive outcomes for children. However, higher stress levels result in more (+) parental depression and more (+) household conflict. Parents might feel stressed due to job or educational demands, financial difficulties, concerns about neighborhood safety, or feeling stigmatized as a result of racial/ethnic discrimination or negative stereotyping associated with nonmarital family forms. Increased parental depression and household conflict result in diminished (–) use of positive parenting practices. Meanwhile, higher levels of perceived social support—due to high levels of family cohesion, private safety nets, or policies and programs that support parents—are positively related (+) to effective parenting practices, hence to positive child outcomes.

Sources: This figure was designed by Agnes Riedmann and derived from research findings from the following: Baxter 1989; Benner and Kim 2010; Broman, Li, and Reckase 2008; Bronte-Tinkew, Horowitz, and Scott 2009; Brush 2008; Burrell and Roosa 2009; Gassman-Pines 2011; Goodman et al. 2011; Goosby 2007; Jackson, Choi, and Bentler 2009; Joshi and Bogen 2007; Lee et al. 2009.

depression and household conflict, in turn, lead to poorer parenting practices—inconsistent discipline, limited parental warmth or involvement, and lower levels of parent-child trust and communication. Poorer child outcomes result (Gassman-Pines 2011; Goodman et al. 2011; Turney 2011). Having social support (from relatives, friends, Internet sites, parenting classes and groups, and social activities that include both parents and children) diminishes this adverse relationship (Lee et al. 2009).

The Transition to Parenthood

Forty-five years ago, in what has become a classic analysis, social scientist Alice Rossi asserted that the **transition to parenthood** is difficult for several reasons. With little experience, many first-time parents abruptly assume twenty-four-hour duty, caring for a fragile dependent (Rossi 1968). Often surprised at how disruptive an infant can be, new parents report being bothered by the baby interrupting their sleep, work, and leisure time. As one new mother said, "[Y]ou have to run to bathe yourself because the child is going to wake up" (in Ornelas et al. 2009, p. 1564).

Moreover, parents may be geographically distant from their own parents and other relatives who might give advice and help. For example, a Mexican immigrant mother told an interviewer:

> In Mexico, when you have a baby, well your mother is always there, or your family is there.... They help you ... like ... to pick him up, to hold him, to change him. All of that. So, you come here, and you find yourself alone and with a little baby that you don't even know how to pick up. (in Ornelas et al. 2009, p. 1,568)

More disconnected from others than before the baby's arrival, nonimmigrant new mothers also report feeling isolated (Paris and Dubus 2005). Employed mothers of infants, especially those who have jobs with inflexible hours and little opportunity for advancement, are more likely to feel stressed (Marshall and Tracy 2009).

It helps to know that babies differ even at birth; the fact that a baby cries a lot does not necessarily mean that she or he is receiving the wrong kind of care (Rankin 2005). Some are "easy," responding positively to new foods, people, and situations, and transmitting consistent cues (such as tired cry or hungry cry). Other infants are more "difficult." They have irregular sleeping or eating habits, adapt slowly to new situations, and may cry for extended periods for no apparent reason (Thomas, Chess, and Birch 1968; Komsi et al. 2006).

Meanwhile, some studies show that women who are more pleased about their pregnancy are less likely later to view parenting as burdensome (East, Chien, and Barber 2012). Becoming a parent typically involves what one researcher has called the *paradox of parenting:* New parents feel overwhelmed, but the motivation to overcome their stress and do their best proceeds from the stressor itself—the child as a source of love, joy, and satisfaction (Coles 2009).

For couples, the transition to parenthood means less time spent relaxing together and declines in their emotional and sexual relationship (Clayton and Perry-Jenkins 2008). A mother's working irregular or night shifts rather than regularly scheduled day shifts increases her stress and adds to a couple's relationship conflict (Grzywacz et al. 2011; Han and Fox 2011). Employed mothers who have established fairly egalitarian relationships with their husbands may find their role becoming more traditional, particularly if they quit working to become full-time homemakers (Fox 2009). Moreover, employed parents face the challenge of finding quality child care. (See "As We Make Choices: Selecting a

Transition to parenthood can be difficult for many reasons, including upset schedules and lack of sleep. It's a paradox that (1) new parents feel overwhelmed while (2) inspiration to overcome their stress and do their best is provided by the stressor itself—the child as a source of profound delight.

As We Make Choices

Selecting a Child Care Facility—Ten Considerations

"Overall, child care is quite safe," and center care is safer than care in private homes (Wrigley and Dreby 2005, p. 729). The general conclusion drawn from research organized by the National Institute of Child Health and Human Development (NICHD) is that children in nonrelative *quality* day care may actually benefit in terms of cognitive and linguistic skills when compared with low-income children cared for by their parents. Research suggests that, in general, family background factors, parenting quality, and parental sensitivity are more important in impacting children's adjustment than time spent in day care (Burchinal et al. 2010; Vandell et al. 2010).

Favorable outcomes are associated with day care providers talking to children, encouraging them to ask questions and responding, reading to the children, and challenging them to pay attention to others' feelings and to different ways of thinking (U.S. National Institute of Child Health and Human Development 1999; Belsky 2008). Other research suggests that children placed in low-quality day care, where the caregivers practice more authoritarian discipline, may be more inclined to take impulsive risks later in adolescence (Fox 2010; Stein 2010; Vandell et al. 2010).

So how does a parent go about choosing a day care facility? Although some parents have access to child care facilities through government programs or their employers, many parents are on their own in selecting a child care facility. Here we make some suggestions to parents who are choosing from commercially available child care.

State laws establish minimal standards, and professional organizations like the American Academy of Pediatrics have guidelines for quality child care. The National Association for the Education of Young Children (NAEYC) is an accrediting agency for child care centers. Check the NAEYC website for accredited centers near you. Although not all good child care centers have taken this step, accreditation by the NAEYC is a good sign.

Here are some things to consider. The facility should:

1. be state licensed;
2. have a low child-to-staff ratio;
3. have a well-trained staff with warm personalities and little turnover;
4. demonstrate cultural sensitivity and knowledge about the diverse race/ethnic, religious, and social class cultures, as well as the diversity of family forms in this society and potentially at the day care facility;
5. give age-appropriate attention to all the children, including the babies;
6. provide age-appropriate and stimulating activities and play spaces without depending on television watching;
7. use time-outs for discipline rather than physical punishment;
8. welcome parental involvement, even in the form of unannounced visits;
9. accurately respond to questions about practical and financial considerations, such as what happens when the child does not attend as usual because of illness or travel, for example; and
10. provide names of other parents for references.

Critical Thinking

What would you like researchers to find out about children in child care? What qualities on this list do you think are most important in choosing a child care center? Could you rank order the ten suggestions, according to your own opinions?

Sources: National Association of Child Care Resource and Referral Agencies (NACCRRA) 2006.

Child Care Facility—Ten Considerations.") Couples who had been more focused on the romantic quality of their relationship may find the transition more difficult (Doss et al. 2009).

On the other hand, one study found that, among parents who rated their relationship high in quality prior to becoming parents, the transition was easier, even with an unusually fussy baby (Schoppe-Sullivan et al. 2007). Coworkers' support helps employed mothers (Perry-Jenkins et al. 2011). Partly because it is a source of community support, religious attendance reduces decline in marital satisfaction after a baby's birth (Wilcox et al. 2011b). When a new mother's expectations are met concerning how much support she will receive or how much her partner will be involved with the baby, the transition to parenthood is easier (Meadows et al. 2007).

Gender and Parenting

According to tradition, mothers assume primary responsibility for raising children. Employed or not, a mother is expected to be the child's primary **psychological parent**, assuming—with self-sacrifice when necessary—major emotional responsibility for her children's upbringing (DeMaris, Mahoney and Pargament 2011). An obvious exception involves gay men who choose to parent, who "must cope with the fact that they will be challenging societal notions regarding the absence of a woman as the primary caregiver" (Berkowitz and Marsiglio 2007, p. 367).

Historically, fathers have been expected to be breadwinners and not necessarily competent in day-to-day child care. Today, however, our culture prescribes that "good" fathers not only assume considerable (often primary) financial responsibility but also actively participate in the child's care. How do cultural expectations regarding mother- and fatherhood correspond with the daily experiences of mothers and fathers? Put another way, how does each gender "do" parenthood?

Doing Motherhood

Mothers typically engage in more hands-on parenting than do fathers, and they take primary responsibility for their children's upbringing (Hook and Chalasani 2008; Newport 2008). In heterosexual families, adolescents of both genders are likely to name their mother as their closest family confidant (Nomaguchi 2008). A study of employed, coupled heterosexual parents found that mothers often define "quality time" differently from fathers. Fathers are more likely to see quality time with their children as being home and available if needed. Mothers more often see quality time as having heart-to-heart talks with their children or engaging in child-centered activities (Snyder 2007).

Some women quit successful careers to accommodate their mothering role (Slaughter 2012). For women of color who have made this choice, an organization called Mocha Moms provides social support (Crowley and Curenton 2011). As mothers have entered the labor force in greater numbers, many men have been encouraged by the redefinition of male roles to play a larger part in the day-to-day care of their families (Bass et al. 2009). When mothers see fathers as competent parents—and when fathers believe that their child's mother has confidence in them—fathers are more likely to be highly involved (Fagan and Barnett 2003). Though stressful for virtually all mothers, mothering as a single parent is generally even more so.

Single Mothers

About 40 percent of all births occur to unmarried women (Martin et al. 2012). As discussed in Chapter 6, some of these births are to women in same-sex couples. In addition, up to one-half of all nonmarital births occur to cohabiting heterosexual couples (Rackin and Gibson-Davis 2012). We can therefore conclude that between 15 percent and 20 percent of all births occur to uncoupled, single mothers. The intersection of gender with family form is evident in the fact that single women, dramatically more often than single men, assume responsibility for child raising (Brush 2008). Especially when there is only one adult in the household, interaction in single-mother families can be highly intense emotionally as parent and child(ren) work to negotiate their respective roles—child or listener/caregiver/problem-solver? Parent or friend? (Nixon, Greene, and Hogan 2012).

The category *single mother* is diverse by race/ethnicity, immigration experience, education, and socioeconomic class. Some women have purposefully decided to raise a child as a single parent. However, the vast majority of single mothers never intended to raise their children without a partner. Some women, becoming mothers while married or cohabiting, believed that their relationship would last but have since divorced or separated. Others realized that their relationship was not permanent, and the pregnancy may have been unplanned, but they chose to bear the child rather than have an abortion (Carter 2009). As one never-married college student and single mother told an interviewer, "My heart felt ready for the baby. I knew it was something I could do with or without a spouse" (in Holland 2009, p. 173).

As a category, single mothers' median family incomes are considerably lower than either married mothers' or single fathers', and single mothers are more likely to live in poverty, as shown in Tables 7.2 and 7.3. Some single mothers fantasize about marrying a responsible partner: "Wouldn't that Prince Charming dream be nice just once in awhile?" (in Gemelli 2008, p. 112). Others may have wanted to wed but did not find a man whom they considered acceptable (Holland 2009). Some single mothers keep the fathers of their children at a distance due to poor relationships with them, safety concerns for their children, apprehension about the father's illegal activities, or their seeing him as generally unreliable (England and Edin 2007).

Nevertheless, single mothers are well aware that to be married is the cultural ideal. One told an interviewer, with reference to her never-married status:

> Even though I have come so far, I have a car, I have a house, not just an apartment, and I'm a CNA [certified nursing assistant] and I will never have to work a minimum wage job again ever. My kids aren't in want for anything really important, but... I'm nowhere. I'm at the bottom, you know. (in Gemelli 2008, p. 112)

In a society that strongly advocates a two-parent model, single mothers report feeling stigmatized (LaRossa 2009; Thornton 2009; Usdansky 2009a, 2009b). Disgrace attached to unwed motherhood has diminished since the mid-twentieth century, but negative attitudes about unmarried parenthood still encourage society-wide reluctance to provide resources for single mothers and their children (Mollborn 2009).

Meanwhile, single mothers evidence creativity and resilience as they construct support networks to help with finances, housing, child care, and other needs (King, Mitchell, and Hawkins 2010). For instance, an association called CoAbode facilitates house sharing for single mothers. In addition, single mothers maintain web-based support groups that offer practical assistance, such as providing business clothes for those in need.

Single mothers may rely on brothers, brothers-in-law, grandfathers, uncles, or male cousins to serve as father figures for their children (Richardson 2009).

A **private safety net**, or social support from family and friends, is associated with children's better adjustment (Ryan, Kalil, and Leininger 2009). Ironically, however, those most in need of financial or other practical assistance are often in neighborhood, family, or friendship networks that are least able to help because their own circumstances preclude doing so (Harknett and Hartnett 2011; Phan, Blumer, and Demaiter 2009). Moreover, social support from extended family is not always without cost. For example, a single mother told of her father's offer to pay for her family's medical insurance but only on the unspoken condition that she listen while he regularly criticized her (Brush 2008, p. 128).

To improve life for themselves and their children, many single mothers choose further education. This decision is not without added stress, however. Given work and parenting obligations, finding time to attend assigned off-campus activities or meetings to plan group projects poses problems. Moreover, some instructors make unappreciated, stereotypic assumptions. As one student explained, "We were going to be starting to talk about welfare laws and programs the following week and [my instructor] wanted to know if I would be comfortable sharing my experiences with the rest of the class. I never even got welfare" (in Duquaine-Watson 2007, p. 234).

For single mothers who do receive welfare, legislative changes have meant added stress (Gemelli 2008). The 1996 "welfare reform bill" dismantled the federal Aid to Families with Dependent Children (AFDC) program. A new federal program, Temporary Assistance for Needy Families (TANF), replaced AFDC. Intended to encourage both marriage and the work ethic, TANF limits to five years what had previously been open-ended assistance and requires that most recipients be employed. As one woman said in response to welfare changes meant to encourage the work ethic, "It's funny that low-income women or women living in poverty can stay at home and be a bad role model, but a middle class mother or wealthy mother is not a bad role model when they stay home. It's like America has two standards and it's based on class" (in Gemelli 2008, p. 101). For many single mothers, TANF has meant unsatisfying work at poverty-level wages, new day care struggles, and less time with their children (Cook et al. 2009; Neblett 2007).

In accordance with the stress model of parental effectiveness, single mothers' time constraints, generally poorer economic resources, and resultant higher depression levels—not family structure per se—result, on average, in less effective parenting behaviors (Teachman 2009). Instead of between-group comparisons of children raised by single mothers with those raised in two-parent homes, researchers sometimes make within-group comparisons of children raised in various single-mother families. In these latter cases, variations in income, education, stress, and depression levels largely explain child-outcome differences (Crawford and Novak 2008). Not surprisingly, stress is less pronounced among single mothers with relatively higher education, fewer children, better jobs, and more personal resources (Lleras 2008).

Toni Axelrod Photography

Given work and other demands, a majority of mothers tell pollsters that they wish they had more time with their children. One way that these middle-class mothers cope with time pressures is by taking their toddlers to their yoga class. One thing that may be sacrificed, however, is mom's time for personal relaxation.

Doing Fatherhood

In general, research shows that a father's involvement in his child's upbringing is related to positive cognitive, emotional, and behavioral outcomes from infancy into adolescence (e.g., Mandara, Rogers, and Zinbarg 2011). Father absence has generally been associated with adverse effects on children's cognitive, moral, and social development (Bronte-Tinkew et al. 2008; Mitchell, Booth, and King 2009). We note that "father absence" can be other than simply residential absence and can include psychological absence, or indifference with minimal positive father-child interaction. Of course, residential fathers have more opportunity to develop psychological presence than do nonresident fathers (Krampe 2009).

However, this situation is more complicated than suggested by these generalizations (Pan and Farrell 2006). Incidences of a father's substance abuse and of father-perpetrated partner violence and child abuse remind us that encouraging father contact is not always best for children (Osborne and Berger 2009;

Salisbury, Henning, and Holdford 2009). Furthermore, **social fathers** (nonbiological fathers in the role of father, such as stepfathers) do not seem to improve adolescents' outcomes when compared with living in a single-mother household (Bzostek 2008). Indeed, a mother's male relatives—for example, the child's uncle or grandfather—may be better and more reliable parent figures than a romantic partner (Jayakody and Kalil 2002). Nevertheless, research does show that, compared to growing up in a single-mother home, younger children benefit economically from living with a social father, provided that he shares his financial resources with the family (Manning and Brown 2006).

Married fathers are increasingly invested in their children's daily lives as they engage in breadwinning, planning, sharing activities, and teaching their children (Livingston and Parker 2011; Wilcox et al. 2011a). Children's interaction with fathers often differs from that with their mothers; fathers more typically play with or engage in leisure activities with their children than do mothers (Cancian and Oliker 2000). Although we tend to stereotype low-income fathers of color as unmarried and absent, interviews with young African American and Hispanic fathers in New York City uncovered married fathers who were actively involved with their children. Better educated fathers with more satisfying jobs showed higher levels of parental engagement (Wilkinson et al. 2009). Experiencing high levels of workplace stressors, including low levels of employee self-direction, adds to fathers' stress, resulting in less effective parenting (Goodman et al. 2008).

Married-Couple Families with a Stay-at-Home Father

In 2010, about 154,000 married-couple families had a stay-at-home father (U.S. Census Bureau 2012c, Table 68)—a situation that has since increased with men's job losses during the recession that began in 2007 (Chesley 2011). Some of these men have been laid off or closed family-owned businesses and remain out of the workforce (Aasen 2009; Kershaw 2009a). Others have wives who earn more than they could and question sending their children to day care when the father could stay home (St. George 2010; J. A. Smith 2009). Despite gender role changes (see Chapter 3), couples with breadwinning mothers and full-time caregiver fathers continue to be viewed as norm violators (Gaunt 2013).

Fathers point out the relative lack of status associated with full-time parenting, at least in some circles, as well as in the media where fathers caring for babies is often the butt of jokes (Poniewozik 2012). One stay-at-home father, recently laid off from a Fortune 500 company, told a *New York Times* reporter, "To go from the 24-7, high-end, deal-making prestige of working for places that are written about in newspapers to this, it took a long time to get comfortable .... It's humbling" (Kershaw 2009a).

Linda Coan O'Kresik/The New York Times/Redux Pictures

A parent's deployment overseas means not only temporary loss of the family member's daily presence but also considerable changes in the family system. Responsibility for household chores may change; the family may move to be with extended kin; the nondeployed parent may experience higher stress and anxiety levels; children may develop new behavior problems (Schlomer 2012). In some communities, "Flat Daddies," cardboard cutouts of a parent deployed overseas, serve as symbolic placeholders in the families left behind. Flat Daddies and Flat Mommies may go to school, sports events, holiday dinner tables. They remind family members of their ongoing connection during deployment (Zezima 2006).

Single Fathers

Compared to mothers, the proportion of fathers who serve as the principal parent is dramatically small. Among families with children under age 18, about 5 percent are single-father families—5 percent for blacks, Hispanics, and nonHispanic whites, and 3 percent for Asians. The majority of single fathers care for just one child, but some are parenting three or more (U.S. Census Bureau 2012c, Tables 64, 67).

Single fathers typically assumed their role because they "stepped up" in difficult and unforeseen circumstances (Coles 2009). In some cases, single fathers have considered their relationship with their own fathers, learned from their past, and want to do things differently. As one single, black father said, "A lot of people take their father not being there when they were young as a bad thing. But I just took the good out of it and took what he did do and took what I'm not going to do like him" (in Coles 2009, p. 1,328). Single fathers often know that extended family support is available but may not rely on it: "I call my sister occasionally for advice, but I have a strong autonomous streak in me. I'd rather do it myself" (in Coles 2009, p. 1,329).

Whether single or married, poor or financially better off, fathers as primary parents report fighting stereotypes that regard them as odd, unmasculine, or weak (Troilo and Coleman 2008). Watched and evaluated as parents, they feel that they have to prove themselves capable (Coles 2009). In response, single fathers have organized various support groups (e.g., National At-Home Dad Network).

Nonresident Fathers

Nonresident fathers are biological or, much less often, adoptive fathers who do not live with one or more of their children. Less true for divorced fathers, never-married nonresidential fathers move in and out of their children's lives (Cross-Barnet, Cherlin, and Burton 2011). Due to multipartnered fatherhood, a father may be living with one or more of his biological children but be "nonresident" with regard to others. Then, too, a nonresident father may be serving as a social father to one or more children whom he did not conceive, usually because he lives with a woman who had at least one child from a previous relationship (McMahon et al. 2007).

Research shows that, in fact, "cooperative coparenting does not occur in most nonresident father families" (McGene and King 2012). However, although often stereotyped as "absent" and disinterested (Troilo and Coleman 2008), many nonresident fathers express love and genuine concern for their children: "These two guys ... are the reason I live, you know" (in Coles 2009, p. 1,334; also see Hohmann-Marriott 2011; Karre and Mounts 2012). Two studies of nonresident fathers who had previously been arrested for drug problems found that their criminal behavior and substance use sometimes declined after the birth of their baby, and many men saw their children daily, several times a week, or weekly (Kerr et al. 2011). Indeed, the majority of nonresident fathers maintain some presence in their children's lives and provide them with various kinds of practical support, at least while the children are young. Some economically disadvantaged fathers take on significant child care responsibility as their contribution to the family's well-being (Amato, Meyers, and Every 2009; England and Edin 2007). Researchers have found that whether a nonresident father is involved depends on his employment status, age, education, religious participation, and substance abuse history as well as on his family background (Goldscheider et al. 2009; Knoester, Petts, and Eggebeen 2007). One recent study shows that a nonresident father is more involved when his child is male (Bronte-Tinkew and Horowitz 2010).

In addition, a nonresident father's involvement largely depends on his relationship with his child's mother and, to a lesser extent, her extended family (Guzzo 2009b; Ryan, Kalil, and Ziol-Guest 2008). He tends to be more highly involved when his relationship with his child's mother is generally without conflict and when his co-parenting relationship is bolstered by social support (Carlson et al. 2011; Fagan and Lee 2011). Involvement is enhanced when the father has been involved prenatally, possibly because he assumes an identity as father during the prenatal period (Cabrera, Fagan, and Farrie 2008a; Marsiglio 2008).

Nevertheless, given limited resources, involvement of nonresident fathers typically declines over their child's life time, especially for daughters (Mitchell, Booth, and King 2009). Many researchers encourage policy support for these "fragile families," especially in communities where father absence is commonplace (Whitehead and Popenoe 2008; Robbers 2009). The next section explores what children need, regardless of the family form in which they reside.

What Do Children Need?

Children of all ages need encouragement, adequate nutrition and shelter, parental interest in their schooling, and consistency in rules and expectations. Parental guidance should be congruent with the child's age or development level (Barnes 2006; Mental Health America 2009).

Children's Needs Differ According to Age

Children's needs differ according to age. Infants need to bond with a consistent and dependable caregiver. To develop emotionally and intellectually, they need affectionate, intimate relationships as well as conversation and variety in their environment. Discipline is never appropriate for babies because they cannot understand its purpose and are unable to change their behavior in response (Brazelton and Greenspan 2000).

Preschool children need opportunities to practice motor development as well as wide exposure to language, especially when people talk directly to them. They also need consistent, clear definitions of what behavior is unacceptable (Del Vecchio and O'Leary 2006; Dorman 2006). School-age children need to practice accomplishing goals appropriate to their abilities and to learn how to get along with others. To better accept criticism as they get older, they need realistic feedback regarding task performance—neither exaggerated praise nor aggressive criticism. They also need to feel that they are contributing family members by being assigned tasks and taught how to do them (Hall 2008; Jayson 2005).

Although the majority of teenagers do not cause familial "storm and stress" (Kantrowitz and Springen 2005), the teen years do have special potential for reducing marital quality and sparking parent-child conflict (Cui and Donnellan 2009). As they begin to define who they are and will be as adults, adolescents need firm guidance coupled with parental accessibility and emotional support (Guilamo-Ramos et al. 2006; Walsh 2007). Teens also need to learn effective methods for resolving conflict (Tucker, McHale, and Crouter 2003).

Karan Kapoor/Riser/Getty Images

The children in this family have different needs that correspond with their varied ages. Meanwhile, all children need encouragement along with consistent parental expectations and rules. Authoritative parents are emotionally involved with their children, setting limits while encouraging them to develop and practice their talents. The children in this extended family will no doubt benefit from the love and attention of aunts and uncles, neglected family members in most research (Bolick 2011a; Milardo 2010).

Despite stereotypical ideas to the contrary, influences from teens' peers are not necessarily negative and can, in fact, be positive (Hall 2008). Furthermore, parents can and do influence their teenagers' behavior (Longmore et al. 2009). It's important for parents to remember "the obvious fact that most adolescents make it to adulthood relatively unscathed and prepared to accept and assume adult roles" (Furstenberg 2000, p. 903). Regardless of age, children have been shown to benefit from an authoritative parenting style (Junn and Boyatzis 2005).

Experts Advise Authoritative Parenting

Parents gradually establish a *parenting style*—a general manner of relating to and disciplining their children. Parenting styles combine two dimensions—(1) parental warmth and (2) parental expectations—coupled with monitoring of their children. As shown in Table 9.1, we can distinguish among authoritarian, permissive, and authoritative parenting styles using these two dimensions (Baumrind 1978; Maccoby and Martin 1983).

The **authoritarian parenting style** is low on emotional warmth and nurturing but high on parental direction and control. The authoritarian parent's attitude is, "I am in charge and set/enforce the rules, no matter what" (Gaertner et al. 2007). Parents who employ this style are more likely to spank their children or use otherwise harsh punishment (Grogan-Kaylor and Otis 2007). Unnecessarily high parental direction or control has been associated with a child's decreased sense of personal effectiveness or mastery over a situation, even among children as young as 4 (Moorman and Pomerantz 2008).

The **permissive parenting style** gives children little parental guidance. Although low on parental direction or control, permissive parenting may be high on emotional nurturing—a situation, characterized as *indulgent*, that leads to the classic "spoiled child." A second variant of the permissive style is low on *both* parental direction *and* emotional support—a situation of *emotional neglect*. Authoritarian and permissive parenting styles are associated with

Table 9.1 Parenting Styles

		Parental Warmth	
		Low	High
Parental Monitoring	High	Authoritarian	Authoritative/Positive
	Low	Permissive-Emotional Neglect	Permissive-Indulgent

children's and adolescents' depression and otherwise poor mental health, low school performance, behavior problems, high rates of teen sexuality and pregnancy, and juvenile delinquency (Hall 2008; Waldfogel 2006).

When both warmth and monitoring are high, parents are said to exhibit an **authoritative parenting style** that is sometimes called *positive parenting*. At least for white, middle-class children, research consistently shows that an authoritative parenting style is the most effective of the four possible styles. Moreover, a recent study of Mexican-origin families found that adolescents did better in school when their parents combined messages about hard work and school success with maternal warmth (Suizzo et al. 2012). This style, characterized as warm, firm, and fair, combines emotional nurturing and support with conscientious parental direction (although not excessive control). Authoritative parents would agree with the statements, "I consider my child's wishes and opinions along with my own when making decisions," "I value my child's school achievement and support my child's efforts," and "I expect my child to act independently at an age-appropriate level" (Manisses Communications Group 2000, p. S1).

Authoritative parenting involves encouraging the child's individuality, talents, and emerging independence, while also consciously setting limits and clearly communicating and enforcing rules (Brooks and Goldstein 2001; Ginott, Ginott, and Goddard 2003). Limits are best set as house rules and stated objectively in third-person terms. A parent may say, for example, "The time to be home is 10 o'clock." With preschoolers, limits need to be set and stated very clearly: A parent who says, "Don't go too far from home" leaves "too far" to the child's interpretation. "Don't go out of the yard at all" is a wiser rule. Authoritative parents monitor their children's activities and whereabouts while giving appropriate consequences for misbehavior when warranted (Waldfogel 2006).

Regardless of family structure, authoritative parents are more likely than others to have children who do better in school and are socially competent, with relatively high self-esteem and cooperative yet independent personalities (Crawford and Novak 2008; Fivush et al. 2009; Jackson-Newsom, Buchanan, and McDonald 2008). Positive effects of authoritative parenting last into adulthood (Schwartz et al. 2009). "A Closer Look at Diversity: Straight Parents and LGBT Children" asks you to consider how an authoritative parenting style would apply to this situation.

When two parents are involved, their collaboration or working together renders them more effective, especially when both parents use the authoritative parenting style (Kjobli and Hagen 2009; Simons and Conger 2009). Studies of two-parent Mexican American families have found that many parents strongly agree with and support each other; children and their parents are happier when they do (Formoso et al. 2007; Updegraff et al. 2012). Collaborative parenting reduces stress and enhances parents' feelings of competence (Jackson, Choi, and Franke 2009), as well as partners' relationship satisfaction (Kluwer, Heesink, and Van de Vliert 2002). Research on married and cohabitating working-class parents of first graders found that families in which both parents practiced an authoritative parenting style were most effective in raising well-adjusted children. Families in which only one parent used an authoritative parenting style were more effective than those in which neither parent did (Meteyer and Perry-Jenkins 2009).

Psychological Control versus Authoritative Parenting

As opposed to direct forms of parental control, such as asking direct questions, explicitly stating expectations, and giving time-outs or denying privileges, **psychological control** involves using manipulative strategies such as inducing guilt or withdrawing signs of affection. Rather than conveying unconditional love for the child while correcting misbehavior, psychological control relies on negative cues such as refusing to acknowledge the child.

The underlying message is that the child's behavior has hurt the parent's feelings. To reduce tension, the child is expected to comply with the parent's wishes. At least among Latino and European American adolescents, and among African American girls, psychological control has been found to hinder children's emergent sense of agency or mastery over their behavior and future goal attainment. Depression can result (Bean and Northrup 2009; Soenens, Vansteenkiste, and Sierens 2009).

Before leaving this discussion, we need to note that some scholars view the authoritarian/permissive/authoritative model as biased. For instance, one study shows that authoritative parenting is a more important prediction of behavior for children of European descent than for the Hmong (Supple and Small 2006). Some scholars argue that the entire model is ethnocentric, or Eurocentric—that is, it uses European, white, middle-class beliefs about parenting as the standard to which all others are compared—often unfavorably (Farver et al. 2007). This point is developed throughout the section "Race/Ethnic Diversity and Parenting," later in this chapter. The next section focuses on the question of whether spanking is ever appropriate.

Is Spanking Ever Appropriate?

Spanking refers to hitting a child with an open hand without causing physical injury. Whether spanking children is ever a good idea is controversial (Coombs-Orme and Cain 2008; Larzelere 2008). Analysis of data from the 13,000 respondents in the National Survey of Families and Households found that about one-third of fathers and 44 percent of mothers had spanked their children during the week before they were interviewed. Boys, especially those under

A Closer Look at Diversity

Straight Parents and LGBT Children

Parenting a child who is lesbian, gay, bisexual, or transsexual (LGBT) can be not only pleasurable and broadening but also emotionally challenging. When a lesbian, gay, or bisexual child "comes out," or discloses his or her orientation to family members, parents may feel confused, ambivalent, alone, embarrassed, or angry (Martin et al. 2009). Even those who view themselves as progressive, readily accepting LGBT friends and acquaintances, may be surprised at their feelings of disappointment and grief at finding out that their child is lesbian, gay, or bisexual. If homosexuality is against the parent's religion, a child's coming out can be even more disconcerting (Schwartz 2012). It helps to recognize the following facts:

- All cultures and historical periods include individuals who have identified themselves as lesbian, gay, bisexual, or transgender.
- The American Psychological Association and the American Pediatric Association do not consider being LGBT as a psychological disorder.
- LGBT adolescents and young adults may feel guilty about their sexual orientation, worried about responses from their families and friends, and fearful of discrimination in clubs, sports, college, or the workplace ("Gay and Lesbian Adolescents" 2006).
- Hiding one's sexual orientation can be extremely stressful and isolating. Adolescents and adult children come out because they want to live their lives openly and honestly without deception ("Questions and Answers" n. d.; Wright and Perry 2006).
- The physical and mental health of LGBT youth is better when they feel social support (Wright and Perry 2006).
- Parents' acceptance of their child's sexual identity allows for open discussions about the child's dating relationships and related issues, such as ways to deal with prejudice and discrimination or how to reduce risks associated with HIV/AIDS.

Experts advise that, whatever a parent's feelings when a child comes out, the child needs assurance that she or he is loved just as much as before:

> Can you imagine the feelings of a youngster who bravely tells his or her parents that they are gay only to be confronted by an anger which may be so severe that they are put out of the house, or told that he or she has brought shame to the family? Yet this happens and it is a fact that gay people occasionally commit suicide because they have been so badly ostracized and made to feel alienated. ("If Your Child Is Gay or Lesbian" 2008)

As parents work through their feelings, they may want to talk their situation over with others. Professionals urge parents not to share the information without their child's consent. An exception involves talking with a counselor. Other resources are available as well. Parents, Families, and Friends of Lesbians and Gays (PFLAG) is a national organization comprised of local educational and support groups, as well as Internet resources.

Critical Thinking

How might the stress model of effective parenting be applied to the situation of an LGBT child's coming out to her or his parents? How might an authoritarian parent's reaction to their child coming out differ from the response of an authoritative parent?

---

age 2, are spanked the most often. Children over age 6 are spanked less often, but some parents spank their children during early adolescence (Guzzo and Lee 2008).

Mothers spank more often than fathers. Younger, less-educated parents in households with more children and less social support, parents who argue a lot with their children, sociopolitical conservatives, those with a fundamentalist religious orientation, and parents who live in relatively violent neighborhoods are more likely to spank (Ellison and Bradshaw 2009). The stress of first-time parenthood is associated with spanking (Guzzo and Lee 2008). One study found that single mothers who become more seriously involved in a romantic relationship, whether with the biological father of their child or not, are more likely to spank their children. The authors speculated that the mothers experienced increased strains as they incorporated the child into their developing relationship (Guzzo and Lee 2008).

A leading domestic violence researcher, sociologist Murray Straus (2007, 2010) advises parents never to hit children of any age under any circumstances. At least among nonHispanic white American youngsters, being frequently spanked in childhood is linked to later behavior problems, as well as to depression, suicide, alcohol or drug abuse, physical aggression against one's parents in adolescence, and later abuse of one's own children and intimate partner violence (Gromoske and Maguire-Jack 2012). Physical punishment can be "harsh" (involving pushing, grabbing, shoving, slapping, or hitting), and disciplining in these ways has been linked to mood and anxiety disorders in adulthood (Afifi et al. 2012).

Furthermore, Straus has argued that spanking teaches children a "hidden agenda"—that it is all right to hit someone and that those who love you hit you. This confusion of love with violence sets the stage for domestic violence. Then, too, especially when a parent spanks in anger—which is never advised—"spanking can escalate and apparently does mix in with more severe hitting" (Kazdin and Benjet 2003, p. 102; Roberto, Carlyle, and Goodall 2007). Infants and babies less than 2 years old should never be spanked (Coombs-Orme and Cain 2008). Spanking or vigorously shaking an infant can lead to permanent damage and even death.

Meanwhile, some researchers and pediatricians contend that Straus and his colleagues may be overstating the case (Saadeh, Rizzo, and Roberts 2002). For one thing, we have become an increasingly diverse society of immigrants from cultures that believe in spanking. One study found that some West African immigrant parents hope to send their children back to relatives in West Africa where they can be effectively disciplined (spanked) because they feel that their inadequately disciplined children have become unruly "mad cows" in the United States (Bledsoe and Sow 2011). Psychologist Marjorie Gunnoe (cited in Gilbert 1997) theorizes that spanking is most likely to have negative results for children only when they perceive being spanked as an aggressive act. She hypothesizes that children under age 8 tend to think it is their parents' right to spank them. Children of conservative Protestant parents who espouse spanking may not see being spanked as demeaning, because their religious environment defines spanking as normal and necessary (Ellison, Musick, and Holden 2011). However, the majority of research does show that corporal punishment has generally negative effects (Christie-Mizell, Pryor, and Grossman 2008; Mulvaney and Mebert 2007).

The American Academy of Pediatrics advises that children less than 2 years old and adolescents should *never* be spanked and recommends that parents learn disciplinary methods other than spanking (Afifi et al. 2012). Straus has argued that spanking usually accompanies other, more effective discipline methods such as explaining or denying privileges. These nonspanking discipline methods are effective by themselves, and parents should be encouraged to follow the principle "just leave out the spanking part" (Straus 1999a, p. 8; Straus 2007).

In general, middle- and upper-middle-class parents are less likely than low-income parents to spank. Low-income parents "may be unaware of current American Academy of Pediatrics policy recommendations about spanking. Or they may consciously disagree with them" (Guzzo and Lee 2008). For parents who disagree with Straus and the American Academy of Pediatrics, a separate organization, the American College of Pediatricians, distinguishes between disciplinary spanking and physical abuse (see Table 9.2) and offers guidelines for the former.

Table 9.2 The American Academy of Pediatrics Position Against Spanking, Versus the American College of Pediatricians' Distinction between Disciplinary Spanking and Corporal Punishment

	American Academy of Pediatrics	American College of Pediatricians	
	Spanking	**Disciplinary Spanking**	**Physical Abuse**
The Act	Hitting a child with an open hand without causing physical injury	Spanking: one or two swats to the buttocks of a child. Spanking should never cause physical injury and "leave only transient redness of the skin."[*]	Physical assault, including to beat, kick, punch, choke, etc.
The Intent	Modify behavior	Training: to modify behavior	Violence: physical force intended to injure, intimidate, get revenge, or abuse
The Attitude	Often, parental frustration	Love and concern	Anger and malice
The Effects	Later emotional and behavior problems: depression, suicide, substance abuse, physical aggression against parents in adolescence, intimate partner violence in adulthood.	Mild to moderate discomfort; behavioral correction	Physical and emotional injury
Is spanking ever appropriate?	No, never.	Yes, if : 1. used for clear, deliberate misbehavior, 2. milder forms of discipline have resulted in noncompliance, 3. not administered on impulse or when out of control—always planned, 4. the child is forewarned and the reasons for the spanking explained.	

[*]American College of Pediatricians, 2013.

Sources: Table created by Agnes Riedmann with data from the following: American College of Pediatricians 2007; Berlin et al. 2009; Gromoske and Maguire-Jack 2012.

Social Class and Parenting

There are effective and ineffective parents in all social classes (Jackson, Choi, and Bentler 2009). Meanwhile, this section examines some ways that social class impacts parental alternatives and choices. You'll recall a theme of this text: Decisions are influenced by social conditions that expand or limit one's options. Virtually all opportunities and experiences, or *life chances*, are influenced by **socioeconomic status (SES)**—one's position in society, measured by educational achievement, occupation, or income. Parenting is no exception (Furstenberg 2006; Lareau 2006).

Research shows that family education and income have more influence on parenting behaviors and children's outcomes than do race/ethnicity or family structure in and of itself (Gibson-Davis 2008). We have seen that parents who are less stressed and relatively content practice more positive child-raising behaviors (Burrell and Roosa 2009; Gibson-Davis 2008). Reduced stress and emotional well-being, in turn, are statistically correlated with higher socioeconomic status.

Middle- and Upper-Middle-Class Parents

In this climate of economic uncertainty, even middle- and upper-middle-class parents with relatively high education have suffered layoffs, salary reductions, and reduced health and retirement benefits (Coy, Conlin, and Herbst 2010). Having already tightened their belts, some have trouble paying their bills (Pew Research Center 2009b). Nevertheless, compared with lower-SES parents, those with higher income can better afford to provide for their children's needs and wants. Then, too, higher-SES parents have the resources to hire household help or purchase devices such as baby-monitoring equipment that might help with parenting (Knoester, Haynie, and Stephens 2006; Nelson 2008). Furthermore, they reside in neighborhoods conducive to successfully raising and educating their children.

More highly educated parents have fewer children on average, show less anxiety regarding their parenting skills, and are likely to emphasize **concerted cultivation** of their child's talents and overall development (Lareau 2003b, 2011). According to this parenting model, they more often praise their children; play with or talk to them "just for fun"; read to them; create and enforce rules about watching television; engage their children in extracurricular lessons, clubs, and sports; take them on outings; enroll them in private or charter schools; and say that there are people in the neighborhood whom they can count on (U.S. Census Bureau 2009, Tables D5-D30).

More so than in low-income neighborhoods, middle- and upper-class children are likely to have neighborhood and school friends who share their parents' values and can therefore serve as parallel socialization agents (Hall 2008). Then, too, volunteering at their children's schools and monitoring other students' and even teachers' behavior, highly educated parents secure educational advantages for their children. Should problems arise at school, higher-SES parents are likely to have network contacts with community professionals who can help and may even challenge school officials' decisions (Hassrick and Schneider 2009).

Higher-SES parents are likely to get parenting information from professional sources such as books or the Internet (Radey and Randolph 2009). Often using the authoritative parenting style, they negotiate with their children in ways meant to foster language and

David Young-Wolff/Alamy

Parents' and children's life experiences are significantly related to their socioeconomic status. Middle- and upper-middle-class parents tend to emphasize concerted cultivation of their children's development and talents. Ironically, although possibly enhancing parental freedom, baby-monitoring devices can also increase anxiety levels because they encourage defining the infant as extremely fragile (Nelson 2008). Can you think of other parental aids that might have similar effects?

critical thinking skills, self-direction, initiative, and self-advocacy (Lareau 2006; Shinn and O'Brien 2008). This parenting model well prepares children for success in the broader society because schools and professions value the self-direction, critical thinking, and self-advocacy that these children learn at home. But can parents take concerted cultivation too far?

Hyperparenting: The "Hurried Child" and "Helicopter Parents" According to some observers, many higher-SES parents engage in **hyperparenting**. Dubbed "helicopter parents," they hover over and meddle excessively in their children's lives (Warner 2006). Some parents have become vulnerable to marketing that prods "good" parents to spend unnecessarily large amounts of money on baby gear, toys, or reading programs for babies, among other things (Purcell 2007; "Your Baby Can Read Set" 2011). For many, excessive spending on their children cuts into their ability to save for emergencies, retirement, or college (Paul 2008). A related issue involves parents' and schools' "hurrying" of children with sports training for toddlers and with the increased use of standardized testing and decreased emphasis on music, art, or time for spontaneity and play (Hyman 2010; Tugend 2011).

Critics have warned that many higher-income parents not only give their children too much but also engage them in too many scheduled activities—private lessons, extracurricular activities associated with school or church, and organized recreational programs. Expecting achievement in these endeavors, parents may place too many demands on their children, even encouraging them to compete for places in the most preferred preschools, for example (Saulny 2006; also see Zernike 2011).

Perhaps filling their own needs while believing that they are acting in the best interest of their children, these parents may be determined to raise "trophy kids" (Perrow 2009). Although there is some evidence that tightly scheduling a child may not be harmful (Cloud 2007), developmental psychologist David Elkind (1988) warned nearly twenty years ago that "scheduled hyperactivity" (Kantrowitz 2000) can produce the "hurried child," not to mention frazzled parents who may feel trapped by parenting demands (Elkind 2007a,b; Nomaguchi and Brown 2011). The overscheduled or "hurried child" is denied free playtime while encouraged to assume too many challenges and responsibilities too soon (Stout 2011; Tugend 2011).

As the hurried child enters young adulthood, some helicopter parents—with the aid of cell phones—meddle inappropriately in their college student's educational experiences or attempt to negotiate job offers, salary, and benefits for their offspring (Blanck 2007). "Constant cellphone connection means parents jump in too quickly," in the opinion of one university housing director. "It surprises me when students say, 'My roommate's mother called and yelled at me,' and I think, 'Are you kidding me?'" (Moore 2010).

Some individuals who see themselves in negative descriptions of helicopter parents find the criticism unfair, especially in this unsettling economy and because they pay considerably for their children's education. Despite media emphasis on possible adverse effects of helicopter parenting, research shows that many young adults enjoy daily (or more) cellphone contact with parents and benefit from intense parental attention in terms of better psychological adjustment and life satisfaction than grown children who did not receive intense support (Fingerman et al. 2012; Khrais 2012; Settersten and Ray 2010).

Meanwhile, compared with life in higher-SES families, "childhood looks different" in families of lower socioeconomic status where children grow up with less supervised play, fewer scheduled activities, and more unplanned interaction with extended family members and friends (Lareau 2003a). Sociologists Elliot Weininger and Annette Lareau (2009) have noted an inconsistency. Higher-SES parents determine to promote self-direction in their children but exercise subtle forms of control that can undermine their intent. Conversely, lower-SES parents tend to espouse obedience but often grant children considerable autonomy (as they play outside unsupervised, for example), thereby limiting emphasis on conformity.

Working-Class Parents

Working-class parents work at construction, manufacturing, repair, installation, and service jobs such as health care assistants that require at least a high school education and pay higher than minimum wage. More so than with higher-SES parents, working-class parents have suffered the negative effects of declining factory work and union power, decreased wages and benefits, insecurities associated with temporary work, and escalating housing, utilities, and transportation costs. Although economic strains affect daily life, working-class families can be less harried by tight scheduling, and children are able to live at a slower pace. They can relax and have free time to create their own play. They can interact equally with cousins, for instance, and spend time with other relatives instead of being on a constant forced march to do well academically and rack up trophies or ribbons in extracurricular activities. They can be children.

Working-class parents do not necessarily view the concerted development parenting model as good parenting. In fact, they may view this model as negative, creating demanding children (Guzzo and Lee 2008). Instead, they tend to follow the **accomplishment of natural growth parenting model**, according to which children's abilities are allowed to develop naturally (Lareau 2006). Among other things, this means that

Issues for Thought

How Would *You* Operationally Define a Quality Home Environment?

What does it mean to say that a low-income single mother's home environment is poor quality? You'll remember from Chapter 2 that *operational definitions* measure researchers' concepts or variables. Among other indicators, the National Longitudinal Survey of Youth (NLSY) measures *quality home environment* for 3- to 6-year-olds by whether the house appears safe, clean, well lighted, and uncluttered; whether the parent reads to the child at least three times a week; whether the child has ten or more books; whether the family gets at least one magazine regularly; and whether a parent talks to the child, kisses or hugs the child, and verbally answers the child's questions (Lleras 2008).

Critical Thinking

In your opinion, are these operational definitions appropriate? Why or why not? Which one(s) depend on finances? Which ones depend on parental literacy? If you were on the NLSY research team, what might you add to or delete from these operational definitions?

working-class children spend more time watching television and playing video games than do children of highly educated parents (Richtel 2012).

Some working-class parents employ the authoritative parenting style. Nevertheless, much parent-child communication tends to be authoritarian, emphasizing obedience and conformity and less often eliciting children's feelings or opinions (Lareau 2003a, 2006). Working-class parents are more likely to tell their children what to do rather than trying to persuade them with reasoning. When dealing with professionals (e.g., doctors, religious clergy, teachers, public officials), working-class parents are likely to encourage their children to keep their thoughts and questions to themselves (Shinn and O'Brien 2008).

Many working-class parents are involved in their children's schools and do promote academic success in their children (Cooper, Crosnoe, et al. 2009). However, the natural growth parenting model, coupled with the authoritarian parenting style, does not correspond well with the middle-class culture and expectations of schools and professions. Although higher-SES children appear to gain a sense of entitlement, working-class children are likely to grow up with feelings of discomfort, constraint, and distrust regarding their school and work experiences (Lucas 2007). For children from working-class families who embark on professional careers, this sense of not fitting in can persist (Lubrano 2003).

Low-Income and Poverty-Level Parents

The rate of child poverty in the United States exceeds the rate of the nation as a whole and is considerably higher than in other wealthy industrialized nations (Portero 2012; U.S. Census Bureau 2012c, Table 713). The majority of low-income and poverty-level parents work at minimum- or less-than-minimum-wage jobs with irregular and unpredictable hours and no employer-subsidized medical insurance or other benefits. Because more and more low-income jobs are part-time and because working even full-time at minimum wage does not pay enough to live above the poverty level, many low-income parents have two or three jobs (Ames, Brosi, and Damiano-Teixeira 2006). Analysis of data from the National Longitudinal Survey of Youth (NLSY) shows that single mothers who work part-time in low-wage jobs with nonstandard hours generally raise their children in poorer-quality home environments (Lleras 2008). "Issues for Thought: How Would *You* Operationally Define a Quality Home Environment?" gives a bit more detail on this study.

Irregular work schedules in low-wage jobs with little autonomy and high supervisor surveillance, coupled with housing or neighborhood troubles as well as financial worries, cause stress. Living in rented homes, apartments, or motel rooms, they struggle with rent burdens, utility payments, and housing instability (Berger et al. 2008). Moreover, many poor families move from city to city to live with relatives or to search for jobs. In sharp contrast to higher-SES parents, many low-income parents struggle to give their children a few "extras," such as a "respectable" birthday party, school field trips, or a high school class ring (Lee, Katras, and Bauer 2009; Mistry et al. 2008).

With fewer resources, lower-income parents are less likely to live in neighborhoods that value education or encourage high achievement (Coles 2009; Henry et al. 2008). In fact, items that middle-class Americans take for granted, such as relatively safe, gang-free neighborhoods, are often unavailable (Frech and Kimbro 2011; Zimmerman and Pogarsky 2011), and parental control is more difficult to achieve in neighborhoods characterized by antisocial behavior (Gayles et al. 2009; Moore et al. 2009). Furthermore, poverty-level families are more likely than others to live with air pollution and to have poorer nutrition, more illnesses such as asthma, schools that are less safe, and limited access to quality medical care (Children's Defense Fund 2012). Children living in poverty—more often disabled or chronically ill than other children—have expensive health care needs that welfare or other social services do not always or completely cover (Cohen and Bloom 2005; Levine 2009).

About 8 percent of children who are raised in poverty (compared with about 5 percent of children raised in families that are not poor) have emotional or behavioral difficulties (Simons et al. 2006, p. 3; Teachman 2008b). Economic hardship in childhood, even in a two-parent married family, particularly when it lasts for a long time or occurs in adolescence, is related to lowered emotional well-being in early adulthood (Sobolewski and Amato 2005; Vandewater and Lansford 2005). Not having enough money causes stress, which often leads to mothers' depression, parental conflict, and general household turbulence. Household turbulence and mothers' depression, in turn, are associated with lower parent-child (especially teen) relationship quality, a situation that results in a child's lower psychological well-being (Jackson, Choi, and Bentler 2009; Teachman 2008b).

Homeless Families Over the past three decades, extreme poverty, a shortage of affordable housing, job erosion, home foreclosures, declining public assistance, lack of affordable health care, domestic violence, substance addiction, and mental illness have helped to create a significant number of homeless families—a phenomenon that would have been unthinkable forty years ago (National Coalition for the Homeless 2009c).

Families with children are among the fastest-growing segments of the homeless population, a situation that has become more pronounced since the beginning of the recession that began in 2007. Partly as a result of changes to welfare laws, described earlier in this chapter, approximately 40 percent of the homeless are mothers and children, with children under age 18 making up about one-quarter of the homeless (Seccombe 2007). Fathers, some of whom are single parents, are also found among the homeless (National Coalition for the Homeless 2009a, 2009c).

Homeless parents, especially those who have been without housing for a longer time, move often and have little in the way of a helpful social network. Getting children to school and supervising their homework—not to mention being actively involved in a child's classroom—become extremely difficult tasks for homeless parents. Although families benefit from entering shelters, life in a homeless shelter is itself stressful. Some shelters require that the family leave during the day, regardless of the weather: "How can this mother go out and look for a job or even look for a place to live when she's got three kids, and it's raining, or it's cold?" Problematic rules involve bedtimes, mealtimes, keeping children quiet, and the requirement that children be with their parents at all times. Other stressors occur as well. For example, one mother told of a single male resident's "getting fresh with my older girl" (Lindsey 1998, p. 248).

Social class may be more important than race/ethnicity in terms of parental values and interactions with children (Lareau 2003b, 2006). Middle-class parents of all race/ethnic groups are more alike than different, and so are poverty-level parents. Upper-middle-class black parents perform their role differently than do working-class black parents or those living below poverty level (Peters 2007). At the same time, social scientists do look at how various ethnic groups evidence culturally specific parenting behaviors (Cohen, Tran, and Rhee 2007). The major focus of the following section is on parenting behaviors and challenges that are specific to various racial, ethnic, and religious minorities.

Chris Hondros/Getty Images News/Getty Images

As with other race/ethnic minorities, African Americans' parental attitudes and behaviors are similar to other parents in their socioeconomic status (SES). Nevertheless, the intersection of gender and race with SES means that this father, culturally expected to be an effective breadwinner, also risks race discrimination as he navigates a job search in this recession economy. He is pictured here at a New York state employment services office in Brooklyn.

Race/Ethnic Diversity and Parenting

As a beginning, we need to note two factors. First, there is considerable overlap among class and race/ethnic categories. For instance, although the upper middle class now includes substantial numbers of people of color, particularly Asians, it is still largely nonHispanic white. Many African American and Hispanic families are now solidly middle class, but these race/ethnic groups remain overrepresented in low-income and poverty categories. A second factor to note is that, as pointed out in Chapter 1, there is considerable ethnic diversity *within* the following groups. For instance, Asian Americans include a broad range of ethnicities, including Chinese, Japanese, Korean, Vietnamese, and Asian Indians, among others. Similarly, the category "black" or "African American" includes not only descendants of African slaves historically brought to the United States but also recent immigrants from Africa and the Caribbean.

African American Parents

Evidence suggests that African American parents' attitudes, behaviors, and hopes for their children are similar to those of other parents in their social class (Peters 2007). Nevertheless, the impact of race remains important. For instance, even when social class is taken into account, it appears that African American mothers (but not fathers) are more likely than European Americans to spank their children. However, spanking may not have the same negative effects on black children as it does on European American children. Among blacks, physical punishment is more acceptable and hence more likely to be viewed by both parent and child as an appropriate display of maternal warmth and positive parenting (Jackson-Newsom, Buchanan, and McDonald 2008). It follows that comparing African American parents to other ethnic groups can be seen as Eurocentric (Dodson 2007).

Besides putting up with research findings that are possibly biased against them, even higher-SES African Americans remain vulnerable to discrimination (Lacy 2007; Welborn 2006). A mother in the previously mentioned Mocha Moms support group described a situation in which her young son tried to get on a tire swing at his school and another boy told him that the swing was only for people with light skin (Crowley and Curenton 2011, p. 8). In addition to incidences of direct racism, other things are difficult. So simple a matter as buying toys becomes problematic. Black dolls only? Should the child choose? What if the choice is a white Barbie doll? For middle-class African American parents, forging a unique *black middle-class* identity and then instilling this identity into their children is a major undertaking (Lacy 2007).

Native American Parents

Native American parents have been described as exercising a permissive parenting style that some critics have viewed as bordering on neglectful. However, describing Native American parenting in this way smacks of Eurocentrism (Seideman et al. 1994). Traditionally, Native American culture has emphasized personal autonomy and individual choice for children as well as for adults. Before the arrival of Europeans and for some time thereafter, Native Americans successfully raised their children by using nonverbal teaching examples and "light discipline," possibly coupled with "persuasion, ridicule, or shaming in opposition to corporal punishment or coercion." Native Americans continue to respond with warmth to their children's needs and also to "respect children enough to allow them to work things out in their own manner" (John 1998, p. 400).

Valuing their cultural heritage, many Native Americans have been reluctant to assimilate into the broader society, and this reluctance may mean a rejection of the authoritative parenting style advised by European American psychologists. The use of tribal elders "to help mitigate the loss of parental involvement and early nurturant figures in the lives of Native American adolescents" is important to many tribes (John 1998, p. 404).

Meanwhile, researchers have noted that Native American parents and children demonstrate resilience. For example, a longitudinal study of twenty-nine Navajo Native American mothers who as teenagers bore infants found that, twelve to fifteen years later, many had completed or gone beyond high school (Dalla et al. 2009). Although single-parent households do occur in Native American communities, the extended family serves as an instrument of group solidarity by reinforcing cultural standards and expectations, and lending practical assistance. With symbolic and actual leadership status in family communities, Native American grandparents often monitor grandchildren and may fully adopt the parenting role when necessary (Letiecq, Bailey, and Kurtz 2008).

Hispanic Parents

A recent study of Mexican-origin parents found that, especially among recent immigrants, parents exhibited "complementary" roles: Fathers were authority figures, and mothers served as principal caretakers (Updegraff et al. 2012). Hispanic parents have been described as more authoritarian than their nonHispanic counterparts. However, as with other race/ethnic minorities, it may be that this description is Eurocentric and therefore inaccurate. The concept of **hierarchical parenting**, which combines warm emotional support for children with a demand for significant respect for parents, older extended-family members, and other authority figures, may more aptly apply to Hispanic parents. Hierarchical parenting is designed to instill in children a more collective value system rather than the relatively high individualism that European Americans favor (McLoyd et al. 2000).

This collectivism has been found to be functional. For instance, a study that compared Mexican American with European American parents in Southern California found that family cohesion (*familismo*) lessened the relationship between economic stress and negative parenting (Behnke et al. 2008; see also Martyn et al. 2009). (Research shows similar positive effects of family cohesion for Vietnamese Americans as well as for families of other ethnicities, including European Americans [Lam 2005; Vandeleur et al. 2009].)

Hispanic parents teach their children the traditions and values of their cultures of origin while often coping with a generation gap that includes differential fluency and different attitudes toward speaking Spanish (Knight et al. 2011; Pew Research Center 2009a). As in other bicultural families, intergenerational conflicts may extend into many matters of everyday life as the younger generation becomes more assimilated into U.S. culture. For example, "My mother would give me these silly dresses to wear to school, not jeans," complained a 15-year-old Mexican American female (Suro 1992, p. A-11). As Hispanic immigrant parents adjust to U.S. culture, they are likely to place less emphasis on *familismo* and become increasingly permissive, the consequences of which can be detrimental to adolescents' behavior (Baer and Schmitz 2007; Driscoll, Russell, and Crockett 2008).

Asian American Parents

Despite the fact that, as a category, Asian Americans have the highest average income of any race/ethnic group measured by the U.S. Census, Asian Americans are also found in lower socioeconomic strata and may experience economic hardship (Ishii-Kuntz et al. 2009). Nevertheless, compared to the average of 29.9 percent for all Americans older than 24, 52.4 percent of Asian Americans are college graduates or have advanced degrees (U.S. Census Bureau 2012c, Table 229).

The Asian American parenting style is often characterized as authoritarian, emphasizing obedience and possibly using physical punishment (but coupled with more praise and hugs than in mainstream American society) (Lindner Gunnoe, Hetherington, and Reiss 2006). Research findings are mixed regarding the success of the expert-preferred authoritative parenting style among Asian Americans, again suggesting that the model may be Eurocentric (Pong, Johnston, and Chen 2010).

Social scientists have offered an alternative parenting concept, the **Confucian training doctrine**. This parenting model is named after the sixth-century Chinese social philosopher Confucius, who stressed (among other things) honesty, sacrifice, familial loyalty, and respect for parents and all elders. The Confucian training doctrine blends parental love, concern, involvement, and physical closeness with strict and firm control, or "training" (Chao 1994; McBride-Chang and Chang 1998). "Training" may involve parents' use of guilt, shame, and moral obligation (defined negatively as "shaming" in mainstream culture) to control their children's behavior (Farver et al. 2007).

> By their own lights, Asian Americans sometimes go overboard in stressing hard work. Nearly four in ten (39 percent) say that Asian-American parents from their country of origin subgroup put too much pressure on their children to do well in school. Just 9 percent say the same about all American parents. On the flip-side of the same coin, about six-in-ten Asian Americans say American parents put too little pressure on their children to succeed in school while just 9 percent say the same about Asian American parents. (Pew Research Center 2012c, p. 4)

However, the extremely strong ties between Asian American parents and their children have been found to lessen parent-child conflict, along with potential negative effects of shaming (Benner and Kim 2009; Park, Vo, and Tsong 2009).

Like other ethnic minorities, Asian Americans have suffered from discrimination (Lau, Takeuchi, and Alegria 2006; Tong 2004). Moreover, Asian immigrant parents may face conflicts with their children when expecting traditional behavior characteristic of the homeland while their children assimilate into the American culture and no longer adhere to traditional expectations regarding dating, for example, or marital monogamy (Ahn, Kim, and Park 2008; Farver et al. 2007).

Parents of Multiracial Children

About 9 million Americans define themselves as more than one race. This figure shows considerable growth from the 6.8 million reported in the 2000 census, the first to offer citizens the option of identifying themselves as multiracial (Jones and Bullock 2012). Of mixed race Americans, 83 percent are white in combination with at least one other race, 34 percent are black in combination with another race, 25 percent are American Indian in combination with some other race, and 30 percent are Asian in combination with another race (Jones and Bullock 2012, Figure 3). As more and more multiracial individuals reach childbearing age and as racial heterogamy loses its taboo, the number of multiracial births is expected to continue to climb (Jones and Bullock 2012).

Raising biracial or multiracial children has unique rewards and challenges (Rockquemore and Laszloffy 2005). One challenge involves insensitive remarks from strangers: "How come she's so white and you're so dark?" (Saulny 2011b). Another challenge may be tension between parents and children—and between the parents and extended family members—over cultural values and attitudes.

A psychologist surveyed multiracial adults and asked whether they thought their parents had been prepared to raise children of mixed race. The majority did not believe so (Dunnewind 2003). Today, however, many schools are more sensitive to the needs of multiracial children, and more resources are available for parents raising multiracial children. According to one recent study, multiracial and multi-ethnic families that consciously foster a family identity as multicultural, multiracial, or multiethnic have happier, better-adjusted children (Soliz, Thorson, and Rittenour 2009).

Parents in Transnational Families

As pointed out in Chapter 1, more and more American families are *transnational*. Due to emigration of one or more individuals, family members in different countries maintain relationships across national borders. For instance, a single mother from Puerto Rico explained her family relationships:

> I have a brother and a sister [in Puerto Rico] and I have a sister in Miami, [but I am closer to] my sister in Puerto Rico even though we are far away. She was here ... when my baby was a few months old. I have plans to go to Puerto Rico for Christmas. (in Dominguez and Lubitow 2008)

This respondent, among others, emphasized her efforts to keep her home culture alive for her children and dreams of taking them to visit her home country so that they might better understand their native culture (Dominguez and Lubitow 2008).

Research is just beginning on transnational families, but we do have other interesting—sometimes disturbing—findings to report. For one thing, according to a recent study published in the *Harvard Educational Review*, "fear and vigilance" characterize the home lives of undocumented parents, with the result that they are less likely to engage with their children's teachers or be active in their communities (Preston 2011). U.S. immigration policy, particularly in states such as Arizona that have passed legislation hostile to undocumented immigrants, has resulted in deportation of undocumented parents and hence their possible separation from their American-citizen children born in the United States.

In the first half of 2011, according to the Department of Homeland Security, the United States deported approximately 46,500 parents with at least one child who is an American citizen (Yoshikawa and Suarez-Orozco 2012). Research speaks to "deep and irreversible harm" done to the children: "Having a parent ripped away permanently, without warning, is one of the most devastating and traumatic experiences in human development" (Yoshikawa and Suarez-Orozco 2012; Yoshikawa and Kholoptseva 2013). How these children fare, who cares for them, and what will be their outcomes as they grow and mature are questions for future research.

Meanwhile, journalists have featured children who were born and raised in the United States, then accompanied their deported parents to Mexico where they miss their American lifestyle and friends, face adjusting to a different culture, and often feel like outsiders. "These kinds of changes are really traumatic for kids," according to a Marta Tienda, a Princeton sociologist born to migrant Mexican workers. "It's going to stick with them" (Cave 2012). Among other research questions that transnational families prompt, studies are beginning to emerge on parent-child relations between emigrating parents who have left their home country and their "left-behind children" who did not accompany them (Graham and Jordan 2011).

Religious Minority Parents

Recent studies suggest that regardless of the particular religion, children in families who adhere to a religious belief system tend to be better adjusted. The thinking is that it is less about religious beliefs and more about being a member of a community that is important for family life (Good and Willoughby 2006; Lees and Horwath 2009). We might think of the religious benefits as *spiritual capital*—coping "resources of faith and values derived from commitment to a religious tradition" (Grace 2002, p. 236).

Ethnicity is often associated with religious belief. For example, Chinese Americans are likely to be Buddhists; Asian Indian Americans are likely to be Hindu or Sikh. In a Christian dominant culture, diverse ethnoreligious affiliations affect parenting for many Americans. For instance, Muslims have their own holy days, such as Ramadan, which may not be taken into account in public schools' scheduling. Wearing flowing robes and, more often, head scarves or veils (called *hijab*), Muslims report that they fear ridicule and face discrimination from employers and others (American Moslem Society 2010; "Muslim Parents Seek Cooperation" 2005).

Those religions such as Islam or Judaism that depart from a Christian tradition have the added burden of raising children in a society that does not support their faith; the same could be said of evangelical Protestant groups as well. American parents of religious minorities generally hope that their children will remain true to their religious heritage amid a majority culture that seldom understands and is sometimes threatening ("Muslim Parents Seek Cooperation" 2005; Lee 2001). One solution has been the emergence of religion-based

Bob Daemmrich/The Image Works

At this festival marking Eid, the end of Ramadan, this Muslim community in central Texas gathers for afternoon prayers. Muslim parents hope that their children will remain true to their religious tradition. Meanwhile, like parents of other minority religions in the United States, they must help their children face fear of ridicule and actual discrimination.

summer camps for children of Baha'i, Buddhist, Catholic, Hindu, Jewish, Mormon, Mennonite, Muslim, and Sikh parents, among others.

Raising Children of Minority Race/Ethnic Identity in a Racist and Discriminatory Society

A parent's feeling victimized by racism adds stress to an already stressful parenting process (Benner and Kim 2009; Brody et al. 2008). Families of color (or religious minorities) attempt to serve as an insulating environment, shielding children from or confronting injustices (Brody et al. 2008; Brown et al. 2007; Cohen, Tran, and Rhee 2007). Most engage in **race socialization**—developing children's pride in their cultural heritage while warning and preparing them about the possibility of encountering discrimination (Cooper and McLoyd 2011; Caughy 2011; Hansen 2012). Higher levels of race socialization are associated with parents' having been discriminated against, their higher sense of personal efficacy, and greater concern that their children will actually encounter racism (Benner and Kim 2009; Brody et al. 2008; Crouter et al. 2008).

Valuing one's cultural heritage while simultaneously being required to deny or "rise above" it in order to advance in mainstream society poses problems for individuals and between parents and their children. For instance, Native Americans must often choose between the reservation and an urban life that is perhaps alienating but may present greater economic opportunity (Seccombe 2007). Hispanics may see a threat to deeply cherished values of family and community in the competitive individualism of the mainstream American achievement path (McLoyd et al. 2000). Asian Americans may follow the "model minority" route to success but experience emotional estrangement from their culturally traditional parents (Kibria 2000).

As one response, minority parents may encourage their children to participate successfully in the larger society with regard to occupation and education while maintaining their original cultural values with regard to religion and family norms (Portes and Zhou 1993). The desire of ethnic minority parents to preserve their culture tends to be sharpened when they feel discriminated against, feel their culture is devalued in mainstream society, or feel that mainstream culture encourages negative behaviors in their children (Kalmijn and van Tubergen 2010). We turn now to a discussion of another variation in the parenting experience—grandparents as parents.

Grandparents As Parents

About 11 percent of U.S. grandparents are raising grandchildren (Lumpkin 2008). More than 3.6 million children under age 18 are living in a grandparent's household, a few with only their grandfather, many with two grandparents, and many more with only their grandmother (King, Mitchell, and Hawkins 2010). Having risen dramatically over the past ten years, the number of grandparents residing with grandchildren (about 7 million) and the number both living with and responsible for raising children (2.7 million) is expected to rise, largely due to recession (U.S. Census Bureau 2012c Table 70; U.S. Census Bureau 2012e, Table CP02).

Taken together, unmarried parenthood, divorce or separation, poverty, substance abuse, HIV/AIDS, domestic violence, abandonment, and incarceration account for a very large majority of grandparent families, or **grandfamilies** (Bailey, Letiecq, and Vannatta 2011; Henderson et al. 2009). "The transition to parenting a second time around requires many grandparents to reconsider their work-family roles and make significant, unexpected changes" (Bailey, Letiecq, and Vannatta 2011, p. F18). One study found that grandparents' coping strategies involved relying on their religious faith as well as imagining that the situation would somehow "just go away" (Lumpkin 2008).

Facts about Families

Foster Parenting

Every state government has a department that monitors parents' treatment of their children. An example is California's Department of Child Protective Services. When government officials determine that an individual under age 18 is being abused or neglected, they can take temporary or permanent custody of the child and remove him or her from the parental home for placement in **foster care**. As wards of the court, foster children are financially supported by the state.

About 400,500 children are in foster care in the United States. There would be more, but there are not enough foster parents or other facilities to fill current needs. Seventy-four percent of foster care takes place in a licensed foster parent's home—47 percent with nonrelatives, and 27 percent with relatives. The remainder of children in foster care live in various arrangements, including group homes (6 percent) or institutional settings (9 percent) such as Nebraska's Boystown (which also accepts girls) (U.S. Department of Health and Human Services 2012a).

The mean age of children in foster care is about eight years. Children stay in foster care for an average of about two years, but 20 percent stay for only one to five months, and nearly 10 percent remain until age 18 when they "age out." Although the rate of children in foster care is higher for African Americans than for other race/ethnic groups, the highest percentage of children in foster care are white (44 percent), followed by 23 percent for blacks, 21 percent for Hispanics, 5 percent for mixed-race children, 2 percent for Native Americans, and 1 percent for Asian children (U.S. Department of Health and Human Services 2012a).

Very often without family support, those who are neither reunited with family members nor adopted "age out" of the system. For the most part, these youth were older when they became foster children and were the developmentally neediest; they face serious challenges as they work toward assuming adult roles. They are more likely than other young adults to become imprisoned, homeless, unemployed, or pregnant outside marriage (Koch 2009; Nunn 2012).

Among others, motivations for becoming a foster parent include fulfilling religious principles, wanting to help fill the community's need for foster homes, enjoying children and hoping to help them, providing a companion for one's only child or for oneself, and earning money. Technically not salaried, foster parents are "reimbursed" in regular monthly stipends by the government.

Although the ultimate goal in half of the cases is reunification of foster children with their parents or principal caretakers, about one-quarter of foster children are available for adoption (U.S. Department of Health and Human Services 2012a). Some foster parents see fostering as a step toward adopting (Baum, Crase, and Crase 2001). The National Foster Parent Association and the Foster Care and Adoptive Community provide online education and support.

In the words of Jo Ann Wentzel, senior editor of the magazine *Parenting Today's Teen* and foster mother to more than seventy-five children over the course of her career,

> I don't regret anything I've ever done for any of my [foster] kids.... Every once in a while, a kid will track me down and leave a cryptic message on my answering machine, which says, I know I was a pain-in-the-butt when I lived with you but I really learned a lot from you.... Or maybe they will tell me about their successes and claim it was because of something we did or said. (Wentzel 2001, p. 2)

Critical Thinking

From the structure-functional perspective discussed in Chapter 2, foster parents are functional alternatives to biological or adoptive parents. What are some ways, do you think, that the foster parent system is functional? What are some instances in which it could be dysfunctional?

Sylvie de Toledo is a social worker whose nephew was raised by her mother after her sister's suicide. Influenced by this experience, Toledo founded a support group called Grandparents As Parents (GAP) "to meet the urgent and ongoing needs of grandparents and other relative caregivers raising at-risk children" (Toledo and Brown 1995). Becoming a primary parent requires considerable adjustment for grandparents (Bailey, Letiecq, and Vannatta 2011). Living with children in the house is a significant change after years of not doing so (Dolbin-MacNab 2006). A grandparent's circle of friends and work life may change. He or she may retire early, reduce work hours, or try to negotiate more flexible ones. On the other hand, a grandparent may return to work to finance raising the child(ren). In either case, a grandparent's finances may suffer while paying for items such as additional beds, food, and clothing (Bailey, Letiecq, and Vannatta 2011).

To help, under the **formal kinship care** system, some states offer financial compensation to grandparents (or other relatives, such as aunts) who raise their grandchildren as state-licensed foster parents. However, some grandparents report having trouble navigating their state's kinship care system due to their fear and distrust of the child welfare system and daunting bureaucratic regulations, among other reasons (Letiecq, Bailey, and Porterfield 2008). "Facts about Families: Foster Parenting" further discusses foster parenting.

The Center for Law and Social Policy finds formal kinship care to be generally good for children. Compared with children in nonrelative foster care, those fostered by relatives are less likely to have tried to run away and more likely to say that they feel loved, like those with whom they live, and want their current placement to be their permanent home (Conway and Hutson 2007). However, a few critics argue that, at least with regard to black children, formal kinship care is overused and detrimental to families in the long run because it fails to emphasize reunification with the children's parents (Harris and Skyles 2008).

Grandparents' raising grandchildren is characterized by ambivalence (Bailey, Letiecq, and Vannatta 2011). Unsure whether or when their grandchildren will return to the parental home, grandparents may "learn a . . . stance of detachment to cope with the shifts they are sure to experience and probably even applaud" (Nelson 2006, p. 822). Furthermore, there are often questions about the possible legal termination of the parent's parental rights (McWey, Henderson, and Alexander 2008). When parental rights are not terminated, grandparents who are responsible for the children in their care lack legal rights over them (LaPierre 2011).

Some grandchildren being raised by a grandparent see one or both parents either regularly or sporadically; but generally these relationships are complicated and often marked with difficulties. In a qualitative study with white, black, and mixed-race children being raised by grandparents, some children hoped for reunification with their parents, but the majority had accepted their situations (Dolbin-MacNab and Keiley 2009).

A grandparent's living in the home of a poor single mother is advantageous inasmuch as it adds income, from Social Security benefits, for example. In addition, researchers and social workers generally maintain that grandparents provide stability, family cohesiveness, and solidarity while often enhancing young children's cognitive development (Spratling 2009). However, not all grandparents raising grandchildren employ effective parenting practices (Barnett 2008). Grandmothers have been found to be most sensitive and beneficial to infants (Dunifon and Kowaleski-Jones 2007).

Research on the responses and feelings of adults who were raised by their grandparents shows that some adult children were grateful and felt a strong bond with their grandmothers whereas others evidenced distance and distrust (Dolbin-MacNab and Keiley 2009). Social service agencies have initiated educational and coping programs for grandfamilies (Dolbin-MacNab 2006; Ross and Aday 2006). The National Center on Grandfamilies promotes awareness of grandfamilies and gives advice on how to help grandparents meet their various needs.

Parenting Young Adult Children

Children benefit from parents' emotional support and practical guidance throughout their twenties and after (Kantrowitz and Tyre 2006). As grown children make the transition to adult roles, parent-child relations often grow closer and less conflicted (Arnett 2004; Straus 2009). At the same time, both parents and young adults may be angry or depressed over student loans due to rising college costs, lingering childhood issues, or difficulties with assuming adult roles (Arnett 2004; Geewax 2012a; Galambos and Krahn 2008). Meanwhile, concerns over the young adult's delayed transition to adulthood can cause parental ambivalence and parent-child conflict, and parents who see their grown children as needing too much support report poorer life satisfaction when compared with other parents (Fingerman et al. 2012; Hay, Fingerman, and Lefkowitz 2007).

A significant majority of higher-SES parents lend or give their children money—about $7,500 annually—to repay student loans, buy a car, help with rent or credit card debt, or to put a down payment on a house (Ray 2012; Wightman, Schoeni, and Robinson 2012). One interesting study found that parents tend to provide money not only to their neediest but also to their most successful children, the latter in anticipation of help from the child as the parent grows older (Fingerman et al. 2009).

As discussed in Chapters 1 and 6, recession, unemployment, and underemployment, along with a decline in affordable housing, make launching oneself into independent adulthood especially difficult today. As one result, more and more young adult children either do not leave the family home or return to it after college, divorce, or on finding their first jobs unsatisfactory (Parker 2012). As reported in Table 6.1, 59 percent of young men and half of women age 18 through 24 live with one or both parents, and a significant fraction of older adults do, too (U.S. Census Bureau 2012b, Table AD-1). Some counselors advise parents to *expect* children to move back home—but not to micromanage their offspring's career or to sacrifice too much. "I see too many parents, especially mothers, helping out grown children when they should be squirreling away more money for their own retirement," said the president of the nonprofit Women's Institute for a Secure Retirement (Kobliner 2010). Meanwhile, there is evidence that grown children's being allowed to live with parents while they get on their occupational feet is good for them (Settersten and Ray 2010).

Parents should feel comfortable in setting reasonable household expectations. One way to do this is to negotiate a parent-adult child residence-sharing agreement (Wadler 2009). The following are typically issues to negotiate:

1. How much money will the adult child be expected to contribute to the household?
2. What are the standards for neatness?
3. Who is responsible for cleaning what and when?
4. Who will cook what and when?
5. How will laundry tasks be divided?
6. What about noise levels?
7. When are guests welcome and in what rooms of the house? Will the home be used for parties?
8. What are the expectations about informing fellow family members of one's whereabouts?
9. If the grown child has returned home with one or more children, who is responsible for their care?

White Packert/The Image Bank/Getty Images

Good parenting involves having adequate economic resources, being involved with the child, using supportive communication and having support from family and/or friends, along with workplace and broader social policies that bolster all families.

More detailed residence-sharing agreements are available on the Internet, some for sale (e.g., "Boomerang Kids Contract" n. d.). Although a residence-sharing agreement can help temporarily, the goal of the majority of parents is for their adult children to move on. Accomplishing this may be complicated by differing ideas on just what a parent owes an adult child. Our culture offers few guidelines about when parental responsibility ends or how to withdraw it.

Toward Better Parenting

One way to improve parenting would be to encourage more research attention to some relatively neglected topics. Two examples are research on siblings, including parent's differential treatment of them (McHale, Updegraff and Whiteman 2012), and how the denial of legal marriage for same-sex couples affects parenting (Goldberg and Kuvalaska 2012). In one piece of recent qualitative research on this latter topic, a young adult explained that permitting her parents to marry legally

> would have affected my life with my mom and Karen [her stepmother] very tangibly and specifically. I couldn't drive Karen's car, because I couldn't be on her car insurance. Karen had dental insurance ... and my mom and I couldn't be on her dental insurance. (in Goldberg and Kuvalaska 2012, p. 44)

In addition to more research, what are some steps that we can take to improve parenting in the United States? Studies show that optimal parenting involves the following factors:

- supportive family communication (Leidy et al. 2009; Lindsey et al. 2009);
- involvement in a child's life and school (Cooper, Crosnoe, et al. 2009);
- private safety nets—that is, support from family or friends (Lee et al. 2009; Ryan, Kalil, and Leininger 2009);
- adequate economic resources (Guzzo and Lee 2008);
- workplace policies that facilitate a healthy work–family balance and support parenting in other ways as well (Aber 2007; Bass et al. 2009);
- safe and healthy neighborhoods that encourage positive parenting, school achievement, and reciprocal social support (Byrnes and Miller 2012); and
- society-wide policies that bolster all parents (Marshall and Tracy 2009).

Chapter 11 explores the first factor listed: supportive family communication. Here we note some existing programs designed to improve child raising, and then we discuss further ways to promote better parenting.

Over the past several decades, many national organizations have emerged to help parents (Lam and Kwong 2012). Some programs serve parents in general.

One of these is Thomas Gordon's Parent Effectiveness Training (PET) (Gordon 2000). Another is Systematic Training for Effective Parenting (STEP) (Center for the Improvement of Child Caring n. d.). Both STEP and PET combine instruction on effective communication techniques with emotional support for parents. Some other parent education classes incorporate anger management training (Fetsch, Yang, and Pettit 2008).

Other programs are designed to improve parenting in specific situations. A variety of intervention programs aim to help teenage and/or substance-abusing parents (Tolan, Szapocznik, and Sambrano 2007). Some curricula are designed for particular race/ethnic groups (Center for the Improvement of Child Caring n. d.; Kumpfer and Tait 2000; Mandara et al. 2012).

Because still more children could benefit from fathers' economic support and additional contact, policy makers advise further interventions that facilitate father involvement (Amato, Meyers, and Emery 2009; Teachman 2009). Then, too, "higher levels of father involvement and a positive coparenting relationship may keep couples together, which allows children to spend their early years with both biological parents in the household" (McClain 2011, p. 889). Hence, there are programs intended to increase fathers' involvement, some fairly successful (Cowan et al. 2009; Hawkins et al. 2008). Some programs that target adolescent fathers aim to teach skills that improve co-parent relations with the child's mother, whether he continues to be romantically involved with her or not (Fagan and Lee 2011; McHale, Waller, and Pearson 2012).

Child care researchers consider the policy implications of the research. Belsky (2002) argues for tax or other policies to support full-time parental care in the home, especially during the first year. Research by the National Institute of Child Health and Human Development (NICHD) suggests that intervention programs that might be effective in enhancing parenting are more important for development than whether a child spends time in day care. This finding was especially true for children who did not have high-quality child care in their first years. "Experimental studies of high-quality early intervention programs have demonstrated that these programs can enhance social, cognitive, and academic development of economically disadvantaged children" (Belsky et al. 2007; Vandell et al. 2010, p. 738).

Other child care scholars agree with the NICHD researchers that subsidies should be available to permit parents to cut back work hours. At the same time, they urge attention to improving the quality of out-of-home care. They also believe child care can make a positive contribution to social development if done well (Maccoby and Lewis 2003).

In line with the family decline perspective, discussed in Chapters 1 and 7, some policy makers promote marriage as the most effective way to enhance father involvement (Wilcox et al. 2011a). According to this perspective,

> [I]t is the institution of marriage that helps men to "sign on" to fatherhood. By choosing to make a legal, social and public commitment to a spouse, a man voluntarily agrees—often well ahead of the actual arrival of a child—to take on the legal and social role of a father. (Whitehead and Popenoe 2008)

From the family change perspective, on the other hand, there is need for "shifting the paradigm in support of multiple family forms" (Jones 2008, p. 208). What are some ways that social policy could better support all parents, regardless of family structure? The stress model of effective parenting suggests that reducing parents' stress would improve parenting. Accordingly, "improving the socioeconomic conditions of parents, particularly among the most vulnerable, might improve parenting outcomes across all relationship types" (Guzzo and Lee 2008, p. 58).

Moreover, because free time to engage in leisurely social interaction and activities is crucial to psychological well-being (Harter and Arora 2008), work and society-wide policies aimed at freeing up time for mothers and fathers would improve parenting. More concerted attention on the part of employers to parental child-raising needs and responsibilities would help, along with their greater recognition that good parenting is essential to a civil society.

Then, too, elementary school administrators might schedule children's performances and parent–teacher conferences so that they are less likely to conflict with parents' work schedules (Barnett et al. 2009). And single parents in particular would benefit from more—and more affordable—day care services (Ornelas et al. 2009). Also, college regulations might be widened to better accommodate student parents. An example would be an institution's explicitly stating that exams can be made up in the case of a child's illness (Duquaine-Watson 2007).

We've seen that informal social support has been found to mitigate stress and hence to be related to more positive parenting (Lee et al. 2009; Ryan, Kalil, and Leininger 2009). But policy makers also urge greater civic and community activism on the part of parents (Dudley 2007; Whitehead and Popenoe 2008). Some parent-education programs include instruction on how to become more civically involved or engage in community activism (Doherty, Jacob, and Cutting 2009). "Pediatrics is politics," the late pediatrician Benjamin Spock once said (quoted in Maier 1998). He meant that good parenting involves working for better neighborhoods, communities, and family-centered social policies—and these, in turn, result in better parenting.

Summary

- Parenting can be difficult today for several reasons, one of which is that work and parent roles often conflict.
- The family ecology theoretical perspective reminds us that society-wide conditions influence the parent-child relationship, and these factors can place emotional and financial strains on parents.
- The stress model of effective parenting posits that stressors of various sorts lead to parental depression and household conflict, which in turn result in less-positive parenting practices and ultimately in poorer child outcomes.
- Although more fathers are involved in child care today, mothers are the primary parent in the vast majority of cases and continue to do the majority of day-to-day child care.
- Child psychologists prefer the authoritative parenting style, although some scholars describe the authoritarian/permissive/authoritative parenting style typology as ethnocentric or Eurocentric.
- Family form, gender, socioeconomic class, and race/ethnicity intersect to result in parenting experiences for individuals.
- Higher-SES parents tend to follow the concerted cultivation parenting model, whereas working-class parents are more likely to adhere to the accomplishment of natural growth model.
- A trend over the past several decades has been for an increasing number of grandparents to serve as primary parents, often as a result of some crisis in the child's immediate family.
- More than 400,500 children are in foster care today, many of them in formal kinship care.
- To have better relationships with their children, parents are encouraged to accept help from others (friends and the community at large as well as professional caregivers), to build and maintain supportive family relationships (the subject matter for Chapter 11), and to engage in community or civic activism.

Questions for Review and Reflection

1. Describe reasons why parenting can be difficult today. Can you think of others besides those presented in this chapter?
2. Compare these three parenting styles: authoritarian, authoritative, and permissive. What are some empirical outcomes of each? Which one is recommended by most experts? Why?
3. How does parenting differ according to social class? Use the family ecology theoretical perspective to explain some of these differences.
4. What unique challenges do African American, Native American, Hispanic, and Asian American parents face today, regardless of their social class? How would *you* prepare an immigrant child or a child of color to face possible discrimination?
5. **Policy Question.** Describe some social policies that could benefit all low-income parents, regardless of their gender, race/ethnicity, or family structure.

Key Terms

accomplishment of natural growth parenting model 236
authoritarian parenting style 231
authoritative parenting style (positive parenting) 232
concerted cultivation parenting model 235
Confucian training doctrine 240
formal kinship care 243
foster care 243
grandfamilies 242
hierarchical parenting 239
hyperparenting 236
multipartnered fertility 222
permissive parenting style 231
private safety net 228
psychological control 232
psychological parent 226
race socialization 242
resilient 223
social fathers 229
socioeconomic status (SES) 235
stress model of parental effectiveness 224
transition to parenthood 224

10

Work and Family

Ariel Skelley/Taxi/Getty Images

Learning Objectives

1. Summarize men's traditional family role and how expectations for husbands have changed over the past several decades.
2. Summarize women's traditional family role and how expectations for wives have changed over the past several decades.
3. List some options and alternatives available to two-earner couples.
4. Discuss division of labor regarding unpaid family work.
5. Give reasons for why women do more unpaid family work than do men.
6. Describe the causal feedback loop regarding work and family living.
7. Discuss family policy regarding work-family conflict.
8. Explain how feelings of fairness regarding work-family roles impact relationship satisfaction.

The three-generation Lee family occupies a four-story building in New York City. With his two grown children, the widowed Chinese-immigrant grandfather, Gung Gung, 86, bought the building fifteen years ago. Now with Gung Gung, two married couples and seven children in the household, the family splits the mortgage, food and repair bills. Gung Gung's son, Warren, makes dough and fillings for the family bakery. Warren also cooks for the family, lights his father's Buddhist altar each morning, and gets him to medical appointments. Warren's wife, Jen is a gym teacher. Gung Gung's daughter, May is assistant principal at a nearby school. May's husband, Ben drives for Federal Express. The family rents out the building's lowest floor to a Mexican restaurant. The Lees employ two Cantonese-speaking babysitters (S. Kramer 2011).

Every Monday, financial executive Karen Cangas leaves home for the airport. She'll be back Thursday night. Until then she'll work with clients in various cities. Her husband drops off their two children at day care on his way to work. Evenings, the family visits with Karen on Skype (Cullen 2007).

Although in different ways, these two contemporary American families (like all families) inhabit the intersection, or *interface* between work and family. In both scenarios, family members work to provide resources; they organize their family lives according to available economic options and opportunities. Until fairly recently in human history, cooperative labor for survival was the dominant purpose of marriages and families, as discussed in Chapter 7. Now we also expect love and emotional support from our families.

In addition to expecting love in marriage, a second development began with the eighteenth- and nineteenth-century Industrial Revolution in Europe and the United States when economic production moved to factories and offices. Wage earners and a **labor force**—people who are employed or who are looking for paid work—emerged. Given these two historical developments—(1) wanting emotional satisfaction from family life and (2) leaving home for employment—combining work and family has become characterized by multiple options and decisions to make.

This chapter focuses on work and family in (heterosexual) marriages rather than in other family forms. There are two reasons for this. First, the work-family interface in cohabiting and in single-parent households is addressed in Chapters 6 and 9, respectively. Second, with few exceptions (e.g., Coughlin and Wade 2012),

The *family ecology perspective* tells us that workplace requirements impact family living. This mom's military career certainly affects her family life. Four in ten military women have children. Even if breast-feeding, new mothers are required to deploy as soon as four months after a birth. Some mothers return after deployment to toddlers who don't remember them. Spokespersons with the National Military Families Association say that military moms should not be deployed overseas until at least twelve months after a birth (Browder 2010; Mann 2008).

the vast majority of research on this topic concerns heterosexually married partners. Domestic relations in same-sex partner households are explored in Chapter 6.

In this chapter we'll examine husbands' and wives' traditional and changing roles as gender is performed in the workplace and family. We'll look at how couples juggle—and struggle—to balance employment with unpaid family labor. We'll see that, as argued in Chapter 3, some aspects of traditional gender roles persist despite social change. The persisting gendered divergence in earnings, as well as the trend toward earnings *con*vergence that we have seen over the past several decades, have had unprecedented implications for the division of household labor. We'll investigate conflict between work and family, then consider what is needed to resolve work-family conflict. To begin, we will explore the interface of work and family life—that space where the social institutions of family and the economy meet.

The Interface of Work and Family Life

The concepts *sociological imagination* and the *family ecology perspective*, discussed in Chapters 1 and 2, hold that family life is influenced by cultural expectations and social structures external to it. The workplace is one such influence (Geist and Cohen 2011). For instance, workplace policies, such as the availability of maternity leave, impact women's fertility decisions and new mothers' labor force participation (Barnes 2013; Laughlin 2011). As another example, in a recession (an economic situation), many families relocate to pursue job opportunities or be nearer to extended kin (Allen 2009).

Meanwhile, family members can influence workplace policies and conditions. In 2010, the Wal-Mart Corporation settled the largest sex-discrimination lawsuit in history when it agreed to pay $86 million in damages to over 200,000 California female employees and family members. Wal-Mart had paid them less than it paid its male employees, and the female employees sued for damages (Stempel 2010). The above are *macro*, or society-wide examples. We can also think in terms of *micro*, or smaller group examples that are closer to home.

On the micro level, family researchers often look at the **spillover** from work situations into family life—how pleasures or stresses associated with work affect interaction within the family (e.g., Jang, Zippay, and Park 2012). One study found significant within-day spillover from the quality of mothers' interactions with supervisors to interactions with their children. Supervisor criticism was positively correlated with harsh and withdrawn mother-child interactions; supervisor praise was positively associated with warm mother-child interactions (Gassman-Pines 2011).

Spillover can be studied in the other direction too. A Singapore study found that workers with distressed marriages were depressed at work and had lower motivation and output (Sandberg et al. 2012). A study of economically disadvantaged mothers found that an adolescent child's delinquency negatively impacted the mother's work stability and performance (Coley, Ribar, and Votruba-Drzal 2011).

Gender and the Work-Family Interface

You may recall from Chapter 3 that men have traditionally been expected to show *instrumental* character traits, such as self-reliance and ambition, while women have been supposed to be *expressive*, relationship-oriented, and supportive helpmates (Parsons and Bales 1955; Sallee 2011). At least in the American middle-class, husbands have been more likely to be in the labor force, an instrumental role. Wives have been more likely to be mother-homemakers, an expressive role. As long as middle-class nonHispanic white women remained out of the labor force, this division of labor was taken for granted by most researchers and policy makers.

Indeed, the family ecology perspective (which suggests that structural forces outside families influence what goes on inside them) operates within a context of gender expectations and structures that influence options and decisions (Garey and Hansen 2011). You'll recall from Chapter 3 that *gender structures* involve the five basic social institutions: family, religion, government, education, and the economy. This chapter focuses on the gendered interface between the family and economic institutions.

When social institutions are not well integrated (that is, they do not work together), individuals who play roles in both institutions experience **role conflict**: meeting the demands of one institution conflict with meeting the simultaneous but different demands of another institution. For example, an employed parent needs (1) to be home to monitor a teenager's after-school behavior *and* (2) to be at work. The career world tends to view someone who takes time from work for family as less than professional, yet family norms often encourage mothers—and increasingly, fathers—to do exactly that (Hochschild 1997).

Work-family conflict on the macro, or society-wide, level often results in personal experiences of role conflict. A study of medical students in the United States and Great Britain found that female students anticipated role conflict. They expected family demands to hamper their careers, while males seemed less influenced by family concerns (Riska 2011).

But the times, they are a-changin'. A 2012 Pew Research survey found for the first time that women age 18 to 34 topped young men in desiring a high-paying career. Nearly two-thirds (66 percent) of young women said that "being successful in a high-paying career or profession" was "very important" or "one of the most important things" in their lives. Among men in the same age group, 59 percent said so. These figures compare to 56 percent of women and 58 percent of men in 1997 (Pew Research Center 2012h).

In addition, a 1977 survey found 43 percent of Americans saying that marriages work better when the husband is breadwinner and the wife takes care of the house and children. By 2010, the percentage thinking this way had dropped to 30. Similarly, 1977 surveys found 48 percent saying that marriages work better when both spouses have jobs and care for the house and children. By 2010, 62 percent thought so (Pew Research Center 2010a, pp. 26, 30).

Despite social change, evidence of traditional expectations and behaviors persists. In 2012, when asked what makes a good husband, 41 percent of Americans said a good husband needs to provide a good income, while only 19 percent said providing a good income is necessary to being a good wife. As another example, a longitudinal study of 61 husbands and 48 wives who began a small business between 2004 and 2006 found that husbands were much more likely than wives to receive help from their spouse—keeping the books, for example. The authors suggest that this difference in spousal support (called *spousal capital*) helps to explain why, among married entrepreneurs, husbands succeed more often than wives do (Matzek, Gudmunson, and Danes 2010).

The Gallup Poll has been asking about U.S. men's and women's work-family preferences for many years. In 1978, for the first time, a decisive majority of women favored employment over the full-time homemaker role. "Since then, no clear consensus in either direction has emerged, with small majorities of women sometimes opting for working outside the home and, ... [at other times] small majorities favoring the traditional role of family caretaker" (Moore 2005). In 2012, the poll showed that a slight majority of women (51 percent) preferred paid employment. A significantly larger proportion of men preferred the provider role. Asked, "If you were free to do either, would you prefer to have a job outside the home, or would you prefer to stay at home and take care of the house and family?" about 75 percent of men have consistently over the past several decades said they preferred a job outside the home. About 22 percent of men would prefer to stay home (Saad 2012a).

Men's Work and Family Roles

Whether single or married, parent or not, men are employed more hours than women and are more likely (86.6 percent) to work full time than are women (73.5 percent) (U.S. Bureau of Labor Statistics 2013 Tables 20). Historically and today, the male **provider role**, in which family men are expected to supply resources (shelter and food, for instance) is evident in all social classes. In this country the male provider role emerged during the 1830s. Before then, a man was expected to be "a good steady worker," but "the idea that he was *the* provider would hardly ring true," because in a farm economy husband and wife together produced family income (Bernard 1986, p. 126).

The male provider role (and its counterpart, the female homemaker, or housewife role) predominated into the late1970s. Since then the proportion of married-couple families in which only the husband is employed has declined steadily—from 42 percent in 1960 to 19 percent in 2009 (U.S. Bureau of Labor Statistics 2013, Table 23). As Figure 10.1 shows, both spouses are employed in the majority (54 percent) of married couples.

In 2013, the Pew Research Center announced that in 40 percent of households with children under age 18 mothers were the sole or primary breadwinner. Single-mother households make up about two-thirds of the 40 percent. Meanwhile, in 15 percent of married households with children under age 18, mothers are the sole or primary breadwinner (Wang, Parker, and Taylor 2013). Although wage earning is certainly no longer reserved for husbands, some still believe that the man should be the sole, or only, family provider, and it works out that way in practice in about one-fifth of married couples.

Meanwhile, men continue to be *primary* breadwinners in the majority of families, and most men in all race/ethnic groups identify with this role (Coltrane 2000). In fact, men's employment and earnings remain important to "facilitating marriage and enhancing marital

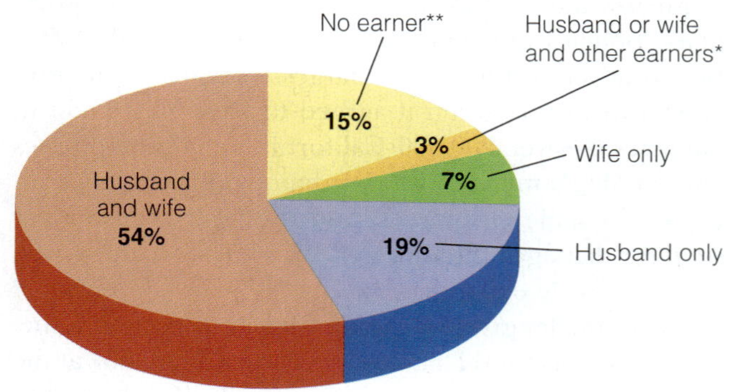

* Includes husband and other family member(s); wife and other family member(s); other earners, neither husband nor wife.

** The spouses may be unemployed, retired, disabled, institutionalized, or imprisoned.

FIGURE 10.1 Married-couple families by percentage of family-member earners, 2012.
Percentages do not add to exactly 100 due to rounding.

Source: U.S. Bureau of Labor Statistics 2013, Table 23.

stability" (Bianchi and Casper 2000, p. 31). Anthropologist Nicholas Townsend interviewed thirty-nine nonHispanic white, Hispanic, and Asian-American men who graduated high school in the 1970s. Regardless of ethnicity, the men described their life goals as "the package deal": marriage, children, home ownership, and a steady job. Work was viewed as essential to being a good father: "Everybody has a purpose in life.... Your purpose is to provide for your family" (Skip, quoted in Townsend 2002, p. 117). Although these men desired to spend more time with their children, in reality their time was devoted to paid work. Many had two jobs or put in extensive overtime.

Other husbands, choosing to work less and spend more time with their families, reject the idea that occupational dedication and achievement comprise their primary purpose. Indeed, a 2007 Harris poll showed that 37 percent of fathers with minor children would leave their job if their partner made enough to support the family, and 38 percent would accept less pay in exchange for more time to spend with their children (Careerbuilder.com 2007).

"Good Providers" versus "Involved Fathers"

It appears that there are two models for the husband-father role. More traditional fathers—researchers have named them **good providers**—emphasize the provider role as defining their merit, and they work *more* hours than do childless men (Glauber and Gozjolko 2011). On the other hand, in order to spend more time with their children, **involved fathers** work fewer hours than do childless men (Kaufman and Uhlenberg 2000). Put another way, some fathers try to decrease the demands of the workplace in order to participate more at home. Indeed, more fathers are taking off work following the birth of a child, and they are more visible in parenting classes, in pediatricians' offices, and dropping off and picking up children at day care centers (Livingston and Parker 2011). However, being an involved father is not easy.

Workplace Obstacles to Being an Involved Father High powered careers often expect "professional dedication," which requires an employee to work long hours and to be available electronically even in officially off hours. Many men who would prefer to work less than they do, "may feel that they have no real option if their jobs demand longer than their desired hours" (Shafer 2011, p. 261). Also, a man who gives priority to family may deal with coworkers' resentment or challenges to his masculinity. Then too, employers may not believe that employees, especially males, should allow family responsibilities to interfere with work (Hochschild 1997).

As one Texas wife, pregnant with the couple's first baby, explained to an interviewer, "Where he works, the men don't take off to take care of sick children. I hope to split time missed from work with my husband, but that battle is yet to be fought.... He's flexible; it's his boss that may not be" (in Stanley-Stevens and Kaiser 2011, p. 121). As a result, some men are reluctant to take advantage of available family benefits. The resistance men encounter, however, may be partly a self-imposed perception that they will be viewed as less committed employees if they access options such as paternity leave (Neal and Hammer 2007).

Some husbands report having lied to bosses or taken other evasive steps at work to hide conflicts between job and family. One man tells his boss that he has "another meeting" so that he can leave the office each day at 6 p.m.: "I never say it's a meeting with my family" (Jacobs and Gerson 2004, p. 6). This situation illustrates the *family ecology perspective*: The greediness of work, especially for men, "has real implications for gender inequality, in that wives may exit the labor force because the husbands' careers make their attempts to combine work and family too difficult" (Shafer 2011, p. 261). We should note, however, that, provided employers allow it, communication technology facilitates working from home and negotiating flexible hours for at least some fathers (Jarrell 2007; Shellenbarger 2007).

Economic Change and the Good Provider Role Many couples need the husband to be employed in the interests of the family's overall financial well-being (Lewin 2009; Smith 2009). Although males are now being re-employed at faster rates than females, men

David Sacks /Lifesize/Getty Images

Many men today expect—and are expected by their partners—to be involved fathers, working at home doing child care or domestic labor, as well as holding a job.

bore the largest brunt of job losses during the recession that began in 2007 (Hartmann, English, and Hayes 2010; Wessel 2010). Some forecasters predict that long-term, at least partly due to globalization and changes in manufacturing, American men, especially those without college educations, will have difficulty finding and keeping jobs (Elsby, Hobijn, and Sahin 2010, Table 1). Women seem to be adjusting more readily to a changing economy (Rosin 2012b). This situation has important consequences for the work-family interface. During the recession many wives became their family's sole breadwinner by default. Then too, with or without recession, some couples have sized up their situation, found that the woman had higher potential earning power or was more desirous of pursuing a career, and decided to reverse traditional roles.

Stay-at-Home Dads As depicted in Figure 10.1, in 7 percent of married couples, only the wife is employed (U.S. Bureau of Labor Statistics 2013, Table 23). A small minority of men have relinquished breadwinning to become **stay-at-home dads**: men who stay home to care for the house and family while their wives work. About 20 percent of fathers of preschool children whose mothers are employed are primary parents (Laughlin and Rukus 2009). In 26 percent of gay male couples with children, one parent stays at home: "To some gay men, the idea of entrusting the care of a hard-won child to someone else seems to defeat the purpose of parenthood" (Bellafante 2004).

Despite gender role changes, families with breadwinning mothers and full-time caregiver fathers continue to be viewed as norm violators to some extent (Gaunt 2013; Poniewozik 2012). Although fathers who are primary parents may experience the loss of a career-based identity and express more sense of isolation than do stay-at-home mothers, being a househusband is not the lonely choice it once was. Local groups, national organizations, and Internet chat rooms bring househusbands together, and stay-at-home mothers are more welcoming of their male counterparts than they used to be. As with many aspects of family life, choice is the key to a man's satisfaction with the househusband role, as is the couple's mutual understanding about the specifics of their division of labor (Gerson 2010; Lewin 2009; Tyre 2009). Is dad responsible for all the laundry or just some of it? Stay-at-home fathers are also discussed in Chapter 9.

Women's Work and Family Roles

As the Industrial Revolution got under way, middle-class white women generally remained at home and engaged in domestic labor and "homemaking." Women in lower social classes, immigrant women, and women of color often supported themselves and their families by taking in laundry, marketing baked goods, working as domestic labor in other people's homes, and housing boarders. Some women worked in factories, but it was largely men who held formal "jobs" and were visible in economic production.

Women in the Labor Force

Gradually, beginning around 1890, more women began to enter the labor force. Industrialization gave rise to bureaucratic corporations, which depended heavily on paperwork. Clerical workers were needed, and not enough men were available. Textile industries sought workers with a dexterity thought to be possessed by women. The expanding economy needed more workers, and women were drawn into the labor force in significant numbers. As Figure 10.2 shows, women's participation in the labor force has increased greatly since the beginning of the nineteenth century.

This trend accelerated during World War I, the Great Depression that followed, and World War II. The trend slowed for a time following the end of World War II in 1945. As soldiers came home, the government encouraged women to return to their kitchens. Despite these cultural pressures, the number of wage-earning women would rise once more. As material expectations increased for housing and consumer goods and as more families began to imagine financing a child's college education, wives' wages grew more important.

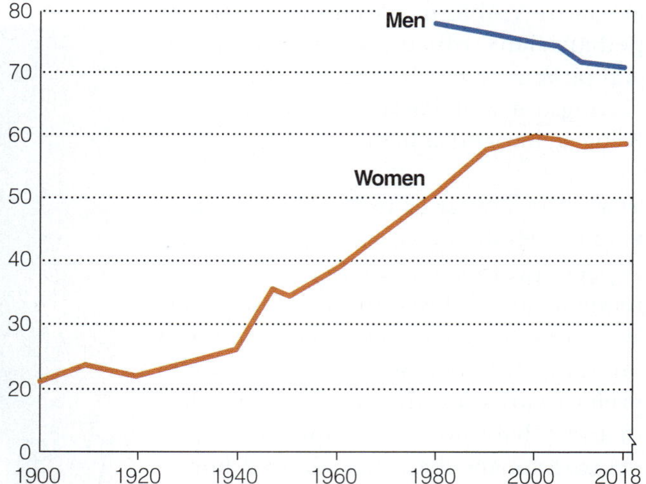

FIGURE 10.2 The participation of women and men over age 16 in the labor force, 1900–2010, and projections to 2018. Women's and men's labor force participation declined between 2005 and 2010 due to the recession that began in 2007.

Source: Thornton and Freedman 1983; U.S. Census Bureau 2012c, Table 587.

Beginning in about 1960, the number of employed nonHispanic white women began to increase rapidly. Stagnant and declining earnings for men led more families to rely on a second earner. The growth in the divorce rate left women uncertain about the wisdom of remaining out of the labor force and dependent on a husband's earnings. The Women's Movement emerged and was a strong force for anti sex-discrimination laws that opened occupations formerly closed to females. The feminist movement also altered societal attitudes regarding women's careers with the result that they began to seem normative. By 1979 a majority of married women were employed outside the home.

Historically, black women have been more likely than nonHispanic whites to work for wages (England, Garcia-Beaulieu, and Ross 2004). Today white women—with labor force participation of 58.5 percent—are catching up to black women at 59.9 percent. Fifty-seven percent of Asian women and 56.5 percent of Hispanic women are employed (U.S. Census Bureau 2012c Table 587).

Mothers in the Labor Force Mothers of young children were the last women to move into the labor force. Although many mothers remained at home while their children were small, by 1970 half of wives with children between ages 6 and 17 earned wages, and that figure increased to 71 percent in 2010 (U.S. Bureau of Labor Statistics 2012b, Table 5).

Today, 61 percent of mothers are employed by the time her child is three years old, the majority working full-time (U.S. Bureau of Labor Statistics 2012b, Table 2). In fact, 58 percent of married women with children under age 1 have joined the labor force. Even larger proportions of single mothers are employed: 79 percent of those with children age 6 to 17, and 68 percent of those with children under age 6 (U.S. Census Bureau 2012c, Tables 599, 600).

> The recent economic recession, the steady decline in men's earnings [see Chapter 1], the growth in single-mother families, and the increase in women's educational levels are some of the factors likely contributing to the growth in the proportion of mothers of young children in the workforce. (Lavery 2012)

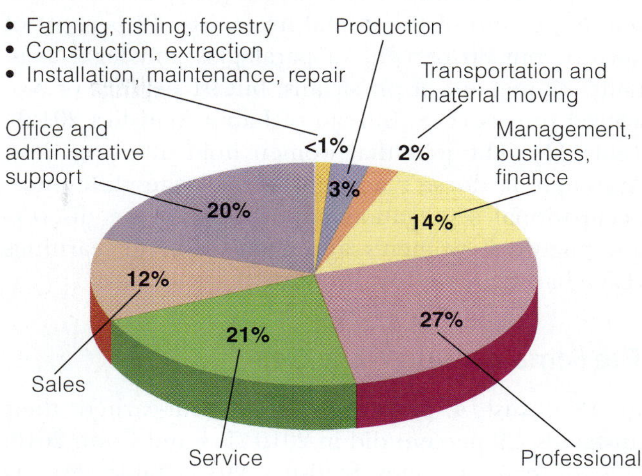

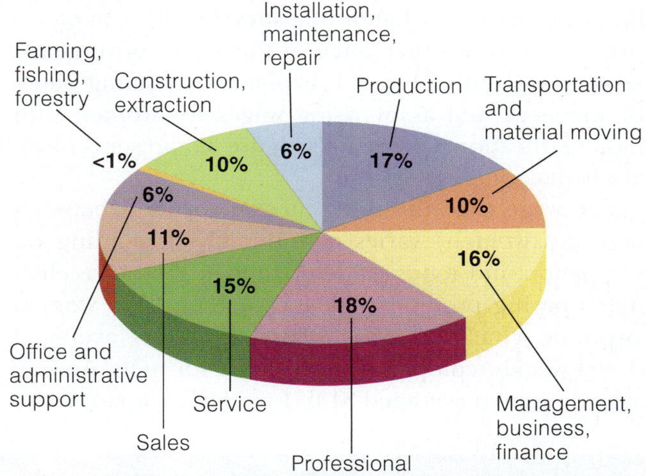

FIGURE 10.3 The jobs held by women and men, 2010. The percentages in each sector of the pie charts indicate the percentages of women and of men in certain job categories. Percentages may not add to exactly 100 percent due to rounding.

Source: U.S. Bureau of Labor Statistics 2011. Women in the Labor Force: A Databook, Table 10.

Women's Occupations

Jobs that women more often hold differ from those of men, as Chapter 3 and Figure 10.3 indicate. The tendency for men and women to be employed in different job types is termed **occupational segregation**. Occupational segregation has declined somewhat since 1960. Women are physicians, attorneys, astronauts, engineers, and military officers. Men have become nurses, dental hygienists, and elementary school teachers. However, as you can see in Figure 10.3, 32 percent of employed women in 2010 were office or sales workers. Only 14 percent were in management, business, or finance positions, while 27 percent were in professional work. Asian American (46.1 percent) and white women (41.5 percent) were the most likely to hold managerial or professional jobs, compared to black (33.8 percent) or Hispanic women (24.1 percent) (U.S. Bureau of Labor Statistics 2012b, Table 12).

Furthermore, jobs typically held by men and women differ *within* major occupational categories, with men more likely to hold the upper-level jobs within each sector. Although women are proportionately more likely to be professionals than men, they occupy the lower-paying ranks. Women are 26 percent of dentists, but 96 percent of the dental hygienists; 32 percent of lawyers, but 86 percent of paralegals and legal assistants; 32 percent of physicians, but 91 percent of registered nurses (U.S. Bureau of Labor Statistics 2012b, Table 11). The jobs that women hold often pay less. Although it doesn't account for all the difference, occupational segregation contributes to the discrepancy between women's and men's average earnings (Gauchat, Kelly, and Wallace 2012).

The Female–Male Wage Gap

In 1970 just 4 percent of wives out-earned their husbands; 29 percent did in 2010 (Fry and Cohn 2010; U.S. Bureau of Labor Statistics 2013, Table 25). As shown in Chapter 3's Figure 3.2, in 2011 women on average earned 82 percent of what men earned, with younger women earning up to 92 percent of what men did (U.S. Bureau of Labor Statistics 2012b). Chapter 3 further points out that some of the wage convergence between women and men is explained by falling wages for men—as well as by rising wages for women with higher education. Chapter 3's Figure 3.3 presents race/ethnic data on the wage gap.

The **wage gap** (the difference in earnings between men and women) varies considerably depending on occupation and tends to be greater in the more elite, higher-paying positions. For instance, in 2011 women corporate chief executive officers (CEOs) averaged $1,464 weekly, compared with $2,122 for men. Among lawyers, women averaged $1,631 weekly, compared with $1,884 for men. Female physicians earned $1,527 weekly, compared with $1,936 for men (U.S. Bureau of Labor Statistics 2013, Table 18).

Men continue to dominate corporate America. In 2010, less than 3 percent of the highest-earning executives in Fortune 500 companies were women ("Women CEOs" 2010). Although racism blocks the path to management for nonwhite and Hispanic men, both racism and sexism block the path for nonwhite and Hispanic women, who hold relatively few executive and board-of-director positions in major corporations. In 2010, the first African American woman assumed a CEO position in the Fortune 500 list of top companies (Angelo 2010).

Some experts have argued that men's earnings are higher because jobs more often held by males are more difficult, require more training, or have less favorable working conditions. A woman also contends with employers' *assumptions* that she will opt out of the labor force to take care of her children or other family members. As a result, an employer may be less likely to select even highly ambitious and fully committed women employees for further training or for positions with advancement potential (Pinto 2009). Also, as explored in Chapter 3, negative stereotyping and discrimination against female employees account for about one-fifth of the male-female wage gap (Bergmann 2011; Bobbitt-Zeher 2011).

The concept of **motherhood penalty** describes the fact that motherhood has a significant negative lifetime impact on female income—a situation that creates a long-term gender-earnings gap. "Women still earn a small proportion of what men earn [over a lifetime] ... and remain financially dependent on men for income during the child-rearing years and indeed throughout much of their adult lives" (Hartmann, Rose, and Lovell 2009, p. 125). The motherhood penalty persists despite women's increasing education and dedicated labor force involvement (Hegewisch and Liepmann 2010; National Women's Law Center 2010). Afraid that impending motherhood will jeopardize workplace status or promotion, some professional women hide their pregnancies for as long as possible (Quart 2012).

James Marshall/The Image Works

Steadily more women have entered the labor force since the 1960s. This woman may be mothering children. Women in blue-collar jobs are still a minority, although more women are entering these jobs, which tend to pay better than traditional women's jobs in service or clerical work.

Stay-at-Home Moms

In 2011 there were about five million stay-at-home mothers. That number represents 23 percent of married-couple families with children under age 15. Compared with employed mothers, stay-at-home moms are more likely to be younger, Hispanic, foreign-born, and mothering a child under age 5 (U.S. Census Bureau 2012a, Table SHP-1). Despite media interest in professional stay-at-home moms, the majority of stay-at-home mothers have no more than a high school diploma and have relatively low household incomes (Saad 2012b). More highly educated women in better-paying jobs or careers are less likely to leave the workforce after having children—and when they do, it's usually not for very long (Goldin 2006).

Many stay-at-home mothers are immigrants who are following a traditional model espoused by their home country's culture. For instance, Arab-American women are employed at lower-than-average rates. Scholars suggest the reasons for this are varied, but traditional gender roles are emphasized in Arab-American families, and women are encouraged to stay out of the labor force because they are considered "bases of security and stability" for other family members (Gold and Bozorgmehr 2007, p. 52). For upper middle class nonHispanic white women, we might think of stay-at-home mothers as comprising **neotraditional families**—that is, families reminiscent of 1950s norms and values, but with the new (neo) aspect that the model is consciously selected from several options—options that the vast majority of 1950s wives did not have.

> This [neotraditional] order is appealing to men and women who are discontented with ... family modernization, the lack of clarity in gender roles ..., and the pressures associated with combining two full-time careers. It is also appealing to women who continue to identify with the domestic sphere, who wish to see homemaking and nurturing accorded high value, and who wish to have husbands who share their commitment to family life. (Wilcox 2004, p. 209)

This family model is often associated with evangelical Christianity, as well as with Orthodox Judaism, traditional Catholicism, and Mormonism (Wilcox 2004). When economic concerns pressure into the labor force a woman who would prefer a neotraditional marriage, she is likely to organize her employment (as much as possible) around part-time or in-home work (Cordes 2009). Although the majority of stay-at-home moms are happy and satisfied with their arrangement, they are somewhat more likely than employed mothers to say that they worry and are sad, angry, or depressed (Mendes, Saad, and McGeeney 2012).

Two-Earner Unions and Work/Family Options

About forty-five years ago, in 1968, there were equal proportions of two-earner and provider–housewife couples: 45 percent of each (Hayghe 1982). Today we've seen that **two-earner unions** in which both partners work are the majority. Partners display considerable flexibility in how they design their two-earner unions. Arrangements are ever-changing and flexible, varying with the arrival and ages of children and with partners' job options and preferences. In this section, we examine four ways that two-earner couples navigate the work-family interface: two-career marriage, part-time employment, shift work, and working at home.

Noah Berger/The New York Times/Redux

Some women have chosen to opt out of the labor force to raise their children at home. This former executive may return to the labor force when her children are older.

Two-Career Marriages

Careers differ from *jobs* in that careers hold the promise of advancement, are considered important in themselves—not just a source of money—and demand a high degree of commitment. Career men and women work in occupations that typically require education beyond a bachelor's degree—for example, medicine, law, academia, financial services, and corporate management. Career workers spend more hours on the job than do the less educated. One third of professional or managerial men work fifty or more hours per week, and one in six women do (Jacobs and Gerson 2004; Usdansky 2011).

The vast majority of two-earner marriages would not be classified as *two career* because the wife's or the husband's employment does not have the features of a *career*. Nevertheless, the two-career couple is a powerful image. Most of today's college students view the **two-career relationship** as an available and workable option. For two-career couples with children, however, family life can be hectic as partners juggle schedules, chores, and child care.

Working Part-Time

About 26 percent of women and 13 percent of men worked part-time in 2011 (U.S. Bureau of Labor Statistics 2013, Table 20). Of part-time workers who had worked full-time, 70 percent say that they changed their employment situation due to child-care problems (U.S. Census Bureau 2012c, Table 613). Less work-family conflict and more family and personal time are clear benefits of part-time employment. Part-time mothers of preschoolers evidence better health and more sensitive and involved parenting than do other mothers, both employed and stay-at-home moms (Buehler and O'Brien 2011).

However, part-time employment has its costs. As it exists now, part-time work seldom offers job security or benefits such as health insurance (U.S. Census Bureau 2012c, Table 655). And part-time pay is rarely proportionate to that of full-time jobs. For example, a part-time teacher or secretary usually earns well below the equivalent hourly wage paid to full-time staff. In higher-level professional and managerial jobs, a different problem appears. To work "part-time" as an attorney, accountant, or aspiring manager is to forgo the salary, status, security, and promotions of a full-time position and still be working a high number of hours.

Shift Work

Sometimes one or both partners engage in **shift work**, defined by the Bureau of Labor Statistics as any work schedule in which more than half an employee's hours are before 8 a.m. or after 4 p.m. It has been estimated that in one-quarter of all two-earner couples, at least one partner does shift work; one in three if they have young children (Presser 2000). With about 20 percent of workers engaged in shift work, men comprise a slight majority (McMenamin 2007). Some couples use shift work for higher wages or to ease child care arrangements, the latter addressed later in this chapter.

Shift workers face physical stress with night work or frequently changing schedules. Also, shift work reduces the overlap of family members' leisure time, which can negatively affect the relationship: "To the extent that social interaction among family members provides the 'glue' that binds them together, we would expect that the more time partners have with one another, the more likely they are to develop a strong commitment to their marriage and feel happy with it" (Presser 2000, p. 94). Unsurprisingly, then, shift work is associated with a decrease in marital stability.

Shift work is also associated with negative spillover effects on mother-child interactions and—probably due to fatigue—with less positive parental mood, poorer maternal sensitivity, and consequent lower children's academic achievement and more negative child behaviors (Gassman-Pines 2011; Grzywacz et al. 2011; Han and Fox 2011). An alternative for some parents involves working at home.

Doing Paid Work at Home

Home-based work—working from home, either for oneself or for an employer—increased significantly over the past few decades (Tozzi 2010). Historically, home-based work involved *piecework*—sewing or making artificial flowers, for example. This mode of home production is declining due to competition from low-wage workers overseas. It still exists, particularly in the assembly of medical kits, circuit boards, jewelry, and some textile work, but many home-based workers are educated and are engaged in professional services such as law, accounting, computer programming, consulting, marketing, finance, and so on (Tozzi 2010). Other home-based businesses include selling cosmetics, kitchenware, or other products.

Home-based work also includes working from home for an employer by connecting to the office, customers, clients, or others via the Internet—that is, telecommuting. About half of telecommuters are women (Mateyka,

Jeffery Salter/Redux

Working from home is one way to manage the work-family interface. What do you see as some positive consequences of this choice? What might be some less positive consequences?

Rapino, and Landivar 2012). Many of them, especially those with children under age 6, say that they work from home to "coordinate work with personal/family needs" (Wight and Raley 2009, Table 1; Guynn 2013).

Illustrating work-family conflict, Yahoo CEO Marissa Mayer caused "an uproar" among telecommuters (most of them mothers) in February 2013 when she announced that Yahoo employees could no longer telecommute and would be expected to be in the Yahoo offices five days a week. Mayer argued that getting workers together in the same workspace encourages the kind of collaboration necessary for Yahoo's ongoing creativity (Guynn 2013; Pofeldt 2013).

Meanwhile, telecommuting does help parents coordinate work and family obligations, but it may not be a panacea. As the author of a study of women in a home-based direct-selling business noted, however, "many women soon discovered... that they had exchanged one set of challenges for another. Indeed, work–family flexibility may be a double-edged sword as boundaries between work and family blur" (Gudmunson et al. 2009). Mothers employed at home "report problems with interruptions...; they are often asked to... run errands for relatives, to watch neighbors' children when bad weather closes the school, or to keep an eye out for the older kids" (Kutner 1988). Put another way, mothers (and others) employed at home may simultaneously be expected to do unpaid family work.

Unpaid Family Work

Social scientists interested in the division of household labor have used many of the theoretical perspectives discussed in Chapter 2 to study unpaid family work. To illustrate this, Table 10.1 applies several of these to the topic and gives examples from the Lee family whose story opened this chapter. **Unpaid family work**

Table 10.1 Theoretical Perspectives Applied to Unpaid Household Labor

Theoretical Perspective	Application to Unpaid Household Labor	Examples from the Lee Family
Family Ecology	Distribution of household labor is influenced by the environment that surrounds the family, particularly the workplace and work-family policy.	The Lee family's urban work options include employment opportunities in local schools, running a family business, and renting out family-owned property, among others.
Family Life Course Development Framework	In our society, adults are assigned more unpaid household labor than are children or the elderly.	• The Lee family's teenagers are expected to be good students, but not expected to work outside the household. • Gung Gung, the Lee family's 86-year-old patriarch, does less work now than he once did.
Structure-Functionalism	What might be some *functional alternatives* to the traditional male breadwinner-female homemaker roles?	Functional alternatives to the traditional division of labor include: • May works outside the home as a school principal. • Warren's wife, Jen, works outside the home as a gym teacher. • Warren does considerable unpaid family work—for example, cooking and getting his father to doctor appointments.
Interaction-Constructionist	Individuals, couples, groups, and societies invest housework with meanings (Sallee 2011). According to the *gender performance* hypothesis, women "do gender" by doing housework, whereas men "do gender" by avoiding housework (Schneider 2011).	The family has defined Warren's primary family role as caregiver to his father.
Exchange or Bargaining	Individuals who provide more family income can exchange resources brought into the household for the right to be excused from more undesirable household tasks.	May and Jen work outside the home, one as a school principal and the other as a gym teacher. They exchange the earnings they bring to the household for not having to prepare meals.
Family Systems Theory	Family members design and perform their respective roles in response to how others in the family system play their roles.	May's husband, Ben, who drives for Federal Express, is out of the family building from 7 a.m. until after 8 p.m. His wife adjusts to this situation by assuming the majority of the discipline for the couple's children.

involves caring for dependent family members, such as the elderly or children, as well as maintaining the family domicile. Chapters 13 and 16 address meeting the needs of ill, disabled, or elderly family members. Chapter 9 explores the different, but changing, expectations and behaviors associated with mother and father roles. Here we focus primarily on household labor associated with maintaining the family domicile.

Household Labor

It may be hard to believe now, but utopians and social engineers once shared a hope that advancing technology and changed social arrangements would make the need for families to cook and clean obsolete (Hayden 1981). But collective arrangements proposed by utopians and early feminists never caught on. As opportunities opened up in the labor force, servants, who had done much of the work for earlier middle-class housewives, entered factory work or took other jobs. Middle-class women were left to do their own housework (Cowan 1983). Then, especially after World War II and during the 1950s, technology raised standards even as it made some housework less time-consuming. With automatic washers, we began to change clothes daily, for example, a practice that creates lots more laundry to do (Newport 2008).

Less Housework Is Being Done Now Than in the Past Today, assisted by microwaves, fast food, sometimes by paid services, and perhaps "living with a little more dust," twenty-first century couples often adjust to women's labor force participation by scaling down what was considered necessary housework a few decades ago (Achen and Stafford 2005, p. 12). Slightly over fifty hours weekly went to housework in 1965, compared with thirty to forty hours in 2005 (Galinsky, Aumann, and Bond 2009; Swanbrow 2008). The whole of this decline is due to women's, not men's doing less household labor.

Husbands' Share of Housework Is More Than in the Past While the trend has been for women to do less household work over the past several decades, men on average have done more (Swanbrow 2008). One typical time study found that women decreased housework time from twenty-six hours per week in 1976 to seventeen hours weekly in 2005. During this same time period, men's housework more than doubled, from six to thirteen hours per week. Husbands' time spent on child care has tripled since the mid-1960s (Galinsky, Aumann, and Bond 2009). However, although husbands are doing more household labor than they did in the past, wives continue to do more of it than husbands do.

Wives Still Average More Household-Labor Hours Than Husbands Do According to the U.S. Bureau of Labor Statistics American Time Use Survey, women average about two hours daily doing domestic chores, while men spend about half that. On average, mothers also spend about two hours daily feeding, bathing, reading to, and playing with children, while husbands spend less than half that (Hartmann, English, and Hayes 2010). Although the gap has lessened (Artis and Pavalko 2003), data from about 8,500 participants in a University of Michigan study showed that women, on average, spend twenty-seven hours a week on household labor (including child care), compared to about thirteen hours for men (Swanbrow 2008).

Moreover, a good deal of women's unpaid family labor goes unnoted and unmeasured. For instance, "health behavior work"—promoting family member's healthy behaviors, such as making family members' dental appointments or monitoring their eating habits—is a virtually invisible aspect of caregiving that is usually assigned to women (Reczek and Umberson 2012). More often than do men, women assume the work of coordinating paid services when a family hires them (Coontz 2007). Women disproportionately organize family activities and care for or "keep" kin—**kin keeping**—maintaining contact, remembering anniversaries and birthdays, sending cards, shopping for gifts (Coontz 2007; Hagestad 1986). Of course, it's possible that, "the work men do to maintain relations as fathers, brothers, and uncles may not be captured by questionnaires or analyses that focus on emotional support or caregiving tasks" (McCann 2012, p. 254). Nevertheless, it's safe to conclude that, overall, women do significantly more unpaid household labor than do men. Why might this be?

Why Do Women Do More of the Household Labor?

Wives do more unpaid family labor whether they are employed outside the home or not (Galinsky, Aumann, and Bond 2009). However, it's important to note that the imbalance may disappear when *hours* employed are counted, rather than simply employment *status*. Husbands spend fewer hours in housework than wives do, but husbands spend more hours in paid employment than do wives (Konigsberg 2011; Treas and Drobni? 2010; van der Lippe 2010). Wives with husbands who earn more and work more hours do more housework. Also wives do more housework when they have husbands who bring work home or are preoccupied with work-related problems when at home (Sullivan and Coltrane 2008). Social scientists have proposed alternative hypotheses for why women do more unpaid household labor.

Hypothesis 1. Partners' Relative Earnings Influence Their Division of Household Labor Men's participation in household labor had long been thought to be related to their relative earnings and the proportionate share of household income produced by each partner (Hartmann, English, and Hayes 2010). On average, employed wives contribute significantly to financial resources—over a third (37.6 percent) of a family's income today and up from 26.6 percent in 1970 (U.S. Bureau of Labor Statistics 2013, Table 24). Nevertheless, on average, wives contribute less than half of family incomes, and this may be why they put in more hours of unpaid household labor. "[A]s married women increase their earnings share toward equality with their husbands, they reduce the amount of time they spend each day on routine housework tasks" (Schneider 2011, p. 857). Wives who outearn their partners have husbands who contribute more to household labor (Crary 2008; Sullivan and Coltrane 2008).

Hypothesis 2. Gender Roles Influence Partners' Division of Household Labor Polls show that even with increased equality in partners' division of household labor, they tend to split household tasks according to traditional gender expectations. For example, men say their household tasks include car upkeep, yard work, and investment decisions, whereas women tend to do more meal preparation, dish washing, grocery shopping, house cleaning, laundry, and child care (Newport 2008). The persistence of "male" and "female" household tasks suggests that household labor can have meaning beyond simple maintenance. "Housework is not just the performance of basic household tasks but it is also a symbolic expression of gender relations, particularly between wives and husbands" (Artis and Pavalko 2003, p. 748; Gilbert 2008).

Virtually all of us judge our own and others' behavior according to whether it agrees with gender norms (normative behavior) or does not (gender deviance) (Gaunt 2013). Doing tasks that are considered traditionally feminine or masculine can reinforce—or threaten—a partner's feminine or masculine gender identity. A man's cleaning the garage may reinforce his masculine gender identity. His doing the dishes may threaten his gender identity; hence he's likely to try to avoid this activity.

Furthermore, if a husband feels threatened by his wife's earnings, he may do *less* housework in an effort to psychologically neutralize his feelings of gender deviance. It has also been proposed that wives who earn more than their husbands may neutralize their gender role deviance by doing more of the housework (Kluwer 2011; Sullivan 2011a, 2011b). This hypothesis, although interesting, lacks empirical support (Schneider 2011). We should note too that the meaning of unpaid family labor may have changed somewhat, as have gender expectations. That is, over the past several decades, it has become more socially acceptable for men to be involved in child care, cooking, and cleaning (Galinsky, Aumann, and Bond 2009).

Hypothesis 3. The Partner with More Power in the Relationship Can Escape Undesirable Household Labor Social scientists studying this topic have long associated power in relationships with household work, particularly with undesirable tasks like cleaning the toilet. Some research has found that the partner with more control over the other does less around the house (Lam, McHale, and Crouter 2012; Sullivan and Coltrane 2008). Marital power is the subject of Chapter 12.

Hypothesis 4. The Relative Time Available to Each Partner Influences the Division of Household Labor The more hours men work, the less housework they do. Men who work fewer hours do more housework. Unemployed men, presumably with more available time, spend almost double what employed men do on household labor (della Cava 2009, Eckel 2010).

> This evidence seems to provide support for the idea that, rather than avoiding housework to mitigate any gender deviance created through unemployment, these men are responding to reduced work time by increasing housework, behavior that appears more consonant with the predictions of time availability theory. (Schneider 2011, p. 857)

At least one writer concludes that, "The widespread belief that working mothers have it the worst... is simply not the open-and-shut case it once was" (Konigsberg 2011). Then too, sharing unpaid family labor varies by race/ethnicity and by sexual identity.

Diversity and Household Labor

In some ethnic groups, such as Vietnamese and Laotian, housework is significantly shared, if not by husbands, by household members other than the child's mother (P. Johnson 1998). Among African Americans, adult children living at home, extended kin, and nonresident fathers are likely to share housework and child care and to help with repairs (Gerson 2010).

Research on race/ethnic differences finds that the pattern of men's spending less time than women in housework occurs in white, black, Asian Indian, and Hispanic families. However, black men spend more time in unpaid family work than do nonHispanic white men (Bhalla 2008; Barajas and Ramirez 2007; Gerson 2010). One explanation is that black men have more egalitarian attitudes regarding unpaid labor, and that black wives are more likely than other race/ethnic categories to have high earnings relative to their husbands' earnings (Gerson 2010; Glauber and Gozjolko 2011). Also, the differences among nonHispanic white, black, and Hispanic men's household labor time may reflect other (as of yet, unknown) differences among them as well (Sullivan and Coltrane 2008).

Bill Aron/Photo Edit

Do you think this man cleans the toilets at home? Changes diapers? Folds laundry? Social scientists have proposed four hypotheses to explain why, on average, husbands do less household labor than wives do. First, husbands who earn more than their wives may exchange their financial input for doing less housework. Second, some tasks may seem like "women's work," and, protecting their gender identity, men are reluctant to do them. Third, the partner with more power in the relationship does less around the house. Finally, the one who spends more hours at paid employment has less time available and therefore does less household labor. Which of these hypotheses do you think is or are more likely to be supported by empirical evidence?

Lesbians, Gay Men, and Household Labor Given that gay couples are comprised of people of the same gender, how does their household division of labor work out and what impact does it have on the relationship? A small, qualitative study of lesbian and gay male couples explored these questions. Each partner was employed full-time, and there were no children in the household. The study looked at who performed some traditionally female tasks. Generally, lesbian couples' division of labor was more egalitarian than that of gay male couples. Perceived equality was closely tied to relationship satisfaction and that, in turn, to relationship stability (Kurdek 2007). As with heterosexual couples, research shows that the number of hours individual members of a couple work in paid labor has the greatest influence on the division of household labor (Sutphin 2006). The next section investigates how partners juggle employment and unpaid family labor.

Juggling Employment and Family Work

The term *juggling* implies hectic and stressful situations. Typically two-earner and single-parent families are hectic indeed. Ways that families manage vary. Many two-earner couples, especially in upper-income families, hire household help and may purchase the services of immigrant, race/ethnic minority, and working-class people for cleaning, child care—and to a lesser extent, cooking (Kelleher 2007; Killewald 2011).

Virtually every researcher studying the work–family interface hears expressions of time pressures and feeling rushed and stressed (Bianchi, Robinson, and Milkie 2006). Stress is most pronounced for parents during children's early years, especially for single mothers and those with demanding careers (Jacobs and Gerson 2004). A Gallup poll asked individuals whether they have enough time to do what they want, and between 55 and 58 percent of adults under age 54 said no (Carroll 2008).

Employed parents put in a **second shift** of unpaid family work that amounts to an extra month of work each year (Hochschild 1989). Nearly twenty-five years ago when sociologist Arlie Hochschild's book by that title was published, she applied the term, "second shift" to wives, not husbands. The women Hochschild interviewed talked about being overtired, sick, and "emotionally drained," as well as how much sleep they could "get by on." (Hochschild 1989, p. 9). Today, husbands feel pressured too (Konigsberg 2011). The second shift means forfeiting leisure time and sleep to accomplish unpaid family work. For a majority of Americans, rest and relaxation time, and time for friends, hobbies, and sleep is not what they would like it to be (Saad 2004).

Work-Family Conflict in the Twenty-First Century

Work demands have increased over the past several decades. Based on the U.S. Census Bureau's Current Population Survey data, sociologists Jerry Jacobs and Kathleen Gerson note that Americans increasingly spend more and more time at work. The authors point to an "increasing mismatch between our economic system and the needs of American families" (Jacobs and Gerson 2004, back cover). American workers lead the industrial world in the number of hours worked (Mishel, Bernstein, and Shierholz 2009). One quarter of U.S. employees work more than forty hours per week, with 9 percent working between fifty and sixty hours a week and another 6 percent working more than that. About 5 percent of the labor force holds two or more paid jobs, with slightly higher proportions of women doing so (U.S. Bureau of Labor Statistics 2013, Table 32; U.S. Census Bureau 2012c, Tables 603, 610). The implications of this situation are important because the inability to escape relentless work demands, even for a short time, correlates with depression, anxiety, and couple conflict.

Even as Americans spend more time at work, younger employees want more family time than did their predecessors. A 2002 study surveyed 2,800 adults from different generations: baby boomers (born between 1946 and 1964), Generation X (born between 1965 and 1979), and Generation Y (born between 1980 and 1994). Respondents were asked whether they put work before family or family before work. Baby boomers were more work- and less family-oriented than today's workers. A majority of the two younger generations said they put family before work (Families and Work Institute 2004). "Gen X and Gen Y men are demanding to have the ability to play a larger role in family life than their fathers did," according to Joan Williams, director of Work/Life Law at American University ("More New Dads Seek Time Off" 2005, p. Bus. 1). The next section looks at stresses peculiar to *two-career* marriages—keeping in mind the distinction between *two-earner* couples and *two-career* couples made earlier.

Managing Two-Career Unions

Some decades ago when the two-career marriage emerged as an ideal lifestyle available to all young couples, Hunt and Hunt (1977) noted that two-career families require a support system of child care providers and household help that depends heavily on ability to pay. The success of today's two-career union is premised on the existence of a labor pool of low-paid, but highly dependable, household help. The vast majority of such help is provided by women, many of whom have their own families to worry about (Romero 1992). "Issues for Thought: When One Woman's Workplace Is Another's Family" examines the intersection of class and gender in these relationships.

The Geography of Two Careers Because career advancement often requires geographic mobility—and even international transfers—juggling two careers may prove difficult. A career move for one may make the other a **trailing partner** who relocates to accommodate the other's career. Provided they can afford it, family-centered couples may turn down transfers because of two-career issues. As a result, some large companies now offer career-opportunity assistance to a trailing partner, such as hiring a job search firm, facilitating intercompany networking, attempting to locate a position for the partner in the same institution, or providing career counseling (Jio 2008).

Although wives still move for their husband's career more often than the reverse, the number of trailing husbands has increased (Shauman and Noonan 2007). More two-career marriages today are based on a conscious mutuality to which partners have become accustomed by the time a career move presents itself. Such couples are less likely to have problems with a female-led relocation than are more traditional marrieds. For many partners, trailing is preferable to commuting, another solution to the problem of career opportunities in two locations.

To Commute or Not? Social scientists have called marriages in which partners live apart for significant periods of time due to work responsibilities **commuter marriages**. The vast majority of commuting couples would rather not do so, but endure the separation for the sake of career or other goals, such as caregiving for a member of the extended family. Since research began on commuter marriages in the early 1970s, social scientists have drawn different conclusions. Some studies suggest that the benefits of such marriages—greater economic and psychological equality between partners—counter their drawbacks. Other research focuses on difficulties in managing the lifestyle. One conclusion to be drawn is that commuters who are able to have frequent reunions are happier than those who cannot.

Commuter marriages are not new. American men have long worked in transient professions where they are gone for extended periods of time (that is, truck drivers, traveling sales, soldiers, and so on). What is different about today's commuter couples is that more and more wives are commuting. The introduction of children increases the stress in commuter marriages—especially for a partner who is the primary caregiver, because that partner becomes a de facto single parent when the commuting partner is absent.

Commuting has its stressors, but strains may also develop when commuting ends and the household reunites (Tessina 2008). Couples married for shorter periods seem to have more difficulties with commuter marriages. Perhaps because of their history of shared time, more established couples in commuter marriages have a greater commitment to the marriage (Rhodes 2002). And, whether two-earner or two-career, whether commuting or not—how are the children doing?

Two-Earner Families and Children's Well-Being

Before women with children entered the work force in large numbers, mothers' employment outside the home was considered problematic by child development experts and the public. However, a 2001 survey

Issues for Thought

When One Woman's Workplace Is Another's Family

"Every weekday evening in affluent homes across America, two groups of women trade places. Mothers who follow careers come home and the women who are paid to care for their children prepare to depart or step aside" (Rimer 1988). Corporate and professional mothers (and some fathers) change places with less financially fortunate women, many of them immigrants, some undocumented. A Manhattan public relations executive says she is "completely dependent on" her nanny from Trinidad: "We couldn't earn a living without her" (Rimer 1988).

In other scenarios, employed parents come home to houses that were left dirty and are now clean (Ehrenreich 2001). Or to aging or ill family members receiving paid, round-the-clock care (Bufkin 2012).

A minority of paid caregivers immigrate to the United States from poorer countries, work for several years, then return to their home countries with money sufficient to join the middle class there (Barrionuevo 2011). From the employers' perspective, paid caregivers have "tremendous power"—"the tyranny of the nannies," as one executive put it (Rimer 1988). For the employer, the nanny is all powerful: A caregiver's not arriving on time (or at all) is out of her employer's control but will assuredly impact the employer's work day.

Tyranny of the Nannies

Many employers say they worry constantly that their nanny might quit. Or arrive late. Or take time off to care for their own sick child. As one executive mother said, "When child care breaks down, everything else breaks down" (Rimer 1988). Sometimes nanny arrangements do fall apart. Some arrangements last just a few days. Even in ongoing arrangements, things can come up. For instance, a middle-aged Iranian immigrant, having been with an Oakland, California, family for ten years, quit in order to return to Iran and care for her aging mother.

Employers say they anxiously strive to keep their nannies happy: "You think, if the nanny is happy the baby is happy. If the baby's happy, you're happy" (in Rimer 1988). But when upper middle-class women hire poorer women, many of them of different race/ethnic or national background, conflict is possible over different mothering ideals, as discussed in Chapter 9 (Macdonald 2011). What about the professional nutritionist mom whose nanny takes the three-year-old to McDonald's weekly? "You have to constantly make compromises," says one parent-employer. "This is one of the most important relationships I'll have in my entire life. I work at it all the time" (in Rimer 1988). But there's another side to the story as well.

Tyranny of Caregivers' Employers

In-home workers are at the low end of the pay scale, many making minimum wage. Typically they work long hours and may be exploited regarding their days off. "When you travel with them, it's 24 hours," explained a Jamaican nanny in Aspen, Colorado. "Sometimes it's two or three weeks before you get time off. There's no [higher paid] overtime" (Rimer 1988). Undocumented immigrant domestic workers are reluctant to complain because they fear being reported to U.S. immigration (INS) and deported.

Tyranny of the Neighborhood

On mothers' Internet forums, such as UrbanBaby, stay-at-home mothers complain that they dislike having to socialize with employed parents' nannies. Public spaces, like playgrounds, have to be shared. But does a nanny belong at a child's birthday party in a neighborhood home? Many nannies say that, paid to be there, they try to be civil while feeling unwelcome (Bindley 2011).

Sometimes the nannies are the casualties of this conflict. If they can't get along in the neighborhood, they may be let go for another paid caregiver who (hopefully) can (Bindley 2011).

Intersectionality

As pointed out in Chapter 3, contemporary feminism stresses *intersectionality*—structural connections among race, class, and gender (Baca Zinn, Hondagneu-Sotelo, and Messner 2007). "Housework is ascribed on the basis of gender, and it is further divided along class lines and, in most cases, by race and ethnicity. Domestic service accentuates the contradiction of race and class in feminism, with privileged women of one class using the labor of another woman to escape aspects of sexism" (Romero 2002, p. 45).

Solutions

In 2011 New York state passed a Domestic Workers' Bill of Rights that allows disability benefits for full-time home workers and legislates the possibility of compensation for workplace sexual harassment or discrimination. The legislation requires employers to pay time and a half for overtime work, and to provide three vacation days annually as well as twenty-four hours off for every seven days worked. A legal workday is defined as eight hours and a workweek as forty hours—or forty-four for live-in workers (Bufkin 2012).

An advocacy organization, Domestic Workers United, organized home workers and lobbied the state legislature in California to pass a similar bill in 2012. California Governor Jerry Brown vetoed the bill, however, explaining that it would place a "burden on working families, who are struggling, I'm sure, to already afford a nanny" (in Bufkin 2012). Meanwhile, policy activists argue that employers and paid home caregivers need to change the definition of their situations from family-like ("She's like family!")—which encourages the exploitation of caregiving laborers—to legally defined and government regulated tradesperson-client relationships (Romero 2002), similar to New York State's Domestic Workers' Bill of Rights.

Critical Thinking

How does the employer-nanny relationship illustrate *intersectionality*? What might be some aspects of work-family conflict in this relationship that are not discussed here? In your opinion, is regulating the employer-nanny relationship as a tradesperson-client agreement a good solution to the problems raised here?

found more than 90 percent of women agreeing that one can be a good mother and have a successful career (Center for the Advancement of Women 2003). Research supports women's opinions in this regard, concluding that maternal employment does not cause behavior problems or other significant differences in children an(Agee, Atkinson, and Crocker 2008). What *is* notable in the research is the correlation between low family income and childhood problems (Jackson, Choi, and Bentler 2009)—a point noted in several places throughout this text.

The economic benefit to children of working mothers cannot be overlooked. Family income tends to be favorably associated with various child outcomes, as pointed out in Chapters 6 and 9. Important for parents, though, is keeping their child's needs in the forefront in the face of daily pressures. Recent studies have found that mothers who work part-time are better at this than those who work full-time—and may indeed spend more time helping their children with homework than even full-time homemakers.

Before the era of working mothers, mothers not employed outside the home did not spend all their time interacting with their children. They devoted more time than today's mothers to housework (including home decorating), leisure activities, or volunteer work. A variety of studies indicate that parents today spend as much or more time with children as in the past. Figure 10.4 presents the time parents spend caring for their children doing routine care and child enrichment activities. Time spent was measured in time diary studies conducted by various universities using samples ranging from 1,200 to over 5,000. In this data-gathering method, respondents keep daily diaries of time spent in certain activities. Consistent with the finding that younger generations are more family-oriented, mothers and fathers spent more time in child care in 2000 than did previous parents going back to 1965 (Bianchi, Robinson, and Milkie 2006).

How did parents, especially mothers, accomplish an increase in time caring for their children and also increase their employment hours? They cut back on housework and spent less time doing things by themselves or only with their partners or friends. Parents combined time in children's activities and time spent with each other. The "thing about the reallocation of mothers' time to employment is that it appears to have been accomplished with little effect on children's well-being," noted sociologist Suzanne Bianchi in her presidential address to the Population Association of America (Bianchi 2000). These results present an optimistic view of how employed mothers' children are faring. Concerns remain, however. For instance, the schedules of parents and their children

Primary child care, average weekly hours

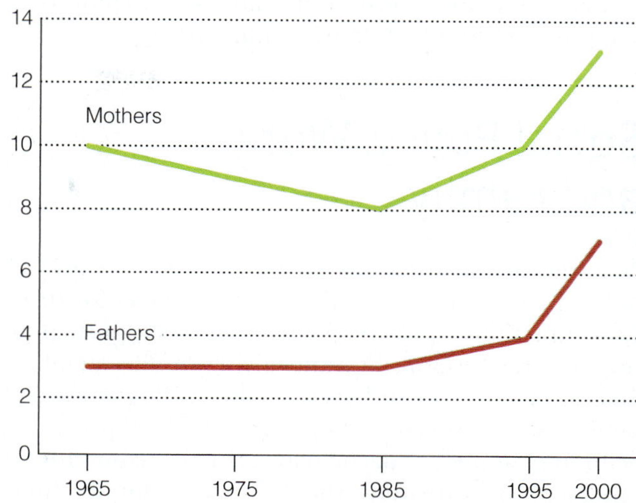

FIGURE 10.4 Average weekly hours that mothers and fathers spend directly caring for their children, 1965–2000.

Source: From Bianchi, Robinson, and Milkie, The Changing Rhythms of American Family Life, 2006, Figure 4.3, p. 72.

© wavebreakmedia/Shutterstock.com

Although increasingly spouses share the provider role today, they actually spend more time with their children than parents did several decades ago when mothers were less likely to be in the labor force. Today's parents, perhaps more family-centered than their grandparents were, spend less leisure time on their own and share more family activities with their children.

have become increasingly frantic with decreasing unstructured or "down" time, an issue explored in Chapter 9 (Chudacoff 2007; Elkind 2010).

Social Policy, Work, and Family

According to the structure–functionalist model, explored in Chapters 1 and 2, families are accountable for three vital functions: providing economic support, raising children responsibly, and giving family members emotional support and security. Regarding work and family issues, social policy asks whether providing economic support (participating in the labor force) enhances or threatens the remaining family functions. Furthermore, this perspective encourages us to ask about functional alternatives to ways that work-family life is structured. Policy issues center on three questions:

1. What are the issues?
2. What is needed to address the issues?
3. Who will provide what's needed to address the issues?

We'll look at these three questions now.

What Are the Issues?

We address three work-family issues in this section—undesirable financial conditions for many Americans, parental stress due to work-family conflict, and persistent gender inequality in today's work-family division of labor.

Inadequate Economic Resources Even with two or more jobs, many Americans do not make enough money to support a family. Low-income and poverty-level parents often work at minimum- or less-than-minimum-wage jobs with unpredictable hours and no benefits. They struggle with housing and transportation instability due to unaffordable rents and utility, gasoline, and car repair costs (Cook et al. 2009).

Not just the poor but also the middle class has suffered as a result of a globalizing economy, the recession that began in 2007, and a current "economic recovery stuck in low gear" (Schwartz 2012; and see Frank 2011; Pew Research Center 2012g). Laid off and still unemployed, some have resorted to temporary ("temp") work. With few if any benefits, temp work is often fraught with the additional stress of landing another placement when the current one ends (Weeks 2010).

Parental Stress Due to Work-Family Conflict Many parents struggle with high stress levels due to work-family conflict and spillover—stress that can damage parents' health and lead to less effective parenting (Gordon et al. 2011; Teachman 2009). Having multiple roles (such as parent, partner, employee, student) does not necessarily increase stress and in fact may enhance personal happiness—provided there's enough time to accomplish everything. However, coupled with financial worries, parents' irregular work schedules in low-wage jobs with little autonomy and high supervisor surveillance increases anxiety and stress (Jacobs and Gerson 2004).

Persistent Gender Inequality Workplace characteristics impact within-family interactions and decision making—and within-family interactions and decision-making impact the workplace. We can think of these reciprocal situations as comprising a **causal feedback loop** (see Figure 10.5). Regarding the work-family interface, economic and workplace expectations, demands, and practices influence decisions made within families, and vice versa. The ideal American career path has required giving priority to one's profession while making family caregiving secondary (Shafer 2011; Slaughter 2012). Meanwhile, women are expected to be the principal parents and family caregivers. Unlike her husband, a wife who is faced with work-family conflict is likely to minimize her work rather than her family role.

When women temporarily leave the labor force to care for children, elderly parents, or other relatives, they forego opportunities for career development and reduce their lifetime earnings (Liu and Hynes 2012; Usdansky 2011). Similarly, when women choose

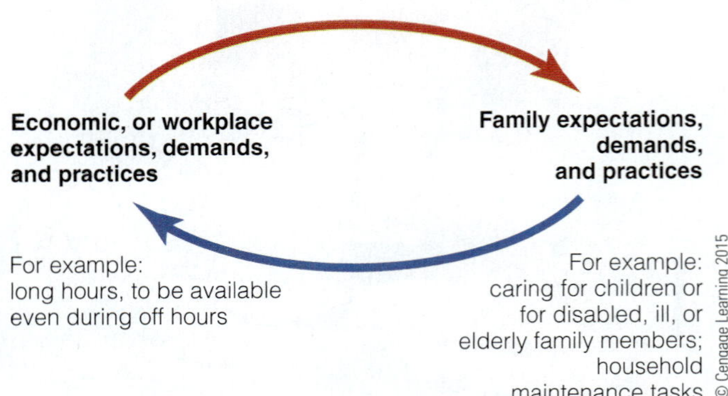

FIGURE 10.5 Work-family causal feedback loop. Workplace expectations, demands, and practices influence family expectations, demands, and practices. Family expectations, demands, and practices, in turn, impact workplace expectations, demands, and practices as the causation continues to loop and feedback.

occupations that are more "family friendly"—shorter or more flexible hours—they choose employment that tends to pay less. As a result, wives risk lower pensions and inadequate support should the marriage break up, and they may not have the careers they might have had otherwise.

Largely because they are paid less on average than their husbands, employed wives are the more likely than husbands to adjust their work schedules to accommodate caregiving (Coltrane 2000). Then, because wives more often than husbands adjust their employment time to accommodate family needs—taking off to care for a sick child, for example—employers feel justified in offering women lower wages and fewer promotions. As a result,

> Women are still paid less than men at every educational level and in every job category. They are less likely than men to hold jobs that offer flexibility or family-friendly benefits. When they become mothers, they face more scrutiny and prejudice on the job than fathers do.
>
> So, especially when women are married to men who work long hours, it often seems to both parties that they have no choice. Female professionals are twice as likely to quit work as other married mothers when their husbands word 50 hours or more a week and more than three times more likely to quit when their husbands work 60 hours or more. (Coontz 2013)

What's Needed to Address the Issues?

Work–family researchers and policy makers are in general agreement that families benefit from(1) adequate wages, (2) quality and affordable elder and child care, (3) family leave, and (4) flexible employment scheduling (Ruppanner 2013).

1. Adequate Wages If the federal minimum wage had kept pace with the rise in executive salaries since 1990, America's poorest paid workers would be making more than $23 an hour (National Employment Law Project, NELP 2013). As he began his first term, President Obama's transition team promised to raise the federal minimum wage from the current $7.25 per hour to $9.50 by 2011. That didn't happen, and in his 2013 State of the Union address President Obama proposed raising the hourly federal minimum wage from $7.25 to $9.00. In his words,

> Today, a full-time worker making the minimum wage earns $14,500 a year.... [A] family that earns the minimum wage still lives below the poverty line. That's wrong. That's why, since the last time this congress raised the minimum wage, nineteen states have chosen to bump theirs even higher. Tonight, let's declare that in the wealthiest nation on Earth, no one who works full-time should have to live in poverty. (President Barack Obama, quoted in Franke-Ruta 2013)

There is evidence from cities and states that have raised the minimum wage above federal mandates that doing so not only improves workers' lives somewhat but also does not cost jobs (Wasikowska 2011). Nevertheless, responding to President Obama's call for a wage hike, Republicans argued that raising the minimum wage would cost many Americans their jobs (Franke-Ruta 2013). Given Congressional deadlock, an increase in the federal minimum wage shouldn't be expected in the near future. A second condition that would benefit many struggling families is quality, affordable child care.

2. Quality, Affordable Elder- and Child Care Many employees have elderly parents in need of their attention and care. Some workers have retired early or just quit to care for parents, whereas others have turned down promotions, switched to part-time work, taken leaves of absence, or simply taken time off from work (Piercy 2007). Many employees expect to provide elder care for five years or more (Council on Contemporary Families 2011). The need for companies to offer employees help with elderly dependents is becoming more recognized, and about 25 percent of U.S. companies now offer elder care benefits as a recruiting tool (McNamara 2009; Woldt 2010). Caring for elderly parents, as well as issues pertaining to the "sandwich generation"—sandwiched between elder- and child care responsibilities—are explored in Chapter 16.

Policy researchers define **child care** as the full-time care and education of children under age 6, care before and after school and during school vacations for older children, and overnight care when employed parents must travel. Child care may be unpaid or paid and provided by parents, relatives, or paid child care workers. Many couples, even two-earner couples, prefer that the mother care for the children. Sometimes influenced by strong conservative religious beliefs, the couple maintains as traditional division of labor as much as possible, with an employed mother working while the children are in school or working nights while the children are sleeping (Hertz 1997).

Tables 10.2 and 10.3 show child care arrangements for young children of employed mothers. As you can see in Table 10.2, 36 percent of children under age 5 whose mothers are employed are cared for by a parent—either mom or dad. Often this arrangement involves a "tag team" approach.

Tag-Team Handoffs "Tag team" is a somewhat whimsical term used when two-earner couples exchange child

Table 10.2 Percent of Employed Mothers' Children Under Age 5 in Various Child Care Arrangements, 2010

Child Care Arrangement	Percent in Arrangement
A parent	36
Grandparent	30
Sibling	3
Other Relative	10
Day care center, nursery school, or Head Start	31
Paid nonrelative in the child's home	5
Family day care—paid daycare in another household	14
Multiple Arrangements	29

Source: U.S. Census Bureau 2011d, Table 1B. Percentages add to more than 100 because some children (29 percent) are in multiple arrangements.

care and work roles daily or even more often than that. The term is also sometimes applied to single mothers who juggle multiple friend and family child care resources in a given day (Gornick, Presser, and Batzdorf 2009). The term suggests the clockwork timing necessary to accomplish exchanges.

Tag team handoffs are tightly scheduled. For example, one parent drives to a predetermined location with the children he's taken care of during the morning hours; he gets out and takes his wife's car to work, while she gets into his car and takes the children home where she will stay with them for the afternoon (Shellenbarger 2009). Other couples, determined to share child care, structure their work to this end. If they can afford it, partners may reduce their paid work hours. In blue-collar and in lower-income families, shift work or periodic male unemployment enables a shared parenting approach.

Child Care by Relatives, Especially Grandparents In another approach, employed parents arrange for relatives, often grandmothers, to care for their children. As Table 10.2 shows, about 30 percent of children under age 5 whose mothers are employed are cared for by grandparents, with another 13 percent cared for by an older sibling or another relative. As depicted in Table 10.3, never-married, Hispanic, Black, and low-income mothers are more likely to ask grandparents and other relatives to help with child care. Despite the prevalence of African-American grandparent caregivers, black mothers—who still rely heavily on kin networks—saw that option decline in the 1990s, as grandmothers themselves were often in the labor force (Brewster and Padavic 2002). As a result, black children are now more likely than Hispanics to be in center care (Capizzano, Adams, and Ost 2006). Chapter 9 addresses grandparents raising their grandchildren; Chapter 16 discusses grandparents helping with child care.

Paid, Nonrelative Child Care Arrangements By the time they enter school, an estimated 44 percent of children have been in a paid, nonrelative child care arrangement. There are essentially three types of paid, nonrelative child care: (1) a paid *in-home caregiver* who

Table 10.3 Percent of Employed Mothers' Children Under Age 5 in Certain Child Care Arrangements, by Characteristics of Mother, 2010

Mother Characteristic	A Parent	Grandparent or Other Relative	Day Care Center	Family Day Care	Multiple Arrangements
Married	39	34	21	8	28
Never Married	29	44	20	8	30
NonHispanic white	37	37	22	9	31
Black	28	42	23	7	20
Asian	38	30	19	6	26
Hispanic*	36	51	15	4	29
High school graduate	36	44	18	4	26
Bachelor's degree or higher	33	32	26	11	29
Less than $1,500 monthly income	33	47	17	4	28
$4,500 and over monthly income	34	37	25	9	29

*Hispanics may be of any race.

Source: U.S. Census Bureau 2011d, Table 1B.

Optimally for families and for the society as a whole, quality daycare would be available to all parents. However, for many American families finding quality daycare is challenging and expensive.

lives in or comes to the house daily; (2) *family child care,* in which care is provided in the caregiver's home, often by a woman who has chosen to remain out of the labor force to care for her own children; and (3) *center care,* which provides group care in child care centers.

Parents who prefer family care seem to be seeking a family-like atmosphere, with a smaller-scale, less routinized setting. They may also desire social similarity of caregiver and parent to better ensure that their children are socialized according to their own values. Center care serves more children per setting, tends to be less expensive, is typically more structured, and can feel more bureaucratized. As indicated in Table 10.3, family day care tends to be relatively expensive and hence used by more highly educated and wealthier parents.

You may want to see "As We Make Choices: Selecting a Child Care Facility—Ten Considerations," in Chapter 9.

Paid Child Care Costs and Concerns As you can see in Table 10.3, paid care is more common for children in higher-income families (U.S. Census Bureau 2011d, Tables1B and 3B). Indeed, paid child care costs are high. On average, married mothers who use paid child care services for children under age 5 spend 22 percent—more than one-fifth—of their income on child care. For never-married single mothers, the comparable figure is 25 percent—one-fourth of a woman's earnings. For mothers with less than a high school education, the comparable statistic is 36 percent—more than one-third of earnings spent on child care. For those living below poverty level, child care costs can run as much as 61 percent—between one-half and two-thirds—of a parent's income (U.S. Census Bureau 2011d, Table 6). Indeed, average monthly child care costs exceed what a low-income family might spend on food.

Other parental concerns involve child care facilities' scheduling that does not meet parents' needs, given today's workforce demands. Most family day care and many child care centers are open weekdays only and close by 7 p.m. Some parents, such as single parents in shift work or those who travel, need access to twenty-four-hour care centers, and some (although not yet enough) are opening (Tavernise 2012). Child care is also difficult to find for mildly ill youngsters too sick to go to their regular day care facility, although there are now centers beginning to fill this need (National Association for Sick Child Day Care 2010).

As they struggle to find quality, affordable child care, many parents make more than one arrangement for each child. A system of patched together child care arrangements becomes increasingly unpredictable and stressful ("Parents and the High Price" 2009). Meanwhile, adequate and affordable child care accommodations are important not only to parents but to the long-term welfare of children.

Self-Care In 2010, of children between ages 5 and 14 whose mothers were employed, nearly one-third

(31.6 percent) were cared for by a parent. Another 17 percent were being watched by a grandparent, 7 percent were in after-school programs, and another 5 percent were in day care centers. In addition, about 10 percent of children ages 9 to 11 and 36 percent of those ages 12 to 14 whose mothers were employed were in **self-care**—that is, without adult supervision—for an average of seven hours per week (U.S. Census Bureau 2011d, Tables 3B, 4).

Self-care is more common in nonHispanic white upper-middle- and middle-class families than in black, Latino, or low-income settings, perhaps due to different cultural attitudes concerning children's capabilities and also due to differences in neighborhood safety (U.S. Census Bureau 2011d, Table 3B). Also, nonHispanic white parents are less inclined to use relative child care arrangements, as shown in Table 10.2. Self-care is also evident in poverty-level families where parents simply cannot afford paid child care arrangements and may not have access to this type of support from family members. Probably due to the high cost of child care, the proportion of school-age children home alone after school has risen since prior to the recession that began in 2007 (McClure 2010).

Compared with those in self-care after school, children with adult supervision are less likely to skip school, use alcohol or marijuana, steal something, or hurt someone (Aizer 2004). Nevertheless, what most children who are alone after school typically do seems innocent enough—eat, get dressed, change clothes, watch TV, study, and play (Swanbrow 2000). And use social media, of course. Sandra Hofferth, a sociologist at the University of Michigan Institute for Social Research has suggested that while unsupervised after-school time creates the potential for trouble, it may have potential positives too. "Spending a little time alone may not be bad, given the hectic, highly scheduled quality of contemporary family life," according to Hofferth. "It gives children a chance to relax..." (Hofferth, in Swanbrow 2000). "As We Make Choices: Self-Care (Home Alone) Kids" gives suggestions for parents considering self-care.

3. Family Leave **Family leave** involves an employee's being able to take an extended period of time from work, either paid or unpaid, for the purpose of caring for their own health needs or for a newborn, a newly adopted or seriously ill child, or for an elderly parent, with the guarantee of a job upon returning. The concept of family leave incorporates maternity, paternity, ill-child, and elder care leaves.

The 1993 U.S. Family and Medical Leave Act mandates up to twelve weeks of unpaid family leave for workers in companies with at least fifty employees. But unpaid leave will not solve the problem for a vast majority of employees, because most working parents need the income. Meanwhile, a majority of the top 100 American companies offer one or more weeks of *paid* maternity leave of between five and eight weeks—although this is seldom available to working-class or minimum-wage workers (Lovell, O'Neill, and Olsen 2007). Not surprisingly, the workers with the highest rates of available paid family leave are in management, professional, and related fields ("National Compensation Survey" 2007, Table 19).

4. Flexible Scheduling In a 2009 poll, nearly half of American men said that companies should provide more flexible work schedules to both men and women (Halpin and Teixeira 2009). **Flexible scheduling** involves such options as **job sharing** (two people share one position), working at home or telecommuting, compressed workweeks, flextime, and personal days (days off for the purpose of attending to a personal matter such as a doctor's appointment or a child's school program). Compressed workweeks allow an employee to concentrate the workweek into three or four or sometimes slightly longer days. **Flextime** involves flexible starting and ending times, with required core hours.

Flexible scheduling, although not a panacea, can help parents share child care or be at home before and after an older child's school hours. In addition, flexible scheduling seems to alleviate negative work-to-family spillover. This finding is particularly true for women, single parents, and employees with more children or perhaps an elderly parent to care for (Jang, Zippay, and Park 2012). Some types of work do not lend themselves to flexible scheduling, but the practice has been adopted by some companies and the federal government because it offers employee-recruiting advantages, reduces turnover, and frees up office space when some employees work at home.

Although telecommuting had been considered the unquestioned way of the future—actually the way of the *present*—and although workers who have flexible hours report enhanced job satisfaction and loyalty to their employer, telecommuting came under corporate scrutiny and renewed national debate when Yahoo changed its policy to prohibit the practice. Several corporate executives outside Yahoo argued that the move was "a backward step in an age when remote working is easier and more effective than ever" (Guynn 2013; Pofeldt 2013). The fact that Marissa Mayer, Yahoo's CEO, is a married mother of a young child was not lost in the debate. "When a working mother is standing behind this [denying employees telecommuting privileges], you know we are a long way from a culture

As We Make Choices

Self-Care (Home Alone) Kids

With parents' difficulties finding available and affordable child care, many children care for themselves after school or at other times. According to the American Academy of Pediatricians, children should not be self in self-care until they are in fourth or fifth grade. Some states have laws regulating the minimum age that a child can be left home alone, and parents considering this option should become familiar with their state's legislation. Here are some things for parents to consider before choosing this option:

1. Are there neighborhood resources offering care or support for children in self-care?
2. How does the child feel about being home alone?
3. Is the child able to follow directions and solve problems on her or his own?
4. How available will a parent or other trusted adult be if needed?
5. How safe is the neighborhood?

Here are some things to consider doing after choosing this option:

1. Set specific rules for the child to follow.
2. Give the child specific instructions about how to reach a parent or other trusted adult at all times.
3. Show the child what to do in certain potential situations, such as if the phone rings or the electricity goes out.
4. Make sure the child knows his or her full name, address, and phone numbers—as well as yours and how to help a third party reach you, if need be.
5. Make sure the child knows how to get help in an emergency, including how to call 911 or other relevant emergency numbers.
6. Show the child appropriate ways to carry a house key so that it is hidden and secure.
7. Provide a daily or weekly schedule of homework, activities, and activities for your child to follow.
8. Have a first aid kit available and teach the child how to use it.
9. Teach the child key safety tips, such as:
   a. Don't walk or play outside alone,
   b. Don't enter the house if there's an open door or window,
   c. Lock the door after entering the house,
   d. Call a parent or parent figure immediately upon getting home to let a responsible adult know,
   e. Always tell callers that a parent is unavailable at the moment and offer to take a message, rather than saying that no one is home,
   f. Don't open the door for anyone the child does not know.
10. Make time to discuss the day's events with your child when you get home.

For more information, visit the National Center for Missing and Exploited Children website.

Critical Thinking

Under what circumstances do you think it is reasonable to leave a child home alone? Under what circumstances would it not be reasonable? How might family finances affect the decision to let a child stay home alone? How might the availability of extended kin affect the decision to let a child stay home alone?

Sources: National Center for Missing and Exploited Children and U.S. Department of Justice. 2013. "Know the Rules ... For Children Who Are Home Alone." Washington, DC: Office of Justice and Delinquency Prevention. Retrieved February 15, 2013 (http://www.missingkids.com).

that will honor the thankless sacrifices that women too often make," read one e-mail sent to a technology blog (Guynn 2013).

Who Will Provide What's Needed to Meet the Challenges?

Policy experts, lawmakers, employers, parents, and citizens disagree over who has the responsibility to provide what is needed regarding work–family solutions. A principal conflict concerns whether such solutions as child care or family leave should be government policy or constitute privileges for which a worker negotiates with an employer.

Even amid global recession, many European countries remain committed to *paid* family leave policies (Human Rights Watch 2011). These nations have a more pronatalist and social-welfare orientation than the United States and view family benefits as a right belonging to all citizens (Lewis 2008). Accustomed to a lack of family policy at the federal level, American parents sometimes turn their attention to local schools as a source of help for the care of older children in after-school programs and younger children in preschool programs and all-day kindergarten (Swanbrow 2000). Evidence that children do better when supervised after school than when left alone has led policy makers to call for expansion of after-school programs (Aizer 2004; McClure 2010).

In the private sector, some large corporations demonstrate interest in effecting **family-friendly workplace policies** that are supportive of employee efforts to combine family and work commitments. Such policies include on-site child care centers, sick-child care, subsidies for child care services or child care locator services, flexible schedules, parental or family leaves, workplace seminars and counseling programs, and support groups for employed parents. These policies may help in recruitment, reduce employee stress and turnover, enhance morale, and thus increase productivity ("Balancing Work and Life" 2009).

But family-friendly policies are hardly available to all American workers (Heymann, Earle, and Hayes 2007). As mentioned above, professionals and managers are much more likely than technical and clerical workers to have access to leave policies, telecommuting, or flexible scheduling ("National Compensation Survey" 2007, Table 19). "At the high end, the big corporations are stepping up to provide benefits to help families, and at the lower end, as women leave welfare, there's now much more support for the idea that they deserve help with child care. But the blue-collar families, the K-Mart cashier, get nothing" (work–family policy expert Kathleen Sylvester, quoted in Lewin 2001).

In another wrinkle, childless workers have begun to argue for a better "work-life balance." Some complain about what they see as the privileging of parents when they themselves may have family caregiving needs: for elderly parents, siblings, or friends with whom they maintain caregiving relationships. Furthermore, they may find it onerous to cover for parent coworkers who are on leave or out of the office. They may feel that fairness should permit schedule flexibility for other than specifically caregiving needs. Some companies have begun to accommodate these workers by redefining policies previously characterized as "family-friendly" and instituting sabbaticals, "flexible culture," and "employee-friendly" policies (Joyce 2006; 2007a).

We would like to think that family- or employee-friendly companies represent the future of work. After all, "children … are 'public goods'; society profits greatly from future generations as stable, well-adjusted adults, as well as future employees and tax payers" (Avellar and Smock 2003, p. 605). Nevertheless, these voluntary programs and benefits do depend on cost constraints and corporate self-interest and are likely to be less available during economic downturns or restructuring. Although family-friendly programs need to be more widely available to all workers, the family life course theoretical framework, described in Chapter 2, directs us to keep in mind that most workers need extensive family support only while parenting young children. From that perspective, the challenge looks less daunting.

In general, multiple roles can be negatively conflicting or positively synergistic, complementing each other. Work and family roles can conflict—when work demands are unreasonable, when child care is difficult to organize or afford, or when family demands are too high as when one is sandwiched between caring for children and also for aging parents. Work and family roles can also be (positively) synergistic—when money earned enhances family life or when what happens at work is uplifting or enhances self-esteem and good feelings spill over into family interactions.

Entering the political arena to work toward the kinds of changes families want is one aspect of creating satisfying marriages and families. We need to focus on "the ways that macrolevel discourses convey the priorities of the powerful from top down and become objects of struggle from the bottom up" (Ferree 2010, p. 433). We have historical evidence that social activism, coupled with government intervention in the service of families, can be effective. European countries evidence ways to successfully alleviate work-family conflict. In the United States, the War on Poverty of the 1960s, which created Head Start programs, among other things, successfully reduced poverty levels (DeNavas-Walt, Proctor, and Smith 2012, Figure 5). But employed couples also want to know what *they* can do themselves to maintain happy marriages. We now turn to that topic.

The Two-Earner Couple's Relationship

Two-earner couples can have heightened satisfaction, excitement, and vitality because partners are more likely to have common experiences and shared worldviews than other couples where the two may lead very different everyday lives. At the same time, conflict may arise as they negotiate the division of household labor.

How a couple allocates paid and unpaid family labor and justifies that allocation has been termed a **gender strategy**, a way of working through everyday situations that takes into account an individual's beliefs and deep feelings about gender roles, as well as her or his employment commitments (Hochschild 1989). Sometimes a couple's gender strategies please both individuals; sometimes they don't. Moreover, even when partners share similar gender attitudes, circumstances may not allow them to act accordingly. In one couple interviewed by Arlie Hochschild (1989), both partners held the traditional belief that a wife should be a full-time homemaker. Yet because the couple needed the wife's income, she was employed and they shared housework on a nearly equal basis.

DEBBIE NODA/MCT/Landov

After a long day on the job, Cabral and Denys get some sleep on the seventeen-mile shuttle bus trip from the plant to Moline, Illinois, where they live. One effect of the trend toward two-earner families is increasing inequality between families with high-status, high-paying *careers* and those with two poorly paid jobs (Paul 2006). The tendency of college-educated professionals to marry other college-educated professions enforces this difference (Usdansky 2011). However, for two-earner couples of all social classes, longer workdays can mean lack of sleep.

Policy analyst Margaret Usdansky has noted a paradoxical incongruity regarding social class and the work-family interface. Many spouses at the upper end of the social-class continuum desire to share paid and unpaid labor equally, whereas less educated couples with less money may prefer traditionally specialized breadwinner-homemaker roles. But the behavior of many couples does not fully align with their values or desires (Usdansky 2011, p. 163). Finding it difficult to manage two careers, housecare, and child care, middle-class couples may decide that the less well paid partner (usually the wife) will leave the labor force, or reduce her hours to part-time: "So I decided to quit, and this was a really, really big deal ... because I never envisioned myself not working" (a female marketing executive in Usdansky 2011, p. 163). On the other hand, many working-class wives, who may not have wanted to, have joined the labor force, either because the family needs more income than what the husband alone can provide or because the husband has been laid off or cannot find work (Usdansky 2011).

Relatedly, human development scholar Jing Zhang and colleagues investigated what they term "gender role disruption"—conflict between ideas about how gender should be performed and how it is actually enacted. This research was conducted among forty university student couples, about half of them with children. They had come to the United States from China for the husband's further education. In China, wives expect to be employed outside the home and to earn wages equal to their husbands'. After giving up their employment to follow their student husband to the United States, they found themselves unable to speak English, unemployed, at home all day, and feeling "useless." Their situations and self-esteem were hardly improved by their husbands' culturally conditioned negative attitudes about stay-at-home wives (Zhang et al. 2011).

Fairness and Couple Happiness

Two kinds of changes are involved today's developing work-family interface: Women are increasingly sharing the provider role, and men take more responsibility for unpaid household work. In considering the provider role, we turn to the notion of *meaning* again: Is women's sharing of the provider role a *threat,* so that men fear losing masculine identity, women's domestic services, and power? Or is a woman's sharing the provider role a *benefit,* because men benefit materially from wives' employment and earnings and from a partner's enthusiasm for the wider world?

Many traditional men view a wife's earning more than they do as a threat. Nontraditional husbands are likely to see this situation as a non-issue (Coughlin and Wade 2012). Generally, whether partners think their work-family arrangement is fair is strongly associated with the couple's happiness (Frisco and Williams 2003; Wilcox et al. 2011a). For instance, in households where the husband is more religious than the wife, particularly in Islam and evangelical Protestantism, strict, traditionally gendered divisions of labor can lead to family tension and decreased marital satisfaction, particularly for the wife (Duba and Watts 2009; Vaaler, Ellison, and Powers 2009; Zink 2008). For the most part, however, social scientists see a "gender convergence" in current attitudes regarding work and family roles. Both men and women want a balance of work and family in their lives (Cohen 2007b, p. A13; Monahan Lang, and Risman 2007). Furthermore, to a wife or female partner, a man's taking up *some,* if not an absolutely equal share, of household tasks may signify caring.

Then too, one's "fair share" of household labor may depend on gendered expectations or on hours spent in the labor force (Konigsberg 2011). Men are doing more than they did in the past, and women are unloading some of their former responsibilities. Today we see relatively comparable men's and women's total hours of paid and unpaid household labor (Bianchi, Robinson, and Milkie 2006). Partners may mentally combine housework and employment to conclude that the total burden of family responsibility is fair (Lavee and Katz 2002). Both husbands and wives are significantly more likely to say that they are "very happy" when both feel that the housework is shared equitably (Wilcox et al. 2011a, p. 38; see also Hall and MacDermid 2009).

Husbands may now carry a greater share of the family work than in the past, and when the transition proceeds from a desire to achieve a more equitable relationship, the result can be greater intimacy. Chapter 11 examines family communication skills.

Summary

- Traditionally, the husband's role was as provider, the wife's as homemaker. These roles have changed and continue to change as more and more women have entered the workforce.
- Nevertheless, women remain segregated occupationally, continue to average lower incomes than men, and do more unpaid household labor than husbands do.
- Paid work is not often structured to allow time for household responsibilities and women, more than men, continue to adjust their time to accomplish both paid and unpaid work.
- In recent years, men have been increasing their share of the housework, and men and women now have a balance in total work hours—that is, time spent in employment plus time spent on unpaid family labor.
- Household work and child care are pressure points as women enter the labor force and the two-earner marriage becomes the norm. To make it work, either the structure of work must be changed, social policy must support working families, or women and men must change their household role patterns—very probably all three.
- Regarding the work-family interface, a causal feedback loop operates, according to which workplace policies affect within-family decision-making. For example, lower average pay for women may lead to deciding that the (lower paid) wife will reduce employment hours when faced with work-family conflict.
- Cultural expectations and public policy affect people's options. Provided that individuals increasingly come to realize this, we can hope for pressure on public officials and corporations to meet the needs of working families by providing supportive policies: parental leave, child care, and flextime.
- To be successful, two-earner marriages will require social policy support and workplace flexibility.

Questions for Review and Reflection

1. Discuss to what extent distinctions between husbands' and wives' work are disappearing.
2. What do you see as the advantages and disadvantages of men being househusbands? Discuss this from the points of view of both men and women.
3. What are some advantages and disadvantages of home-based work?
4. What work–family conflicts do you see around you? Interview some married or single-parent friends of yours for concrete examples and for some suggestions for resolving such conflicts.
5. **Policy Question.** What family-friendly workplace policies would you like to see instituted? Which would you be likely to take advantage of?

Key Terms

causal feedback loop 266
child care 267
commuter marriage 263
family-friendly workplace policies 272
family leave 270
flexible scheduling 270
flextime 270
gender strategy 272
good provider role 253
involved fathers 253
job sharing 270
kin keeping 260
labor force 250
motherhood penalty 256
neotraditional families 257
occupational segregation 255
provider role 252
role conflict 251
second shift 262
self-care 270
shift work 258
spillover 251
stay-at-home dad 254
trailing partner 263
two-career relationship 258
two-earner unions 257
unpaid family work 259
wage gap 256

11

Communication in Relationships, Marriages, and Families

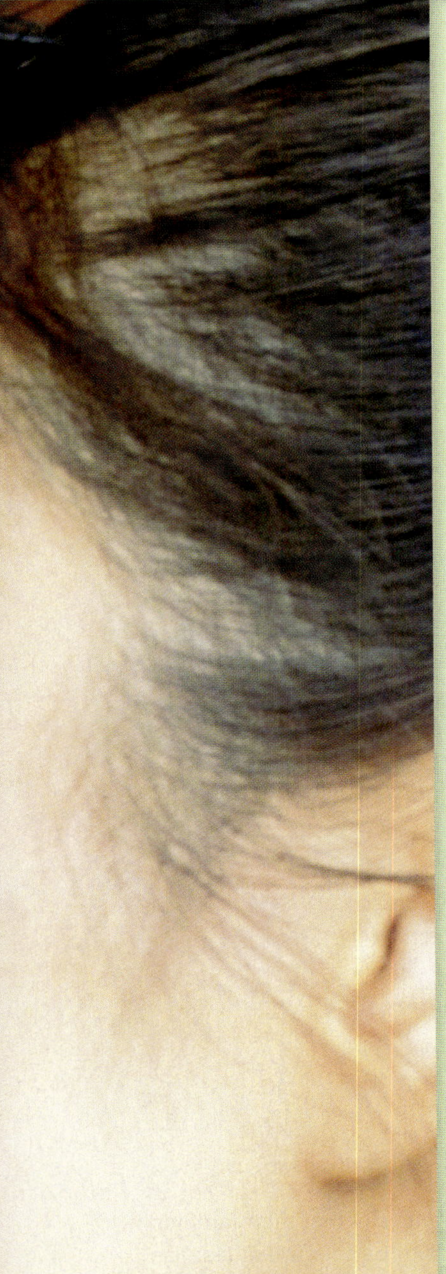
Ron Chapple/Taxi/Getty Images

Learning Objectives

1. Define and list six characteristics of *cohesive families*.
2. Explain how the macro-environment (family ecology theoretical perspective) joins with the micro family communication perspective to help create cohesive families.
3. Describe the Four Horsemen of the Apocalypse and explain why they should be avoided.
4. Discuss gender differences in couple and family communication.
5. Give ten guidelines for effective interpersonal communication.
6. Describe the myth of conflict-free conflict.
7. Explain the dyadic approach to studying family cohesiveness and communication.

We all want good relationships, ones that we can trust to continue in supportive ways (Gottman 2011). Young adults hope to develop loving, ongoing relationships. Middle-aged family members hope to maintain positive relations while faced with the often difficult challenges of putting food on the table or getting their children through school. Older adults focus on maintaining already established bonds with partners, friends, or other family members (Villar and Villamizar 2012).

Families are powerful environments. Virtually nowhere else in our society is there such capacity to support, hurt, comfort, denigrate, reassure, ridicule, hate, and love. Research from a variety of samples and pertaining to a variety of family situations overwhelmingly supports what may be intuitively obvious: Conveying affection for one's partner and other family members is a very important determinant of relationship and family happiness, as well as of each family member's emotional well-being (Fagan 2009; Soliz, Thorson, and Rittenour 2009). And although conflict is a natural part of every relationship, developing positive communication skills can help family members to resolve conflicts in constructive ways.

Some scholars have criticized family communication research. They argue that most family communication studies are designed as if only the couple or family members are involved. However, we need to look at the couple's or family's social environment as well (Trail and Karney 2012). Incorporating the family ecology model (see Chapter 2), some communication researchers have begun to take a "dyadic approach" in which "elements of the macroenvironment, such as [economic and] cultural background [are recognized as interacting] directly and indirectly with spouses' individual characteristics and marital behavior" (Helms, Supple, and Proulx 2011; and see Trail and Karney 2012). We return to this point at the end of this chapter.

This chapter emphasizes the importance of communicating affection as well as addressing conflict in positive ways. We'll examine connections between communication and relationship satisfaction. We'll discuss gender differences regarding couple and family communication. We will review ten guidelines suggested for addressing family and couple conflicts. To begin, we'll look at characteristics of cohesive families.

Characteristics of Cohesive Families

As depicted in Figure 11.1, many Americans believe that their marriage and family relationships are closer today than in the past. Fifty-one percent of marrieds feel that their spousal relationship is closer than their parents' was, and 40 percent say that the families they have today feel closer than their childhood families did (Pew Research Center 2010a). **Family cohesion**, **closeness**, or togetherness, is defined as "the emotional bonding that couples and family members have toward one another" (Olson and Gorall 2003, p. 516). A couple or family can have too much cohesion (an *enmeshed* couple or family) or too little (a *disengaged* or *disconnected* couple or family).

Experts advise a *balanced level of cohesion*—one that combines a reasonable and mutually satisfying degree of emotional bonding with individual family members'

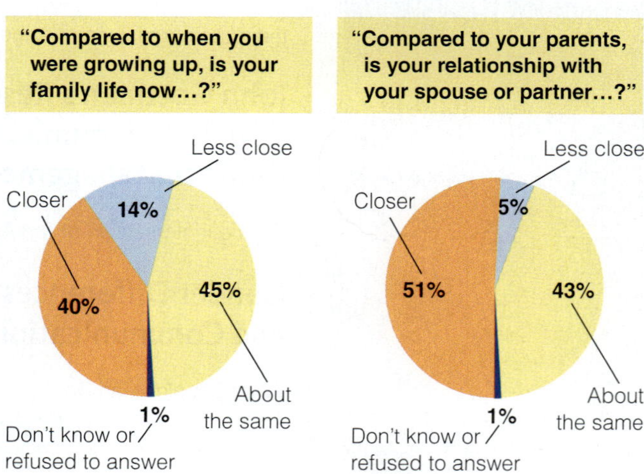

Figure 11.1 Survey respondents say that their families and relationships with a spouse or partner are at least as satisfying as the families they experienced as children. Forty percent say that their families are closer now than when they were growing up. Fifty-one percent of those who are in a relationship say that their relationship is closer compared to their parents'.

Source: Pew Research Center, 2010a. p. 5.

Pew Research Center. 2010a. "The Decline of Marriage and Rise of New Families." *Pew Social & Demographic Trends*. November 18, p. 5.

need for autonomy. In this chapter, we will use the term *family cohesion* to refer to a balanced degree of cohesion—neither enmeshed nor disengaged. Before going further, we should recognize that for different families—and for families of different race/ethnicities—the definition of *balance* with regard to family cohesion varies. "If a couple's/family's expectations or subcultural group norms support more extreme [cohesion levels], families can function well as long as all family members desire the family to function [at that level]" (Olson and Gorall 2003, p. 522). For instance, in Mexican American families, a relatively high level of family cohesion has been related to positive outcomes for adolescents (Behnke et al. 2008; Martyn et al. 2009).

To find out what makes families cohesive, social scientist Nick Stinnett researched 130 "strong families" in rural and urban areas throughout Oklahoma (Stinnett 1985, 2008). Obviously, this limited sample, selected with help from home economics extension agents, has no claim to representativeness. Furthermore, the concept *strong family* is subjective. Various individuals or groups have their own ideas about just what a strong family is. But Stinnett's research helped to advance ideas about what makes for couple and/or family cohesion. In general, Stinnett's families constructed their lives in ways that enhance family relationships. Instead of drifting into relationship habits by default, they made knowledgeable choices, each member playing an active part in carrying out family commitments. When Stinnett made his observations, the following six qualities stood out. More recent research has corroborated Stinnett's findings and is included in the following discussion. Stinnett's observations:

1. Both verbally and nonverbally, family members often openly expressed their *appreciation for one another.* Intent on finding the best in one another, they "built each other up psychologically" (Stinnett 2008; and see Niehuis et al. 2011; Samek and Reuter 2011).
2. Members of cohesive families had a *high degree of commitment* to the family group as a whole (Stinnett 2008). Families like this create a shared family identity and reality (Offer 2013; Rueter and Koerner 2008). From a symbolic interactionist perspective, families create a shared reality through frequent, spontaneous, and unconstrained conversations that allow family members to participate together in defining family beliefs, values, situations, events, and rituals. Social media and video Internet communication technologies, such as Skype, allow for developing or ongoing family connectedness across distances, even across continents (Furukawa and Driessnack 2013).
3. Dinner time together is important to family cohesiveness (and to positive adolescent behaviors) (Musick and Meier 2012). Leisure time together is important for family members (Clayton and Perry-Jenkins 2008; Smith, Freeman, and Zabriskie 2009). Stinnett found that, on a regular basis, family members *arranged their personal schedules* so that they could do things together. They invested in their family. When life got so hectic that members didn't have enough time for their families, they listed the activities they were involved in, found those that weren't worth their time, and scratched them off their lists (Stinnett 2008). In a college-student sample, those who reported more shared time with their grandparents had more satisfying relationships with them (Mansson, Myers, and Turner 2010).
4. Stinnett found that strong families were able to *deal positively with crises.* Family members were able to see something good in bad situations, even if it was just gratitude that they had each other and could face the crisis together (Stinnett 2008). Chapter 13 addresses dealing creatively with stress and crises.
5. Many of the families that Stinnett studied had a *spiritual orientation.* Although they were not necessarily members of any organized religion, some of these families had a sense that their union and family had some "sanctified" purpose and power greater than themselves (Ellison et al. 2011). Typically these families had a "hopeful attitude toward life" (DeFrain 2002).
6. These families had positive communication patterns. Members of families like these talk with and listen to one another, conveying respect and interest (Schrodt 2009). They confirm, validate, and accept one another (Dailey 2009). "Issues for Thought: A Postdivorce Family Communicates As a Child-Raising Institution" describes one family's efforts to improve their communication patterns.

In addition to the previously mentioned characteristics, cohesive families and supportive couple relationships involve fairness, prudence, humility, tolerance, gratitude, justice, charity, and forgiveness (Day and Acock 2013; Fincham, Stanley, and Beach 2007). How do children benefit from family cohesiveness?

Children, Family Cohesion, and Unresolved Conflict

Regardless of family structure, a family characterized by warmth, cohesion, and generally supportive communication is better for children (Froyen et al. 2013; Lindsey et al. 2009; Matiasko, Grunden, and Ernst 2007; Offer 2013). If a child has one or more siblings, feeling close to them is positively associated with healthy development (Samek and Reuter 2011). Parental values are more readily passed on to children when the family atmosphere is generally cohesive (Roest, Dubas, and Gerris 2009).

Issues for Thought

A Postdivorce Family Communicates As a Child-Raising Institution

Jo Ann is 38 and has been divorced for six years. She has five children. Gary, nineteen, her oldest, lives with his father, Richard. At the time of this interview, Jo Ann and Richard and their children have recently begun family counseling. The purpose, Jo Ann explained, is to create a more cooperative and supportive atmosphere for their children. Jo Ann and Richard do not want to revisit their spousal relationship, but they and their children are still in many ways a family fulfilling traditional family functions and working to improve communication patterns. In Jo Ann's words:

We've been going to family counseling about twice a month now. The whole family goes—all five kids, Richard, and me. The counselor wants to have a videotaping session. He says it would help us gain insights into how we act together. The two older girls don't want any part of it, but the rest of us decided it might be really good for Joey to see how he acts. [Joey's] the reason we're going in the first place. At the first counseling sessions, he sat with his coat over his head....

Joey's always been a problem. He's used to getting his own way. Some people want all the attention. They will do anything to get it. I guess I never knew how to deal with this.... He drives us nuts at home. He calls me and the girls names.... He was disrupting class and yelling at the teacher. And finally they expelled him.... Joey gets anger and frustration built up in him.

So I took him to a psychologist, and the psychologist said he'd like the whole family to come in, including Richard. Well, Richard still lives in this city and sees all the kids, so I asked him about it. And he said okay....

In between sessions, the counselor wants us to have family conferences with the seven of us together. One day I called Richard and asked him over for supper. In the back of my mind I thought maybe we could get this family conferencing started.

Well, after dinner Joey, our eleven-year-old, started acting up. So I went for a walk with him. We must have walked a mile and a half, and Joey was angry the whole time. He told me I never listen to him; I never spend time with him. Then he started telling me about how he was mad at his dad because his dad won't listen to him.

He said his dad tells all these dumb jokes that are just so old, but he just keeps telling them and telling them. So when we got home, I saw Richard was still there, and I asked Joey, "Would you like to have a family conference? Maybe tell your dad some of the things that are bothering you?" And he said, "Could we?"... [During the conference] Joey talked first. Then everybody had a chance to say something. There was one time I was afraid it was going to get out of hand. Everybody was interrupting everybody else. But the counselor had told me you have to set up ground rules. This is where we learned that we got some neat kids because when I said "Let somebody else talk," everybody did! So it went real well.... And then finally Richard said, "I think it's time for us to come to a conclusion." I said, "Well, you're right."

Critical Thinking

How do communication patterns in this family influence the functions that family members are performing? We usually think of family cohesion in terms of intact families, but how do communication patterns in this family affect this family's cohesion?

Conversely, a home characterized by significant, unresolved, and ongoing conflict negatively impacts children (Schoppe-Sullivan, Schermerhorn, and Cummings 2007). However, contrary to the idea that all couple conflicts in the home are necessarily detrimental to children, conflicts can end in constructive ways from the children's perspective. "[C]hildren [can] feel an increased sense of well-being resulting from the confidence of knowing that although their parents disagree, their relationship is safe and will endure" (Goeke-Morey, Cummings, and Papp 2007, p. 751).

However, a climate of *unresolved* marital conflict, especially when accompanied by parental depression, which it often is, correlates with children's emotional insecurity (Kouros, Merrilees, and Cummings 2008). A Hong Kong study of children's responses to ongoing parental conflict found that the children felt anxious over the future of their parents' relationship as well as feeling that they had to mediate the conflict (Lee et al. 2010).

Research shows a link between unresolved parental conflict and children's behavior problems (Feinberg, Kan, and Hetherington 2007; Teachman 2009). One study (Buehler et al. 1998) sampled 337 sixth through eighth graders. Three-quarters were nonHispanic white, 12 percent were Hispanic, and 13 percent represented other race/ethnic groups. The parents of 87 percent of the children in the sample were married. The parents' average education level was somewhere between high school graduate and some college.

The students were asked to fill out questionnaires that assessed their behavior and any conflict between their parents. *Externalizing behavior problems* (associated with "acting out," aggression toward others, or rule breaking) were measured by students' agreeing or disagreeing with statements such as "I cheat a lot"

Blend Images/Alamy

Among other things, cohesive families have high levels of commitment and positive communication patterns. Making time to be together, they build one another up psychologically.

showed that the interparental conflict influenced a child's behavior *indirectly:* Marital discord negatively affected parental discipline and the parent-child relationship more generally. This situation then negatively affected the child's behavior.

Interestingly, a study from the biosocial theoretical perspective (see Chapter 2) sampled mothers and found that a mother's high level of cortisol, a chemical associated with being stressed, "spilled over" after having been secreted into her bloodstream during parental conflict to subsequently have a negative effect on her parenting behaviors (Sturge-Apple et al. 2009).

or "I tease others a lot." *Internalizing behavior problems* (those associated with emotional or psychological problems) were measured by students' agreeing or disagreeing with statements such as "I am unhappy a lot" or "I worry a lot." The children were also asked about their parents' conflict. Negative *overt parental conflict styles* involved such things as the parents' calling each other names, telling each other to shut up, or threatening each other in front of the child. Negative *covert parental conflict styles* included such things as trying to get the child to side with one parent and asking the child to relay a message from one parent to the other because the parents refused to speak to each other.

The researchers found that conflict between parents was not the only cause of children's behavior problems. Nevertheless, for both girls and boys, a strong correlation existed between interparental conflict and behavior problems. When parents used an overtly negative style, the youth were more likely to report externalizing behavior problems. When parents used a covert, negative style, the youth were more likely to report internalizing behavior problems.

In another study of fifty-five NonHispanic white middle- and upper-middle-class five-year-olds and their married mothers, the mothers completed questionnaires on parent-child relations and interparental arguing that were mailed to them at home. Later, the mothers took their children to be observed in a university laboratory setting. The researchers found that marital discord was positively related to children's externalizing and internalizing behavior problems. However, this research also

Researchers conclude that "If parents are able to maintain good relations with children in the face of marital conflict, the children may be buffered from the potential emotional fallout of the conflict" (Harrist and Ainslie 1998, p. 156; see also Lindsey, Caldera, and Tankersley 2009). Recent research has investigated the effects of parental and general family conflict on sibling relationships.

Unresolved Family Conflict and Sibling Relationships

Close sibling relationships can provide helpful emotional and practical support over the life course. Cohesive families foster sibling closeness (Samek and Reuter 2011). But is there another aspect of this story?

A study of mothers, fathers, and adolescents from 200 middle- and working-class, mostly European American families found parental conflict to be causally associated with parents' differential treatment of their children. A possible reason is because ongoing, unresolved parental conflict encourages a parent to form a parent-child alliance with one sibling—a situation that leaves other siblings out. Perceived differential treatment among siblings leads to sibling conflict and underlying resentments that can linger into adulthood (Kan, McHale, and Crouter 2008).

Another study looked at a racially and ethnically diverse sample of elderly mothers with at least two adult children and found that a mother's perceived favoritism while the children were growing up reduced siblings' closeness in adulthood. "Further, mothers' favoritism appeared to reduce closeness regardless of which child was favored, suggesting that siblings' relationships are

As We Make Choices

Communicating with Children—How to Talk So Kids Will Listen and Listen So Kids Will Talk

There are more and less effective ways to communicate with children, and a knowledgeable choice would involve more effective ways. What are some of these methods?

Helping Children Deal with Their Feelings

Children—including adult children—need to have their feelings accepted and respected.

1. *You can listen quietly and attentively.*
2. *You can acknowledge their feelings with a word.* "Oh ... mmm ... I see ...."
3. *You can give the feeling a name.* "That sounds frustrating!"
4. *You can note that all feelings are accepted, but certain actions must be limited.* "I can see how angry you are at your brother. Tell him what you want with words, not fists."

Engaging a Child's Cooperation

1. *Describe what you see, or describe the problem.* "There's a wet towel on the bed."
2. *Give information.* "The towel is getting my blanket wet."
3. *Describe what you feel.* "I don't like sleeping in a wet bed!"
4. *Write a note* (above towel rack): Please put me back so I can dry. Thanks! Your Towel

Instead of Punishment

1. *Express your feelings strongly—without attacking character.* "I'm furious that my saw was left outside to rust in the rain!"
2. *State your expectations.* "I expect my tools to be returned after they've been borrowed."
3. *Show the child how to make amends.* "What this saw needs now is a little steel wool and a lot of elbow grease."
4. *Give the child a choice.* "You can borrow my tools and return them, or you can give up the privilege of using them. You decide."

Encouraging Autonomy

1. *Let children make choices.* "Are you in the mood for your gray pants today or your red pants?"
2. *Show respect for a child's struggle.* "A jar can be hard to open."
3. *Don't ask too many questions.* "Glad to see you. Welcome home."
4. *Don't rush to answer questions.* "That's an interesting question. What do you think?"
5. *Encourage children to use sources outside the home.* "Maybe the pet shop owner would have a suggestion."
6. *Don't take away hope.* "So you're thinking of trying out for the play! That should be an experience."

Praise and Self-Esteem

Instead of evaluating, describe.

1. *Describe what you see.* "I see your car parked exactly where we agreed it would be."
2. *Describe what you feel.* "It's a pleasure to walk into this room!"
3. *Sum up the child's praiseworthy behavior with a word.* "You sorted out your pencils, crayons, and pens, and put them in separate boxes. That's what I call *organization!*"

Freeing Children from Playing Roles

1. *Look for opportunities to show the child a new picture of himself or herself.* "You've had that toy since you were three, and it looks almost like new!"
2. *Put children in situations in which they can see themselves differently.* "Sara, would you take the screwdriver and tighten the pulls on these drawers?"
3. *Let children overhear you say something positive about them.* "He held his arm steady even though the shot hurt."
4. *Model the behavior you'd like to see.* "It's hard to lose, but I'll try to be a sport about it. Congratulations!"
5. *Be a storehouse for your child's special moments.* "I remember the time you ...."

Critical Thinking

What bit of advice given here might you choose to practice when communicating with the child(ren) in your life? Why is it important to encourage children to talk? Why is it important to listen to children? Why does how we talk to children matter?

Source: Excerpts from Rawson Associates/Scribner, an imprint of Simon & Schuster, from *How to Talk So Kids Will Listen and Listen So Kids Will Talk*, by Adele Faber and Elaine Mazlish. Copyright © 1980 by Adele Faber and Elaine Mazlish. Also see Faber and Mazlish (2006) as well as the Faber Mazlish website and the Mental Health America website.

shaped ... by principles of equity" (Suitor et al. 2009, p. 1032). According to linguist Deborah Tannen (2006), many grown daughters continue to feel rejected because their mothers persist in showing preference to their brothers.

However, a different study found that, in childhood and adolescence, the sibling who felt slighted was likely to be depressed (Shanahan et al. 2008). Meanwhile a study of 246 two-parent Mexican American families found that adolescents in families with a solution-oriented conflict management style, rather than ongoing unresolved conflict, had better sibling relationships (Killoren, Thayer, and Updegraff 2008).

Of course, children's behavior also depends on how the parents communicate with the children themselves, even when family conflict exists (Schrodt et al. 2009). "As We Make Choices: Communicating with Children—How to Talk So Kids Will Listen and Listen So Kids Will Talk" describes some effective ways to communicate with children.

By now you may have surmised that in families headed by couples, family communication tends to be influenced by the degree of supportiveness or negativity in the couple relationship itself (Doohan et al. 2009). Distressed couples tend toward negative exchanges that put family relationships on a downward spiral (Driver and Gottman 2004). Partners who communicate mutual affection create a positive "spiraling effect" so that the atmosphere becomes one of emotional support (White 1999). Meanwhile, as we'll see in the following section, not all supportive couples and families are alike.

Communication and Couple Satisfaction

Couples demonstrate different **relationship ideologies**—expectations for closeness and/or distance as well as ideas about how partners should play their roles. A pivotal task for all couples is to balance each partner's need for autonomy with the simultaneous need for intimacy, togetherness, and support (Brock and Lawrence 2009; Lavy et al. 2009).

Couples also differ in their attitudes toward conflict. Some expect to engage in conflict only over big issues. Others argue more often. Still others expect a relationship that largely avoids not only conflict but also demonstrations of affection (Fitzpatrick 1995). All of these couple types can be happy with their relationship. What matters is whether the partners' actual interaction matches their ideology.

Meanwhile, unhappy relationships have some common features: less positive and more negative verbal and nonverbal communication (for instance, refusing to smile or glaring at the partner during conflict), together with more reciprocity of negative—but not of positive—messages (Gottman and Levenson 2000; Noller and Fitzpatrick 1991; and see Patterson et al. 2012).

Having gathered data on married couples, researchers Ted Huston and Heidi Melz (2004) classified relationships into four types: *warm*, or friendly; *tempestuous*, or stormy; *bland*, or empty shell; and *hostile*, or distressed (p. 951). This research by Huston and Metz (2004) studied heterosexual, married couples only. The extent to which their findings and conclusions apply to otherwise committed couples, such as cohabiting or same-sex partners, is unknown. Increasingly, researchers are making the point that correlates of relationship satisfaction need to be studied among other than heterosexually married couples, and researchers are beginning to do this (e.g., Lincoln, Taylor, and Jackson 2008).

Warm relationships are high on showing signs of love and affection while low on antagonism. Tempestuous unions are high on both affection and antagonism. Bland marriages are low on showing signs of affection as well as on antagonism. Hostile marriages are low on love and affection but high on antagonism.

We can assume that warm and friendly relationships best fill the family function of providing emotional security. We can also conclude that hostile ones are undesirable. Huston and Melz called both bland and tempestuous unions "mixed blessing" relationships because these two types evidenced only one of two desirable attributes. Although bland relationships have little antagonism, they lack displays of affection. And although tempestuous couples intermittently show affection, they deal with conflicts in aggressive, or antagonistic, ways.

Regarding married couples, we note that some maintain flirtatious and playful communication patterns throughout their marriage (Frisby and Booth-Butterfield 2012). However, a more common finding is that after the honeymoon stage, there is a "coming down to earth" stage in a marriage (Huston and

Tetra Images/Getty Images

An important characteristic of happy couples involves disclosure of feelings and showing affection for one another. Meanwhile, even the happiest couples experience conflicts. Whether and how a couple resolves interpersonal conflicts creates a "spiraling effect" that positively or negatively influences communication throughout the family.

Melz 2004). Interestingly, however, in the years after "the honeymoon's over," couples did not necessarily argue more. Instead, their marriages showed a decline in signs of love and affection. "One year into marriage, the average spouse says 'I love you,' hugs and kisses their partner, makes their partner laugh, and has sexual intercourse about half as often as when they were newly wed." Although marriages do not necessarily "become more antagonistic as time passes, the unpleasant exchanges that *do* occur are embedded in a less affectionate context, and thus, the spouses are likely to come to feel that their marriage is less of 'a haven in a heartless world'" (p. 951).

> If our goal is to identify the early signs of a marital rupture, our research suggests that we look to the loss of love and affection early in marriage as symptomatic.... This loss of good feelings, rather than the emergence of conflict early in marriage, seems to be what sends relationships into a downward spiral, no doubt eventually leading to increased bickering and fighting and, ultimately, to the collapse of the union. (Huston and Melz 2004, pp. 951–52)

This situation helps to explain the general finding, discussed in Chapter 7, that for many couples the early years of marriage are the happiest. Of course, partners can change this by making informed decisions (see Chapter 1) about communicating intimacy.

A recently completed ten-year longitudinal study of newly married couples shows that early support in a young couple's life enhanced their marital satisfaction and helped reduce conflict and marital dissolution (Verhofstadt, Ickes, and Buysse 2010; Sullivan et al. 2010). The researchers noted that "how spouses respond to one another's everyday disclosures and requests for support may be more consequential than how they negotiate their differences of opinion in producing behavioral changes that foreshadow later marital satisfaction and stability" (Sullivan et al. 2010, p. 640).

This research found that the types of support needed for long-term union satisfaction involved asking for and offering validation of feelings, and asking for and offering understanding and compassion rather than anger and contempt when disagreements arise (Sullivan et al. 2010)—or when roles change. As an example, traditional role expectations—although relatively rigid and limiting—can be a way of expressing love and caring. When a wife cooks her husband's favorite meal or a husband can pay for family travel, each feels cared about. Many people have noted the potential of shared work and of shared provider and caregiving roles for enriching a marriage (Galinsky, Aumann, and Bond 2009; van der Lippe 2010). As partners relinquish some traditional behaviors, they need to create new ways of letting each other know they care.

Communicate Positive Feelings

Other research, conducted by widely recognized communication psychologist John Gottman and his colleagues, found that "[t]he absence of positive affect and not the presence of negative affect...was most predictive of later divorcing" (Gottman and Levenson 2000, p. 743). **Positive affect** involves the verbal or nonverbal expression of affection. Affection can be expressed verbally, nonverbally (a smile, a touch), or symbolically. A husband's sharing in housework carries a symbolic meaning for a wife, indicating that her work is recognized and appreciated (Wilcox et al. 2011b).

Gottman further argued that, at least for the middle-class couples in his sample, he could predict a married couple's later divorce by examining how well the spouses showed that they were interested in each other:

> In a careful viewing of the videotapes, we noticed that there were critical moments during the events-of-the-day conversation that could be called either "requited" [returned, acknowledged, or reciprocated] or "unrequited" interest and excitement. For example, in one couple, the wife reported excitedly about something their young son had done that day, but she was met with her husband's disinterest. After a time of talking about errands that needed doing, he talked excitedly about something important that happened to him that day at work, but she responded with disinterest and irritation. No doubt this kind of interaction pattern carried over into the rest of their interaction, forming a pattern for "turning away" from one another. (Gottman and Levenson 2000, p. 744)

Relatedly, UCLA psychologist Shelly Gable studied how one partner responds when something positive happens to the other one, such as a promotion at work (Gable et al. 2004). A partner might respond enthusiastically ("That's wonderful, and it's because you've had so many good ideas in the past few months."). But he or she could instead respond in a less-than-enthusiastic manner ("Hmmm, that's nice."), seem uninterested ("Did you see the score of the Yankees game?"), or point out the downsides ("I suppose it's good news, but it wasn't much of a raise."). According to Gable's research, the only "correct" reaction—the response that's correlated with intimacy, satisfaction, trust, and continued commitment—is the first response: the enthusiastic, active one (Lawson 2004a). "Facts about Families: Ten Rules for Successful Relationships" presents ideas on how to show positive affect. But even the happiest couples have conflicts, and how they are addressed has much to do with maintaining supportive relationships.

Facts about Families

Ten Rules for Successful Relationships

Psychologists Nathaniel Branden and Robert Sternberg have developed some rules for nourishing relationships. Here are ten. The first seven can be applied to all family relationships. The final three pertain to romantic couple relationships.

In All Family Relationships

1. *Express your love verbally.* Say "I love you."
2. *Be physically affectionate.* Offer (and accept) a touch or hug that says, "I care," "I'm sorry," or "I understand."
3. *Express your appreciation.* Tell your loved ones what you like, enjoy, and admire about one another. Listen with interest.
4. *Help the relationship or family to become an emotional support system.* Be there for each other in times of illness, difficulty, and crisis; be generally helpful and nurturing—devoted to each other's well-being.
5. *Express your affection in material ways.* Send cards or give presents on more than just expected occasions. Lighten a family member's burden once in a while by doing more than your agreed-upon share of the chores.
6. *Accept your family members' shortcomings.* We are not talking about putting up with physical or verbal abuse here. But harmless shortcomings are part of every relationship. Love your family members, not an unattainable idealization of them.
7. *Do unto each other as you would have the other do unto you.* Unconsciously, we sometimes want to give less than we get, or to be treated in positive ways that we fail to offer our family members. Try to see things from another family member's viewpoint.

In Romantic Couple Relationships

1. *Share more about yourself with your partner than you do with any other person.* In other words, keep each other primary (see Chapter 7).
2. *Make time to be alone together.* This time should be exclusively devoted to the two of you as a couple. Understand that love requires attention and leisure.
3. *Do not take your relationship for granted.* Make your relationship your first priority and actively seek to meet each other's needs.

Critical Thinking

Often, we read a list like the previous one and think about whether our partner or other family members are doing them, not whether we ourselves are. How many of the items on this list do you yourself do? Which two or three items might you begin to incorporate into a relationship?

Sources: Branden 1988, pp. 225–228; Mayo Clinic Health Letter 2012; Sternberg 1988b, pp. 272–277; see also Gottman and Silver 1999; Gottman and DeClaire 2001; Mackey, Diemer, and O'Brien 2000; Markman, Stanley, and Blumberg 2001.

Stress, Coping, and Conflict in Relationships

Daily stresses make for communication challenges; they call for coping on the part of family members as individuals and also as parts in the family system (Bernard 2012b; Duba et al. 2012; Falconier and Epstein 2011; Ledermann et al. 2010). Stresses might involve work-family spillover (see Chapter 10), economic worries, conflict between parents over child raising practices, illness or health concerns about a family member, or neighborhood issues. As a further example, communication challenges can surface in same-sex unions as partners may have to negotiate about one's wanting to come out to relatives while the other does not (Lannutti 2013). Sometimes stress involves how to talk with a family member about necessary lifestyle changes after a heart attack or how to discuss with an older family member the need to give up driving privileges (Goldsmith, Bute, and Lindholm 2012). As addressed in Chapter 10, couples often negotiate (and renegotiate) work and family roles.

Couples have to negotiate differences and challenges about their children, money, household chores, in-laws and other relatives, how to allocate their time, and irritating little habits that one of them has. Another challenging topic involves family communication itself, often with one family member feeling that another is not paying attention or understanding (Papp, Cummings, and Goeke-Morey 2009). Meanwhile, daily stresses interact with communication patterns to make the stressful situation feel either better or worse (Ledermann et al. 2010). Partners can help or hinder one another in dealing with stresses and strain. Experts advise **relationship-focused coping**—that is, coping in which "each partner attempts to cope with his or her own strain in ways that do not harm the relationship and also [attend] to the other's emotional needs" (Falconier and Epstein 2011, p. 307).

Through it all, anger and conflict should be viewed as challenges to be met rather than avoided (Schechtman and Schechtman 2003). Sociologist Judith Wallerstein (Wallerstein and Blakeslee 1995) conducted lengthy interviews with fifty predominantly nonHispanic white, middle-class married couples in northern California. The shortest marriage was ten years and the longest forty years. To participate, both husband and wife had to define their marriage as happy. When discussing what she found, Wallerstein wrote this:

> [E]very married person knows that "conflict-free marriage" is an oxymoron. In reality it is neither possible nor desirable.... [I]n a contemporary marriage it is expected that husbands and wives will have different opinions. More important, they can't avoid having serious collisions on big issues that defy compromise. (p. 143)

The couples in Wallerstein's research quarreled:

> In one marriage the husband and wife sat in the car to argue, to avoid upsetting the children. She told him that passive smoke was a proven carcinogen, and while the children were young he could not smoke in their home. He could do what he wanted outside. The man admitted that the request was reasonable, but he was furious. He punished her by not talking to her except when absolutely necessary for three months. Then he accepted the injunction on his smoking and they resumed their customary relationship. (p. 148)

Wallerstein concluded that

> [t]he happily married couples I spoke with were frank in acknowledging their serious differences over the years.... What emerged from these interviews was not only that conflict is ubiquitous but that these couples considered learning to disagree and to stand one's ground one of the gifts of a good marriage. (p. 144)

Counselors generally advise that an important aspect of learning to disagree involves expressing anger directly, a skill explored later in this chapter. Here we turn to an examination of *indirect* expressions of anger.

Indirect Expressions of Anger

Many of us may feel uncomfortable about expressing our anger directly. As a result, we can find ourselves engaging in **passive-aggression**—that is, expressing anger indirectly. Chronic criticism, nagging, nitpicking, and sarcasm are all forms of passive-aggression. Procrastination, especially when you have promised a partner that you will do something, may be a form of passive-aggression (Ferrari and Emmons 1994). These behaviors create unnecessary distance and pain in relationships. Most people who use sarcasm do so unthinkingly, unaware of its hurtful consequences. But being the target of sarcastic or otherwise hurtful remarks can result in partners feeling alienated from each other (Murphy and Oberlin 2006). Then too, sex and other expressions of intimacy become arenas for ongoing conflict when partners passive-aggressively withhold them. For example, a partner makes a disparaging comment in front of company. The hurt spouse says nothing at the time but rejects the other's sexual advances later.

Other forms of indirect anger include sabotage and displacement. **Sabotage**, a means of getting revenge, or "payback" (Boon, Deveau, and Alibhal 2009), involves one partner's attempts to spoil or undermine some activity that the other has planned. For example, the partner who is angry because the other invited friends over when he or she wanted to relax may sabotage the evening by acting bored. In **displacement**, a person directs anger at people or things that the other cherishes. An individual who is angry with a partner for spending too much time on a career may hate the partner's expensive car.

Typically, individuals express anger indirectly because they are afraid of conflict, either generally or with reference to a specific person or persons (Murphy and Oberlin 2006; Oyamot, Fuglestad, and Snyder 2010). Advising partners to express anger directly rests on assumptions of equitable power and feelings of security in a relationship (Knobloch and Knobloch-Fedders 2010; Knudson-Martin and Mahoney 2009). Nevertheless, partners and family members who do not express their anger directly risk emotional and/or sexual detachment (Gottman and Levenson 2000). Of course, it is important to recognize that direct expressions of anger can go too far, resulting in domestic violence

Myrleen Pearson/PhotoEdit

Learning to express anger and dealing with conflict early in a relationship are challenges to be met rather than avoided. Acknowledging and resolving conflict is painful, but it often strengthens the couple's union in the long run. Meanwhile, feeling uncertain about how invested one's relationship partner is can cause psychological turmoil and pain (Knobloch and Theiss 2013). A key to keeping a relationship healthy is often to share positive events and feelings so that angry arguments occur within an overall context of couple satisfaction and mutual trust.

(Rehman et al. 2009), discussed in Chapter 12. We turn now to what an important research team has to say about conflict management.

John Gottman's Research on Couple Communication and Conflict Management

Social psychologist John Gottman (1979, 1994, 1996; Gottman et al. 1998; Gottman and DeClaire 2001; Gottman and Notarius 2000, 2003) has made his reputation in the field of marital communication. The extent to which Gottman and colleagues' findings apply to other than heterosexually married couples has been questioned (Kim, Capaldi, and Crosby 2007).

In the 1970s, applying an interactionist perspective to partner communication, he began studying newly married couples in a university lab while they talked casually, discussed issues that they disagreed about, or tried to solve problems. Video cameras recorded the spouses' gestures, facial expressions, and verbal pitch and tone. After he began this research, Gottman kept in contact with more than 650 of the couples, some for as many as fourteen years. Typically, the couples were videotaped intermittently. Some couples volunteered for laboratory observation that monitored shifts in their heart rate and chemical stress indicators in their blood and urine as a result of their communicating with each other (Gottman 1996). Still later, Gottman and his colleagues identified similar patterns among gay and lesbian couples (Gottman et al. 2003; see also Houts and Horne 2008).

Studying marital communication in this detail, Gottman and his colleagues were able to chart the effects of small gestures. For example, early in his career he reported that when a spouse—particularly the wife—rolled her eyes while the other was talking, divorce was likely to follow sometime in the future, even if the couple was not thinking about divorce at the time (Gottman and Krotkoff 1989). Recognizing the need for continued research in this area, we present Gottman's highly influential findings here.

The Four Horsemen of the Apocalypse

Gottman's research (1994) showed that conflict and anger themselves did not predict divorce, but processes that he called the **Four Horsemen of the Apocalypse** did. The word *apocalypse* refers to the biblical idea that the world is soon to end, being destroyed by fire. The Four Horsemen are allegorical figures representing war, famine, and death, with the fourth uncertain (*Concise Columbia Encyclopedia* 1994, p. 309). Gottman used the phrase to indicate attitudes and behaviors that foreshadow impending divorce.

The Four Horsemen of the Apocalypse are contempt, criticism, defensiveness, and stonewalling. Rolling one's eyes indicates **contempt**, a feeling that one's spouse is inferior or undesirable. **Criticism** involves making disapproving judgments or evaluations of one's partner. **Defensiveness** means preparing to defend oneself against what one presumes is an upcoming attack. **Stonewalling** involves resistance—refusing to take a partner's complaints seriously. Stonewallers react to their partner's attempts to raise tension-producing issues by refusing to entertain them. Avoiding or evading an argument is an example of stonewalling. Argument evaders use several tactics to avoid fighting, such as vacating the scene when an argument threatens; turning sullen and refusing to talk; declaring, "I can't take it when you yell at me"; using the hit-and-run tactic of filing a complaint, then leaving no time for an answer or resolution; and saying "OK, you win" without meaning it.

In several of Gottman's studies, these behaviors identified those who would divorce, with an unusually high accuracy of about 90 percent. Later, after more research, Gottman added **belligerence**, a behavior—as in the above example—that challenges the other's power or authority (for example, "What can you do if I do go drinking with Dave? What are you going to do about it?") (Gottman et al. 1998, p. 6). To illustrate how Gottman's horsemen make their ways into communication, consider the following exchange:

Partner A: I can't find my phone.

Partner B: It's never on when I call you anyway.

Partner A: What's that got to do with anything?

Partner B: So look for it.

Partner A: So help me.

Partner B: You're just like your dad—always expecting somebody to do things for you.

Partner A: "#!@#*!"

In this scenario, *Partner A* mentions having misplaced a cell phone, and an argument develops, illustrating contempt, defensiveness, and belligerence. When *Partner A* announces the need for the phone, *Partner B* raises a complaint: "It's never on when I try to call you anyway." In a less-distressed couple, *A* might respond to *B*'s complaint. However, *A* fails to de-escalate the interchange and does not acknowledge *B*'s complaint. Less distressed couples might stop this negative spiral with shared humor or some sign of affection. However, *Partner A* subsequently requests that *B* help look for the phone. *Partner B*'s reply is contemptuous and critical: "You're just like your dad—always expecting somebody to do things for you." Again, the couple fails to de-escalate the negative affect. This time *A* calls *B* a name, maybe with an expletive. Name-calling is contemptuous.

It appears the couple has forgotten what the fight is about. In fact, one wonders whether they ever knew what the fight was about. Counselors point out that distressed couples, like the couple depicted here, may unconsciously allow trivial issues to become decoys so that they evade the real area of conflict and leave it unresolved. In sum, contempt, criticism, defensiveness, stonewalling, and belligerence characterize unhappy marriages and may signal impending divorce (Gottman and Levenson 2002). Gottman, like other researchers, found that communicating positive feelings for a partner characterized happier, more stable unions.

Positive versus Negative Affect Gottman and his colleagues videotaped 130 newlywed couples as they discussed a problem that caused ongoing disagreement in their marriage for fifteen minutes (Gottman et al. 1998). Each couple's communication was coded in one-second sequences, and then synchronized with each spouse's heart rate data, which was being collected at the same time. The heart rate data would indicate each partner's physiological stress.

The researchers examined all the interaction sequences in which one partner first expressed **negative affect**: anger, sadness, whining, disgust, tension and fear, belligerence, contempt, or defensiveness. Belligerence, contempt, and defensiveness (three of Gottman's indicators of impending divorce) were coded as *high-intensity, negative affect.* The other emotions listed previously (anger, sadness, whining, and so on) were coded as *low-intensity negative affect.*

Next, the researchers watched what happened immediately after a partner had expressed negative affect or raised a complaint. Sometimes the partner reciprocated with negative affect in kind, either low or high intensity. As examples, *Partner A* whines, and *Partner B* whines back; *A* expresses anger, and *B* responds with tension and fear; or *A* is contemptuous, and *B* immediately becomes defensive.

At other times, one partner's first negative expression was reciprocated with an escalation of the negativity. As examples, *Partner A* whines, and *Partner B* grows belligerent; or *A* expresses anger, and *B* becomes defensive. Gottman and his colleagues called this kind of interchange *refusing-to-accept influence,* because the spouse on the receiving end of the other's complaint refuses to consider it and, instead, escalates the fight. Here's an example:

> A thirty-eight-year-old woman from a Chicago suburb had a husband who was addicted to porn. She said, "He would come home from work, slide food around on his plate during dinner, play for maybe half an hour with the kids, and then go into his home office, shut the door, and surf Internet porn for hours.... I would continually confront him on this. There were times I would be so angry I would cry and cry and tell him how much it hurt.... It got to the point where he stopped even making excuses. It was more or less 'I know you know and I don't really care. What are you going to do about it?'" (Paul 2005, p. 3)

Negative escalation is also evidenced in the cell phone conflict, above.

In Gottman's study, many couples were likely to communicate with positive affect, responding to each other warmly with interest, affection, or shared (not mean or contemptuous) humor. Positive affect typically de-escalated conflict (Gottman and Levenson 2000, 2002). Gottman and his colleagues found that "[t]he only variable that predicted both marital stability and marital happiness among stable couples was the amount of positive affect in the conflict" (1998, p. 17). In stable, happy couples, shared humor and expressions of warmth, interest, and affection were apparent even in conflict situations and, therefore, de-escalated the argument.

The researchers "found no evidence... to support the [idea that] anger is the destructive emotion in marriages" (1998, p. 16). Instead, they found that contempt, belligerence, and defensiveness were the destructive attitudes and behaviors. Furthermore, Gottman and his colleagues concluded that the interaction pattern best predicting (heterosexual) divorce was a wife's raising a complaint, followed by her husband's refusing-to-accept influence, followed, in turn, by the wife's reciprocating her husband's escalated negativity, and the absence of any de-escalation by means of positive affect. Despite changing gender roles (see Chapter 4), researchers continue to observe gender differences in communication patterns.

Gender Differences and Communication

Before the nineteenth century, men's and women's domestic activities involved economic production, not personal intimacy. With the development of separate gender spheres in industrializing societies during the nineteenth century, expressions of emotion became the domain of middle-class women, whereas work was defined as more appropriate to masculinity. As a result of this historical legacy, we see men as less well equipped than women for emotional relatedness (Real 2002). Empirical evidence shows that, on average, females in heterosexual relationships do more than males to keep the relationship in existence and satisfying (Malinen, Tolvanen, and Ronka 2012). More often than a man, a woman acts as a "relationship barometer"—taking the pressure of the relationship (Faulkner, Davey, and Davey 2005; Fleming and Cordova 2012).

Sociologist Francesca Cancian (1987) expanded on these points to argue that men are equally loving, but women, not men, are made to feel primarily responsible for love's endurance or success. Furthermore, expressions of love are defined and perceived mostly on feminine terms—that is, verbally—and women are the more verbal sex. Expressions of love that men may make, such as doing favors or reducing their partners' burdens, are not credited as love (Cancian 1985). Recent research appears to support Cancian's analysis. For instance, a study with 453 heterosexual couples drawn from a national representative survey looked at changes that women would like in their partners, compared with changes that men would like. The women were more likely than the men to want increases in a partner's demonstrations of positive emotion (Heyman et al. 2009).

In Cancian's analysis, "The consequences of love would be more positive if love were the responsibility of men as well as women and if love were defined more broadly to include instrumental help as well as emotional expression" (1985, p. 262). Cancian has also argued that a more balanced view of how love is to be expressed—one that includes masculine as well as feminine elements—would find men equally loving and emotionally profound.

Meanwhile, Deborah Tannen's book *You Just Don't Understand* (1990) suggests that men typically engage in **report talk**, conversation aimed mainly at conveying information. Women, on the other hand, are likely to engage in **rapport talk**, speaking to gain or reinforce intimacy or connection with others. Men are likely to bring up problems, for instance, only when hoping to trigger suggestions for solution. Women, on the other hand, are likely to talk about problems simply to share or foster rapport. These gendered differences

> lead to an imbalance in many families. If the mother is telling about troubles she confronted during her day but the father is not, the result is that mothers come across as more problem-ridden and insecure than fathers. And many men, because they don't tend to talk in this way, understandably assume that a woman who recounts a problem must be seeking help solving it; why else would she talk about it? That's why they generously provide solutions. So the woman's conversational gambit ends up being refracted through the man's point of view. This misunderstanding of women's rapport-talk often results in mothers appearing to their families as less confident, or even less competent, than their husbands. (Tannen 2006, pp. 83–84)

Moreover, some researchers speculate that women, being more expressive and attuned to the emotional quality of a relationship, are more likely than men to suggest marriage counseling or a marriage enrichment program (Fleming and Cordova 2012). Women are also more likely to bring conflict into the open, sometimes in an attention-getting negative tone (Cui et al. 2005). Men try to minimize the impending conflict either by conciliatory gestures or by stonewalling. The male's minimization of conflict may appear to the female as failure to recognize her emotional needs (Noller and Fitzpatrick 1991; Canary and Dindia 1998). In the following, a husband describes this situation:

> The more I try to be cool and calm her the worse it gets. I swear, I can't figure her out, I'll keep trying to tell her not to get so excited, but there's nothing I can do. Anything I say just makes it worse. So then I try to keep quiet, but... wow the explosion is like crazy, just nuts. (in Rubin 2007, p. 323)

We might compare this to a wife, who told her interviewers that,

> I can't stand that he's so damned unemotional and expects me to be the same. He lives in his head all of [the] time, and he acts like anything that's emotional isn't worth dealing with. (in Rubin 2007, p. 322)

Reviews of research on couple communication in the 1990s (Gottman and Notarius 2000; Bradbury, Fincham, and Beach 2000) concluded that men and women differ in their responses to negative affect in close relationships. When faced with a complaint from a partner, men tend to withdraw emotionally whereas women do not (Green and Addis 2012). Researchers have found this pattern to be common enough that some call it the *female-demand/male-withdraw communication pattern* (Gottman and Levenson 2000). In distressed unions, this pattern becomes a repeated cycle of negative verbal expression by one partner and withdrawal by the other (Bradbury, Fincham, and Beach 2000).

Many researchers and therapists agree that generally there is a female-demand/male-withdraw pattern—at least among middle-class white heterosexually married couples like the ones Gottman studied (Miller and Roloff 2005; Weger 2005). However, an alternative view argues that "it is not gender per se but the nature of the marital discussion—for example, whether it is the wife or the husband who desires a change—that may determine who is demanding and who is withdrawing" (Roberts 2000, p. 702; also see Kim, Capaldi, and Crosby 2007). Research on same-sex couples has found the same pattern—that is, one partner demands while the other withdraws (Parker-Pope 2008). And research in stepfamilies suggests that neither partner is likely to demand as much as in many first marriages, while both partners are more likely to withdraw from a conflict (Halford, Nicholson, and Sanders 2007).

Obviously, the **female-demand/male-withdraw interaction pattern** leads to both partners feeling misunderstood, thereby decreasing marital satisfaction (Weger 2005). Gottman and his colleagues (1998) concluded that wives and husbands have different goals when they disagree:

> The wife wants to resolve the disagreement so that she feels closer to the husband and respected by him. The husband, though, just wants to avoid a blowup. The husband doesn't see the disagreement as an opportunity for closeness, but for trouble. (p. 17)

In one husband's words, "I just got mad and I'd take off—go out with the guys and have a few beers or something. When I'd get back, things would be even worse." From his wife's perspective, "The more I screamed, the more he'd withdraw, until finally I'd go kind of crazy. Then he'd leave and not come back until two or three in the morning sometimes" (Rubin 1976, pp. 77, 79).

Gottman and his colleagues sought to better understand this pattern. You'll recall that the researchers monitored spouses' heart rates as indicators of physiological stress during conflict. They hypothesized that, "it is likely that the biological, stress-related response of men is more rapid and recovery is slower than that of women, and that this response is related to the greater emotional withdrawal of men than women in distressed families" (Gottman et al. 1998, p. 19). That is, when confronted with conflict from an intimate, men may experience more intense and uncomfortable physical symptoms of stress than women do. Therefore, men are more likely than women to withdraw emotionally and/or physically.

An alternative—or complementary—view is that men have been socialized to withdraw (Green and Addis 2012). The cultural options for masculinity include "no 'sissy' stuff," according to which men are expected to distance themselves from anything considered feminine. We guess this could include taking a wife's complaints seriously. In two books on men and communication, the first called *I Don't Want to Talk About It,* therapist Terrence Real (1997, 2002) attributes males' withdrawal to a "secret legacy of depression," brought on by men's traditional socialization, particularly society's refusal to let them grieve over losses (e.g., "Don't cry over nothing"). It is possible that physiology and culture interact to create a female-demand/male-withdraw pattern.

What Couples Can Do

The general conclusion of Gottman's research on couple communication and conflict management is as follows. Both partners:

1. need to try to be gentle when they raise complaints;
2. can help to reduce anxiety in their mate by communicating care and affection, hence reducing physical stress symptoms;
3. can learn techniques for reducing anxiety in oneself—for example, taking a time-out or saying that "I just can't talk about it now, but I will later" (and meaning it);
4. need to be willing to accept influence from each other;
5. need to do what they can—perhaps using authentic, shared humor, kindness, and other signs of affection—to de-escalate the argument. It is important to recognize that this does not mean avoiding the issue altogether.

Finally, Gottman and his colleagues (1998) suggest that, as we have already seen, it is important for couples to think about communicating with positive affect more often in their daily living and not just during times of conflict (Gottman and DeClaire 2001). "Facts about Families: Ten Rules for Successful Relationships" suggests ways to do this. Then too, as we have seen, cohesive families have arguments. But the argument ends; conflicts are resolved. We turn to some guidelines for accomplishing this.

Working Through Conflicts in Positive Ways—Ten Guidelines

Different cultural groups vary in their endorsement of openly expressing emotion and directly expressing conflicts (see, for example, Hirsch 2003). We need to recognize that preferred standards for communication vary culturally (Matsunaga and Imahori 2009) and that the guidelines suggested in this chapter, which accent direct communication styles, may be ethno- or Eurocentric. Deborah Tannen's (1990) book *You Just Don't Understand,* which drew wide attention for its comparison of men's and women's communication styles, also points out cultural communication differences among, for example, New Yorkers, Californians, New Englanders, and Midwesterners, and among Scandinavians, Canada's native peoples, and Greeks. Interpersonal communication differences are evidenced by other races/ethnicities as well (Matsunaga and Imahori 2009). Nevertheless, counselors do advise that there are better (and not-so-good) ways that virtually all couples and family members can resolve differences.

Before going further, we want to point out that not all negative facts and feelings need to be communicated. Before voicing a complaint, we might ask ourselves, "How important is it?" (Sanford 2006). Counselors suggest that if, after giving it some time and thought, we believe that raising a particular grievance is important, then we should do so. Similarly, when offering negative information, it is important to ask ourselves why we want to do so and whether the other person really needs to know. We turn now to ten specific guidelines for constructive conflict management.

Noel Hendrickson/Digital Vision/Getty Images

Chronic stonewallers may fear rejection or retaliation and therefore hesitate to acknowledge their own or their partner's angry emotions. Examples of stonewalling include saying things like, "I can't take it when you yell at me," or turning sullen and refusing to talk. It may sound impossible to fight more fairly when you're angry, but "practice makes better." Using "I" statements, avoiding mixed messages, focusing your anger on specific issues, and being willing to change are some guidelines worth trying.

Guideline 1: Express Anger Directly and with Kindness

Family members may have the false belief that their intimates automatically know—or should know—what they think and how they feel. This incorrect idea is detrimental to relationships (Hamamci 2005). When complaints are not addressed directly, conflict goes unresolved, with lingering grievances sparked again and again by "subtle triggers." Consider the following family situation:

> An ongoing point of contention in this family is the mother's belief that her teenage daughter, Joyce, spends too much money on clothes and makeup, which she buys in upscale stores rather than more economical stores, like Wal-Mart. So when the father, who is scanning a newspaper, remarks, "I see Wal-Mart set a record for sales yesterday," the seed is planted for an argument to sprout. (Tannen 2006, p. 123)

The underlying conflict is voiced as follows:

Mom: So? We don't shop at Wal-Mart, so what's the point?

Dad: Okay.

Joyce: What does that have to do with anything?

Mom: Okay, I'm just saying—

Joyce: Saying what?

Mom: Yeah, so what's the point?

Joyce: What point, Mom? You don't shop there either, Mom.

Mom: Yes, I do. You could shop there for toiletries.

Joyce: For clothes you shop there, Mom?

Mom: No.

Joyce: See, so why should we go shopping there for toiletries?...I don't go shopping for toiletries anywhere because you buy them for me.

Mom: No, but you buy makeup.

Dad: Well, this year we can do all our Christmas shopping at Wal-Mart. (Tannen 2006, pp. 123–25)

Tension and conflict go unresolved.

Counselors advise expressing anger directly because doing so makes way for resolution (Bernstein and Magee 2004). For example, the mother might say, "I feel that you've been spending more than we can afford on makeup." Counselors further advise that a grievance will be less threatening to the receiver when positive feelings are conveyed at the same time that the grievance is voiced.

> If you're angry and resentful, requests for change will be met with resistance and countercharge efforts: "It's not my problem; it's your problem." But if you learn to approach each other with acceptance and empathy, you can create a collaborative context, and often people will make spontaneous changes. ("Loving Your Partner as a Package Deal" 2000)

So, even better, the mother might say, "You always look nice, and I like the way that you choose to wear your makeup, but I feel that you're spending more than we can afford on it." Being direct is not the same as being unnecessarily critical. In fact, it's possible to be simultaneously direct, sensitive, and kind. ("Issues for Thought: Biosociology, Love, and Communication," adds a somewhat different perspective on this point.)

Issues for Thought

Biosociology, Love, and Communication

An intriguing area of research from the biosocial theoretical perspective (see Chapter 2) suggests that brain chemistry helps to explain why shared new activities enhance romantic relationships (Gottlieb 2006; Slater 2006). This research points to two hormones. The first, dopamine, is a chemical naturally produced in our brains. Although dopamine has many functions, its importance to love is that it acts upon the pleasure center in our brains, giving us powerful feelings of enjoyment and motivating us to do whatever we're doing that is so pleasurable over and over again (Berridge and Robinson 1998).

Dopamine helps to explain why we have a second helping of a really tasty dessert, for example. Furthermore, dopamine is associated with new or novel pleasurable experiences and activities. Research shows that when people are newly in love, they tend to have higher brain levels of dopamine (Slater 2006). Dopamine makes you "high on" your partner. The second hormone relevant here is oxytocin, also produced naturally in our brains. Some researchers have nicknamed oxytocin the "love" or "cuddle" hormone (Barker n. d.; Bosse 1999).

Like dopamine, oxytocin has several functions (inducing labor and stimulating breast milk production in females, for example). Research in mammals has long demonstrated that oxytocin facilitates more general maternal, nurturing behaviors ("Oxytocin" 1997). In addition, oxytocin seems to be related to human feelings of deep friendship, trust, sexuality, love, bonding, and commitment, and it may help to reduce feelings of stress during conflicts and thereby increase positive communication behaviors ("Biology of Social Bonds" 1999; Bosse 1999; Ditzen et al. 2009; and see Priem, McLaren, and Solomon 2010).

Hormones affect feelings and behavior, but the reverse is also true: Behaviors can stimulate hormone production. Doing novel things together and engaging in supportive touch, including sex, stimulate the production of dopamine and oxytocin, respectively (Slater 2006; see also Ellison and Gray 2009). All this gives a somewhat new slant to the phrase "making love," doesn't it?

Critical Thinking

The hormones dopamine and oxytocin may influence communication behaviors. What else can you think of that influences couple or family communication? Recall the idea from Chapter 2 that "nature" (genetics, hormones) and "nurture" (culture and social relations) interact to influence attitudes and behaviors (Horwitz and Neiderhiser 2011). Can you think of an example of how dopamine or oxytocin ("nature") might interact with "nurture"?

Guideline 2: Check Out Your Interpretation of Others' Behaviors

Because family members and partners in distressed relationships seldom understand each other as well as they think they do, a good habit is to ask for feedback by a process of *checking it out:* asking the other person whether your perception of her or his feelings or of the present situation is accurate. Checking it out often helps to avoid unnecessary hurt feelings or imagining trouble that may not exist, as the following example illustrates:

> **Family Member A:** I think you're mad about something. (*checking it out*) Is it because it's my class night and I haven't made dinner?
>
> **Family Member B:** No, I'm irritated because I was tied up in traffic an extra half hour on my way home.

Guideline 3: To Avoid Attacks, Use "I" Statements

Attacks, sometimes interpreted as blame, involve insults or assaults on another's character or self-esteem, which should be considered a "shared relationship resource'" that is, both partners are happier when each one's self-esteem is high, rather than low (Robinson and Cameron 2011). Needless to say, attacks do not help either to enhance self-esteem or to bond a couple (Sinclair and Monk 2004). A rule in avoiding attack is to use the word, *I* rather than *you* or *why*. For example, instead of declaring, "You're late," or asking "Why are you late?"—both of which can smack of blame—a statement such as, "I was worried because you hadn't arrived" may allow for more positive dialogue. "And while comments like 'Are you trying to put us in the poorhouse?' may be emotionally satisfying in the moment, they're ineffective in the long run" (Mangla 2013). The receiver is more likely to perceive "I" statements as an attempt to recognize and communicate feelings; "you" and "why" statements are more likely to be perceived as attacks, even when not intended as such.

Of course, making "I" statements may be too much to ask in the heat of an argument. One social psychologist has admitted what many of us may have experienced: "It is impossible to make an 'I-statement' when you are in the ... 'wanting-revenge, feeling-stung-and-needing-to-sting-back' state of mind" (quoted in Gottman et al. 1998, p. 18). Of course, this is partly the point. Keeping in mind the possibility of expressing a complaint—at least *beginning* a confrontation—with an "I" statement can discourage family members from getting to that wanting-revenge state of mind in the first place.

Guideline 4: Avoid Mixed, or Double Messages

Mixed, or double, messages contradict each other. Contradictory messages may be verbal, or one may be verbal and one nonverbal. For example, a family member offers to take the family to a movie yet sighs and says that he or she is exhausted after a really hard day at work. Or a partner insists, "Of course I love you" while picking an invisible speck from his or her sleeve in a gesture of indifference.

Senders of mixed messages may not be aware of what they are doing, and mixed messages can be very subtle. They sometimes result from simultaneously wanting to recognize and to deny conflict or tension. A classic example is the *silent treatment.* One partner becomes aware that she or he has said or done something upsetting and asks what's wrong. "Oh, nothing," the other replies without much feeling, but everything about the partner's face, body, attitude, and posture suggests that something is indeed wrong (Lerner 2001).

Communication scholars and counselors point out that there are two major aspects of any communication: *what* is said (the verbal message) and *how* it is said or interpreted (the nonverbal "meta-message"). The meta-message depends on tone of voice, inflection, and body language, as well as on the receiver (Nierenberg and Calero 1973). In a mixed message, the verbal message does not correspond with the meta-message.

Moreover, communication involves both a sender and a receiver. Just as the sender gives both an overt message and an underlying *meta-message,* so also does a receiver give cues about how seriously she or he is taking the message. For example, listening while continuing to do chores sends the nonverbal message that what is being heard is not very important.

Guideline 5: When You Can, Choose the Time and Place Carefully

Arguments are less likely to be constructive if the complainant raises grievances at the wrong time. One partner may be ready to argue about an issue when the other is almost asleep or working on an important assignment, for instance. At such times, the person who picked the fight may get more—or less—than he or she had expected.

Family members might negotiate a time and place for addressing issues. Arguing "by appointment" may sound silly and be difficult to arrange, but doing so has advantages. For one thing, complainants can organize their thoughts and feelings more calmly and deliberately, increasing the likelihood that they will be heard. Also, recipients of complaints have time before the argument to prepare themselves to hear some criticism.

A qualitative British study of teenagers and their parents found that some parents and teens use mobile phones to raise sensitive issues that they intend to later pursue face to face. One teen said, "Well, I think mobiles can be really good if you've got something you don't wanna tell straight away," and another young respondent said, "I'd maybe text if it's something that I can't, I dunno, something I can't get across [face to face] and stuff" (in Devitt and Roker 2009, p. 192; see also Lasen and Casado 2012). [On the other hand, reliance on mobile phones and other social media may result in people's not learning and consequent inability to communicate *face-to-face* about touchy issues (Moore 2010).]

Guideline 6: Address a Specific Issue, Ask for a Specific Change, and Be Open to Compromise

Constructive relationships aim at resolving current, specific problems. Recipients of complaints need to feel that they can do something specific to help resolve the problem raised. This will be difficult if they feel overwhelmed by old gripes. Furthermore, complainants should be ready to propose one or more solutions. Recipients might come up with possible solutions themselves. When family members can entertain potential solutions to a definite problem at hand, they are better able to negotiate alternatives.

John Gottman found that happily married couples reached agreement rather quickly. Either one partner gave in to the other without resentment, or the two compromised. Unhappily married couples continued in a cycle of stubbornness and hostility (Gottman and Krotkoff 1989; see also Busby and Holman 2009).

Guideline 7: Be Willing to Change Yourself

The principle that couples or family members should accept each other as they are sometimes merges with the idea that individuals should be exactly what they choose to be. The result is an erroneous assumption that if someone loves you, he or she will accept you just as you are and not ask for even minor changes. In truth, partners need to be willing to be influenced by their loved ones and to change themselves (Lerner 2001).

Therapists note that, in some relationships, each person expects the other one to do the changing: "You have to understand, she's [or he's] impossible to live with" (Ball and Kivisto 2006, p. 155). One counselor team (Christensen and Jacobson 1999) has suggested "acceptance therapy," helping individuals accept their partners and other family members as they are instead of demanding change—although these counselors also suggest that, paradoxically, showing acceptance can lead to a partner's changing behavior. We need to balance acceptance of another against not being a doormat, but being willing to change ourselves is key.

Guideline 8: Don't Try to Win

Counselors encourage us to recognize that there are probably several ways to solve a particular problem, and backing others into a corner with ultimatums and counter-ultimatums is not negotiation but attack. Moreover, wanting to win a dispute with a loved one typically encourages us to use unnecessarily hurtful language, which nonproductively increases the recipient's stress (Priem, McLaren, and Solomon 2010). We're reminded that recipients of painful messages typically see them as more hurtful than do the senders (Zhang 2009). How we say things impacts how others perceive them (Young 2010). Even hurtful information can be conveyed with sensitivity.

Societies that emphasize competition, such as ours does, encourage people to see almost everything they do in terms of winning or losing (Fromm 1956). Yet research clearly indicates that for same-sex and heterosexual couples, the tactics associated with winning in a particular conflict are also those associated with lower relationship satisfaction (Clunis and Green 2005; Heene, Buysse, and Van Oost 2007; Houts and Horne 2008). Losing lessens a person's self-esteem, increases resentment, and adds strain to the relationship. On the other hand, everyone wins when family members mutually agree on solutions to their differences (Carroll, Badger, and Yang 2006).

Guideline 9: Be Willing to Forgive

A growing number of therapists suggest that being willing to forgive is critical to ongoing happy relationships (Fincham, Hall, and Beach 2006). Forgiveness "is the idea of a change whereby one becomes less motivated to think, feel, and behave negatively (e.g., retaliate, withdraw) in regard to the offender." Forgiveness is not something to which the offender is necessarily entitled, but it is granted nevertheless.

Contrary to what many individuals believe, however, forgiveness does not require that the offended partner minimize or condone the offense. Rather, "an individual forgives despite the wrongful nature of the offense and the fact that the offender is not entitled to forgiveness." Further, "forgiveness is distinct from denial (an unwillingness to perceive the injury) ... or forgetting (removes awareness of offence from consciousness)" (Fincham, Hall, and Beach 2006, p. 416).

Forgiveness is often a process that takes time, rather than one specific decision or act of the will. Being willing to forgive has been associated in research with marital satisfaction, lessened ambivalence toward a partner, conflict resolution, enhanced commitment, and greater empathy (Fincham, Hall, and Beach 2006).

Guideline 10: End the Argument

Ending the argument is important. Sometimes when individuals are too hurt to continue, they need to stop arguing before they reach a resolution. A family member may signal that he or she feels too distressed to go on by calling for a time-out. Or it could help to bargain about whether the fight should continue at all.

As pointed out earlier in this chapter, the happily married couples that Gottman and his colleagues, as well as Wallerstein, interviewed knew how and when to stop fighting. Arguments can end with compromise, apology, submission, or agreement to disagree (Goeke-Morey, Cummings, and Papp 2007). Ideally, a fight ends when there has been a mutually satisfactory airing of each partner's views.

Toward Better Couple and Family Communication

As we saw earlier in this chapter, many Americans see their current couple and family relationships as closer than what they experienced in childhood. Due to communication research over the past several decades, people know more about how to nurture supportive relationships today than they did in the past, and they may more actively value doing so. Put another way, they realize that keeping a loving relationship or creating a cohesive family is not automatic. Meanwhile, the quality of a relationship has become pivotal to a union's survival (Cherlin 2009a; Ledermann et al. 2010).

Thinking of the family ecology model, we might say creating a supportive and cohesive family involves community initiatives to strengthen family cohesion. With regard to same-sex families, for instance,

> Community initiatives to strengthen families could emphasize the importance of staying in touch and getting along despite disagreements among family members; these campaigns could identify sexual orientation as a topic where adult family members can learn to disagree with each other while agreeing to provide a loving environment for children that is free from conflicts between adults. (Oswald and Lazarevic 2011)

Doing so requires working on ourselves as well as on our relationships. A first step involves consciously recognizing how important the relationship is to us. A second step is to set realistic expectations about the relationship (Cloud and Townsend 2005). As one married woman put it,

> You just have this idealized version of getting married, you know, everybody plays it up as so romantic and so wonderful and sweet. Now that I am married and now that I have gotten older and hit the real world I'm kind

of like . . . It's a lot more hands-on, you know, getting stuff done . . . than it is that idealized romantic notion that you get as a girl. (in Fairchild 2006, p. 13)

A third step involves improving our own (1) **emotional intelligence**—awareness of what we're feeling so that we can express our feelings more authentically; (2) ability and willingness to repair our moods, not unnecessarily nursing our hurt feelings; (3) healthy balance between controlling rash impulses and being candid and spontaneous; and (4) sensitivity to the feelings and needs of others (Keaten and Kelly 2008). We can develop greater flexibility of thought, learning to think of several alternative workable solutions to problems and to have several ways of responding to a situation, not just one that habitually comes up by default (Koesten, Schrodt, and Ford 2009). Support is mutually reinforcing. When we can support others, they are more likely to be supportive (Priem, Solomon, and Steuber 2009).

Regarding the relationship itself, counselors encourage making time for play and incorporating new activities into relationships (Lawson 2004b; Smith, Freeman, and Zabriskie 2009). Social psychologist John Crosby points out that people may misinterpret the idea of "working at" committed relationships: Instead of working *at* relationships, "we may, with all good intentions, end up making work *of*" them (Crosby 1991, p. 287).

Meanwhile, some of us have grown up with poor role modeling on the part of our parents (Ledbetter 2009; Rovers 2006; Schrodt et al. 2009). Regardless of how our parents behaved, we can choose to change how we communicate (Braithwaite and Baxter 2006; Turner and West 2006; Wright 2006). Training programs in couple and family communication, often conducted by counseling psychologists and sometimes designed for specific races/ethnicities, life course stages (new parents), or family forms (stepfamilies), have proven effective in helping to change negative communication patterns (Blanchard et al. 2009; Lucier-Greer and Adler-Baeder 2012; Trillingsgaard et al. 2012). Some family life education and communication education programs are available online (Hughes et al. 2012).

One program for married and cohabiting couples is ENRICH, originally developed by social psychologist David Olson at the University of Minnesota. A similar program is PREP (the Prevention and Relationship Enhancement Program), developed by marital communication psychologists Scott Stanley and Howard Markman, with the overall aim of strengthening marriages and preventing divorce (Markman, Stanley, and Blumberg 2001; Schilling et al. 2003). Marriage Encounter and similar organizations offer weekend workshops, designed for mostly satisfied marrieds who want to improve their relationship (Yalcin and Karaban 2007). Advertising "psychological care for the whole family," the Family Success Consortium offers programs for all couples, whether or not married.

Men's groups aimed at encouraging their expressions of emotional intimacy have been shown to enhance couple and family relationships (Garfield 2010). Some conflict management programs have been developed for child or adolescent siblings (Kennedy and Kramer 2008; Thomas and Roberts 2009) and/or for families of particular races/ethnicities (e.g., Soll, McHale, and Feinberg 2009). Some programs have been designed for same-sex couples (Heffner 2003; Unitarian Universalist Association n. d.). As mentioned in Chapter 9, some parenting enhancement programs incorporate anger management components (Dixon et al. 2012; Fleming and Cordova 2012).

Couples or family members who want to work for change on their own might practice the previously mentioned guidelines for conflict resolution. As partners and family members grow accustomed to voicing grievances regularly and in more respectful or caring ways, their disagreements less often become full-fledged fights: Family members gradually learn to incorporate many irritations and requests into their normal conversations, arguing in normal tones of voice and even with humor. Although these suggestions may help, learning

Ronnie Kaufman/Cusp/Corbis

Couples *can* change their fighting habits. The key to staying happily together is to make knowledgeable choices—about not avoiding conflict but dealing with it openly, or directly, and in supportive ways. Doing so involves listening—without judgment, without formulating a response while the other talks, and without interrupting. The goal isn't necessarily agreement, but acknowledgment, insight, and understanding.

to fight fair is not easy. Sometimes couples and families feel that they need outside help, and they may decide to engage a counselor, discussed later in this chapter.

Then too, a number of books and Internet resources are available that could help. As examples, there are books on overcoming passive-aggressive behavior (Murphy and Oberlin 2006), recognizing how we sabotage our relationships (Matta 2006), and changing habits that can thwart a satisfying life in general (Kagan and Einbund 2008). Some books, such as *Person to Person: Positive Relationships Don't Just Happen* (Hanna, Suggett, and Radtke 2008) focus on both individual self-improvement and couple communication. Susan Halpern's *Finding the Words: Candid Conversations with Loved Ones* (2009) covers topics such as cultivating conscious conversations as a couple, communicating in ways that might lessen the disruptive effects of divorce, and improving communication between parents and their adult children. John Gottman's research, described in this chapter, comes to life in his readable *10 Lessons to Transform Your Marriage* (Gottman, Gottman, and DeClaire 2006). There are books on communication designed specifically for same-sex couples (Clunis and Green 2005). And, of course, there are university courses and textbooks on interpersonal communication and relationships (e.g., Knapp and Vangelisti 2009; Verderber, Verdeber, and Berryman-Fink 2010).

M Stock/Stock Connection

Families are made up of individuals, each one seeking not only a unique identity but also a cohesive place to belong. Supportive communication facilitates family togetherness, or cohesiveness. These sisters are sharing memories recorded in old family photos. They're good friends, but the main reason that the majority of adult siblings in one small study gave for continuing their relationship was because "we're family" (Myers 2011).

In addition, there is a vast number of good (and perhaps not-so-good, so be selective) Internet resources on communication of all sorts. Among others, topics range from managing unresolved family conflict, parent-child communication, opening the door to renewed communication between estranged siblings, and—of course—communication between partners (e.g., Gilkey, Carey, and Wade 2009; Gottman and Gottman 2011; Robinson 2009; South-DeRose n. d.; Von Rosenvinge n. d.; Wakeland n. d.). Family relationships are dynamic and can change for the better. For example, an adult woman told this story about her improving relationship with her sister:

> [We] spent some time together.... We hadn't done that in 3 or 4 years.... It was... getting to the point where ... we could just continue to stick our head in the sand or we could... try this again. Because this is the only family.... So [now, after beginning to repair the relationship], it's sort of inching along like that. A little better, a little better. (in Connidis 2007, p. 489)

Finally, there is the option of relationship or family counseling.

Relationship and Family Counseling

Relationship and family counseling is a professional service having two goals: (1) helping individuals, couples, and families gain insight into the actually or potentially troublesome dynamics of their relationship(s); and (2) teaching clients more effective and supportive communication techniques. According to the American Association for Marriage and Family Therapy (AAMFT), this type of counseling is meant to be "solution-focused; specific, with attainable therapeutic goals; [and] designed with the 'end in mind'" ("What Is Marriage and Family Therapy?" 2005; Clinton and Trent 2009).

Experts advise couples or families to visit a counselor when communication is typically hostile or conflict goes unresolved, when they cannot figure out how to resolve a family problem themselves, or when a partner is thinking of leaving a committed relationship. However, counseling is also appropriate—and perhaps more effective—as a preventive technique, undertaken at the onset of family stress or when a couple or family sees a potentially troublesome transition ahead.

People go to counselors for help in working through premarital and engagement issues, as well as cultural clashes, same-sex couples, cohabitation, infidelity, divorce, substance abuse, finances, unemployment, co-parenting conflict, infertility, sexual difficulties, and changing roles such as with retirement, remarriage, and stepfamily issues, among others (Clinton and Laaser 2010; Mayo Clinic Staff 2005).

Qualifications of Counselors The qualifications of counselors vary. A counselor who is a member of the American Association for Marriage and Family Therapy (AAMFT) has a graduate degree and at least three years of clinical training under a senior counselor's supervision. The safest way to choose a qualified counselor is to select one who belongs to the AAMFT. To do so, check the organization's website. Personal recommendations from family members or friends or both may also be helpful.

It is important to have a counselor whom you like and trust and who empathizes with you. It is also important that the counselor respect your religious and personal values. Even well-trained counselors can be capable of unintentional bias that may get in the way of productive therapy (Charles, Thomas, and Thornton 2005; Knudson-Martin and Laughlin 2005). If after three or four sessions you do not feel comfortable with the counselor or don't believe the counselor is effective, it might be a good idea to try someone else. Experts advise interviewing a prospective counselor before beginning therapy. The Mayo Clinic Staff (2005) advises asking lots of questions, including the following:

- Are you a clinical member of the AAMFT or licensed by the state, or both?
- What is your educational and training background?
- What is your experience with my type of problem?
- How much do you charge?
- Are your services covered by my health insurance?
- Where is your office and what are your hours?
- How long is each session?
- How often are sessions scheduled?
- How many sessions should I expect to have?
- What is your policy on canceled sessions?
- How can I contact you if I have an emergency?

Will Counseling Save a Relationship? Despite its substantiated benefits, the extent to which counseling "saves" a relationship is difficult to measure (Corliss and Steptoe 2004). Counseling is based on the presumption that partners are willing to cooperate, and it is possible that one's partner may not be willing. No counselor can or will attempt to change a person to a partner's liking without active cooperation from all involved (Rasheed, Rasheed, and Marley 2010). To read—or participate in—an online discussion of ordinary people's opinions on whether counseling "saved" their relationship, you might want to visit the Berkeley Parents Network webpage, "Does Couples Counseling Work?" on the University of California, Berkeley website (parents .berkeley.edu).

The Myth of Conflict-Free Conflict

"Generally, marriages that have built up positive emotional bank accounts through respect, mutual support, and affirmation of each person's worth are more likely to survive" (Hetherington 2003, p. 322). Relationships require attention (Cole 2011). It is the *rewards* of a long-term relationship—love, respect, friendship, and good communication—that are most effective in keeping marriages and other relationship unions together. It follows that partners need to keep their relationship rewarding for one another.

Meanwhile, we've devoted enough attention to conflict resolution techniques that it may seem as if conflict itself can be free of conflict. It can't. Even the fairest fighters hit below the belt once in a while, and just about all fighting involves some degree of frustration and hurt feelings. Moreover, some individuals have a partner who chooses not to learn to face conflict positively. Sometimes attending a relationship enhancement program can end in one or both partners' disappointment when one partner doesn't seem to be motivated or trying during the program itself or when what seemed to be progress during the program itself is not carried out or followed through. Disappointed partners may feel hurt or anguished, try to make their partner feel guilty, or ignore the problem—these are coping mechanisms (Dixon et al. 2012).

In relationships where one wants to change and the other doesn't, sometimes much can be gained if just one partner begins to communicate more positively. Other times, however, positive changes in one individual do not spur growth in the other. Situations like this may end in alienation, separation, or divorce. We need to note that communication patterns can convey messages of dominance and disaffiliation (McLaren, Solomon, and Priem 2012). The next chapter examines issues of dominance and power in relationships.

Then too, even when both partners develop constructive habits, all their problems will not necessarily be resolved (Booth, Crouter, and Clements 2001; Driver and Gottman 2004). Although a complainant may feel that he or she is being fair in bringing up a grievance and discussing it openly and calmly, the recipient may view the complaint as critical and punitive, and may not want to bargain about the issue.

Family Well-Being Depends on Positive Communication Habits Together with the Family's External Social Environment

The introduction to this chapter pointed out that family communication takes places within and is influenced by the family's external social environment (Trail and

Karney 2012). A "dyadic approach" to discussing couple or family communication involves incorporating a sociological imagination or the family ecology theoretical perspective (Helms, Supple, and Proulx 2011). Here's something to think about from this point of view:

> To improve or further enhance marital functioning, people should try to reduce and cope—individually or dyadically—with both high levels of external stress that tends to spill over into the relationship and high levels of relationship stress. To reduce the level of external stress, employers are required to provide safe working conditions and fair wages. In addition, governmental and other social service programs should pay special attention to the needs of low-income couples and help them to overcome external strains, as they often experience more stress and face greater problems in building and maintaining a healthy intimate relationship than better off couples.
>
> [Meanwhile, finding] effective ways to deal with stress occurring inside the relationship is important to stave off deterioration of marital functioning on both the individual and dyadic levels. Couple programs that teach coping skills ... have demonstrated promising results in improving aspects of marital functioning. (Ledermann et al. 2010, p. 204)

The above quote addresses families' social (work) environments in general. Regarding same-sex couples,

> Community initiatives to strengthen families could emphasize the importance of staying in touch and getting along despite disagreements among family members; these campaigns could identify sexual orientation as a topic where adult family members can learn to disagree with each other while agreeing to provide a loving environment for children that is free from conflicts between adults. (Oswald and Lazarevic 2011, p. 383)

Ben Glass/Warner Bros. Pictures/Everett Collection

Many observers strongly criticize the way that American culture tends to equate love with infatuation, or chemistry. "Every pop-cultural medium portrays the heights of adult intimacy as the moment when two attractive people who don't know a thing about each other tumble into bed and have passionate sex." But infatuation "merely brings the players together.... Relationships live on time" (Lewis, Amini, and Lannon 2000, pp. 205–207). We need to move from infatuation to "the deep connection that is the hallmark and destination of true love" (Love 2001, p. xi). Positive communication is critical to this process.

Meanwhile, to close this chapter, we note that not every conflict can be resolved, even between the fairest and most mature individuals. If an unresolved conflict is not crucial, then the two may simply have to accept their inability to resolve that particular issue and let it go. Family cohesiveness, as well as supportive couple relationships, has much to do with commitment, gentleness, and humor, and on letting our loved ones know how much we care about and appreciate them—a task largely accomplished by little gestures such as a touch or hug, and also by sharing ourselves and listening with genuine interest (Love and Stosny 2007).

Summary

- Members of cohesive families express their appreciation for each other, have a high level of commitment to the family group as a whole, do things together, know how to deal positively with stress or crises, and evidence positive communication patterns.
- From a family ecology perspective, cohesive families depend on supportive external social environments (the economy and work environments, for instance), together with positive communication patterns.
- There is evidence that a family's having a spiritual orientation is positively related to cohesiveness.
- Research on couple communication indicates the importance to relationships of both positive communication and the avoidance of a spiral of negativity.
- Although some family communication patterns may reach the point of pathology, family conflict itself is an inevitable part of normal family life.
- Although arguing is a normal part of the most loving relationships, there are better and worse ways of managing conflict.
- Alienating practices, such as belligerence and the Four Horsemen of the Apocalypse—contempt, criticism, defensiveness, and stonewalling—should be avoided.
- Constructive arguing habits may not only resolve issues but also bring participants closer together.
- Constructive arguments are characterized by efforts to be gentle and by de-escalation of negativity. No one loses.
- There is no such thing as conflict-free conflict.

Questions for Review and Reflection

1. We often hear that communication is important to maintaining family relationships. Can you discuss some specific reasons why this is true?
2. Explain the interactionist theoretical perspective on families, and show how John Gottman's research illustrates this perspective.
3. Describe the Four Horsemen of the Apocalypse. If someone you care for treated you this way in a disagreement, how would you feel? Do you ever treat others with one or more of the "Four Horsemen"?
4. Discuss your reactions to each of the ten guidelines proposed in this chapter for constructive arguing. What would you add—or subtract?
5. **Policy Question.** Besides the suggestions in "Facts about Families: Ten Rules for Successful Relationships," what *society-wide* ideas might you offer for maintaining loving relationships?

Key Terms

belligerence 287
contempt 287
criticism 287
defensiveness 287
displacement 286
emotional intelligence 295
family cohesion or closeness 278
female-demand/male-withdraw interaction pattern 290
Four Horsemen of the Apocalypse 287
mixed, or double, messages 293
negative affect 288
passive-aggression 286
positive affect 284
rapport talk 289
relationship-focused coping 285
relationship ideologies 283
report talk 289
sabotage 286
stonewalling 287

12

Power and Violence in Families

Learning Objectives

1. Define power and distinguish between personal and social power.
2. Describe the social psychological bases of relationship power.
3. Define the resource hypothesis, then discuss how cultural context impacts it.
4. Summarize the incidence of family violence, including trends and reasons for the trends.
5. Distinguish between intimate terrorism and situational couple violence.
6. Distinguish between child abuse, child neglect, and willful child neglect.
7. Describe the following three approaches to stopping family violence: separating victim from perpetrator, engaging the criminal justice system, and psychological therapy.

- Sarah applies for a promotion, and accepting it will mean moving to another city; Sarah's partner does not want to relocate.
- Antonio wants a new stereo for his truck; his partner would prefer to spend the money on ski equipment.
- Marietta would like to talk to her partner about what each of them does (and doesn't do) around the house, but her partner is too busy to discuss it.
- Greg feels that he gives more and is more committed to developing a satisfying, cooperative marriage than his wife is.
- Nicole and Paul go to counseling and parenting education classes in efforts to curtail violence in their family.

Each of these scenarios illustrates power in relationships. This chapter examines family power. We will look at a classic study on marital decision making, one indicator of marital, or **conjugal power**, then explore contemporary social science research on conjugal power. We will discuss why playing power politics is harmful to intimacy and look at alternatives. We will then explore family violence as an unfortunate, too often tragic, consequence of abuse of power. We will explore intimate partner violence (IPV), then address child maltreatment—a form of family violence that is often present along with IPV. We close this chapter with a discussion of ways that family violence can be addressed and lessened. We begin by defining power.

What Is Power?

Power is the ability to exercise one's will. Power exercised over oneself is *personal power*, or autonomy. Having a comfortable degree of personal power is important to self-development. *Social power* is the ability to exercise one's will over others. Social power may be exerted in different realms, including within relationships. Parental power, for instance, operates between parents and children.

Until fairly recently analysis of power in couples focused on marriage, but research has now been extended to include unmarried couples, both heterosexual cohabitors and same-sex. However, the majority of studies that we have now still focus on power dynamics in heterosexually married couples. The extent to which research findings on conjugal power can be generalized to same-sex or to nonmarried heterosexual couples is generally untested and therefore open to speculation—including yours.

Relationship power involves (a) *objective measures of power*—who makes more, or more important, decisions, or who does more housework, for example—and (b) *subjective measures of fairness*—whether each partner feels that their arrangement is a fair, or equitable one. Objective and subjective measures of couple power often yield similar results. This is not always the case, however. For example, a career-invested spouse who makes most of the important family decisions and rarely does housework may perceive the relationship as fair—although from an objective viewpoint the distribution of power is not equal. Feelings of fairness can be grounded in an **equality** standard whereby partners equally share rights and responsibilities in the relationship. For example, partners buy a new TV set and divide the cost in half. On the other hand, fairness can be thought of in terms of **equity**—whether the rewards of the relationship feel subjectively proportional to each partner's contributions, which may not necessarily be equal. In this case, partners buy a new TV set, and the one earning more money foots more of the bill.

Objective measures of equality, together with partners' subjective *perceptions* of fairness, influence relationship satisfaction and commitment, but the perception of fairness is generally more significant. Moreover, when partners perceive themselves as reciprocally respected, listened to, and supported, they are more apt to define themselves as equal partners. They are also less depressed, generally happier, and more satisfied with their relationship (Corra et al. 2009; Greenstein 2009; Sullivan and Coltrane 2008). Another way to think about relationship or family power is to consider the concept, *power bases.*

Power Bases

Two psychologists (French and Raven 1959) developed a typology of six bases, or sources, of social power: coercive, reward, expert, informational, referent, and legitimate power. These bases of social power can be applied to the family, and we use them in this chapter in analyzing couple power relationships (see Table 12.1).

Table 12.1 Bases of Social Power As Applied to the Family

Type of Power	Source of Power	Example
Coercive Power	Ability and willingness to punish the partner	Partner sulks, refuses to talk, and withholds sex; physical violence
Reward Power	Ability and willingness to give partner material or nonmaterial gifts and favors	Partner gives affection, attention, praise, and respect to partner, and assists him or her in realizing goals; takes over unpleasant tasks; gives material gifts
Expert Power	Knowledge, ability, judgment	Savings and investment decisions shaped by partner with more education or experience in financial matters
Informational Power	Knows more about a consumer item, child rearing, travel destination, housing market, health issue	Persuades other parent about most effective mode of child discipline, citing experts' books
Referent Power	Emotional identification with partner	Partner agrees to purchase of house or travel plans preferred by the other because she or he wants to make partner happy
Legitimate Power	Society and culture authorize the power of one or the other partner, or both	In traditional marriage, husband has final authority as "head" of household; current ideal is that of equal partners

Source: Typology of power concepts from French and Raven (1959). Specific wording of definitions and the examples are by this textbook's authors.

Coercive power is based on the dominant person's ability and willingness to punish the partner with psychological–emotional abuse or physical violence or, more subtly, by withholding favors or affection (Davies, Ford-Gilboe, and Hammerton 2009, p. 28). Slapping a mate and spanking a child are examples of *coercive power*; so is refusing to talk to the other person—the silent treatment. **Reward power** is based on an individual's ability to give material or nonmaterial gifts and favors, ranging from emotional support—for instance, eye contact, a smile, a gentle hand on a shoulder, listening—and and attention to financial support or recreational travel.

Expert power stems from the dominant person's superior judgment, knowledge, or ability. Although this is certainly changing, our society traditionally attributed expertise in such important matters as finances to men, while women were attributed special knowledge of children and expertise in the domestic sphere. **Informational power** is based on the persuasive content of what the dominant person tells another individual. A partner may be persuaded to charge less on a credit card when the other shares information on the card's high interest rate.

Referent power is based on a person's emotional identification with the partner. A partner who attends a social function when he or she would rather not "because my loved one wanted to go and so I wanted to go too" has been swayed by *referent power*. In happy relationships, *referent power* increases as partners grow older together (Raven, Centers, and Rodrigues 1975; Pyke and Adams 2010).

Finally, **legitimate power** stems from the dominant individual's ability to claim authority, or the right to request compliance. *Legitimate power* in traditional marriages involves acceptance by both partners of the husband's role as head of the family. Although this is not the case for all families in the United States, the current ideal in mainstream culture is an egalitarian couple partnership (Gerson 2010).

Throughout this chapter, we will see the various power bases at work. For instance, the consistent research finding that economic dependence of one partner on the other results in the dependent partner's being less powerful may be explained by understanding the interplay of both *reward power* and *coercive power*. If I can reward you with financial support—or threaten to take it away—then I am more able to exert power over you. We turn now to look at research on marital power and the theoretical perspectives used to explain couple power relationships.

DonSmetzer/Stone/Getty Images

Although an older generation may hold to traditionally legitimated patriarchal power, the next generation may renegotiate and consciously change those roles, especially as women assume more autonomy and make gains in the workplace. In this photo, the classic card game of Hwoa-tu and the Chinese vase and screen in the background suggest a world of traditional authority, a man's legitimate power as head of the family. The posture and clothing of the younger family members suggest more casual and democratic family relations.

Classical Perspectives on Marital Power

Research on marital power began in the 1950s. At that time—before the feminist movement of the 1970s—interest in marital power was more academic than political. Today research on marital power, particularly that on violence, is highly politicized, a point that we will return to later in this chapter. In the 1950s social scientists Robert Blood and Donald Wolfe were curious about how married couples made decisions. Their research resulted in *Husbands and Wives: The Dynamics of Married Living* (1960), an important book that significantly shaped thinking on couple power.

The Resource Hypothesis

Blood and Wolfe's research popularized the resource hypothesis, derived from the exchange theoretical perspective, described in Chapter 2. The **resource hypothesis** posits that the partner with more resources can exchange them for greater power in the relationship. Resources primarily include earnings and education, the latter resulting in informational and expert power. In Blood and Wolfe's research, the resource hypothesis was supported by the finding that the spouse with higher earnings and educational attainment made more decisions. Relatedly, Blood and Wolfe found that a wife's power was greater when she had no young children and when she worked outside the home—situations that made her less financially dependent on her husband. Today's research continues to show that wives' relative income increases their say in important decisions (Amato et al. 2003, 2007).

Resources and Gender The Blood and Wolfe study made a major contribution by encouraging researchers to see couple power as based on each partner's relative resources. However, the study was strongly criticized, primarily because it ignored sources of power other than individual resources—namely, the power of gender expectations, norms, and socialization (Safilios-Rothschild 1970; Tichenor 2010). Accordingly, Blood and Wolfe came under heavy fire for their naïve conclusion that a patriarchal power structure had been replaced by egalitarian marriages.

Feminist Dair Gillespie (1971) pointed to a situation that persists today, although to a lesser extent: Power-granting resources are socially structured by gender and hence unevenly distributed in heterosexual relationships. Despite social change, the majority of husbands earn more than their wives and hence have greater access to economic *resources*. Better educated than wives at the time of the Blood and Wolfe study in the 1950s, husbands were also likely to have more *expert* and *informational power* than wives. Even men's greater physical strength can be an important resource, granting actual or potential coercive power (Collins and Coltrane 1995). According to Gillespie, the resource hypothesis, which presents resources and power as gender-free, was simply "rationalizing the preponderance of the male sex" (1971, p. 449). Although more couples today are moving toward equal, or **egalitarian relationships**, research continues to support Gillespie's insight that American marriages are not yet fully equal (Gerson 2010; Spade and Valentine 2010).

Resources in Cultural Context In some cases patriarchal norms may be strong enough to override personal resources and legitimate greater conjugal power for husbands (Safilios-Rothschild 1967; Blumberg and Coleman 1989). The concept of **resources in cultural context** stresses that society-wide gender structures (see Chapter 3) influence conjugal power, tempering the impact of relative individual resources. Individual resources fully influence conjugal power only when there is no cultural norm for conjugal power—either an **egalitarian norm** or a **patriarchal norm**. When traditional norms of male authority are strong, husbands will likely dominate regardless of the partners' personal resources. Also, if an egalitarian norm were to be thoroughly accepted society-wide, partners would share equal power regardless of their relative economic, educational, or other resources. Only in societies or situations where neither patriarchal nor egalitarian norms are firmly entrenched is power negotiated by couples according to their relative resources (Stevenson and Wolfers 2006; Wight, Bianchi, and Hunt 2012).

Current Research on Couple Power

Current research measures couple power in the following four ways:

1. *Decision making:* Who gets to make decisions about everything from where the couple will live to how they will spend their leisure time?
2. *Division of labor:* Who provides income? Who does the household labor? Who takes primary responsibility for child care, if there are children?
3. *Allocation of money* earned by either or both partners: Who controls household spending? Who has personal spending money?
4. *Ability to influence* the other partner and feeling comfortable in raising complaints about the relationship.

We will review the research on decision-making, household labor, control over money, and the ability to influence one's partner.

Decision Making

A national survey in 2000 compared conjugal decision making in 1980 with that in 2000 and found that in 2000, "respondents were significantly more likely to report equal-decision-making" (Amato et al. 2003, p. 9). Even wives in evangelical families often have more decision-making power than their formal submission to the male family head would indicate. In fact, some research shows that "co-parenting and joint decision-making are more common in evangelical homes than in secular and mainline religious households" (Bartkowski and Read 2003, p. 88; Vaaler, Ellison, and Powers 2009).

On the other hand, a study of wives who earn more than their husbands suggests that "the gender structure exerts an influence that is independent of breadwinning or relative financial contributions" (Tichenor 2005, p. 117). A residual sense of the propriety of traditional male privilege—that is, *legitimate power*—ascribed more authority to men even in situations where they had fewer relative resources. "Just as women's income does not buy them either relief from domestic labor or greater financial power..., it does not give them dominion in decision making" (p. 117; see also Treas and Tai 2012).

In an interesting twist, women can sometimes gain power from their greater knowledge of the household. They can use this *informational power* to shape decisions about purchases and household arrangements, as we noted earlier.

Division of Household Labor

Social scientists use housework as one criterion of power on the assumption that no one really wants to do it. As discussed in Chapter 10, women's satisfaction with the fairness of how household labor is allocated is strongly associated with relationship happiness and commitment. Conversely, when a wife has more egalitarian expectations than her spouse fulfills, depression and marital conflict likely follows (Gerson 2010; Greenstein 2009; Sullivan and Coltrane 2008).

Meanwhile, women whose husbands work more hours than they do are apt to see the division of household labor as fair (Parker and Wang 2013). So are women who espouse more traditional gender values. Perhaps their expectations are shaped by a religious doctrine of separate spheres and male headship (Myers 2006; Nock, Sanchez, and Wright 2008). Still, even in evangelical couples, there is an implicit acknowledgment of a norm of equality in the attention given by evangelical men to expressing great appreciation for their wives' doing the preponderant share of housework. In the context of a societal egalitarian ideal, the additional domestic work of evangelical wives becomes a "gift," which is reciprocated by the husband's emotional work of expressed appreciation in an "economy of gratitude" (Vaaler, Ellison, and Powers 2009; Wilcox 2004, p. 154, referencing Hochschild 1989).

Recent decades have seen a significant increase in men's share of housework (Eckel 2010). Nevertheless, the fact that women continue to do more housework than men do is seen as an objective indicator of their relatively less conjugal power (Gaunt 2013). As shown in Figure 12.1, when the three categories are calculated together, mothers and fathers now spend about the same number of hours in paid work, housework and child care. In fact, fathers spend an average of one more hour per week—54 hours for fathers, compared with 53 for mothers. We might note here, though, that what is labeled "leisure" for women is often indirectly child care and household management; taking the kids to the park would be an example (Gerson 2010).

Partners may be inclined to see a division of labor similar to the one depicted in Figure 12.1 as fair. It is important to note, however, that such a division of labor does not readily facilitate a woman's maximizing her career potential or advancing into top management positions (Sandberg 2013a). Being saddled with more than half the housework and child care responsibilities takes time from women's career participation and development. Accepting this situation at home renders wives and mothers—even potential wives and mothers—less effective in the labor force (Sandberg 2013b).

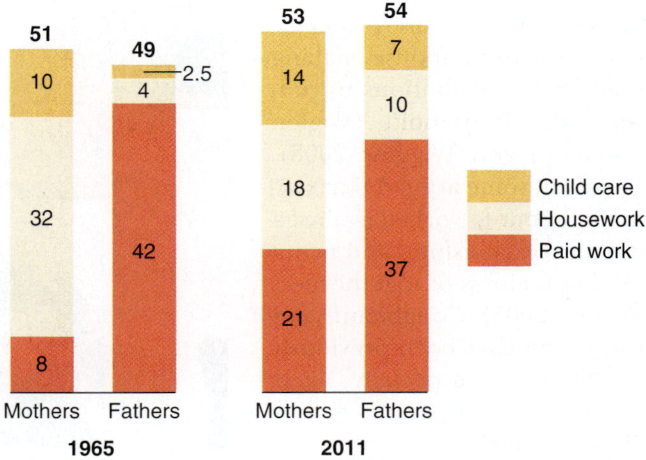

FIGURE 12.1 How mothers and fathers spend their time, 1965 and 2011: Average number of hours per week spent on...

Source: Pew Research Center, "Modern Parenthood." Available at http://www.pewsocialtrends.org/2013/03/14/modern-parenthood-roles-of-moms-and-dads-converge-as-they-balance-work-and-family/

Money Allocation

Research on couples' **money allocation systems**—whether they pool their money and who controls pooled or separate money—is relatively recent. British social policy scholar Jan Pahl (1989) described historical money allocation systems that subsequent researchers have used or adapted (e.g., Kenney 2006; Vogler 2005). According to Pahl, in the industrial era—from about the 1800s through about the 1950s—a family's allocation system was typically one of complete control of his earnings by the male breadwinner, who doled out a housekeeping allowance to his homemaker wife. The allowance was often rather skimpy, while the husband was privileged to take money "off the top" for personal spending and recreation.

Even before the second wave of feminism emerged in the 1970s, husbands' complete control of the money began to be viewed as inappropriate to a companionate model of marriage. Accordingly, the typical conjugal money allocation system became one of pooled resources whereby the husband's earnings were deposited into an account maintained in both names and theoretically controlled jointly by the spouses. However, given male dominance in decision making in this era, the husband usually controlled the pooled account. Moreover, nonearning women typically felt uncomfortable making decisions about "his" earnings. Hence in reality the joint account was not jointly controlled.

By the 1970s, feminists had begun to criticize the husband-controlled joint pool system. With more wives now earning income, separate financial accounts and control began to be seen as a favored alternative, with each spouse making equitable contributions to running the household (Vogler, Lyonette, and Wiggins 2008). However, some argued that separate accounts worked against a couple's establishing and maintaining feelings of togetherness (Vogler 2005). Cohabitants and those who have been previously divorced are especially likely to maintain separate money (Kenney 2006).

At present, a variety of allocation systems operate in American marriages, involving two dimensions: whether to pool and who controls—husband, wife, or both? Meanwhile, gender continues to play a strong role, and men seem to retain more control over the family's income no matter who earns it. Men are especially likely to retain personal spending money and/or to feel free to spend the family's income on personal and recreational desires without consulting their partner. Moreover, women may spend some of their "personal" money on household and children's needs, such as new linens, kids' clothing, or family recreation. And even otherwise egalitarian-oriented men may assume they have "veto power" over major decisions (Vogler, Lyonette, and Wiggins 2008). This would seem to be an example of the ongoing influence of traditional *legitimate power* ascribed to a male, overriding the wife's *resource power*.

Ability to Influence the Other

In supportive and satisfying relationships, each partner feels confident that he or she can be heard, even when raising uncomfortable issues. Partners can air concerns without fear of being dismissed or otherwise treated badly. Each has power to influence the other. Where this is not the case, research suggests that husbands wield more power. Men's power may not always be visible, but they may be able to suppress—or ignore—issues raised by their partners (Tichenor 2005). Meanwhile, Rather than risking confrontation, "The spouse with less power [usually the wife]

Pixland/Jupiter Images

This family is having breakfast in a household where roles may be somewhat differentiated by gender. Working for wages or even outearning a husband are resources that enhance a wife's conjugal power. However, she still may not be a fully equal partner (Teachman 2010). In this photo, is mom standing so that she can more readily serve whoever might need something?

A Closer Look at Diversity

Mobile Phones, Migrant Mothers, and Conjugal Power

Many times in this text we say that technology has dramatically changed—and will continue to change—family interaction. But we had not thought of this example until we came across this study in the journal, *New Media & Society*. The study looks at conjugal power in transnational Filipino families, power between "left-behind" fathers and migrant mothers who leave the Philippines to get better paying jobs (often in health care fields) in wealthier nations.

Researchers visited with the fathers and their children in ten Filipino homes. The researchers concluded that, "while the mobile phone can lead to increasing cooperation between left-behind fathers and migrant mothers, it has mostly resulted in exacerbating the already tremendous chasms that divide them" (Cabanes and Acedera 2012, p. 916). How did the authors come to this conclusion? For one thing, the mobile phone can facilitate ongoing conflict. Said one husband,

> [My wife and I] always fight because of texting. She keeps on saying that every time I text her, there is nothing else I talk about except financial problems ... But sometimes, the only thing I want is for her to show some sympathy. It can be very overwhelming when you're alone, you know. (Cabanes and Acedera 2012, p. 923)

The authors explain that mobile phones and related technologies, such as Skype and facetime, allow family members to be virtually in two places at once. This is a dramatically new development in history. Not so long ago communication between family members in different countries would have depended on letters, handwritten on paper and sent via "snail mail"—a process that could take weeks, even months. Today family members who are nations apart can readily phone or text each other, a development that allows them to communicate daily, hourly, or even more often.

Meanwhile, a migrating mother reverses traditional Filipino conjugal power, because she becomes the family breadwinner, sending money back to her family. Her left-behind husband becomes responsible for homemaking and child care.

> The mothers' crossing over to the role of provider tends to be very bruising to the fathers' egos. And the mobile phone appears to be central to this experience. Many fathers share that their wives "abuse" mobile phone calls .... Said one father, "When that thing rings, it does not matter what time it is or where I am. I have to answer. And I have to answer well." (Cabanes and Acedera 2012, p. 923)

How does a mother abuse a mobile phone call, in her husband's opinion? The mothers call regularly to help ensure that the money they're sending home isn't going to waste. One husband's wife accused him of throwing away money on gambling. Another husband explained a somewhat different situation. He resisted in small ways, such as by delaying answering his wife's phone calls and text messages.

Despite the authors' conclusion that the mobile phone primarily increased the power struggles between these spouses, the mothers also often called to ask their children to help their fathers with the household chores. And fathers texted their wives for help with children's homework or discipline (Cabanes and Acedera 2012).

Critical Thinking

Has the mobile or iPhone changed the power dynamics in any of your relationships? Why or why not? If you do see changes, how would you describe them? In your opinion, are these changes for the better, for the worse, or both?

typically spends more time aligning emotions with [the spouse's] expectations" (Coltrane 1998, p. 201).

On the other hand, a wife may wield considerable conjugal power by means of referent power, her husband's love for her (but see Lois 2010). Then too, men have been much influenced by the cultural expectation that they should respond favorably to their wives' raising concerns (Connell 2005). As discussed in Chapters 3 and 11, women have traditionally been socialized to be more attuned to relationship dynamics. However, men today *are* encouraged to engage in "emotion work," "to express emotion to their wives, to be attentive to the dynamics of their relationship and the needs of their wives, [and] ... to set aside time for activities focused especially on the relationship" (Wilcox and Nock 2006, p. 1322).

Diversity and Marital Power

Because the United States is a pluralistic society, we expect to find varied visions and enactments of marital power. One example of diversity and conjugal power is described in "A Closer Look at Diversity: Mobile Phones, Migrant Mothers, and Conjugal Power." We'll look at three general types of conjugal power now: egalitarian, neotraditional, and gender-modified egalitarian unions. If we were to place these on a continuum from most to least equal, egalitarian unions would be most equal and neotraditional unions, least so. Gender-modified egalitarian unions would fall somewhere in between.

Egalitarian Unions In egalitarian unions partners share equally in the four components of couple power: decision making; division of household labor, particularly housework; money allocation; and ability to raise relationship issues. Whether egalitarianism is fully realized or not, social scientists have generally assumed that egalitarian unions are the most sought after by U.S. couples. Sociologist Steven Nock has described *mutually economically dependent spouses* as dual-earner couples in which each spouse earns between 40 and 59 percent of the family income and thus are "equally dependent on one another's earnings" and hence—due to equalization of earnings resources—more egalitarian (2001, p. 755; and see also Amato and Hohmann-Marriott 2007).

Marriage would be grounded in "extensive dependencies by both partners," as was the case in traditional society. As Nock sees it, the decline in divorce rates since the early 1980s may reflect "the gradual working out of the gender issues first confronted in the 1960s. If so, this implies that young men and women are forming new types of marriages that are based on a new understanding of gender ideals" (Nock 2001, p. 774).

Another possibility is that society-wide equality *norms* could become so strong that spouses would share power equally regardless of their respective individual resources. This situation would be a new form of legitimate power—one that justifies and supports absolute objective equality regardless of resources. Pepper Schwartz's research on *peer marriage* (also referred to in the literature as "post-gender" or "equal sharer" unions) offers an example of a strong equality norm at work (1994, 2001; see also Hall and MacDermid 2009). Schwartz identified couples who work to fashion their marriage according to an egalitarian ideal and continually monitored and negotiated family responsibilities, perhaps forgoing male career advancement in order to equally pursue female workplace success. Presently lesbian couples seem to have come closest to attaining this ideal. For example, the resources that each brings to the relationship do not affect each person's power (Jeong and Horne 2009).

As gender norms move from traditional to egalitarian, all family members' interests and preferences gain legitimacy, not only or primarily those of the husband. This situation means that more and more family decisions must now be consciously negotiated. Although a possible outcome of such conscious negotiating can be increased conflict, another is greater intimacy.

Neotraditional Unions A second model favors a traditional division of labor and male family leadership. What makes this model different from the traditional models of the nineteenth and early twentieth centuries is the melding of traditional ideals with a new (neo) egalitarian spirit. Evangelical Christians and other conservative religious groups tend to embrace this model, characterized by a gendered division of labor, formal male dominance in decision making, and an "egalitarian spirit."

Although a husband's dominant power is legitimated in this milieu, marital power in practice is often negotiated. First "articulated by evangelical feminists," the "mutual submission" (of husband and wife to each other) has become increasingly popular because it justifies the shared decision making that characterizes many evangelical marriages (Dolan 2008, p. 32; Vaaler, Ellison, and Powers 2009). Another way in which a norm of equality is represented in these ostensibly husband-dominant marriages is in an "economy of gratitude" (Pugh 2009, p. 6; Wilcox 2004, p. 154), as husbands display appreciation for their wives' "gift" of household work. When compared to traditional conjugal power, the sharp edges of male dominance are softened in the **neotraditional family**, as the title of sociologist Bradford Wilcox's book—*Soft Patriarchs* (2004)—suggests.

Gender-Modified Egalitarian Unions The strictly egalitarian marriage model might not represent what most couples want—at least not now (Wilcox and Nock 2006). In the **gender-modified egalitarian model**, absolute equality is diminished by the symbolic importance of maintaining fairly traditional, comfortable, and familiar gender roles. Compromise with the fully egalitarian ideal occurs "as spouses work together to construct appropriate gender identities and maintain viable marriages" (Tichenor 2005, p. 32). In this view,

> Because wives—even wives with egalitarian attitudes—have been socialized to value gender-typical patterns of behavior, wives will be happier in marriages with gender-typical practices in the division of household labor, work outside the home, and earnings. Because husbands—even husbands with egalitarian attitudes—have been socialized to value gender-typical patterns of behavior, husbands will be happier in marriages that produce gender-typical patterns. (Wilcox and Nock 2006, p. 1328)

In their study of over 5,000 couples drawn from the 1992-94 National Survey of Families and Households, Wilcox and Nock found support for the hypothesis that "the gendered character of marriage seems to remain sufficiently powerful as a tacit ideal" (2006, pp. 1339–40).

Immigration and Marital Power Recent immigrants from traditional cultures, such as in Eastern Europe, Asia, or Central and South America, typically arrive in

Comstock Images/Getty Images

It seems from these partners' body language that they share various aspects of couple power on a fairly equal basis. What conditions might influence which one does more housework? Makes more of the important couple decisions? Has more say in how their money is spent?

the United States having fairly traditional gender roles and conjugal power relations. As they assimilate, subsequent generations tend to adopt less traditional patterns according to which "husband–wife relationships are more flexible and negotiated ... [and] socioeconomic achievements become the basis for negotiation within the family" (Cooney et al. 1982, p. 622).

Even among native-born Americans, we must recognize the continuing salience of tradition and the assumption that to some degree it is legitimate for husbands to wield family authority (Nock, Sanchez, and Wright 2008).

Power Politics versus Freely Cooperative Relationships

We've seen that relationships perceived as fair and equitable are more apt to be stable and satisfying. Power asymmetry often characterizes dissatisfied couples (Gottman et al. 1998). Not seeking to "win" an argument and highly respecting each other facilitates satisfying relationships (see Chapter 11). Put another way, supportive partners avoid **power politics**.

In a union characterized by playing power politics, partners lock into a relationship-damaging cycle of behaviors. Sometimes (or often) hinting at leaving the marriage, partners alternate in acting sulky, critical, or distant. The sulking, critical, or distant partner carries on this behavior until she or he fears that the mate will "stop dancing" if it goes on much longer. Then it's the other's turn. This seesawing creates alienation and loneliness for both partners (Blumberg and Coleman 1989; Chafetz 1989).

Developing a Freely Cooperative Relationship

There are alternatives to this kind of power struggle. Blood and Wolfe (1960) observed one alternative in which partners grew increasingly separate in their decision making as they took charge in separate domains—one person buying the family car perhaps and the other, managing the kitchen. This alternative can be a poor one for partners wanting intimacy, because it enforces separateness rather than the sharing.

There may be better options, however. As discussed in Chapter 11, communication expert John Gottman and his colleagues advise partners to share power if they want to be happy together (Coan and Gottman 2007; Gottman et al. 1998). "As We Make Choices: Domination and Submission in Couple Communication Patterns" illustrates communication patterns of dominance and submission.

Partners who see themselves as mutually respected, equally committed, and listened to when they raise concerns are more likely to see their relationship as egalitarian and are more satisfied overall with their relationship. Changing power patterns can be difficult, because usually power patterns have been established from the earliest days of the relationship. Certain behaviors not only become expected but also come to have symbolic meaning: "She always grocery shops and cooks for me, and therefore I know she loves me," for example. Sociologist William Goode offers the following insight:

> Men have .... occupied the center of the stage, and women's attention was focused on them .... [But] the center of attention shifts to women more now than in the past. I believe that this shift troubles men far more, and creates more of their resistance, than the women's demand for equal opportunity and pay in employment. (Goode 1982, p. 140; see also Herrera 2012)

Nevertheless, the best way to work through power changes is not to deny them but to openly negotiate, using communication techniques described in Chapter 11. At the same time, it's good to remember that, although it promises a more rewarding relationship in the long run, changing a power relationship is a challenge to any relationship and can be painful for both partners. One option for handing power change is to pursue family or relationship counseling. Whether on

As We Make Choices

Domination and Submission in Couple Communication Patterns

Our communication patterns reflect the power relationships with whomever we are communicating. We can classify communication patterns as illustrating (1) dominance, (2) submission, or (3) mutual cooperation. Below are a few questions that researchers use to measure dominance or submission.

Measuring Dominance—Agree or Disagree

1. When we disagree, my goal is to convince my partner that I am right.
2. When we argue or fight, I try to win.
3. I try to take control when we argue.
4. I rarely let my partner win an argument.
5. When we argue, I let my partner know I am in charge.

Measuring Submission—Agree or Disagree

1. I give in to my partner's wishes to settle arguments on my partners' terms.
2. When we have conflict, I usually give in to my partner.
3. I surrender to my partner when we disagree on an issue. Sometimes I agree with my partner just so the conflict will end.
4. When we argue, I usually try to satisfy my partner's needs rather than my own (Zacchilli, Hendrick, and Hendrick 2010, p. 1081).

Critical Thinking

Do you recognize yourself in any of these questions? What might be some questions designed to measure mutual cooperation in couple communication? What are some ways that partners change their dominant or submissive communication patterns?

their own or with help, partners can choose cooperation over playing power politics. Unfortunately, when freely cooperatively relationships do not develop, one result can be family violence. We turn to that topic now.

Family Violence

Using emotional or physical violence to manifest power in families has occurred throughout history, but only in the last fifty years has family violence been labeled a social problem. First, child abuse was recognized as a social problem in the 1960s (Kempe et al. 1962). That recognition was followed in the 1970s—in conjunction with the Second Wave of the Women's Movement (see Chapter 3)—by attention to wife abuse, then typically called "wife beating" or "wife battering." Gradually researchers and policy makers expanded the concept, *wife abuse,* to **intimate partner violence (IPV)**—the physical or emotional abuse of spouses of either gender, cohabiting or noncohabiting heterosexual or same-sex relationship partners, or former spouses or intimate partners. The federal government includes all these forms of couple violence in its reports on IPV.

With the 1980s came concern about elder abuse; and later, about husband abuse. More recently, researchers have begun to address sibling violence and child-to-parent violence. Dating violence is discussed in Chapter 5; elder abuse and neglect, in Chapter 16. We discuss many of the remaining forms of family violence here. We'll look first at data sources for intimate partner violence.

Skjold Photographs/PhotoEdit

Gay and lesbian couples are more likely to share power and domestic duties than are heterosexual couples. At the same time, some same-sex couples do experience family violence, a situation that went ignored for many decades but has now begun to be researched and otherwise addressed.

IPV Data Sources

Data on physical intimate partner violence come from four types of sources: (1) violent crime reports filed in police departments, then reported to the FBI; (2) national

Facts about Families

Major Sources of Family Violence Data

There are several sources of family violence data. Here are the major ones:

1. The work of Murray Straus, Richard Gelles, and their colleagues in their National Family Violence Surveys pioneered and shaped the scientific study of family violence. Having developed the Conflict Tactics Scale, the research group undertook a household survey in 1975, followed by another ten years later. Together, the surveys produced data from more than 8,000 husbands, wives, and cohabiting individuals (Straus, Gelles, and Steinmetz 1980; Straus and Gelles 1986, 1988, 1995; Gelles and Straus 1988).
2. The National Violence Against Women Survey (NVAWS), commissioned by the National Institute of Justice and the Centers for Disease Control and Prevention, was conducted between 1995 and 1996 and employed a modified version of the Conflict Tactics Scale. Despite some methodological limitations (for instance, the survey used random digit dialing and thus did not include women without telephones), NVAWS is an important resource on violence against women (Chen and Ullman 2010).
3. The National Crime Victimization Survey (NCVS), conducted every two years by the Federal Bureau of Justice Statistics (BJS) is a national survey that asks respondents about any physical violence they have experienced as victims, their relationship to the perpetrator, and whether the violence was reported to the police. Violent acts covered in this survey include assault and rape, or sexual assault. The Bureau of Justice Statistics also reports on intimate partner violence. Spouses, ex-spouses, and current or former boyfriends or girlfriends, including same-sex partners, are considered *intimate partners.*
4. The Uniform Crime Reports issued by the Federal Bureau of Investigation (FBI) provide data on criminal incidents reported to the police by many (though not all) local law enforcement agencies. A weakness of these data is that many crimes are not reported to the police, including an estimated half of intimate partner violent crimes (Groves and Cork 2008). However, because most homicides are reported, the homicide data are more valid (Loftin, McDowall, and Fetzer 2008).
5. The National Child Abuse and Neglect Data System, directed by the U.S. Department of Health and Human Services, collects data from government child protection agencies and other community agencies voluntarily submitted by states. The collected data results in an annual report, called *Child Maltreatment.*

Critical Thinking

Why might a survey that asks a nationally representative sample of U.S. couples about whether they ever get violent with each other—including throwing something or shoving—yield significantly different results regarding IPV than a survey asking whether the respondent had even been a crime victim?

surveys of the general population; (3) smaller research studies, often qualitative, that relate persons' experience with family violence; and (4) reports from social workers, counselors, or volunteers at hospital emergency rooms or in other settings. "Facts About Families: Major Sources of Family Violence Data" gives more detail on major family violence data sources. These four IPV data types give divergent pictures of IPV, a point that we will return to for further exploration later in this chapter. Researchers continue to work at explaining the differences in findings and at getting an accurate picture of family violence. Here we turn attention to pioneering survey research that helped bring family violence to public attention and remains highly influential.

The National Family Violence Surveys The 1970s Women's Movement raised attention to wife abuse, particularly to serious injuries observed in hospital emergency rooms. Meanwhile, the early and continuing research of Murray Straus, Richard Gelles, and their colleagues in their **National Family Violence Surveys** shaped scientific research on family violence among the general population. The research group undertook a national household survey in 1975, followed by another one ten years later (Straus, Gelles, and Steinmetz 1980; Straus and Gelles 1986, 1988, 1995; Gelles and Straus 1988).

Interested in *physical* domestic violence, the research team defined family violence as "an act carried out with the intention of causing physical pain or injury to another person." The researchers developed a measure of family violence called the **Conflict Tactics Scale**. Respondents were asked whether they had done the following and how often:

- threw something at the other
- pushed, grabbed, or shoved
- slapped or spanked (a child)
- kicked, bit, or hit with a fist

- hit or tried to hit with something
- beat up the other
- burned or scalded (children) or choked (spouses)
- threatened with a knife or gun
- used a knife or gun (Straus and Gelles 1988, p. 152).

Severe violence was defined as acts that have a relatively high probability of causing an injury: kicking, biting, punching, hitting with an object, choking, beating, threatening with a knife or gun, using a knife or gun—and, for violence by parents against children, burning or scalding the child (Straus and Gelles 1988, p. 16). To this day many family violence researchers use the Conflict Tactics Scale, later modified to add sexual assault (Straus et al. 1996). It's important to note that the Conflict Tactics Scale is different from and broader than the crime categories of assault and homicide that form the basis of criminal justice system statistics. For instance, behaviors that tend to be less severe—throwing things, pushing, grabbing, or shoving—are counted in this research as violent acts. We return to this point later in this chapter. At this point we turn to the statistics on violence between intimate partners.

The Incidence of Intimate Partner Violence (IPV)

First, the good news: Intimate partner violence (IPV) has declined significantly over the past several decades. The incidence of child maltreatment, discussed later in this chapter, has declined as well (U.S. Department of Health and Human Services 2012b). These welcome declines show that efforts to combat family violence are paying off (Catalano 2007). The overall rate of nonfatal IPV declined by 64 percent between 1994 and 2010—from 9.8 victimizations per 1,000 persons age 12 or older to 3.6 per 1,000 (Catalano 2012). Figure 12.2 shows declining victimization rates for both women and men. Intimate partner homicides are down as well (Catalano et al. 2009).

Nevertheless, despite the fact that it has declined significantly, family violence makes up about 11 percent of all violence and one third of all police recorded violence (Catalano 2012). According to a report based on the Federal Bureau of Justice Statistics National Crime Victimization Survey, there were about 907,000 victimizations by intimate partners in 2010. One-third of these were serious violent crimes: rapes, sexual assaults, aggravated assaults, and crimes involving serious injuries or weapons. The other two-thirds were lesser offenses, mostly simple assaults (Catalano 2012).

Although rates of intimate partner violence have dropped for all race/ethnic categories, IPV rates vary greatly by race/ethnicity as Figure 12.3 indicates. Victimization rates are relatively high (11.1 per 1,000) for Native American women. Black women's rates are somewhat high (5.0). White and Hispanic females have moderate rates (4.0 and 4.3, respectively), whereas Asians have very low IPV rates. White, black, and Hispanic male victimization rates are low, whereas those of Native American men are relatively high (Catalano 2007). We note here that family violence crosses all generational, social, cultural, class, and religious groups. A Study comparing IPV rates in black-white couples with those in nonHispanic white and in black couples found that interracial couples have a

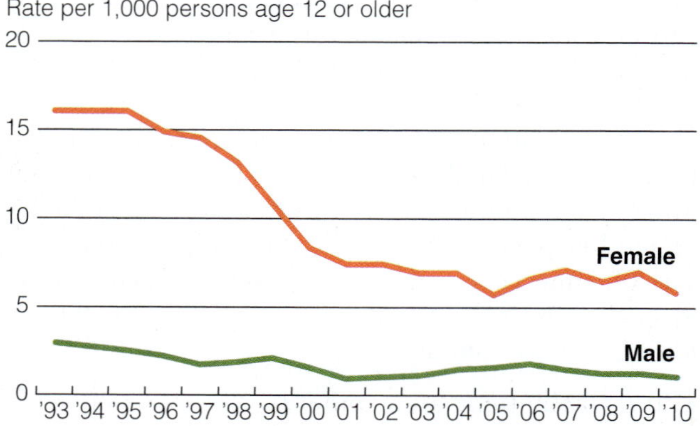

FIGURE 12.2 Intimate partner violence nonfatal victimization rates, by sex, 1993–2010.

Source: Catalano 2012, Figure 2.

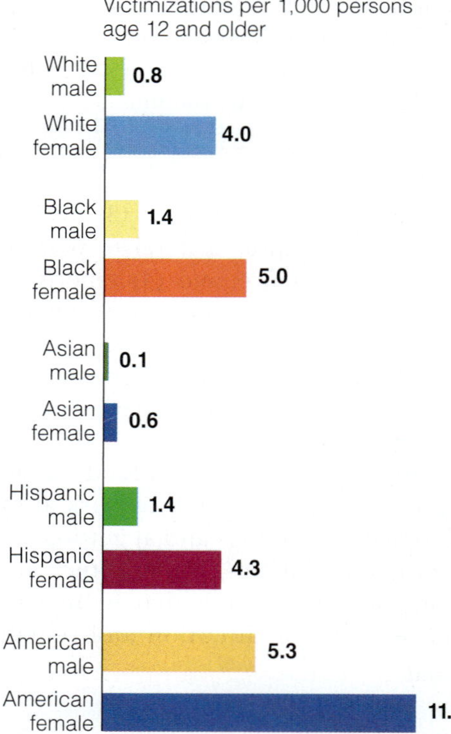

FIGURE 12.3 Intimate partner victimization rate by gender, race, and Hispanic origin, 1993–2005.

Source: Catalano 2007.

rate similar to black couples (and thereby higher than that of nonHispanic whites) (Martin et al. 2013).

Correlates of Family Violence

IPV perpetrators and victims tend to be young adults (Shortt et al. 2013). As discussed in Chapter 6, the rate of violence between cohabiting partners is significantly higher than that of marrieds (Anderson 2010; Dush 2011; Martin et al. 2013). Overall, cohabitors are younger, less integrated into family and community, and more likely to have psychobehavioral problems such as depression and alcohol abuse—all factors associated with family violence (Stets 1991). Another possibility is that there is less institutional control over cohabiting relationships than over marriage (Ellis 2006). Still another thesis is that the less-violent cohabiting couples end up getting married whereas more-violent married couples get divorced, sharpening the difference between the two groups (Kenney and McLanahan 2006).

Another correlate of family violence—for victims and perpetrators—involves having experienced or witnessed domestic abuse while growing up (Hendy et al. 2012). Children who grow up with family violence may come to regard it as normal. Both men and women are more likely to be IPV victims if they were abused as children or witnessed violence between parents (Lee et al. 2012; Pettit et al. 2010). Unfortunately, children reside in an estimated 35 percent of households where IPV takes place (Catalano 2007).

Whether victims of child abuse or witnesses of adult interpersonal violence, those who are exposed to family violence in childhood are more likely than others to abuse their own children and their spouses as well (Heyman and Slep 2002; Milner et al. 2010; Zielinski 2009). This does not mean that abused children are predestined to be abusive parents or partners. Gelles and Cavanaugh (2005) report an intergenerational transmission rate of 30 percent. That is much higher than the general average of 2 to 4 percent; nevertheless, "[t]he most typical outcome for individuals exposed to violence in their families of origin is to be nonviolent in their adult families. This is the case for both men and women" (Heyman and Slep 2002, p. 870).

Stress is another correlate of family violence. Financial troubles "pervade the typical violent home" (Gelles and Straus 1988, p. 85; Zielinski 2009). Overload—that is, having too much to deal with or to worry about—creates stress that can lead to family violence, including child abuse. We note here that African Americans and Latinos living in socially disorganized communities encounter more social stressors, on average, than non Hispanic whites, including work-related problems and financial strain, as well as interpersonal and institutional racism (e.g., at the hands of police, schools, public officials, lenders and landlords, and others) (Ellison et al. 2007).

Other causes of family stress are changing lifestyles and standards of living, children's misbehavior, and a parent's feeling pressure to do a good job but being perplexed about how to do it. Poverty and economic stressors, neighborhoods with higher concentrations of drug offense arrests and bars and liquor stores and higher rates of violence in general are correlated with IPV (Anderson 2010). Tending to be concentrated in disadvantaged neighborhoods and lack affordable housing options are also issues confronting more vulnerable segments of our population. These factors help to explain race/ethnic disparities in children's exposure to violence as well (Zimmerman and Messner 2013). Dealing with family stressors is the subject matter of Chapter 13.

In addition to stress, substance abuse, including heavy alcohol use and binge drinking—often resulting in impulsivity, together with the tendency to feel put down by another's fairly innocent remarks—is a factor in IPV (Kilmer et al. 2013; Wiersma et al. 2010). Drinking may also serve as a rationalization and excuse for violence that would have occurred in any case (Gelles 1974). Depression and other mental disorders are also associated with being either victim or perpetrator of family violence (Trevillion et al. 2012). Some studies have identified a pattern of assortative partnering, described in Chapter 5, whereby individuals at higher risk for family violence are mutually attracted and match risk factors for IPV—with high-risk individuals assortative selecting other high-risk individuals while low-risk individuals are more likely to choose low-risk individuals (Anderson 2010).

Gender and Intimate Partner Violence (IPV)

As indicated in figures 12.1 and 12.2, crime victimization survey data show an overwhelming victimization of women when compared with men in every race/ethnic category (Catalano 2012, Table 1). Separated or divorced women are more likely to be IPV victims (Catalano 2012, Figures 3 and 6). However, the National Family Violence Surveys and large-scale studies using the Conflict Tactics Scale find approximately equal amounts of IPV perpetrated by men and women (Cook 2009; Straus 2008; see also Fountain et al. 2009).

These conflicting findings have set off an unresolved dispute about whether IPV is *asymmetrical*—with women primarily the victims of male aggression—or whether couple violence is *symmetrical*—with men and women perpetrating IPV at about the same rate (Anderson 2010, 2013; Fincham et al. 2013; Hines and Douglas 2009). The assertion that IPV is asymmetrical and that males are far more often

Local advocacy groups draw attention to efforts to prevent domestic violence and to the need for more resources. In many localities, there are still not enough shelters to meet the needs of battered women and their children. Moreover, nationwide there are very few shelters or other services for male victims of family violence.

perpetrators than victims is a strongly held feminist position, going back to the 1970s when feminists first raised the issue of wife abuse. The question whether women's violence toward men is mostly in self-defense, as feminist violence researchers argue, is part of the debate about gender differences in domestic violence perpetration.

Self-Defense? Demie Kurz (1993) offers evidence that intimate partner violence by women is largely in self-defense or at least retaliation rather than the initiation of a violent attack, especially in the case of husband homicides committed by women (Belknap 2012; see also Loseke and Kurz 2005). Sociologist Murray Straus, on the other hand, claims that better data indicate that wives often strike out first and that the data do "not support the hypothesis that assaults by wives are primarily acts of self-defense or retaliation" (Straus 1993, p. 76; 2008; Straus and Ramirez 2007). Moreover, he acknowledges that assuredly women's violence produces far fewer serious injuries and deaths than does men's violence. Nevertheless, injuries to men should not be dismissed (Straus 2008).

Some researchers suggest that a principal reason for contradictory findings is that questions on the Conflict Tactics Scale include a broader range of actions that may be characterized by the survey respondent as violent, although not reported to crime authorities (Dutton and Nicholls 2005; Medeiros and Straus 2006; Straus 2008). These researchers propose that there are indeed two distinct forms of IPV—*intimate terrorism* (formerly termed *patriarchal terrorism*) and *situational couple violence* (Johnson 2008).

Situational Couple Violence

Situational couple violence refers to symmetrical (mutual, perpetrated by women as well as by men) violence between partners that occurs in conjunction with a specific argument, tends to be less severe in terms of injuries, and is not likely to escalate as the relationship progresses (Johnson 2008; Johnson and Ferraro 2000). Among couples, situational couple violence appears to be relatively more common than intimate terrorism—especially among young, cohabiting couples (Johnson 2008; Whitaker et al. 2007). Situational couple violence typically erupts during a fight and is often accompanied by heavy drinking (Anderson 2010; Smith et al. 2011). The following admission by one young woman, speaking to a researcher, may shed some light on this situation:

> I was like quick-tempered. It was little things that set me off. I used to get in a lot of fights with my brother. And my mom, we used to argue a lot… My temper is still bad and I take things out on him [her partner].
> (in Smith et al. 2011)

Situational couple violence can indeed result in serious injury. Nevertheless, intimate terrorism is the more

Table 12.2 Two Forms of IPV—Intimate Terrorism versus Situational Couple Violence

Intimate Terrorism	Situational Couple Violence
Explained by feminist theory (see Chapter 2); also called "patriarchal violence"	Explained by family systems theory (see Chapter 2)
Less commonly found among couples	More commonly found among couples
More often (but not exclusively) found among marrieds	More often (but not exclusively) found among cohabitors
Almost always perpetrated by men (asymmetrical violence)	Perpetrated by both women and men fairly equally (symmetrical violence)
May occur frequently in the relationship and follows a pattern of occurring more and more often	Probably occurs less frequently in a relationship
Motivated by the need to dominate or possess the partner. Aimed at controlling the partner through intimidation and physical, emotional, and/or sexual abuse	Sparked by frustration and anger and aimed at winning a particular fight through physical or emotional abuse
Follows a "cycle of violence"	Does not necessarily follow a predictable pattern or cycle
Severe injuries probably, even homicide	Less severe injuries, on average

Source: Hench 2004; Johnson 2008; Johnson and Ferraro 2000; Loseke and Kurz 2005; Straus 2008.

dangerous form of IPV—even leading in some cases to homicide. Table 12.2 compares situational couple violence with intimate terrorism.

Intimate Terrorism

Intimate terrorism refers to abuse that is decisively oriented to controlling one's partner through fear and intimidation. Feminists have vehemently maintained—and research evidence consistently shows—that intimate terrorism is almost entirely perpetrated by males (Anderson 2010; Johnson 2008). Not focused on a particular matter of dispute between the partners, intimate terrorism is intended to establish an overall pattern of dominance and seems to occur more often in marriage than among cohabitors. It is believed that intimate terrorism includes more incidents within couples than does situational couple violence. Furthermore, intimate terrorism is likely to escalate and, especially in more advanced stages, is more likely than situational couple violence to produce serious injury.

Contrary to what many victims think, intimate terrorism is not about the perpetrator's loss of control. Nor is intimate terrorism necessarily precipitated by

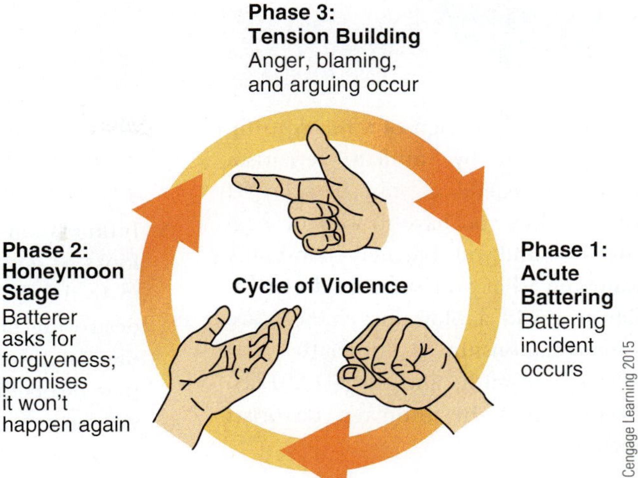

FIGURE 12.4 Intimate terrorism and the cycle of violence.

an outburst of anger. Rather, intimate terrorism is a method of establishing and maintaining power. As depicted in Figure 12.4, intimate terrorism follows a cycle often identified as the **cycle of violence**, consisting of three consecutive phases:

Phase 1—the violent episode itself, including, physical, emotional, and/or sexual abuse, and typically growing more violent over the course of the relationship;

Phase 2—a calm period, sometimes termed the "honeymoon" phase during which the abuser may ignore or deny the violence; blame the episode the victim; or act genuinely sorry, sending cards and flowers, for instance—hence the expression "honeymoon" phase;

Phase 3—tension buildup during which the victim feels increasingly disappointed and intimidated while the abuser's behavior is unpredictable and threatening.

In the absence of therapy—and perhaps even despite therapy—the cycle of violence repeats itself, usually with stage 2 shortening while stages 1 and 3 lengthen. An Internet search for "cycle of violence" yields several graphic depictions of this cycle. Here we note that physical abuse is just one of the tools an intimate terrorist uses; emotional abuse is frequent as well. "Facts about Families: Signs of Intimate Terrorism" gives specific examples of intimate partner terrorism.

Emotional Abuse **Emotional abuse** includes often making verbal threats or routinely making comments that damage a partner's self esteem (Meyer 2013). Emotional abuse includes verbal abuse, such as name-calling or demeaning verbal attacks; threats (to take away the victim's children, for instance, or somehow to harm a victim's family or friends); and attacks on pets or the victim's property. Emotional abuse includes name calling,

Facts about Families

Signs of Intimate Terrorism

The most telling sign of a relationship characterized by intimate terrorism is feeling fearful of your partner. "If you feel like you have to walk on eggshells around your partner—constantly watching what you say and do in order to avoid a blow-up—chances are your relationship is unhealthy and abusive" (Smith and Segal 2012a). Tactics used by intimate terrorists include:

1. dominance—making unilateral decisions for other family members, telling them what to do and expecting unquestioned obedience;
2. humiliation—making remarks or gestures that are likely to encourage others to feel bad about themselves or defective in some way; includes insults, name-calling, shaming, and public put-downs;
3. isolation—cutting other family members off from the outside world, a situation that increases their dependence on the abuser; keeping family members from seeing relatives or friends;
4. threats—threatening to hurt other family members or pets; threatening to commit suicide if challenged;
5. Intimidation—making frightening looks or gestures, smashing things in front of other family members, destroying property, putting weapons on display with the message that disobedience could result in violent consequences (Smith and Segal 2012a,b).

An intimate terrorist is not changing if he or she:

1. minimizes the abuse or denies the seriousness of its consequences;
2. continues to blame others;
3. insists that the victim *owes* him or her another chance;
4. resists staying in treatment or threatens to quit;
5. says he or she can't make it if the partner leaves;
6. expects something from the victim in exchange for promises to get help (Smith and Segal 2012b).

- You can find out more about intimate terrorism at http://helpguide.org.
- If you believe that you are in an abusive relationship, information at this website can help: http://au.reachout.com/What-to-do-if-youre-in-an-abusive-relationship.
- If you believe that someone you know is in an abusive relationship, a first step, experts advise, is to say something like this: "I just want to be there for you. How can I help?" Expressing to the victim your anger with the abuser will most likely not be productive (Welch 2011).
- If you believe that you are acting abusively in one or more of your relationships, there are steps that you can take. See http://loveisrespect.org.

Critical Thinking

If you believe that you yourself or someone you know is in an abusive relationship, what first, or next step will you take now to address the situation?

constant criticism, isolating the victim, intimidation, causing a partner's sleep deprivation, and withholding money or other basic necessities. Often part of a pattern of control and domination, emotional abuse virtually always accompanies physical violence in intimate terrorism (Johnson 2008; McCue 2008).

Marital Rape and Reproductive Coercion Estimates are that between 10 and 14 percent of women experience marital rape (Ferro, Cermele, and Saltzman 2008, p. 765; Bergen 2006). These sexual assaults often involve other violence as well. The feminist movement defined and publicized **marital rape** in the 1970s. Under traditional common law, a husband's sexual assault or forceful coercion of his wife was not considered rape because marriage meant the husband was entitled to unlimited sexual access to his wife. However, as a result of feminist political activity, all states have outlawed marital rape since 1993 (McMahon-Howard, Clay-Warner, and Renzulli 2009; National Clearinghouse on Marital and Date Rape, http://www.ncmdr.org, n. d.).

Several small studies have reported that pregnancy increased the likelihood of male-to-female IPV (Brownridge et al. 2011; Cox 2008; Martin et al. 2004). On the other hand, research using national samples has found that when age was controlled, there was no increased risk of being abused with pregnancy. Still, the many studies that have found an association between pregnancy and violence have kept this hypothesis alive, along with a possible explanation—the male partner's jealousy, specifically that the new baby would interfere with the wife's attention to and care of the man (Sarkar 2008).

The American College of Obstetricians and Gynecologists recognizes the phenomenon of **reproductive coercion**—behavior related to reproductive health that is used to maintain power and control in a relationship. For instance,

> A partner may sabotage efforts at contraception, refuse to practice safe sex, intentionally expose a partner to a sexually transmitted infection (STI) or human immunodeficiency virus (HIV), control the outcome

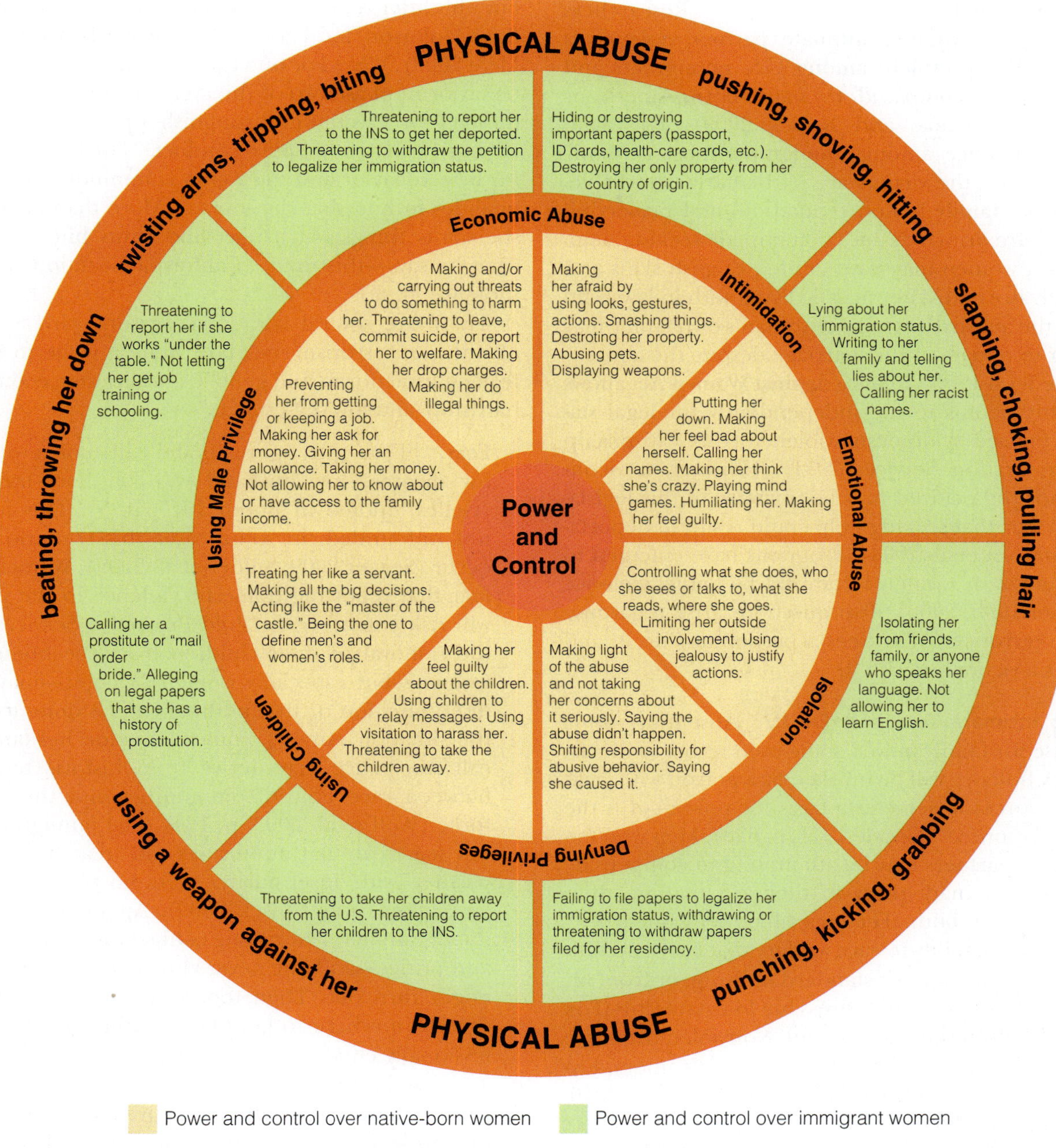

FIGURE 12.5 The power and control wheel for native born and for immigrant women: behaviors that some male partners use for coercive power and control of immigrant women.

Source: Immigrant Battered Women Power and Control Wheel, produced and distributed by National Center on Domestic and Sexual Violence, Austin, TX, available at www.endingviolence.org/files/uploads/ImmigrantWomenPCwheel.pdf and adapted from the original wheel by Domestic Abuse Intervention Project, Duluth, MN.

> of a pregnancy (by forcing the woman to continue the pregnancy or to have an abortion or to injure her in a way to cause a miscarriage), forbid sterilization, or control access to other reproductive health services. (American College of Obstetricians and Gynecologists 2012)

About 20 percent of women who seek care in family planning clinics and also report a history of IPV had experienced reproductive coercion (with the majority reporting birth control sabotage) (American College of Obstetricians and Gynecologists 2012).

Figure 12.5, the "Control Wheel," further illustrates how an intimate terrorist's need for control can result in emotional and physical violence. The inner circle describes intimate terrorism generally. The outer circle applies to intimate terrorism to immigrant couples specifically.

Immigrants and Intimate Terrorism Researchers disagree about whether intimate partner violence is greater or less prevalent among U.S. immigrants and refugees when compared to native-born Americans (Runner, Yoshihama, and Novick 2009). This said, domestic violence among immigrants most certainly does exist, and the victims are particularly vulnerable. Sometimes, "family honor, reputation, and preserving harmony" are primary values that impede seeking help. Immigrant women may have limited English skills and be socially isolated. They may be living with in-laws who support the abusive husband; a woman's own family may urge her to remain in the marriage despite the abuse.

The federal 1996 **Violence Against Women Act** allows immigrant victims to file independently for legal status if victimized by domestic violence (U.S. Citizenship and Immigration Services 2011). However, a victim who is not yet a citizen may be unaware of this legislation and fear that seeking help could result in deportation (Yoshioka et al. 2003). Programs have emerged to assist immigrant women victimized by family violence (Abraham 1995, 2000). Two questions arise regarding intimate terrorism: Why does a perpetrator do it, and why does a victim live with it?

Why Do Intimate Terrorists Do It? Absent a *reward power* base for family power, some men resort to *coercive power:* "[V]iolence will be invoked by a person who lacks other resources to serve as a basis for power—it is the 'ultimate resource' (Goode 1971, p. 628). Men who terrorize their partners may be attempting to compensate for feelings of inadequacy in their occupation, their relationship, or both. Feelings of powerlessness may stem from an inability to earn a salary that keeps up with inflation and the family's standard of living, or from living with the irony of a high-stress but low-status job (Fox et al. 2002; "Promoting Respectful, Nonviolent Intimate Partner Relationships" 2009). A husband's unemployment is associated with family violence (Condon 2010; Lauby and Else 2008).

In terms of *relative* status, a woman's risk of experiencing severe violence is greatest when she is employed and her husband is not. Considerable research has found violence associated with status reversal, where the woman is superior the man in terms of employment, earnings, or education (Kaukinen 2004). Among immigrants, a husband's loss of status upon immigration could be associated with intimate terrorism when jobs commensurate with education or expectations do not measure up, economic hardship is the family's lot, and wives, children, and people in general do not accord a male the respect he is accustomed to in his country of origin (Min 2002).

Why Do Victims Continue to Live with It? For the most part, battered women do leave an abusive relationship, but only after repeated violent episodes and reconciliations (Johnson and Ferraro 2000; Roberts, Wolfer, and Mele 2008). Why? For one thing, an intimate terrorist's behavior in phase two of the cycle of violence makes it difficult for a victim to distinguish a partner's genuine change from manipulative conduct. "The batterer is on his best behavior and the victim is reminded of all the qualities in him that she loves.... More than anything, she wants things to change. She wants him to mean what he says—this time" (California Coalition Against Violence, n. d.).

There are several additional reasons why a victim of intimate terrorism may take quite a while to leave: fear, cultural norms and gender socialization, economic hardship, and low self-esteem.

- **Fear.** "[T]he wife figures if she calls police or files for divorce, her husband will kill her—literally" (Gelles, quoted in Booth 1977, p. 7). This fear is not unfounded; in some cases this scenario does occur (Seager 2009, Snider et al. 2009). Women also fear that reporting domestic violence to the police will risk contact with Child Protective Services and the removal of their children from the home. All women—but especially women of color—may fear discrimination or that nothing can be done to ease the situation. Among women of color, hesitancy to call the police may derive from historic tensions between race/ethnic communities and the police force (Wolf et al. 2003, p. 124). The immigrant victim may fear that authorities will treat her or her partner with insensitivity, even hostility (Runner, Yoshihama, and Novick 2009). An undocumented immigrant may fear risking contact with immigration authorities—and one never knows how that situation might turn out. Then too, a victim may fear that she has nowhere to go. On the other hand, a professionally employed wife may fear potential stigma associated with being a victim of family violence, a situation that could result in negative consequences to her professional reputation and career (Kaukinen, Meyer, and Akers 2013).
- **Cultural norms and gender socialization.** Historically, women were encouraged to put up with abuse. English common law, the basis of the American legal structure, asserted that a husband had the right to physically "chastise" an errant wife. Although the legal right to physically abuse women has long since disappeared, our cultural heritage continues to have an influence on seeking help and getting it (Ellison et al. 2007). For one thing, cultural norms suggest that it is primarily a woman's responsibility to keep a marriage intact. Believing this, wives are often convinced that their emotional support may lead husbands to reform (Roberts, Wolfer, and Mele 2008).

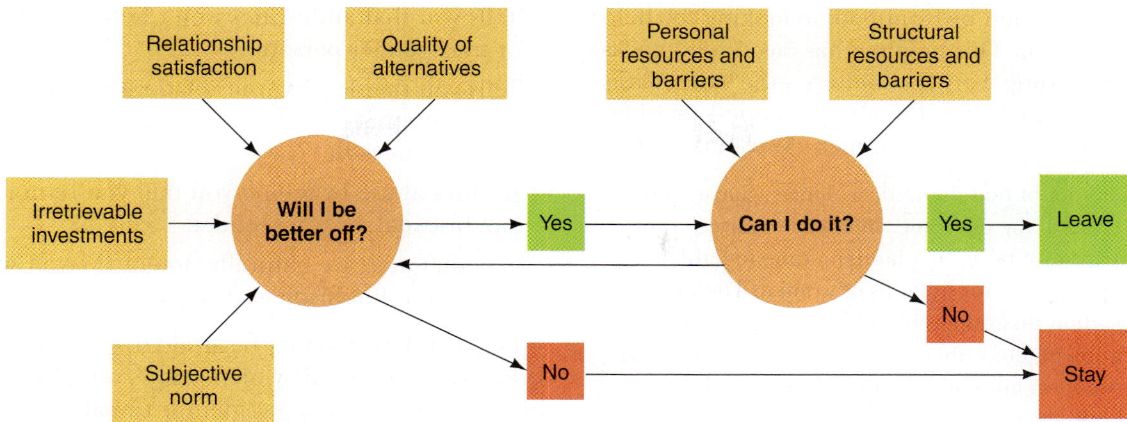

FIGURE 12.6 Conceptual model of abused women's stay/leave decision-making process.

Source: Choice and Lamke 1997, p. 295.

- **Economic hardship.** Economic uncertainty and concerns about future financial troubles can discourage victims from leaving (Anderson et al. 2003; see also Enander and Holmberg 2008). Leaving—or just pressing charges—could mean loss of a husband's income or damage to his professional reputation. The prospect of financial troubles is heightened when children are involved and a woman is likely looking ahead to single parenthood (Kaukinen, Meyer, and Akers 2013). Then too, an abusive husband may have inhibited his wife's economic independence in several ways—for example, by preventing her from getting to work, showing up at her workplace to harass her, contacting her supervisor or coworkers with intimidating messages or threats, or destroying her work clothes (Kimerling et al. 2009).
- **Low self-esteem.** Low self-esteem interacts with fear, depression, confusion, anxiety, feelings of self-blame, and loss of a sense of personal control (Umberson et al. 1998) to create the **battered woman syndrome**, in which a wife cannot see a way out of her situation (Walker 2009).

Social scientists have applied *exchange theory* to an abused woman's decision to stay or leave (McDonough 2010). As Figure 12.6 illustrates, an abused partner weighs such things as investment in the relationship, her (dis)satisfaction with the relationship, the possibility and quality of alternatives, and beliefs about whether it is appropriate to leave ("subjective norm") against such questions as whether the victim will be better off after leaving (might the perpetrator retaliate, for example?) and whether leaving is actually possible. The victim's personal resources along with available community, or structural resources, such as whether shelters or other forms of assistance are available, further affect the decision. Personal barriers might involve not having a job with adequate pay or an extended family that could help. Structural barriers might include the lack of community systems for practical help.

Much of this discussion has assumed that in a heterosexual relationship, the male is the perpetrator of family violence. However, this is not always the case (Fiebert 2012).

Male Victims of Heterosexual Terrorism

The butt of jokes, rather than social policy, male IPV victims are often considered unmanly. Some social scientists question the extent to which males, socialized to be the strong ones, would acknowledge to researchers that they had been battered by a female (Brown 2008; Dutton and Nicholls 2005). If males can be suspected of minimizing IPV victimization, so also can some family violence researchers. Often, for instance,

> female-perpetrated abuse is minimized and understood as either defensive or situational in nature, an isolated expression of frustration in communicating with an unsympathetic partner, in contrast to the presumably intentional, pervasive, and generally controlling behaviors exhibited by men. (Hamel and Nicholls 2006, p. xxxix; see also Johnson and Leone 2005)

If male victims do call police, they may themselves be arrested as suspected perpetrators. The police appear ill-equipped to deal with female abuse of males, going so far as to downplay violence against men even when they're called to the scene (Brown 2008).

The position that domestic violence is asymmetrical impacts resources available to male victims. One result is that funding to combat IPV has overwhelmingly been spent on programs designed to support only women victims (Watson 2010). As another example, 80 percent of the calls to the Domestic Abuse Helpline for Men and

Women are from men or from people looking for help on behalf of a man. This helpline has faced roadblocks in search for funding. Very few shelters—the Valley Oasis shelter in Antelope Valley, California, is one—provides a full range of shelter services to men (Watson 2010).

> There is no doubt that domestic violence against men can be reduced; the domestic violence initiatives of the past 40 years have brought a hidden crime to light and provided protection for millions of women. The next step is to admit that domestic violence is not a male or female problem, but rather a human problem, and that a lasting solution must address the cruelty—and suffering—of both sexes. (Watson 2010)

Men can find help online at Help for Abused Men.

Abuse among Same Gender, Bisexual, and Transgender Couples

Rates of same-sex intimate partner violence (SSIPV) are "comparable to rates of heterosexual domestic violence, with approximately one quarter to one half of all same-sex intimate relationships demonstrating abusive dynamics," and upwards of half of all transgendered people reporting intimate partner victimization (Murray and Mobley 2009, p. 361; see also Little and Terrance 2010).

Some of the relationship dynamics in same-sex abusive partnerships are similar to those in abusive straight relationships (Kurdek 1994). The 1996 Violence Against Women Act was expanded in 2010 to include same-gender couples, then expanded again in 2013 to include some categories, including transgender individuals (Barron 2010; Calms 2013). As with heterosexual IPV, batterers may abuse drugs or alcohol or have a history of childhood violence. Also like heterosexuals involved in domestic violence, the gay or lesbian couple is likely to deny or minimize the violence, along with believing that the violence is at least partly the victim's fault (Kulkin et al. 2008).

Although sexual minorities experience all of the same threats as do heterosexual victims, they have an additional concern—the abusive partner can threaten to "out" them to employers, family members, and friends. Hate crimes, discrimination, internalized homophobia, fear of being "outed," are all stressors that can impact homosexual and bisexual relationships and SSIPV (Brown 2008; Fountain et al. 2009). According to a Mayo Clinic publication, if you're lesbian, gay, bisexual, or transgender, you can experience domestic violence if you're in a relationship with someone who:

- Threatens to tell friends, family, colleagues, or community members your sexual orientation or gender identity,
- Tells you that authorities won't help a gay, bisexual, or transgender person,
- Tells you that leaving the relationship means you're admitting that gay, bisexual, or transgender relationships are deviant,
- Justifies abuse by telling you that you're not "really" gay, bisexual, or transgender,
- Says that men are naturally violent (Mayo Clinic, referenced in Robinson and Segal 2012).

LGBT individuals may be afraid to go to the police—or to use any domestic violence intervention services—for fear of having their gay identity revealed or receiving a hostile response (Simpson and Helfrich, 2005). Domestic violence services oriented to LGBT individuals or couples are now somewhat available in large cities with substantial gay/lesbian communities (Fountain et al. 2009). Nevertheless, as with male victims of heterosexual IPV, few resources exist to serve same-sex couples.

Generally, research shows a substantial overlap in households between IPV and child abuse (Anderson 2010). Unrealistic or rigid expectations of other family members and low levels of positive family interactions are related to both IPV and to child abuse, discussed next.

Violence Against Children—Child Maltreatment

The federal government and researchers use the umbrella term *child maltreatment* to cover both abuse and neglect. Perceptions of what constitutes child abuse or neglect have differed throughout history and in various cultures. Practices that we now consider abusive were accepted in the past as the normal exercise of parental rights or as appropriate discipline. Today standards of acceptable child care vary according to culture and social class. What some groups consider mild abuse, others consider right and proper discipline. In 1974, however, Congress provided a legal definition of *child maltreatment* in the Child Abuse Prevention and Treatment Act.

The act defines **child maltreatment** as the "physical or mental injury, sexual abuse, or negligent treatment of a child under the age of 18 by a person who is responsible for the child's welfare under circumstances that indicate that the child's health or welfare is harmed or threatened" (U.S. Department of Health, Education, and Welfare 1975, p. 3; and see Sedlak et al. 2010). According to this definition, nearly 3 million children—about one in twenty-five—experience maltreatment annually (Sedlak et al. 2010).

Current estimates from the 2012 U.S. Department of Health and Human Services report, *Child Maltreatment,* are based on state reports of child abuse. Of the

Where does discipline end and child abuse begin? For one thing, physical discipline should not be undertaken when a parent or parent figure is angry, because it is then too easy to go beyond reasonable limits.

reported cases of child maltreatment in 2011, approximately 78 percent are of neglect; 18 percent, of physical abuse; 9 percent, of sexual abuse; 7 percent, of psychological mistreatment; and 2 percent, of medical neglect—failure to provide a child with necessary medical services and treatment (U.S. Department of Health and Human Services 2012b).

Neglect and Abuse

Child neglect, by far the most common form of child maltreatment, involves failing to provide adequate physical or emotional care. Often "neglect" is grounded in parents' or guardians' economic problems, mental health issues, lack of parenting skills, or a history of themselves having been a victim of childhood neglect or abuse (Currie and Widom 2010; Daniel, Taylor, and Scott 2010). Less common, *willful* neglect involves purposeful failure to provide care even when resources are available. Physically neglected children often show signs of malnutrition, lack immunization against childhood disease, lack proper clothing, attend school irregularly, and need medical attention for such conditions as poor eyesight or bad teeth.

Emotional neglect involves a parent's or guardian's often being overly harsh and critical, failing to provide guidance, or being uninterested in a child's needs. Recently some activists have begun to label more specific forms of emotional neglect, such as *educational neglect* (failure to see that a school-age child gets to school regularly), and *medical neglect* (failure to attain necessary medical care for a child).

The term **child abuse** refers to overt acts of aggression—excessive verbal derogation (emotional child abuse) or physical child abuse such as beating, whipping, punching, kicking, hitting with a heavy object, burning or scalding, or threatening with or using a knife or gun. Data collected by the U.S. Department of Health and Human Services between 1993 and 2007 show a 29 percent decline in the rate of physical child abuse. The incidence of children victimized by sexual abuse decreased by 47 percent during that time. The incidence of emotionally abused children decreased by 48 percent.

However, the estimated number of emotionally *neglected* children more than doubled during that interval, rising from about 500,000 in 1993 to more than 1 million in 2006—an 83 percent increase in the rate (Sedlak et al. 2010, p. 7). An estimated 1,570 children died from abuse or neglect in 2011, with 80 percent under the age of 4 and 78 percent of the deaths caused by a parent (U.S. Department of Health and Human Services 2013, p. 10).

Child Sexual Abuse A form of child abuse is **sexual abuse**: a child's being forced, tricked, or coerced, by an older person, into sexual behavior—exposure, unwanted kissing, fondling of sexual organs, intercourse, rape, incest, prostitution, and pornography—for purposes of sexual gratification or financial gain (U.S. Department of Health and Human Services 2012b). Of abused children, between 10 and 20 percent are sexually abused (Sedlak 2010; U.S. Department of Health and Human Services 2012b). As with other forms of family violence, data collected between 1992 and 2007 show that child sexual abuse has declined—by a significant 53 percent (U.S. Department of Health and Human Services 2010).

Incest involves sexual relations between related individuals. The most common forms are sibling incest followed by father–daughter incest. The definition of child sexual abuse *excludes* mutually desired sex play between or among siblings close in age, but coerced sex by strong and/or older siblings is sexual abuse and is more widespread than parent-child incest (Kiselica and Morrill-Richards 2007; Thompson 2009).

Incest is the most emotionally charged form of sexual abuse and also the most difficult to detect. Having been an incest victim in childhood appears in the background of a variety of sexual, emotional, and physical problems among adults (Banyard et al. 2009; Carlson, Maciol, and Schneider 2006; Sachs-Ericsson et al. 2010).

We see occasional media stories about female sex abusers, but research indicates that the vast majority of sexual abusers are male (Peter 2009). About 47 percent of sexual assaults on children are by relatives; and 49 percent by others such as clergy, teachers, coaches, or neighbors; only 4 percent by strangers (Hernandez 2001).

How Extensive Is Child Maltreatment?

Abused children live in families of all socioeconomic levels, races, nationalities, and religious groups. Nevertheless, child maltreatment is reported much more frequently among poor and nonwhite families than among middle- and upper- class white families. Reporting on 2009 data, a federal government author concluded that, "Children in low socioeconomic status households... experienced some type of maltreatment at more than 5 times the rate of other children; they were more than 3 times as likely to be abused and about 7 times as likely to be neglected" (Sedlak et al. 2010, p. 12). According to a subsequent federal document, "Analyzing 5 years of race and ethnicity data reveals that the percentage and rate per 1,000 distributions have remained stable for several years" (U.S. Health and Human Services 2012b, p. 20).

Differences in neglect rates are largely attributed to insufficient income. Meanwhile, differences in all maltreatment rates may be partly due to reporting differences with poor children more likely to be seen by social welfare authorities (Gelles and Cavanaugh 2005). Another reason may be unconscious race or class-based discrimination on the part of medical personnel and others who report abuse and neglect (Lane et al. 2002). There are also real differences, however, and the fact that low-income or race/ethnic minority parents can be highly stressed is often offered as one explanation (Gelles and Cavanaugh 2005).

The percentages of male (48.3 percent) and female (51.3 percent) victims are not very different. The youngest children (through age 3) are more vulnerable than older children (U.S. Department of Health and Human Services 2010, p. 26). A child faces the greatest risk of becoming a victim of homicide during the first year of life (Collymore 2002).

About 54 percent of perpetrators are women. Approximately 80 percent of abused or neglected children are mistreated by at least one parent: by mother only in 38.3 percent of cases; by father only in 18 percent of cases; and by both mother and father in nearly 18 percent of cases. In 10 percent of cases, children were mistreated by other caregivers: foster parents or legal guardians, day care workers, or unmarried partners of a parent (U.S. Department of Health and Human Services 2013).

Risk Factors for Child Abuse When exaggerated, the following factors can encourage even well-intentioned parents to mistreat their children:

- Having learned to view children as requiring physical punishment in order to develop properly (Milner et al. 2010, p. 335).
- Having unrealistic expectations about what a child is capable of and unknowledgeable the children's physical and emotional abilities and needs (Letarte, Normandeau, and Allard 2010). For example, slapping a bawling toddler to stop her or his crying is completely unrealistic, as is too-early toilet training.
- Feeling highly stressed or helpless in the parent or provider role (Milner et al. 2010).
- Being a young adult and inexperienced with child care or managing stress (Milner et al. 2010).
- Experiencing marital discord, divorce, especially when coupled with children perceived as unusually demanding or otherwise difficult (Milner et al. 2010).
- Abusing alcohol or other substances (Milner et al. 2010).
- A mother's cohabiting with a male partner, who could potentially abuse the child—and is significantly more likely to do so than the child's biological father (Crary 2007b; Sedlak et al. 2010).
- Having a stepfather—because stepfathers are more likely than biological fathers to abuse children (Crary 2007b; Sedlak et al. 2010). Still, it is important to remember that 80 percent of child maltreatment perpetrators are the child's biological parents, whereas cohabitants and stepparents make up just 4 percent each (U.S. Department of Health and Human Services 2013).

Abuse versus "Normal" Child Raising Issues surrounding spanking children are addressed at length in Chapter 9. Here we note that it is too easy for parents to go beyond reasonable limits when angry or distraught or to include as "discipline" what most observers would define as abuse (Baumrind, Larzelere, and Cowan 2002; Feigelman et al. 2009). Hence, child abuse must be seen as a potential behavior in many families (Feigelman et al. 2009). Moreover, immigrant families may come from cultures where rather severe physical punishment is considered necessary for good child rearing. Those parents may not be aware that what they are doing by way of parental discipline is illegal in this country. They may instead view themselves as very responsible parents (Renteln 2004, pp. 54–57).

Then too, immigrant parents may be mistakenly identified as having abused children because of certain cultural practices not initially understood in this country. There are healing practices in certain cultures that can produce what looks like evidence of injuries to an American doctor or social service worker. Southeast Asians employ a practice known as "coining" whereby they rub the edge of a coin along the skin. This leaves marks that can appear to be those of a whip (Child Abuse Prevention Council of Sacramento, n. d.). Similarly, Asian children may have "Mongolian spots" on their skin, a natural phenomenon, but one that appears as bruising to an unaware health practitioner (Families with Children from China 1999). The issue whether spanking children should be considered abusive is addressed in Chapter 9. We turn now to two more recently recognized forms of family violence, sibling violence and child-to-parent violence.

Sibling Violence

Sibling violence—violent acts perpetrated by one sibling against another—is often overlooked and rarely studied (Butler 2006a; Finkelhor et al. 2005), even though the early National Family Violence Survey found it to be the most pervasive form of family violence (Straus, Gelles, and Steinmetz 1980). Furthermore, sibling violence is not only of the "harmless" teasing variety ("UF Study" 2004). A fairly recent national study found that 35 percent of children had been hit or attacked by siblings in the previous year. Fourteen percent were repeatedly attacked, 5 percent hard enough to have injuries such as bruises, cuts, chipped teeth, and sometimes broken bones. Two percent were hit with rocks, toys, broom handles, shovels, or knives (Butler 2006a).

Child psychologist John Caffaro (in Butler 2006a) sees sibling abuse as situational, not personality driven. When parents are frequently physically or emotionally absent from the home, or when they have their own problems, sibling violence is more apt to occur. Failure to intervene effectively also plays a part, as does parental favoritism of one child over another. Trauma, anxiety, and depression are likely to result from experiencing sibling violence, as well as an increased likelihood to perpetrate violence as an adult and to have relationship problems (Butler 2006a; Hoffman and Edwards 2004; Noland et al. 2004).

Perpetrators of sibling violence are more likely than others to become perpetrators of dating violence, according to a study of more than 500 men and women at a Florida community college. "Siblings learn violence as a form of sibling manipulation and control as they compete with each other for family resources.... They carry these bullying behaviors into dating, the next peer relationship in which they have an emotional investment" (researcher Virginia Noland in "UF Study" 2004). Furthermore, adolescent dating violence tends to continue at least into young adulthood (Cui et al. 2013). Yet sibling violence has received comparatively little research attention and even less attention has been given to preventive or therapeutic responses. Noland et al. (2004) recommends that sibling violence be taken more seriously and that anger management programs be implemented while potentially violent individuals are still children ("UF Study" 2004).

Child-to-Parent Violence

Relatively little research has been done on **child-to-parent violence** (Cottrell and Monk 2004; Holt 2013). Yet, like other forms of family violence, child-to-parent abuse has been there all along.

This section relies heavily on a review article by Cottrell and Monk (2004) and on the more recent book *Adolescent-to-Parent Abuse* (2013) by criminal justice professor Amanda Holt. Data suggest that 9 to 14 percent of parents have been abused by adolescent children, with injuries that include bruises, cuts, and broken bones. Types of assaults have included kicking, punching, biting, and weapons. Mothers, especially single mothers, and elderly parents of youth are the most frequent victims (Holt 2013).

Adolescent boys are the most frequent perpetrators, with their growth in size and strength associated with increases in violence. Although there are no clear findings of differences in race/ethnicity or social class, poverty and other family stressors are related to child-to-parent violence. Abusive children may exhibit diminished emotional attachments to parents. The child may have been abused by the parent or witnessed intimate partner abuse in the household. Overly permissive parents and those who abandon their authority in response to the violence tend to see more of it. Parents whose child-raising styles contradict each other are also at risk. Drug use by the adolescent may play a role (Holt 2013). Parents who are victims of assaults by their adolescent children often engage in denial. Unfortunately, at the

© prudkov/Shutterstock.com

An estimated 300,000 to 400,000 children are used annually in pornography or prostitution (National Institute of Justice 2007; Svevo-Cianci 2010). Child victims of organized sexual exploitation are typically runaways from troubled homes. Many have left home to escape family violence (Domestic Sex Trafficking of Minors n. d.; Harris 2009).

moment, few if any support services exist, and the criminal justice system has not responded systematically (Cottrell and Monk 2004).

Stopping Family Violence

Stopping family violence involves policy action on both the micro (relationship or personal) and the macro (structural) levels (Larrivée, Brabant, and Lessard 2012). First, we will explore micro, or relationship approaches. On the relationship level, three major approaches to combating family violence involve (1) separating victim from perpetrator, (2) the criminal justice approach, and (3) the therapeutic approach.

Micro, or Relationship Approaches

On the relationship level, three major approaches to combating family violence involve(1) separating victim from perpetrator,(2) the criminal justice approach, and (3) the therapeutic approach.

Separating Victim from Perpetrator With regard to both IPV and child maltreatment, separation of the victim from the perpetrator is one approach to stopping family violence. Separating victims from perpetrators involves escaping to shelters for IPV victims and removing an abused child from the home in the case of child abuse.

First established in the 1970s by feminists for female abuse victims, a network of **shelters** now provides a woman (and, often, her children) with temporary housing, food, and clothing to alleviate the problems of economic dependency and physical safety. Shelter staff also provide counseling to encourage a stronger self-concept so that the woman can view herself as worthy of better treatment and capable of making her way alone if need be. Finally, shelters may provide guidance in obtaining employment, legal assistance, or family counseling. In two studies on correlates of domestic homicide in large U.S. cities, researchers found that the availability of shelter and hotline services for domestic violence and more aggressive arrest and prosecution lowered homicide rates between intimate partners (Anderson 2010).

Protecting abused or neglected children may involve removing them from their family homes and placing them in foster care, discussed at greater length in Chapter 9. This practice is controversial, as foster parents have been abusive in some cases, and there are not enough foster parents to go around in many regions of the country. Moreover, removal from the home can be traumatic to children, who are very often attached to their parents despite the abuse (Kaufman 2006). They may blame themselves for the breaking up of the family (Gelles and Cavanaugh 2005).

An alternative to removal of the child is **family preservation**, whereby a Child Protective Services worker is able to "leave the child with [the offending] family and provide support in the form of housekeeping help or drug treatment, and then visit frequently to monitor progress" (Kaufman 2006, p. A12). The family preservation approach would not be appropriate if harm to the child appears imminent. Family preservation is a controversial strategy (Gelles 2005; Wexler, 2005), but both removal of the child from the home and a family preservation approach carry risk. Sometimes separating victim from perpetrator is accompanied by a criminal justice response.

The Criminal Justice Response Until the 1970s there was little legal protection for battered women (see Dugan, Nagin, and Rosenfeld 2004 and Leisenring 2008 for more detail on this point). Today experts believe that arrest can deter future intimate partner violence, at least by perpetrators who are employed and married—spouses with a "stake in conformity." Those who are unemployed and/or not married to the person they abused may react to arrest by *increased* violence (Dugan, Nagin, and Rosenfeld 2004).

Another problem with the arrest strategy has involved the fact that a literal reading of a mandatory arrest law has resulted in the arrest of victims, along with perpetrators, when the victim has resisted with violent force. In some cases, women claimed that the exchange between the perpetrator and the police officer was characterized by "male bonding," in which the perpetrator's story overrode the woman's complaint of violence. Today more victims report positive and protective experiences to researchers (Wolf et al. 2003).

Several interventions to address partner violence have been implemented through the criminal justice system including specialized domestic violence courts, coordinated community responses, mandatory arrest and prosecution policies, and court-ordered batterer treatment. Civil protective orders have also been used, more as a preventive measure, the goal being to prevent future violence rather than punish for past acts of violence (Logan, Walker, and Hoyt 2012).

Meanwhile, some advocates favor the **criminal justice response**—that is, the punitive approach—for perpetrators of child maltreatment. These advocates believe that one or both parents should be held legally responsible for child abuse. Although all states have criminalized child abuse, the approach to child protection has gradually shifted from punitive to therapeutic. Not all who work with abused children are happy with this shift. They reject the family system approach to therapy because it implies distribution of responsibility for change to all family members (Stewart 1984). Nevertheless, social workers and clinicians—rather than the police and the court system—increasingly investigate and treat abusive or neglectful parents.

A complicated issue related to this approach involves holding battered women criminally responsible for "failing to act" to prevent such abuse at the hands of their male partners. Fathers are typically *not* held accountable for child abuse committed by female partners (Liptak 2002). Meanwhile, feminist legal advocates question whether the law should hold a battered woman responsible for failing to prevent harm to her children when, as a battered woman, she cannot even defend herself:

> When the law punishes a battered woman for failing to protect her child against a batterer, it may be punishing her for failing to do something she was incapable of doing.... She is then being punished for the crime of the person who has victimized her. (Erickson 1991, pp. 208–9)

Some courts have recognized this paradox. The Illinois Supreme Court overturned such a mother's conviction in 2002, and courts have ruled in favor of mothers who lost custody or had children removed from the home, citing the mothers' domestic violence victimization (Liptak 2002; *Nicholson v. Scopetta* 2004; Nordwall and Leavitt 2004). An alternative—or, in some cases, a complement—to the criminal justice approach is the therapeutic approach.

The Therapeutic Approach The therapeutic approach involves establishing counseling and educational programs designed to define and treat offenders as in need of educational and psychological guidance, rather than as criminals. As an example, preventive programs that address relationship skills among teen mothers have reduced IPV levels (Langhinrichsen-Rohling and Capaldi 2012). Programs designed to prevent child maltreatment might include participation of perpetrators' mothers and siblings "to enhance development of more peaceful conflict resolution patterns within and outside the family" (Hendy et al. 2012).

Although therapy was earlier thought to be ineffective, a number of therapy programs for perpetrators of family violence have now emerged. Many IPV perpetrators have difficulty controlling anger and frustration, dealing with stress, and relinquishing control over others. Some therapists' data now suggest that supportive group counseling can sometimes help create a setting in which abusers learn constructive ways to cope with anger and recognize their partner's right to autonomy and respect (Carney, Buttell, and Dutton 2006; Scott and Straus 2007; Soliz, Thorson, and Rittenour 2009).

Another tactic involves teaching empathy—the ability to experience the condition of another person vicariously and to care about how the other person feels in that situation (Barlinska, Szuster, and Winiewski 2013). This solution has been used to counter bullying. Violence has been associated with lack of empathy, but we now know that in many cases individuals can be encouraged or taught to experience empathy.

Then too, because research shows that spirituality can be important to many people who have experienced IPV, religious communities have strengthened their family violence counseling (Austin and Falconier 2012; Ellison et al. 2007). In some cases, couple therapy or individual counseling programs may help women to

> assert themselves in an appropriate but determined way early in relationships, rather than being passive in the face of mounting assaults on their integrity and autonomy. The hope is that if women can create interpersonal limits, for both their own and their partner's speech and actions, they will be less likely to overlook incremental escalations of offending behavior. (Stein 2013, p. 190)

We note, however, that couple therapy programs designed to stop IPV remain controversial because they proceed from the premise that a couple's staying together without violence after an abusive past is possible. Feminist scholars have argued that this assumption is dangerous and that the danger women face in violent relationships is highly underestimated (Helfrich and Simpson 2005; Kulkin et al. 2008).

The therapeutic approach with regard to child maltreatment involves increasing parents' self-esteem and their knowledge about children (Goldstein, Keller, and Erne 1985). Programs attempt to reach stressed parents before they hurt their children, and many operate a twenty-four-hour hotline for parents under stress. High school classes on family life, child development, and parenting are now virtually universal and are thought to have helped reduce child abuse by teaching parents what to expect from children at different ages.

Parent education directed toward new immigrant parents might decrease the number of situations that end in removal of children from their home. For example, if some immigrant families do not realize that their traditional disciplinary practices constitute criminal child abuse in the United States, parent education offered through refugee service centers could anticipate that problem (Gonzalez and O'Connor 2002; Renteln 2004, pp. 54–58).

Macro, or Structural Approaches

The family ecology model, introduced and described at length in Chapter 2, argues that family dynamics are influenced by all that goes around outside families, from neighborhood to workplace, to state and federal legislation—and even to global developments, such as migration patterns. When looking at stopping family violence, a family ecology approach urges us to consider how more macro, or social structural programs and efforts impact more micro, relationship solutions.

A macro approach notes the social, cultural, and economic context of family violence, then provides programs and services to help reduce or otherwise address it. For instance, housing assistance and subsidized child care might dissuade a low-income or socially isolated parent from neglectfully leaving children alone while working (U.S. Department of Health and Human Services 2013). After-school programs have been known to positively impact adolescents who live in violent homes (Gardner, Browning, and Brooks-Gunn 2012).

It is worth noting again here that family violence, both involving intimate partners and children, has declined since the 1990s. This decline may well be a consequence of the support and treatment programs developed since the 1970s. Shelter options, women's increased employment and ability to support themselves, cultural change that takes domestic violence seriously and endorses women's taking self-protective actions, and increased interest in and understanding of domestic violence on the part of law enforcement agencies are all developments that may account for the decrease in fatal and nonfatal violence against women by their intimate partners. Although a dramatic rise in women's employment and earnings may prove threatening to low-earning husbands in the short run, in the long run, mutual awareness of a woman's potential economic independence may deter wife abuse by changing the family power dynamic (Yakushko and Espin 2010).

The 1994 and subsequently broadened Violence Against Women Act has resulted in greater perpetrator accountability for rape and stalking; in increasing rates of prosecution, conviction, and sentencing of offenders; in more easily granting a victim's protection order; and in ensuring that police respond to crisis calls. The Act also established the National Domestic Violence Hotline, which has answered more than 3 million calls since it was established. ("Factsheet: The Violence Against Women Act" n. d.).

Information, advice, and referral help for IPV and other forms of family violence is available on the Internet—for example from the National Coalition Against Domestic Violence, Helpguide.org, and from the National Domestic Violence Hotline. The costs of all anti-family violence programs may be high, but the estimated costs of IPV are up to $6 billion annually, with $4 billion of that for direct medical and mental health services (National Coalition Against Domestic Violence 2009).

A 2012 Rape Crisis Center Survey found that up to 80 percent of rape victims suffer from PTSD, and a significant percent miss time at work or school. Studies show that when victims receive advocate-assisted services, the do better short-and long-term recovery. Unfortunately, however, statewide, and regional budget cuts, along with increasing difficulty of competition for donor funding have resulted in program closings. Today 65 percent of programs have a waiting list for counseling services and 30 percent have a waiting list for support groups. Meanwhile, half of programs nationwide have laid off staff and two thirds have reduced the hours they spend on prevention (National Alliance to End Sexual Violence 2012). The consequences for some abuse victims may be increasingly serious as money for shelters and related services dries up (Lauby and Else 2008; "U.S. Recession Causing Increase ..." 2009). A qualitative study of thirteen women who had entered shelters for abuse victims found that, in a follow-up two years later, "long after women leave the shelter and their abusive relationships they will need the continued support of vocational psychologists to provide career assessment and counseling with focus on long-term career and educational opportunities" (Brown, Trangsrud, and Linnemeyer 2009).

Structural solutions or help to combat child maltreatment would involve more energetic national measures to raise living standards, keep children out of poverty, and include children of immigrants (Svevo-Cianci 2010). Poverty makes things worse for children growing up in violent homes, and yet these children can and do show resilience, especially when aided by community resources and support (Yoo and Huang 2012). Encouraging greater neighborhood cohesiveness would help. In neighborhoods that have support systems and tight social networks of community-related friends—where other adults are somewhat involved in the activities of the family—child abuse and neglect are much more likely to be noticed and stopped (Falconer et al. 2008; U.S. Department of Health and Human Services 2013).

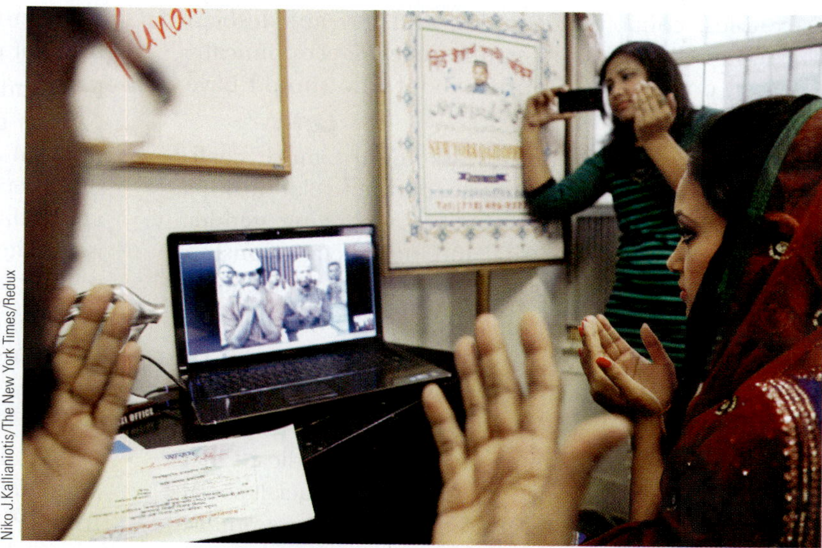

Niko J.Kallianiotis/The New York Times/Redux

In this photo Bangladeshi Punam Chowdhury, in a Queens, New York, mosque is marrying Tanvir Ahmmed, in Bangladesh, via the Internet. Tanvir will move to the United States later to apply for citizenship (see Nir 2013). Future research will no doubt investigate power relations between Internet spouses, such as Punam and Tanvir. What are some hypotheses that would be worth testing, do you think?

Social policy and programs to combat child maltreatment, first established in the 1960s and 1970s, have contributed to the decline in the incidence of child abuse and willful neglect. For instance, there are community support options such as volunteers available to babysit with potentially abused children in order to give a stressed parent a break. Another community resource is the *crisis nursery*, where parents may take their children when they need to get away for a few hours. Ideally, crisis nurseries are open twenty-four hours a day and accept children at any hour without prearrangement. We need programmatic support for male victims of spouse abuse: "Compassion for victims of violence is not a zero sum game .... Reasonable people would rationally want to extend compassion, support, and intervention to all victims of violence" (Kimmel 2002, p. 1,354). Indeed, the male victim of violence has few resources and often little sympathy (Cook 2009, p. 41).

Power disparities discourage intimacy, which is based on honesty, sharing, and mutual respect. For most, therefore, attainment of the American ideal of equality in marriage would seem to support the development of intimacy in marital relationships (Gudmunson et al. 2009; Rudman and Glick 2008; Soliz, Thorson, and Rittenour 2009). We close this chapter with a reminder of everyone's basic right to be respected—and not to be physically, emotionally, or sexually abused—in any relationship. And we end on an optimistic note, as most forms of family violence show evidence of declining rather than increasing.

Summary

- Power, the ability to exercise one's will, may rest on cultural authority, on economic and personal resources that are gender-based and/or involve love and emotional dependence, on interpersonal dynamics, or on emotional or physical violence.
- Marital, or conjugal power, or power in other intimate partner relationships, includes decision making, control over money, the division of household labor, and a sense of empowerment in the relationship. American marriages experience tension between egalitarianism on the one hand and, on the other, gender identities that in effect preserve male authority.
- The relative power of a husband and wife within a marriage or other intimate partnership varies by education, social class, religion, race/ethnicity, age, immigration status, and other factors. It varies by whether or not the woman works and with the presence and age of children. Studies of married

couples, cohabiting couples, and gay and lesbian couples illustrate the significance of economically based power and of norms about who should have power.

- Couples can consciously work toward more egalitarian marriages or intimate partner relationships and relinquish "power politics." Changing gender roles, as they affect marital and intimate relationship power, necessitate negotiation and communication.
- Researchers do not agree on whether intimate partner violence is primarily perpetrated by males or whether males and females are equally likely to abuse their partners.
- The effects of intimate partner violence indicate that victimization of women is the more crucial social problem, and it has received the most programmatic attention.
- Some programs have now been developed for male abusers. Studies suggesting that arrest is sometimes a deterrent to further wife abuse illustrate the importance of public policies in this area.
- Economic hardships and other stress factors (among parents of all social classes and races) can lead to physical and/or emotional child abuse as can lack of understanding of children's developmental needs and abilities. One difficulty in eliminating child abuse is drawing a clear distinction between "normal" child rearing and abuse.
- Physical, verbal, and emotional abuse, as well as sexual abuse and child neglect, are forms of violence against children. Sibling violence is an often overlooked form of child abuse.
- Criminal justice, therapeutic, and social welfare approaches are ways of addressing family violence, from IPV to child maltreatment.
- Child-to-parent abuse is a recently "discovered" form of family violence. It may grow out of previous abuse of a child.

Questions for Review and Reflection

1. How is gender related to power in marriage? How do you think ongoing social change will affect power in marriage?
2. Do you think that power in a marriage or other couple relationship depends on who earns how much money? Or does it depend on emotions? Is it possible for a couple to develop a no-power relationship?
3. Looking at domestic violence, why might women remain with the men who batter them? Do you think that shelters provide an adequate way out for these women? What about arresting the abuser? Should intimate partner violence against men receive more attention in the form of social programs? Why or why not?
4. What factors might play a role when well-intentioned parents abuse their children?
5. **Policy Question.** What can we as a society do to combat child neglect that is really due to family poverty?

Key Terms

battered woman syndrome 319
child abuse 321
child maltreatment 320
child neglect 321
child-to-parent violence 323
coercive power 303
Conflict Tactics Scale 311
conjugal power 302
criminal justice response (to family violence) 325
cycle of violence 315
egalitarian norm 304
egalitarian relationships 304
emotional abuse 315
emotional neglect 321
equality 302
equity 302
expert power 303
family preservation 324
gender-modified egalitarian model 308
incest 321
informational power 303
intimate partner violence (IPV) 310
intimate terrorism 315
legitimate power 303

marital rape 316
money allocation systems 306
National Family Violence Surveys 311
neotraditional family 308
patriarchal norm 304
power 302
power politics 309
referent power 303
reproductive coercion 316
resource hypothesis 304
resources in cultural context 304
reward power 303
sexual abuse 321
shelters 324
sibling violence 323
situational couple violence 314
Violence Against Women Act 318

13

Family Stress, Crisis, and Resilience

Gideon Mendel/Corbis News/Corbis

Learning Objectives

1. Define *family stress* and *family crisis* and distinguish between them.
2. Describe several types of stressors.
3. Explain stressor overload and stressor pile-up.
4. Describe the course of a family crisis.
5. Discuss how family members' appraisal of the stressor situation affects the outcome, or course of the crisis.
6. List family characteristics associated with resilience.
7. List family resources associated with meeting crises creatively.

Americans are stressed. In a national poll taken just before the onset of the recession that began in 2009, just 16 percent of 18- to 29-year-olds and 12 percent of 30- to 49-year-olds said that they rarely experience stress (Carroll 2007c). Since the recession's onset, Americans' stress levels have climbed as more and more couples have experienced financial strain (Elias 2009; Falconier and Epstein 2011; Shugrue and Robison 2009). In a recent poll by the American Psychological Association, more than half of U.S. adults reported that they had lain awake at night with worry at least once during the previous month (American Psychological Association 2013).

Children are worried as well. A poll by the national organization, KidsHealth, asked children ages 9 to 13 whether they worry and, if so, what about. Results showed that children as young as nine worry about the health of family members and getting into automobile accidents, among other concerns ("KidsPoll" 2008). Unfortunately, but a sign of the times, even some children in kindergarten worry about their body image ("Body Image" 2008). Children with a deployed military parent worry about whether their parent will come home and when, or about whether the parent will be injured: "The worst time is when the phone rings because you don't know who is calling. They could be calling, telling you that he got shot or something" (in Huebner et al. 2007, p. 117; see also Lane et al. 2012). "Facts about Families: Shielding Children from Stress Associated with Frightening Events" lists resources for parents in this regard.

We can think of families as continually balancing the demands put upon them against their capacity to meet those demands. This chapter addresses family stress, crisis, and resilience. We will review various theoretical perspectives on the family and discuss how these can be applied to family stress and crises. We'll discuss what precipitates family stress or crisis, and then look at how families define or interpret stressful situations and how their definitions affect the course of a family crisis.

In several places throughout this text, we point out that families are more likely to be happy when they work toward mutually supportive relationships—and when they have the resources to do so. Nowhere does this become more apparent than in a discussion of how families manage stress and crises. To begin, we'll define the concepts of family *stress*, *crisis*, and *resilience*.

Defining Family Stress, Crisis, and Resilience

Stress is a state of tension that results from the need to respond to change. Individual stress is physically and emotionally experienced in reaction to a change that requires a response or adjustment. Stresses can come from any situation or concern—even positive changes—that make one feel anxious, frustrated, nervous, or excited. Having your best friend as a house guest for a period of time can be stressful. Finally landing the job you've been working toward can be stressful.

The Holmes-Rahe Life Stress Inventory Scale provides a way to measure an individual adult's or child's stress level. Many of the scale items, such as death of a spouse, death of a parent, divorce, divorce of parents, marital separation, death of a close family member, marriage, marital reconciliation, change in health of a family member, pregnancy, fathering a pregnancy out of marriage, gaining a new family member, a child leaving home, and trouble with in-laws, are actually family stressors and result not only in individual stress but also in family stress or crisis.

The authors rank various life events according to how difficult they are to cope with. For instance, death of a spouse, the most stressful life event on the adult scale, is equivalent to 100 *life change units*. Divorce is equivalent to 73 life change units, while trouble with in-laws is equivalent to 29. For children, divorce of parents is equivalent to 77 life change units, while hospitalization of a parent is equivalent to 55 life change units and loss of a job by a parent is equivalent to 46. You can find the Holmes–Rahe Stress Inventory at The American Institute of Stress website.

Family stress is a state of tension that arises when demands test or tax a family's capabilities. As with individual stress, situations that we think of as good, as well as those that we think of as bad, are all capable of creating stress in our families. Moving to a different neighborhood, taking a new job, getting a promotion, or bringing a baby home might be examples of "good" situations that create family stress. As sociologist Pauline Boss (1997) reminds us,

> Perhaps the first thing to realize about stress is that it's not always a bad thing to have in families. In fact it can make family life exciting—being busy, working, playing hard, competing in contests, being involved in community activities, and even arguing when you don't agree with other family members. Stress means change. It is the force exerted on a family by demands. (p. 1)

Facts about Families

Shielding Children from Stress Associated with Frightening Events

We can never completely shield our children from natural disasters, news of terrorism, or other dangers. And it's appropriate for children to feel anxious about parents and other relatives serving in the military, being hospitalized, imprisoned, addicted to drugs or alcohol, or going through anxiety-provoking situations. Various websites offer information addressed to children and stress, including:

- American Academy of Pediatrics has developed a guide for pediatricians;
- Department of Defense Education Activity on school safety, children of military personnel, and other programs;
- FEMA (Federal Emergency Management Agency), on preparing children for natural disasters and national security emergencies;
- The KidsHealth organization has material designed for parents and children regarding things that children worry about and how parents can help;
- The National Child Traumatic Stress Network, sponsored by UCLA and Duke University, is another good resource.

Critical Thinking

When or how might you make use of these resources, or of resources like them, in your family life? How might family dynamics in the face of a stressor event such as an earthquake or explosion mitigate, or lessen a child's stress? How might family dynamics exacerbate, or worsen a child's stress?

Family stress might be also be caused by potentially harmful, ambiguous, or difficult situations such as trying to find adequate housing on a poverty budget, financing children's education on a middle-class income, being laid off in a recession economy, or losing one's home due to inability to pay the mortgage. A family member's injury or death is a source of family stress. Responding to the needs of aging parents is stressful for a family (see Chapter 16). Undergoing infertility treatments or losing a pregnancy are stressors (see Chapter 8) (Shreffler, Greil, and McQuillan 2011). Living as a cancer survivor is a stressor (Marshall 2010). In fact, many of the topics addressed throughout this text are family stressors: weddings, moving in together as cohabitors, forming multigenerational households, immigrating, transitioning to parenthood, parenting, separation or divorce, juggling family needs against workplace obligations, family violence, poverty—can you think of others?

Family stress calls for family adjustment. As an example of adjustment, more and more older parents are moving into the homes of their grown children in response to recession-related financial pressures (M. Alvarez 2009). These and other family households that expand temporarily to include more family members—often young adults returning home, as discussed in Chapter 6—then contract when the additional family members leave, are sometimes called *accordion families* and demonstrate often remarkable adjustments (Newman 2012; Newman and Knapp 2013).

When adjustments are not easy to come by, family stress can lead to a **family crisis**: "a situation in which the usual behavior patterns are ineffective and new ones are called for immediately" (National Ag Safety Database n. d., p. 1; Patterson 2002b). We can think of a family crisis as a sharper jolt to a family than more ordinary family stress. The definition of *crisis* encompasses three interrelated ideas:

1. Crises necessarily involve change.
2. A crisis is a turning point with the potential for positive effects, negative effects, or both.
3. A crisis is a time of relative instability.

Family crises are turning points that require some change in the way family members think and act to meet a new situation (Hansen and Hill 1964; McCubbin and McCubbin 1991; Patterson 2002b). In the words of social worker and crisis researcher Ronald Pitzer:

> *Crisis* occurs when you or your family face an important problem or task that you cannot easily solve. A crisis consists of the problem and your reaction to it. It's a turning point for better or worse. Things will never be quite the same again. They may not necessarily be worse; perhaps they will be better, but they will definitely be different. (Pitzer 1997a, p. 1)

In part, what makes the difference between whether things get better depends on a family's level of resilience—the ability to recover from challenging situations (McCubbin et al. 2001). We return to the topic of resilience later in this chapter. Meanwhile, we note several ways that social science theory gives insight into family stress, crisis, and resilience.

Theoretical Perspectives on Family Stress and Crises

We saw in Chapter 2 that there are various theoretical perspectives concerning marriages and families. Throughout this chapter we will apply several of these theoretical perspectives to family stress and crises. Here we give a brief

Peggy Peattie/U-T San Diego/ZUMA Press

For decades some family members, many of them U.S. citizens, have been divided by the U.S.-Mexico border. In the 1990s the United States built a 17-foot wall along the border. Weekends, hoping to catch glimpses of each through cracks, divided family members gather on either side of the wall. In this photo, Jimena Angulo, age 5, meets her father for the first time in a rare, brief, U.S.-monitored opening of one spot in the wall (Hamilton 2013). A family's being divided in this way constitutes a family crisis. The family ecology perspective focuses on how factors external to the family, such as economic conditions and immigration policies in both countries, can result in family crisis. From a family systems perspective, all the members in this family have to adapt to being separated as positively as they can.

review of several theoretical perspectives that are typically used when examining family stress and crises.

You may recall that the *structure–functional* perspective views the family as a social institution that performs essential functions for society—raising children responsibly and providing economic and emotional security to family members. From this point of view, a family crisis threatens to disrupt the family's ability to perform these critical functions (Patterson 2002b).

The *family development*, or *family life course*, perspective sees a family as changing in predictable ways over time. This perspective typically analyzes **family transitions**—expected or *predictable* changes in the course of family life—as family stressors that can precipitate a family crisis (Carter and McGoldrick 1988). For example, having a baby or sending the youngest child off to college taxes a family's resources and brings about significant changes in family relationships and expectations. Over the course of family living, people may form cohabiting relationships, marry, become parents, break up or divorce, remarry, and make transitions to retirement and widow- or widowerhood. All these transitions are stressors (Cooper, McLanahan, et al. 2009).

In addition, the family development perspective focuses on the fact that predictable family transitions, such as an adult child's becoming financially independent, are expected to occur within an appropriate time period, although the window of acceptable time has lengthened over that past several decades (Arnett 2004; Furstenberg et al. 2004). As discussed in Chapter 2, transitions that are "outside of expected time" create greater stress than those that are "on time" (Hagestad 1996; Rogers and Hogan 2003). Partly for this reason, teenage pregnancy is often a family stressor. Another example, explored in Chapter 9, involves a grandparent's or, less often but not uncommon, an aunt's filling the parent role (Davis-Sowers 2012). "A Closer Look at Diversity: Young Caregivers" provides a third example of assuming a role outside of expected time.

The *family ecology* perspective explores how a family is influenced by the environments that surround it. From this point of view, many causes of family stress originate outside the family—in the neighborhood, workplace, and national or international environment (Boss 2002; Socha and Stamp 2009). Living in a violent neighborhood causes family stress and has potential for sparking family crises (Bertram and Dartt 2009). Conflict between work and family roles, largely created by workplace demands, is another example of an environmental factor that can cause family stress (Barnett et al. 2009; Bass et al. 2009). Natural disasters, such as tornadoes, hurricanes, or earthquakes, create family stress and crises (Sattler 2006; Taft et al. 2009). Moreover, as we'll see later in this chapter, our family's external environment offers or denies us resources for dealing with stressors.

The *family system* theoretical framework looks at the family as a system—like a computer system or an organic system, such as a living plant or the human body. In a system, each component or part influences all the other parts. When one family member changes a role, all the family members must adapt and change as well. As an example, when a family member becomes addicted to steroids, alcohol, or other mind-altering drugs, the entire family system is affected (El-Sheikh and Flanagan 2001; Foster and Brooks-Gunn 2009). As another example, when one sibling has a disability, other siblings—as well as parents and other relatives—have to adjust (Neely-Barnes and Graff 2011).

Finally, exploring the discussions, gestures, and actions that go on in families, the *interactionist perspective*

A Closer Look at Diversity

Young Caregivers

We tend to think of caregivers as middle age or older, and a large majority of them are. But caregivers are diverse in age, some of them children. For young caregivers, the role involves additional stress because it does not take place "on time." Here's what one thirty-something caregiver wrote in her blog:

> Being thrust into a caregiver role at a younger age, when my mom at the age of 57 had a debilitating stroke, I was faced with all the "common" caregiver challenges but at a time in my life when it was least expected and with absolutely no warning. I immediately left my career, my home, my friends to move back home (2,000 miles away) to do everything that was humanly and sometimes inhumanly possible to help my mom .... Being a caregiver, especially at a young age, is a huge sacrifice. I don't regret it, but sometimes I can't help but feel that I am missing out on some of the best years of my life.
>
> During my 20s, I mostly focused on my career. I was always a very driven person and while I had one or two serious relationships during that time, I was not ready to "settle down." In my mind, I felt like that's what my 30s would be for. Had I been able to predict the future, I would've married my college boyfriend and started having babies immediately. OK, maybe not, but the idea of it sure sounds good now (laugh). So, here I am, one year into caregiving and I just started working again (my career had to be redefined too) ....
>
> Meanwhile, my friends and acquaintances are getting married, having babies, buying houses, etc. Sometimes I feel like everyone is moving forward, and I am frozen in time .... I wonder if and when will I have the opportunity to fulfill my own hopes and dreams. As a young caregiver, and in my particular situation, this is my biggest challenge and fear .... So to all the young caregivers out there—whether you are caring for your spouse/significant other, a sibling, or a parent—You are not alone. (Caregiver Support Blog 2008)

In addition to caregivers in their twenties and thirties, an estimated 1.5 million U.S. children under age 18 serve as caregiver to a family member (Shifren 2009; "Young Caregivers" 2009). Experts expect the numbers to grow as chronically ill patients leave hospitals sooner and live longer, as the recession compels patients to forgo paid help, and as more returning veterans need home care (Belluck 2009).

Child and teen caregivers are often responsible for keeping the care recipient company. In addition, they shop, do household chores, and help with meal preparation. Some assist the care recipient with eating, getting in and out of bed, getting dressed, taking a bath, or going to the bathroom. Some administer medications; help the care recipient communicate with doctors, nurses, or other medical professionals; make appointments; or arrange for others to help the care recipient (Hunt, Levine, and Naiditch 2005; see also Champion et al. 2009).

Caretaking can give purpose to a young person's life (Shapiro 2006a). Although some child caregivers do well, others grow depressed and/or angry as they sacrifice social and extracurricular activities. Some miss—or even quit—school (Belluck 2009). Policy makers urge further research on questions, such as how to improve support groups for young caregivers, how teachers and schools can assist them, and how educational, social, and career opportunities can be fostered within the context of caregiving (Hunt, Levine, and Naiditch 2005).

Support organizations for young caregivers include the American Association of Caregiving Youth (AACY), the National Alliance for Caregiving (NAC), the National Family Caregivers Association (NFCA), the Family Caregiver Alliance (FCA), the Children of Aging Parents (CAPS), and the Caregiving Youth Project, sponsored by AACY (Belluck 2009).

Critical Thinking

From a policy point of view, what might be done to assist young caregivers? What might your local community do to help?

views families as shaping family traditions and family members' self-concepts and identities. By interacting with one another, family members struggle to create shared family meanings that define stressful or potentially stressful situations—for example, as good or bad, disaster or challenge, someone's fault or no one's fault. As we will explore later in this chapter, "a family's shared meanings about the demands they are experiencing can render them more or less vulnerable in how they respond" (Patterson 2002b, p. 355).

What Precipitates a Family Crisis?

Demands put upon a family cause stress and sometimes precipitate a family crisis. Social scientists call such demands **stressors**—a precipitating event or events that create stress. Stressors vary in kind and degree, and their nature is one factor that affects how a family responds.

In general, stressors are less difficult to cope with when they are (1) expected, (2) brief, (3) seen as not so serious, and (4) gradually improve over time.

Types of Stressors

There are several types of stressors (Figure 13.1). We will briefly examine nine of the most common here.

1. Addition of a Family Member Adding a member to the family—for example, through birth, adoption (Bird, Peterson, and Miller 2002), marriage, remarriage, or the onset of cohabitation—is a stressor. You may recall Chapter 9's discussion on why the transition to parenthood is stressful. As another example, the addition of adult family members may bring people who are very different from one another in values and life experience into intimate social contact. Furthermore, not only are in-laws (and increasingly stepparents, stepgrandparents, and step-siblings) added through marriage or cohabitation but also a whole array of their kin come into the family. Then too, having an adult child return home or having a member of one's extended family move into the household because of financial problems are stressors. Transition to a multi-generation household is a stressor but can have positive outcomes for some children of unmarried mothers who move into parental homes (Augustine and Raley 2012).

Like any system, a family has boundaries. Family members need to know "who is in and who is outside the family" (Boss 1997, p. 4). Adding a family member is stressful because doing so involves family boundary changes; that is, family boundaries have to shift to include or "make room for" new people or to adapt to the loss of a family member (Boss 1980). This situation applies to the addition of a cohabiting partner to the family, as well as his or her departure from the household (Cherlin 2009a).

2. Loss of a Family Member The death of a family member is, of course, a stressor. Family systems theory reminds us that children as well as adults grieve the absence of a family member, and their grief needs to be recognized and addressed (Boss 1980; Monroe and Kraus 2010; Saint Louis 2012). The same is true for other family members who are sometimes "forgotten grievers," such as grandparents on occasion, or aunts and uncles, among others (Gilrane-McGarry and O'Grady 2012).

Interestingly, the likelihood of death in a society can influence how people define a death in the family. For instance, under the mortality conditions that existed in this country in 1900, half of all families with three children could expect to have one die before reaching age 15. Social historians have argued that parents defined the loss of a child as almost natural or predictable and, consequently, may have suffered less emotionally than do parents today (Wells 1985, pp. 1–2). In contrast, family members who lose a child of any age today do so "outside of expected time," a situation that exacerbates, or adds to, their grief. The long-term effects of grieving such a loss may negatively affect a couple's intimacy (Gottlieb, Lang, and Amsel 1996).

Loss of potential children through miscarriage or stillbirth has the possible added strain of family disorientation (Shreffler, Greil, and McQuillan 2011). Attachment to the fetus may vary substantially so that the loss may be grieved greatly or little. Add to that the generally minimal display of bereavement customary in the United States and the omission of funerals or support rituals for perinatal (birth process) loss, and "all these ambiguities mean that a family may have to cope with sharply different feelings among family members … [and] the family as a whole may have to cope with the fact that they as a family have a very different reaction to loss than do the people around them" (Rosenblatt and Burns 1986, p. 238).

In addition to permanent loss, the temporary loss of a family member, such as through an older sibling's going away to college or a parent's leaving for long periods due to work demands, is a stressor. Temporary losses that not only create change in family structure but also introduce fear of the unknown are a form of ambiguous loss (Huebner et al. 2007; Whealin and Pivar 2006).

3. Ambiguous Loss The loss of a family member is ambiguous when it is uncertain whether the family member is "really" gone (Boss 2007):

> Ambiguous loss is a loss that remains unclear…. [U]ncertainty or a lack of information about the

Addition of a family member

Loss of a family member

Ambiguous loss

Sudden, unexpected change

Ongoing family conflict

Caring for a dependent, ill, or disabled family member

Demoralizing event

Daily family hassles

Anxieties about children in a culture of fear

FIGURE 13.1 Types of stressors.

> whereabouts or status of a loved one as absent or present, as dead or alive, is traumatizing for most individuals, couples, and families. The ambiguity freezes the grief process and prevents cognition, thus blocking coping and decision-making processes. Closure is impossible. (Boss 2007, p. 105)

Having a family member who has been called to war or who is missing in action are situations of ambiguous loss (Pittman, Kerpelman, and McFadyen 2004; Boss 2007). Even having a family member in the military reserves during a period in history when she or he is likely to be called into active duty is a significant stressor (Lane et al. 2012).

In addition, a family member may be physically present but psychologically absent, as in the case of family members with alcoholism or mental illness, those suffering from Alzheimer's disease or who have experienced brain injury, or children with cognitive impairment or severe disabilities (Blieszner et al. 2007; Roper and Jackson 2007). The ambiguity of post-separation or postdivorce family boundaries can be stressful. A nonresident father whose relationship with the child's mother—and hence with his future child—is uncertain can experience ambiguous loss (Leite 2007).

From the family systems perspective, ambiguous loss is uniquely difficult to deal with because it creates family **boundary ambiguity** (see Figure 13.2)—"confused

Situations of physical absence & psychological presence
There is a preoccupation with thinking of the absent member. The process of grieving and restructuring cannot begin because the facts surrounding the loss of the person are not clear.

Catastophic and unexpected situations:
- war (missing soldiers)
- natural disaster (missing persons)
- kidnapping, hostage-taking, terrorism
- incarceration
- desertion, mysterious disappearance
- missing body (murder, plane crash, etc.)

More common situations:
- divorce
- military deployment
- young adults leaving home
- elderly mate moving to a nursing home

Situations of physical presence & psychological absence
Families where a member is physically there but not emotionally available to the system. The family is intact, but a member is psychologically preoccupied with something outside the system.

Catastophic and unexpected situations:
- Alzheimer's disease and other dementias
- chronic mental illness
- addictions (alcohol, drugs, gambling, etc.)
- traumatic head injury, brain injury
- coma, unconsciousness

More common situations:
- preoccupation with work
- obsession with computer games, Internet, TV

FIGURE 13.2 Two common forms of boundary ambiguity—(1) a family member's physical absence coupled with psychological presence, and (2) a family member's physical presence coupled with psychological absence. "Sometimes a family experiences an event or situation that makes it difficult—or even impossible—for them to determine precisely who is in their family system" (Boss 1997, pp. 2–3).

perceptions about who is in or out of a particular family" (Boss 2004, p. 553; Carroll, Olson, and Buckmiller 2007):

> With a clear-cut loss, there is more clarity—a death certificate, mourning rituals, and the opportunity to honor and dispose [of the] remains. With ambiguous loss, none of these markers exists. The clarity needed for boundary maintenance (in the sociological sense) or closure (in the psychological sense) is unattainable.... [P]arenting roles are ignored, decisions are put on hold, daily tasks are undone, family members are ignored or cut off, and rituals and celebrations are canceled even though they are the glue of family life. (Boss 2004, p. 553)

4. Sudden, Unexpected Change A sudden, unexpected change in the family's income or social status may also be a stressor. A family member's having a heart attack, or having a child run away are examples (Cohen 2008; Goldsmith, Bute, and Lindholm 2012). Sudden job loss is another example. Natural disasters, mentioned earlier, cause sudden change. The bombing at the 2013 Boston marathon caused sudden change for injured individuals and their families, as well as for all Bostonians who may take safety less for granted now. Most people think of stressors as being negative, and some sudden changes are. But positive changes, such as winning the lottery (don't you wish?) or getting a significant promotion, can cause stress too.

5. Ongoing Family Conflict Ongoing, unresolved conflict among family members is a stressor (Hammen, Brennan, and Shih 2004). Deciding how children should be disciplined can bring to the surface divisive differences over parenting roles, for example. The role of an adult child living with parents is often unclear and can be a source of unresolved conflict. Watching an adult grandchild go through family conflict can be a stressor for a grandparent. If children of teenagers or of divorced adult children are involved, the situation becomes even more challenging (Hall and Cummings 1997).

6. Caring for a Dependent, Ill, or Disabled Family Member Caring for a dependent or disabled family member is a stressor (Berge and Holm 2007; Patterson 2002a). This is especially true when work demands conflict with those of family care (Stewart 2013). Examples involve being responsible for an adult child or a sibling with mental illness and/or physical or developmental disabilities (Fields 2010; Levine 2009; Lowe and Cohen 2010). Due mainly to advancing medical technology, the number of dependent people and the severity of their disabilities have steadily increased over recent decades. For instance, more babies today survive low birth weight and birth defects. Caring for a disabled child can be stressful enough for parents that some decide not to have another. Analysis of national data found that mothers of firstborn children with a disability were statistically less likely to have a second child (MacInnes 2008).

Also, more people now survive serious accidents, and many seriously injured soldiers deployed abroad have survived to require extensive ongoing care and medical attention. "Issues for Thought: Caring for Patients at Home—A Family Stressor" discusses how recent technological advances, coupled with the goal of containing medical care costs, have created new stressors for families who are increasingly expected to care for very ill patients at home.

In addition, parents may be raising children with chronic physical conditions, such as asthma, diabetes, epilepsy, or autism (O'Brien 2007; Rao and Beidel 2009). Families may need to see their children through bone marrow, kidney, or liver transplants, sometimes requiring several months' residence at a medical center away from home (LoBiondo-Wood, Williams, and McGhee 2004).

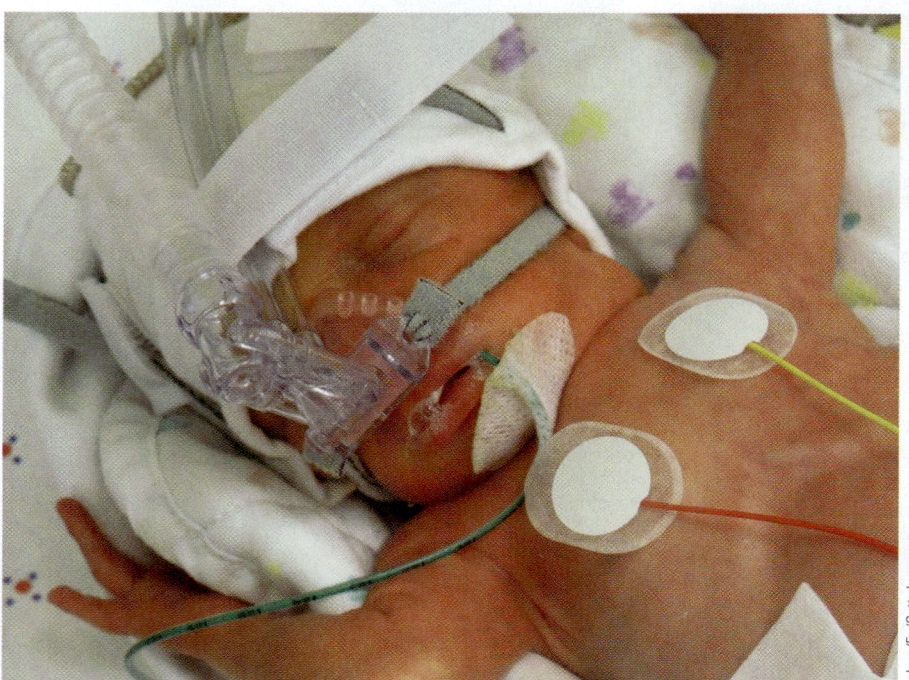

tioloco/E+/Getty Images

Sometimes a situation may be classified as more than one type of stressor. Due to advancing medical technology, for instance, more newborns today survive low birth weight or birth defects but may need ongoing remedial attention. Therefore, adding a baby to the family may also mean caring for a medically fragile child.

Issues for Thought

Caring for Patients at Home—A Family Stressor

Between 20 million and 50 million family members in the United States today are providing care that medical professionals once performed in hospitals. Family caregivers—mostly wives and daughters, but also partners, siblings, husbands, sons, grandchildren, and grandparents—provide about 80 percent of all care for ill or disabled relatives, which represents an estimated $13 billion in unpaid caregiving services annually (Brody 2008; Guberman et al. 2005). We can expect need for family caregiving to increase as the population ages; the incidence of chronic disease such as diabetes rises; the number of day surgeries grows; and modern medicine is able to save more and more lives, resulting in more special-needs infants and returning veterans who need care, among others (Brody 2008; Guberman et al. 2005).

> It is somewhat disconcerting to imagine an activity taking place in the hospital and then displacing this same activity to the home. In the hospital...the patient is in a supposedly sterile environment. Diet and medications are completely controlled by hospital staff. Indeed, a patient who asks to keep and self-administer his or her medications is refused. An interdisciplinary professional team is present, and when any one member is confronted with a problem (leaking IV tube, patient discomfort, apparatus malfunction), the members of the team are backed up by specialists (IV technicians, specialized doctors, technicians, and so on) and by a team of people responsible for the organization of the instrumental activities of daily living (meals, toileting, and so on)....
>
> Now transfer this to the home setting.... The IV pole is squeezed in between the bed and the night table, and there is almost no room to move because of the addition of a small table that is used to lay out equipment. The patient frequently gets caught in the line and loosens the catheter in his or her vein. When it starts bleeding at the site, the home care nurse has already come and gone. What to do? The caregiver makes an adjustment. He or she is abused for hurting the patient, but the IV starts to flow again and the bleeding stops. There is another dispute between patient and caregiver concerning hygiene around the IV. What does keeping a sterile area mean? Can the dog sit on the bed? Does the caregiver have to wear gloves? Are these questions important enough to disturb medical personnel for answers? Who should be called—the hospital, the home care nurse, or the 24-hour medical-information line? ...
>
> [P]erhaps the most unsettling aspect of the transfer of care responsibilities to patients and their families is the anxiety and insecurity of assuming this care without sufficient supervision and emergency backup. In the hospital, you have an emergency call button if something goes wrong. But what replaces this button when you are being cared for at home? Indeed, the home is psychologically, and sometimes physically, very far from immediate help in the case of an emergency or an unforeseen development. [In this study, the] majority of patients and caregivers assuming complex care felt alone and abandoned, causing high levels of stress and anguish and conflicts within couples and within families....
>
> Based on our study, we raise serious questions about the legitimacy of the transfer of high-tech care to the family.

Critical Thinking

Can you apply the family ecology theoretical perspective to this situation? What are some creative ways that a family might deal with high-tech caregiving at home? In what ways might community activism play a part in addressing this situation?

Source: Largely Excerpted from Guberman et al. 2005, pp. 247–72; also Brody 2008).

Adults with advanced AIDS may return home to be taken care of by family members. A family's caring for a disabled, or terminally ill member can be a stressor for young children, who may exhibit behavior problems as a response, as well as for the adults in the household (Seltzer and Heller 1997; Toot et al. 2013). Chapter 16 addresses the issue of the sandwich generation of caring for one's own children as well as for aging parents.

7. Demoralizing Events Stressors may be demoralizing events—those that signal some loss of family morale (Early, Gregoire, and McDonald 2002; Wildeman 2012). Demoralization can accompany the stressors already described. But, among other things, this category also includes financial troubles, poverty, homelessness, having one's child placed in foster care, juvenile delinquency or criminal prosecution, scandal, family violence, mental illness, incarceration, or suicide (Bricker et al. 2012; McNamara 2008; Taft et al. 2009; Wharff, Ginnis, and Ross 2012). Being the brunt of racist treatment can assuredly be demoralizing (Murry et al. 2001). Grandparents' raising grandchildren is a situation that is often—although not always—associated with demoralizing events (Henderson et al. 2009).

Physical, mental, or emotional illnesses or disorders can be demoralizing. Alzheimer's disease or brain injury, in which a beloved family member seems to have become a different person, can be heartbreaking.

Some illnesses can be especially demoralizing when they are associated with the possibility of being socially stigmatized. Autism, HIV/AIDS, attention deficit/hyperactivity disorder (ADHD), anorexia nervosa, and bulimia are examples (Hall and Graff 2012; Odom 2009; Rumney 2009; Shannon 2009). In military personnel who have served during wartime, post-traumatic stress disorder (PTSD) can be demoralizing, causing "family members [to] feel hurt, alienated, or discouraged, and then become angry or distant toward the partner" ("PTSD and Relationships" 2006; and see Lane et al. 2012; Schlomer et al. 2012).

8. Daily Family Hassles Daily family hassles are stressors. Examples involve balancing work against family demands, working odd hours, being regularly stuck in traffic on long commutes to work, or arranging child care or transportation (Adamo 2013; Evans and Wachs 2010). Another example involves protecting children from danger, especially in neighborhoods characterized by violence (Bertram and Dartt 2009; Foster and Brooks-Gunn 2009). A recent study found that, compared to married mothers, single mothers were more likely to feel greater stress and feelings of inadequacy, evidenced by migraines or chronic back pain, when experiencing daily hassles associated with their children's allergies or frequent colds (Ontai et al. 2008). "Facts about Families: ADHD, Autism, Stigma, and Stress" explores this point further.

Some scholars have investigated everyday stressors that are unique to certain professions. For instance, especially in recent years, military families "are subjected to unique stressors, such as repeated relocations that often include international sites, frequent separations of service members from families, and subsequent reorganizations of family life during reunions" (Drummet, Coleman, and Cable 2003, p. 279; Hawkins et al. 2012; Lowe et al. 2012; Schlomer et al. 2012; Wadsworth 2012). Families of Protestant clergy experience not only the stressors of ministry demands but also family criticism and situations in which members of the congregation "intrusively assume that the minister will fulfill their expectations without due consideration of the minister's priorities" (Lee and Iverson-Gilbert 2003, p. 251).

9. Anxieties about Children in a "Culture of Fear" A final stressor involves living in a situation of chronic anxiety with regard to children's safety. Increased media portrayal of various dangers seems to have led to a general "culture of fear" (Glassner 1999), which makes "anxiety about children ... a central matter in twentieth-century American culture" (Fass 2003; Stearns 2003). High-profile kidnappings and school shootings certainly inspire worry, but parental fear may exceed the reality of the risk. Misrepresented by the media as high and/or rising, many perceived threats to children are statistically low or have actually declined. In 2008, the U.S. murder rate was at its lowest level since the 1960s (Von Drehle 2010). Crimes against children at school have declined in the past several years (Dinkes, Kemp, and Baum 2009). And although one wouldn't know it by watching televised news, the odds that a stranger will kidnap a child are extremely low.

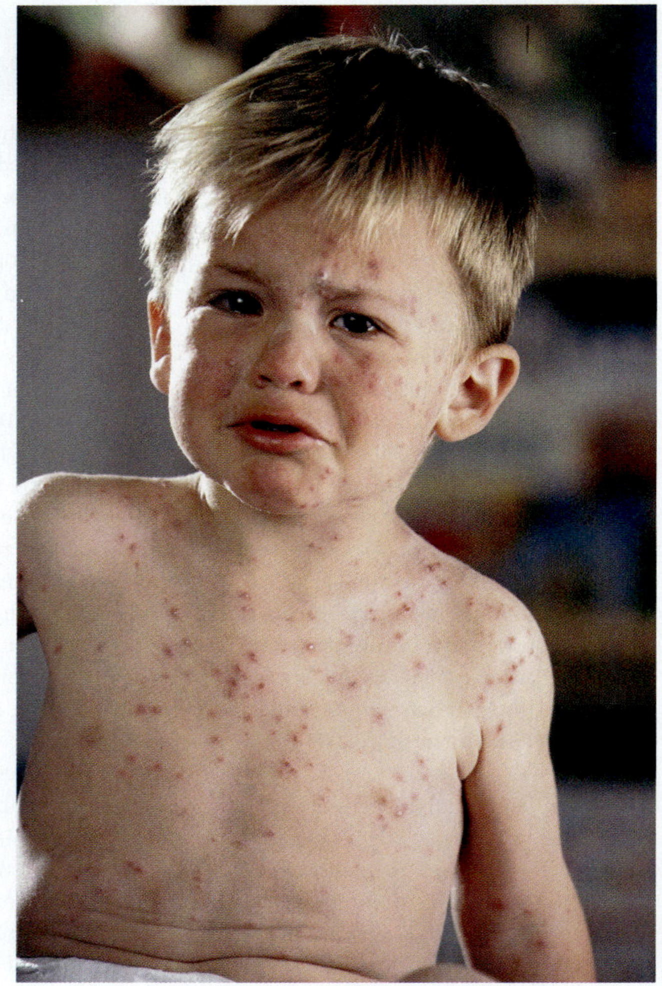

Daily family hassles, such as a child's coming down with chicken pox, put demands on a family. Sometimes everyday hassles pile up to result in what social scientists call "stressor overload." This is especially true when a new stressor is added to already difficult daily family life.

We don't mean to imply that kidnappings, crimes at school, or other feared threats to children never occur. A realistic analysis of dangers, collecting information on strategies appropriate to living safely in our neighborhoods, a plan for talking with children about protection from actual risks, and a "check it out" attitude toward frightening media stories are good parental approaches.

As you read this section, you may have noted that sometimes a single event can be classified as more than one type of stressor or combine stressor types. Adopting

Facts about Families

ADHD, Autism, Stigma, and Stress

The Centers for disease Control and Prevention estimate that about one in 110 children in the United States has some degree of autism (Ramisch 2012). According to estimates by the American Psychiatric Association, attention deficit/hyperactivity disorder (ADHD) has been diagnosed in 3 to 7 percent of U.S. school children, more often in boys than in girls (Firmin and Phillips 2009). ADHD can also be an adult diagnosis (Retz and Klein 2010). Autism and ADHD are similar inasmuch as children with either disease risk being stigmatized.

Families with children diagnosed with autism or ADHD face ongoing stressors that can be understood as daily hassles, some severe and demoralizing. Parents report that, due to the child's behavior, they often feel interrupted; miss social events because they are hesitant to leave their child with a babysitter; are anxious about taking the child out in public; must deal with other parents', neighbors', teachers', and/or school bus drivers' complaints; spend excessive amounts of time with the child's homework; worry that the child will get into trouble or be injured; have difficulty finding adequate after-school placement for their child and are unable to find or afford other professional or school services for the child; lose patience due to especially trying morning routines; must address siblings' resentment of parents' extra time and attention spent with the child diagnosed with ADHD; miss work; face lack of sleep due to disrupted bedtimes; and do not have enough time for themselves (Firmin and Phillips 2009; Ramisch 2012; Hall and Graff 2012; Reader, Stewart, and Johnson 2009). Regarding airplane travel, parents face potential troubles with boarding or in-flight regulations when taking an autistic child who may not readily follow the rules (Swarns 2012).

Raising a child diagnosed with autism or ADHD can involve feeling embarrassed as a result of specific instances of misbehavior and also as a consequence of being stigmatized. Stigmatizing others involves prejudice or discrimination based on others' perceived negative characteristics, status, or behaviors (Goffman 1963). **Courtesy stigma** refers to a situation in which not only the initially stigmatized individual but also her or his intimates are stigmatized by association (Goffman 1963; Koro-Ljungberg and Bussing 2009).

Focus groups (see Chapter 2) with thirty parents of children diagnosed with ADHD revealed that the parents often received unsolicited advice and felt negatively judged by both extended family members and strangers. Indeed, public debate over whether the diagnosis itself is legitimate or simply a convenient label for bad behavior increases the possibility of stigma (Koro-Ljungberg and Bussing 2009).

Some parents informed others of the diagnosis to ward off potential criticism of their child's behavior. Parents who were able to resist or shrug off negativity from neighbors, extended kin, or community were better able to cope. Some parents mentioned spirituality as a resource. As one mother said, "I just leave it in God's hand because the only thing I can do is just pray for him. Just pray and ask God to shield and protect him" (in Koro-Ljungberg and Bussing 2009, p. 1192).

For many parents, however, management of courtesy stigma may involve withdrawal on the one hand, coupled with activism on the other hand. The majority of the parents in the focus group research managed stigma mainly by avoiding potentially stressful situations. They kept the diagnosis to themselves, interacting primarily or only with families of children who demonstrated behaviors similar to their child's. Many admitted doing homework and school projects for their children to reduce the possibility of being stigmatized by their child's teacher or by other parents (Koro-Ljungberg and Bussing 2009).

Parents also engaged in activism. Some volunteered at school to advocate for their child. They pressed for special school services for children diagnosed with ADHD, and they politicized ADHD by demanding more public education and increased awareness that could help reduce the stigma associated with ADHD (Koro-Ljungberg and Bussing 2009).

A variation of the STEPP (Strategies to Enhance Positive Parenting) program has been developed specifically to address parenting children diagnosed with ADHD (Chacko et al. 2008). Although sponsored by the pharmaceutical company Shire, the website ADHDaction guide.com includes nonpharmaceutically based tips for managing adult ADHD (see also Manning, Wainwright, and Bennett 2011; Retz and Klein 2010).

Critical Thinking

Do you know an adult, child, or parents of a child who has been diagnosed with autism or ADHD? Could you have added to their feelings of being stigmatized? If you are parenting a child diagnosed with autism or ADHD, have you experienced courtesy stigma? If so, how have you handled it? How might you handle it in the future?

a child with special needs often involves both adding a family member and caring for a disabled child (Schweiger and O'Brien 2005). As another example, raising a child with emotional or behavior problems adds to everyday family hassles, can be demoralizing, and may precipitate family conflict (Chacko et al. 2008; Talan 2009). Having immigrated to the United States means having lost the physical companionship of family members in one's home country as well as struggling with daily hassles and challenges, perhaps regarding a new language, perhaps, documentation, or financial issues (Raffaeli et al. 2012). From another point of view, having parents who emigrate, leaving children or elderly parents in the home country, causes stress for those left at home (Bodrug-Lungu and Kostina-Ritchey 2013; Cabanes and Acedera 2012).

The September 11, 2001, attack on New York City and Washington, DC, and more recently the Boston marathon bombing were events that can be classified as a sudden changes in our family environments—changes that, among other things, sparked parents' need to consider how to talk with their children about terrorism (Boss 2004; Myers-Walls 2002; Walsh 2002). For many, the attacks also constituted demoralizing events. For others, the events were not only sudden and demoralizing but also sadly marked the loss of a family member.

Stressor Overload

A family may be stressed not just by one serious, chronic problem but also by a series of large or small, related or unrelated stressors that build on one another too rapidly for the family members to cope effectively (McCubbin, Thompson, and McCubbin 1996). This situation is called **stressor overload**, or *pileup*:

> Even small events, not enough by themselves to cause any real stress, can take a toll when they come one after another. First an unplanned pregnancy, then a move, then a financial problem that results in having to borrow several thousand dollars, then the big row with the new neighbors over keeping the dog tied up, and finally little Jimmy breaking his arm in a bicycle accident, all in three months, finally becomes too much. (Broderick 1979b, p. 352)

In some cases stressor overload characterizes the primary stressor event itself. For instance, experiencing a natural disaster such as a wild fire, hurricane, or tornado may require evacuation, finding shelter, getting children to school from a new location, getting appraisals for home damage, worrying about whether belongings left at home are secure, and going without familiar clothing, cosmetics, or toys (McDermott and Cobham 2012).

Then too, stressor overload can creep up on people without their realizing it. Even though it may be difficult to point to any single precipitating factor, an unrelenting series of relatively small stressors can add up to a crisis. In today's economy, characterized by longer working hours, a family's need to rely on more than one paycheck, fewer high-paying jobs, fewer benefits, and little job security, stressor overload may be more common than in the past. A second example of stressor overload is the addition of psychological depression to an earlier stressor, such as chronic poverty or an adolescent family member's living with epilepsy (Seaton and Taylor 2003).

A third example might involve the ambiguous loss of a family member deployed overseas, followed by the stressors associated with the family member's return home, possibly compounded by the soldier's serious physical injuries and/or post-traumatic stress disorder (England 2009; Hoge 2010; Johnson 2010). As a final example, a parent's promotion and consequent relocation to another part of the country involves packing, moving, finding new doctors and pharmacies, locating new schools for the children, and the need to make friends and forge an informal support system in the new locale. We'll return to the idea of stressor pileup shortly. Now, however, with an understanding of the various kinds of events that cause family stress and can precipitate a family crisis, we turn to a discussion of the course of a family crisis.

The Course of a Family Crisis

Although family stress simply involves "pressure put on the family," a family *crisis* results from an "imbalance between pressure and supports" (Boss 1997, p. 1). A family crisis ordinarily follows a fairly predictable course, similar to the truncated roller coaster shown in Figure 13.3. Three distinct phases can be identified: the event that causes the crisis, the period of disorganization that follows, and the reorganizing or recovery phase after the family reaches a low point. Families have a certain level of organization before a crisis; that is, they function at a certain level of effectiveness—higher for some families, lower for others.

Families that are having difficulties or functioning less than effectively before the onset of additional stressors or demands are said to be **vulnerable families**; families capable of "doing well in the face of adversity" are called **resilient families** (Patterson 2002b, p. 350). It's important to note here that the definition of a family's "doing well" (that is, family well-being) may differ

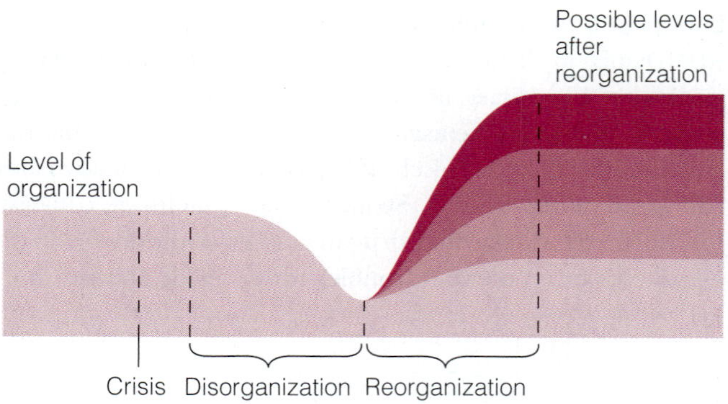

FIGURE 13.3 Patterns of family adaptation to crisis
When families reorganize after the onset of a crisis, they may do so at a level that is less, the same as, or more effective than that preceding the crisis.

Source: Adapted from Hansen and Hill 1964, p. 810.

Spencer Grant/age fotostock

Resilient families do well in the face of adversity. Greater financial resources are advantageous in coping with family stress and crises, but low income families are often creatively resilient in locating resources.

according to whether the family views its own well-being from a more individualistic or from a more collectivist set of values (see Chapter 1). From a more individualistic perspective, family well-being may emphasize economic, educational, and occupational resilience. From a more collectivist perspective, however, family well-being may emphasize ability of family members to get together and a return to practicing cultural traditions (McCubbin et al. 2013).

In the period of disorganization following the crisis, family functioning declines from its initial level. Families reorganize, and after the reorganization is complete, (1) they may function at about the same level as before; (2) they may have been so weakened by the crisis that they function only at a reduced level—more often the case with vulnerable families; or (3) they may have been stimulated by the crisis to reorganize in a way that makes them more effective—a characteristic of resilient families.

At the onset of a crisis, it may seem that no adjustment is required at all. A family may be confused by a member's alcoholism or numbed by the new or sudden stress and, in a process of denial, go about their business as if the event had not occurred. Gradually, however, the family begins to assimilate the reality of the crisis and to appraise the situation. Then the **period of family disorganization** sets in.

The Period of Disorganization

At this time, family organization slumps, habitual roles and routines become nebulous and confused, and members carry out their responsibilities with less enthusiasm. Although not always, this period of disorganization may be "so severe that the family structure collapses and is immobilized for a time. The family can no longer function. For a time no one goes to work; no one cooks or even wants to eat; and no one performs the usual family tasks" (Boss 1997, p. 1). Typically, and legitimately, family members may begin to feel angry and resentful.

Expressive relationships within the family change, some growing stronger and more supportive perhaps, and others more distant. In the words of social psychologist Benjamin Karney,

> [S]tressful events occurring outside of a relationship interfere with couples' ability to maintain an intimate bond within the relationship. First, stress outside the relationship changes what couples need to talk about and the time available to talk about it.... Time that couples spend deciding how they are going to cut back to get their bills paid, or negotiating who is going to take off work to care for a sick relative, is time that is not spent on other activities, like having sex or participating in shared interests that are more likely to promote closeness. (2011, p. F2)

Other research shows that sexual activity, one of the most sensitive aspects of a relationship, often changes sharply and may temporarily cease. Parent-child relations may also change. As one example, a child described life with his mother during the period of disorganization after his father was deployed:

> I could tell my mom was getting like really depressed and since she wouldn't talk, I wouldn't talk. And so around the house everyone was just kind of depressed for a little while and you could tell because they didn't speak a lot. (in Huebner et al. 2007, p. 117)

Relations between family members and their outside friends, as well as the extended kin network, may also change during this phase. Some families withdraw from all outside activities until the crisis is over; as a result, they may become more private or isolated than before the crisis began. As we shall see, withdrawing from friends and kin often weakens rather than strengthens a family's ability to meet a crisis.

At the **nadir**, or low point, of family disorganization, conflicts may develop over how the situation should be handled. For example, in families with a seriously ill member, the healthy members are likely either to overestimate or to underestimate the sick person's incapacitation and, accordingly, to act either more sympathetically or less tolerantly than the ill member wants (Conner 2000; Pyke and Bengston 1996). Reaching the optimal balance between nurturance and encouragement of the ill person's self-sufficiency may take time, sensitivity, and judgment.

During the period of disorganization, family members face the decision of whether to express or to smother any angry feelings they may have. Expressing anger as blame will usually sharpen hostilities; laying blame on a family member for the difficulties being faced will not help to solve the problem and will only make things worse (Stratton 2003). At the same time, when family members opt to repress their anger, they risk allowing it to smolder, thus creating tension and increasingly strained relations. How members cope with conflict at this point—for instance, whether they use communication techniques that foster bonding or cause hurtful conflict or distancing—greatly influence the family's overall level of recovery (McDermott and Cobham 2012).

Recovery

Once the crisis hits bottom, things often begin to improve. Either by trial and error or by thoughtful planning, family members usually arrive at new routines and reciprocal expectations. They are able to look past the time of crisis to envision a return to some state of normalcy and to reach some agreements about the future. Some families do not recover intact, as today's high divorce and separation rates illustrate. Divorce or the separation of a cohabiting relationship can be seen as an adjustment to family crisis and as a family crisis in itself (see Figure 13.4).

Other families stay together, although at lower levels of organization or mutual support than before the crisis. As Figure 13.3 shows, some families remain at a low level of recovery, with members continuing to interact much as they did at the low point of disorganization. This interaction often involves a series of circles in which one member is viewed as deliberately causing the trouble and the others blame that individual and nag him or her to stop. This is true of many families in which one member is alcoholic or is otherwise chemically dependent or a chronic gambler, for example. Rather than directly expressing anger about being blamed and nagged, the offending member persists in the unwanted behavior.

In some instances, social structural, or environmental conditions limit a family's odds of recovery. For instance, former prisoners are often denied work opportunities from employers who ask situations like these on employment applications. Although this remains the situation for many who have been incarcerated, in 2012 the Equal Employment Opportunity Commission "approved an updated policy making it more difficult for employers to use background checks to systematically rule out hiring anyone with a criminal conviction" (Greenhouse 2012).

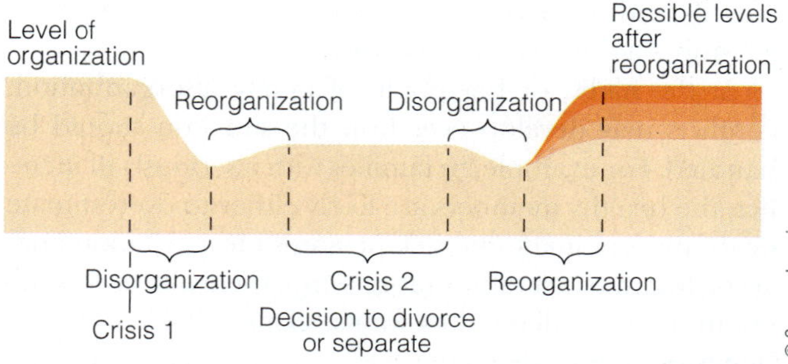

FIGURE 13.4 Divorce or separation as a family adjustment to crisis and as a crisis in itself.

Some families match the level of organization they had maintained before the onset of the crisis, whereas others rise to levels above what they experienced before the crisis (McCubbin 1995). For example, a family member's attempted suicide might motivate all family members to reexamine their relationships.

Reorganization at higher levels of mutual support may also result from less dramatic crises. For instance, partners in midlife might view boredom with their relationship as a challenge and revise their lifestyle to add some zest—by traveling more or planning to spend more time together rather than in activities with the whole family, for example. Research on family reorganization after a young child has been diagnosed with asthma found that families that were more cohesive and operated on a higher level of supportive functioning prior to the diagnoses recovered more positively than did families that were less supportive of one another prior to the diagnosis (Spagnola and Fiese 2010).

We have suggested that separation or divorce is a family crisis, and one study suggests that former partners' recovery trajectories evidence a similar range of possibilities. A longitudinal study of 216 men's and 238 women's recovery "pathways" over the course of ten years after divorce found that about one-fifth of the sample "grew more competent, well adjusted, and fulfilled" than they had been before the divorce. Meanwhile, the largest group had fashioned lives about as satisfying as what they'd had before divorce. Others, about one-tenth of the sample, evidenced low social responsibility and self-esteem and were more depressed and defeated than they had been before they divorced (Hetherington 2003, pp. 324–325). Now that we have examined the course of family crises, we will turn our attention to a theoretical model specifically designed to explain family stress, crisis, adjustment, and adaptation.

Family Stress, Crisis, Adjustment, and Adaptation: A Theoretical Model

Some decades ago, sociologist Reuben Hill proposed the ABC-X family crisis model, and much of what we've already noted about stressors is based on the research of Hill, his colleagues, and his successors (Hill 1958;

Hansen and Hill 1964). The **ABC-X model** states that **A** (the stressor event) interacting with **B** (the family's ability to cope with a crisis; their crisis-meeting resources) interacting with **C** (the family's appraisal of the stressor event) produces **X** (the crisis) (see Sussman, Steinmetz, and Peterson 1999). In Figure 13.5, *A* would be the demands put upon a family, *B* would be the family's capabilities—resources and coping behaviors—and *C* would be the meanings that the family creates to explain the demands.

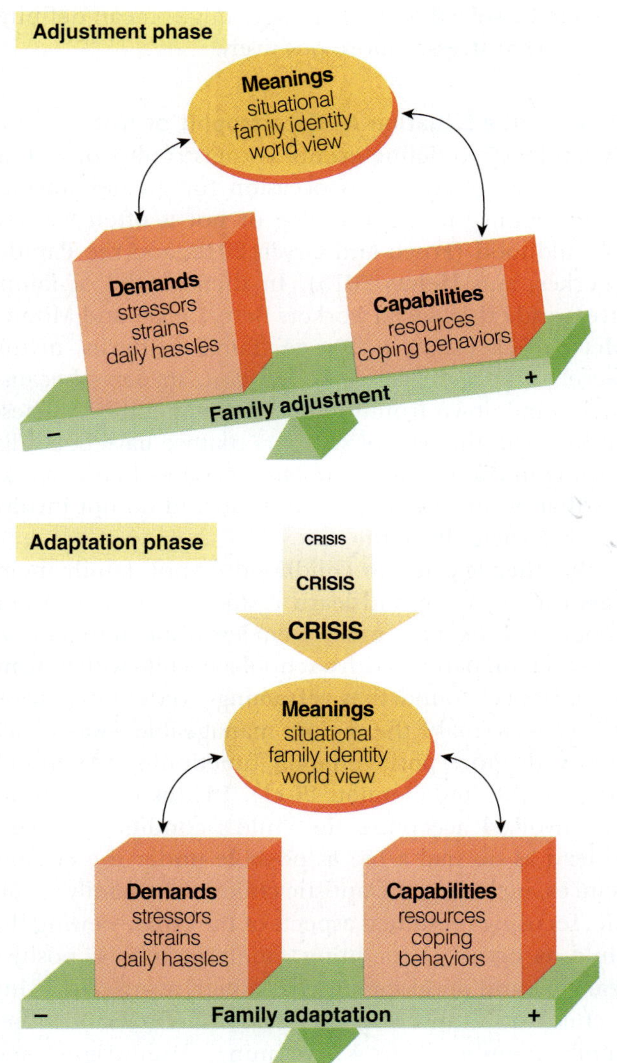

FIGURE 13.5 Family stress, crisis, adjustment, and adaptation. Families continuously balance the demands put upon them against their capabilities to meet those demands. When demands become heavy, families engage their resources to meet them while also appraising their situation—that is, they create meanings to explain and address their demands. When demands outweigh resources, family adjustment is in jeopardy, and a family crisis may develop. Through the course of a family crisis, some level of adaptation occurs (Patterson 2002b).

Source: Patterson 2002b, p. 351.

As Figure 13.5 illustrates, families continuously balance the demands put upon them against their capabilities to meet those demands. When demands become heavy, families engage their resources to meet them while also appraising their situation—that is, they create meanings to explain and address their demands. When demands outweigh resources, family adjustment is in jeopardy, and a family crisis may develop. Through the course of a family crisis, some level of adaptation occurs (Patterson 2002b).

Stressor Pileup

Building on the ABC-X model, Hamilton McCubbin and Joan Patterson (1983) advanced the *double* ABC-X model to better describe family adjustment to crises. In Hill's original model, the *A* factor was the stressor event; in the double ABC-X model, *A* becomes *Aa,* or "family pileup." *Pileup* includes not just the stressor but also previously existing family strains and future hardships induced by the stressor event.

When a family experiences a new stressor, prior strains that may have gone unnoticed—or been barely managed—come to the fore. Prior strains might be any residual family tensions that linger from unresolved stressors or are inherent in ongoing family roles, such as being a single parent or a partner in a two-career family. For example, ongoing but ignored family conflict may intensify when parents or stepparents must deal with a child who is underachieving in school, has joined a criminal gang, or is abusing drugs. As another example, financial and time constraints typical of single-parent families may assume crisis-inducing importance with the addition of a stressor, such as caring for an injured child.

An example of future demands precipitated by the stressor event would be a parent's losing a job, a stressor followed by unpaid bills, possibly even loss of the family domicile. A study about parenting a disabled child found that the child's rehabilitation often led to parental job changes, severe financial problems, and sleep deprivation (Rogers and Hogan 2003; Porterfield 2002). A somewhat related study about being an aging parent of a disabled adult found that the stress and uncertainty of aging is exacerbated, or worsened when an elderly parent is worried about who will care for a disabled offspring in the future (Fujiura 2010).

The pileup concept of family-life demands, or stressors (similar to the concept of stressor overload described earlier), is important in predicting family adjustment over the course of family life. Social scientists believe that, generally, an excessive number of life changes and strains occurring within a brief time, perhaps a year, are more likely to disrupt a family. Pileup renders a family more vulnerable to emerging from a crisis at a lower level of effectiveness (McCubbin and

McCubbin 1989). We have examined various characteristics of stressor demands put upon a family. Next we will look at how the family makes meaning of, defines, or appraises those demands. We'll look at crisis-meeting resources and coping behaviors after that.

Appraising the Situation

From an interactionist perspective, the meaning that a family gives to a situation—how family members appraise, define, or interpret a crisis-precipitating event—can have as much or more to do with the family's ability to cope as does the character of the event itself (McCubbin and McCubbin 1991; Patterson 2002a, 2002b). For example, a study of families faced with caring for an aging family member found that some families felt ambivalent, even negative, about having to provide care; other families saw caregiving as one more chance to bring the family together (Pyke and Bengston 1996; Roscoe et al. 2009).

Several factors influence how family members define a stressful situation. One is the *nature of the stressor itself.* How serious is it? How long can it be expected to last? Is the stressor event likely to improve or get worse? A study of families coping with a preschooler who had asthma found that the severity of the child's asthma attacks affected how smoothly the family adjusted (Spagnola and Fiese 2010). Sometimes in the case of ambiguous loss, families do not know whether a missing family member will ever return or whether a chronically ill or a chemically dependent family member will ever recover. A stressor event that puts a family in limbo this way is very difficult to manage.

In addition to the nature of the stressor itself, a second factor that influences a family's appraisal of the situation is *the degree of hardship or the kind of problems the stressor creates.* Temporary unemployment at age 16 is less a hardship than not finding a position upon college graduation or being laid off at age 45. Being victimized by a crime is always a stressful event, but coming home to find one's house burglarized may be less traumatic than being robbed at gunpoint. Caregiving for a short time after a family member's surgery differs from the long-term caregiving of indefinite length associated with serious chronic illness.

A third factor is *the family's previous successful experience with crises,* particularly those of a similar nature. If family members have had experience in nursing a sick member back to health, they will feel less bewildered and more capable of handling a new, similar situation. Believing from the start that demands are surmountable, and that the family has the ability to cope collectively, may make adjustment somewhat easier (Pitzer 1997b; Wells, Widmer, and McCoy 2004). Family members' interpretations of a crisis event shape their responses in subsequent stages of the crisis. Meanwhile, the family's crisis-meeting resources affect its appraisal of the situation.

A fourth, related factor that influences a family's appraisal of a stressor involves the *adult family members' legacies from their childhoods* (Carter and McGoldrick 1988; Fosco and Grych 2012). For example, growing up in a family that tended to define anything that went wrong as a catastrophe or a "punishment" from God might lead the family to define the current stressor more negatively. On the other hand, growing up in a family that tended to define demands simply as problems to be solved or as challenges might mean defining the current stressor more positively.

Defining the Situation As Catastrophic or Not

A family tendency to define events as catastrophic or not, as only negative or not, as occasion for greater warmth and demonstrations of caring or not is often learned in childhood (Fosco and Grych 2012; and see Parade, Leerkes, and Helms 2013). In their model of family stress and crisis, social workers Betty Carter and Monica McGoldrick (1988, p. 9) see "family patterns, myths, secrets [and] legacies" as *vertical stressors*—because they come down from the previous generations. These authors call the type of stressors that we have been discussing in this chapter *horizontal stressors*—that is, stressors that occur in real, present time and do not involve family legacies from the past.

Whether learned in childhood or not, family members can focus on and learn resilience—ways to think about stressors that render them less difficult to handle. Research on parents with a school-age child with autism, for instance, found that **reframing**—redefining stressful events to make them more manageable—was associated with more positive family functioning (Manning, Wainwright, and Bennett 2011). In this case, reframing involved accepting the child's condition, striving to lead as normal a life as possible under the circumstances, including the autistic child in the family's daily life, focusing on valued aspects of the child, viewing the child as occasion for other family members' positive learning and development—although assuredly having a child with autism does increase parental and other family members' stress (Manning, Wainwright, and Bennett 2011).

Not All Family Members Necessarily Agree in Their Appraisal of the Situation

Although we are discussing the family's appraisal, or definition of the situation, it is important to remember the possibility that each family member experiences a stressful event in a unique way:

> These unique meanings may enable family members to work together toward crisis resolution or they may prevent resolution from being achieved. That is, an individual's response to a stressor may enhance or impede the

LOREN HOLMES/The New York Times/Redux

A positive outlook, spiritual values, supportive communication, adaptability, public services, and informal social support—all these, along with extended family and community resources, are factors in family resilience, or meeting a crisis creatively. When living in rural Alaska got too difficult and expensive, this Yupik Eskimo family moved from their rural hometown of Kongiganak, with a population of 439, to urban Anchorage with a population of about 300,000. The family has extended kin in Anchorage. With a basement freezer stocked with bottles of seal oil, sea duck, and spotted seal meat, the family wards off homesickness by working to maintain their rural traditions and rituals (Severson 2011).

> family's progress toward common goals, may embellish or reduce family cohesion, may encourage or interfere with collective efficacy. (Walker 1985, pp. 832–33)

A dramatic example involves a family member's needing for a sibling to donate an organ, such as a kidney, or stem cells to fight a fatal illness (Begley and Piggott 2012). Some family members will see the donation, which requires surgical risk and time off work, as a taken-for-granted requirement to save a brother's or sister's life. However, the sibling who is expected to donate may see things differently and feel ambivalent. The same may be true for other family members. For instance,

> Sarah has three children, and her husband is unhappy about her becoming a donor. He is going for promotion, the children have important school years ahead of them and he does not want a disruption to family life. Sarah feels trapped. (Begley and Piggott 2012, p. 184)

Feeling trapped is a stressor. Resolving situations like these involve honest family communication, facilitated when possible by counseling.

Crisis-Meeting Resources

A family's crisis-meeting capabilities—resources and coping behaviors—constitute its ability to prevent a stressor from creating severe disharmony or disruption. We categorize a family's crisis-meeting resources into three types: personal/individual, family, and community.

The personal resources of each family member (for example, intelligence, problem-solving skills, and physical and emotional health) are important. At the same time, the family *as family* or family system has a level of resources, including bonds of trust, appreciation, and support (family harmony); sound finances and financial management and health practices; positive communication patterns; healthy leisure activities; and overall satisfaction with the family and quality of life (Boss 2002; Patterson 2002b).

Family rituals (see Chapter 1) are resources (Boss 2004; Oswald and Masciadrelli 2008). A study of families with alcoholism found that adult children of alcoholics who came from families that had maintained family dinner and other rituals (or who married into families that did) were less likely to become alcoholics themselves (Bennett, Wolin, and Reiss 1988; Goleman 1992).

And, of course, *money* is a family resource. For instance, a breadwinner's losing his or her job is less difficult to deal with when the family has substantial savings. In a qualitative study among U.S. working-poor rural families, one respondent explained that "I had absolutely nothing after I paid my bills to feed my kids. I scrounged just so that they could eat something, and I had to shortchange my landlord so that I could feed them, too, which put me behind in rent." Another said, "I felt overwhelmed and stressed because every time I get paid, I just don't have money for everything... because I have two... children [with medical problems]" (Dolan, Braun, and Murphy 2003, p. F14). At the other end of the financial spectrum are families who can afford to send troubled adolescents to costly "wilderness camps," for example, or other residential treatment facilities for illegal drug use or otherwise negative behaviors. Parents have told evaluation researchers that facilities such as these help to abate or relieve a family crisis and also to stabilize the family (Harper 2009).

The family ecology perspective alerts us to the fact that *community resources* are consequential as well (Socha and Stamp 2009). Increasingly aware of this, medical and family practice professionals have in recent years designed a wide variety of community-based programs to help families adapt to medically related family demands, such as a partner's cancer or a child's diabetes, congenital heart disease, and other illnesses (Marshall 2010; Tak and McCubbin 2002). In fact, in many instances, family members have become community activists, working to create community-based resources to aid them in dealing with a particular family stressor or crisis. Parents have been a driving force in shaping services and laws related to individuals with mental challenges (Lustig 1999). As a second example, parent groups and adults with disabilities worked together to help pass the Americans with Disabilities Act (Bryan 2010; Turnbull and Turnbull 1997).

Vulnerable versus Resilient Families Ultimately, the family either successfully adapts or becomes exhausted and vulnerable to continuing crises. Family systems may be high or low in vulnerability, a situation that affects how positively the family faces demands; this enables us to predict or explain the family's poor or good adjustment to stressor events (Patterson 2002b).

More prone to poor adjustment from crisis-provoking events, *vulnerable families* evidence a lower sense of common purpose and feel less in control of what happens to them. They may cope with problems by showing diminished respect or understanding for one another. Vulnerable families are also less experienced in shifting responsibilities among family members and are more resistant to compromise. There is little emphasis on family routines or predictable time together (McCubbin and McCubbin 1991).

From a social psychological point of view, *resilient families* tend to emphasize mutual acceptance, respect, and shared values. Family members rely on one another for support. Generally accepting difficulties, they work together to solve problems with members, feeling that they have input into major decisions (McCubbin et al. 2001). It may be apparent that these behaviors are less difficult to foster when a family has sufficient economic resources. The next section discusses factors that help families to meet crises creatively.

Meeting Crises Creatively

Meeting crises creatively means that after reaching the nadir in the course of the crisis, the family rises to a level of reorganization and emotional support that is equal to or higher than that which preceded the crisis. For some families—for example, those experiencing the crisis of domestic violence—breaking up may be the most beneficial (and perhaps the only workable) way to reorganize. Other families stay together and find ways to meet crises effectively. What factors differentiate resilient families that reorganize creatively from those that do not?

A Positive Outlook

In times of crisis, family members make many choices, one of the most significant of which is whether to blame one member for the hardship. Casting blame, even when it is deserved, is less productive than viewing the crisis primarily as a challenge (Stratton 2003).

Put another way, the more that family members can strive to maintain a positive outlook, the more it helps a person or a family to meet a crisis constructively (Burns 2010; Thomason 2005). Electing to work toward developing more open, supportive family communication—especially in times of conflict—also helps individuals and families meet crises constructively (Stinnett, Hilliard, and Stinnett 2000). Families that meet a crisis with an accepting attitude, focusing on the positive aspects of their lives, do better than those that feel they have been singled out for misfortune (Burns 2010). For example, many chronic illnesses have downward trajectories, so both partners may realistically expect that the ill mate's health will only grow worse (Marshall 2010). Some couples are remarkably able to adjust to this, "either because of immense closeness to each other or because they are grateful for what little life and relationship remains" (Strauss and Glaser 1975, p. 64).

Spiritual Values and Support Groups

"Spirituality, however the family defines it, can be a strong comfort during crisis" (Thomason 2005, p. F11). Many authors have argued that strong religious faith is related to high family cohesiveness and helps people manage demands or crises, partly because it provides a positive way of looking at suffering (Ellison et al. 2011; Lepper 2009; Wiley, Warren, and Montanelli 2002). A spiritual outlook may be fostered in many ways, including through Buddhist, Christian, Muslim, and other religious or philosophical traditions. However, a sense of spirituality—that is, a conviction that there is some power or entity greater than oneself—need not be associated with membership in any organized religion. Self-help groups, such as Alcoholics Anonymous or Al-Anon for families of alcoholics, incorporate a "higher power" and can help people take a positive, spiritual approach to family crises.

Open, Supportive Communication

Families whose members interact openly and supportively meet crises more creatively (Olson and Gorall 2009; Orthner, Jones-Sanpei, and Williamson 2004). For one thing, free-flowing communication opens the way to understanding (Thomason 2005). As an example, research shows that expressions of support from parents help children to cope with daily stress (Valiente et al. 2004). As another example, the better-adjusted husbands with multiple sclerosis believed that even though they were embarrassed when they fell in public or were incontinent, they could freely discuss these situations with their families and feel confident that their families understood (Power 1979). And as a final example, talking openly and supportively with an elderly parent who is dying about what that parent wants—in terms of medical treatment, hospice, and burial—can help (Fein 1997).

Knowing how to indicate the specific kind of support that one needs is important at stressful times. For example, differentiating between—and knowing how to request—just listening as opposed to problem-solving discussion can help reduce misunderstandings among family members—and between family members and others as well (Stinnett, Hilliard, and Stinnett 2000; Tannen 1990). Families whose communication is characterized by a sense of humor, as well as a sense of family history, togetherness, and common values, evidence greater resilience in the face of stress or crisis (Thomason 2005). Interestingly, research with women experiencing breast cancer showed that support from others was more likely to be offered and more often from the healthy spouse or other family members when the woman suffering was able to feel positive enough to express appreciative feelings toward her caregivers (Sheridan et al. 2010).

Adaptability

Adaptable families are better able to respond effectively to crises (Boss 2002; Uruk, Sayger, and Cogdal 2007). And families are more adaptable when they are more democratic and when conjugal power is fairly egalitarian (see Chapter 12). In families in which one member wields authoritarian power, the whole family suffers if the authoritarian leader does not make effective decisions during a crisis—and allows no one else to move into a position of leadership (McCubbin and McCubbin 1994). A partner who feels comfortable only as the family leader may resent his or her loss of power, and this resentment may continue to cause problems when the crisis is over.

Family adaptability in aspects other than leadership is also important (Burr, Klein, and McCubbin 1995). Families that can adapt their schedules and use of space, their family activities and rituals, and their connections with the outside world to the limitations and possibilities posed by the crisis will cope more effectively than families that are committed to preserving sameness. For example, a study of mothers of children with developmental disabilities found that mothers who worked part-time had less stress than those who worked full-time or were not employed at all (Gottlieb 1997). As another example, a study of married parents caring for a disabled adult child found that when their division of labor was adapted to feel fair, both parents experienced greater marital satisfaction and less stress (Essex and Hong 2005).

Informal Social Support

It's easier to cope with crises when a person doesn't feel alone (Tak and McCubbin 2002; Wickersham 2008). In fact, polls show that time spent with others is necessary to individuals' emotional well-being (Harter and Arora 2008). The caregiver needs support too. Spouse caregivers of elderly, disabled, or chronically ill themselves need social support (Fujiura 2010). Family members in hospitals need support from nurses and other hospital staff (Lind et al. 2012). Caregivers of burn patients showed symptoms of anxiety and need of counselling and support up to a year after the burn patient was injured (Backstrom, Armstrong, and Puentes 2013). We need support services of all types, including for parents of disabled or emotionally or behaviorally disturbed children (Vaughan et al. 2012).

Families may find helpful support in times of crisis from kin, good friends, neighbors, and even acquaintances such as work colleagues (Johnson 2010). Analysis of data from the National Survey of Black Americans found that many of them in times of crisis received support from fellow church members (Taylor, Lincoln, and Chatters 2005). These various relationships provide a wide array of help—from lending money in financial emergencies to helping with child care to just being there for emotional support. Research on families in poverty shows that, although the informal social support that they receive rarely helps to lift them out of poverty, it does help them to cope with their economic circumstances (Henly, Danziger, and Offer 2005).

Even continued contact with more casual acquaintances may be helpful, as they often offer useful information, along with enhancing one's sense of community (Orthner, Jones-Sanpei, and Williamson 2004). And, of course, the Internet offers information, social media options, and support for many, many stressors (Gilkey, Carey, and Wade 2009). A qualitative study that recruited participants by means of Web pages asked the seventy-seven respondents who answered an Internet-based survey about the advantages and disadvantages of Internet support, compared with face-to-face social networks (Colvin et al. 2004). Respondents mentioned two main Internet advantages—anonymity and the ease of connecting with others in the same situation despite geographical distance. Disadvantages related to lack of physical contact: "No one can hold your hand or give you a Kleenex when the tears are flowing" (p. 53).

An Extended Family

Sibling relationships and other kin networks can be a valuable source of support in times of crisis (Ryan, Kalil, and Leininger 2009). Grandparents, aunts, uncles, or other relatives may help with health crises or with more common family stressors, like running

Fuse/Jupiter Images

Many—although not all—turn to their extended family for social support in times of stress. Kin may provide emotional support, monetary support, and practical help. We have to note that this is not always the case, however. Sometimes extended family do not support an individual family member's wants or needs. Can you think of instances when this might be the case?

errands or helping with child care (Milardo 2005). Families going through divorce often fall back on relatives for practical help and financial assistance. In other crises, kin provide a shoulder to lean on—someone who may be asked for help without causing embarrassment—which can make a crucial difference in a family's ability to recover.

Although extended families as residential groupings represent a small proportion of family households, kin ties remain salient (Furstenberg 2005). One aspect of all this that is beginning to get more research attention involves reciprocal friendship and support among adult siblings (White and Riedmann 1992; Kluger 2006; Spitze and Trent 2006). In times of family stress or crisis, new immigrants (as well as African Americans) may rely on **fictive kin**—relationships based not on blood or marriage but rather on "close friendship ties that replicate many of the rights and obligations usually associated with family ties" (Ebaugh and Curry 2000).

We need to be cautious, though, not to overestimate or romanticize the extended family as a resource. For instance, a study that compared mothers who had children with more than one father found that the women received less support from their kin networks than did single mothers who did not have multipartnered births. The researchers concluded that "smaller and denser kin networks seem to be superior to broader but weaker kin ties in terms of perceived instrumental support" (Harknett and Knab 2007). Among the poor, extended kin may not have the resources to offer much practical help—and when they do, they are more likely to offer it to female than to male family members (Henly, Danziger, and Offer 2005; Mazelis and Mykyta 2011).

Moreover, along with some previous research, a small study of low-income families living in two trailer parks along the mid-Atlantic coast concluded that "low-income families do not share housing and other resources within a flexible and fluctuating network of extended and fictive kin as regularly as previously assumed." Extended family members may not get along, or individuals may be too embarrassed to ask their kin for help. One woman explained that neither her parents nor any one of her five siblings could help her because "they all have problems of their own." A Hispanic mother told the interviewer, "I know you've probably heard that Hispanic families are close-knit, well, hmmph! [It's not necessarily true.]" (Edwards 2004, p. 523). Then, too, among some recent immigrant groups, such as Asians or Hispanics, expectations of the extended family may clash with the more individualistic values of a more Americanized family member who needs help. We saw in Chapter 12, for instance, that some immigrant extended family members might discourage a battered wife from leaving her abusive husband or reporting him to police.

Community Resources

In 2008, Nebraska became the last state in the United States to adopt a safe-haven law whereby parents can abandon their children at hospitals without fear of prosecution. Intended as a way to save unwanted newborns from being murdered or left in dumpsters or motel rooms, Nebraska's safe-haven law failed to limit the ages of children who could be legally abandoned. During the month after the legislation passed, more than thirty youngsters were left at Nebraska hospitals. Most of them were older than eleven. Some had extremely severe mental and behavioral problems (Hansen and Spenser 2009). Some had been transported to Nebraska by overwhelmed parents from outside the state. A month later, Nebraska amended its law to require that legally abandoned children had to be younger than thirty days old

(Italie 2008; Jenkins 2008). The story became fuel for jokes on late-night television and afternoon talk shows.

> But… something is wrong when so many parents are so eager to abandon so many children…. Because what happened in Nebraska constitutes a message from overstressed parents, one we ignore at our own peril. It is not a complicated message. On the contrary it is as simple and succinct as a word: Help. (Pitts 2008)

Upon calling a special session of the Nebraska legislature to address this issue, more than one legislator indicated that the state would have to examine the accessibility of social services for older children and their families (Eckholm 2008). Subsequent reviews showed that in some cases the state had failed to help desperate parents who did not receive necessary services for their children until after the children had been dropped off. In other cases, the parents themselves did not seem to know where to turn—although appropriate services were available—until they heard about the safe-haven law (Hansen and Spenser 2009).

The success with which families meet the demands placed upon them depends at least partly on the availability of community resources, coupled with families' knowledge of and ability to access the community resources available to help (Karney 2011; Odom 2009; Trask et al. 2005).

> **Community-based resources** are defined as all of those characteristics, competencies and means of persons, groups and institutions outside the family which the family may call upon, access, and use to meet their demands. This includes a whole range of services, such as medical and health care services. The services of other institutions in the family's… environment, such as schools, churches, employers, etc.[,] are also resources to the family. At the more macro level, government policies that enhance and support families can be viewed as community resources. (McCubbin and McCubbin 1991, p. 19, boldface added)

Among others, community-based resources include schools and school personnel; social workers and family welfare agencies; foster child care; church programs that provide food, clothing, or shelter to poor or homeless families; twelve-step and other support programs for substance abusers and their families; programs for crime or abuse victims and their families; support groups for people with serious diseases such as cancer or AIDS, for parents and other relatives of disabled or terminally ill children, or for caregivers of disabled family members or those with cancer or Alzheimer's disease; and community pregnancy prevention and/or parent education programs. An Oregon study of non-Hispanic white and Hispanic teen mothers found that a government-funded home-visitation program increased family functioning, especially for the Hispanics in this sample (Middlemiss and McGuigan 2005).

Strength-Based Programs As an example of community-based resources, family empowerment, or *strength-based* programs recognize and build on a family's positive attributes (strengths) to foster resilience (Cleek et al. 2012). A unique example of family empowerment and parent education programs, mandated by the U.S. government in 1995, involves federal prison inmates. Parent inmates learn general skills, such as how to talk to their child. They also learn ways to create positive parent-child interaction from prison—such as games they can play with a child through the mail—as well as suggestions on what to do when returning home upon release (Coffman and Markstrom-Adams 1995; see also Comfort 2008; Kohl 2012). "Issues for Thought: When a Parent Is in Prison" further describes some of these programs.

Family Counseling Another community resource, family counseling (see Chapter 11) can help families after a crisis occurs, such as a family member's suffering from post-traumatic stress disorder (PTSD) (England 2009). Counseling can also help when families foresee a family change or future new demands (Clinton and Trent 2009; Rasheed, Rasheed, and Marley 2010). For instance, a couple might visit a counselor when expecting or adopting a baby, when deciding about work commitments and family needs, when the youngest child is about to leave home, or when a partner is about to retire. Family counseling is not just for relationships that are in trouble but is also a resource that can help to enhance family dynamics. Increasingly, counselors and social workers emphasize empowering families toward the goal of *enhanced resilience*—that is, emphasizing and building upon a family's strengths (Burns 2010; Power 2004).

Books and Online Resources In addition to counseling, resources include books on various subjects related to family stress and crises. Some examples: Barbara Monroe and Frances Kraus's *Brief Interventions with Bereaved Children* (2010); Avis Rumney's *Dying to Please* (2009) on eating disorders (see also Siegel, Brisman, and Weinshel 2009); Lynn Adams's (2010) "survival guide" for parenting autistic children (see also Roth and Barson 2010); Wes Burgess's *The Bipolar Handbook for Children, Teens, and Families: Real-Life Questions with Up-to-Date Answers* (2008); Kenneth Talan's *Help Your Child or Teen Get Back on Track* (2009), which addresses emotional and behavior problems (see also Kearney 2010); Chelsea Lowe and Bruce Cohen's *Living with Someone Who's Living with Bipolar Disorder* (2010); and Diane England's *The Post Traumatic Stress Disorder Relationship* (2009). Some books on topics of family stress or crisis are written specifically for children. Julianna Fields's *Families Living with Mental and Physical Challenges* (2010) is one example.

We also note the many resources available online. Resources on an enormous variety of stressors—from

Issues for Thought

When a Parent Is in Prison

More than 2 million children have a parent who is in jail or prison (Sabol and West 2009)—a demoralizing family stressor event, coupled with boundary ambiguity (Wildeman 2012). Incarceration rates rose sharply during the 1990s (Arditti 2003). Although rates are not increasing as rapidly today as ten years ago, they do continue to rise (McCarthy 2009). A child's risk of parental incarceration increased more than 60 percent between 1978 and 2000 (Roberts 2012). Although the risk remains low for non Hispanic white children, calculations show that 14 to 15 percent of black children born in 1978 and 25 to 28 percent "of black children born in 1990 had a parent imprisoned by the time the child was 14" (Wildeman 2009, p. 271).

Prior to their imprisonment, 79 percent of mothers and 53 percent of fathers were living with their children (National Resource Center on Children and Families of the Incarcerated 2009a). While mothers are in prison, about one-quarter of their children live with their fathers. Grandparents care for about half of all children with incarcerated mothers. Non Hispanic white children are more likely to be in nonfamily foster care (see Chapter 10) than are African American or Hispanic children (Lee, Genty, and Laver 2005). This may be because black and Hispanic communities have had more of a tradition of shared care of children (e.g., Stack 1974), a situation facilitating making arrangements that place children with adult relatives, often grandparents (Enos 2001; Poehlmann 2005). The children's caregivers often

> feel compelled to lie about their loved one's whereabouts. If the children are young, their mother may explain the father's absence by saying that "Daddy's away on a long trip" or "He's working on a job in another state." One caregiver ... explained to her nephews that their father was away at "super-hero school." Older children who know the truth may feel that they need to be careful not to discuss it at school or with friends. (Arditti 2003, p. F15)

Children's visiting an incarcerated parent can be expensive and otherwise difficult to arrange, because prisons are often far from their homes (McManus 2006; National Resource Center on Children and Families of the Incarcerated 2009b). One study found that half of children of women prisoners did not visit at all during their mother's incarceration. However, including phone calls and letters, 78 percent of mothers and 62 percent of fathers had at least monthly contact with children (Mumola 2000).

More and more, policy makers have realized that disrupted family ties have a severe and negative impact on the next generation (Arditti 2003; Cho 2011; Kohl 2012; Poehlmann 2005). Consequently, a number of correctional systems, including the Federal Bureau of Prisons, have developed visitation programs to facilitate parent-child contact. Many correctional facilities have returned to an earlier practice of permitting babies born in prison to remain with their mothers for a time (Rutgers University School of Criminal Justice and the New Jersey Institute for Social Justice 2006; Comfort 2008). Although visitation programs were initially oriented solely to mothers, prisons have more recently developed programs for fathers as well (Enos 2001; McManus 2006).

---

involuntary infertility (Resolve, a website of the National Infertility Association); to having a disabled child (Support for Families of Children with Disabilities); to experiencing the death of a child (The Compassionate Friends); to having a family member in prison (The Prison Talk Online Community)—offer Web-based virtual communities and information from experts as well as from others who are experiencing similar family demands.

The other side of the community-based-resource story, however, is that there just aren't enough of them. For instance, families struggling with caring for a disabled family member often face workplace insensitivity, career challenges, and financial difficulties (Fujiura 2010; Stewart 2013). Moreover, federal and state budget cuts have resulted in reduction of services and programs for families struggling with difficult stressors, such as caring for a disabled child, among many other stressors (Streitfeld 2010). In the words of social psychology professor Benjamin Karney,

> [F]amilies will benefit from policies that make their lives easier. Higher wages, more job security, and access to health care—to the extent that these policies would reduce the stress of modern life—would also promote the stability and quality of relationships directly. [Furthermore], even in the absence of serious changes to their lives, couples might be encouraged to recognize the ways that stress affects their relationships and assisted in developing communication patterns and concrete resources to help them cope with stress effectively when it arises. (Karney 2011)

As discussed in Chapter 1, family policy analyst Karen Bogenschneider has argued that political decisions (such as cutting family-centered federal, state, county,

Joel Gordon

Having a family member in prison or jail is a crisis that a small but growing number of families face today. Family stress and adjustment experts tell us that virtually all family crises have some potential for positive as well as negative effects. Can you think of any possible positive effects in this case? What community supports might help this family? What might be some alternatives to incarcerating parents who have been actively involved in raising their children?

We focus here on children's needs, but imprisonment demoralizes other family members as well and usually has a negative economic impact on the family system, not only when the prisoner has been an essential breadwinner but also due to costs associated with visiting the prisoner and making long-distance family telephone calls, among others (Arditti, Lambert-Shute, and Joest 2003). Moreover, because of stigma associated with incarceration, prisoners' families receive little community support (Arditti 2003, p. F15).

Strong family bonds appear to reduce children's negative behaviors (Poehlmann 2005), although "incarceration can undermine social bonds, [and] strain marital and other family relationships" (Western and McLanahan 2000, p. 323). Policy analysts argue that "[a]n over reliance on incarceration as punishment, particularly for nonviolent offenders, is not good family policy" (Arditti 2003, p. F17; Wildeman 2009; and see Dyer, Pleck, and McBride 2012). They propose alternatives to incarceration, such as home confinement with work release (Comfort 2008; see also sentencingproject.org).

Critical Thinking

Can you apply the ABC-X model to this situation of having an incarcerated family member?

or city programs) need to be evaluated through a *family impact lens* (Bogenschneider et al. 2012). From this point of view, most family policy scholars are apt to argue that program cutting—even such apparently small things as closing city swimming pools—unfortunately adds to family stress.

Crisis: Disaster or Opportunity?

A family crisis is a turning point in the course of family living that requires members to change how they have been thinking and acting (McCubbin and McCubbin 1991, 1994). We tend to think of *crisis* as synonymous with *disaster*, but the word comes from the Greek for *decision*. Although we cannot control the occurrence of many crises, we can decide how to cope with them.

Most crises—even the most unfortunate ones—have the potential for some positive as well as negative effects. For example, Professor Joan Patterson, a recognized expert in the field of family stress, has observed that many parents who are raising children with "complex and intense" medical needs

> seem to find new meaning for their life. Having a child with such severe medical needs and such a tenuous hold on life shatters the expectations of most parents for how life is supposed to be. It leads to a search for meaning as a way to accept their circumstances. When families get to this place, they not only accept their child and their family's life, but they often experience a kind of gratitude that those of us who have never faced this level of hardship can't really understand. (2002a, p. F7)

It's important to remember that not all stressors are unhappy ones. Happy events, such as moving into a new house, can be family stressors too.

Tom & Dee Ann McCarthy/Cusp/Corbis

Whether a family emerges from a crisis with a greater capacity for supportive family interaction depends at least partly on how family members choose to define the crisis (e.g., Manning, Wainwright, and Bennett 2011). A major theme of this text is that, given the opportunities and limitations posed by society, people create their families and relationships based on the choices they make. Families whose members choose to be flexible in roles and leadership meet crises creatively.

However, although they have options and choices, family members do not have absolute control over their lives (Coontz 1997; Kleber et al. 1997). Many family troubles are really the results of public issues. For example, the serious family disorganization that results from poverty is as much a social as a private problem (Trask et al. 2005). Also, most American families have some handicaps in meeting crises creatively. The typical American family is under a high level of stress at all times. Providing family members with emotional security in an impersonal and unpredictable society is difficult even when things are running smoothly. Family members are trying to do this while holding jobs and managing other activities and relationships.

Moreover, many family crises are more difficult to bear when communities lack adequate resources to help families meet them (Byrnes and Miller 2012; Pitts 2008). One response to this situation is to engage in community activism (Bryan 2010). For example, one couple, frustrated by the lack of organized community support available to them and their autistic child, founded Autism Speaks. A project of Autism Speaks is "to develop a central database of 10,000-plus children with autism that will provide, for the first time, the standardized medical records that researchers need to conduct accurate clinical trials" (Wright 2005, p. 47; see also autismandactivism.com; Roth and Barson 2010). A second example is the grassroots Disability Rights Movement (Bryan 2010). When families act collectively toward the goal of obtaining needed resources for effectively meeting the demands placed upon them, family adjustment can be expected to improve overall.

Summary

- Throughout the course of family living, *all* families are faced with demands, transitions, and stress.
- Family stress is a state of tension that arises when demands test, or tax, a family's resources.
- A sharper jolt to a family than more ordinary family stress, a family crisis encompasses three interrelated factors: (1) family change, (2) a turning point with the potential for positive and/or negative effects, and (3) a time of relative instability.
- Demands, or stressors, are of various types and have varied characteristics. Generally, stressors that are expected, brief, and improving are less difficult to cope with.
- The predictable changes of individuals and families—parenthood, midlife transitions, post-parenthood, retirement, and widowhood and widowerhood—are all family transitions that may be viewed as stressors.
- A common pattern can be traced in families that are experiencing family crisis. Three distinct phases can be identified: (1) the stressor event that causes the crisis, (2) the period of disorganization that follows, and (3) the reorganizing or recovery phase after the family reaches a low point.
- The eventual level of reorganization a family reaches depends on a number of factors, including the type of stressor, the degree of stress it imposes, whether it is accompanied by other stressors, the family's appraisal or definition of the crisis situation, and the family's available resources.
- Meeting crises creatively means resuming daily functioning at or above the level that existed before the crisis.
- Several factors can help families meet family stress and/or crises more creatively: a positive outlook, spiritual values, the presence of support groups, high self-esteem, open and supportive communication within the family, adaptability, counseling, and the presence of a kin network.

Questions for Review and Reflection

1. Compare the concepts *family stress* and *family crisis*, giving examples and explaining how a family crisis differs from family stress.
2. Differentiate among the types of stressors. How are these single events different from stressor overload? How might economic recession cause stressor overload?
3. Discuss issues addressed in other chapters of this text (e.g., work–family issues, parenting, separation, divorce, and remarriage) in terms of the ABC-X model of family crisis.
4. What factors help some families recover from crisis while others remain in the disorganization phase?
5. **Policy Question.** In your opinion, what more, if anything, could/should government do to help families experiencing stress? Families experiencing crisis?

Key Terms

ABC-X model 345
boundary ambiguity 337
community-based resources 351
courtesy stigma 341
family crisis 333
family stress 332
family transitions 334
fictive kin 350
nadir (of family disorganization) 343
period of family disorganization 343
reframing 346
resilient families 342
stress 332
stressors 335
stressor overload (pileup) 342
vulnerable families 342

14

Divorce and Relationship Dissolution

Howard Grey/STONE/Getty Images

Learning Objectives

1. Describe historical and current trends in divorce and relationship dissolution.
2. Identify economic, sociocultural, legal, and demographic factors associated with divorce and relationship dissolution.
3. Contrast the economic, social, and emotional consequences of divorce for women, men, and children.
4. Discuss the process by which couples decide to divorce and decide on custody arrangements for their children.
5. Describe the rules for successful co-parenting.

Divorce is a common experience in the United States for all social classes, but divorce is significantly less common among the more highly educated. Between 40 percent and 50 percent of all first marriages will likely end in divorce (National Marriage Project and the Institute for American Values 2012). In this chapter, we'll examine factors that affect people's decisions to divorce, the experience itself, and ways to make the experience less painful. We'll also analyze why so many couples in our society decide to divorce and examine the debate over whether a divorce should be harder to get. Divorce is associated with lower well-being for children, so we will end this chapter by discussing what parents can do to reduce these negative effects.

Although the research on divorce is mostly specific to marriage, many of the dynamics apply to breakups of committed nonmarital relationships (Kamp Dush 2011). As discussed in Chapter 6, the rate of relationship dissolution for cohabitors is considerably higher than that of marrieds. Divorce research may also apply to some extent to same-sex couples, although much less is known about their process of breaking up. We'll begin by looking at divorce rates in the United States, which are among the highest in the world.

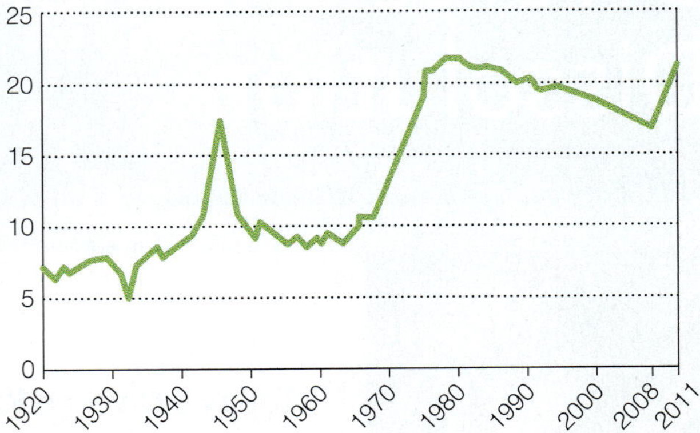

FIGURE 14.1 Divorces per 1,000 married women age 15 and older in the United States, 1920–2011. This includes the latest data available for the refined divorce rate.

Source: U.S. National Center for Health Statistics 1990a, 1998, p. 3; Wilcox and Marquardt 2009, p. 76, Figure 5. National Marriage Project and the Institute for American Values 2012, p. 70, Figure 5.

Today's Divorce Rate

Even though divorce was much less common in past centuries than today, divorce is not new. The divorce rate started its upward swing in the late nineteenth century (Amato and Irving 2006; Teachman, Tedrow, and Hall 2006). The frequency of divorce increased throughout most of the twentieth century, as Figure 14.1 shows, with dips and upswings surrounding historical events such as the Great Depression, Great Recession of 2007, and major wars. There was an unprecedented rise in divorce between 1960 and 1980 when the **refined divorce rate** (the number of divorces per 1,000 married women) more than doubled (Wilcox and Marquardt 2009, p. 75, Table 5). The divorce rate declined throughout the 1990s and for the most part leveled off, though at a high level (Pew Research Center 2010a).

Another way to look at divorce trends is by using the **crude divorce rate**, the number of divorces per 1,000 population (see Figure 14.2). This rate includes portions of the population—children and the unmarried—who are not at risk for divorce. Despite its limitations, the crude divorce rate is sometimes used for comparisons over time because these data are the only long-term annual data available. The crude divorce rate has declined almost 30 percent since 1979, and it has not been so low since around 1970 (U.S. Census Bureau 2012c, Table 133).

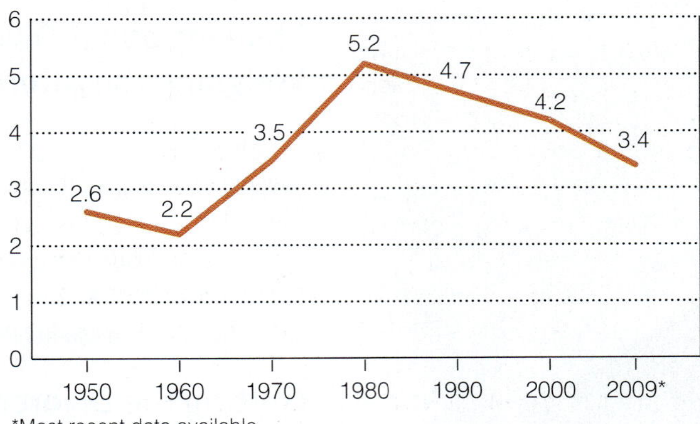

FIGURE 14.2 Divorces per 1,000 population, 1950 to 2009 (crude divorce rate).

Source: Adapted from U.S. Census Bureau 2010b, Table 78; U.S. National Center for Health Statistics 2006, Table A; U.S. Census Bureau 2012c, Table 133.

The Divorce Divide

Despite the widely held notion that "fifty percent of all marriages will end in divorce," the risk of divorce varies significantly by social class, as measured by education. For example, since 1990 the crude divorce rate declined among college graduates, but has remained stable for men and women with less education (Amato 2010; Elliot and Simmons 2011). This holds true for African Americans and whites (Kim 2012). According to the National Marriage Project and the Institute for American Values (2012), a woman with a college degree has a 25-percent lower risk of divorce than one who did not complete high school. This has produced what sociologist Steven Martin calls the **divorce divide**, the large disparity in divorce rates between those with and without a college degree (Ono 2009). This divide is similar to the marriage gap discussed in Chapter 7, which describes the disparity in marriage rates between the poor and nonpoor. Divorce rates also vary by income. Compared with people who earn under $25,000 annually, people who earn over $50,000 have a 30-percent lower risk of divorce (National Marriage Project and the Institute for American Values 2012).

Starter Marriages and Silver Divorces

Most divorces occur relatively early in marriage—a large percentage end within five years, do not involve children, and are followed by remarriage—situations that led journalist Pamela Paul to coin the term **starter marriage** (Paul 2002). A *starter marriage* is a first marriage that ends in divorce within the first few years, typically before the couple has children. In her interviews, young divorced men and women listed marrying too soon, the desire to move out of their parents' house, to have a fairytale wedding, and the inevitability of marriage after a long period of dating as reasons for why their marriage did not last.

Despite the early demise of some marriages, the median length of a first marriage that ends in divorce is about eight years (Kreider and Ellis 2011b). Marriages lasting weeks or months (as opposed to years) are rare. For example, in Iowa in there were 74 marriages from 2000 to 2009 that lasted 72 days or fewer. Two lasted only seven days (Kilen 2011).

Meanwhile, the proportion of divorces among older couples and those in long-term marriages has recently shown an increase. This came as a surprise to many academics and social commentators, who refer to this phenomenon as a **silver divorce** (Wingert and Kantrowitz 2010). This is particularly true for the baby boom generation, born between 1946 and 1964. Sociologists Susan Brown and I-Fen Lin, in their analysis of the American Community Survey, found that the divorce rate among adults aged 50 and older doubled between 1990 and 2010. Divorces among this age group accounted for one-fourth of all divorces in 2010 and is the only age group for which the rate of divorce is currently growing (Brown and Lin 2012).

What explains the jump in divorce among baby boomers? First, the baby boom generation is one of the largest in history with the largest number of marriages and divorces. Second, the high rate of remarriage among this age group also contributed to the high rate of divorce. The divorce rate of those aged 50 and older was 2.5 times higher in remarriages than first marriages (Brown and Lin 2012). Sociologist Dr. Susan Brown (Brown 2013), the lead author of the silver divorce study, points to the "complex marital biographies" of baby boomers, which increase the probability of divorce later in life. She identifies several life transitions that may cause older adults to "take stock," such as an "empty nest" (adult children leaving home) and impending retirement. Although married couples have been making such life transitions for decades, divorce was heavily stigmatized and therefore was not considered an option, which is less the case today. Moreover, longer life expectancies among more recent generations of older couples translate into many more years in an unhappy marriage Another difference between previous and current generations of older couples is society's increased focus on individual self-fulfillment, especially among baby boomers (Thomas 2012).

Divorce Among Gay and Lesbian Couples

In 2005, Carolyn Conrad filed for dissolution of her civil union from Kathleen Peterson. What is noteworthy about this couple is that they were in the first same-sex marriage in the state of Vermont (Barlow 2005). Like heterosexual couples, gay and lesbian couples sometimes break up. Now that some states are allowing same-sex marriage, some of those relationships will end in divorce. Same-sex divorce is one of the benefits of same-sex marriage (where it exists) in that it provides for a formal, clearly recognized way to address couple breakup issues regarding property, child custody, and so on (Allen 2007). However, there are many legal barriers for gay and lesbian married couples who want a divorce, for example if the couple should move to a state that doesn't recognize the marriage (Van Eeden-Moorefield et al. 2011).

Research on divorce among same-sex couples is just getting started, and it is too early to tell how divorce among same-sex couples will affect overall divorce rates. The number or percentage of same-sex marriages that end in divorce is currently difficult to estimate. However, research conducted thus far show either minor or no difference in divorce rates between gay and lesbian and heterosexual couples (Badgett and Herman 2011; Balsam et al. 2008; Joyner, Manning, and Bogle 2013).

Redivorce

In summing up the statistics, we need to note that a high divorce rate does not mean that Americans have given up on marriage. It means that they find an unhappy marriage intolerable and hope to replace it with a happier one. But a consequence of remarriages—which have higher divorce rates than first marriages—is an emerging trend of redivorce, or experiencing multiple divorces. [Although not unheard of, divorced couples remarrying each other a second time is extremely uncommon (National Public Radio 2011).] As sociologist Andrew Cherlin explains in his book *The Marriage Go-Round* (2009a), many who divorce—and their children—can expect several emotionally significant transitions in family structure and lifestyle. The American Community Survey in 2009 indicated that one-third of men and women who divorced in the last 12 months had been previously divorced; 26 percent had one prior divorce and 9 percent had two or more prior divorces (Elliot and Simmons 2011).

In a context of a high though declining divorce rate, along with a positive view of marriage, why is it that married couples get divorced? The following section identifies economic, sociocultural, legal, interpersonal, and demographic factors associated with divorce.

Why Did the Divorce Rate Rise Throughout the Twentieth Century?

Various factors can bind married couples together: economic interdependence between spouses; spouses' having similar social and demographic characteristics; legal, social, and moral constraints; and the spouses' relationship itself. Yet the binding strength of these factors has lessened.

Demographic Factors

Certain demographic factors that have been shown to be major risk factors for divorce. Many of these same factors are associated with the break-up of cohabiting union (Kamp Dush 2011). These include:

- Remarriage. Remarried couples are more likely than first marrieds to divorce (Bumpass and Raley 2007).
- Young age at first marriage, especially marrying as a teenager (Amato et al. 2007; Copen et al. 2012, p. 7).
- Marrying someone of a different race, ethnicity, or religion (Amato 2010).
- Cohabitation before marriage increases the likelihood of divorce (see Chapter 5), although the increased risk of divorce for cohabiting couples has become smaller and depends on the couples' education, income, and number of prior cohabitations (Miller, Sassler, and Kusi-Appouh 2011; Teachman, Tedrow, and Hall 2006).
- Premarital sex, premarital pregnancy, and premarital childbearing usually increase the risk of divorce in a subsequent marriage (Heaton 2002; Teachman 2002), unless birth occurred during cohabitation (Teachman, Tedrow, and Hall 2006).
- Having no or older children, compared to young children, increases divorce risk (Hetherington 2003), except in the case of remarriages involving stepchildren (see Chapter 15 for further discussion).
- Having parents and grandparents who divorced increases one's own likelihood of divorcing (Amato 2010 Teachman, Tedrow, and Hall 2006).
- Black and Hispanics have higher divorce rates compared to whites and Asians (see Table 14.1), but "racial differences in dissolution are not well understood... and we know little about the underlying processes that may generate differences in divorce rates among racial and ethnic groups" (Teachman, Tedrow, and Hall 2006, p. 76). For example, among Hispanics, immigration status is important. Mexican Americans born outside the United States have very low rates of divorce whereas those born in the U.S. have rates of divorce similar to nonHispanic whites (Amato 2010).
- It is widely thought that military couples have higher divorce rates. Despite the stress and demands of military service and long deployments, military marriages are not more prone to divorce than civilian marriages (Karney, Loughran, and Pollard 2012).

Economic Factors

Traditionally, as we've seen, the family was a self-sufficient productive unit. Survival was far more difficult outside of families, so members remained economically

Table 14.1 Percentage of Ever-Married Women Age 50 to 59 Ever-Divorced, by Race/Ethnicity, 2009

Race/Ethnicity	Ever-Divorced (%)
White, nonHispanic	42.4
Black	48.2
Asian	18.5
Hispanic	30.2
Total	41.1

Source: Kreider, Rose M. and Renee Ellis. 2011b. *Number, Timing, and Duration of Marriages and Divorces: 2009.* Current Population Reports P70–125. Washington, DC: U.S. Census Bureau, Table 2.

bound to one another. But today, because family members no longer need one another for basic necessities, they are freer to divorce than they once were (Davis 2010).

Families are still somewhat interdependent economically. Even though marriage "has become less economically necessary... it remains economically advantageous in most cases" (Wilcox and Marquardt 2009, p. 42). As long as marriage continues to offer practical benefits, economic interdependence will help hold marriages together. The economic practicality of marriages varies according to several conditions.

The higher the social class as defined in terms of education, employment, income, wealth (e.g., home ownership), the less likely a couple is to divorce or separate (Kreider and Ellis 2011b). Moreover, the loss of income has been found to increase the likelihood of divorce, especially when it is the male who loses his income (Wilcox and Marquardt 2009, pp. 19, 34). What about women's earnings? The answer is complicated. The upward trend of divorce and the upward trend of women in the labor force have accompanied each other historically. A woman's earning a wage might give an unhappily married woman the economic power, the increased independence, and the self-confidence to help her decide on divorce. For instance, in unhappy marriages (but not happy ones) women's employment may increase divorce risk (Sayer et al. 2011; Schoen et al. 2002), and Amato et al. (2007) found wives' work hours increased their spouses' perceptions of marital problems. However, this **independence effect** may hold for white women only (Teachman 2010), and women's relative contribution to the household income and divorce is inconclusive (Sayer 2006).

Another perspective is that a wife's earnings may actually help to hold the marriage together by counteracting the negative effects of poverty and economic insecurity on marital stability (Schoen, Rogers, and Amato 2006). Moreover, women's educational gains seem to be a stabilizing factor in marriage (Heaton 2002). Other research shows that the couple's financial well-being is less predictive of divorce than financial disagreements between spouses (Dew, Britt, and Huston 2012).

Weakening Social, Moral, and Legal Constraints

There are far fewer social, moral, and legal constraints on divorce than in the past. Only about one-third of Americans today see divorce as morally unacceptable (Gallup Poll 2012b). In a study of marital cohesion, virtually no respondents mentioned stigma or disapproval as a barrier to divorce (Amato and Hohmann-Marriott 2007).

These changes in views have become widespread. Nationally, conservative Protestant and Catholic churches, traditionally strongly opposed to divorce, have become more tolerant of divorced people and often explicitly welcome them into the congregation (Cherlin 2009a). Rural and nonmetropolitan areas, known for their more conservative family values and religious beliefs, had always had lower divorce rates. Yet now, divorce rates in those areas have grown to such an extent that the rural–urban distinction in divorce has completely disappeared (Tavernise and Gebeloff 2011). Southern states, sometimes referred to as the "Bible Belt," are also known for their traditional cultural values and many young adults feel pressure to marry young (Hetter 2011). It may come as a surprise that divorce rates are higher there than the more liberal Northeast. The high marriage rates in the South combined with younger age at marriage has resulted in more divorces and higher divorce rates in that region (Elliot and Simmons 2011; Population Reference Bureau 2013).

To say that societal constraints against divorce no longer exist would be an overstatement. Religious views and differences in culture still affect peoples' attitudes toward divorce. For example, Hispanics have more negative attitudes toward divorce, despite their higher divorce rates (Ellison, Wolfinger, and Ramos-Wada 2012). Most states have waiting periods of various lengths before a divorce can be granted, and it may take many more months for divorcing couples to come to an agreement about how to divide assets (and debts), financial support, and child custody. Despite greater societal acceptance of divorce, social pressure against divorce among affluent groups has recently been growing, leading sociologist Andrew Cherlin to remark, "The condemnation of divorce is also coming from the group that is most confident it can make its marriages succeed, and that allows them to be dismissive of divorce" (cited in Paul 2011).

With greater societal acceptance of divorce came lessening legal restrictions on divorce, such as no-fault divorce laws, which were widely adopted throughout the United States in the 1970s and 1980s. This revision of divorce law was intended to reduce the hostility of the partners and to permit an individual to end a failed marriage readily. Before the 1970s, the fault system predominated. A fault divorce required a legal determination that one party was guilty and the other innocent. Parties seeking divorce had to prove that they had "grounds" for divorce, such as the spouse's adultery, mental cruelty, or desertion. Obtaining a divorce might require falsifying these facts. The one judged guilty rarely received custody of the children, and the judgment largely influenced property settlement and alimony awards, as well as the opinions of friends and family (Coontz 2010a; Stevenson and Wolfers 2007). Such a protracted legal battle of adversaries increased hostility and diminished

Issues for Thought

Should Divorce Be Harder to Get?

In reviewing the process of divorce and its effects, this question arises: Should divorce be harder to get? Some Americans and some family scholars and policy makers think so. At the same time, the American public holds somewhat ambivalent attitudes about divorce. In one poll, one-half of those surveyed thought divorce should be harder to get (Stokes and Ellison 2010, p. 13, Table 1). It also appears that some unhappily married individuals postponed divorce until their children were older (Foster 2006; P. Schwartz 2010).

To address this situation, some policy makers have proposed changes in state divorce laws so that divorces would be more difficult to get. As noted earlier, with no-fault divorce laws, a marriage can be dissolved simply by one spouse's testifying in court that the couple has "irreconcilable differences" or that the marriage has suffered an "irretrievable breakdown."

Concerned about the sanctity of marriage, the impact on children of marital impermanence, and what seems to some a lack of fairness toward the spouse who would like to preserve the marriage, laws and policies have been proposed that would make divorce harder to obtain. These so-called **covenant marriages**, have included the restoration of fault for all divorces; a waiting period of as long as five years; a two-tier divorce process, with a more extensive process for divorces involving children; prioritization of children's needs in postdivorce financial arrangements; requirement of a "parenting plan" to be negotiated prior to granting a divorce; and publicizing research that would convince the public of the risks of divorce (Stokes and Ellison 2012; Wilcox 2009). Many states and cities have established premarital counseling, marriage education, marriage counseling, or some combination of the three as either required or elective for couples planning to marry. Research is lacking on their effectiveness in preventing divorce. However, research shows these kinds of programs are useful in addressing the needs of separating and divorcing couples, improving the post-divorce lives of families (Pollet and Lombreglia 2008).

Noting that divorce is "an American tradition" that began in colonial times and has grown more prevalent with industrialization and urbanization, historian Glenda Riley, among others, argues that making divorce harder to get will not change this trend (Riley 1991; Wolfers 2006).

Critical Thinking

Do you think we should make it harder to get a divorce? What are the reasons for your position? If you lived in a state with a covenant marriage option, would you choose it?

chances for a civil postdivorce relationship and successful co-parenting. With no-fault divorce, a marriage became legally dissolvable when one or both partners declared it to be "irretrievably broken" or characterized by "irreconcilable differences." No-fault divorce is sometimes termed **unilateral divorce** because one partner can secure the divorce even if the other wants to continue the marriage. Researchers have studied whether no-fault divorce laws are responsible for the increase in the divorce rate, but have found little evidence (Wolfers 2006).

High Expectations for Marriage

As discussed in Chapter 7, the focus of institutional marriages is on meeting obligations to one's spouse, family, and community—love, especially romantic love, was not required or even necessarily expected. In companionate marriage, couples are expected to be "in love." But love, especially romantic love, tends to wane over time making it a rather risky basis for marriage (Cherlin 2009a; Demo and Fine 2010). The subsequent move from companionate to individualistic marriage in the mid-twentieth century put added pressure on marriage to meet each spouse's need for love but also personal happiness and life satisfaction (Cherlin 2009a; Demo and Fine 2010).

If pressure from extended families, communities, churches, and the legal system can no longer be counted on to preserve marital stability, the quality of the relationship becomes central to the survival of a marriage (Bodenmann, Ledermann, and Bradbury 2007; Ledermann et al. 2010). Research has found that couples whose expectations are more practical are more satisfied with their marriages than are those who expect completely loving and expressive relationships (Demo and Fine 2010; see also Oberlander et al. 2010 for a discussion of racial differences and marital expectation). Although many couples part for serious and specific reasons, others may do so because of unrealized expectations and general discontent. Sociologist Andrew Cherlin (2009a) suggests that the U.S.'s high divorce rates may be in part a reflection of a broad sense of "restlessness" unique to American culture.

Interpersonal Dynamics

Along with larger social forces, researchers have examined aspects of marital relationships that increase divorce risk. Marital complaints associated with divorce are a partner's infidelity, alcoholism, drug abuse, jealousy, moodiness, violence, low levels of trust, and, much less often, homosexuality, as well as perceived incompatibility and growing apart (Amato 2010; Fincham and Beach 2010). Counselors suggest that some common complaints—about money, sex, and in-laws, for example—are really arenas for acting out deeper conflicts, such as who will be the more powerful partner, how much autonomy each partner should have, and how emotions are expressed (Amato 2010; Amato and Hohmann-Marriot 2007). Amato cautions that "not all couples display a pattern of relationship dysfunction prior to divorce (Amato 2010, p. 653). Even couples who are moderately happy and who have low levels of conflict can be at risk of divorce in the presence of other risk factors. Yet, a number of social trends in recent years are helping to decrease the divorce rate. Some of these are discussed below.

Why the Divorce Rate Stabilized over the Past Three Decades

Most observers conclude that the divorce rate has stabilized, and even declined, for the time being (National Marriage Project and the Institute for American Values 2012). Social scientists propose several reasons for this development:

Fewer people are marrying at younger ages. Those who wait are likely to make better choices and to have the maturity and commitment to work through problems (National Marriage Project and the Institute for American Values 2012).

The standard of living has improved over the past few decades for two-earner families with good jobs, a situation that leads to less tension at home and lower probability of divorce (Wilcox and Marquardt 2009).

Federally funded marriage education programs may have had an impact. For poorer families, who could not afford family counseling on their own, the programs may have helped couples to better manage their relationships (Crary 2007b; "Divorce Rate" 2007).

There is an increased determination on the part of children of divorced parents to make their own marriages work (Crary 2007b; Teachman, Tedrow, and Hall 2006).

Ironically, divorce rates may have declined because cohabitation has increased—if riskier relationships never become marriages, they never become divorces either. Cohabiting men and women are often leery of marriage, expressing their desire to "do it right," meaning marrying only once (Miller, Sassler, and Kusi-Appouh 2011).

More recently, the bad economy and real estate market and the high costs of obtaining a legal divorce keeps people married (Bilefsky 2012).

Image Source/Corbis

Deciding to divorce is difficult. Couples struggle with feelings about their past hopes, current unhappiness, and an uncertain future.

Thinking about Divorce: Weighing the Alternatives

Not everyone who thinks about divorce actually gets one. Because divorce is now an available option, spouses may compare the benefits of their current union to the projected consequences of not being married. One model of deciding about divorce, derived from exchange theory (see Chapter 2) is social psychologist George **Levinger's model of divorce decisions**. He posits that spouses assess their marriage in terms of the *barriers* to divorce, *alternatives* to marriage (possibilities for remarriage or satisfying), and *rewards* of marriage (Levinger 1965, 1976).

"What's Stopping Me?" Barriers to Divorce

Respondents to the Marital Instability Over the Life Course surveys named children, along with religion and lack of financial resources, as barriers to divorce

When parents consider divorce, they often think about the potential impact on their children—and that is a barrier to divorce.

(Previti and Amato 2003). Indeed, another study found that both mothers and fathers anticipated that "divorce would worsen their economic situation and their abilities to fulfill the responsibilities of being a parent" (Poortman and Seltzer 2007, p. 265). However, when researchers Chris Knoester and Alan Booth (2000) examined these data, they found that only three of nine barriers studied were associated with a lower likelihood of divorce: (1) when the wife's income was a smaller percentage of the family income, (2) when church attendance was high, or (3) when the couple had a new child. Affection for their children and concern about the children's welfare after divorce discouraged some parents from dissolving their marriage. This concern sometimes leads to delaying an intended divorce (Furstenberg and Kiernan 2001; Heaton 2002; Poortman and Seltzer 2007).

Long marriages are less likely to end in divorce. One reason for this, in addition to the marital bond itself, is that common economic interests and friendship networks increase over time and help stabilize the marriage at times of tension (Brown, Orbuch, and Maharaj 2010).

"Would I Be Happier?" Alternatives to the Marriage

Alternatives, another of Levinger's concepts, were found to be the least important in decisions to divorce (Previti and Amato 2003). Yet, some married people may ask themselves whether they would be happier if they were to divorce. Some people may prefer to stay single after divorce, but many partners probably weigh their chances for a remarriage.

Leaving a bad marriage may have a positive outcome regardless of whether the individual remarries. A British study found people to be less happy one year after separation, but by one year after the divorce, both men and women were happier than they had been while married (Gardner and Oswald 2006). Nevertheless, marriage can and often does provide emotional support, sexual gratification, companionship, and economic and practical benefits, including better health. But unhappy marriages do not provide these benefits and may be a factor in poorer health (Elias 2004).

"Can This Marriage Be Saved?" Rewards of the Current Marriage

Some partners respond to marital distress by trying to improve their relationship and by focusing on the rewards of their current marriage (the third component of Levinger's theory). A recently completed ten-year longitudinal study of newly married couples indicates that marital satisfaction does not necessarily have to decline and may even increase (Sullivan et al. 2010). Improvements in those marriages came about through the passage of time (children got older, jobs or other problems improved); because partners' efforts to work on problems, make changes, and communicate better were effective; or because individual partners made personal changes (travel, work, hobbies, or emotional disengagement) that enabled them to live relatively happily despite an unsatisfying marriage.

One must decide whether divorce represents a healthy step away from an unhappy relationship that cannot be improved. A marriage counselor may help partners become aware of the consequences of divorce so that people can make informed decisions.

Other Solutions to Marital Distress

Not all unhappy couples divorce. Below are some alternative solutions to marital distress.

Marital Separation In nearly 5 percent of married couple families in the United States, roughly 2.7 million families, the main householder reported that the couple was "separated" (U.S. Census Bureau 2012f, Table 66). According to the 2009 American Community Survey, the average length of separation among men and women who eventually divorce is about 10 months (Kreider and Ellis 2011). The duration between separation and divorce for first marriages varies by race and ethnicity, with longer separations among Hispanics and blacks than among whites (Copen et al. 2012). These groups are also more likely than whites to dissolve their marriage through permanent marital separation rather than divorce (Bramlett and Mosher 2002).

Aside from the few government statistics cited above, very little research covers the period between marital separation and divorce. Some marital partners who have separated do make efforts to reconcile. In one of the few studies of marital separation, Howard Wineberg used a sample of white women from the 1987 National Survey of Families and Households and found that 44 percent of the separated women attempted reconciliation. Half of the resumptions of marriage that followed took place within a month, suggesting that those separations may have been impulsive and soon regretted. Virtually no marriages were resumed after eight months of separation. Only one-third of the reconciliations "took"—that is, resulted in a continued marriage (Wineberg 1996). The author cautions that "not all separated couples should be encouraged to reconcile because a reconciliation does not ensure a happy marriage or that the couple will be married for very long" (p. 308).

Stable Unhappy Marriages From time to time, researchers have taken up the question of what happens to couples who are distanced, unhappy, or in conflict if they don't divorce (e.g., Waite, Luo, and Lewin 2009). It is, however, "surprising" that "long-term low-quality marriage ... has received relatively little attention" (Hawkins and Booth 2005, p. 451). Hawkins and Booth (2005) followed unhappy marriages for twelve years and compared people in unhappy marriages to divorced single and remarried individuals. "Divorced individuals who remarry have greater overall happiness, and those who divorce and remain unmarried have greater levels of life satisfaction, self-esteem, and overall health than unhappily married people.... We suggest that unhappily married people who dissolve low-quality marriages likely have greater odds of improving their well-being than those remaining in such unions" (p. 468). Yet, another study suggests that people who remain in an unsatisfying marriage are less depressed than those who leave their marriages (Waite et al. 2002; Waite, Luo, and Lewin 2009).

Getting the Divorce

One of the scariest things about divorce is uncertainty and lack of knowledge about what the future may hold in terms of one's finances, living situation, and children. A much less mentioned stressor is the divorce process itself.

The "Black Box" of Divorce

Whereas we know a great deal about how to get married—buy a ring, propose, have an engagement party, plan the wedding, etc.—how to get divorced is a "black box." We know quite a bit about relationship dynamics before divorce (inputs) and after the divorce (outputs), but not about the inner workings of the divorce process itself. Hearing "I want a divorce" from one's spouse involves entering a world where no one wants to go and learning a whole new language of petitioners and respondents, of financial disclosures and waiting periods, of parenting plans and spousal support. Use caution before taking the advice of friends and relatives. What to do when one spouse wants a divorce? Consult an attorney who has expertise in divorce law.

Initiating a Divorce

Not all divorced people wanted to or were ready to end their marriage. It may have been their spouse's choice. Initiating and noninitiating partners tend to talk about their reasons in different terms. The initiator of the divorce typically invokes a "vocabulary of individual needs" while the noninitiating partner speaks in terms of "familial commitment" (Hopper 1993). Indeed, partners contemplating divorce must grapple with the contradictory pull of commitment to one's family and one's personal happiness and individual goals (see Chapter 1; Amato 2004; Cherlin 2009a).

Women more often initiate a divorce (Coontz 2010b; Thomas 2012). Not surprisingly, research shows that the degree of trauma a divorcing person suffers usually depends on whether that person or the spouse wanted the dissolution. The one "left" experiences a greater loss of control and has much mourning yet to do—the divorce-seeking spouse may have already worked through his or her sadness and distress (Amato 2000; Braver, Shapiro, and Goodman 2006). These negative effects of one's spouse initiating a separation (as opposed to oneself or jointly) extend to both physical and mental health and affect both men and women (Hewitt and Turrell 2011). Even for those who actively choose to divorce, however, divorce and its aftermath may be unexpectedly painful.

Legal Aspects of Divorce

A legal divorce is the dissolution of the marriage by the state through a court order terminating the marriage. The principal purpose of the legal divorce is to dissolve the marriage contract so that emotionally divorced spouses can conduct economically separate lives and be free to remarry.

Two aspects of the legal divorce make marital breakup painful. First, divorce, like death, creates the need to grieve. But the usual divorce in court is a rational, unceremonial exchange that takes only a few minutes. Divorcing individuals may feel frustrated by their lack of control over a process in which the lawyers are the principals.

A second aspect of the legal divorce that aggravates conflict and misery is the adversary system. Under our judicial system, lawyers advocate their client's interest only and are eager to "get the most for my client" and "protect my client's rights." Opposing attorneys are not trained to and ethically are not even supposed to balance the interests of the parties and strive for the outcome that promises most mutual benefit.

A final note on the legal divorce is that, by definition, it applies only to marriage. There is no legal forum in which cohabitants, whether heterosexual or gay/lesbian, may obtain a divorce. Some couples may be cohabiting precisely to avoid the prospect of going to court should their relationship sour, especially couples who have been divorced once already. However, they are likely to find that the absence of a venue in which to resolve separation-related disputes in a standardized way is also a problem. Many cohabiting couples must go to small claims court to sort out belongings, assets, and debts (the legal side of living together is discussed in Chapter 6).

Divorce Mediation

Divorce mediation is an alternative, nonadversarial means of dispute resolution by which a couple, with the assistance of a mediator or mediators (frequently a lawyer–therapist team), negotiate the settlement of their custody, support, property, and visitation issues. In the process, they hope to learn a pattern of dealing with each other that will enable them to resolve future disputes. Mediation is recommended or mandatory in all states for child custody and visitation disputes before litigation can be commenced (Comerford 2006).

Couples who utilize divorce mediation have less relitigation, feel more satisfied with the process and the results, and report better relationships with ex-spouses and children (Bailey and McCarty 2009; Holtzworth-Munroe, Applegate, and D'Onofrio 2009). Mediation can vary in quality, however, with agreements that are "clear, fair, comprehensive, and "tailor-fit" and resulting in higher post-divorce well-being among adults (Baitar et al. 2012, p. 65). Having a mediation professional who displayed "problem-solving behaviors" including (1) structuring the mediation process, (2) noticing relevant arrangement details, and (3) was sensitive to the emotional reactions of conflicting parties was also associated with more positive outcomes (Baitar et al. 2012, p. 71).

There are arguments for and against mediation in child custody. Women's advocacy groups have claimed that mediation may be biased against females in that they may be less assertive in negotiations. Some scholars have argued that mediators take insufficient account of prior domestic violence (Comerford 2006; Freeman 2008, note 16). Judith Wallerstein believes that the positive effects of divorce mediation for children may be overstated, but new practitioners are attempting to level the field (2003, p. 80; see also Freeman 2008 for a discussion on positive and negative qualities of mediation). Yet it does seem that "[m]ediation produces higher levels of compliance [with court decisions] and lower relitigation rates than litigation or attorney-negotiated settlement." It is less costly and generally less time-consuming than litigation (Comerford 2006; Crary 2007a).

Divorce "Fallout"

Marriage is a public announcement to the community that two individuals have joined their lives. Marriage usually also joins extended families and friendship networks and simultaneously removes individuals from the world of dating and mate seeking. **Divorce fallout** refers to ruptures of relationships and changes in social networks that come about as a result of divorce. At the same time, divorce provides the opportunity for forming new ties. (Divorce fallout also involves financial fallout or the economic consequences of divorce, which will be discussed in the next section.)

Kin No More? When a couple divorces, relationships with family members can be "reinterpreted" in the following ways: (1) kin promotion, redefinition of a distant relative to a close relative, (2) kin exchange, reclassifying a family relationship (such as a sibling relationship becoming more like a parent-child relationship), (3) non-kin conversion, turning friends and colleagues into family members, and (4) kin retention, in which ex-in laws are kept as family members (Allen, Blieszner, and Roberto 2011). What to do about "ex" family members is a vexing problem, especially given that these remain the family members of one's children. Another problem is how to incorporate new family members.

The Divorce-Extended Family A surprising phenomenon encountered by those who do research on divorced families is the expansion of the kinship system that is produced by links between ex-spouses and their new spouses and significant others and beyond to their extended kin, including grandparents, aunts, uncles, cousins, and friends. This kinship system is referred to as the *divorce-extended family*. Ahrons (1994) believes it is important for parents to "accept that your child's family will expand to include nonbiological kin" (1994, p. 252) and that the "relationships formed when a parent remarries also tend to be more rewarding for the children as their kinship system expands rather than contracts" (Ahrons 2007, p. 64).

Some postdivorce extended families find they can enjoy and benefit from connections to one another

Lori Waselchuk/The New York Times/Redux

These eight grandparents, all connected to the young basketball player by marriage, divorce, and remarriage, come together to cheer him on and enjoy his game.

even when there had earlier been conflict and old tensions sometimes resurface (Kleinfield 2003). "Letting go of old resentments, whether between ex-spouses, parent and child, or stepparent and child, is the most challenging part of" creating divorce-extended families, but doing so presents additional resources, friendship, and love to individuals and families (A. Bernstein 2007, p. 74).

Grandparenting After Divorce Given the frequency of divorce, most grandparents find themselves touched by it. Grandparents fear losing touch with grandchildren, and this does happen. In response, all fifty states have passed grandparent visitation laws. However, a Supreme Court decision struck down Washington's law (*Troxel v. Granville* 2000) because it was considered to interfere with parents' rights to determine how their children are to be raised. The status of other states' laws is uncertain; some courts have allowed grandparents visitation rights in certain circumstances (Dao 2005; Henderson and Moran 2001; Hsia 2002; Stoddard 2006).

In favorable circumstances, grandparents become closer to grandchildren, as adult children turn to grandparents for help or grandchildren seek emotional support (Henderson et al.2009; Ruiz and Silverstein 2007; Timonen, Doyle, and O'Dwyer 2011). Researchers and therapists have concluded that

> these relationships work best when family members do not take sides in the divorce and make their primary commitment to the children. Grandparents can play a particular role, especially if their marriages are intact: symbolic generational continuity and living proof to children that relationships can be lasting, reliable, and dependable. Grandparents also convey a sense of tradition and a special commitment to the young.... Their encouragement, friendship, and affection has special meaning for children of divorce; it specifically counteracts the children's sense that all relationships are unhappy and transient. (Wallerstein and Blakeslee 1989, p. 111; see also Henderson et al. 2009)

Indeed, children who were close to their grandparents had fewer problems adjusting to their parents' divorce (Connidis 2009; Ruiz and Silverstein 2007). Of course, more and more grandparents' own marriages are not intact today. Nevertheless, one can assume that

Monkey Business Image/Kalium/Age Fotostock

Divorce affects the extended family as well as the nuclear one. In some families, grandparents may lose touch with grandchildren, whereas in others, they may become more central figures of support and stability.

even a loving, divorced grandparent could add to the support system of a grandchild of divorce.

Women are more likely than men to retain in-law relationships after divorce, particularly if they had been in close contact before the divorce and if the in-law approves of the divorce (Connidis 2009). Relationships between former in-laws are more likely to continue when children are involved. In any event, grandchildren were most likely to remain closest to maternal grandparents, as mothers typically grew closer to and relied more on their parents after divorce (Connidis 2009; Henderson et al. 2009).

Friends No More? A change in marital status is likely to mean changes in one's community of friends. Divorced people may feel uncomfortable with their friends who are still married because activities are done in pairs; the newly single person may also feel awkward. Couple friends may fear becoming involved in a conflict over allegiances, and they may experience their own sense of loss. Moreover, if married friends have some ambivalence about their own marriages, a divorce in their social circle may cause them to feel anxious and uncomfortable. A common outcome is a mutual withdrawal.

Like many newly married people, those who are newly divorced must find new communities to replace old friendships that are no longer mutually satisfying. The initiative for change may in fact come not only from rejection or awkwardness in old friendships but also from the divorced person's finding friends who share with him or her the new concerns and emotions of the divorce experience. Priority may also go to new relationships with people of the opposite sex; for the majority of divorced and widowed people, building a new community involves dating again.

Deciding knowledgeably whether to divorce means weighing what we know about the consequences of divorce. The next section examines the economic consequences of divorce.

The Economic Consequences of Divorce

Upon divorce, a couple must become distinct economic units, each with its own property, income, control of expenditures, and responsibility for taxes, debts, and so on. The economic fallout of divorce can be severe. No one "wins" financially in a divorce. Everyone's standard of living suffers, especially children's. The economic consequences of divorce also vary for husbands and wives, with wives experiencing greater, and more enduring, losses (Avellar and Smock 2005; Sayer 2006). Although many parents help out financially when their adult child gets divorced, these are generally "one time" gifts and not enough to prevent subsequent economic decline (Leopold and Schneider 2011; Mazelis and Mykyta 2011; Timonen, Doyle, and O'Dwyer 2011). Older couples who divorce also face substantial losses. As one financial planner says, "When you divorce in your 50s and 60s, you lose the luxury of time to recover from any financial shortfalls or mistakes" (Yip 2012).

Consequences for Children: Single-Parent Families and Poverty

The continuation of a high incidence of divorce contributes to the increased prevalence of single-parent families. Children's living arrangements vary greatly by race and ethnicity, as Figure 14.3 indicates. Based on the most recent available data, we find that nonHispanic white children are most apt (75 percent) to be living in two-parent families (with biological parents or a parent and stepparent). However, whereas about two-thirds (67 percent) of Hispanic children live with two parents, only a little more than a third (37 percent) of black children are living in a two-parent household (Kreider and Ellis 2011a). There are many reasons why so many African American children live in single-mother households. We have discussed these reasons in previous chapters and will continue to shed light on the issue in the next chapters of this book.

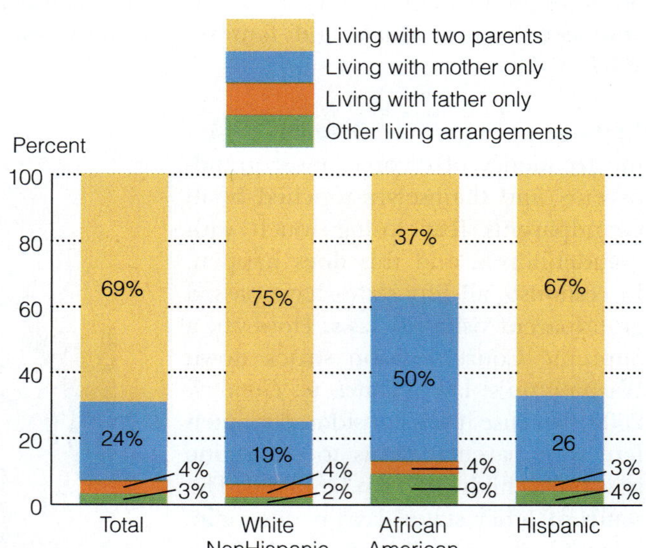

FIGURE 14.3 Living arrangements of children under age 18 by race/ethnicity, 2009. *Other living arrangements* include children who are living in the homes of relatives, in foster homes, or with other nonrelatives, or who are heads of their own households.

Source: Kreider, Rose M. and Renee Ellis. 2011a. *Living Arrangements of Children: 2009.* Current Population Reports, P70-126, U.S. Census Bureau, Washington, DC. Table 1.

Family structure and poverty are interrelated. Figure 14.4 shows the proportions of children who were living in poverty in 2010, for the largest racial/ethnic groups, comparing poverty rates by family type. As you can see, 47 percent of all children who reside in mother-only, single-parent families live in poverty. This compares to 11 percent of those living in married-couple families. The relationship between family type and poverty is consistent across racial/ethnic categories. Children residing in single father families are about half as likely to live in poverty as children living with single mothers (Kreider and Ellis 2011a).

Economic Losses for Women

Sociologist Lenore Weitzman researched the financial plight of divorced women and their children in her landmark book *The Divorce Revolution* (1985). She found that wives' (and children's) income declined considerably after divorce while husbands' disposable incomes increased. Although Weitzman's figures were later questioned, newer data continue to show that women's income declines after divorce—from 27 percent to 51 percent (Peterson 1996; Sayer 2006). Another statistic used to examine divorced women's economic decline is the **income-to-needs ratio**—that is, how well income meets financial needs. Women and their children experience a decline of 20 percent to 36 percent in their income-to-needs ratio (Meadows, McLanahan, and Knab 2009; Sayer 2006). Research on the economic consequences of cohabiters' breaking up finds similarities to getting divorced.

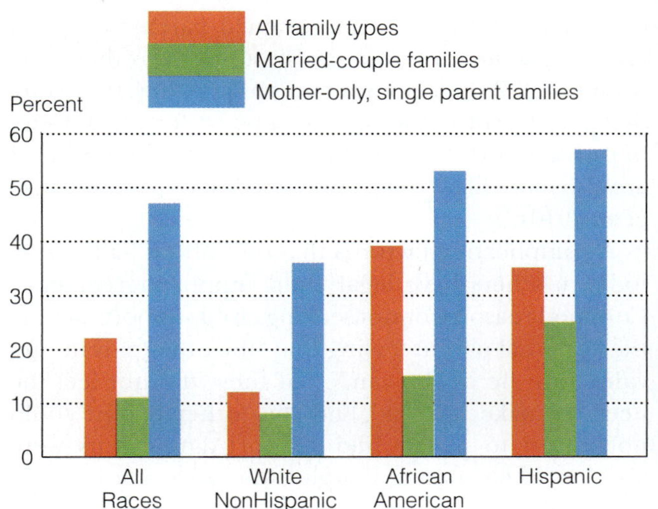

FIGURE 14.4 Poverty status of families with children under 18 by race/ethnicity and type of family, 2010 (percentage of families with incomes below the poverty level).

Source: Federal Interagency Forum on Child and Family Statistics 2012, Figure 4.

A fundamental reason for the income disparity between ex-husbands and their former wives is men's and women's unequal wages and different work patterns (Hartmann, English, and Hayes 2010; Oldham 2008a). As discussed in Chapter 10, women on average work in less well paying occupations and earn less money overall than men. However, as women make gains in education and earnings, and with increased enforcement in child support orders, the income disparity between women and men post-divorce has been getting smaller. Men's post-divorce economic recovery is discussed in the next section.

Even if tangible assets, such as houses and savings accounts, are divided equitably, that does not usually put her on an equal footing with her former husband for the future (Oldham 2008b). That reason for this disparity is that, for most couples, assets are the "non-tangible" kind, such as a professional degree, a business or managerial position, work experience, or a skilled trade. "Most women would have to make heroic

Patrick Pihl/Interpix/Alamy

Women and their children experience a substantial decline in their standard of living after a divorce. They may need to move to less-expensive—and less-desirable—housing and away from their former neighborhood, school, and friends. Many men also experience a decline in standard of living.

leaps in the labor or marriage market to keep their losses as small as the losses experienced by the men from whom they separate" (McManus and DiPrete 2001, p. 266).

Returning to the labor force and increasing work hours does not allow most women to fully recoup the economic losses from divorce, especially women with children, because of the lower pay of occupations in which women more often work. Remarriage or cohabitation with a new partner remain a surer strategy to recovering the family's standard of living (Jansen, Mortelmans, and Snoeckx 2009), even though the pool of eligibles available to divorced women tend to have lower incomes. Women who are custodial parents must also depend on child support from the other parent to meet their new single-parent family's expenses. Child support amounts are set relatively low, and much child support remains unpaid (as discussed in a later section of this chapter).

Economic Losses for Men

Although men's economic losses after divorce are not as great as those of women, men's incomes, overall, also decline. Depending on the study, married men experience a decline in family income of from 8 percent to 41 percent; cohabiting men's drop in income is comparable (Sayer 2006). The chief reason for their declining standard of living is the loss of the partner's income but also because the noncustodial parent is now taxed at a much higher rate because (usually) he has no dependents to claim (Braver, Shapiro, and Goodman 2006). It's also possible that divorced men may be discriminated against in the workplace. Killewald (2013) found that that the earnings of men not in traditional family structures are paid significantly less than married men.

However, though family income drops for a man when he is no longer part of a couple, so do his expenses; his household is now smaller. When the smaller number of family members in men's households are taken into account, studies show an increase in men's income postdivorce anywhere between 18 percent and 93 percent. In fact, child support, discussed in the next section, is a major source of tension between men and women after divorce.

Child Support

Child support involves money paid by the noncustodial to the custodial parent to support the children of a now-ended marital, cohabiting, or sexual relationship. Because mothers retain custody in the preponderance of cases, the vast majority of those ordered to pay child support, about 85 percent, are fathers (U.S. Census Bureau 2012g).

For many years, the child support awarded to the custodial parent was often not paid, and states made little effort to collect it on the parent's behalf. Policy makers' concerns about poverty, the economic consequences of divorce for women, and in particular, welfare and social services costs led to a series of federal laws that have changed that situation considerably. In recent years, government authorities have been more successful at securing payment and at standardized amounts that are often higher than previously provided. With better child support enforcement, the number of child support payments have increased and the poverty rate of custodial parents and their children dropped (Grall 2011; Huang 2009).

Despite the fact that child support amounts are increasing overall, child support awards continue to be small, given what it costs to raise a child (see Chapter 8). Even when fully paid, child support amounts are not very impressive, averaging between $300 and $500 per month (Grall 2011; U.S. Census Bureau 2012c). Moreover, declining child support amounts may be associated with the increase in fathers having joint physical custody (Cancian, Meyer, and Han 2012).

Although the majority of custodial parents are women, about one in six custodial parents are men. This figure should increase as men's and women's work and parenting roles continue to evolve. Custodial fathers receive about 82 percent of the child support amounts that custodial mothers receive (Grall 2011; U.S. Census Bureau 2012g) given that, due to their lower incomes, noncustodial mothers pay less in child support than noncustodial fathers (Stewart 2010).

Some noncustodial parents do make additional "in kind" contributions in the form of gifts, clothes, food, medical costs (beyond health insurance), and child care. About 60 percent of custodial parents reported receiving some form of noncash support for their children (Grall 2011). However, it is a myth that noncustodial parents substitute in kind support for the amount of cash support due. Parents who receive child support are also more likely to receive noncash support (Garasky et al. 2010).

A complicating factor is that only about half of custodial mothers have legal child support agreements. Common reasons for not seeking child support include the custodial parent feeling that "the other parent provides what he or she can," that they "did not feel the need to make it legal," and the "other parent could not afford to pay." Child support agreements were also less likely among couples who were not married (Grall 2011).

Why Do Some Parents Not Pay Their Child Support Obligations? Some research suggests that the principal reason for a noncustodial parent's failure to pay is unemployment or underemployment (Sorensen 2010).

Among families in which the absent parent has been employed during the entire previous year, payment rates are 80 percent or more. Not so when unemployment is involved. "The key to reducing poverty [among single-parent families] thus appears to be the old and unglamorous one, of solving un- and underemployment, both for the fathers and the mothers" (Braver, Fitzpatrick, and Bay 1991, pp. 184–185). Compliance with count-ordered child support decrees may also be related to the noncustodial parent's involvement in the child's life. Seventy-eight percent are in compliance when they have either joint custody or visitation arrangements; only 67 percent are in compliance when the parent has neither (Garasky et al. 2007). Sometimes fathers withhold child support from the mother as a "power play" to see their children more often, although mothers may prevent visitation in order to ensure the child support due them is paid (Moore 2012). Another problem is that mothers and fathers often go on to have more children with a new spouse or partner, a situation associated with lower child support payments to "first families" (Manning and Smock 2000; Meyer and Cancian 2012).

Two suggested solutions to the problem of nonpayment of child support are guaranteed child support and a children's allowance. Both are based on the principle of society-wide responsibility for all children. With guaranteed child support, a policy adopted in France and Sweden, the government sends to the custodial parent the full amount of support awarded to the child. It then becomes the government's task to collect the money from the parent who owes it. A second alternative, a children's allowance, provides a government grant to all families—married or single-parent, regardless of income—based on the number of children they have. All industrialized countries except the United States have some version of a children's allowance. In the present political and economic climate in the United States, such measures seem unlikely to be adopted.

As another approach to securing payment of child support, some states have begun experimenting with "responsible fatherhood" programs, often supported by government grants. Low-income fathers are typically expected to pay a higher portion of their income in child support than are middle-class fathers, resulting in a spiral of expanded debt and often withdrawal from their children (Huang, Mincy, and Garfinkle 2005). Recognizing that there are low-income fathers who want to provide support for their children but lack the income to do so, these multifaceted programs provide employment services, family support, and mediation services. Results thus far are only modest. An increase in child support payments has occurred, though not yet enough to produce the dramatic improvement in the lives of men and their children that program designers had hoped for (Sorensen 2010).

We have been speaking of child support in the context of marriage and divorce and heterosexual couples, but some courts have awarded child support when same-sex couples who have been raising children together break up (Graham 2008; Kravets 2005; "State Court Orders" 2005). We turn now from the economics of postdivorce family support to a broader examination of the aftermath of divorce for children.

The Social and Emotional Consequences of Divorce

For obvious reasons, most research has focused on how divorce affects children. However, divorce has important effects on women and men as well. We will examine those effects first.

Consequences for Women and Men

Numerous studies over several decades indicate that compared to married individuals, people who have divorced have more health problems, more symptoms of anxiety and depression, more substance use, and greater overall mortality (Amato 2010). Researchers have been interested in to what extent the negative outcomes are caused by the divorce itself or whether it is the result of selection. The selection perspective posits that individuals have pre-existing traits, called selection factors, which sort them into groups with higher divorce risks as well as negative outcomes after divorce (Amato 2000, 2010).

Although several studies provide evidence in favor of selection (for example that divorce is more common among those with poor mental health), there is strong evidence that divorce has a negative effect on the health and well-being of adults despite such predispositions. Researchers have identified a number of contributors including the as loss of social support and companionship, decline in standard of living, having to change residences, feelings of anger and sadness, and discontinuing measures associated with a healthy lifestyle, such as yearly doctor's visits (Amato 2010). Older Americans who divorce may have to prolong one's working life and re-enter the labor force and marriage market. Nevertheless, according to a study of divorced men and women ages 40 to 79 conducted by the American Association of Retired Persons (AARP), 80 percent considered themselves happy, scoring above five on a ten point happiness scale with 56 percent scoring an 8, 9, or 10 (Thomas 2012).

Rameshwar Das

Divorced mothers, most of whom are the residential parent, often face financial worries and emotional overload as they try to be the complete parent for the children.

Divorce and Stress-Related Growth Scholars and clinicians have begun to talk about **stress-related growth** (for children as well as adults). There is now more emphasis on the diversity of outcomes of divorce. Stress-related growth can take different paths. A crisis-related pathway is marked when a traumatic event generates an ultimate result that makes the person stronger. (See Chapter 13 for a more general discussion of these ideas as applied to the family.) A stress-relief pathway occurs when, for example, the end of a marriage and its problems brings relief to one or both of the partners. Kinds of growth include growth in the self, growth in interpersonal relationships (closer to family and friends), and growth or change in philosophy of life. The specifics are still a little vague, and more research is needed. Yet, scholars reviewing the literature conclude that

> [r]esearch on stress-related growth indicates that most individuals who have experienced traumatic events report positive life changes.... One thing that is clear from the existing research ... is that it is at least as common to experience positive outcomes following divorce as negative one[s], and that positive outcomes can coexist with even substantial pain and stress. (Tashiro, Frazier, and Berman 2006, pp. 362, 364)

Divorce provides an escape from marital behaviors that may be more harmful than divorce itself, such as a parent's alcoholism or drug abuse (Coontz 1997). An interesting study by economists Betsey Stevenson and Justin Wolfers (2004, 2007) found that no-fault divorce was associated with declines in suicide rates for women, domestic violence against both men and women, and intimate partner homicides of women. The existence of an escape route seems to change the balance of power and reduce violence.

Another way of looking at stress-related growth—as well as less happy outcomes—comes from E. Mavis Hetherington's Virginia Longitudinal Study of Divorce and Remarriage. She and her colleagues followed 144 couples for twenty years; half were divorced initially, and half not. Additional families were added as time went by. Families were interviewed at various points, but our interest here is in the ten-year point. A previously developed typology of postdivorce adaptive patterns was used to assess the adjustment of divorced adults. Hetherington found 20 percent of those studied to have "enhanced" lives, while 40 percent had "good enough adjustment" (Hetherington and Kelly 2002; Hetherington 2003).

Perhaps the best overall assessment of the outcomes of divorce is from Paul Amato:

> On one side are those who see divorce as an important contributor to many social problems. On the other side are those who see divorce as a largely benign force that provides adults with a second chance for happiness and rescues children from dysfunctional and aversive home environments.... Based on ... research ... it is reasonable to conclude that ... [d]ivorce benefits some individuals, leads others to experience temporary decrements in well-being that improve over time, and forces others on a downward cycle from which they might never fully recover. (Amato 2000, p. 1282)

Whether a divorce can have a positive effect on adult adjustment depends on a range of moderating factors, such as income, the level of distress in the marriage, and the presence of children. Social and emotional support from family and friends, as well positive religious beliefs have been shown to reduce the negative effects of divorce (Krumrei, Mahoney, and Pargament 2011; Webb et al. 2010). Among divorced (and noncohabiting) parents with infants, maintaining a positive relationship with one's ex-spouse or partner is associated with less depression (Paulson, Dauber, and Leiferman 2011). A divorce might evoke feelings of elation, at least at first. "Divorce parties" have become a booming business as are "divorce rings," which some newly divorced women buy for themselves to replace their missing wedding bands (Fishman 2013; White 2012). For most however, divorce is an overwhelming and draining experience often involving enormous changes in family life for parents and children.

How Divorce Affects Children

More than half of all divorces involve children under 18, and about 40 percent of children born to married parents will experience marital disruption (Amato 2000). How do separation and divorce affect children? Outcomes for children depend a great deal on the circumstances before and after the divorce (Amato 2010). Although the divorce experience is psychologically stressful and, in most cases, financially disadvantageous for children, children in high-conflict marriages seem to benefit from a divorce. Living in an intact family characterized by unresolved tension and alienating conflict can cause as great or greater emotional stress and a lower sense of self-worth in children than living in a supportive single-parent family (Barber and Demo 2006). When the conflict level in the home has been low, however, children have poorer postdivorce outcomes. They are likely surprised by a divorce and seem to suffer more emotional damage. Among other things, it is difficult for them to see the divorce as necessary (Stevenson and Wolfers 2007).

The "Child of Divorce" Perspective The research of Judith Wallerstein and her colleagues has been very influential in defining the situation of children of divorce for both professionals and the public. In their longitudinal study of children's postdivorce adjustment, psychologists Judith Wallerstein and Joan Kelly interviewed all of the members of some sixty families with one or more children who had entered counseling at the time of the parents' separation in 1971. Wallerstein and her colleagues reinterviewed children at one year, two years, five years, ten years, and, in some cases, fifteen years later and finally again at the twenty-five-year point (Wallerstein and Lewis 2007, 2008).

In the initial aftermath of the divorce, children appeared worst in terms of their psychological adjustment at one year after separation. By two years post-divorce, households had generally stabilized. At five years, many of the 131 children seemed to have come through the experience fairly well: 34 percent "coped well"; 29 percent were in a middle range of adequate, though uneven, functioning; and 37 percent were not coping well, with anger playing a significant part in the emotional life of many of them (Wallerstein and Kelly 1980). In their twenty-five-year follow-up, they found a substantial proportion of the children had no or limited contact with their fathers (Wallerstein and Lewis 2007). According to the authors, the "twists and turns of the postdivorce father–child relationship, its high vulnerability to change, satisfactions laced with disappointments, and undercurrent of love, longing, and anxiety all call attention to the complexity of building a lasting father-child relationship outside the marriage" (Wallerstein and Lewis 2008).

Their research and that of other researchers indicates that children of divorced parents have less money available for their needs, which is especially significant because some of the negative impact of divorce can be attributed to economic deprivation (Strohschein 2005; Wallerstein and Lewis 2007). Data from the Health and Retirement Study indicates that, compared with continuously married parents, divorced parents give less money to their adult children (Shapiro and Remle 2010). In the Wallerstein study, financing college was found to be especially problematic, even among those who were middle-class. Sixty percent of those children were likely to receive less education than their fathers; 45 percent were likely to receive less than their mothers. Even divorced fathers who had retained close ties, who had the money or could save it, and who ascribed importance to education seemed to feel less obligated to support their children through college (Wallerstein and Blakeslee 1989; Wallerstein and Lewis 2007, 2008).

The Wallerstein study is well-known for its methodological problems: it was a small, unrepresentative sample recruited by offering free counseling to the family; it lacked a control group; there was difficulty separating family troubles and mental health concerns that predate the separation and divorce from those that might be effects of divorce; and it looked at a time period when divorce was still a new trend. Still, other research supports Wallerstein's conclusion that divorce has negative and long-term effects (Amato 2010; Conger, Conger, and Martin 2010; Frisco, Muller, and Frank 2007). Divorce is a "risk factor for multiple problems in adulthood" (Amato 2000, p. 1,279). Children of divorce continue to have lower outcomes than children from intact families in the areas of academic success, conduct, psychological adjustment, social competence, and self-concept, and they have more troubled marriages and weaker ties to parents, especially fathers, that can extend into adulthood (Bouchard and Doucet 2011; Krampe and Newton 2012: Magnuson and Berger 2009; Mandara, Rogers, and Zinbarg 2011).

Race/Ethnic Differences in Effects of Divorce on Children Research based on a national representative sample of more than 10,000 U.S. teens found that the negative effects of divorce and living with a single mother were less serious for black and Hispanic adolescents than they were for whites (Mandara, Rogers, and Zinbarg 2011). Why might this be? First, although nonmarried parenthood remains somewhat stigmatized in the United States (Usdansky 2009a, 2009b), it is noteworthy that this stigma may be minimal within the black community where single-parent families are more normative (Heard 2007; Liu and Reczek 2012). Accordingly, "The common history among blacks allows for the emergence and primacy of social supports, such as women-centered kinship networks, coresidence with extended family, and strong ties to the church, which can buffer the negative effects of stress caused by family instability" (Heard 2007, p. 336). Analyzing data on 867 African American children from the Family and Community Health Study, Simons and her colleagues found that "child behavior problems were no greater in either mother-grandmother or mother-relative families than in those in intact nuclear families." At least among blacks, these researchers found mother-grandmother families to be "functionally equivalent" (Simons et al. 2006, p. 818). Then too, other extended kin in black families—uncles, for example—may be involved in child care which may offset some of the negative effect of father absence (Richardson 2009).

Reasons for Negative Effects of Divorce on Children

Researchers and theorists offer a variety of explanations for why and how divorce could adversely affect children. Divorce researcher Paul Amato (1993) has summarized seven theoretical perspectives concerning the reasons for negative outcomes.

1. The **life stress perspective** assumes that, just as divorce is known to be a stressful life event for adults, it must also be so for children. Furthermore, divorce is not one single event but a process of associated events that may include moving—often to a poorer neighborhood—changing schools, giving up pets, and losing contact with grandparents and other relatives (Benner and Kim 2010; Burrell and Roosa 2009; Jackson, Choi, and Bentler 2009; Teachman 2009; White et al. 2009). This perspective holds that an accumulation of negative stressors results in problems for children of divorce.
2. The **parental loss perspective** assumes that a family with both parents living in the same household is the optimal environment for children's development. Both parents are important resources, providing children love, emotional support, practical assistance, information, guidance, and supervision, as well as modeling social skills such as cooperation, negotiation, and compromise. Accordingly, the absence of a parent from the household is problematic for children's socialization.
3. The **parental adjustment perspective** notes the importance of the custodial parent's psychological adjustment and the quality of parenting. Supportive and appropriately disciplining parents facilitate their children's well-being. However, the stress of divorce and related problems and adjustments may impair a parent's child-raising skills, with probably negative consequences for children. Divorced parents do spend less time with children. Compared to married parents, divorced parents are "less supportive, have fewer rules, dispense harsher discipline, provide less supervision, and engage in more conflict with their children" (Amato 2000, p. 1,279; see also Schoppe-Sullivan, Schermerhorn, and Cummings 2007). On the other hand, in their recent review of the literature, Thomson and McLanahan (2012) conclude that differences in parenting between divorced versus married parents probably play a small role in child outcomes relative to other factors.
4. The **economic hardship perspective** assumes that economic hardship brought about by marital dissolution is primarily responsible for the problems faced by children whose parents divorce (Amato 2010; Sun and Yuanzhang 2011; Thomson and McLanahan 2012). Indeed, economic circumstances do condition diverse outcomes for children—perhaps accounting for one-half the differences between children in divorced compared to intact two-parent families. Nevertheless, children in better-off remarried or single-parent families still lag behind children from two-parent families on various outcome indicators (Amato 2010; Kreider and Elliott 2009b; Magnuson and Berger 2009; Waite, Luo, and Lewin 2009).
5. The **interparental conflict perspective** holds that conflict between parents is responsible for the lowered well-being of children of divorce. Many studies, including that of Wallerstein, indicate that some negative results for children may not be simply the result of divorce per se, but are also generated by exposure to parental conflict prior to, during, and subsequent to the divorce (Barber and Demo 2006; Benner and Kim 2010; Burrell and Roosa 2009; Jackson, Choi, and Bentler 2009; Teachman 2009; White et al. 2009).
6. The **selection perspective**, discussed previously in relation to adults, says that at least some of the child's problems after the divorce were present before the marriage (Thomson and McLanahan 2012). Some of these might be the result of the forthcoming marital breakdown, but some may be due to earlier dysfunctional family patterns, behaviors, or personality traits on the part of the child's parents (e.g., Uecker and Ellison 2012). Yet other research indicates that the negative effects of divorce on children persist even after accounting for pre-divorce factors (Kim 2011).
7. The **family instability perspective** stresses that the number of transitions in and out of various family settings are the key to children's adjustment (Amato 2012; Bures 2009; Magnuson and Berger 2009). David Demo and Mark Fine in their book *Beyond the Average Divorce* (2010) refer to this as

family fluidity, which refers to "the frequency and rate of changes in family related experiences and outcomes" (p. 7). The logic of the family fluidity or instability hypothesis is this:

> Transitions may include parents' separation; a cohabiting romantic partner's move into, or out of, the home of a single parent; the remarriage of a single (noncohabiting) parent; or the disruption of a remarriage. The underlying assumption is that children and their parents, whether single or partnered, form a functioning family system and that repeated disruption of this system may be more distressing than its long-term continuation.... Stable single-parent households or stepfamilies, in contrast, do not require that children readjust repeatedly to the loss of coresident parents and parent-figures or the introduction of cohabiting parents and stepparents. (Fomby and Cherlin 2007, p. 182)

Similar to the effect of living with a single parent, multiple transitions did not seem to impact black children as much. The instability perspective is continuing to gain support as a result of advanced statistical techniques that allow researchers to track children's movements over time (Demo and Fine 2010; Sun and Yuanzhang 2011).

Divorce and Sibling Relationships With so much focus on how divorce affects relationships between parents and children, we sometimes forget how divorce affects relationships between other family members, such as siblings. Siblings may respond to their parents' divorce in two ways. One hypothesis is that sibling rivalry and conflict increases as a result of scarcity in the mother's time and her attention. A second hypothesis is that siblings become closer and more supportive in the face of unstable and unreliable relationships with adults. Most available research unfortunately supports the first hypothesis. Biological sibling relationships have been found to be more conflictual, negative, and ambivalent in divorced and remarried homes than in original two-parent families (McHale, Updegraff, and Whiteman 2012).

A More Optimistic Look at Outcomes for Children of Divorce Having considered reasons for the negative effects of divorce on children, we now try to assess just how important divorce is in the lives of affected children. We've given considerable attention to the research of Wallerstein and her colleagues because it has been very influential. "Judith Wallerstein's research on the long-term effects of divorce on children has had a profound effect on scholarly work, clinical practice, social policy, and the general public's views of divorce" (Amato 2003, p. 332).

Nevertheless, concern that children of divorce are disadvantaged does not rest solely on Wallerstein's research and what many see as her exaggerated presentation of the dangers of divorce (Cherlin 1999). A "persuasive body of evidence supports a moderate version of [her] thesis" (Amato 2003, pp. 338–339; Hetherington 2005). The remaining question, then, is this: How much does divorce affect children?

E. Mavis Hetherington has been studying divorcing families for about the same length of time as Judith Wallerstein, and she has a much more optimistic view of the outcomes for children—and adults. Starting in 1974 in Virginia with parents of 4-year-olds, forty-eight divorced and forty-eight married-couple families, her research ultimately included 1,400 stable and dissolved marriages and the children of those marriages, some followed for almost thirty years. Hetherington found that 25 percent of these children of divorced parents had long-term social, emotional, or psychological problems, compared to 10 percent of those whose parents had not divorced. However, in assessing the impact of divorce, she would emphasize the 75 to 80 percent of children who are coping reasonably well (Hetherington and Kelly 2002).

> Researchers have clearly demonstrated that, on average, children benefit from being raised in two biological or adoptive parent families rather than separated, divorced, or never-married single-parent households... but... there is considerable variability, and the differences between groups, while significant, are relatively small. Indeed, despite the well-documented risks associated with separation and divorce, the majority of divorced children as young adults enjoy average or better social and emotional adjustment. (Kelly and Lamb 2003, p. 195, citations omitted)

Constance Ahrons (2004) also followed children of divorce into adulthood. Now averaging 31 years of age, they had been 6 through 15 at the time of the marital separation. At present—twenty years later—79 percent think that their parents' decision to divorce was a good one, and 78 percent feel that they are either better off than they would have been, or else not that affected. Twenty percent, however, did not do so well, having "emotional scars that didn't heal" (p. 44). Ahrons judges that the prime factor affecting outcomes was how the parents related to each other in terms of avoiding conflict.

Clinical Problems Versus Psychological Pain In his recent review of the divorce literature, Amato (2010), based on the work of Laumann-Billings and Emery (2000), draws a distinction between clinical problems versus psychological pain. That is, although children of divorced parents have an increased risk of social and emotional problems, they are not more likely than children with continuously married parents to exhibit "serious" issues, such as being in the clinical range for depression or anxiety.

Other scholars agree: "On average, parental divorce and remarriage have only a small negative impact on the well-being of children" (Barber and Demo 2006, p. 291; see also Demo, Aquilino, and Fine 2005, p. 125). Experiencing adversity in childhood, such as parental divorce, is associated with stronger sense of control and more positive expectations for the future in adulthood (Kim and Woo 2011; Schafer, Ferraro, and Mustillo 2011).

This does not mean, however, that children of divorce do not experience substantial stress emanating from the divorce such as sadness over not seeing one of their parents, nervousness about being left home alone, or worry about where to spend the holidays. Children of divorce also are more likely to have been "parentified," meaning they were forced to take on adult responsibilities before they were developmentally mature enough to handle them. **Parentification** has been linked to worse child outcomes (Shaffer and Egeland 2011). These types of horizontal relationships happen even in households where parents work hard to keep the distinction between "parent" and "child" (Nixon, Greene, and Hogan 2012). A 13-year old girl living alone with a single mother explains, "I pretend that I'm not sad sometimes because it just worries her so much.... I don't like to worry her. I think if I'm sad that's mine to deal with; she doesn't have to worry about it all the time." A 12-year-old boy from the same study says, "If she's stressed at work, if she has to meet a deadline or something, she can come back and I can help her out with stuff around the house and everything.... I'd bring her a cup of tea or whatever and put my brother to bed" (Nixon, Greene, and Hogan 2012, p. 149).

All in all, divorce researchers seem to be moving to a middle ground in which they acknowledge that children of divorce are disadvantaged compared to those of married parents—and that those whose parents were not engaged in serious marital conflict have especially lost the advantage of an intact parental home. But many have moved away from simplistic or overly negative views of the outcomes of divorce. A theme that runs through virtually all studies on the impact of marriage and divorce on the well-being of children rests on the behavior of the parents, toward the children and each other (Ahrons 2004; Barber and Demo 2006; Freeman 2008). If a divorced couple continues to have a relationship that is fraught with conflict, the outcomes for their children will be negative. A good mother–child (or custodial parent–child) bond and competent parenting by the custodial parent seem to be the most significant factors (Amato and Cheadle 2008).

A good nonresident parent-child relationship also has a positive influence on child outcomes (Demo and Fine 2010; Guzzo 2009b). If children of married, cohabitating, divorced, remarried, or single parents feel nurtured, loved, and supported by parents and families who engage in conflict resolution and work hard at getting along with one another, the outcomes for those children tend to be positive.

We now turn to issues of custody, the setting in which children will live after the divorce.

Child Custody Issues

A basic issue in a divorce of parents is the determination of which parent will take custody—that is, assume primary responsibility for caring for the children and making decisions about their upbringing and general welfare. **Legal custody** refers to who has the right to make decisions with respect to a child's upbringing (e.g., health, religion, education) and **physical custody** refers to where a child will live (Stewart 2007). In **joint custody**, both divorced parents continue to take equal responsibility for important decisions regarding the child's general upbringing. Joint custody may be legal, physical, or both. In joint physical custody parents or children move periodically so that the child resides with each parent in turn on a substantially equal basis. Arrangements can vary from the child spending one week with one parent and the next with the other, cycles of several days with one parent followed by several days with the other, or other arrangements such as weekdays with one parent and weekends with the other. The second variation is joint legal custody—in which both parents have an equal right to participate in important decisions and retain a symbolically important legal authority.

Custody patterns and preferences in law have changed over time. As part of a patriarchal legal system, fathers were automatically given custody until the mid-nineteenth century. Then the first wave of the women's movement made mothers' parental rights an issue. Emerging theories of child development also lent support to a presumption that mother custody was virtually always in the child's best interest, the so-called tender years doctrine (Artis 2004).

In the 1970s, states' reforms of divorce law incorporated new ideas about men, women, and parenthood; custody criteria were made gender neutral. Under current laws, a father and a mother who want to retain custody have theoretically equal chances. Judges try to assess the relationship between parents and children on a case-by-case basis. Other approaches, such as the friendly parent concept, the idea that custody should favor the parent who is more likely to foster the child's relationship with the other parent, have also been tried (Dore 2004).

Because mothers are typically the ones who have physically cared for the child, and because many judges still have traditional attitudes about gender, some courts continue to give preference to mothers (Demo and Fine 2010). Mothers still receive sole physical custody of their children about 80 percent of the time, although, as discussed below, this might be declining (Kreider and Fields 2005). Eighty-two percent of custodial parents are mothers; 18 percent are fathers. Not all custodial mothers in government statistics have been divorced however; 35 percent were never married (Grall 2011).

However, there has been a dramatic increase in fathers who have joint physical custody. Cancian, Meyer, and Han (2012) analyzed court records from Wisconsin over the past twelve years and found that although the proportion of fathers awarded sole physical custody remained stable at about 8 percent, the proportion awarded joint physical custody increased from 24 percent to 44 percent. Among lesbian couples with children, the majority opt for joint custody (Gartrell et al. 2011). Split custody, in which each parent has physical custody of at least one child, remains uncommon and only occurs in about 2 percent to 4 percent of divorce cases. Joint physical custody among never married couples also increased over the period, from 1 percent to 10 percent. A very small minority of joint custody couples set up bird's nests for their children, in which the parents, rather than the children, move between households (Luscombe 2011). Another arrangement, seen recently among low-income families, is living together apart, or LTA, which is when couples who are no longer romantically involved reside in the same households to share the day-to-day care of the children (Cross-Barnet, Cherlin, and Burton 2011).

Growth in father custody (whether joint or sole) is likely the result of men's changing social roles and their increasing involvement with children, in terms of direct care and emotional closeness. It also may be the case that mothers have become less inclined to insist on sole custody. Fatherhood scholar James Levine thinks that "[w]e're seeing some weakening of the constraints on women to feel they can only be successful if they are successful mothers," so they are more willing to concede custody to willing fathers (quoted in Fritsch 2001, p. 4). Other research paints a different picture of father custody, with some men utilizing gender-neutral laws about custody to continue to exert power and control over their ex-spouse and to avoid paying child support (Elizabeth, Gavey, and Tolmie 2012). Several studies show that residence patterns of children in father custody are more unstable than those of children in mother custody (Stewart 2007). Children in father custody are more likely than children in mother custody "drift" to their other parent's home. Because returning to court is costly, most mothers whose children have returned to them do not seek an adjustment in child support awards or other changes in divorce decrees, leaving them at a financial disadvantage.

Studies have found few or no difference in well-being between children living with only their fathers (apart from their mothers) versus only with their mothers (apart from their fathers; Stewart 2007). Overall, children greatly benefit from contact and closeness with both biological parents (Bastaits, Ponnet, and Mortelmans 2012; King and Sobolewski 2006; Sobolewski and King 2005).

The Residential Parent

As discussed above, most custodial parents are women. Single parenthood is a role associated with stressor overload, or stress pileup (see Chapter 13 for a more detailed discussion of these concepts). Custodial parents have the challenge of being responsible for their children's care on a full-time basis. They often feel overwhelmed and exhausted, and have little time for meeting their own needs for rest, socializing, exercise, and hobbies which negatively affects their mental and physical health. Given that child support awards are typically low, they must grapple with "making ends meet" financially. As discussed in Chapter 15, they must also balance the needs of new spouses and partners with their children, and often must play "mediator" between the two.

The Visiting Parent

Most noncustodial parents are fathers. They often find it difficult to construct a satisfying parent–child relationship. During the marriage, a father's authority in the family gave weight to his parental role, but this vanishes in a nonresidential situation. Custodial mothers are effectively gatekeepers, facilitating, or not, the noncustodial father's relationship with his children (Adamsons and Pasley 2006; Moore 2012; Sano, Richards, and Zvonkovic 2008).

About a third of children with a nonresident father have no contact him (Huang 2009; Stewart 2007). Visitation with children also declines over time in most cases (Manning and Smock 1999). Factors associated with low visitation between nonresident fathers and their children include being Nonwhite, not having been married to the child's mother, low level of education, geographical distance, conflict with the mother, and nonpayment of child support (Huang 2009; Paulson, Dauber, and Leiferman 2011; Troilo and Coleman 2012). Swiss and Le Bourdais (2009) discuss the material conditions that impact the noncustodial father's relationship negatively. For example, they point out that working-class fathers, especially those who earn "lower incomes are more likely to be working in low-paying, part-time, or shift-oriented work where they may not be available to their children when the latter are free, that is, in the evenings and weekends" (p. 644).

A new marriage or cohabiting relationship is not by itself a factor in decreasing visitation, but the presence of children in a new family, particularly biological children, does lead to a decline (Manning and Smock 2000). Fathers seemed to find it difficult to parent their children across two families (Guzzo 2009b; Juby et al. 2007; Swiss and Le Bourdais 2009).

The situation of noncustodial fathers has improved since earlier research found that many had detached from their children. Data from the National Survey of America's Families (NSAF) indicates that a third of children had at least weekly contact with their fathers, although overnight and extended visits were still rare (Stewart 2010). Many nonresident fathers embrace their new parenting role, if only on a part-time basis. Some fathers felt they were even more involved with their children than they had been, for example, by taking their children to doctor's appointments. One father with a 5-year-old daughter explains, "I make breakfast for her. She likes that, she sits at the counter while I'm cooking and we talk" (Troilo and Coleman 2012, p. 607). Other fathers expressed frustration with their new "part-time full-time role,"

> I pick her up around 5:30 [on Wednesday evening], then before you know it, it's almost 8, and I'm telling her, "Ok, let's start getting toward going to bed. Do you know what you're going to wear tomorrow?" If not, I wash clothes if needed, make sure her teeth are brushed, and see if she needs a bath, read stories, and you know 8:30, 9 is gone before you know it. You're talking 2.5 hours on a weeknight. The mornings are pretty much getting her up and out the door. There's no quality time there. (Troilo and Coleman 2012, p. 610)

Email, texting, and social media can be used to increase communication between nonresident parents and children, but they may come with difficultiess (such as children "defriending" a parent). Nevertheless, one father, who communicates with his daughter daily says "with email and text, it's just like you're there is some ways" (Troilo and Coleman 2012, p. 606).

How Nonresident Father Involvement Affects Children

When the divorced or separated father uses an authoritative parenting style (see Chapter 9) and when the visit does not lead to conflict between the parents, it has a favorable impact on the child's adjustment. Authoritative parenting on the part of the nonresident father is especially important to children's well-being, at least among adolescent boys, when the mother is low on this parenting style (Karre and Mounts 2012). The quality of the father-child relationship is more important to children's well-being than how often the child sees his or her father or whether the father lives in the household (Booth, Scott, and King 2010; Stewart 2007). Among adolescents, relationship quality, closeness, and responsive parenting (that fathers consider the child's point of view and explain decisions) were found to reduce adolescents' "internalizing" (depression, self-esteem) and "externalizing" (aggression, antisocial behavior, drug use) problems and was associated with higher grades (Booth, Scott, and King 2010; Stewart 2003).

Nonresident fathers are not always the best role models, for example, if he abuses drugs or alcohol or if is violent (see Chapter 9). In those cases, it may be best if the father is not involved (Osborne and Berger 2009; Salisbury, Henning, and Holdford 2009). In such cases, the courts may step in to limit or prevent fathers from having contact with the mother and children.

Divorced fathers, most of whom are the nonresidential parent, face the loss of time with children, as well as a more general loneliness. Being the "visiting parent" is often difficult, but maintaining the father–child bond is significant in a child's adjustment to divorce.

Noncustodial Mothers According to the most recent data available, there are more than two million noncustodial mothers in the United States (Sousa and Sorensen 2006). The popular stereotype is that noncustodial mothers are deviant in some way. Whereas some of those mothers have lost custody of children due to their abuse, neglect, or problems with substance abuse or their mental health, most noncustodial mothers voluntarily surrender custody. The most common reason for doing so is worries about how they will financially support their children (Stewart 2007). It is important to point out too that about half of children with a nonresident mother are not living with their fathers, but are living with grandparents, relatives, or are in foster care (Stewart 2010). Many of these placements are temporary, and children are returned to living with their mothers when their mothers are financially stable.

Most research and discussion on visiting parents has been about fathers, but one study did compare the two sexes. Using two national datasets, the NSFH and NSAF, Stewart (1999a, 1999b, 2010) compared visitation and child support payments between children with noncustodial mothers versus fathers. Whereas noncustodial mothers exhibited more day-to-day visitation, phone calls, and extended visits, noncustodial fathers were more likely to pay child support and paid higher amounts.

Being a noncustodial mother is stressful parenting role. Reasons include the stigma of not being the primary caretaker of one's children, the involvement of stepmothers in decisions about the children, and loneliness from being separated from the children. Deborah Eicher-Catt's qualitative study of noncustodial mothers (2004) found that the minority of mothers whose custody was abrogated by the courts were restricted in their contact, perhaps permitted only supervised visitation with their children. The larger group who voluntarily relinquished custody found it hard to achieve a workable relationship with the child as well. They were unable to be traditional mothers, but found a "mother-as-friend" role insufficient and uncomfortable. Eicher-Catt advises noncustodial mothers to focus on building a relationship, rather than thinking in terms of the traditional maternal role.

Joint Custody

Recall that joint custody means that both divorced parents continue to take equal responsibility for important decisions regarding the child's general upbringing. When parents live close to each other and when both are committed, joint custody can bring the experiences of the two parents closer together, providing advantages to each. Both parents may feel they have the opportunity to pass their own beliefs and values on to their children. In addition, neither parent is overloaded with sole custodial responsibility. Joint custody gives each parent some downtime from parenting (M-Y. Lee 2002). Table 14.2 lists advantages and

Table 14.2 Joint Custody from a Father's Perspective

John is an engineer who shares physical custody of his 13-year-old son, Dustin. Dustin lives with his dad during the week and with his mom and stepdad on weekends, except for Wednesday nights, when he stays overnight with his mom. John and Dustin's mom, who owns a hair salon with her husband, live in the same town, and Dustin transitions to and from each parent's house in the afternoons by way of the school bus. Below John talks about the advantages and disadvantages of joint custody.

Advantages	Disadvantages
Shared custody schedule gives me much more time with my son than a traditional, limited-visitation schedule.	Dustin sometimes uses the rotating schedule to avoid following rules he does not like, such as consistently doing his homework.
I get to be a "real dad," instead of a weekend guest in my child's life.	I feel sometimes "out of the loop," regarding events going on in Dustin's life, such as knowing who his friends are and who he is taking to the 8th grade dance.
I think the time we spend physically together could not be replaced with phone conversations and texting.	Every time Dustin comes back from his mom's, he has to spend a bit of time re-learn my house rules.
Our time together happens at a "normal" pace, instead of being crammed into a short visitation.	Bedtime and chores are often a struggle—I probably let Dustin stay up later than he should because I often work late and want to spend time with him.
Me and his mom each get to have "child-free" time. I have more time with my girlfriend and with friends. I can take weekend trips on my motorcycle.	I worry that Dustin never gets fully settled in either house.
Dustin gets to have two regular households, instead of a regular house and a "foreign" house.	With joint custody, I am tied to a small geographic area which affects my career. I would go up for a promotion at my company except most opportunities for advancement would take me out of state.
Dustin has a room, clothes, and his things at each house and doesn't have to "pack up" to move between households.	His mom, her husband, and I have to work harder to get along.
Dustin has become accustomed to following two sets of rules, one for each house. For example, he goes to church and youth group when at his mom's.	
Both households have equal weight, so there's no power-play between me and his mom.	
Direct expenses for Dustin (clothes, lunch money, etc.) are more evenly shared.	

disadvantages of joint custody from a father's perspective. Shared custody gives children the chance for a more realistic and normal relationship with each parent (Arditti and Keith 1993). It results in more father involvement and in closer relationships with both parents (Kelly 2007).

The high rate of geographic mobility in the United States can make joint physical custody difficult. Joint custody can also be expensive. Each parent must maintain housing, equipment, toys, and often a separate set of clothes for the children and must sometimes pay for travel between homes if they are geographically distant. Mothers, more than fathers, would find it difficult to maintain a family household without child support, which is often not awarded when custody is shared. There may be situations—an abusive parent, other domestic violence, or extremely high levels of parental conflict, for example—where sole custody is preferable (Hardesty and Chung 2006).

Research does not consistently support the presumption that joint custody is always best for children or their parents. Melinda Stafford Markham and Marilyn Coleman (2012) in their interviews with mothers who shared custody reveal that not all joint custody parents get along—parents have disagreements about child support, parenting style, and often have trouble communicating. An early study on joint custody, The Stanford Child Custody Study (Maccoby and Mnookin 1992), found that children in joint custody and mother custody did similarly well: "'[T]he welfare of kids following a divorce did not depend on who got custody, but on how the household was managed and how the parents cooperated'" (psychologist Eleanor Maccoby in Kimmel 2000, p. 141; Barber and Demo 2006; Freeman 2008). One reviewer of the research literature concluded that there is "no consistent evidence of the superiority of one arrangement over another" (M-Y. Lee 2002, p. 673). Yet, another review of thirty-three studies of joint and sole custody (Bauserman 2002) did find that children in joint custody arrangements had superior adjustment. What about the effect of joint custody on parents? A recent meta-analysis of parents found better father-child relationships, less parenting stress, lower parental conflict, and better overall adjustment among parents with joint custody than sole custody, but were less satisfied with custody arrangements (Bauserman 2012).

Among lesbian couples with children, co-parent adoption was associated with a greater likelihood of shared custody than sole custody, and children who had been adopted by their "other" mother (the nonbirth/nonlegal mother) had higher well-being, regardless of custody status (Gartrell et al. 2011).

Styles of Parental Relationships After Divorce

Sociologist Constance Ahrons (1994) led one of the first explorations of relationships between parents after divorce, the Binuclear Family Study. In a **binuclear family** the child is the "nucleus" in two households within one family. The study is based on interviews with ninety-eight divorced couples approximately one year after their divorce Ninety percent of them were followed to the five-year point, and each couple was interviewed three times. These were primarily white, middle-class couples from one Wisconsin county in which the mother had primary custody of the children.

At the one-year point, 50 percent of the ex-spouses had amicable relations, whereas the other 50 percent did not. In half of those cases, the divorce was a bad one and harmful to family members; in the other half, the divorcing spouses had "preserved family ties and provided children with two parents and healthy families" (p. 16).

The ninety-eight couples represented a broad range of postdivorce relationships. Half of the couples exhibited what she considered "cooperative" parenting. These include two styles, one of which she termed *perfect pals* and the other of which she termed *cooperative colleagues.* Parents who are perfect pals (12 percent) were friends who called each other often and brought their common children and new families together on holidays or for outings or other activities. This pattern occurred in a minority of the "good divorces." More often (38 percent), the couples were cooperative colleagues, who worked well together but did not attempt to share holidays or be in constant touch—occasionally, they might share children's important events such as birthdays. Ex-spouses might talk about extended family, friends, or work. They still had areas of conflict but were able to compartmentalize them and keep them out of the collaboration that they wanted to maintain for their children (Ahrons 1994; 2007). Cooperative parenting is associated with the higher social and emotional well-being for children, although the effect may be small and may only apply to certain outcomes (Amato, Kane, and James 2011).

Although not mentioned in the Ahrons study, other research (Amato, Kane, and James 2011; Maccoby and Mnookin 1992) has identified a post-divorce parental relationship referred to as **parallel parents**, parents who parented alongside each other but with minimal contact or communication or conflict. Relative to both cooperative parenting and parenting without any involvement from the ex-spouse, parallel parenting is

BananaStock/JupiterImages

When divorcing parents continue to engage in conflict and especially when children are drawn into it, a child's adjustment is poorer. Interparental conflict does tend to diminish with the passage of time.

associated with worse child outcomes (Amato, Kane, and James 2011; Pryor 2011). Returning to the Ahrons study, other divorcing couples were the angry associates (25 percent) or fiery foes (25 percent) that we may think of in conjunction with divorce.

Co-Parenting

Co-parenting is a "team" approach to raising children after divorce. In this model of parenting, parents strive to work together as "colleagues," with the goal of meeting their children's emotional and financial needs (this is similar to Ahron's idea of "cooperative" parenting discussed above). Co-parenting should not be confused with joint custody, which refers to couples' legal arrangements for where the children should live and how decisions about the children are made. That is, parents with sole or joint custody can be co-parents. Although divorced parents for the most part are allowed to spend time and interact with their children as they want, co-parenting is considered the best model for enhancing children's well-being and future success. As discussed below, many states require couples with children to complete co-parenting classes before a divorce is granted.

Most states offer or require parent education for divorcing parents (Kelly 2007; McGough 2005; Pollet and Lombreglia 2008, p. 375). In some communities, the children also meet in groups with a teacher or mental health professional (Pollet and Lombreglia 2008). The idea is that parents will continue to raise their children as co-parents, and they are likely to need help in meeting this new challenge. Evaluation forms completed after the sessions have shown predominantly positive responses, but more research is needed on the effect of co-parenting on children's well-being (Amato, Kane, and James 2011).

The Our Family Wizard website can help facilitate co-parenting relationships by minimizing "face-time" between parents. Randy Kessler, chair of the American Bar Association's Family Law Section explains, "People don't want to talk to their exes because the sounds of their voice is irritating. But they can email. The can share and on-line calendar. They can use any number of resources on the Internet. There are even divorce apps" (Paul 2012).

Communication technology does not make things easier if the relationship between ex-spouses is a contentious one (Ganong et al. 2012). Overall, however, the effect has been positive. Andrew Cherlin explains, "Before this electronic media, noncustodial parents had very formalized, appointment-driven communication with their kids," he said. "Electronic media may help noncustodial parents by informalizing the process of communicating with their children. They become less dependent on schedules and therefore more consistent with an easy, informal flow of information, which may be what teenagers like" (Meyers 2011).

Yet, data based on the National Survey of Family Households indicates that most divorced parents are not engaging in co-parenting. Most have low (70 percent) or medium (17 percent) levels of co-parenting (defined as at least monthly discussions about the children and "a great deal" or "some" influence from the father in childrearing decisions) as opposed to high (12 percent) levels and that co-parenting declines over time (McGene and King 2012).

The factors that seem to affect co-parenting success are rather straightforward: a previous good co-parenting relationship during the marriage, a mediated rather than hostile divorce process, a reasonably good postdivorce relationship between ex-spouses, and length of time since the divorce (Adamsons and Pasley 2006). Parents who share physical and/or legal custody and who are thought to get along better than other divorced parents are not always consistently nice and easy with one another. They, too, display various patterns of co-parenting, including continuously contentious, always amicable, and bad to better (Markham and Coleman 2012). One bad to better mother explains:

> I think always keeping [my son] at the focus of what I was trying to do as a person, as a parent you have to let some of that go. And since it's been working so far, I have no reason to rock the boat. Anything that I would do to be vengeful against my ex wouldn't do anything but damage my son.... I've let a lot of anger go. (Markham and Coleman 2012, p. 593)

This discussion of mothers' and fathers' custody, visitation, and child support issues suggests that being divorced is in many ways a very different experience for men and women. The fact is that both men's and women's postdivorce situations would be somewhat alleviated by eliminating the economic discrimination faced by women, especially women reentering the labor force, by strong child support enforcement, and by constructing co-parenting relationships that give fathers the sense of continuing involvement as parents that most would like. "As We Make Choices: Rules for Successful Co-Parenting" provides some general guidelines for divorcing

Michael Newman/ Photo Edit

These divorced parents have both come to meet with their child's teacher. When parents work together to co-parent their children, they continue to have a sense of "family."

As We Make Choices

Rules for Successful Co-Parenting

Children in the Middle is a national organization that is "committed to supporting adults in raising children between two homes and to offering tools to the professionals assisting these families ... dealing with transitions such as separation, divorce, or other family matters involving two homes" (Children in the Middle 2013). They provide co-parenting classes and information for divorcing couples with children. Such classes are mandatory in many states—a couple's divorce will not be granted without it. Here are their guidelines for parents:

1. Do not talk negatively, or allow others to talk negatively, about the other parent, their family and friends, or their home in hearing range of the child. This would include belittling remarks, ridicules, or bringing up allegations that are valid or invalid about adult issues.
2. Do not question the children about the other parent or the activities of the other parent regarding their personal lives. In specific terms, do not use the child to spy on the other parent.
3. Do not argue or have heated conversations between the parents when the children are present or during exchanges
4. Do not make promises to the children to try and win them over at the expense of the other parent.
5. Do communicate with the other parent and make similar rules in reference to discipline, bedtime routines, sleeping arrangements, and schedules. Appropriate discipline should be exercised by mutually agreed upon adults.
6. At all times, the decisions made by the parents will be for the child's psychological, spiritual, and physical well being and safety.
7. Schedule changes will be made and confirmed beforehand between the parents without involving the child, in order to avoid any false hopes and cause any disappointments or resentments toward the other parent.
8. Do notify each other in a timely manner of need to deviate from the order including canceling time with the child, rescheduling, and promptness.
9. Do not schedule activities for the child during your child's time with their other home without the other parent's consent. However, both parents will work together to allow the child to be involved in extracurricular activities.
10. Do keep the other parent informed of any scholastic, medical, psychiatric, or extracurricular activities or appointments of the child.
11. Do keep the other parent informed at all times of your address and telephone number. If you are out of town with the child, do provide the other parent the address and phone where the children may be reached in case of an emergency.
12. Do refer to the other parent as the child's Mother or Father in conversation, rather than using the parent's first or last name.
13. Do not bring the child into adult issues and adult conversations about custody, the court, or about the other party.
14. Do not ask the child where he or she wants to live. But do encourage the child to understand they have two homes.
15. Do not attempt to alienate the other parent from the child's life.
16. Do not allow stepparents or others to negatively alter or modify your relationship with the other parent.
17. Do not use phrases that draw the children into your issues, or make the children feel guilty about the time spent with the other parent. For example, rather than saying "I miss you!" say "I love you!"

Critical Thinking

Do you know any divorced parents who follow, or do not follow, the rules of successful co-parenting? What do you think it would be like to be a "child of divorce"? How well do you think your parents would get along?

Source: Children in the Middle 2013.

parents who want to cooperate in parenting their children.

The title of Ahrons's most recent book—*We're Still Family* (2004)—characterizes the positive outcomes of divorce that she has found among many of the families she studied. She argues that we must "recognize families of divorce as legitimate." To encourage more "good divorces," it is important to dispel the "myth that only in a nuclear family can we raise healthy children" (Ahrons 1994, p. 4). People often find what they expect, and social models of a functional postdivorce family have been lacking. In Chapter 15, we turn to a consideration of that common step for many divorced people: remarrying.

Summary

- Divorce rates rose sharply in the twentieth century, and divorce rates in the United States are now among the highest in the world. Since around 1980, however, they have declined substantially and have since stabilized.
- Among the reasons divorce rates have increased to the present level are changes in society. Economic interdependence and legal, moral, and social constraints are lessening. Expectations for intimacy have risen, while expectations of permanence are declining.
- People's personal decisions to divorce involve weighing the advantages of the marriage against marital complaints in a context of weakening barriers to divorce and an assessment of the possible consequences of divorce.
- Two consequences that receive a great deal of consideration are how a divorce will affect any children and whether it will cause serious financial difficulties.
- The economic divorce is typically more damaging for women than for men, and this is especially so for custodial mothers. Over the past twenty-five years, child support policies have undergone sweeping changes. The results appear to be positive, with more child support being collected. Fathers' chief concern is maintaining a relationship with their children, so visitation, joint custody, and the moving away of custodial mothers are their chief legal and policy issues.
- Researchers have proposed a number of possible theories to explain negative effects of divorce on children. These include the life stress perspective, the parental loss perspective, the parental adjustment perspective, the economic hardship perspective, the interparental conflict perspective, the family instability perspective, and selection perspective.
- Debate continues among family scholars and policy makers concerning how important a threat of divorce is to children today. Some call for return to a fault system of divorce or other restrictions on divorce. Others see divorce as part of a set of broad social changes, the implications of which must be addressed in ways other than turning back the clock. Now there is also a centrist view of the impact of divorce on children: Yes, there is some disadvantage; no, divorce is not the most powerful influence on children's lives.
- New norms and new forms of the postdivorce family seem to be developing. Some postdivorce families can share family occasions and attachments and work together civilly and realistically to foster a "good divorce" and a binuclear family.

Questions for Review and Reflection

1. What factors bind marriages and families together? How have these factors changed, and how has the divorce rate been affected?
2. How is the experience of divorce different for men and women? How are these differences related to society's gender expectations?
3. In what situation(s), in your opinion, would divorce be the best option for a family and its children?
4. Do you think couples are too quick to divorce? What are your reasons for thinking so?
5. **Policy Question.** Should divorced parents with children be required to remain in the same community? Permitted to move only by court authorization? Be free to choose whether to be geographically mobile?

Key Terms

binuclear family 380
child support 370
co-parenting 381
covenant marriage 362
crude divorce rate 358
divorce divide 359
divorce fallout 366
divorce mediation 366

economic hardship perspective 374
family fluidity 375
family instability perspective 374
income-to-needs ratio 369
independence effect 361
interparental conflict perspective 374
joint custody 376
legal custody 376
Levinger's model of divorce decisions 363
life stress perspective 374
parallel parents 380
parental adjustment perspective 374
parental loss perspective 374
parentification 376
physical custody 376
refined divorce rate 358
selection perspective 374
silver divorce 359
starter marriage 359
stress-related growth 372
unilateral divorce 362

15

Remarriages and Stepfamilies

Karin Dreyer/Blend Images/Getty Images

Defining and Measuring Stepfamilies

What Makes a Stepfamily?

Issues for Thought: What Makes a Stepfamily?

Various Types of Stepfamilies

Perceptions of Stepfamilies: Stereotypes and Stigma

Choosing Partners the Next Time

Dating with Children

What Kinds of People Become Stepparents?

Second Weddings

Happiness Satisfaction, and Stability in Remarriage

Happiness and Satisfaction in Remarriage

The Stability of Remarriages

Day-to-Day Living in Stepfamilies

Challenges to Developing a Stepfamily Identity

A Closer Look at Diversity: Do You Speak Stepfamily?

The Stepfamily System

Stepfamily Roles

Stepfamily Relationships

Financial and Legal Issues

Well-Being in Stepfamilies

The Well-Being of Parents and Stepparents

The Well-Being of Children

Creating Supportive Stepfamilies

Learning Objectives

1. Define what makes a stepfamily and describe various stepfamily types.
2. Describe the process through which adults and children become members of stepfamilies.
3. Compare happiness, satisfaction, and stability in remarriages and first marriages.
4. Identify the challenges of stepfamily roles and relationships.
5. Compare adult and child well-being in stepfamilies relative to that in two-biological/adoptive parent families.
6. Explain the characteristics of a supportive and healthy stepfamily environment.

- "[My stepdad] is very listening. He's very open to communication. He's willing to talk to you about anything. He's always been there" (in Baxter, Braithwaite, and Bryant 2006, p. 393).
- "I wasn't very happy about [the remarriage] because before this, you know, we got to spend a lot of time with my dad.... But then when she came into the picture...we kind of got left out" (in Baxter et al. 2009, p. 479).

Today, approximately 30 percent of all marriages are remarriages for one or both partners (Cruz 2012). Stepfamilies are also increasingly being formed through cohabitation as well as through legal marriage. In addition to remarriage and cohabitation, stepfamilies can be formed when a woman or man who has had a child prior to marriage marries for the first time. All in all, 42 percent of adults have at least one step relative, with 30 percent reporting having a step- or half-sibling (National Stepfamily Resource Center 2013). Incidentally, stepfamilies are more common in the United States than in any other industrialized nation (Andersson and Philipov 2002).

Most remarried people are happy with their relationships and lives. Much of what we have said throughout this book—for example, about good parenting practices, supportive communication, and how best to handle family stress—applies to remarriages and stepfamilies. At the same time, stepfamily members experience unique challenges, because stepfamilies are different from original two-parent families in a number of important ways. Foremost, forming a cohesive and satisfying stepfamily requires "leaving behind traditional views of [family] belonging" (Sky 2009).

In this chapter, we will discuss how new family patterns have created various types of stepfamilies and complicated stepfamily dynamics. We'll look at negative stereotypes of stepfamilies and their ongoing stigmatization. We will talk about the delicate process of stepfamily formation; ambiguity in norms, roles, and family boundaries in stepfamilies; and what is known about how stepfamily relationships change over time. We'll discuss happiness and stability in remarriages and among first-married and cohabiting couples with stepchildren and how living in a stepfamily affects the well-being of its members. We'll explore some challenges typically associated with stepfamily life and ideas for creating supportive remarriages and stepfamilies. Throughout this chapter, we will point out how stepfamily processes vary for different types of stepfamilies. We'll begin with a discussion of how stepfamilies are currently defined and measured.

Defining and Measuring Stepfamilies

This section describes changes in how stepfamilies are defined and sources of diversity in stepfamilies, providing statistics on the prevalence of remarriages and stepfamilies in the United States.

What Makes a Stepfamily?

The seemingly straightforward question of "what is a stepfamily?" is not as easy to answer as you might think (see "Issues for Thought: What Makes a Stepfamily?"). You may be surprised to find out that even among family scholars there is disagreement about what constitutes a stepfamily (Stewart 2007). Traditionally, a stepfamily has been defined as a household containing a remarried couple and one or both spouses' children from a previous marriage. However, there have been important social and demographic shifts in recent decades that have altered family and stepfamily life alike. These shifts include growth in the number of women having children outside of marriage, an increase in unmarried couples with children living together, increasing involvement of nonresident parents in their children's lives (including growth in joint custody), increasing awareness and tolerance of lesbian and gay relationships, and other factors (Stewart 2007). The result is that today there are many different types of stepfamilies. Although life can be very different within each, they have one thing in common—the absence of a biological relationship between one's children and one's romantic partner.

Pathways to Stepfamily Living Sociologist Kathryn Tillman (2007) has described the various pathways that lead to stepfamily living (see Figure 15.1). For example, a stepfamily can originate with a birth to a married or cohabiting couple (House A). Stepfamilies

Issues for Thought

What Makes a Stepfamily?

What comes to mind when you hear the word *stepfamily*? Perhaps *The Brady Bunch*, the iconic 1960s TV family in which a widower with three sons and a widow with three daughters marry and merge households. Below are the stories of five real-life families from this century.

Rhonda and her boyfriend, Al, have a child, Emily. The couple's relationship doesn't work out and Al moves out of state. He has no contact with his daughter and pays no child support. Rhonda meets Peter and they marry. Peter adopts Emily.

Carol and Randy are divorced. Carol has custody of their two children, Emma and Sophia. After a few months of dating, Carol's boyfriend, Roger, moves in with Carol and the girls.

Bobby lives with his mother, Elaine. Bobby's father, Doug, has remarried Leslie. The couple lives with Leslie's children from her first marriage, Teddy, Austin, and Abbey. Bobby sees his father every other weekend, usually for a movie and a bite to eat. Sometimes Leslie and her children go along (depending on what's showing).

Janet is married to Ron, and the couple has two sons, Billy and Justin. Janet falls in love with Ann, a coworker, and divorces Ron. Ann moves in with Janet, Billy, and Justin. The boys refer to her as "Aunt Ann."

Rosemary is divorced and has three college-aged children, Sarah, Ben, and Lily. Rosemary meets James, a widower, at a retirement party and after a two-week courtship, they fly off to Las Vegas and get married. James has a grown son, Todd. The kids get together and plan a reception for the couple.

Critical Thinking

Do you know any families that are similar to these?

Would you consider any of these a stepfamily? Why or why not?

can also originate from a birth to a single mother who is neither married nor cohabiting (House B). Children born to married or cohabiting parents may later experience the death of a parent (D) or their parents' divorce or union dissolution (C), after which a parent may marry or remarry, forming a married stepfamily (F). Alternatively, rather than marrying, a parent may go on to form a cohabiting union with a new partner (G). A cohabiting stepfamily may be a permanent arrangement or it may transition to a married or remarried stepfamily, depending upon whether either partner has been married previously. Children born outside of a union may later experience their mother's marriage, remarriage, or cohabitation, resulting in either a married or a cohabiting stepfamily. Multipartnered parenthood (having children with more than one partner) and subsequent union dissolutions and formations, adds even greater complexity to these pathways (Cherlin 2010).

The various pathways stepfamilies follow can result in vastly different experiences for stepfamily members. For instance, stepfamilies formed by previously married spouses are likely to include relationships with ex-spouses and relatives of ex-spouses. Although stepfamilies that form after the death of one parent do not have relationships with ex-partners, they may have ongoing ties with relatives of the deceased parent. The complexity that results from various stepfamily-formation pathways is just one factor that influences the diversity of the stepfamily experience. Other factors will be discussed later.

Various Types of Stepfamilies

The family formation patterns described above have created many different types of stepfamilies, and each type varies in its frequency. For example, whereas stepfamilies formed through divorce and remarriage remain very common, stepfamilies with gay and lesbian parents, even though becoming more prevalent in recent years, represent only a tiny fraction of all stepfamilies. Below we describe different kinds of stepfamilies and provide the most recent statistics on each. Statistics on families, and stepfamilies in particular, should be viewed with caution, however. First, estimates are produced or funded by government agencies and their availability is dependent upon yearly budgets. Second, although our figures are the most recent available, many are based on older surveys. Third, remarriages and cohabitations are less durable than first marriages. As a result, membership in stepfamilies tends to ebb and flow, making stepfamilies difficult to capture in social surveys. Statistics on stepfamilies are therefore probably underestimates.

Stepfamilies Created by Widowhood or Divorce Followed by Remarriage Historically, well into the twentieth century, almost all stepfamilies were formed after the death of a spouse (Strow and Strow 2006). The term stepparent originally meant a person who replaces a dead parent, not an additional parent figure (Bray 1999). Remarriage after the death of a spouse was common throughout American history (Ihinger-Tallman and Pasley 1987). Analyzing 1689 census data from the

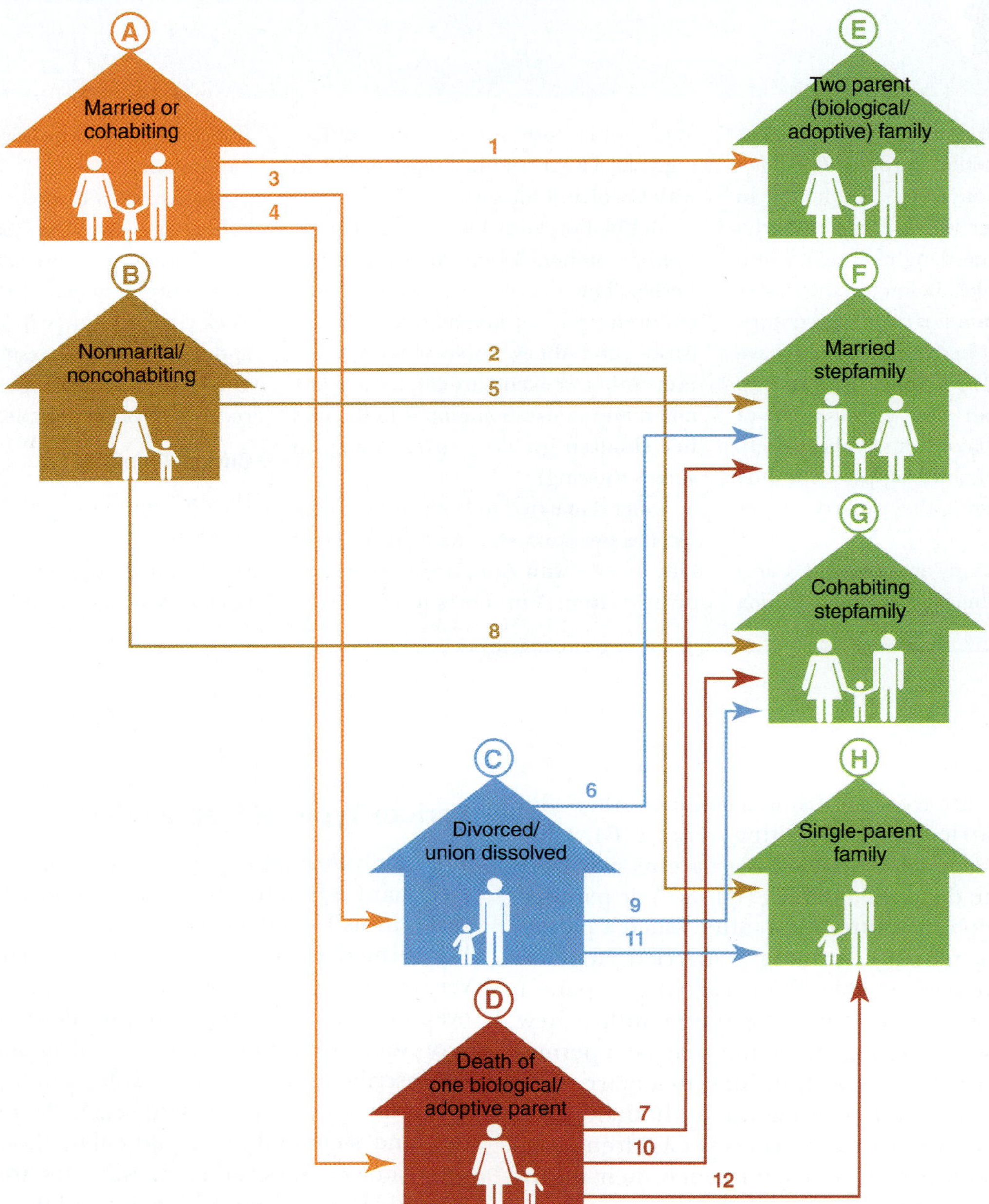

FIGURE 15.1 Pathways to stepfamily living and other family types. The various possible pathways to stepfamily living begin with a baby being born (or being adopted into) to a married or cohabiting couple (House A) or to a single (unmarried, not cohabiting) parent (House B). The couple in House A may take Path 1 and stay together (House E). The parent residing in House B may take Path 2 and choose to not marry or repartner (House H). The couple from House A may take Path 3 and divorce or break up (House C) or they may take Path 4 as a result of one of the spouses/partners passing away (House D). The single parents in House B, House C, House D may choose to take Path 5, Path 6, or Path 7 and marry (or remarry), resulting in a married stepfamily (House F). Alternatively, these parents may take Path 8, Path 9, or Path 10 choose to cohabit, resulting in a cohabiting stepfamily (House G). The previously coupled parents in House C and House D may also take Path 11 or Path 12 and not marry or repartner, resulting in a single parent family (House H). There is further complexity, based on the gender of the biological/adoptive parent, the continued involvement of ex-spouses and partners, and subsequent dissolutions and unions.

Source: Adapted from Tillman 2007, Figure 1, 383–424.

Plymouth Colony town of Bristol, John Demos (1966) estimated that, among people at least 50 years old, 40 percent of men and 26 percent of women remarried at least once. These remarriages took place quickly, usually within a year of a spouse's death. Data from the English village of Clayworth (Nottinghamshire) in 1688 indicates that roughly one in six households was a stepfamily (Phillips 1997).

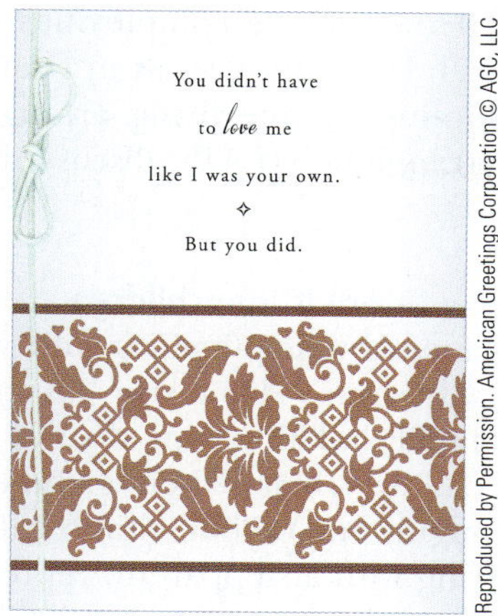

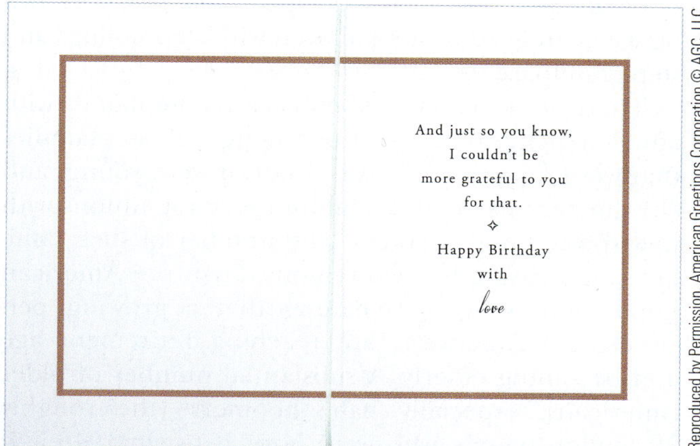

About half of all marriages today are remarriages, a fact that greeting card companies acknowledge.

Today, a number of websites are dedicated to remarital wedding planning and services (e.g., I Do! Take Two)—an indication that remarriages remain common. However, the vast majority of remarried couples today have been divorced rather than widowed. Although the divorce rate has declined in the last two decades (U.S. Census Bureau, 2012c), roughly half of all marriages will still end in divorce (Amato 2010; Bumpass and Raley 2007).

The majority of divorced people remarry, and remarriage rates are much higher among men than women (43 per 1,000 for men versus 23 per 1,000 for women). The gender gap in remarriage increases with age (Cruz 2012). Widowed men also remarry at higher rates than widowed women (Sweeney 2010). The 2009 American Community Survey found that among currently married couples, 20 percent of one or both spouses had been previously married, 16 percent had been married twice, and 4 percent had been married three or more times (Kreider and Ellis 2011b). Like first marriage, the likelihood of remarriage is substantially higher for men and women with a college degree (Cruz 2012). Remarriage rates also vary by race and ethnicity. Hispanic and Asian men have the highest remarriage rates and nonHispanic white and black women have the lowest (Cruz 2012).

About half of remarriages involve children (Kreider 2006). About 10 percent of all children (5.3 million) are living in a married stepfamily with one biological parent and one stepparent (Kreider and Ellis 2011a). Remarried mothers are more likely to have their children living with them than remarried fathers. Although joint custody is increasing, mothers receive sole physical custody of their children in about 80 percent of divorce cases (Cancian and Meyer 1998; Kreider and Fields 2005). However, the stepfamily is no longer merely the product of divorce or the death of a spouse. Some new kinds of stepfamilies are discussed below.

Stepfamilies Created by Nonmarital Childbearing Stepfamilies can originate from a nonmarital birth, followed by first marriage, remarriage, or cohabitation. Forty percent of all babies in the U.S. are born to unmarried women (Hamilton, Martin, and Ventura 2009). This statistic varies dramatically by race and ethnicity: 28 percent among nonHispanic whites, 51 percent among Hispanics, 72 percent among African Americans, 65 percent among American Indians/Alaska Natives, and 17 percent among Asians/Pacific Islanders. Thus, the stepfamilies of some racial and ethnic groups are more likely than others to be formed in this manner.

Data based on the Fragile Families and Child Wellbeing Study found that more than a third of women who had a child with an unmarried partner formed a married or cohabiting union with a new partner within 5 years (Bzostek, McLanahan, and Carlson 2012). It is important to note that childbearing outside of marriage can occur to previously married women, too (Downs 2003; Teachman 2008a).

Stepfamilies Created by Cohabitation As discussed in Chapter 6, growth in cohabitation has transformed American families, including stepfamilies. The majority of young men and women today will cohabit at some point in their lives (Cherlin 2010). Cohabitation is particularly common among the previously married. In 2002, more than two thirds of women under age 45 in the U.S. had cohabited with a partner between their first and second marriages (Teachman 2008a). At least two-thirds of children enter stepfamilies through cohabitation rather than marriage, and cohabiting stepfamilies make up one-quarter of all stepfamilies (Bumpass, Raley, and Sweet 1995; Kreider and Fields 2005). Many children in cohabiting stepfamilies are a product of a nonmarital birth.

Starting with the 2000 census, unmarried men and women can now report stepchildren, reflecting these trends (Kreider 2003).

Children and Households Living in Stepfamilies The 2009 Survey of Income and Program Participation estimates that 13.3 percent of all households with children contain a stepparent, stepchild, step-sibling, or half-sibling (Kreider and Ellis 2011a). These same data indicate that 7.5 percent of children under age 18 reside with a married or cohabiting stepparent. Among stepfamilies, 72.7 percent of children under age 18 reside in first married and remarried stepfamilies (Figure 15.2). Another 23.0 percent of children live with their biological or adoptive mother and her cohabiting partner. An additional 5.2 percent of children live in cohabiting households with their biological father and his partner. These statistics describe a situation at one point in time. However, "it is important to keep in mind that as children age, they may spend time in several [living] arrangements" (Kreider and Elliott 2009a, p. 16), especially among minority racial and ethnic groups because they tend to have greater relationship instability.

Multiple-Household Stepfamilies It is not always easy to determine where people in stepfamilies live. We tend think of families and households as synonymous, but in stepfamilies, it is not unusual for members of the same family to live in different households or for family members to move back and forth between households. One of the most common scenarios is for children living with their mothers to visit their fathers on holidays, weekends, and during the summer months. About one-third of children with a nonresident parent see him or her weekly (Stewart 2010). Because most nonresident parents eventually cohabit and/or marry, these visits often include stepparents, an issue discussed later in the chapter. However, the complexities of combining households can sometimes result in couples with stepchildren adopting unique living arrangements, such as living apart together (LAT), discussed in Chapter 6 (Green 2010).

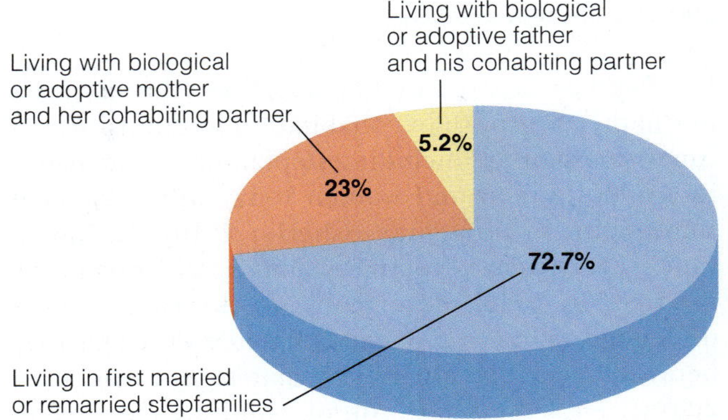

FIGURE 15.2 U.S. children under age 18 living in stepfamilies, 2009.

Source: Calculated from Table 2 and Figure 7 of Kreider and Ellis 2011 based on the Survey of Income and Program Participation (SIPP), 2008 Panel, Wave 2 Topical Module.

Stepfamilies with Adult Stepchildren Most studies of stepfamilies focus on stepfamilies with young and/or adolescent children. However, as discussed in Chapter 9, family relationships do not simply disappear when children reach their eighteenth birthday. There are roughly one million children 18 years of age and older living with a stepparent, representing 6 percent of all adult children (Kreider 2003). This figure vastly underestimates adults in stepfamilies because most do not live with their stepparents, nor does this figure include adult stepchildren with step-siblings and step-grandparents.

There are two modes of entry into a stepfamily with adult stepchildren: (a) the "aging" of stepfamilies that were formed when the children were young, and (b) the parents of adult children forming unions with new spouses and partners. The number of such families is increasing for two reasons. First, the American population is aging, meaning that a growing percentage of Americans are reaching retirement age and becoming elderly. A substantial number of older Americans, especially baby boomers (the roughly 80 million people who were born between 1946 and 1964), have experienced divorce, remarriage, and stepfamily life. The stepfamilies formed by this generation are aging along with the population, resulting in an unprecedented number of older Americans who are stepparents (Stewart 2007). These trends will continue as the children and stepchildren of the "divorce generation" become adults themselves and are in the process of forming (and sometimes, dissolving) their own families. Second, the divorce rate has risen among older adults (Brown and Lin 2012), as has cohabitation (Vespa 2012). The remarriage rate among older Americans remains high (Kreider and Ellis 2011b). These trends are producing an increasing number of stepfamilies formed later in life.

Race/Ethnic Diversity in Stepfamilies Increasing racial and ethnic diversity in American society, and increasing awareness of and attention to this diversity, is also influencing our understanding of stepfamily life. Unfortunately, few studies have examined stepfamilies of races and ethnicities other than nonHispanic white, and to a lesser extent, black. Although minorities such as Hispanics and Asians are increasing rapidly as a proportion of the population, even less is known about stepfamily life among these groups (Grady 2009). One explanation for our lack of knowledge is that race/ethnic minorities tend to be underrepresented in stepfamilies formed through divorce and remarriage (the type of stepfamily researchers have studied most). It is much more common for blacks and Hispanics to enter stepfamilies though nonmarital childbearing and cohabitation than divorce and remarriage (Sassler 2010; Smock and Greenland 2010; Sweeney 2010). In fact, studies that account for this indicate that stepchildren are disproportionately African American (Kreider 2003; Kreider and Fields 2005). One rare study of Hispanic stepfamilies indicated the importance of considering the effects of immigration and intergenerational relationships on stepfamily dynamics (Coltrane, Gutierrez, and Parke 2008).

Stepfamilies with Gay and Lesbian Parents Stepfamilies with gay or lesbian parents are formed when a man or a woman with children from a previous relationship (heterosexual or same-sex) forms a romantic union with a partner of the same sex. The new partner of the children's parent becomes the children's stepparent in the same way that heterosexual partners do. In the case of gays and lesbians, however, it is important to be clear that to be a stepfamily, the children must be from a previous relationship of one of the partners. Gays and lesbians who together adopt or bear a child (who is biologically or legally related to just one partner) is not a stepfamily because that child is a product of the current relationship.

Stepfamilies with gay or lesbian parents are absent from most estimates of stepfamilies. In general, gay and lesbian parents tend to be hidden because many do not disclose their sexual identity, sometimes even to their children (Lynch 2000). According to the 2010 census, 16.7 percent of same-sex couple households contained children from a previous relationship or the current union (Lofquist et al. 2012). Data from the 2010 American Community Survey indicates that among same-sex couples with children, 27.2 had step- or adopted children (stepchildren were not counted separately; Lofquist 2011). These figures underestimate gay and lesbian stepfamilies because many gay and lesbian couples have children and stepchildren who reside elsewhere (Berger 2000). Table 15.1 shows how researchers used to define stepfamilies compared to how they define them today.

Other Types of Stepfamilies There are other variations in the structure of stepfamilies. For example, studies of stepfamilies usually don't include people who *used to be* stepparents or *used to be* stepchildren and who no longer are. This scenario is not infrequent given the instability of remarriage and cohabitation. The 2009 SIPP indicates that nearly 300,000 U.S. children reside with just a stepparent and no biological parent (Kreider and Ellis 2011a). Studies often don't consider the special case of stepchildren who have been legally adopted by

Table 15.1 A New Model of Stepfamilies

The social and demographic trends described in this section have important implications for the way that stepfamilies are defined. This table compares the traditional definition of a stepfamily to a "revised" definition that incorporates these trends.

Assumption	Traditional	Revised
Union type	Remarriage	First marriage, remarriage, cohabitation
Residence of children	Co-resident; static	Co-resident and non-resident; dynamic
Stage of family life cycle	Childrearing; children ages 0–18	Parenting across the life course (includes children ages 18+)
Race/ethnicity	White	White, African American, Hispanic, etc.
Social class	Middle class	All classes (lower, middle, upper)
Sexual orientation	Heterosexual ("straight")	Heterosexual or homosexual (gay/lesbian)

Source: Stewart 2007.

their stepparents (Stewart 2010). Finally, a child may be a member of a stepfamily but not be a stepchild him or herself, if that child is a product of the new union. About five percent of households with children contain both stepchildren and the biological child(ren) of the couple, referred to as a *shared* or *mutual child* (Kreider 2003). These are children whose family is technically "intact" (i.e., they have both biological parents in the home) but who have half-siblings from their parents' previous relationships. The remainder of this chapter explores what family life is like for stepfamily members. We begin with a look at how stepfamilies are perceived in society.

Perceptions of Stepfamilies: Stereotypes and Stigma

Stigmatization refers to people being subjected to negative labels, stereotypes, and cultural myths that portray them as deviant and harmful simply as a result of having certain social characteristics (Coleman, Ganong, and Fine 2000). Stepfamilies are *stigmatized* in that they are perceived as being less functional and desirable than original two-parent families. Professor of social work Irene Levin (1997) has argued that "the nuclear family has a kind of model monopoly when it comes to family forms" (p. 123). According to this **nuclear-family model monopoly**, the first-marriage family is the "real" standard for family living, with all other family forms seen as deficient alternatives.

A study of movie plot summaries showed that stepparents were portrayed negatively more than half (58%) of the time and were otherwise portrayed as neutral rather than positive (Claxton-Oldfield and Butler 1998). The media has perpetuated the idea that stepparents are physically or sexually abusive (Claxton-Oldfield 2008). But data from the UK New Stepfamily Study suggest that mothers in stepfamilies spanked the children much less frequently than did mothers in biological-parent families (Smith 2008). Stigmatization extends to stepchildren as well as stepparents. In one study, 211 university students were asked to examine an eight-year-old boy's report card. All the students saw the same report card, but some were told that the child lived with his biological parents whereas others were told that he lived with his mother and stepfather. Asked about their impressions of the boy, male (but not female) students rated the stepchild less positively than the biological child with respect to social and emotional behaviors (Claxton-Oldfield et al. 2002).

One self-help book, *The Blended Family: Achieving Peace and Harmony in the Christian Home* (Douglas and Douglas 2000), compares stepfamilies to "sinners" when it advises that "blended families be given encouragement, support, and teaching *just as the drug addict, murderer, fornicator, adulterer, and other sinners*" (quoted in Coleman and Nickleberry 2009, p. 556, italics added by Coleman and Nickleberry). Some religions, such as Catholicism, do not recognize a remarriage after divorce unless the first marriage has been annulled (Hornik 2001). Cultural variation in the stepfamily concept can create difficulties for stepfamily members and programs designed to help stepfamilies. For instance, the facilitators of one such program had trouble recruiting Hispanics because "stepfamily" is not a recognized term in the Spanish language (Reck et al. 2012). There was no word for stepfamily in Japanese, until a new Japanese word *suteppufamiri* was created, adopted from the English word "stepfamily" (Nozawa 2008). The word *stepchild* is often used pejoratively to refer to a person who or thing that is treated shabbily. An Internet search of the phrase, "poor stepchild of," by the authors yielded 681,000 hits. For example, in a town hall meeting jointly sponsored by the Canadian Psychiatric and Canadian Medical Associations, mental health was referred to as "the poor stepchild of medical care"(Rich 2011).

Some stepfamilies face multiple stigmas stemming from their race, ethnicity, or sexual identity. One African American stepfamily illustrates the "double stigma" that black stepfamilies experience:

> **Interviewer:** Does living in a black stepfamily make your experience special in any way?
>
> **Donnamae (biological mother):** It certainly does. Being black is difficult enough in this society. Being a stepfamily is not an easy experience either. Being both presents a serious challenge to your skills of dealing with unpleasant reactions from teachers, physicians, and practically everybody else. It, excuse my cynical expression, may color one's life in black. Sometimes it takes a great deal of patience and humor to deal with all the ignorance and attitudes that you have to take.
>
> **Malcolm (stepfather):** Most of the occasions on which I feel the burden occur because of other people's reactions... mostly white people's reactions. Raising other people's kids is no big deal for me because this is not an unusual thing in my family. It is done all the time. When I was very young, my father was placed in Europe with the army and my mother joined him for a year, leaving me to live with my aunt and uncle. It was kind of fun growing up with my cousins. Later on, when I was an adolescent, my older sister moved in with us because she got into some marital and psychological trouble, so in a way I helped raise my niece and nephews. Currently, all of them still live with my parents. So when Lorenzo came into my life it was not an unfamiliar experience... and it is pretty common in my community too. (Berger 1998, p. 145)

Negative stereotypes associated with stepfamilies influence society's appraisal of stepfamily functioning (Jones and Galinski 2003). Negative stereotypes can be held by members of stepfamilies themselves (Planitz and Feeney 2009). However, stepfamily members who believed no or very few myths about stepfamilies had higher optimism about the remarriage and higher family, marital, and personal satisfaction (Kurdek and Fine 1991). Therapist Anne Bernstein has urged that we work toward "deconstructing the stories of failure, insufficiency, and neglect" and, instead, "collaboratively reconstruct stories that liberate steprelationships" from this legacy (1999, p. 415). Ganong and Coleman (1997) identify another issue. They say that in addition to being regarded negatively, stepfamilies are simply disregarded and are ignored by society. The lack of consideration of stepfamilies' special needs and unique concerns might have a negative effect, but the effect needs to be addressed with research.

Choosing Partners the Next Time

For people who have been divorced or who have children from previous relationships, forming new partnerships is serious business. Parents have to consider their children's needs in relation to their own needs for companionship. Moreover, finding a partner who accepts children as part of relationship can be challenging. Finally, if wedding planning weren't difficult enough, couples with children have to consider the expectations and desires of children and in-laws who may be ambivalent about the marriage. As discussed in the next section, parents navigate this unchartered territory with varying degrees of success.

Dating with Children

Courtship among people who have been previously married has not been a major topic of research. Nevertheless, counselors note that people who ended troubled first marriages through divorce may experience personal conflicts that they need to resolve before they can expect to fashion a supportive, stable second marriage (Dupuis 2007). Counselors advise waiting until one has worked through grief and anger over the prior divorce before entering into another serious relationship (Marano 2000; see also Bonach 2007). Nevertheless, most men and women remarry relatively quickly after divorce, usually between three and four years (Kreider 2006).

Meanwhile, dating before remarriage may differ in many respects from traditional dating. Courtship may proceed much more rapidly, with individuals viewing themselves as mature adults who know what they are looking for. Others are more cautious, with the partners feeling wary of repeating an unhappy marital experience. Dating websites, such as Match.com and eHarmony are efficient tools for single moms and dads to identify suitable partners. Members can simply select the characteristics they are looking for in a mate, including their desires for children.

Dating may include outings with one or both partners' children, or couples may prefer to keep their dating relationships and home lives separate. The fact that a couple is dating might not even be known to the children, who may be led to believe that their mother or father's new dating partner is simply a "good friend" who likes to come over and do things with them (Marsiglio 2004), especially with new sexual relationships. Sociologist William Marsiglio's (2004) interviews with stepfathers reveal the great lengths couples will go through to keep their sex life secret. He explains the strategy one couple used to keep the child from finding out the mother's boyfriend had stayed over: "Juan would wake up early and walk outside and knock on the door to wake up the child and signal to him that he had just arrived early that morning. This little trick apparently worked" (p. 36). Marsiglio claims that, "A child's resistance to parents' dating is one of the most difficult scenarios faced by divorced, dating parents" (p. 34). Couples with children who are dating often struggle to determine the "right time" to introduce children to dating partners, and when to introduce children to each other.

Overall, "a common courtship pattern is as follows: (a) male partner spends a few nights per week in the mother's household, followed by (b) a brief period of full-time living together, followed by (c) remarriage" (Coleman, Ganong, and Fine 2000:1,290). The fact that the majority of stepfamilies begin as cohabiting relationships is a complicating factor (Cherlin 2009a). Similar to remarital courtship, not much is known about the process through which individuals with children choose cohabiting partners. Some research suggests that unmarried mothers often "partner up," forming subsequent cohabiting relationships with men who have better financial stability and fewer behavioral problems than a prior partner (Bzostek, Carlson, and McLanahan 2007; Graefe and Lichter 2007). As discussed in Chapter 1, we sometimes "slide" into a family situation, rather than making a conscious decision (Stanley 2009). Couples with previous children can "drift" into cohabitation, with new partners spending increasing amounts of time in the home (Manning and Smock 2005). Marsiglio (2004, p. 33) states, "life together as a stepfamily seldom begins on a given day; it evolves from the bits and pieces of experience woven into the fabric of their dating relationship."

What Kinds of People Become Stepparents?

People who remarry and/or become stepparents are not randomly selected. Foremost, they are either divorced themselves or have children from a previous relationship, or they are willing to form a partnership with someone who is divorced or who already has children. Children are generally a liability in the marriage market, especially for women (Upchurch, Lillard, and Panis, 2001). On the other hand, men with children (either co-resident or nonresident) are significantly more likely than childless men to cohabit or marry a woman with children (Goldscheider and Sassler 2006; Stewart, Manning, and Smock 2003). Research based on the National Survey of Families and Households indicates that for men, marrying "someone with children" is less desirable than marrying "someone of a different race," "someone 5 years older," "someone of a different religion," or "someone who earns much more/much less than you" (Goldscheider and Kaufman 2006). Only marrying "someone unlikely to hold a steady job" ranked lower. Women were less adverse to children than men but this characteristic still ranked low. For both men and women, willingness to marry a partner with children was stronger among people who were older, were less educated, were white, had more egalitarian gender role attitudes, were from a nonintact family, and who had been married or who had children themselves.

Several studies based on nationally representative data indicate that remarried spouses and stepparents have less desirable characteristics than first-married spouses and spouses with no prior children. Different studies indicate that they tend to be less integrated with parents and in-laws; are more willing to leave the marriage; are more likely to exhibit risky and immature behaviors; have lower occupational status, education, and income; have greater disparity in age between partners; are more likely to come from different religions; and are more likely to have been divorced (Killewald 2013; Schramm et al. 2012; Stewart 2007). Yet, with divorce, nonmarital childbearing, and cohabitation now so pervasive in society, the uniqueness of repartnering adults may be declining (Smock 2000).

Like marriage, remarriage has physical, emotional, and financial benefits for both men and women, with the benefits somewhat greater for men (Dewilde and Uunk 2008; Dupre and Meadows 2007; Hughes and Waite 2009; Williams 2003). For women, the advantages are mostly financial. Page and Stevens (2004) used The Panel Study of Income Dynamics to track the financial well-being of children whose parents divorced. Among children whose primary parent did not remarry within six years of divorce, family income was 45 percent lower than it was pre-divorce. Among those who had remarried within six years, family incomes were 9 percent *higher* than their pre-divorce levels. Cohabitation also improves a divorced family's economic status, but not as much as does remarriage (Morrison and Ritualo 2000).

While men to do not, on average, benefit as much financially as do women, there are still many advantages to recoupling for men. In general, single and divorced men have less healthy lifestyles than do married men. On average, they drink more alcohol, are more likely to smoke cigarettes, eat less balanced meals, and go to the doctor less often (Waite and Gallagher 2000). However, the extent to which *re*marriage is associated with healthier behaviors among men is not yet known. It is possible that women with children from previous relationships may have less time and energy to devote to their spouse.

Second Weddings

Americans tend to have rigid notions of what a wedding is supposed to look like. In her book *White Weddings*, sociologist Chrys Ingraham (2008) describes the very homogeneous way that the media portrays weddings—white, middle-class, heterosexual, and the *first* marriage of both partners. The modern wedding is not playing into this stereotype however and is in fact "celebrating" growing diversity in marriage pool. The wedding industry estimates that remarried couples are responsible for about 40 percent of all their revenues (Ingraham 2008). Remarriage weddings are different from first weddings—they tend to be smaller, the couple is older, and most involve children from previous relationships. The presence of children from prior relationships can encourage a couple to marry as opposed to carrying on an ongoing dating or cohabiting relationship (Nozawa 2008). Remarried couples are more likely to get married away from home, and are thought to be fueling the trend toward "destination weddings," which involves flying the wedding party and guests to an exotic location.

Remarried wedding ceremonies are complicated, emotionally-charged, and often awkward affairs that need to be handled delicately. University of Iowa communication studies professor Leslie Baxter and colleagues asked thirty male and fifty female stepchildren from two Midwest colleges to describe their parents' remarriage ceremonies (Baxter et al. 2009). The wedding ceremonies the students described ranged from a downsized version of the traditional wedding to a courthouse civil ceremony to an informal event in a backyard or casual restaurant. Sometimes a couple left the area to remarry, and then informed family and friends later.

Remarrying couples differ from first-marrying couples in their degree of homogamy because choosing a remarriage partner differs from making a marital choice the first time inasmuch as there is a smaller pool of eligible on any given attribute. People in their thirties and forties meet others at work or in other settings that bring people together from more varied backgrounds.

Many of the students were critical of their stepparent's wedding ceremony. Some argued that "the relationship didn't deserve that" degree of celebration. Other students critiqued the ceremony as being too casual: "Well, if you want everyone to take [your remarriage] seriously, it needs to be a little more than um, a barbeque" (Baxter et al. 2009, pp. 476, 477). A male student said he felt that the ceremony had inadvertently insulted his family of origin:

> The only part that upset me was the pastor was talking about how life's events lead you up to this moment and how there's bumps in the road, and blah, blah, blah, but this is where you're supposed to be. And I got pissed, because I was like, was my mom the bump in the road? (p. 480)

The children of remarrying couples are likely to have mixed emotions about the event. Some couples design a family-centered ceremony paying particular attention to their children's feelings. In this case, not only the bride and groom but also all members of the new stepfamily are celebrated. For example, one of Baxter's respondents described how all of the stepchildren as well as the remarrying couple "got little rings to show that we all got married" (Baxter et al. 2009, p. 475). Although the family-centered ceremony was the least common wedding type described by the students, it was the one most appreciated:

> Over and over again, participants told us that they had wanted far greater involvement with the remarriage event... [whether] it was being granted sufficient time to get to know the stepparent, being informed and consulted about the decision to marry, participating in the planning of the ritual, or creating a ceremony and/or artifacts that celebrated the family. (Baxter et al. 2009, pp. 481, 485)

Because the wedding ceremony can influence how children feel about their future stepfamily, involving them may be more important than a couple realizes. Stepchildren's adjustment is a factor in a remarried couple's overall happiness. A 20-year-old respondent offered the following advice: "If people are thinking about starting a stepfamily, they should take their time and, you know, keep everyone informed and pay attention to everyone's feelings.... Like [a mother should] talk to her daughters and see how we, um, feel, and take those feelings into consideration" (in Baxter et al. 2009, p. 481). Indeed, researchers advise that forming stepfamilies when a stepchild-to-be is in adolescence can be especially difficult (Bray and Easling 2005). On the other hand, remarriages formed when the children are older may be viewed quite positively. One adult stepchild from a British study says this about his mother's remarriage:

> Oh yeah, brilliant together. I mean mum never went abroad with my dad because dad wouldn't go anywhere and they didn't have the money anyway. It's been very good. It's been good for each other, you know.... If something had happened to one of them, I think the other one would just roll up because they go to their clubs four or five times a week. (Allan, Crow, and Hawker 2011, p. 136)

Because the re-wedding ceremony can influence how children feel about their future stepfamily, involving them may be more important than a couple realizes. Stepchildren's adjustment is a factor in a remarried couple's overall happiness.

Happiness, Satisfaction, and Stability in Remarriage

As pointed out elsewhere in this text, marital happiness or satisfaction, and marital stability are not the same. *Marital happiness* and *marital satisfaction* are synonymous phrases that refer to the quality of the marital relationship whether or not it is permanent; *marital stability* refers simply to the duration of the union. We'll look at both ways of evaluating remarriages.

Happiness and Satisfaction in Remarriage

Research on remarried partners' happiness and satisfaction was relatively prevalent through the 1980s and 1990s (Shriner 2009). However, since then, scholars have focused on other topics, such as communication in stepfamilies. What research does exist consistently shows no difference in marital happiness, satisfaction, and other dimensions of marital quality between couples in first versus subsequent marriages (Amato et al. 2007), although remarried couples have been found to have higher levels of *perceived* marital instability (Bulanda and Brown 2007).

We know that wives' satisfaction with the division of household labor is important to marital satisfaction. Some research suggests that there may be more equity, or fairness, in remarriages because housework is more equally divided and because remarried wives tend to have more financial resources than first married wives (Clarke 2005; Snoeckx, Dehertogh, and Mortelmans 2008). On the other hand, a small study based on extensive interviews with fifteen adult stepchildren found "the persistence of traditional gender practices in the parenting and stepparenting of children" (Schmeeckle 2007, p.174). And a recent, small qualitative study of wedding planning among remarrying Canadian brides concluded that their gendered division of labor—at least regarding wedding-planning tasks—was much like that for first marriages (Humble 2009). Other factors associated with marital satisfaction operate differently in second marriages than first marriages. For second marriages, premarital cohabitation had a similar negative effect on marital quality whether or not the couple became engaged before living together. For first marriages, couples who cohabited with their spouse without first being engaged had lower marital quality than couples who had gotten engaged before living together (Stanley et al. 2010).

Interestingly, research with thirty-two black, New York City partners in lesbian stepfamilies—those in which at least one child was from a mother's prior heterosexual relationship—found that in some ways the women followed traditional gender norms (Moore 2008). Although both women in virtually all of the couples were employed, biological mothers tended to do more child care and household chores. The author speculated that gendered division of labor is defined differently among black lesbian stepmothers than among heterosexuals. Among the lesbian couples, "control over some forms of household labor [resulted in] greater relationship power. Biological mothers want more control over the household because such authority affects the well-being of children—children who biological mothers see as primarily theirs and not their partner's" (Moore 2008, p. 344).

Whatever the case regarding gender roles in remarriages and stepfamilies, considerable research shows that remarrieds experience more tension and conflict than do first-marrieds, usually on issues related to stepchildren and specifically discipline (Smith 2008; Stewart 2007). There are also stressors unique to stepfamilies, on top of the economic problems and the everyday pressures of married life, that negatively affect marital quality of remarried spouses (Shramm and Adler-Baeder 2012). Remarital satisfaction is influenced by the wider society through negative stereotyping of remarriages and stepfamilies (Ganong and Coleman 2004). Moreover, very few remarried couples prepare for living in a stepfamily, and most relationship education programs and interventions used by clinicians do not address the unique needs of stepfamilies (Whitton, Nicholson, and Markman 2008). Nevertheless, supportive family communication (see Chapter 11) is important to remarital happiness, and many of the same factors that affect marital quality also affect the quality of remarriages, such as religiousness (Schramm et al. 2012).

The Stability of Remarriages

Remarriages are less stable than first marriages. For example, 40 percent of remarriages formed between 1985 and 1994 ended in separation or divorce within a decade, compared with 32 percent of first marriages (Bumpass and Raley 2007). There are several reasons for this difference. You might recall the discussion in Chapter 6 about how cohabiting—at least serial cohabiting—before marriage generally increases the odds of divorce. Research that analyzed more than 3,000 remarried respondents from National Survey of Families and Households (NSFH) data similarly found that post-divorce cohabitation is positively associated with remarital instability (Xu, Hudspeth, and Bartkowski 2006; but see Teachman 2008a).

The researchers suggested that the *selection effect* (the idea that divorced people who "select" themselves into cohabitation are different from those who don't) largely explained this situation. For one thing, people who divorce in the first place—and those who cohabit—are disproportionately from the lower-middle and lower classes, which generally have a higher tendency to divorce or redivorce.

Maybe you've heard people who are about to remarry say that they plan to work harder in their new marriage and not to repeat the mistakes they made in their first. For many couples, this may be the case (Brimhall, Wampler, and Kimball 2008). Many remarried couples may go to great lengths to avoid conflict. Longitudinal data from the National Survey of Families and Households showed that couples with less marital conflict had higher marital quality, which in turn had a lower risk of divorce, especially among remarried couples with stepchildren as opposed to remarried couples who had only shared children (Van Eeden-Moorefield and Pasley 2008). However, remarried couples who are reluctant to directly address problems that arise in their relationship may fare no better. "The experience of destructive conflict that often precedes the breakup of a first marriage can be highly stressful and ... might prompt avoidance of communication about the difficulties that stepfamily

David Young-Wolff/PhotoEdit

Research consistently shows that stepfamilies are less stable than nuclear families and that, in general, children in stepfamilies have more problems than those in intact, nuclear families. However, research also indicates that the quality of the communication and relationships among family members – and the extent to which children are monitored – may be more important to positive child outcomes than is the family structure itself.

couples have to negotiate" (Halford, Nicholson, and Sanders 2007, p. 480). As pointed out in Chapter 11, researchers and counselors advise directly addressing difficult issues. In one recent Canadian study (Saint-Jacques et al. 2011), researchers interviewed two groups of participants–one group was comprised of stepfamilies who had been together for at least 5 years. The other group consisted of stepfamilies that had dissolved within the first 5 years. The key difference between the two was not the nature, number, and intensity of their problems, but was the *way* in which they approached the problems. Whereas families that stayed together tried multiple solutions, families that did not either avoided talking about the problem or gave up entirely by ending the relationship.

A third reason that remarriages are more likely to end in divorce may be that if seemingly irresolvable problems do arise, remarrieds are, as a category, more accepting of divorce; they have already demonstrated their willingness to divorce. As one remarried husband said, "We're not going to tolerate the kind of crap we did the first time around.... I don't need it again. She doesn't either" (Brimhall, Wampler, and Kimball 2008, p. 378).

Fourth, although stepfamilies' increasing numbers and visibility have led to their growing social and cultural acceptance, remarried families continue to be stereotyped as "less than" or other than "normal" (Ganong and Coleman 2004). One indication of this situation involves re-weddings. As discussed in Chapter 7, weddings publicly announce a couple's commitment. Indicating their culturally diminished importance, re-weddings are typically less extravagant than first weddings, as we have seen. As a result of the relative devaluing of remarriage, remarrieds may receive less social support from friends or extended kin and be somewhat less integrated with parents and in-laws, thus not experiencing the encouragement or social pressures that can act as barriers to divorce (Coleman, Ganong, and Fine 2000).

Finally, a significant factor in the relative instability of remarriages is the presence of stepchildren (Stewart 2005a). There are surprisingly few recent studies of specific risk factors of a second divorce (Sweeney 2010). However, one study found increased risk of divorce among women (but not men) under age 45 who brought their own biological children into a second marriage (Teachman 2008a). In an older study, sociologists Lynn White and Alan Booth interviewed a national sample of more than 2,000 married people under age 55 in 1980 and reinterviewed four-fifths of them in 1983. White and Booth found that the quality of the remarital relationship itself did not affect the odds of divorce, but the partners' overall satisfaction with family life did:

> [W]e interpret this as evidence that the stepfamily, rather than the marriage, is stressful.... These data suggest that... if it were not for the children these marriages would be stable. The partners manage to be relatively happy despite the presence of stepchildren, but they nevertheless are more apt to divorce because of child-related problems. (White and Booth 1985, p. 696; see also Schrodt, Soliz, and Braithwaite 2008)

The previous discussion suggests that when neither spouse enters a remarriage with children, the couple's union is usually very much like a first marriage. But when at least one spouse has children from a previous marriage, family life often differs sharply from that of first marriages. These unique family dynamics are discussed in the next section.

Day-to-Day Living in Stepfamilies

Although the formal marriage or remarriage ceremony is typically considered the "start date" of a stepfamily (even though many stepfamilies start out with cohabitation), stepfamily formation is a process that unfolds over time. Stepfamily members do not become an "instant family," no matter how much they want to. Moreover, stepfamily dynamics will always be different from those of original two-parent families. Below we discuss what it is like to live in a stepfamily and identify factors that affect the development of a stepfamily identity, stepfamily roles, relationships between family members, and changes in those relationships over time.

Challenges to Developing a Stepfamily Identity

A **cultural script** is a set of socially prescribed and understood guidelines for defining responsibilities and obligations and hence for relating to each other (Ganong and Coleman 2000). Society offers members of stepfamilies an underdeveloped script. As discussed above, our traditional wedding "script" assumes the couple is marrying for the first time. Lacking a standard set of guidelines, couples planning a second wedding must make things up as they go. Noting the cultural ambiguity of stepfamily relationships, social scientist Andrew Cherlin thirty years ago called the remarried family an **incomplete institution** (1978). He said that, unlike first marriages, remarriage lacks social norms to guide behavior and therefore remarried couples do not have the tools to solve problems and "get along." Moreover, "cohabiting stepfamilies are arguably even less institutionalized than married stepfamilies, which are formed through a tie that is legally binding" (Brown and Manning 2009, p. 88). The result is that, in stepfamilies, everyday living can be very challenging. In terms of evidence, he cites our language, laws, and customs (see "A Closer Look at Diversity: Do You Speak Stepfamily?").

A Closer Look at Diversity

Do You Speak Stepfamily?

Whereas children in traditional families call their parents "Mom" and "Dad," there is no standard way for stepchildren to refer to their stepparents. Communication researchers interviewed thirty-nine stepchildren at a large Midwestern university (Kellas, LeClaire-Underberg, and Normand 2008). The students described how they purposefully choose language to clarify their family form for others:

> Whenever I talk about [my stepfamily] with people it's always my stepdad, my stepmom, stepsister.... I always put those terms in there because I do have a biological, real sister and so I guess, I try to help people out because obviously my family's really confusing. (p. 251)

Note this respondent's reference to her "real" sister.

In addition to clarifying their family situation for others, the students deliberately used language that normalized stepfamily living for outsiders: "If I am outside the family and people ask me where I am going I say I am going to my mom and dad's house. So face-to-face, I call [my stepmother by her name], but with everybody else, it's just my mom" (p. 249).

The students in this study also described the way that language—interestingly, language associated with the nuclear-family model—symbolized and communicated stepfamily members' closeness and solidarity. One participant reported that as a young child she referred to her stepfather as "Daddy" to acknowledge that she felt close to him. Another student reported overhearing his younger stepbrother talking with his friends about how happy he was to have a new big brother: "[H]e called me his brother.... After I heard that it went from being a stepfamily... to being an actual family" (p. 249). Similarly, in a different study, a stepfather told an interviewer: "I have never, ever, thought of these two girls as my stepchildren. They're just my daughters, and I've always referred to them as such" (in Hans and Coleman 2009, p. 611).

On the other hand, some students reported strategically using language that communicated separateness: "At the very beginning I wouldn't even call her my stepmom. I would call her my dad's wife.... I didn't want that connection" (in Kellas, LeClaire-Underberg, and Normand 2008, p. 250). Other students pointed to consciously negotiated terms by which they referred to or addressed stepfamily members as they balanced relationships in an ambiguous family environment. Fairly common was the decision to call a stepfather by a different term than that reserved for the biological father—referring to a biological father as dad, for example, while calling a stepfather by his first name. One stepdaughter reported that when with her biological father, she "always has to be really careful" to refer to her stepfather as "Paul" although she usually calls him "Dad."

Another student confessed that he consistently addressed his stepfather as "Bill," although he wished he could have called him "Dad," but "it just never came"—even though "he really is the one that raised me" (Kellas, LeClaire-Underberg, and Normand 2008, p. 251). Finally, some members of stepfamilies may not call their step-relatives anything, engage in a language pattern called *no-naming* or *zero forms of address* (Schneider 1968, p. 84; Duvall 1954, p. 7).

Critical Thinking

Have you ever considered how you refer to your family members? What makes a family member "not real" as opposed to "real"?

Boundary Ambiguity in Stepfamilies In a stepfamily, "Whose picture goes on the mantle?" may be a difficult question to answer (Munroe 2009, p. 168). You may recall that family **boundary ambiguity** is a "state when family members are uncertain in their perception of who is in or out of the family or who is performing what roles and tasks within the family system" (Boss 1987, p. 709). Boundary ambiguity is common in stepfamilies. One stepfamily member, a wife and mother who complained about never knowing how much to fix for dinner on any given day, describes her family as an "accordion" that "shrinks and expands alternately" (Berger 1998).

Interviews with stepfamily members reveal that definitions of family often differ between parents and children and between siblings. Susan Stewart (2005a), one of the authors of this textbook, analyzed data from the National Survey of Families and Households. The data included 2,313 stepfamilies, defined as "married or cohabiting couples in which at least one partner has a biological or adopted child from a previous union living inside or outside the household." Stewart defined *boundary ambiguity* as "any discrepancy in partners' reports of shared children (the biological or adopted children of both partners) and/or stepchildren (biological or adopted children from previous unions)" (p. 1,009).

Boundary ambiguity was present among 25 percent of couples with stepchildren and higher among couples with nonresident stepchildren than with resident stepchildren, especially when they were the biological children of the wife. Boundary ambiguity was also more common among cohabitors. Couples with greater boundary ambiguity were found to have significantly lower relationship quality and stability.

Table 15.2 Boundary Ambiguity in Four Family Forms
No boundary ambiguity means that a mother's report of who is in the family coincided with that of her adolescent child.

Family Form	Boundary Ambiguity (%)
Two-biological-parent family	0.6
Single-mother family	11.6
Married stepparent family	30.2
Cohabiting stepparent family	65.9

Source: Based on analysis of data from the National Longitudinal Study of Adolescent Health (N = 14,047), Adapted from Brown and Manning 2009, p. 92.

In a more recent study by sociologists Susan Brown and Wendy Manning (2009), 14,047 adolescents and their mothers were asked to list the members of their families. Brown and Manning compared boundary ambiguity in four family types: (1) families headed by two biological parents, (2) single-mother families,(3) married families in which one parent is a stepparent, and (4) cohabiting families in which one parent is a stepparent. As shown in Table 15.2, cohabiting stepparent families evidenced greatest boundary ambiguity whereas families with two biological parents evidenced the least.

In a rare study of siblings, evidence of boundary ambiguity has been found among step-and half siblings as well (White 1998). Relationships with kin outside the immediate stepfamily are also complex and uncharted (Ganong and Coleman 2004). There are few mutually accepted ways of defining and relating to new extended and ex-kin relationships, but it appears that new relatives do not so much *replace* as *add* kin (White and Riedmann 1992).

The Stepfamily System

All families, including stepfamilies, are part of a *system* of relationships. In other words, relationships between two stepfamily members are affected by relationships between all other stepfamily members.

Family Systems Theory A family systems approach (see Chapter 2) is especially useful for understanding stepfamily relationship dynamics. Family systems theory emphasizes *interdependence* in family relationships, the idea that the relationship between two family members is influenced by each one's relationship with all the other members of the family. This interplay is particularly important in stepfamilies (Afifi 2008). A study of British children found that conflict between the child's biological parent and stepparent was associated with children's negative appraisals of their parenting which in turn was associated with internalizing problems (Shelton, Walters, and Harold 2008).

Because stepfamily members are often uncertain about how they should behave toward one another, they tend to look to other family members for cues. For example, a new stepmother may look to her husband for guidance about how best to get the children to eat their vegetables. The strategy she employs will depend on her husband's relationships with the children and their relationships with each other. Perhaps they will decide that the husband will be the one to say something to them. As the children grow older and as the family becomes more comfortable in their relationships, she will need her husband's guidance less. One of the biggest mistakes stepparents make is to try to actively parent, especially with respect to discipline, too soon (Stewart 2007). Having the biological parent act as a "mediator" between the stepparent and stepchildren, at least initially, is a wise choice.

Triadic Communication Family dynamics can sometimes become set. In one study, researchers asked fifty university student stepchildren to describe the communication patterns in their stepfamilies (Baxter, Braithwaite, and Bryant 2006). As illustrated in Figure 15.3, the results showed four different relationship/communication patterns among a biological parent, stepparent, and child. In a *linked triad,* a child's interaction is connected with the stepparent through the child's biological parent: "A lot of stuff I communicate with my stepdad goes through my mom" (Baxter, Braithwaite, and Bryant 2006, p. 389). In the *outsider triad,* the child and the biological parent maintain interaction, but the stepparent remains an outsider and pretty much irrelevant to the child's life: "I focus my talking towards my mom. And um like when we're [all] watching the TV, I just always turn to my mom and usually talk to her" (Baxter, Braithwaite, and Bryant 2006, p. 391).

In the *adult-coalition triad,* the child views the biological parent and the stepparent as maintaining a couple relationship that ignores the child's concerns. As an example:

> My sister and I have some difficulties with my stepmom .... And [my dad will] say, "Well, honey, you have to understand this and this and this," and like makes excuses for her, and I kind of want to say to him, "Well, we're your *daughters,*" you know. (Baxter, Braithwaite, and Bryant 2006, p. 392, italics in original)

In the *complete triad,* communication flows freely, involving all stepfamily members equally: "I'll call home and I'll talk to my stepfather and I'll talk to my mom .... He's like a father to me" (Baxter, Braithwaite, and Bryant 2006, p. 393).

Unanimously, the student respondents regarded the adult-coalition pattern as negative and saw the complete triad pattern as ideal. However, the researchers hastened to add that "the outsider triad can be functional

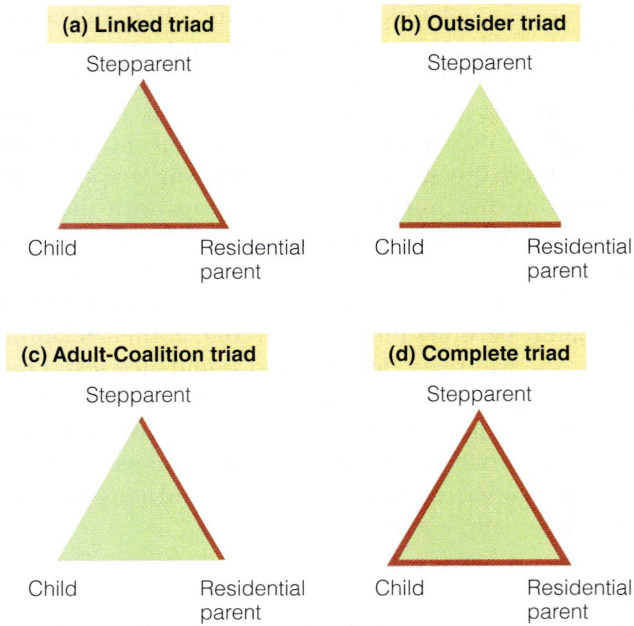

Figure 15.3 Perceived types of triadic communication structures in stepfamilies. "A darker line represents the presence of a direct, positive ... line of / in a given stepfamily dyad, and a lighter line represents the absence of a direct, positive ... line of communication" (Baxter, Braithwaite, and Bryant 2006, p. 388). A=Linked Triad. Communication between child and stepparent is linked though the residential biological/adoptive parent. B=Outsider Triad. Child's communication takes place primarily with the residential biological/adoptive parent; child feels little interdependence with the stepparent. In this triad, the stepparent is an outsider. C=Adult-Coalition Triad. Stepparent and residential biological/adoptive parent are viewed as forming a coalition to the relative exclusion of the child. D=Complete Triad. Equal communication and interaction along all three sides of the triad, incorporating all three triadic components equally.

Source: Baxter, Braithwaite, and Bryant 2006, p. 388

for the stepfamily so long as it is compatible with expectations of the stepparent" (Baxter, Braithwaite, and Bryant 2006, p. 395). Nonetheless, creating stepfamily cohesion is important, although it can be difficult. One obstacle involves the fact that stepparent, stepchild, and stepsibling relationships are involuntary. Those involved may feel that they had little choice in the matter and hence be disinclined to cooperate (Schrodt, Soliz, and Braithwaite 2008).

Then too, disruptions associated with one or more stepchildren's comings and goings according to a visitation schedule may be stressful (Kheshgi-Genovese and Genovese 1997). Stepchildren in joint custody arrangements (see Chapter 14) may regularly move back and forth between two households with two sets of rules. A child's ties with the noncustodial parent can make the pre-divorced or separated family seem "more real" than the stepfamily. After returning from visits with the noncustodial parent, a stepchild may unintentionally undermine stepfamily definitions. As an example, a five-year-old visited her biological mother's house where she mentioned a "brother" in her stepfamily. When the girl returned to her stepmother's home, she announced, "My mother says he's not my brother, he's my half-brother" (Bernstein 1997).

Furthermore, stepsiblings may not get along well. Especially in the case of multipartnered parenthood, the lives of stepchildren living in one household can vary greatly:

> one child may have a devoted nonresident father who sees her regularly, another child who has no contact with her father jealously watches her half sister go away for weekends with her dad, and a third child—from the new partnership—has both of her parents in the household .... The inequalities among children in the same household can be stark. (Cherlin 2009a, p. 195)

In some cases stepsibling rivalry is sparked by actual or perceived inequality of treatment by one parent. As one stepchild explained,

> The way my brothers and I saw it was that my dad treated us the same and he tried to treat her children the same, but we saw a difference in the way she treated her children and the way she treated us .... My brothers and I were being treated differently than her children were. (in Baxter, Braithwaite, and Bryant 2006, p. 390)

Moreover, as noted by psychotherapist Susan Pacey (2005):

> Grandparents are powerful figures in the hierarchy of the stepfamily, and can help or hinder the couple in forming a new life together. For example, gifts or bequests made to the biological grandchildren only, when a long established stepfamily home includes step or half siblings, may prove divisive and detrimental to the stepfamily and the couple. (p. 368)

Another challenge to fashioning cohesive stepfamilies lies in children's lack of desire to see them work out. After age 2 or 3, children often harbor fantasies that their original parents will reunite (Bray 1999; Gamache 1997). This hope can last into young adulthood. As one college student described hearing of her parent's remarriage intentions,

> It didn't hit me until I hung up the phone and then I remember just crying and I couldn't understand why I was crying. You know? And I think it had just hit me that my parents are never going to get back together .... I had never thought that [their] getting divorced was the end of them. (in Baxter et al. 2009, p. 480)

In the case of remarriage after widowhood, children may have idealized, almost sacred, memories of the parent who died and may not want another to take his or her place (Andersen 2002; Barash 2000).

Thus, although it is common to think of children as the "receptors" or even "victims" of family dynamics, children also shape their interactions with adult family members and adult family members' relationships with each other (Menning 2008). Whereas original two-parent families might be thought of as operating from the "top-down" (from parents to children), stepfamilies are thought to operate from the "bottom-up" (from children to parents). Rosenberg and Hajfal (1985) refer to these as **dripolator** and **percolator effects**. Several classic studies suggest that children have a greater impact on the marital relationship and overall family happiness in stepfamilies than in original two-parent families (Bray, Berger, and Beothel 1994; Crosbie-Burnett 1984; Hetherington and Jodl 1994; White and Booth, 1985). One stepparent laments, "We didn't have a very long romance or honeymoon period. In fact, we never had a honeymoon period ... when [stepchild] came home, the romance went out of our marriage" (Felkner et al. 2002, p. 131).

Stepfamily Roles

Incomplete institutionalization of stepfamilies includes stepfamily roles themselves. As we have seen, there is no cultural script to show members of stepfamilies how to play their roles (Ganong and Coleman 2004; Speer and Trees 2007; Wooding 2008). Given this **role ambiguity**—that is, few clear guidelines regarding what responsibilities, behaviors, and emotions stepfamily members are expected to exhibit—it may not be surprising that relationship and communication patterns in stepfamilies are highly variable.

Lower role ambiguity has been associated with higher remarital satisfaction, especially for wives, and with greater parenting satisfaction, especially for stepfathers (Kurdek and Fine 1991; Munroe 2009). However, the role of the stepparent is "precarious; the relationship between a stepparent and stepchild only exists in law as long as the biological parent and stepparent are married" (Beer 1989, p. 11). Although some stepchildren certainly do maintain relations with a stepparent after a stepparental divorce, one must forge personalized ways to do so in the absence of commonly understood norms (Dickinson 2002).

One result of role ambiguity is that society seems to expect stepparents and children to love each other in much the same way as biologically related parents and children do. In reality, this is not often the case, and therapists point out that stepparents and stepchildren should not expect to feel the same as they would if they were biologically related (Barash 2000). For example, it is a good idea for a stepparent to wait awhile before becoming an active disciplinarian (Ganong and Coleman 2000).

Stepparents exhibit wide variation in their relationships with their stepchildren. At one extreme, there are stepparents who legally adopt their children, have their stepchildren legally take their last name, or pretend to be their child's biological parent (Stewart 2010; Filinson 1986). At the other end of the spectrum are stepparents with little or no involvement in their stepchildren's lives, or who are abusive (Daly and Wilson 1994). Among stepfathers living with minor stepchildren who were part of the National Survey of Families and Households, whereas one-third felt that it is *definitely true* or *somewhat true* that a stepparent is "more like a friend than a parent," half of the stepfathers felt that this statement was *somewhat false* or *definitely false* (Marsiglio 1992). A little over half (55%) of stepfathers said that it was *somewhat true* or *definitely true* that "having stepchildren is just as satisfying as having your own children," yet one-quarter (27%) said that it was *somewhat true* or *definitely true* that "it is harder to love stepchildren than it is to love your own children." A more recent study of stepfathers by the same author (Marsiglio 2004) reveals a similar ambivalence with respect to **paternal claiming**, the extent to which to stepfathers see their stepchildren as biological children.

In general, stepparents take a less active role in the lives of their children than do biological parents (Cooksey and Fondell 1996). This pattern, at least early on, is not necessarily bad. In one study, stepfathers and their spouses tended to agree that a low level of involvement of the stepfather was good (Kurdek and Fine 1991) and that too much involvement too soon can result in negative backlash from the stepchild (Hetherington and Kelly 2002). Other research indicates that the effect of stepparental time on stepchildren's well-being (such as in the case of academic achievement) is small, especially for older children (Hofferth 2006). The quality of the stepparent-stepchild relationship is more important than quantity of time together in determining their degree of influence on their stepchildren (e.g., Carlson and Knoester 2011). For example, among adolescents from the NSFH, those who report higher quality stepparent-stepchild relationships had fewer emotional and behavior problems (White and Gilbreth, 2001), and adolescents who report a closer relationship to stepparents (both stepmothers and stepfathers) had significantly higher self-esteem (Berg 2003).

The Stepfather Role

Many children have positive relationships with their stepfathers and men who may eventually become their stepfathers. At one end of the continuum are men who should be considered stepparents because they are deeply involved in the children's lives and likely to marry the mothers after a period of living together. At the other end are mothers' short-term romantic liaisons who may have little to do with the children and may be around only for a few months.

In between is a gray area of men in the household who are not taking on a parental role but who do spend some time with the children and may be present for a year or two (Cherlin 2009a, p. 101).

Jesse, speaking about his girlfriend's seven month old child, said this:

> We played with him for a day at the park and it was a blast! It was fun.... It was just a lot of fun going to the park with Shaun and everything. It was neat.... I was like—oh my goodness, this thing is—he's awesome.... I was still kind of uneasy because I wasn't around kids a lot... but it was a really fun day for both of us.... (Marsiglio 2004, p. 41)

There is more ambiguity in the stepfather role among stepfathers who are cohabiting as opposed to married (Manning, Smock, and Bergstrom-Lynch 2009). These men are less involved with their stepchildren and their relationships are of shorter duration (Stewart 2007). In the absence of marriage, stepfather involvement tends to drop sharply after relationships between romantic partners end (Tach, Mincy, and Edin 2010).

Men who decide to marry women with children come to their new responsibilities with varied emotions, typically quite different from those that motivate a man to assume responsibility for his biological children. "I was really turned on by her," said one stepfather of his second wife. "Then I met her kids." This sequence is a fairly common "situation of many stepparents whose primary focus may be the marriage rather than parenting" (Ceballo et al. 2004, p. 46). Along with feeling positive about what he is undertaking, a new husband may be anxious, fearful, or ambivalent.

To enter a single-mother family, a stepfather must work his way into a closed group—a reason that many stepfathers "tend to be marginalized in households where mothers are regarded as the disciplinarians" (Pacey 2005, p. 366). Furthermore, the mother and children share a common history, one that does not yet include the stepfather. In addition, many stepfathers must construct their stepparenting role parallel to that of the nonresidential, but still involved, biological father (Cheadle, Amato, and King 2010; King 2009).

Children and their stepfathers are more likely to feel positive about their relationship when the stepfather assumes a parental identity, when his parenting behavior meets his own and other family members' expectations, and when his parental demands are not contested by an involved, nonresidential biological father (Coleman, Ganong, and Fine 2000; MacDonald and DeMaris 2002; Marsiglio 2004). For teens in remarried families, close ties to stepfathers are more likely to develop when the adolescent has close ties to his or her mother before the stepfather entered the family. Moreover, stepchildren do not have to "choose" between their stepfather and biological father; they can have close relationships with both (King 2009). Although less is known about stepfather-stepchild relationships in minority stepfamilies, in one study, relationships between Mexican American stepfathers and stepchildren and white stepfathers and stepchildren were found to be remarkably similar (Coltrane, Gutierrez, and Parke 2008).

The Stepmother Role There has been far less research on mothers and stepmothers in stepfamilies than fathers and stepfathers (Sweeney 2010). Decades of research indicates that the stepmother role and their relationships with stepchildren are more problematic than those of stepfathers (Stewart 2007). Stepmothers are given what many consider an impossible job. This has been described as the **stepmother trap**. On the one hand, mothers in our society are expected to be self-sacrificing and completely devoted to their children (Arendell 2000). On the other hand, stepmothers are often stigmatized—seen and portrayed as cruel, vain, selfish, competitive, and even abusive (Schrodt 2008; Whiting, Smith, and Barnett 2007). In reality, the majority of adolescents report feeling close to their stepmothers (King 2007).

Maureen McHugh (2007) is a stepmother who writes for the website Second Wives Café: Online Support for Second Wives and Stepmoms (secondwivescafe.com). The following is an excerpt from her online article "The Evil Stepmother":

> My nine-year-old stepson Adam and I were coming home from Kung Fu. "Maureen," Adam said—he calls me "Maureen" because he was seven when Bob and I got married and that was what he had called me before. "Maureen," Adam said, "are we going to have a Christmas Tree?"
>
> "Yeah," I said, "of course." After thinking a moment, "Adam, why didn't you think we were going to have a Christmas Tree?"
>
> "Because of the new house," he said, rather matter-of-fact. "I thought you might not let us." It is strange to find that you have become the kind of person who might ban Christmas Trees.

A small study asked 265 stepmothers about their expectations about the stepmother role. The researchers found that stepmothers expected to be included in stepfamily activities but certainly do not see themselves as replacing the stepchild's mother. The more time a stepmother spent with her stepchildren, the more she expected to be included in stepfamily functions and decisions, and the more she behaved as concerned parent, rather than as a friend (Orchard and Solberg 2000).

Another important variable is whether or not the stepmother lives with her stepchildren. Many nonresident stepmothers do not think of themselves as a "stepmother" and are resistant to using the term (Allan, Crow, and Hawker 2011). Orchard and Solberg's (2000) survey of stepmothers who were members of the *Stepfamily Association of America* found that stepmothers who had stepchildren who spent the majority of their time in their home had higher expectations for functional inclusion in the family (e.g., feeling welcomed into the stepfamily, share equally in discipline), parental love, household responsibility, and mother replacement (e.g., not competing for affection, not being a wicked stepmother) than stepmothers who spent less time with their stepchildren. Smith (1990) argues that "part-time stepmothering can sometimes be even more difficult and stressful than full-time; the role of the part-time stepmother is extremely ambiguous" (p. 3). Indeed, in another study of full and part-time stepmothers, 30 percent thought that "spending more time together" would help improve their relationship with their stepchildren (Quick, McKenry, and Newman 1994).

Another part-time stepmother tells about her experience:

> In the early stages I [had feelings of hatred toward the children] and I felt ashamed of the hatred in myself. I did not want the children to be hurt by it. I used to go off until I could control the hatred better. I felt everyone hated me. I felt mean! But as time goes by I have learned that I don't need to invest so much.... They have two very good parents already.... I can have closeness and fun without the whole burden. But it is muddling and difficult to get the balance right. (Smith, 1990, p. 20)

Nonresident stepmothers' relationships with their stepchildren may also depend on whether or not they have had a child with their spouse or partner. Stepchildren influence stepparents' intentions to have additional children, even if the stepchildren live elsewhere (Stewart 2002). Adding a new biological child may help nonresident stepmothers feel more secure in their role and allowed them to better tolerate their stepchildren's visits (Ambert 1989). Yet, some nonresidential stepmothers have reported that their partner's visiting children are bad influences on their own, biological children (Henry and McCue 2009).

Stepmothers who live with their stepchildren may face somewhat different challenges. Social worker and stepmother Emily Bouchard tells her story:

> When I moved in with my husband and his two teenage daughters, he had a real "hands off" approach.... Sparks began to fly as soon as I asserted what I needed to be different... For example, when I noticed that my car had been "borrowed" (the odometer was different) without my knowledge or permission, I had to show up as a parent the way I needed to parent—setting limits, confronting the greater issues of lying and sneaking, and asserting the natural consequences for unacceptable behavior. This method was foreign to their family, and there were reactions all the way around! Thankfully, my husband supported me in front of his daughter, and then we discussed our differences privately and came to a mutual understanding about how to handle parenting together from then on. (Bouchard n. d.)

Stepfamily Relationships

Relationships between stepfamily members depend on many factors, such as the age of the children at the time of stepfamily formation, whether the stepchildren are boys or girls, and whether the stepparent is a mother or a father. These relationships are discussed in this section.

Between Stepparents and Stepchildren As discussed above, relationships between stepparents and stepchildren are complicated and diverse. Moreover, they are not static. Stepfamily relationships evolve. *How* they evolve is unclear because there have been few longitudinal studies of stepfamilies. Studies that have utilized longitudinal data, which provide the best assessment of relationship change, indicate that stepparents become less involved with and close to stepchildren over time (Stewart 2005b). However, this decline may have more to do with natural developmental changes in children than the stepparent-stepchild relationship. Similar declines in involvement and closeness occur between biological parents and children, with increasing conflict during adolescence (Cooksey and Fondell 1996; Marsiglio 1991; Rossi and Rossi 1990).

Between Biological Parents and Children Most stepfamily research has been devoted to the relationship between the stepparent and stepchild, but biological parent-child relationships in stepfamilies should not be overlooked. Most of what is known is limited to biological parent-child relationships after divorce (Coleman, Ganong, and Fine 2000). What happens after divorce is important to understand because it is the starting point for most children who enter stepfamilies. For instance, Mavis Hetherington's classic longitudinal study of divorced families in Virginia (Hetherington 1987; Hetherington and Jodl 1994; Hetherington and Stanley-Hagan 1999) indicated that the quality of mothers' parenting declines after divorce, a result of being the sole primary caretaker of children combined with busy work and/or school schedules and a loss of income. In particular, divorced mothers were found to engage in less *authoritative parenting* than other mothers (see Chapter 9). Divorce appears to have a greater negative effect on mothers' relationships with sons than daughters (Orbuch, Thornton, and Cancio 2000). Newer data from the UK New Stepfamily Study found that biological mothers in first families and stepfamilies were similar in their relationship quality, time spent with children, and discipline (Smith 2008). The author said that "there was no evidence that mothers disengaged from parenting after stepfamily formation" (p. 170). At the same time, mothers in stepfamilies had "an additional layer of parenting activities" in the form of defending, mediating, and facilitating the stepparents and stepchildren (Smith 2008, p. 171; Weaver and Coleman 2010).

Most information on biological father-child relationships in stepfamilies focuses on children with nonresident fathers. How do children manage relationships with "two dads"? White and Gilbreth (2001) found that the majority of adolescent stepchildren from the NSFH maintained positive relationships with both and that positive relationships with nonresident biological fathers were more common among children in stepfamilies than among children living with a single mother. There is not a lot of information specifically on resident biological father-child relationships in stepfamilies. It is thought that resident father-child relations undergo a more dramatic change with remarriage than resident mother-child relations because stepmothers take over many of the primary caretaking duties formerly performed by single fathers (Hetherington and Stanley-Hagan 1997). Resident biological fathers after divorce tend to have greater difficulty than mothers in the areas of communication and monitoring (especially with adolescent daughters), but they have fewer problems with control and are as warm and nurturing as mothers (Buchanan, Maccoby, and Dornbusch 1996; Hetherington and Stanley-Hagan 1999).

Between Full, Step- and Half Siblings Relationships between siblings, including stepsiblings, in families has been a neglected area of study (McHale, Updegraff, and Whiteman 2012; Sweeney 2010). Over three-quarters (77.9 percent) of children live with a sibling, and 12.5 percent of children live with a stepsibling, half-sibling, or both (Kreider and Ellis 2011a). The majority of children in stepfamilies (76.8 percent) live with a sibling. According to the 2008 Survey of Income and Program Participation, 6 percent of children in stepfamilies live with half-sibling (and no stepsibling), 8.2 percent live with a stepsibling (and no half-sibling), and 4.0 percent live with both half-siblings and stepsiblings (Kreider and Ellis 2011a). Nonresident stepsibling relationships are much more common, but we lack official estimates. In general, biological siblings are closer and more engaged than stepsiblings, as are siblings (biological, half, or step-) that reside in the same household (Sweeney 2010; Allan, Crow, and Hawker 2011). Stepsibling relationships are less intense, competitive, and conflictual than biological and half-sibling relationships (Ganong and Coleman 1993; Hetherington and Stanley-Hagan 1999). However, there is a great deal of variability in stepsibling relationships. Some children make no distinction between biological and stepsiblings–however this appears to be related to whether the stepparent does so (Allan, Crow, and Hawker 2011).

Stepparents' Decisions about Having Children When people remarry or form a new committed partnership, they have decisions to make about having children together. Does it make a difference whether one or both partners already have children? The answer to that question is yes.

A study of more than two thousand couples drawn from a national sample—the National Survey of Families and Households—found that individuals living with a second spouse or partner were most likely to want to have a child if neither partner had children (Stewart 2002). Desire for another child was lower for cohabiting couples than for married ones. If *both* partners already had children, an intention to have another child was especially low, with one exception. Because of the symbolic importance of joint parenthood, if the couple did not have a biological child, they were very likely to intend to have one.

The biological children of both partners in a stepfamily are called *mutual, shared,* or *joint* children (Stewart 2007). These are the half-siblings of any stepchildren. Research shows that a principal reason for the couple choosing to have a child together involves hope that the mutual child will "cement" the remarriage bond (Ganong and Coleman 2004). However, there is little empirical support for the so called *concrete baby effect* (MacDonald and DeMaris 1995; Stewart 2005b but see Pasley and Lipe 1998).

Believing that they will now be ignored by the stepparent—or seeing the mutual child as having a privileged place in the family—the stepchildren may feel threatened, jealous, or resentful (Cartwright 2008; Munroe 2009). Feelings of jealousy about a future mutual child can be set up as early as a parent's wedding ceremony. For instance, a 19-year-old female whose father had been remarried for two years told an interviewer:

> I think at the wedding it would have been more ideal if like family members had, I don't know, I guess, paid more attention to me and my brother. 'Cause they were kind of like you're married and you're going to have kids and everything, but they kind of forgot that there's already two kids in this family. (in Baxter et al. 2009, p. 482)

Relationships with Grandparents Even if they do not maintain a relationship with one another, ex-spouses and partners do not always make a "clean break" from in-laws (Finch and Mason 1990), usually because children remain the grandchildren of former in-laws. The extent of relationships with former in-laws after divorce depends on many things, such as the quality their relationship during the marriage, amount of reciprocal support, how close the grandparents are to their grandchildren, whether the children's father stays involved with them, whether either partner remarries, and who has custody (Duran-Aydintug 1993; Finch and Mason 1990; Kennedy and Kennedy 1993).

Many older adults find themselves "step-" grandparents to their children's stepchildren. Grandparents can make a positive contribution to stepfamilies. They provide economic and social support to grandchildren, act as intermediaries to keep everyone informed and involved regarding the children, and provide a "neutral zone" for ex-spouses, child-care, and other assistance (Stewart 2007). For example, contact, closeness with, and confiding in grandparents is associated with less depression in grandchildren, including step-grandchildren (Ruiz and Silverstein 2007).

Becoming a stepgrandparent is a *process* not unlike the process of becoming a stepparent. According to Henry, Ceglian, and Matthews (1992), grandparents must (a) accept the loss of the fantasy of a "lifelong happy marriage" for their child, and traditional grandparenthood for themselves, (b) cope with ambiguous family relationships, and (c) accept new family members (p. 28). Allan, Crow, and Hawker's (2011) study of British stepfamilies included stepgrandparents. They found that the stepgrandchild-grandparent relationship was dependent upon the ages of the children, how much time they spend together, as well as the number of biologically related grandchildren in the family. Perhaps the most important factor is the quality of the relationship between the grandparent and their adult child (and with their spouse/partner). As a result, there is a great deal of variability in grandparent-grandchild relationships in stepfamilies. Whereas some stepgrandchildren and stepgrandparents never come to view one another as "kin," others do not make a distinction between biological and step-grandchildren. Common issues that might face stepgrandparents is whether to extend monetary gifts to stepgrandchildren, and whether and to what extent to include stepgrandchildren in family holidays and on vacations.

Financial and Legal Issues

Two important, yet understudied, issues in stepfamilies have to do with how stepfamilies manage their finances and how stepfamilies navigate the laws governing stepfamilies. These issues are often intertwined, as stepfamily finances are impacted by legal decisions about child support, provision of health insurance, and other income transferred between divorced parents living in separate households. Moreover, laws governing stepfamilies are complicated, often ambiguous, and vary by state.

Financial Arrangements in Stepfamilies Financial challenges, like many issues that characterize stepfamilies, often begin with the previous divorce (Gold 2009; Henry and McCue 2009). Remarried couples are generally financially responsible for each spouse's children from previous unions and for any new children they may have together (Manning, Stewart, and Smock 2003). Not surprisingly, some remarrieds believe that prior child support agreements need to be modified better to accommodate the needs of the current family (Hans 2009). Some states, concerned about child support, passed legislation in the 1970s designed to prevent the remarriage of people whose child support was not paid up. But the Supreme Court ruled in *Zablocki v. Redhail* (1978) that marriage—including remarriage—was so fundamental a right that it could not be abridged in this way. Whether legally required to or not, most stepparents help to support the stepchildren with whom they reside, either through direct contributions to the child's personal expenses or through payments toward general household expenses such as food and shelter (Mason et al. 2002; Stewart 2007). Nonresidential stepmothers also spend their own money on stepchildren—usually for incidentals during visitation periods (Engel 2000).

Many remarried husbands report feeling caught between what they see as the impossible financial demands of both their former family and their present one (Hans and Coleman 2009). Some second wives—more often, those without children of their own—feel resentful about the portion of the husband's income that goes to his first partner to help support his children from that union (Hans and Coleman 2009). As an Australian nonresidential stepmother told an interviewer,

> I have always felt very cross that we have to give a large amount of our income [in child support] when I would really like that income to help … [my autistic son] … with therapy…. I am forced to work many hours to pay for [his] therapy. (in Henry and McCue 2009, p. 196)

Other mothers may feel guilty about the burden of support that her own children place on their stepfather (Barash 2000). Mothers with children from a previous relationship may worry about receiving regular child support from an ex-partner. Research shows that child support payments and visitation both decline when nonresident fathers have additional children with new partners (Manning, Stewart, and Smock 2003). Similarly, in a study of Wisconsin mothers receiving welfare, nonresident fathers provided less informal support to those mothers who went on to have an additional child with a new man (Meyer and Cancian 2012).

Despite the fact that financial issues and money problems are the primary issue that all couples argue about, there have been relatively few studies of stepfamilies' financial arrangements (Gold 2009). Although it is 25 years old, the best-known study of how stepfamilies manage their finances is that of Barbara Fishman (1983). Her in-depth interviews with 16 middle-class, resident stepfamilies suggest that stepfamilies tend to adopt one of two models of economic behavior, a common pot or two pot economic system. In the **common pot system**, economic resources are pooled and distributed according to need regardless of biological relatedness. In the **two-pot system**, economic resources are divided and are distributed along biological lines, and are only secondarily distributed according to need. For example, common pot families would put their money together in a joint account and share ownership of assets, such as a home, whereas two pot families would keep their money in separate savings accounts and might have prenuptial agreement. Fishman suggests that a concern for the "common good" (family harmony, trust, and closeness) underlies the common pot system whereas rationality, economic independence, and personal autonomy underlie the two pot system. She feels that whereas the common pot encourages household unity; separate pots discourage it.

Fishman's two system model of financial management is innovative and appealing but it lacks rigorous testing with nationally representative data. A number of studies using nonrepresentative samples indicate that the majority of stepfamilies are common pot economies (e.g., Lowan and Dolan 1994; Mason et al. 2002), as does one national study (e.g., Addo and Sassler 2010). These studies show that, consistent with Fishman's findings (1983), joint banks accounts were, for the most part associated with higher relationship quality. Newer research also indicates that contextual factors play a role, such as income, race/ethnicity, and gender (Addo and Sassler 2010; Higginbotham, Tulane, and Skogrand 2012). For example, in a study of low-income married and cohabiting couples, previous marriages and children had no effect on whether couples had separate or joint bank accounts (Addo and Sassler 2010), and that Hispanic couples were more likely to modify their arrangements to move toward shared management. Whether a common pot or two pot system is adopted, having a clear financial management strategy was associated with greater financial security among remarried couples (Van Eeden-Moorefield et al. 2007).

Legal Issues in Stepfamilies Unfortunately, family law generally assumes that all marriages are first marriages. Therefore, many of the legal provisions that affect stepfamily life proceed from one or both partners' divorce decrees. Meanwhile, few legal provisions exist for the special circumstances of stepfamilies, especially if the remarried union should dissolve. Should a stepparent pass away, assets would go to children who are biologically related as opposed to stepchildren, unless the stepparent has specified otherwise – which many stepparents have not. Stepparents are also faced with the dilemma of how to divide their assets. One stepparent explains:

> No, I haven't made a will but it wouldn't be equal. It is awfully complicated.... On the one hand, you sort of say well Nick's my son I should leave it all to my son, then blow the rest of [them]. But I don't feel that strongly enough to actually say it. I, I feel I owe it to Nick that he should get the lion's share or he should certainly get a bigger proportion, even if you just base it on the technicality that the others will inherit from their blood father (Allan, Crow, and Hawker 2011, p. 45).

Laws governing stepfamilies vary greatly by state, and the legal rights and responsibilities of stepparents are ambiguous (Troilo 2011). In many states, stepparents do not have the authority to see the school records of stepchildren or make medical decisions for them. Studies of childhood cancer patients indicate that although some stepparents voluntarily "step back" or "step away" from the decision making process, others report being "pushed" out of the way, even if they are one of the child's primary caretakers (Kelly and Ganong 2011a,b).

Some states have passed legislation that holds stepparents responsible for the support of stepchildren during the marriage (Hans 2002; Malia 2005). However, "in most states, stepparents are not required to support their spouses' children financially, although most voluntarily choose to provide contributions" (Malia 2005, p. 302). Although a stepfamily may come to rely on such financial support, if a remarriage ends in divorce, stepparents are not legally responsible for child support unless they have formally adopted the stepchildren or signed a written promise to pay child support in the event of divorce. Stepparents who are no longer married to their child's biological parent also have no rights in regard to their stepchildren's custody nor are they allowed provisions for visitation. Even within marriage, most states consider stepparents and stepchildren *legal strangers* (Malia 2008). (Interestingly, college financial aid applications require information on a stepparent's income to calculate student need.)

The preservation of stepparent–stepchild relations when death or divorce severs the marital tie is also a serious issue. Some stepparents continue relationships with their stepchildren after a divorce (Dickinson 2002). Visitation rights (and a corresponding support obligation) of stepparents are beginning to be legally clarified (Hans 2002; Mason, Fine, and Carnochan 2001). But outcomes are unpredictable at this point. It's possible that when a custodial, biological parent dies, the absence of custodial preference for stepparents over extended kin may result in children's being removed from a home in which they had a close psychological tie to a stepparent. Stepchildren whose stepparent and biological parent are not legally married are at even greater risk (Holtzman 2011).

One way to be certain that situations like these do not occur is for a stepparent to legally adopt a stepchild, and this situation may be virtually impossible due to the noncustodial biological parent's objections (Malia 2008; Mason et al. 2002). This is because, in the U.S., children are limited to having two parents, and in most states those parents must be of opposite sex. In other industrialized countries, such as New Zealand, stepparents and stepchildren can establish a legal relationship regardless of whether two biological parents are present in the child's life (Atkin 2008). Because family law is constantly evolving, members of stepfamilies are advised to check with an attorney regarding applicable laws in their state. Another good resource is the National Stepfamily Resource Center's Frequently Asked Questions on law and policy.

Michael Newman/PhotoEdit

Conflicting expectations concerning a stepfather's —or stepmother's—role may make it stressful. When stepparents can ignore the myths and negative images of the role and maintain optimism about the remarriage, they are more likely to have high family, marital, and personal satisfaction.

Well-Being in Stepfamilies

One of the most researched areas concerning stepfamilies has to do with the well-being of their members, especially children. As discussed in previous chapters, members of nontraditional families (not living in a married, two-parent household) generally do not fare as well on a range of economic and social and emotional variables. As you will see in this section, the reasons for these differences are quite complicated.

The Well-Being of Parents and Stepparents

Compared to what is known about the effects of stepfamily living on children, very little research concerns the well-being of adults in stepfamilies (Sweeney 2010). What we do know is that parents have lower levels of happiness and life satisfaction and higher levels of psychological distress than nonparents, especially among women (McLanahan and Adams, 1987). Fortunately, the economic strain and personal demands of parenthood tend to subside as the children grow older.

Several older studies indicate that parenting stepchildren is less satisfying than parenting biological children (Rogers and White 1998), especially for stepmothers (Levin 1997; Thoits 1992). Stepmothers have reported feeling isolated, resentful, guilty, and can suffer from low self-esteem (Smith 1990). There is also evidence that stepmothers have worse mental health than biological mothers (Shapiro and Stewart 2011; Smith 2008). Stepchildren do not generally "substitute" for biological children. For example, stepchildren provide less social support to their aging parents than do biological children, and this support to stepparents is more dependent upon the quality of the relationship than support to biological parents (Ganong and Coleman, 2006a, 2006b). However, stepparenting has many positive benefits for stepparents. Some benefits of stepmothering mentioned by stepmothers are the opportunity to engage in the maternal role, enjoying the challenge of living in a stepfamily, and personal growth (Quick, McKenry, and Newman 1994). Self-esteem is higher among stepmothers who have better relationships with their stepchildren (Quick, McKenry, and Newman 1994).

The Well-Being of Children

How does membership in a stepfamily affect children's well-being? The majority of children in remarried households show few, if any, negative outcomes. However, considerable research has found that, on average, stepchildren of all ages have somewhat higher rates of smoking, alcohol and drug use, and juvenile delinquency; do less well in school; are more likely to have experienced early sexual behavior and childbearing; may experience more family conflict; and are somewhat less well-adjusted than children in first-marriage families (for a review of this research see Sweeney 2010).

"Studies consistently indicate... that children in stepfamilies exhibit more problems than do children with continuously married parents and about the same number of problems as do children with single parents" (Amato 2005, p. 80). This is the case whether or not the stepchild has been adopted by their stepparent (Stewart 2010) and also applies to shared children born into stepfamilies.

Not all children living in stepfamilies are stepchildren. Some children are born into stepfamilies and are the biological children of both parents, and half-siblings to stepchildren. Research suggests that all children raised in stepfamilies, even if both of their biological parents are in the household, experience lower levels of well-being compared to children raised in continuously intact families (Thomson and McLanahan 2012). Moreover, the effect of stepfamily living on children is not straightforward. Acquiring a stepparent may increase some problems but improve other areas of children's lives and these effects may depend on the child's gender, age, race/ethnicity, and income (Sweeney 2010). In one study of 8,008 children in kindergarten through fifth grade, living with a stepparent was associated with higher school performance (Sun and Li 2011). In another, living in a stepfamily was associated with increased use of marijuana use among African American adolescent boys (Mandara, Rogers, and Zinbarg 2011).

Sociologist Andrew Cherlin explains these findings:

> The addition of a stepparent increases stress in the family system at least temporarily, as families adjust to new routines, as the biological parent focuses attention on the new partnership, or as stepchildren come into conflict with the stepparent. This increased stress could cause children to have more emotional problems or to perform worse in school, which could counterbalance the positive effects of having a second adult and a second income in the household. (2009a, p. 22)

Some research suggests that the effect of stepfamilies on child well-being may be less negative among African Americans, whose culture is more likely to support multiparental models. African American children may be more accustomed to life transitions and stressful life conditions and therefore have greater capacity to adjust to changes, compared with nonHispanic white children. Analysis of data from a national sample of African American youth found that the presence of a father, whether biological or stepfather, served to increase the likelihood of positive outcomes (Adler-Baeder et al. 2010).

Further, children in stepfamilies grow up with fewer economic resources than children in original, two-parent families. Stepfamilies have lower earnings, less savings, are less likely to be home owners, and have less equity in their homes (Stewart 2001). The parents of stepchildren have lower aspirations for them in terms of going to college, they are less involved in their schoolwork, and are less likely to help them go to college, establish a business, or buy a home (Astone and McLanahan 1991; Hetherington and Kelly 2002). The lower financial resources of stepfathers relative to biological fathers explain a large portion of stepchildren's lower school achievement (Hofferth 2006). There is also evidence that fewer family resources go to stepchildren than to biological children. Resident stepparents spend less on food, schooling, clothing, and miscellaneous items (gifts, hobbies, and pocket money) for children than do biological parents (e.g., Case, Lin, and McLanahan 2000). Data from the National Postsecondary Student Aid Study shows that remarried parents' and divorced parents' contribution to their children's college costs are similar, despite remarried parents' higher incomes (Turley and Desmond 2011).

Stepparents make fewer investments in their children's health than biological parents. Children with resident stepmothers are significantly less likely to have routine doctor and dentist visits and to have a place for regular medical care, they are less likely to wear seatbelts, and are more likely to be living with a cigarette smoker than children living with a biological mother (Case and Paxson 2001). A longitudinal study of high school students in St. Paul, Minnesota, shows that teenagers in remarried stepfamilies do significantly more routine housework (cooking, cleaning, and laundry) than teens in original two-parent families (Gager, Cooney, and Call 1999).

Some researchers have found that remarriage lessens some negative effects for children of divorce—but only for those who experienced their parents' divorce at an early age and when the subsequent remarriage remains intact (Arendell 1997; Cherlin 2009a). One of the main benefits of a parent's remarriage for children is economic and in the area of school achievement (Sweeney 2010). Additional research shows that younger children adjust better to a parent's remarriage than do older children, especially adolescents (Amato 2005; Carlson 2006). Siblings in the same family system may have different levels of exposure to marital conflict and therefore may have differential levels of well-being in the areas of self-blame, depression, and behavior problems (Richmond and Stocker 2003).

What about the well-being of stepchildren living in cohabiting stepfamilies? Chapters 6 and 7 discuss how the outcomes of children from cohabiting couple households are significantly worse compared to children with two married parents. These findings apply to children who have both of their biological parents in the home and children living with one biological parent and one stepparent. In fact, children living with a mother and her cohabiting partner have outcomes more similar to children living with a single mother than to children living with a stepparent (Thomson and McLanahan 2012). Nevertheless, research also shows that, compared with growing up in a single-parent home, children benefit economically from living with a cohabiting partner (Manning and Brown 2006).

Stepchildren in cohabiting stepfamilies fare less well than do those in remarried stepfamilies (Sweeney 2010). Cherlin suggests reasons why:

> Single parent families, whether after divorce or in the absence of marriage, create a new family system. Then into that system, with its shared history, intensive relationships, and agreed-upon roles, walks a parent's new live-in partner.... Lone mothers may be willing to live with a partner whom they wouldn't necessarily marry.... Their partners, in turn, may not be interested in.... developing a parentlike relationship with their children. They could even be a net drain on children's resources if the parent becomes preoccupied with the intimate relationship.... Moreover, cohabiting partnerships tend to be short-lived, and the departure of a cohabiting partner could once again produce more stress in the household. If the cohabiting stepfamily has established

family routines, these would be disrupted again, and the biological parent and the children would have to adjust to the loss. (2009a, pp. 21–23).

Sequential transitions from one family structure to another also create upheaval and stress for children (Bures 2009; Cavanagh 2008; Magnuson and Berger 2009).

The lower levels of well-being that stepchildren experience during childhood can follow them into adulthood. Stepchildren leave home earlier than biological children. For example, stepdaughters from the National Study of Adolescent Health had an elevated risk of nonmarital childbearing and cohabitation than to those with continuously married biological parents (Amato and Kane 2011). Among adolescents whose parents had a high-distress relationship there was an increased risk of early marriage and marital births. Stepchildren who leave the parental home are more likely to give "friction at home" as the reason for their early departure than are other children (Kiernan 1992). This is referred to as extrusion, "defined as individuals' being 'pushed out' of their households earlier than normal for members of their cultural group, either because they are forced to leave or because remaining in their households is so stressful that they 'choose' to leave" (Crosbie-Burnett et al. 2005, p. 213).

Although we note the preceding findings, we also recognize other aspects of this story. For one thing, most stepchildren do well and appreciate their stepfamily lives. Moreover, numerous studies show that the *instability* of family structure (i.e., children moving in and out of different family types), as opposed to family structure per se that is associated with the worst child outcomes (Sun and Li 2011; Thomson and McLanahan 2012). Similarly, several studies have concluded that stepchildren's well-being and future outcomes largely depend on the quality of the relationships and communication among family members regardless of family structure (Crawford and Novak 2008; Doohan et al. 2009; Schoppe-Sullivan, Schermerhorn, and Cummings 2007). The extent to which parents or stepparents monitor their children's comings and goings is probably more important to positive child outcomes than is family structure itself (Crawford and Novak 2008). Furthermore, recent research shows that a close, nonconflictual relationship with a stepfather enhances the overall well-being of adolescents, and this is especially true when the child has a similar relationship with the biological mother (Yuan and Hamilton 2006; Booth, Scott, and King 2010). Among Hispanics, stepfather involvement was associated with fewer behavior and emotional problems in adolescents (Coltrane, Gutierrez, and Parke 2008). Positive relationships with siblings and half-siblings are correlated with higher levels of adjustment in adolescents in stepfamilies (Baham et al. 2008).

Creating Supportive Stepfamilies

Creating a supportive stepfamily is not automatic. Supportive stepfamily relationships must be forged. Patricia Papernow's (1993) book *Becoming a Stepfamily* uses a *developmental approach* to stepfamilies. Developmentalists focus on the complexity and processes that occur within family systems and how these aspects of family life unfold over time (O'Brien 2005). Papernow, a psychologist, bases her theoretical framework on in-depth interviews with over 50 stepfamilies. The **stepfamily cycle** describes the process by which veritable strangers to one another form "nourishing, reliable relationships," (Papernow 1993, p. 12). She argues that it is critically important for stepfamilies to have what she refers to as a "developmental map." A developmental map helps family members know what is normal and predictable and what is a family crisis. She compares it to the experience of a child throwing a temper tantrum. Once parents understand that tantrums are to be expected of children of a certain age, then the family will be less likely to be "stressed-out" and overreact to the incident. The Cycle therefore offers both guidance and reassurance to stepfamilies as they go about their daily lives, as well as a model of intervention points for professional working with stepfamilies. The stages of the Stepfamily Cycle and the life cycle tasks associated with each stage are outlined in Table 15.3.

Papernow cautions that the Stepfamily Cycle does not unfold in a neat and precise way, and it can take anywhere from 4 to 12 years to complete. Moreover, it is possible for stepfamilies to get "stuck" and never reach maturation. For stepfamilies that do make it, the end result of the process is the **intimate outsider role** for the stepparent, which Papernow (1993) describes as, "intimate enough to be a confidante, and outside enough to provide support and mentoring in areas too threatening to share with biological parents: sex, career choices, drugs, relationships, remaining distress about the divorce" (p. 16–17).

Many stepparent-stepchild relationships improve over time. One stepchild says the following about her stepfather, "He was [trying], and I was rebelling because I wasn't happy with him being there.... I wanted a reason not to like him, but eventually the harder he tried, the wall broke down with me, and I saw what he was doing for my mom" (Ganong, Coleman, and Jamison 2011, p. 406). In fact Ganong, Coleman, and Jamison (2011) interviewed young adults who had had one or more stepparents over the course of their lives. They found six patterns of step-relationship development: accepting as a parent, liking from the start, accepting with ambivalence, changing trajectory, rejecting, and

Table 15.3 Papernow's Stepfamily Cycle[a]

Stage	Description	Life Cycle Tasks
Early		
1. Fantasy	Parents have hopes of being an "instant family." Children harbor the fantasy that their parents will reunite.	Acknowledging fears and fantasies. For children, acknowledge loss of their "real family."
2. Immersion	Reality of the challenges of living in a stepfamily sets in. Increasing sense of unease among family members. Stepparents in particular are likely to have negative emotions.	Bearing disappointment of "instant family" fantasy not coming true. For stepparents, dealing with rejection and new routines. Keeping hopeful about the future.
3. Awareness	Stepparents become aware that they are an "outsider." Increasing pressure on the biological parent as the "insider."	Each stepfamily member gathering data about the family and his or her place in it. Naming feelings, lowering expectations.
Middle		
4. Mobilization	Differences are aired more openly. This state is marked by increased chaos and conflict.	Voicing unheard needs and perceptions. Increasing the focus on the couple.
5. Action	Power struggles between insiders and outsiders diminish.	Negotiating agreements about how the family will function. Drawing new boundaries around step-relationships.
Later		
6. Contact	The "honeymoon" period. The chaos of previous periods has stabilized. Stepparents and stepchildren forge a "real" relationship. The marital relationship improves.	Working on developing a "workable" stepparent role. Letting go of old hurts.
7. Resolution	Stepfamily norms are established. Step-relationships no longer require constant attention. Increased clarity, acceptance, and satisfaction. The stepparent is established as an "intimate outsider."	Continued acceptance of differences between family members. Staying on developmental track as new issues emerge (e.g., new babies, financial problems).

[a]Adapted from Papernow 1993. Reprinted from Stewart 2007.

coexisting. These patterns were dependent upon the age of the stepchild when the stepfamily was formed, the child's gender, and the amount of time the stepparent and stepchild spent together.

It helps to remember that the unrealistic "urge to blend the two biological families as quickly as possible" may lead to disappointment when one or more adult or child members "resist connecting" (Wark and Jobalia 1998, p. 70). You may have noticed that we have not used the once-familiar term *blended family* in this chapter. That's because family therapists and other experts have concluded that stepfamilies do not readily "blend" (Deal, in Kiesbye 2009). Playing with the language, stepmother and online columnist Dawn Miller refers to stepfamily living as "life in a blender" (Miller 2004).

Despite stepfamilies' unique challenges, people can and do create supportive and resilient remarriages and stepfamilies (Ahrons 2004). Counselors remind remarrieds not to forget their couple relationship (Munroe 2009). In many locations, prospective partners can participate in remarriage preparatory courses that alert remarrying couples to expect challenges and help to address them.

In a clever play on words, columnist Dawn Miller (n. d.) titled one of her essays "Don't Go Nuclear—Negotiate." Chapter 7 presents some things for couples to talk about when forging an adaptable, supportive marriage relationship. Many of those questions also apply to remarriages. But there are additional issues to discuss regarding stepfamilies: household rules, expectations for a stepparent's financial support of stepchildren, how emergency medical care will be handled if the biological parent isn't there to sign a release, questions of inheritance, and whether there will be a mutual child or children (Lavin 2003).

Chapter 13 points out that life transitions, such as remarriage or the transition to stepparent, are family stressors. That chapter explains that resilient families deal with family transitions creatively by emphasizing mutual acceptance, respect, and shared values (Ahrons 2004). You may also recall that Chapter 11 presents guidelines for constructive family communication—all applicable in stepfamilies.

It is important for members of stepfamilies (and the clinicians who work with them) to recognize that the "stepfamily architecture" presents special challenges (Papernow 2008). That is, creating supportive and

Michael Newman/PhotoEdit

This family portrait is of a mother and stepfather and her daughter/his stepdaughter, along with a baby son from the new union. The remarried family structure, which is complex and has many unique characteristics, has no accepted cultural script. When all members are able to work thoughtfully together, adjustment to a new family life can be easier.

cohesive stepfamilies involves recognizing and building upon some potential family strengths that are unique to stepfamilies. Relationships with new extended kin may be a potential source of new friendships. Beginning a renewed sense of family history is another strength builder. Although holidays often divide the stepfamily because of visitation agreements with a noncustodial parent, it is possible to create new family holidays when the entire stepfamily is sure to be together (Wurzel n. d.). Well-designed empirically-based relationship interventions for stepfamilies have been shown be effective at improving couple relationships, parenting, and child well-being (Lucier-Greer and Adler-Baeder 2012; Nicholson et al. 2008).

Technology is also continuously transforming family dynamics and communication. Research is still emerging about the extent to which members of stepfamilies use new technologies such as email, Skype, cell phones, texting, Facebook, and photosharing sites, as well as informational and stepfamily support blogs and websites. On the one hand, new technologies such as texting and email allows parents greater control over their communications with ex-spouses and partners (Ganong et al. 2012). On the other hand, technology may introduce new problems, such as disagreements between partners about limiting children's online surfing, texting, or gaming as well as their own use (Hertlein 2012). Technology also allows ex-spouses to communicate more frequently, and how to manage those interactions can create tension between partners. Given the complexity of stepfamily relationships, technology can help stepfamily members to stay connected, especially for family members who reside in different households.

Increasing access to technology means that today's stepfamilies have access to more resources than in the past. For instance, several online websites by stepfamily counselors and well-respected researchers are designed to give advice and report research findings concerning stepfamilies, such as the website of the Stepfamily Foundation. The website, I Do! Take Two offers not only purchasable wedding products but also respected counselors' advice on topics ranging from second wedding etiquette to religious, financial, and legal issues. This website has good suggestions for including children in the wedding ceremony, possibly creating a family-forming ritual.

There are also more and more books written by psychologists and others for remarrieds and stepfamily members. Living up to its title, Erin Munroe's *The Everything Guide to Stepparenting* (2009) addresses topics from dating a parent to the logistics of moving in together to questions about maintaining step-relationships after a second divorce. Increasingly, there are stepfamily books written for children. One of these, directed to teens, is *Stepliving for Teens: Getting along with Step-Parents, Parents, and Siblings* (Block and Bartell 2001). Another book for teens and preteens is *The Step-Tween Survival Guide: How To Deal with Life in a Stepfamily* (Cohn, Glasser, and Mark 2008). Sally Hewitt's *My Stepfamily* (2009) is written for younger children. For further suggestions, see Coleman and Nickleberry (2009).

Stepfamily enrichment programs, support groups, and various other group counseling resources for stepfamilies of various ethnicities are increasingly available and have been found to be helpful (Higginbotham, Miller, and Niehuis 2009; Skogrand, Barrios-Bell, and Higginbotham 2009). One example is the Active Parenting for Stepfamilies program (Popkin and Einstein 2006). Another is the Stepfamily Enrichment Program (Michaels 2006). Such programs can increasingly be found on-line (e.g., Gelatt, Adler-Baeder, and Seeley 2010). In general, researchers and family therapists tend to agree that:

> it is neither the structural complexity nor the presence/absence of children in the home *per se* that impacts the marital relationship. Rather, the ways in which couples interact around these issues are the key to understanding marital relationships in general and marital relationships in remarriages specifically. (Ihinger-Tallman and Pasley 1997, p. 25; see also Halford, Nicholson, and Sanders 2007)

Afifi (2008) emphasizes the importance of examining how stepfamilies cope as a social unit. Interacting in positive ways in remarriages and stepfamilies involves making knowledgeable choices.

We close this chapter with the paragraph that stepfamily scholar Susan Stewart uses to close her significant book *Brave New Stepfamilies* (2007):

> One might conclude that Americans can maximize their well-being by getting married, staying married, reproducing their own biological offspring, and toughing it out. Yet an increasing number of Americans live increasing portions of their lives in increasingly diverse families that do not align with this idea. Perhaps Americans might do better by admitting the emerging normality of stepfamilies and building institutional supports to make their brave new stepfamilies strong. (p. 224)

Summary

- Remarriages have always been fairly common in the United States but are more frequent now than they were early in this century, and they now follow divorce more often than widowhood.
- There is a great deal of diversity among stepfamilies as a result of growing rates of nonmarital childbearing and cohabitation, increasing involvement of noncustodial parents with their children, the aging of the population, growing racial and ethnic diversity, and increasing societal support of same sex couples.
- Remarriages are usually about as happy as first marriages, but they tend to be slightly less stable.
- One reason for greater instability in remarriage is lack of a widely recognized cultural script for living in remarriages or stepfamilies.
- Remarried adults and stepchildren often unconsciously try to approximate the nuclear-family model, but it does not work well for most stepfamilies.
- Relationships within stepfamilies and also with extended kin are often complex, yet there are virtually no social prescriptions and few legal definitions to clarify roles and relationships.
- Stepparents are often troubled by financial strains, role ambiguity, and stepchildren's hostility.
- Stepparents and children raised in stepfamilies, on average, have fewer financial resources and lower levels of socioemotional well-being.
- Marital happiness and stability in remarried families are greater when the couple has strong social support, good communication, a positive attitude about the remarriage, low role ambiguity, and little belief in negative stereotypes and myths about remarriages or stepfamilies.

Questions for Review and Reflection

1. Discuss some structural differences between stepfamilies and original two-parent families. Discuss different types of stepfamilies.
2. The remarried family has been called an incomplete institution. What does this mean? How does this affect remarried couples with stepchildren?
3. What evidence can you gather from observation or your own personal experience or both to show that stepfamilies (a) may be more culturally acceptable today than in the past and (b) remain negatively stereotyped as not as functional or as normal as original two-parent families?
4. What are some challenges that stepparents face? What are some challenges faced particularly by stepfathers? Why might the role of stepmother be more difficult than that of stepfather? How might these challenges be confronted?
5. **Policy Question.** In terms of social policy, what might be done to increase the stability of stepfamilies and the well-being of children in stepfamilies?

Key Terms

boundary ambiguity 400
common-pot system 407
cultural script 399
dripolator effects 403
incomplete institution 399
intimate outsider role 411
nuclear-family model monopoly 393
paternal claiming 403
percolator effects 403
role ambiguity 403
stepfamily cycle 411
stepmother trap 404
stigmatization 393
two-pot system 408

16

Aging and Multigenerational Families

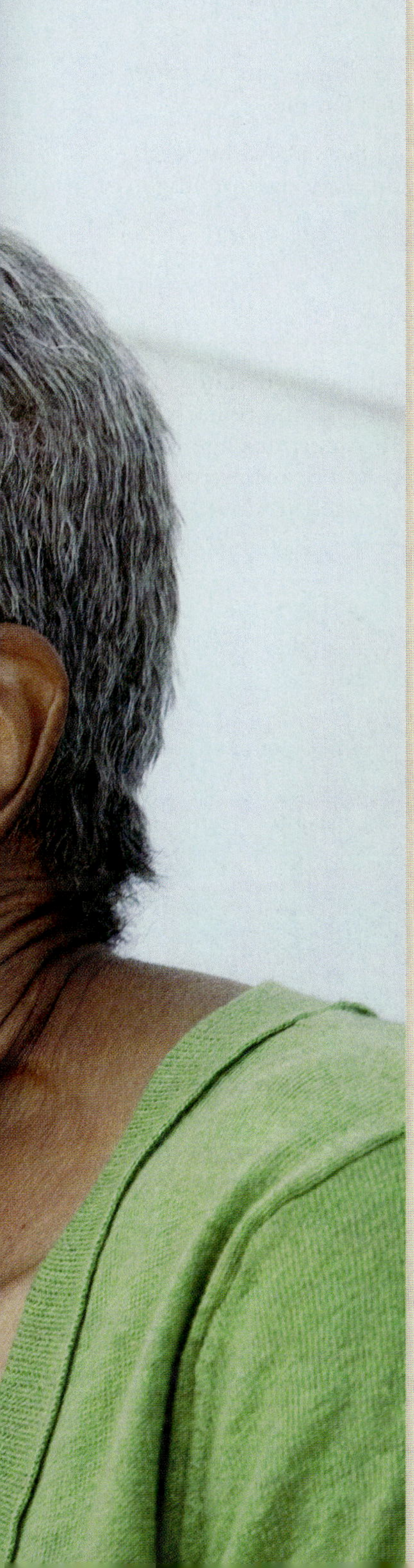

Ronnie Kaufman/Blend Images/Getty Images

Learning Objectives

1. Describe the changing age structure in the United States and other industrialized countries.
2. Discuss how the diversity of family forms among the elderly can be expected to affect caring for them in the future.
3. Describe economic circumstances and issues faced by the elderly.
4. Describe grandparent roles.
5. Explain how gender affects caregiving.
6. Discuss issues surrounding elder abuse and neglect.
7. List some costs and benefits of caring for an elderly family member.

- "I feel very young," boasts an 82-year-old great grandmother who lifts weights weekly. She and her husband regularly take their great grandchildren to car shows (Rosenbloom 2006).
- In her seventies, Martha Stewart joins Match.com and shows off her new guys on television.
- After work Ramona visits her dad at his assisted living facility, then rushes home to help her children with their homework.
- Jeanne intervenes to rescue her grandparents whose caregiver son lives with them and is financially abusing and physically neglecting them (Beidler 2012).

Americans are living longer now than in decades past, a situation making for more and more **multigenerational families**—families that include several generations. Compared with just a few decades ago, many older Americans feel better and behave more youthfully today as well. Americans' happiness level, although highest for those in their early twenties and gradually dropping after that, begins to increase once more at about age 60 and does not drop again until after about age 75. Even into their nineties, 72 percent of Americans tell pollsters that they "experienced happiness, enjoyment, and smiling or laughter during a lot of the day" (Newport and Pelham 2009).

This chapter examines various issues concerning multigenerational families, particularly issues focused on the contributions and needs of aging family members. We will explore older Americans' living arrangements. We'll discuss the grandparent role, and then look at issues concerning caregiving to older family members. The ever-increasing diversity within today's older population—both in race/ethnicity and in family form—provides a backdrop to these discussions. To begin, we note that many of the topics explored elsewhere in this text apply to aging and multigenerational families. For instance:

- Older families, like other families, comprise a diversity of family forms, including LGBT couples.
- Older wives—like younger ones—are concerned about fairness and equity when it comes to power, decision making, housework, and other/caregiving tasks.
- Aging family members may be actively engaged in parenting.
- As today's adults age, more older families are stepfamilies.
- Supportive communication is important in older, younger, and multigenerational families.

To begin, we'll examine some facts about our aging population, the older generations in today's multigenerational families.

Our Aging Population

The number of older people in the United States (and all other industrialized nations) is growing remarkably. In 1980, there were 25.5 million Americans age 65 or older; today about 40.2 million Americans are age 65 or older, and this number is expected to double over

Dave & Les Jacobs/Alamy

Changing the concept of aging itself, seniors are increasingly active into older ages. According to LeRoy Hanneman of Del Webb Retirement Communities, at least some "Boomers should be called Zoomers" (in "The Demographics of Aging" n. d.).

the next 40 years. Americans age 75 and older numbered close to 10 million in 1980; by 2010, there were more than 18.5 million. Of those age 85 and above, there were 2.2 million in 1980, compared to more than 5.8 million today. Projections are that, by the year 2050, there will be nearly 88.5 million Americans age 65 and older, with about 19 million of them age 85 and over (U.S. Census Bureau 2012c, Table 9).

Not just the *number* of elderly has increased but also their *proportion* of the total U.S. population. This is especially true for those in the "older-old" (age 75 through 84) and the "old-old" (85 and over) age groups. The proportion of Americans age 75 and above rose from 4.4 percent in 1980 to 6 percent in 2010, with projections to 11 percent in 2050. The proportion of Americans age 85 and older rose from 1 percent in 1980 to nearly 2 percent in 2010, with projections to 4.3 percent in 2050 (U.S. Census Bureau 2003a, Table 11; 2012c, Table 9). Although the number of centenarians (those age 100 and above) is small at less than 1 percent, that percent is growing (Meyer 2012).

Aging Baby Boomers

Between 1946 and 1964, in the aftermath of World War II, more U.S. women married and had children than ever before. The high birthrate created what is commonly called the **baby boom**. Now baby boomers have begun to retire, and within the next decades, they will generate an unprecedentedly large elderly population (see Figure 16.1). Meanwhile, the number of children under age 18 is about the same today as it has been for several decades (about 75 million).

As a result, children now make up a decreasing proportion—and older Americans a growing proportion—of the population. The proportion of the U.S. population under age 18 was 24 percent in 2012, compared to 36 percent in 1960 (U.S. Census Bureau 2003a, Table 11; 2012c, Table 9). Along with the impact of the baby boomers' aging and the declining proportion of children in the population, longer life expectancy has contributed to the fact that, as a whole, our population is growing older.

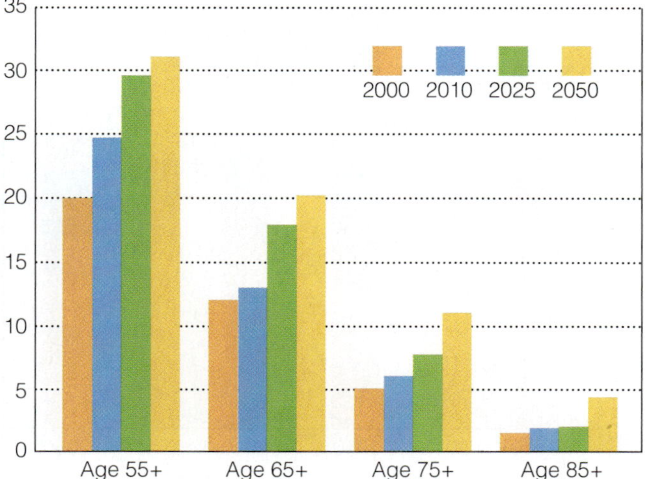

FIGURE 16.1 Older Americans as a percentage of the total U.S. population, 2000 and 2010, with projections for 2025 and 2050. Currently, baby boomers are in their fifties and early sixties. As the baby boom cohort grows older, populations over age 55, 65, 75, and 85 will increase.

Sources: Calculated from U.S. Census Bureau 2002, Table 12; 2012c, Table 9.

Longer Life Expectancy

Americans are now living long enough that demographers divide the aging population into three categories: the "young old" (age 65 through 74), the "older old" (age 75 through 85), and the "old old" (age 85 and over). Life expectancy at birth increased from 70.8 years in 1970 (67.1 years for men and 74.7 years for women) to 78.7 years in 2010 (76.2 years for men and 81.1 years for women) (Minino and Murphy 2012). By the year 2020, life expectancy at birth is projected to reach 79.5 (77.1 for men, and 81.9 for women) (U.S. Census Bureau 2012c, Table 104).

Race/Ethnicity and Life Expectancy Life expectancy differs by race/ethnicity. Life expectancy at birth for nonHispanic whites in 2010 was 78.8, compared with 74.7 for nonHispanic blacks (Minino and Murphy 2012). Much of the difference between blacks and whites has historically been associated with whites' averaging higher incomes and lower poverty rates than blacks. Higher incomes, along with higher education levels, are associated with longer life expectancy, largely because people in higher socioeconomic groups are less likely to work or live in hazardous environments and have access to better health care (Conner 2000, p. 16).

Meanwhile, due to the group's high proportion of immigrants, Hispanics now have a higher average life expectancy—81.3 in 2012—than do nonHispanic whites (Minino and Murphy 2012). Immigrants arrive with more healthy eating habits than are customary here, and they have generally grown used to more walking and other physical activity. Upon attaining but a minimal level of economic success here, immigrants—and more so their American-born children—walk and exercise less while consuming unnecessarily high numbers of calories, situations that result in life-shortening obesity, diabetes, and high blood pressure (Tavernise 2013). Today Hispanic females average the longest life expectancy of any race/ethnic/gender category—83.8 years in 2010 (Minino and Murphy 2012).

Gender and Life Expectancy Women, on average, live about five years longer than men. Because women live longer on average, the makeup of the elderly population differs by gender. In 2010, there were 22.9 million women age 65 and older, compared to 17.3 million men. For Americans over age 84, there are 3.9 million women and about 1.9 million men (U.S. Census Bureau 2012c, Table 9). This gendered difference in life expectancy means that, among other things, elderly women are more likely than men to be widowed. Eighty percent of centenarians today are female (Meyer 2012).

For reasons explained later in this chapter, elderly women are more likely than their male counterparts to be poor (Federal Interagency Forum on Aging-Related Statistics 2013). However, trends show that the life-expectancy gap between women and men is slowly narrowing (U.S. Census Bureau 2012c, Tables 104). Should this trend continue, policy analysts point to at least two implications for women. For one thing, more elderly heterosexual women may have spouses or cohabiting partners to care for them should they need it. In addition, a shorter widowhood could mean that older women will be better off financially (Zernike 2006).

Family Consequences of Longer Life Expectancy Demographers point to at least two family-related consequences of our living longer. First, because more generations are alive at once, as members of multigenerational families we will increasingly have opportunities to maintain ties with grandparents, great-grandparents, and even great-great-grandparents (Bengston 2001). It is estimated that, by 2030, more than two-thirds of 8-year-olds will have a living great-grandparent (Rosenbloom 2006).

A second consequence of longer life expectancy is that, on average, more Americans spend time near the end of their lives with chronic health problems and/or physical disabilities (Scommegna 2013). We can think not just in terms of overall life expectancy but also in terms of **active life expectancy**—the period of life free of disability. After this, a period of being at least partly disabled may follow, partly because today's older Americans, while living longer, are not necessarily healthier and are more likely to be obese, have diabetes or high blood pressure than previous generations of similar ages—situations that can lead to higher odds of stroke and resulting disability (Scommegna 2013). As Americans get older, more and more of us will be called upon to provide care for a parent or other relative who is disabled. We will return to issues surrounding giving care to aging family members later in this chapter. We turn now to the race/ethnic composition of the older American population.

Race/Ethnic Composition of the Older American Population

"While the elderly are often subsumed under the same umbrella as the 'over 65-generation,' it is important... to note that the aging population in the United States comes from a wide range of backgrounds and cultures" (Trask et al. 2009, p. 301). As a category, nonHispanic whites are older than people in other race/ethnic categories. Although the national median age is 36.9 years, the median age for nonHispanic whites is 41.2 (U.S. Census Bureau 2012c, Table 10). Of the total population, 12.9 percent are currently age 65 and older, while 16 percent of nonHispanic whites are 65 and above. These figures compare to just 9.6 percent of Asian Americans, 8.5 percent of blacks, and 5.7 percent of Hispanics (U.S. Census Bureau 2012c, Table 10).

Figure 16.2 looks at the nation's age distribution by race/ethnicity in another way. As you can see from that figure, about 80 percent of today's U.S. population over age 64 is nonHispanic white. Another 8 percent is African American, with another 7 percent Hispanic. But the older population is becoming more ethnically and racially diverse as members of race/ethnic minority

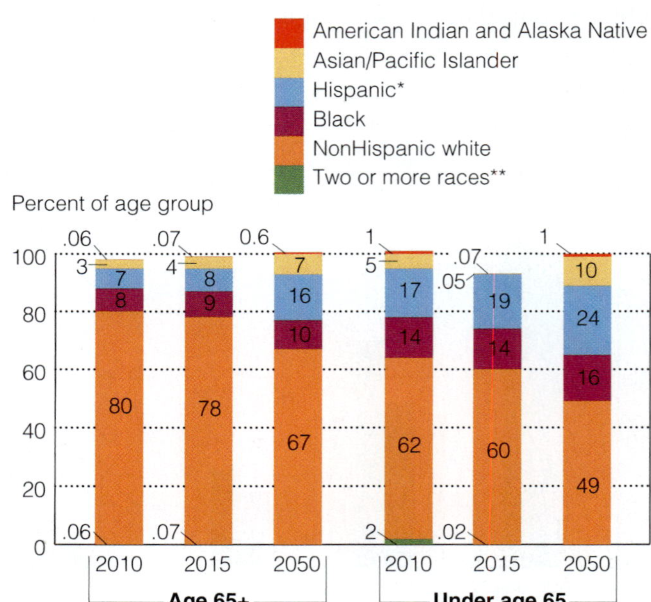

*Hispanic may be of any race.
**Projected figures for 2050 not available.
Percentage totals do not always add up to 100 because of rounding.

FIGURE 16.2 U.S. elderly and nonelderly population by race/ethnicity, 2010, with 2015 and 2050 projections.

Sources: Calculated from U.S. Census Bureau 2010b, Table 9; 2012c, Table 10.

groups grow older (Federal Interagency Forum on Aging-Related Statistics 2013, p. 4).

Due to immigration and to the relatively high birthrates of ethnic and racial minority groups, Hispanic, African American, and Asian populations are growing faster than nonHispanic whites. Members of these ethnic groups will age, of course. Hence by 2050, the nonHispanic white share of the population over age 64 is projected to fall from 80 to 67 percent. While some senior centers "offer *tai chi* exercise classes or serve tamales for lunch, a reflection of greater ethnic diversity," scholars argue for continued and more research on the multicultural needs of aging Americans (Treas 1995, p. 8; Trask et al. 2009).

Older Americans and the Diversity of Family Forms

We have seen throughout this text that today's families are diverse in form. As the postmodern family (see Chapter 1) grows older, we can expect late-life family forms to exhibit increasing diversity. You can see in Figure 16.3 that 72 percent of men and 45 percent of women age 65 and older were married in 2012. Thirteen percent of women and 10 percent of men in that age category were divorced or separated, with another 4 percent never married (U.S. Administration on Aging 2013, Figure 2).

The birthrate in the mid 1970s struck a record low while the divorce rate peaked. Americans who were in their twenties and thirties in the 1970s are now moving into later life. Increasingly, therefore, families will enter into older ages with fewer, if any, children and with histories of cohabitation, separation, divorce, and repartnering. Additionally, now that same-sex families are increasingly visible, we can expect researchers and policy makers to pay more attention to aging same-sex families (Grant 2010). In fact, the Gay and Lesbian Association of Retired Persons (GLARP) formed in 1999 (www.gaylesbianretiring.org). The caregiving implications of the trend toward increased family diversity are addressed later in this chapter. Here we turn to a discussion of living arrangements among older Americans.

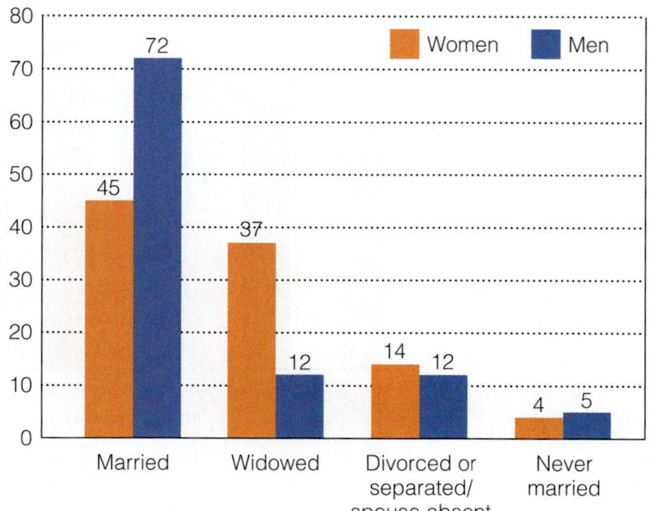

FIGURE 16.3 Marital status of persons age 65 and older, 2012.

Source: U.S. Administration on Aging 2013, Figure 2, p. 4.

Living Arrangements of Older Americans

About 1.3 million or 3.5 percent of Americans age 65 and older live in nursing homes. The likelihood of living in an institutional setting, such as a nursing home, increases with age: 1 percent of Americans between ages 65 and 74 reside in institutional settings. Among those between ages 75 and 84, the percentage is 3. Of those age 85 and older, 11 percent live in institutional settings (U.S. Administration on Aging 2013, p. 5).

Among the vast majority of older adults not residing in institutional settings, some have moved to retirement communities, some of these in the "Sun Belt"—Florida and the Southwest—as well as to communities in countries south of the U.S. border where the cost of living is considerably less (Bjelde and Sanders 2009; Banks 2009). A few retirement communities have begun to emerge specifically for lesbians and gays ("Birds of a Feather" 2010; Grant 2010). Some older Americans are cohabiting, whereas others live in cohousing arrangements, described in Chapter 6. Still others have established "living alone together" (LAT) relationships, also discussed in Chapter 6.

Then too, complementing the term, "boomerang kids," which refers to adult children who move back into their parents' homes, one journalist now writes of "boomerang seniors" (Kluger 2010). Since the onset of the recent economic downturn, a significant and growing number of older Americans—many of them healthy and active but needing to reduce expenses—have moved into their grown children's or grandchildren's homes (Lofquist et al. 2012, p. 15). "It's a return to much closer intergenerational ties than we saw through much of the twentieth century" (Stephanie Coontz, quoted in Brandon 2008). Then too, more frail elders today reside in their children's homes when the family cannot afford assisted living or other caregiving facilities . Recent studies have generally found that for the most part multigenerational households are happy (Donaldson 2012). In some instances a grandparent's moving in benefits the grandchildren's school performance (Augustine and Raley 2013).

Nevertheless, historical trends in family living arrangements in American society show a long-term preference for separate households (Bures 2009). Currently, as shown in Chapter 1's Figure 1.2, about 27 percent of U.S. households are made up of people living alone. Many of them are older people. This situation represents a growing trend since about 1940. Among Americans age 65 and older, approximately 30 percent live alone. Because of the increased likelihood of being widowed as people age, about 37.5 percent of people over age 75 live by themselves (U.S. Census Bureau 2012c, Table 58). As a result, 50 percent of women age 75 and older live alone (U.S. Administration on Aging 2010, p. 7).

Despite the fact that living alone can be a financial strain for elders, demographers project that—due to the increasing number of never-marrieds and to the significant proportion of middle-aged individuals who are divorced—an increasing percentage of older Americans will live alone in the future (Klinenberg, Torres, and Portacolone 2012). Among elderly parents, both they and their adult children generally prefer to live near one another, although not in the same residence (Moody 2006). The elderly who live alone "are usually within a close distance of relatives or only a phone call or email away. Fewer than one out of twenty are socially isolated, and usually are so because they have lived that way most of their lives" (Moody 2006, p. 331). It remains to be seen whether preferences will change as our aging population grows more ethnically diverse with more elderly coming from a heritage of communal, rather than individualistic values.

Table 16.1 Living Arrangements of People 65 Years Old and Over, by Race/Ethnicity, 2010

	Living Arrangement	65–74 Years Old (%)	75 and Older (%)
Total Population	Alone	22	37
	With spouse	64	45
	With other people[b]	14	18
NonHispanic White	Alone	22	39
	With spouse	67	47
	With other people[b]	10	14
Black	Alone	32	39
	With spouse	42	27
	With other people[b]	26	33
Asian American	Alone	14	21
	With spouse	65	51
	With other people[b]	21	28
Hispanic Origin[a]	Alone	19	22
	With spouse	54	41
	With other people[b]	27	37

[a]People of Hispanic origin may be of any race.
[b]The category, with other people, includes relatives other than a spouse, as well as institutional settings, although the proportion of older Americans in institutional settings is relatively small. See footnote 4 in this chapter for further information on aging individuals in institutional settings.

Source: Calculated from U.S. Census Bureau 2012c, Table 58.

Race/Ethnic Differences in Older Americans' Living Arrangements

Table 16.1 compares the living arrangements of nonHispanic white, black, Asian, and Hispanic adults age 65 and older. Due to both economic and cultural differences, the living arrangements of older Americans vary according to race and ethnicity. For instance, as shown in Figure 16.4, Asian Americans, Blacks, Hispanics, American Indians, Alaska Natives, Native Hawaiians, and Pacific Islanders in the United States are far more likely than nonHispanic whites to live in multigenerational households (Lofquist 2012). Older family members may reside in the homes of their grown children, or they may have opened their doors to adult offspring who have moved in with them (Glick and Hook 2002).

One generalization that we can make from the statistics in Table 16.1 and Figure 16.4 is that ethnic groups other than nonHispanic whites are significantly more likely to live in multigenerational households with people other than their spouse—grown children, siblings, or other relatives (Connidis 2007; Lofquist 2012). Partly as a result of economic necessity, coupled with social norms involving family members' obligations to one another, older Asian Americans and Hispanics are less likely than nonHispanic whites

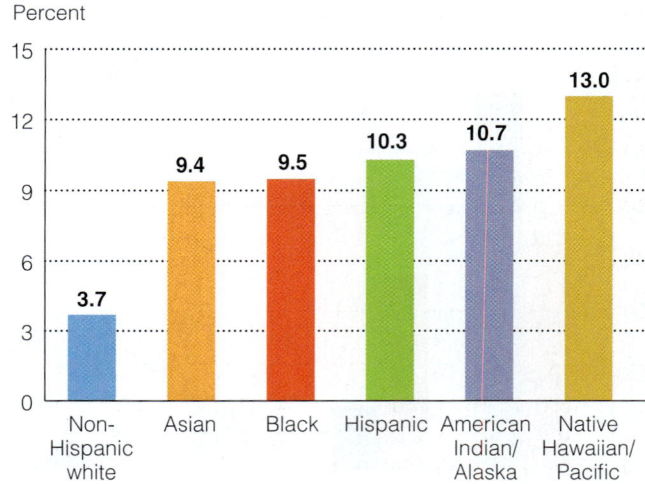

Figure 16.4 Intergenerational households as a percentage of family households in the relevant race/ethnic category, by race of the householder.

Source: Adapted from Lofquist 2012, Table 2.

to live alone. This is true for Hispanics even though they are also less likely than nonHispanic whites to live with a spouse (U.S. Census Bureau 2012c, Table 58).

Older Hispanics and African Americans are less likely than nonHispanic whites to live with a spouse for two reasons. First, due to differences in patterns of marriage and divorce, African Americans are more likely than whites to enter older ages without a spouse. Second, gender differences in life expectancy (with women living longer than men) are slightly higher among blacks and Hispanics (about seven years) than among nonHispanic whites (about five years) (U.S. Census Bureau 2010b, Table 102).

Gender Differences in Older Americans' Living Arrangements

Due mainly to differences in life expectancy, older heterosexual men are much more likely to be living with their spouse than are older heterosexual women (72 percent of men age 65 and older, compared to about 45 percent of women). Older women (40 percent) are far more likely than men (13 percent) to be widowed (U.S. Census Bureau 2012c, Table 34). Of Americans between ages 65 and 74 who are living alone, 66 percent are women. Of those 75 and over who are living alone, 76 percent are women (U.S. Census Bureau 2012b, Table 72). Older women are significantly more likely than older men to live with people other than their spouse—a pattern that persists into old-old age (U.S. Administration on Aging 2010, Figure 3).

Generally, among older Americans without partners, living arrangements depend on a variety of factors, including the status of one's health, the availability of others with whom to reside, social norms regarding obligations of other family members toward their elderly, the history of the family's relationship quality, personal preferences for privacy and independence, and economics (Bianchi and Casper 2000). Older Americans with better health and higher incomes are more likely to live independently, a situation that suggests strong personal preferences—at least among today's mostly nonHispanic white elderly—for privacy and independence. However, those in financial need are more likely to live with relatives.

Aging in Today's Economy

Today's older Americans live on a combination of employment income and/or investment income, Social Security benefits, private pensions from employers, personal savings, and social programs designed to meet the needs of the poor and disabled. About 40 percent of the income of Americans age 65 and older is from Social Security benefits and related federal programs, such as Medicare, Medicaid, and Supplemental Security Income (SSI). During the Great Depression of the 1930s, millions of Americans lost their jobs and savings. In response, the federal government passed the 1935 Social Security Act, a new program designed to assist the elderly. The Social Security Act established the collection of taxes on income from one generation of workers to pay monthly pensions to an older generation who had retired from work. Initially, only those who contributed to Social Security were eligible to receive benefits, but over the years, the U.S. Congress has extended coverage to spouses and to the widowed, as well as to the blind and permanently disabled.

Medicare, begun in 1965, is a compulsory federal program that does not provide money but does offer health care insurance and benefits to the aged, blind, and permanently disabled. Before the program began, just 56 percent of the aged had hospital insurance. In 1992, at least 97 percent of all older people in the United States had coverage because all who qualify for Social Security are eligible for Medicare.

In 1965, intending to provide health care to poor Americans of all ages, Congress created the Medicaid program in conjunction with Medicare. Eligibility for Medicaid is based on having virtually no family assets (saving and checking accounts, stocks, bonds, mutual funds, and any form of property that can be converted to cash) and very little income. In 1972, Congress created the federal Supplemental Security Income program (SSI), a program that provides monthly income checks to poverty-level older Americans and the disabled (Meyer and Bellas 2001).

Social Security benefits are the only source of income for one-fifth of Americans over age 64. The expansion of Social Security benefits to older Americans resulted in dramatic changes in U.S. poverty rates over the last several decades. Before Social Security was initiated, the elderly were disproportionately poor (Federal Interagency Forum on Aging-Related Statistics 2013, pp. 14–15). But poverty has declined sharply for those age 65 and over—from 36 percent in 1959 to about 10 percent today. Due partly to older Americans' lobbying to protect Social Security benefits, the poverty rate for those over age 64 is now less than half that of children (U.S. Federal Interagency Forum on Child and Family Statistics 2012, p. 12). Then too, many of today's older Americans benefit from what was a generally stable or rising economy during most of their working years.

However, having noted that today's older Americans are generally better off than generations preceding

Robert Brenner/PhotoEdit

Social Security and Medicare have raised the incomes of older Americans, beginning in 1940, so that the proportion of elderly in the United States living in poverty today has declined and is less than that of children. Nevertheless, about 3.7 million, or 10 percent of older Americans—disproportionately the unmarried and women—are living in poverty. Another 2.4 million of the elderly are classified as "near poor," with income up to 125 percent of the poverty level (U.S. Administration on Aging 2010).

them, we need to acknowledge that on average their income declines by up to one-half upon retirement (U.S. Census Bureau 2011b, Table S1903)—a situation that can lead to stress and relationship conflict (Dew and Yorgason 2010). Furthermore, the recession that began in 2009 made for unprecedented negative economic changes in the lives of many older Americans (Mossaad 2010).

Moreover, 9 percent of adults age 65 and older are living in poverty—nearly 20 percent of Blacks and/or Hispanics (U.S. Census Bureau 2012c, Table 713). Between 6 and 10 percent of the homeless are age 65 and older, and this proportion may be modestly increasing (National Alliance to End Homelessness 2013; National Coalition for the Homeless 2009b). Along with the "near poor" (those with incomes at or below 125 percent of the poverty level, who make up another 9 percent of the elderly), these poverty-level older Americans are hardly enjoying the comfortable and leisurely lifestyle that we may imagine when we think of retirement (Employee Benefit Research Institute 2009). For instance, Medicaid recipients living in nursing homes are wards of the state. Hence, their entire monthly income, except for a small personal-needs allowance, goes toward nursing home costs (Meyer and Bellas 2001).

Retirement?

From an historical standpoint, widespread retirement only became possible in the twentieth century, when the industrial economy was productive enough to support sizable numbers of nonworking adults. At the same time, the economy no longer needed so many workers in the labor force, and companies believed that older workers were not as quick or productive as the young. Governments, corporations, labor unions, and older workers themselves saw retirement as a desirable policy, and it soon became the normal practice (Moody 2006).

Today, however, an unpredictable economy, together with longer active life expectancy, may render the policy of retiring at about age 65 outdated (Brown et al. 2010; "Perpetually Deferred Retirement" 2012). Although most older people eventually retire, some do not—and many of those who don't are employed beyond age 70 (Purcell 2009, Figure 16). Then too, many people retire gradually by steadily reducing their work hours or intermittently leaving, then returning to the labor force before retiring completely (Kim and DeVaney 2005).

Not wanting to give up the psychological benefits associated with working—that is, feeling that one's life is meaningful and experiencing personal growth—is a reason that people give for not retiring. A less satisfying reason, increasingly relevant in today's economy and particularly applicable to divorced older women, is not being financially able to retire (Taylor 2009). Even before the onset of the recession that began in 2009, there was evidence that the majority of aging baby boomers expected to work at least part-time after retirement age (Brougham and Walsh 2009). By 2009, 38 percent of those over age 62 had delayed retirement due to the recession, according to the Pew Research Center. Among those between ages 50 and 61, 63 percent (54 percent of men and 72 percent of women) said that they might need to delay retirement because of the recession (Taylor 2009).

Gender Issues and Older Women's Finances

A recent demographic study of aging baby boomers shows that, continuing a pattern described in Chapters 6 and 7, among both women and men, those who married and stayed together are the best off financially, with never-marrieds faring worst (Addo and Lichter 2013; Lin and Brown 2013). Compared to widows, never-married, divorced and separated women are worse off financially and many need to work for several years after traditional retirement age (Herd 2009; Ulker 2009). In fact, the marriage gap between U.S. blacks and whites accounts for much of the wealth gap between the races in old age (Addo and Lichter 2013).

Nevertheless, on average older men are considerably better off economically than are older women. In 2011, the median income of American individuals age 65 and older was $27,707 for males and $15,362 for females (U.S. Administration on Aging 2013). This dramatically unequal situation is partly due to the fact that, throughout their employment years, men averaged higher earnings than did women (U.S. Bureau of Labor Statistics 2012b; and see Chapters 3 and 10). Consequently, older women today have smaller, if any, pensions from employers and lower Social Security benefits. Furthermore, older women on average did not begin to save for retirement as early as did men (Even and Macpherson 2004; Herd 2009).

Moreover, although women are much more likely than men to rely on Social Security for at least 90 percent of their income, women's Social Security benefits average only about 76 percent of men's. Maximum Social Security benefits ($2,533 monthly in 2013) are available only to workers with lengthy and continuous labor force participation in higher-paying jobs (U.S. Social Security Administration 2013). This situation works against older women today, who either did not participate in the labor force at all or are likely to have dropped in and out of the labor force while taking lower-paying jobs (Herd 2009). "Thus, women are penalized for conforming to a role that they are strongly encouraged to assume—unpaid household worker—and their disadvantaged economic position is carried into old age" (Meyer and Bellas 2001, p. 193).

Unfortunately the future does not look much better. Today singles of both genders in their twenties and thirties—and particularly single mothers, many of whom cannot afford to put away money—evidence inadequate retirement saving activities (Knoll, Tamborini, and Whitman 2012).

At the time of this writing, the federal Defense of Marriage Act (DOMA) precludes even legally married same-sex couples from receiving federal married-couple benefits, such as Social Security payments to a widowed or disabled spouse (Grant 2010; and see Chapter 6 in this text).

However, an older wife married heterosexually for at least ten years to a now-retired worker can receive a spousal "allowance," equal to one-half of her husband's benefits. Employed women may qualify on the basis of either their own or—if legally married by federal (not just state) definitions—their husband's work records, although they cannot receive benefits under both categories. Ex-spouses also qualify for one-half the amount of their ex's Social Security benefits, provided the marriage lasted at least ten years. We turn to an examination of couple relationships in later life.

Relationship Satisfaction in Later Life

As shown in Figure 16.3, only about 4 percent of men or women 65 and older today have never married (U.S. Administration on Aging 2010, Figure 2). Some later-life marriages are remarriages, and the proportion of repartnered older Americans will increase as those who are now middle-aged grow older. Today, however, the majority of older, heterosexually married couples have been wed for quite some time—either in first or second marriages. Retirement represents an important and, usually temporarily, stressful change for couples (Dew and Yorgason 2010; Wickrama, Walker, and Lorenz 2013).

Chapter 5's discussion of love and mate selection suggests topics that couples are encouraged to discuss early in their relationship—such things as sexual expectations and preferences; partners' roles; extended family obligations; how much time alone each needs (and how much togetherness); and what do intimacy, commitment, and relationship responsibility mean to each partner. Counselors suggest that couples revisit topics such as these as they move into retirement (Mintzer and Taylor 2011).

For both partners, role flexibility is important to successful adjustment (Bulanda 2011). Also, health is an important factor in morale in later life, and it has a substantial impact on marital quality as well (Connidis 2010; Radina 2013). The majority of older married couples place companionship and intimacy as central to their lives and describe their unions as happy (Szinovacz and Schaffer 2000; Villar and Villamizar 2012; Walker et al. 2001).

On average, older couples report having fewer disagreements, and marital happiness often increases in later life when couples have the time, energy, and financial resources to invest in their relationship (Hatch and Bulcroft 2004; see also Story et al. 2007). Mothers' marital satisfaction has been shown to increase with age, a finding that is especially true after their grown children leave home (Association for Psychological Science 2008). As a category, aging marrieds are happier and more satisfied with their lives than are their nonmarried counterparts (LaPierre 2009; Villar and Villamizar 2012).

This is not to say that all older couples are happy together. A spouse's being depressed or otherwise in

Ariel Skelley/Photographer's Choice/Getty Images

Married older couples may be in first unions or in remarriages. Most older couples describe their marriages as happy. A retired husband may choose to spend more time doing homemaking tasks and give increased attention to being a companionate spouse. Role-sharing, feeling that work is fairly shared, and having supportive communication predict good adjustment for retiring couples.

poor mental or physical health can negatively affect couple happiness (Wickrama, O'Neal, and Lorenz 2013). A study using national data found that older spouses who felt unfairly treated by their mates were more distressed than were singles of the same age (Hagedoorn et al. 2006). Other research has found that later-life couples who hold more egalitarian attitudes toward gender roles and who experience high levels of warm mutual interaction report significantly greater marital happiness (Kaufman and Taniguchi 2006; Schmitt, Kliegel, and Shapiro 2007).

Sexuality in Later Life

As discussed in Chapter 4, the frequency of sexual intercourse tends to decline with age (Marshall 2009). Nonetheless, many older Americans continue to be sexually active—even into old-old age and even in nursing homes (Lindau and Gavrilova 2010; Schwartz 2012). A 2009 national survey of Americans age 45 and older found that among respondents age 70 and older, 11 percent of women and 22 percent of men reported having sexual intercourse at least once or twice a month (Fisher 2010, Tables 19 and 20). Whether it involved intercourse or other forms of sexual expression, 80 percent of men and 39 percent of women age 70 and older said that a sexual relationship was important to their quality of life (Fisher 2010, p. 9).

This is not to imply that older adults have no sexual problems. Although aging single women may be interested in sex, lack of a partner can be a problem. We have seen that, as they age, women are far more likely than men to be widowed. Moreover, as they grow older, women are adversely affected by the *double standard of aging*—that is, men aren't considered old or sexually ineligible as soon as women are (England and McClintock 2009). Beauty, "identified, as it is for women, with youthfulness, does not stand up well to age" (Sontag 1976, p. 352). For older single women, this situation can exacerbate more general feelings of loneliness (Narayan 2008).

Moreover, stress, dissatisfaction with one's partner, and health-related issues can inhibit sexual desire and activity for both sexes (Laumann, Das, and Waite 2008). According to psychiatrist Stephen Levine, "Over age 50, the quality of sex depends much more on the overall quality of a relationship than it does for young couples" (quoted in Jacoby 1999, p. 42; see also Elliott and Umberson 2008).

Later-Life Divorce, Widowhood, and Remarriage

The majority of couples who divorce do so before their retirement years. However, some couples do divorce in later life, and baby boomers are divorcing at relatively high rates at older ages (Brown and Lin 2012). As noted in Chapter 14, "silver," or later-life divorces are not necessarily easy on the couple's adult children. Family celebrations and holidays are disrupted. An adult child's graduation or wedding can be difficult when forced to accommodate recently divorced parents.

Furthermore, adult children of divorcing parents may worry about having to become full-time caregivers to an aging parent in the absence of the parent's spouse. "Years after parents split, their children may wind up helping to sustain two households instead of one, and those households can be across town or across the country" (Span 2009a). Nevertheless, although some later-life marriages end in divorce, the vast majority do so with the death of a spouse.

Widowhood and Widowerhood

Adjustment to widowhood or widowerhood is an important common family transition in later life. A spouse's death brings the conjugal unit to an end—often a profoundly painful event. "The stress and emotional trauma of losing a spouse as a confidant might be greater now than in the past, [for those who do not divorce] the average duration of marriage becomes longer with increasing life expectancy" (Liu 2009, p. 1,170).

Because women's life expectancy is longer and older men remarry more often than women do, widowhood is significantly more common than widowerhood. Just over half (51 percent) of women between ages 75 and 84 are widowed, compared with 17 percent of men. Among those age 85 and over, just under three-quarters of women (73 percent) are widowed, compared with over one-third (35 percent) of men (Federal Interagency Forum on Aging-Related Statistics 2013, p. 87).

Typically, widowhood and widowerhood begin with **bereavement**, a period of mourning, followed by gradual adjustment to the loss and to the new, unmarried status. Bereavement manifests itself in physical, emotional, and intellectual symptoms. Recently, widowed people perceive their health as declining and report feeling depressed. The bereaved experience various emotions—anger, guilt, sadness, anxiety, and preoccupation with thoughts of the dead spouse—but these feelings tend to diminish over time (Connidis 2010; Jin and Chrisatakis 2009). Social support, adult children's help with housework and related tasks, and activities with friends, children, and siblings help (Cornwell and Waite 2009; Population Reference Bureau 2009).

There is evidence that being single in old age is more physically and emotionally detrimental for men than for women. Wives, as opposed to husbands, tend to be central "in the household production of health" (Jin and Chrisatakis 2009, p. 605). Put another way, females are more likely than males to concern themselves with the health of all household members, so in a wife's absence, a man's health is more likely to decline than vice versa. Then too, as sources of support, women more often have friends in addition to family members. Men are more often dependent solely on family. For some of the widowed, remarriage promises resumed intimacy and companionship.

Aging and Remarriage

Annually about one-half million Americans over age 65 remarry (Belkin 2010). Remarrying elders, particularly those who are widowed, are likely to choose homogamous partners similar in social class, race/ethnic, and religious identity. Sometimes the new partner is someone who reminds them of their first spouse or is someone whom they've known for years.

However, middle-aged and older people may face considerable opposition to remarriage—from restrictive pension and Social Security regulations (Ebeling 2007) and from their adult children. In general, a widow/widower cannot receive Social Security survivors' benefits if he or she remarries before age 60. Remarrying after age 60 (or 50 if disabled) does not negatively affect receipt of survivor benefits (Social Security Online n. d.).

Chapter 15 fully explores adult children's attitudes about a parent's remarriage. Although some grown children may be supportive of a parent's remarriage, others may find it inappropriate (Sherman and Boss 2007). Adult children may worry about the biological parent's potentially diminished interest in them or about their inheritance (Connidis 2010). Even when a will leaves everything to one's children, a second spouse may have the legal right to claim a share of the estate. Presuming that the remarrying couple desires it, a common option is signing a prenuptial agreement according to which the spouse-to-be relinquishes any claim to the estate (Ebeling 2007; see also Barnes 2009). Older couples with children from previous marriages are encouraged to see a financial adviser before marrying.

Especially for women, age reduces the likelihood of remarrying. Although the pattern of remarriage since 1960 has been similar for all age categories, the remarriage rate for younger women is consistently higher than for older women (Moorman, Booth, and Fingerman 2006). An uneven sex ratio (see Chapter 6) decreases the odds that older heterosexual women will repartner: Among Americans age 55 and older, there are approximately eighty-three men for every 100 women (U.S. Census Bureau 2012c, Table 9). The double standard of aging also works against women in the remarriage market (England and McClintock 2009). Then too, women may be less interested than men in late-life remarriage (Levaro 2009).

Meanwhile, research shows that the myth that widowers are quick to replace a deceased wife is just that—an exaggerated stereotype. "Men with high levels of social support from friends are no more likely than women to report interest in repartnering" (Carr 2004, p. 1,065). During later life, morale and well-being frequently derive from relations with siblings, as well as from friends, neighbors, and other social contacts (Nelson 2011; Voorpostel and Blieszner 2008). Particularly for aging mothers, relationships with children and grandchildren are important.

Multigenerational Ties: Older Parents, Adult Children, and Grandchildren

More often than spousal relationships, those between parents and their biological children last a lifetime (Kaufman and Uhlenberg 1998). Many parents continue to aid their adult offspring virtually as long as possible. Often following traditional gender roles, aging mothers are likely to help with childcare while older fathers may help adult offspring with home repairs, yardwork, or car repair (Kahn, McGill, and Bianchi 2011). In this section, we examine relations between older parents and their adult children and grandchildren.

Older Parents and Adult Children

Adults' relationships with their parents range from *tight-knit*, to *sociable*, to *obligatory*, to *intimate but distant*, or *detached*. Sociologists Merril Silverstein and Vern Bengston (2001) developed six indicators of relationship solidarity, or connection: geographic proximity, contact between members in a relationship, emotional closeness, similarity of opinions, providing care, and receiving care. Based on survey evidence and using these six indicators, Silverstein and Bengston then developed a typology of these five kinds of parent–adult child relations.

Parent–adult child relations vary depending on how family members combine—or, in the case of the detached relationship style, do not combine—the six indicators. For instance, in tight-knit relations, the parent and the adult child live near each other (geographic proximity), feel emotionally close, share similar opinions, and help each other (give and receive assistance). Sociable relations involve all these characteristics except that the parent and adult child do not exchange assistance. Table 16.2 defines all five relationship types.

Research shows that there is no one typical model for parent–adult child relationships (Arnett 2004; Silverstein and Bengston 2001). Furthermore, parent–adult child relations might change over time, moving from one relationship type to another depending on the parent's and the adult child's respective ages, the parent's changed marital status, and the presence or absence of grandchildren, among other factors. For instance, to be nearer to their aging parents, adult children sometimes return to the area in which they grew up, or retired grandparents may decide to relocate to be near their grandchildren (Lee 2007). Both of these situations could move an intimate-but-distant relationship to a tight-knit one. Then, too, a parent–adult child relationship might change depending only on emotional factors, such as when an adult child chooses to forgive an aging parent for some past transgression, or vice versa.

Using national survey data from a sample of 971 adult children who had at least one surviving nonco-resident parent, Silverstein and Bengston (2001) made the following findings (among others):

- The majority of relations were neither tight-knit nor detached, but "variegated"—one of the three relationship styles in between (see Table 16.2). Variegated

Table 16.2 Types of Intergenerational Relations

Class	Definition
Tight-knit	Adult children are engaged with their parents based on geographic proximity, frequency of contact, emotional closeness, similarity of opinions, and providing and receiving assistance.
Sociable	Adult children are engaged with their parents based on geographic proximity, frequency of contact, emotional closeness, and similarity of opinions but not based on providing or receiving assistance.
Obligatory	Adult children are engaged with their parents based on geographic proximity and frequency of contact but not based on emotional closeness and similarity of opinions. Adult children are likely to provide or receive assistance or both.
Intimate but distant	Adult children are engaged with their parents based on emotional closeness and similarity of opinions but not based on geographic proximity, frequency of contact, providing assistance, and receiving assistance.
Detached	Adult children are not engaged with their parents based on any of these six indicators of solidarity.

Source: Adapted from Silverstein and Bengston 2001, p. 55.

relations characterized 62 percent of adult children's interaction with their mothers and 53 percent with their fathers.

- Tight-knit relations are more likely to occur among lower socioeconomic groups and race/ethnic minorities.
- NonHispanic whites were more likely than African Americans to have detached relationships with their parents and more likely than blacks or Hispanics to have obligatory relationships with their mothers.
- The most common relationship between a mother and her adult child was tight-knit. The next most common was sociable, followed by intimate but distant, obligatory, and, finally, detached.
- The most common relationship between a father and his adult child was detached, followed by sociable, tight-knit, obligatory, and intimate but distant. Almost four times as many adult children reported being detached from their fathers as from their mothers.
- Daughters were more likely than sons to have tight-knit relations with their mothers.
- Sons were more likely than daughters to have obligatory relations with their mothers.
- Adult children were more likely to have obligatory or detached relations with divorced or separated mothers than with married mothers.
- Adult children were more likely to have detached relations with divorced or separated fathers than with consistently married fathers.

Mackenzie Stroh Photography

Financially independent adults' relationships with a parent can be of several types: tight-knit, sociable, obligatory, intimate but distant, or detached. Then too, today's parents can find themselves in the "senior sandwich generation"—paying for a child's college tuition, worrying about the financial burden of elder care for aging parents, while trying to save for their own retirement.

From these findings, we can conclude that daughters are more likely than sons to have close relationships with their parents, especially with their mothers. Even for mothers, a parent's divorce or separation often weakens the bond with adult children (Hans, Ganong, and Coleman 2009). However, having detached relations with one's adult children after divorce is nearly five times greater for fathers than for mothers. Partly, at least, this is true because a divorced father is less likely than either a consistently married father or a divorced mother to live with his biological children and more likely to remarry (Pezzin, Pollak, and Schone 2008; Silverstein and Bengston 2001).

In some families, the reality of past abuse, a conflict-filled divorce, or simply fundamental differences in values or lifestyles makes it seem unlikely that parents and children will spend time together (Arnett 2004; and see Winter, Gitlin, and Dennis 2011). Money matters can also cause tension. This is especially true in stepfamilies where "[a]dult children can feel resentful when they see a stepparent spending what they consider as their rightful inheritance" (Sherman 2006, p. F8). Especially if they have children, older persons are encouraged to thoroughly discuss and agree on money matters prior to a remarriage (Yip 2012). Overall, the majority of adult children's relationships with parents, although not necessarily tight-knit, continue to be meaningful.

Grandparenthood

As noted in Chapter 1, technology and the desire for grandchildren can be strong enough that some people finance freezing the eggs of an adult daughter not yet ready to have children (Gootman 2012). Much more commonly, medical technology and grandparenthood meet as grandmothers-to-be study ultrasound images of their grandchild's developing fetus (Harpel and Hertzog 2010).

Thanks to longer life expectancy, which creates more opportunity for the role, grandparenting—and great-grandparenting—became increasingly important to families throughout the twentieth and into the twenty-first centuries (Bengston 2001; Rosenbloom 2006). Many older Americans are raising grandchildren, as discussed in Chapter 9. This section focuses on grandparents who are not primarily responsible for raising their grandchildren.

Younger grandparents are often employed and partnered while older grandparents may have physical disabilities. Hence, a grandchild's experiences with a younger grandparent is typically quite different from those of with an older grandparent (Davey et al. 2009). Married grandparents have different experiences with grandchildren than do single grandparents. Once retired, however, both married and single grandfathers spend more time with grandchildren than they did while they were employed (Kahn, McGill, and Bianchi 2011). Geographical distance affects grandparent-grandchild relationships (Dunifon and Bajracharya 2012).

Grandparents tend to adopt a grandparenting style similar to one that they themselves experienced as a grandchild (Mueller and Elder 2003). Some grandparents (especially when they live far away) have remote relationships with their grandchildren, while other grandparent-grandchild relationships are more companionate, doing things with their grandchildren but exercising little authority in the grandparent-parent-grandchild relationship. Still other grandparents are highly involved (Cherlin and Furstenberg 1986; see also Dunifon and Bajracharya 2012).

Then too, a grandparent may have different relationship styles with different grandchildren. Although grandparents do interact with their teenage grandchildren, they generally are most actively involved with (more available and emotionally responsive) preschoolers (Davey et al. 2009). After a typically uninterested adolescence, adult grandchildren often renew relationships with grandparents (Cherlin and Furstenberg 1986; see also Mansson, Myers, and Turner 2010).

Grandchildren give personal pleasure and a sense of immortality. Some elderly grandfathers see the role as an opportunity to be involved with babies and very young children, an activity that may have been discouraged when their own children were young (Cunningham-Burley 2001). Overall, the grandparent role is mediated by the parent. Not getting along with the parent dampens the grandparent's contact and, hence, the relationship with her or his grandchildren (Dunifon and Bajracharya 2012).

Grandparents may serve as valuable "family watchdogs," ready to provide assistance when needed (Troll 1985; Winerip 2009). For instance, the Interactive Autism Network (2010) conducted an online survey of individuals who had a grandchild with autism. This survey was hardly representative of all grandparents in this category, because to be aware of the survey one would have had to be interested enough in the topic to visit the website. However, it is interesting to note that nearly one-third of the grandparent respondents reported being the first in the family to notice anything out of the ordinary regarding their grandchild's development. Some grandparents (14 percent) had moved closer to their grandchild's family to help out. Many grandparents assisted with treatment-related costs, some dipping into retirement savings (Hamilton 2010; Interactive Autism Network 2010).

A close relationship with a grandparent can help to facilitate a grandchild's adjustment after parental divorce (Henderson et al. 2009). If an adult child of divorced parents becomes divorced, financial assistance and emotional support may be more readily available from a grandparent (Leopold and Schneider 2011; Timonen, Doyle, and O'Dwyer 2011). We might note too that **surrogate grandparents** (a form of fictive kin, not biologically related to the "grandchildren") sometimes attach to families (and vice versa). Surrogate grandparents "fill some of the gaps in our mobile society" for both older people and children (O'Brien n. d.).

In low-income and ethnic-minority families, parents and children readily rely on grandparents for child care and other help. Even among middle- and upper-middle-class families, it is not unusual for grandparents to help with child care—or contribute to the cost of a grandchild's schooling, wedding, or first house. A (pre-recession) *New York Times* article featured several upper-middle-class grandparents who commuted by plane weekly to help with child care (Lee 2007).

Race/Ethnicity and Grandparenting Although there isn't yet much research on the subject—maybe you'll do some in the future—we know some ways that race/ethnicity affects grandparenting (Karasik and Hamon 2007). For instance, a study found that 87 percent of black grandparents felt free to correct a grandchild's behavior, compared to just 43 percent of white grandparents. As one black grandmother said of her fourteen-year-old grandson, "He can get around his mother, but he can't get around me so well" (quoted in Cherlin and Furstenberg 1986, p. 128).

As another example, Native American elders have traditionally served as *cultural conservator grandparents*—actively seeking contact and temporary coresidence with their grandchildren "for the expressed purpose of exposing them to the American Indian way of life" (Weibel-Orlando 2001, p. 143, quoted in Karasik and Hamon 2007, p. 145). Maintaining an ethnic-minority culture into future generations is typically a concern for ethnic-minority grandparents.

As a third example—and one that addresses transnational families—in a study of 112 Asian-Indian grandchildren in the United States, 40 percent had weekly phone conversations with grandparents in India;

Grandparenting styles differ—shaped by the grandparent and grandchild's ages and personalities, as well as by the grandparent's health and employment status. Ties with a grandchild can be remote, companionate, or involved, and a grandparent may have different ties with different grandchildren. Maintaining one's culture into future generations may be of particular concern for an ethnic-minority, *cultural conservator* grandparent.

Walter Hodges/Corbis/Glow Images

another one-third called one or more times a month. Seven percent maintained weekly email contact. Forty-three percent of the grandchildren visited their grandparent(s) in India every two years (Saxena and Sanders 2009).

Divorce, Remarriage, and Grandparenting How does a grown child's divorce affect a grandparent relationship? Evidence suggests that a grandparent may fret over whether to intervene on behalf of the grandchildren. As might be expected, effects of the divorce are different for the **custodial grandparent** (parent of the custodial parent) than for the **noncustodial grandparent** (parent of the noncustodial parent), with noncustodial grandparents significantly less likely to see their grandchildren as often as they had before the divorce (Connidis 2010). Because mothers more often get custody, the most common situation is for maternal grandparent relationships to be maintained or enhanced while paternal ones diminish (Connidis 2009; Henderson et al. 2009).

Due to pressure from noncustodial grandparents, states have passed laws giving grandparents the right to seek legalized visitation rights, but courts are reluctant to do so when parents object. Grandparents who go to court to seek visitation rights are successful between 30 and 40 percent of the time. When courts recognize visitation rights for grandparents in spite of parental objections, the reason usually involves the best interests of the child (Henderson 2005a, 2005b).

Remarriages and re-divorces create step-grandparents and ex-step-grandparents. Little information exists on step-grandparents, but available data suggest that they tend to distinguish their "real" grandchildren from those of remarriages. Asked about his step-grandparents, one young man told this story:

> They were pretty good, but again, there was that line. And we knew.... I mean I remember a couple of Christmases ago... [my grandmother made] a family quilt and everyone was on it except me and my brother.... As we got older the fact that they weren't our grandparents became more predominant. (in Kemp 2007, p. 875)

Younger step-grandchildren and those who live with the grandparent's adult child are more likely to develop ties with the step-grandparent (Coleman, Ganong, and Cable 1997; Pacey 2005). "As We Make Choices: Tips for Step-Grandparents" suggests ways to foster positive relationships with step-grandchildren. We turn now to an examination of caregiving to aging family members.

As We Make Choices

Tips for Step-Grandparents

Stepchildren can benefit from an older adult's genuine concern and support. The following advice to step-grandparents has been excerpted from the University of Florida Extension website.

> You may [have] become an instant grandparent with step-grandchildren. You may have both grandchildren and step-grandchildren in the same family.... You probably have many thoughts and feelings about this role. You may think:
>
> - I'm not old enough or ready to be a grandparent.
> - This interferes with dreams about the birth of my first grandchild.
> - Will my step-grandchild like me? Will I like my step-grandchild?
> - Is it okay to feel differently toward my step-grandchildren than my real grandchildren?
> - Will our family celebrations and traditions have to change?...
>
> Remember that relationships are built over time. Your relationship and role as a step-grandparent will take time to develop. Communicate and spend time together in order to get to know each other.
>
> Recognize the vital role of grandparents and step-grandparents in today's families. You can offer children in busy stepfamilies companionship, time, and a listening ear... children who are exposed to such contact are less fearful of old age and the elderly. They feel more connoted to their families.
>
> Create the grandparenting role that is comfortable to you and rewarding for your stepfamily. Step-grandparenting, like other stepfamily roles, is challenging and undefined. It is up to you to carve a role for yourself that fits your son or daughter's new family. Here are some things to consider doing with step-grandchildren:
>
> - Spend time one-on-one with them.
> - Teach them a game or skill.
> - Listen for their concerns, as well as their joys.
> - Offer companionship for activities they enjoy.
> - Share your history and family traditions.
> - Show them acceptance.

Critical Thinking

How might a step-grandparent benefit from choosing to follow some of these suggestions? How might a step-grandchild benefit?

Source: Excerpted from Ferrer-Chancy 2009.

Aging Families and Caregiving

When we associate caregiving with the elderly, we may tend to think only in terms of older generations as care *recipients.* However, about one-fifth of Americans age 75 and older are engaged in some form of care *giving,* whether child care or caring for other elders (Shapiro 2006b). In addition to assisting adult children and grandchildren financially and otherwise, older Americans give much to their communities. Many are volunteers in their churches, hospitals, schools, and various other settings. In this section, however, we focus on **elder care**—that is, care provided to the elderly.

Elder care involves emotional support, a variety of services, and, sometimes, financial assistance. A growing number of tax-funded, charity, and for-profit services provide elder care. Nevertheless, the persisting social expectation in the United States is that family members will either care for elderly relatives personally or organize and supervise the care provided by others. About 43.5 million adults age 18 and over (nearly 19 percent of all American adults) are engaged in **informal caregiving**—unpaid and personally provided care—to a family member or friend age 50 or older (National Alliance for Caregiving 2009). Adults responsible for aging family members often pay for some of the care recipient's expenses, such as for groceries, drugs, medical co-payments, or transportation (Gross 2007).

Being concerned about elderly family members might involve nothing more than making a daily phone call to make sure they're okay, stopping by for a weekly visit, or accompanying them to appointments to translate or help explain. However, **gerontologists** (social scientists who study aging) specifically define **caregiving** as "assistance provided to persons who cannot, for whatever reason, perform the basic activities or instrumental activities of daily living for themselves" (Uhlenberg 1996, p. 682). Caregiving may be short term (taking care of someone who has recently had joint-replacement surgery, for example) or long term.

About 16 percent of caregivers help out for fewer than six months, but more than one-third provide care for between one and four years, and up to 30 percent assist a disabled family member for five years or more. On average, long-term caregivers spend about twenty hours each week in this role, with another 11 percent spending more than forty hours per week. About 50 percent of caregivers are employed full-time; 11 percent, part-time (National Alliance for Caregiving 2009, Figures 4 and 8, and p. 14).

The majority of the young-old need almost no help at all, but as individuals age, they may increasingly require assistance with tasks such as paying bills and, later,

Facts about Families

Community Resources for Elder Care

More older Americans and their care providers are turning to professional elder care service providers for help. The table in this box defines several of these services and facilities. The following are three considerations regarding making decisions about these options:

1. *Don't wait.* Exploring options before they're needed helps all family members know what to expect and begin to prepare for the future (Greenwald 1999, p. 53).
2. *Seek expert advice.* Geriatric social workers can help assess an elderly person's needs and develop action plans.
3. *Shop around.* To compare cost and quality, families should visit as many facilities as possible on different days of the week and hours of the day. Ask for references (Shapiro 2001).

Critical Thinking

What kinds of community elder care services can you think of that are not included in this table? How might they be useful to caregivers and to the elderly as well?

The Options	What Is It?
Home care	Wide range of services, including shopping and transportation, health aides who give baths, nurses who provide medical care, and physical therapy brought to the home (estimated average cost $20,000–22,000 per year)
Adult day care	A place to get meals and spend the day, usually run by not-for-profit agencies
Congregate housing	A private home within a residential compound, providing shared activities and services; sometimes considered one type of assisted living
Assisted living	Numerous kinds of housing with services for people who do not have severe medical problems but who need help with personal care such as bathing, dressing, grooming, or meals (estimated average cost $42,000 per year)
Continuing-care retirement community (CCRC)	A complex of residences that includes independent living, assisted living and nursing home care, so seniors can stay in the same general location as their housing needs change over time, beginning when they are still healthy and active
Nursing home (skilled nursing facilities)	Residential facilities with twenty-four-hour medical care available for those who need continual attention (estimated average cost $85,000 per year)

Sources: "Choosing Senior Housing" 2007; Geewax 2012b; also see "Eldercare Locator" at www.eldercare.gov, and "Glossary of Senior Housing Terms" (2010) at senioroutlook.com/glossary.asp.

bathing or eating (National Alliance for Caregiving 2009, Figure 5). Severely ill, physically disabled older people, and those with dementia often need a great deal of continuing care. An increasingly common alternative involves continuing care facilities that first offer small apartments for elderly able to care for themselves but with meals, housekeeping services, transportation to community events, and in-house social programs; then assisted living services, such as help with bathing; then nursing home care. For families who can afford them, continuing care facilities ease some eldercare decisions and needs. "Facts about Families: Community Resources for Elder Care" describes elder care options.

A number of elder care givers are relatively young—in their thirties, with some in their twenties, partly because children born to older parents begin elder care at younger ages (Allen 2012; Bigda 2013). Then too, grandchildren may be providing elder care (Fruhauf, Jarrott, and Allen 2006). As explored in Chapter 13's "A Closer Look at Family Diversity: Young Caregivers," some family caregivers are children, many of whom essentially put their life—including their childhood—on hold (Wilson 2012). However, the mean age for caregivers is around 50. Between about 15 percent and one-third of caregivers are age 65 or older—and may themselves be in poor health or beginning to suffer from age-related disabilities (Johnson and Wiener 2006; National Alliance for Caregiving 2009, p. 15).

Which family members provide elder care and how much they provide depend on the family's understanding of who is primarily responsible for giving the care, as well as the care receiver's preference (Connidis and Kemp 2008). The first choice for a caregiver is an available spouse, followed by adult children, siblings, grandchildren, nieces and nephews, friends and neighbors, and, finally, a formal service provider. This system of elderly care receivers' preference for caregivers is termed the **hierarchical compensatory model of caregiving** (Cantor 1979; Horowitz 1985; and see Nelson 2011).

Older LGBT individuals may be more inclined than straights to rely on friends, often fellows in their LGBT communities and more likely found in urban than rural areas (Lee 2013). A small study of seventeen gay men

Issues for Thought

Filial Responsibility Laws

Should filial responsibility be a law? Consider the following argument by a legal scholar.

> A filial responsibility statute is simply a law that "create[s] a statutory duty for adult children to financially support their parents who are unable to provide for themselves." Typically, such support includes an obligation to pay for "food, clothing, shelter, and medical attention." The rationale behind such laws arises from the reciprocal duty that parents have to care for their children; because parents extended voluntary care to their minor children, it is the filial responsibility of children to return that support to their parents....
>
> In 1601, filial responsibility was put into statutory form with the [British] enactment of the Elizabethan Poor Relief Act, from which most modern statutes are derived.... Carrying on the historical tradition of supporting indigent parents, [in] the United States... prior to the 1960s, federal legislation recognized this obligation as well. However, ... the establishment of Medicare in the 1960s led to the repeal of the federal statute. Today, there are twenty-two states with filial responsibility statutes, and few—if any—of the states currently enforce these laws.... [C]hildren are only required to support their parents so long as "they are of sufficient ability".... Therefore, children should be excused from their filial responsibility if they have no economic means of supporting their parents....
>
> Children can also avoid filial responsibility if they can demonstrate the parent abandoned them....
>
> [According to one court decision] "The selection of the adult children is rational on the ground that the parents, who are now in need, supported and cared for their children during their minority and that such children should in return now support their parents to the extent to which they are capable.... [Filial responsibility] statutes would be beneficial to our society and provide desperately needed relief for our strained public treasury."

Critical Thinking

Other experts argue that (1) there is already considerable voluntary assistance from children to parents; (2) filial responsibility legislation can undermine parent–child relationships, creating resentment on the payee's part and guilt on the recipient's part; (3) government programs such as Social Security are preferable. Making Social Security contributions, children are paying for parents, but without the tension created by legislated direct payments (Callahan 1985). In your opinion, what would be some benefits of requiring filial responsibility to families and to society? What might be some drawbacks?

Source: Excerpted from Lundberg 2009, 534–582.

and twenty lesbians over age 65 in Los Angeles concluded that, when relying on neighbors or acquaintances for social support, aging lesbians and gays differ in their approaches (Rosenfeld 1999). Those who formed a gay or lesbian identity before the 1970s, when homosexuality was highly stigmatized, tended to hide their sexual identity. Others—usually baby boomers who formed a lesbian or gay identity during the 1970s or later (in a "gay liberation atmosphere")—celebrate their identity, often becoming activists for LGBT rights regarding elderly housing, health care, and other services.

Whether gay or straight, older Americans provide a considerable amount of elder care to each other. Caregiving and receiving are expected components of marriage for most of today's older couples, who have developed a relationship of mutual exchange over many years (Machir 2003; Roper and Yorgason 2009). Among marrieds, spouses of either gender provide up to 40 percent of elder care (Johnson and Weiner 2006). Caregivers in longer, emotionally close marriages with little ongoing conflict evidenced better overall well-being (investment) (Roper and Yorgason 2009). Caring for a disabled spouse in later years can involve mentally and emotionally reconstructing a new sense of marital closeness based on a shared history (Boylstein and Hayes 2012).

For the uncoupled and childfree, siblings are important in mutual caregiving (Eriksen and Gerstel 2002). After spouses, however, adult children are most likely to be providing elder care.

Adult Children as Elder Care Providers

Grown children caring for elderly parents generally provide most of the care themselves, although they also seek assistance from formal service providers. Motivated by **filial responsibility** (a child's obligation to parents), respect, and affection, adult children care for their folks "because they're my parents" (Anderson, Fields, and Dobb 2013). Highly religious adult children report higher levels of filial responsibility (Gans, Silverstein, and Lowenstein 2009).

As pointed out in "Issues for Thought: Filial Responsibility Laws," principles of reciprocity are at work regarding filial responsibility. As one caregiver explained to a researcher, "My mom's been there for me, helping me out, so I'm always going to be there for her" (Radina 2007, p. 158). And from another caregiver, "For me, it is a joy and a form of self reward to care for my elders, see to it there is a roof over their heads, food in the [fridge]. All the things they did for me as a young one" (in Gitner 2012). One caregiver explained that her mother deserved care now because of all that her mother had previously done for an elderly aunt (Gitner 2012; see also Allen 2012).

Generally, aging parents expect to receive help from offspring in proportion to the aid that the parents had once given them, and offspring who have received more financial help from their parents are more likely than their siblings to be caring for the elderly parent (Bucx, Van Wel, and Knijn 2012; Fingerman et al. 2009). Interestingly, adult stepchildren receive less financial and other help than do biological or adopted children—a situation that may help to explain the lower level of felt filial responsibility among stepchildren (Berry 2008).

On average, parent–adult child ties with stepchildren are not as close as are those with biological or adopted children (Ward, Spitze, and Deane 2009). So it may be no surprise that adult children feel less obligation to aging stepparents than they do to biological parents, especially when the stepparent was acquired later in the adult child's life (Pezzin, Pollak, and Schone 2008). Feeling obligated to help one's remarried *biological* parent is often the principal reason for helping the *step*parent (Ganong, Coleman, and Rothrauff 2009).

Shared Elder Care among Siblings There is some evidence that the oldest sibling in a family feels filial responsibility more strongly and thus provides more care to an aging parent than do younger sibs (Fontaine, Gramain, and Wittwer 2009). Today, it's often the case that siblings have geographically moved away from each other and from their aging parents; geographical distance typically excuses a sibling from hands-on, day-to-day caregiving. Even when siblings live near an elderly parent, however, the burden of elder care does not always fall upon each one equally. A study based on forty focus groups (see Chapter 2) asked eldercare givers with siblings to describe how they felt about the caregiving situation:

> Siblings who described an imbalance in caregiving responsibilities reported feeling considerable distress.... One participant confessed that she was straddling a "real thin line between just taking her [barely participating sister's] head off some day, because I'm so mad at the inequity of it." (Ingersoll-Dayton et al. 2003, p. 205)

However, the majority in this study sought defined the situation as more or less fair (see also Connidis and Kemp 2008; Kuperminc, Jurkovic, and Casey 2009). They took into account a sibling's geographical distance, employment responsibilities, and other obligations. Some participants (both men and women) called upon gendered expectations to help justify women's inequitable elder care responsibilities. "I guess it's my gender," said a woman whose brother did little to help (Ingersoll-Dayton et al. 2003, p. 207). Feeling that the aging parent favored one sibling over another—and perhaps continues to favor one sibling over another—can add to conflict and tension among siblings caring for an aging parent or grandparent (Jensen et al. 2013).

Gender Differences in Providing Elder Care

Men *are* involved in elder care, especially when retired (Kahn, McGill, and Bianchi 2011). About one-third of unpaid caregivers are male. Moreover, the proportion of sons involved in elder care may increase in the future due to the growing number of only-child sons, smaller sibling groups from which to draw care providers, and changing gender roles that could one day make eldercare giving an expectation for adult male behavior (National Alliance for Caregiving 2009).

For now, however, women account for about two-thirds of all unpaid caregivers (National Alliance for Caregiving 2009, p. 14). This statistic partly reflects the fact that women live longer, so more caregivers are wives. Feelings of obligation to care for a disabled spouse are fairly equal between husbands and wives (Roper and Yorgason 2009). However, gender makes a significant difference in adult children's caregiving obligations (Barnett 2013; Hill 2011).

Norms in many Asian American families designate the oldest son as the responsible caregiver to aging parents (Kamo and Zhou 1994; Lin and Liu 1993). Sons of other ethnicities often provide eldercare in the absence of available daughters (Lee, Spitze, and Logan 2003). Except in these cases, the adult child involved in a parent's care is considerably more likely to be a daughter or even a daughter-in-law than a son. This situation—one that raises issues of gender equity similar to those of parenting and other unpaid family labor, discussed in Chapters 9 and 10—is partly due to ongoing employment differences between women and men (Sarkisian and Gerstel 2004).

Then, too, in accordance with the findings, discussed earlier, that relations with daughters are more often tight-knit than those with sons, a parent may prefer a daughter's help or assume that even an employed daughter will help while a son is more often perceived as too busy. A son-in-law explained his wife's caring for her mother this way: "Mom just, I think, calls on her more, so that's the way.... It's not that the brothers wouldn't help at all, but it's just...she gets called on more" (Ingersoll-Dayton et al. 2003, p. 207).

Then too, men and women tend to provide care differently—in ways that reflect socially gendered expectations (Raschick and Ingersoll-Dayton 2004). Sons, grandsons, and other male caregivers (although not husbands) tend to perform occasional tasks, such as cleaning gutters or mowing the lawn, whereas daughters more often provide continually needed services like housekeeping, cooking, or doing laundry (Hequembourg and Brallier 2005). Sons more often serve as financial managers, as well as organizers, negotiators, supervisors, and intermediaries between the care receiver and formal service providers. A son is more likely than a daughter to enlist help from his spouse, the elder care receiver's daughter-in-law (Raschick and Ingersoll-Dayton 2004). "Providing intimate, hands-on care is culturally defined as feminine, [and the] dirty parts of care work are mainly women's work" (Isaksen 2002, pp. 806, 809).

The Sandwich Generation

Many who are caring for aging parents have children at home. About thirty years ago journalists and social scientists took note of an emergent **sandwich generation**: individuals who are sandwiched between the simultaneous responsibilities of caring for their dependent children and aging parents. Between 9 and 13 percent of U.S. households with someone between ages 30 and 60 are members of the sandwich generation (Neal and Hammer 2007). But "[t]he sandwich generation is not really a single generation" (Council on Contemporary Families 2011). The majority (55 percent) of people with elder and child care responsibilities are between ages 28 and 42. However, many are older (38 percent between ages 43 and 61) and 7 percent of "sandwichers" are younger than age 28 (Council on Contemporary Families 2011).

Recently, gerontologist Neal Cutler, who studies the effect of aging on finances, coined the term *senior sandwich generation*. These individuals are between middle- and near-retirement age "and facing the ultimate financial trifecta: college for their kids (either current tuition bills or paying back borrowed money), retirement for themselves and at-home or nursing-home care for one or more parents. All at the same time" (Chatzky 2006). About "one-in-seven middle-aged adults (15 percent) is providing financial support to both an aging parent and a child" (Parker and Patten 2013).

Today's sandwiched caregivers often house financially strapped young adults who have returned home. Sometimes young adults who return home serve as caregivers to their aging parents. Even when the parent is healthy and active, young adults who live with them can be quite helpful—doing the lawn, for instance, cleaning gutters, or carrying groceries in from the car. Nevertheless, providing economically for multigenerations can mean mounting financial burdens for the sandwich generation (Parker and Patten 2013).

The sandwich generation experiences all the hectic task juggling discussed in Chapters 9 and 10. For some in this category, work demands take the majority of their attention. Others emphasize child care responsibilities, whereas still others place greater focus on care for their aging parents (Cullen et al. 2009). Regardless of what corner of this triangle predominates—work, children, or aging parents—stress builds as family members maneuver chaotic everyday life (Williams and Boushey 2010).

Ronnie Kaufman/Corbis

The majority of the young old need almost no help at all, but as they grow older they are likely to need more help, sometimes requiring a great deal of continuing care. Often employed, members of the *sandwich generation*—adults sandwiched between the simultaneous responsibilities of caring for their children and aging parents—feel the strains associated with juggling work, child care, and elder care.

Elder Care—Joy, Ambivalence, Reluctance, and Conflict

The media typically present eldercare as tenderly nursing "loved ones," and this is often the case—but not always. Sometimes individuals resent being called upon to care for aging parents (or other family members) with whom they never felt close (Span 2013). Many caregiver relationships are best characterized as ambivalent (Connidis 2011; Kiecolt, Blieszner, and Savla 2011). Unfortunately, ambivalent or consistently problematic caregiving relationships result in diminished well-being for both caregiver and receiver (Rook et al. 2012). One study of spouse and adult child caregivers found that better quality caregiver-receiver relationship prior to the receiver's onset of dementia resulted in caregivers' being less willing to place the receiver in a nursing home (Winter, Gitlin, and Dennis 2011).

It may help to remember that older people can be stubborn, demanding, and angry about receiving help. Needing assistance may threaten their sense of autonomy and self-esteem (Bottke 2010; Halpern 2009). In the words of one psychotherapist, "For men in particular, their ego and sense of self can be hit very hard when they can no longer do what they used to do. It's safe to take it out on someone you love, not because you don't love her or respect her, but because it's safe" (in Brody 2012). Then too, caregivers may expect a certain amount of deference, or courteous submission to their opinions and decisions, from an aging parent in need—and when this does not occur, "intergenerational relations become strained" (Pyke 1999, p. 661).

Some conflicts between caregiver and recipient are about finances—when the elderly parent seems to be letting go of money unwisely, for example (Duhigg 2007). Whether an aging parent can drive safely or should move into a residential care facility are potential matters of contention (Dyer, Pickens, and Burnett 2007; Frum 2012). Discussing a parent's need for lifestyle changes following a critical health event, such as a heart attack, can be problematic (Goldsmith, Bute, and Lindholm 2012). More routine issues, such as how urgent is an errand that needs to be run, can also cause conflict. Using positive communication skills (see Chapter 11) to set boundaries is important to a caregiver's well-being (Bottke 2010). Susan Halpern's *Finding the Words: Candid Conversations with Loved Ones* (2009) is a good resource.

Disagreements may also emerge among caregivers (Mills and Wilmoth 2002). For instance, conflicts may arise when siblings are "forced to make urgent, complex decisions for loved ones in intensive care units" (Siegel 2004) Adult siblings may have differing feelings about a parent (Flora 2007). Sometimes old sibling rivalries or perceptions of a parent's favoritism emerge (Suitor et al. 2009). All else being equal, families that have developed a shared understanding of the caregiving situation make more effective caregiving teams (Halpern 2009). With regard to the decision to withdraw treatment from a dying parent, the older family member's having previously written an advanced directive on this issue is advised and generally most helpful.

Caregiver Stress Caregiving may enhance one's sense of purpose and overall life satisfaction due not only to enhanced intimacy but also to the hope that helping others will result in assistance with one's own needs when the time arrives (Shapiro 2006b). Nevertheless, caregiving can be overwhelming—physically, financially, and emotionally costly (Christakis and Allison 2006; Johnson and Wiener 2006; Parra-Cardona et al. 2007).

A recent national poll found that among adults with a living parent over age 64, nearly 32 percent contributed to a parent's financial support during the preceding year; the majority were contributing to ongoing expenses (Parker and Patten 2013). Caregivers not only spend their own money but (especially women) may also take extended time from work or pass up promotions (Geewax 2012b). Some family caregivers take early retirement or geographically relocate to provide elder care (Anderson, Fields, and Dobb 2013). Caregivers (usually women) whose jobs are not accommodating—do not allow for flexibility or phone calls from work, for example—are more likely than others to quit working altogether (Lahaie, Earle, and Heymann 2012; Stewart 2013). As discussed in Chapter 3, feminist scholars point to the *intersectionality of race, class, and gender* when analyzing women's lives. As a case in point here, while professional women may relinquish promotions, less affluent women *often* experience work situations hostile enough to their caregiving needs that they quit work altogether (Lahaie, Earle, and Heymann 2012).

A caregiver making choices such as these limits his or her own old-age resources (Hill 2011).

Furthermore, providing elder care can be socially isolating, bring on depression, and further strain one's physical health (Clark and Diamond 2010). Older caregivers report losing contact with friends and other family members (National Alliance for Caregiving 2009). Younger caregivers experience limitations on dating and other relationships. As one 26-year-old explained to a research team, "I would like to go camping with my husband once in a while, but I can't just get up and go away, because of taking care of my grandparents" (in Dellmann-Jenkins, Blankemeyer, and Pinkard 2000, p. 181).

Caregiver stress and depression among Americans result partly from the fact that ours is an individualistic culture in which adult offspring are expected to establish lives apart from their parents and to achieve success as individuals rather than (or as well as) working to benefit the family system (Killian and Ganong 2002). Furthermore, because the chronically ill or disabled of all ages are decreasingly cared for in hospitals, today's informal caregivers are asked to perform complicated care regimens that have traditionally been handled only by health care professionals (Byrne, Orange, and Ward-Griffin 2011; Guberman et al. 2005). This situation places added demands on a family caregiver's time, energy, and emotional stamina (Richards 2009). (See "Issues for Thought: Caring for Patients at Home—A Family Stressor," in Chapter 13.)

On days when a chronically ill care recipient appears to be doing better, caregivers report being in a better mood (Roper and Yorgason 2009). Also, when the caregiver–recipient relationship feels more reciprocal—with the caregiver feeling that he or she is getting something in return—the caregiver is less often depressed (LeBlanc and Wright 2000). Receiving adequate training and learning specific caregiver skills can lessen caregiver stress, as do various forms of social support—phone calls and other contacts with friends as well as finding information and connecting with other caregivers either in face-to-face support groups or on the Internet (Roper and Yorgason 2009; Wilkins, Bruce, and Sirey 2009).

Race/Ethnic Diversity and Family Elder Care

Adult children of all races and ethnicities feel responsible for their aging parents and to siblings with whom they likely share eldercare responsibilities. However, race/ethnic differences do exist regarding elder care. For instance, Asians have traditionally tended to emphasize the centrality of filial obligations over conjugal relationships (Burr and Mutchler 1999). In addition, blacks are more likely than nonHispanic whites to expect their adult children to personally care for them in old age (Lee, Peek, and Coward 1998). As we saw earlier in this chapter, ethnic groups other than nonHispanic whites have been more likely to establish multigenerational households.

Although more than two-thirds (67 percent) of family caregivers today are nonHispanic white, this situation will change as members of other race/ethnic groups grow older. Research is beginning to accumulate on how race and ethnicity influence caregivers' emotional and mental health (Skarupski et al. 2009). One study compared 89 Latina with 96 nonHispanic white female caregivers of older relatives with Alzheimer's disease and found that, regardless of ethnicity, lack of financial resources negatively impacted the caregivers' emotional health. However, stress from lack of resources was mitigated, or lessened by Latinas' relatively strong familistic values (Montoro-Rodriquez and Gallagher-Thompson 2009). A study of 307 caregivers compared blacks with nonHispanic whites and found that when exposed to caregiver stress blacks coped better emotionally (Skarupski et al. 2009; see also Wilkins, Bruce, and Sirey 2009). Prior research with African American wife caregivers found that receiving support from their churches lessened their stress and helped their marital relationship (Chadiha, Rafferty, and Pickard 2003).

Race/ethnic minorities have not been as likely as nonHispanic whites to use community-based services, such as senior centers. This discrepancy occurs due to language barriers; because individuals don't know that they qualify for government-funded services; or because they are reluctant to include paid service providers as members of their caregiving team (Levine 2008). **Fictive kin** (family-like relationships that are not based on blood or marriage but on close friendship ties) are often resources for elder care exchanges among African Americans (sometimes called "going for sisters"), Hispanics (*compadrazgos*), Italians (*compare*), and other ethnic groups as well (Ebaugh and Curry 2000).

However, it is incorrect to assume that ethnic minority families "take care of their own" without needing to rely on community services. Among many immigrant ethnic groups, **acculturation** (the process whereby immigrant groups adopt the beliefs, values, and norms of their new culture) affects norms of filial obligation. Younger generations are more likely than their elders to become acculturated—a situation that creates the need for family members to renegotiate their respective roles (see Chapter 2) with the potential for intergenerational conflict (Rudolph, Cornelius-White, and Quintana 2005; Wong, Yoo, and Stewart 2006). A study of older Puerto Ricans found that filial obligation has declined in the younger generations (Zsembik and Bonilla 2000). And research on Chinese immigrant families in California found that sons often outsourced elder care:

> "I told her that I hire you to help me achieve my filial duty," Paul Wang, a 60-year-old Taiwanese immigrant owning a software company in Silicon Valley, California, described... his conversation with the in-home care worker he employed for his mother suffering from Alzheimer's disease. (Lan 2002, p. 812)

Acculturation has also meant that, increasingly, aging immigrants live in housing designed for the elderly, rather than with their grown children (Kershaw 2003, p. A10). We turn now to a discussion of elder abuse and neglect.

Elder Abuse and Neglect

Parallel to child abuse and neglect, discussed in Chapter 12, **elder abuse** involves overt acts of aggression, whereas **elder neglect** involves failure to give adequate care. Elder abuse includes physical assault; verbal abuse and other forms of emotional humiliation; purposeful social isolation (for example, forbidding use of the telephone); and financial exploitation (Hildreth, Burke, and Glass 2009).

The profile of the abused or neglected elderly person is of a female, 70 years old or above, who has physical, mental, and/or emotional impairments and is dependent on the abuser/caregiver for companionship and help with daily activities (Leisey, Kupstas, and Cooper 2009). Unfortunately, accurate statistics on the prevalence of elder abuse and neglect do not exist, partly because much elder abuse and neglect goes unreported (Federal Interagency Forum on Aging-Related Statistics 2013). About 38,000 cases of elderly financial abuse are reported every year, and experts believe the actual number of occurrences is much higher (Bendix 2009; Miles 2008; see also Hull 2008). Depending on how broadly a researcher defines elder maltreatment, various studies have concluded that as many as 5 million—between 1 percent and about 10 percent—of individuals age 60 and above are abused or neglected annually by professional and family caregivers (Hildreth, Burke, and Glass 2009; Ramnarace 2010).

Data suggest that *non*family members—typically paid caregivers either in the aged person's home or in an institutional setting—are responsible for more than half of all elder verbal, physical, and financial abuse (Laumann, Leitsch, and Waite 2008). Accordingly, education programs for family members and others who work with the elderly—social workers, physicians, nurses, dentists, attorneys—have been designed to facilitate detection and prevention of elder maltreatment in institutional settings (Bendix 2009; Phelan 2009; Rinker 2009; Wagenaar 2009).

Bill Aron/PhotoEdit

The majority of elderly Americans maintain their own homes. However, many of the frail elderly depend on, or live with, family members. Although care for an aging parent—most often by daughters—is often given with fondness and love, it can also bring stress, conflicting emotions, and great demands on time, energy, health, and finances.

Elder Maltreatment by Family Members

In 1996, the federal government sponsored the National Elder Abuse Incidence Study, a national study that combined reports from Adult Protective Services with interviews with people in the community who had contact with the elderly. Estimates from this study—still our best source for statistics—were that about one-half million Americans over 60 and living in family households were abused or neglected (Ramnarace 2010). Neglect was the more common form of elder maltreatment (National Center on Elder Abuse 2013). Another national study found that, among elderly who suffered maltreatment by a family member, 9 percent experienced verbal abuse, and 3.5 percent had suffered financial abuse. Less than 1 percent experienced physical abuse (Laumann, Leitsch, and Waite 2008).

Common to cases of physical elder abuse by family members are shared living arrangements, the abuser's poor emotional health (often including alcohol or drug problems), and a pathological relationship between victim and abuser (Anetzberger, Korbin, and Austin 1994). Abusive acts may be "carried out by abusers to compensate for their perceived lack or loss of power" (Pillemer 1986, p. 244). "In many instances, both the victim and the perpetrator [are] caught in a web of interdependency and disability, which [make] it difficult for them to seek or accept outside help or to consider separation" (Wolf 1986, p. 221).

However, the largest proportion of abuse is from a spouse or a romantic partner (Laumann, Leitsch, and Waite 2008). In fact, "there is reason to believe that a certain proportion of elder abuse is actually spouse abuse grown old" (Phillips 1986, p. 212; see also Leisey, Kupstas, and Cooper 2009; Shannon 2009). In some cases, marital violence among the elderly involves abuse of the caregiving spouse by a partner who is ill with Alzheimer's disease (Pillemer 1986). Adult children are also responsible for elder abuse, and an abuser is often financially dependent on the elderly victim (Laumann, Leitsch, and Waite 2008; Pillemer 1986).

Two Models to Explain Elder Abuse Researching and combating elder abuse generally proceed from either of two models: the *caregiver model* or the *domestic violence model.* The **caregiver model of elder abuse and neglect** views abusive or neglectful caregivers as individuals who are simply overwhelmed by caregiving burdens, a situation that can cause the caregiver to lose control and verbally or physically abuse the receiver (Abbey 2009; Bainbridge et al. 2009). This model also points to ecological factors or environmental factors—or absence of resources, combined with work demands, or one's own parenting demands—that make things harder (Norris et al. 2013).

Abusive or neglectful family caregivers are often stressed by socially structured conditions, such as job conflicts. An interesting recent study attempted to explain intimate partner violence in later life by analyzing readers' comments posted after Internet news stories that reported it (Brossoie, Roberto and Barrow 2012). Typical of one pattern found in the data was a spouse-caregiver frustrated with available social services:

> I have begged and pleaded for help [from community services] and there simply is none. If [my husband] were to die I'm sure there would be charges of neglect against me even though I am not neglecting him.... His doctors know, the courts know, his family knows and no one cares as long as there is someone else to blame. (in Brossoie, Roberto, and Barrow 2012, p. 798)

Professional care providers employed by community agencies are often trained to recognize potentially abusive family situations, and then can work to reduce dangerous levels of caregiver stress that may trigger abuse (Brandl 2007; Thobaben 2008).

A second model, the **domestic violence model of elder abuse and neglect** views elder abuse as a form of unlawful family violence and focuses on negative personal characteristics of abusers and on a possible criminal justice response, such as arrest and prosecution (Wallace 2008; Hagan 2010). The criminal justice response is especially likely in the case of financial abuse (Gross 2006b).

The Changing American Family and Elder Care in the Future

In the future, families may be more and more creative in fashioning caregiving arrangements for elderly members. Multigenerational and communal households are examples, and architects are beginning to design homes that better accommodate a variety of family and friendship living options, such as "granny flats"—additions or renovations to standard dwellings specifically for in-laws, including "Next Door Garage Apartments" or little units on a family property called "backyard living" (Gerace 2012).

Nevertheless, longer life expectancy and more chronically ill Americans in old-old age raise concerns about providing elder care now and in the future (Levine 2008). For one thing, a considerable number of elderly live alone now and will age alone. "One hundred years ago, 70 percent of American widows and widowers moved in with their families. Today nearly the same proportion of widows and widowers live alone" (Klinenberg, Torres, and Portacolone 2012, p. 1).

The American family is changing in form, as we have seen throughout this text. Many of these changes could result in a diminishing caregiver "kin supply" (Bengston 2001, p. 5; Himes 2001). As noted throughout this text, aging baby boomers have had higher cohabitation rates, smaller families, and high divorce rates. Consequently, fewer spouses or grown children—the two most important caregiving resources—will be available for caregiving (Jacobsen, Mather, and Dupuis 2012).

A high sense of filial obligation has been positively related to caregiving (Kuperminc, Jurkovic, and Casey 2009; Anderson, Fields, and Dobb 2013). But parental divorce reduces the younger generation's sense of filial obligation, particularly to divorced fathers and (expectedly) from former daughters-in-law (Ganong, Coleman, and Rothrauff 2009). Research suggests that ties between parents and adult children are generally less close when the parents are divorced (Connidis 2009). Children raised in divorced families may have received fewer resources or less emotional support from a parent and hence feel less filial obligation.

Moreover, the extent to which one's cohabiting partner participates in elder care is a matter for future research, but since the majority of cohabiting relationships are short-lived, we might expect that eldercare is or will not be part of the cohabitation equation.

We are gradually learning more about the ramifications of remarriage for elder care. In one qualitative study of late-life remarried caregivers,

> [w]ives revealed how little support or assistance they received from adult stepchildren for [the children's] ailing father. Often these caregivers endured a kind of amplified stress, isolation, and conflict in their caregiving role, which they attributed to their remarried, stepmother status. By the same token, adult children and stepchildren are not always

STAR TRIBUNE/MINNEAPOLIS–ST. PAUL 2013

After experiencing it herself, Nancy Edwards, left, wrote a thesis on lesbians coming out later in life. Surrounded by family photos from their prior heterosexual marriages, she and her partner sit in their Minneapolis home. Older individuals who formed a LG identity before the 1970s, when homosexuality was highly stigmatized, may tend to hide their sexual orientation. But baby boomers, who formed a LG identity in a "gay liberation atmosphere" after the 1970s, more often become visible activists for elderly LGBT rights (Rosenfeld 1999).

> granted access to the critical decision-making or caretaking. (Sherman 2006, p. F8; see also Sherman and Boss 2007)

Furthermore, females' increased participation in the labor force decreases the time that women, the principal providers of elder care, have available to engage in elder care. Then too, geographical mobility negatively affects hands-on eldercare. Moreover, in an era of more transnational families (see Chapter 1), aging parents may be in the country of origin while their adult offspring struggle to help to care for their aging parents from abroad (Bodrug-Lungu and Kostina-Ritchey 2013; Zhou 2012).

In addition to demographic changes, some have suggested that American society may be becoming increasingly individualistic, a situation that—in this view—threatens to weaken the family.

Today's older Americans are likely to have sisters and brothers who may be able to help them. However, as more younger parents choose to have fewer or only children, more elderly in the future will have fewer or no siblings. Then too, more siblings in a family can ease each one's burden of care for an aging parent, because they can share necessary tasks (Wolf, Freedman, and Soldo 1997; see also Fontaine, Gramain, and Wittwer 2009). However, increasingly, fewer siblings will be available to share in elder care.

Never-married, childfree women are particularly active socially; future childfree and single elderly are likely to fashion fictive-kin families of mutually caring friends (Spencer and Pahl 2006). On the other hand, elderly individuals without children may prove to be "less likely than are parents to have robust network types capable of maintaining independent living" (Wenger et al. 2007, p. 1,419; and see Bures, Koropeckyj-Cox, and Loree 2009).

Same-Sex Families and Elder Care

Same-sex couples and families face elder care challenges. "But along with getting older, they also have to face the prejudices of being gay or lesbian.... Nursing homes and private retirement centers... [often] make assumptions that their residents are heterosexual and structure activities on the basis of these assumptions" (Powell 2004, p. 60). Partly so as not to be separated from their partners by well-meaning relatives who may put them in separate nursing homes, gay men and lesbian couples (who can afford it) have begun to create assisted living communities of their own ("Birds of a Feather" 2010; Grant 2010; Mitchell 2007).

Although many policy analysts express concern about families' ability to effectively provide elder care in the future, some also point to the **latent kin matrix**, defined as "a web of continually shifting linkages that provide the potential for activating and intensifying close kin relationships" (Riley 1983, p. 441; see also Spencer and Pahl 2006). An important feature of the latent matrix is that, although they may remain dormant for long periods, family relations with adult siblings and extended kin emerge as a resource when the need arises (Silverstein and Bengston 2001).

Moreover, while family forms become increasingly diverse, so many ways that members of the postmodern family deal with elder care. For instance, in research involving in-depth interviews with forty-five older respondents, a 67-year-old separated wife, whose husband had become diabetic and asthmatic and had heart trouble, reported that she still loved him very much: "If something happened to him and this gal [his new romantic partner] didn't take care of him, I would go and take care of him myself" (quoted in Allen et al. 1999, p. 154). It is not entirely unusual for a divorced mate to provide assistance to an aging ex-spouse who needs it:

> Hospice workers, academics, and doctors say they are seeing more such cases.... Often a person feels deep ties to a former husband or wife or feels a responsibility born of common experience and child rearing.... "They are acting more like a brother or sister, or cousin, or extended family member." (in Richtel 2005, p. ST1-2)

Toward Better Caregiving

Family sociologist and demographer Andrew Cherlin (1996) distinguishes between the "public" and the "private" faces of families. The **private face of family** "provides individuals with intimacy, emotional support, and love" (p. 19) and is very important to the well-being of aging family members (Lowe and McBride-Henry 2012). The **public face of family** produces public goods and services by educating children and caring for the ill and elderly. "While serving each other, members of the 'public' family also serve the larger community" (Conner 2000, p. 36; Gross 2004).

The Private Face of Family Caregiving

We've seen throughout this text that providing family members with intimacy, emotional support, and love involves positive communication techniques described in Chapter 11. Hopefully making use of these techniques, families need to discuss many topics regarding an aging member—from who will go for groceries to money matters to advanced care planning and end-of-life care to the possibility of organ donation upon death (Boerner, Carr, and Moorman 2013).

With regard to multigenerational living, counselors advise that before moving in together, family members of all ages (even young children) are encouraged to talk about how they expect life to change: what they want, what they are excited about, and what they're nervous about the changing situation. For instance, how much will grandparents help with child care? How will meals be handled? Families are encouraged to schedule fairly regular meetings to discuss issues before they become problems (Alvarez 2009; Gitner 2012).

Discussing things may be particularly important in families where cultural expectations differ widely between generations. Immigrants arrive from more familistic cultures, such as Hispanic and Asian nations, with high emphasis on filial obligation. However, children of immigrants become more individualistic as they assimilate into American society (Ruiz and Ransford 2012; Lee and Smith 2012). For instance, a study of Korean Americans found that older caregivers were much more likely to want to personally care for a spouse with dementia at home, while (next-generation) adult children more often favored hiring professionals (Lee and Smith 2012).

There are many good resources that can help in these discussions. Several Internet examples: American Association of Retired Persons, Center for Retirement Research at Boston College, National Family Caregivers Alliance, National Alliance for Caregiving, National Elder Law Network, National Institute on Aging, Rand Center for the Study of Aging, Social Security Administration, and the U.S. Administration on Aging. For those who still like paper, the National Institute on Aging, a division of the U.S. Department of Health and Human Services, will send you free of charge various pamphlets on topics such as aging in one's home, choosing a nursing home, caregiving from a geographical distance, and elder abuse, among others.

The Public Face of Family Caregiving

More than at any time since the 1935 onset of Social Security, contemporary families serve a **multigenerational safety net function** whereby older family members care for younger ones (even young adults), and vice versa. Ironically, however,

> the Great Economic Recession has increased the need for the family safety net, while at the same time, family members have become less able to act as a safety net because parents and children are often exposed to the same disadvantages, such as unemployment. The public safety net is also fraying, thereby increasing the need for kin support and simultaneously reducing the resources available to share among kin. (Seltzer 2011, F2)

The public face looks to government leaders, who ironically tend to emphasize the "paramount importance" of "personal responsibility and accountability for planning for one's longevity" (White House Conference on Aging 2005, p. 18). However, the financial and emotional costs of providing good care for ill and disabled family members are often too high for family caregivers to manage without help (Parker and Patten 2013; Piercy 2010; Span 2009b). Many point to the increasing need for public services to assist family caregivers—for example, transportation, personal care services, and adult day care (Rosen 2007).

However,

> The media constantly reinforce the conventional wisdom that the care crisis is an individual problem. Books, magazines and newspapers offer American women an endless stream of advice about how to maintain their "balancing act," how to be better organized and more efficient or how to meditate, exercise and pamper themselves to relieve their mounting stress. Missing is the very pragmatic proposal that American society needs new policies that will restructure the workplace and reorganize family life. (Rosen 2007, p. 13)

Social scientists Francesca Cancian and Stacey Oliker (2000) have proposed the following strategies for moving our society toward better elder care coupled with greater gender equity in providing elder care:

- Provide government funds that support more care outside the family, such as government-funded day care centers for the elderly and respite (time off) services for caregivers.
- Increase social recognition of caregiving—both paid and unpaid—as productive and valuable work.
- Make caregiving more economically rewarding or, at least, less economically costly to the caregiver (p. 130).

How these policy changes might be accomplished may be difficult to imagine, but this fact negates the utility neither of the vision nor of the political debate that needs to emerge (Gross 2007). Elder care (as well as child care) is indeed a responsibility not only of individual families but also of an entire society.

Fuse/Getty Images

Many older Americans remain active even into old-old age. Nonetheless, longer life span does mean that more and more Americans will spend the last months or years of their lives with chronic health problems and/or disabilities. The aging of the population raises concerns about how a changing family structure will be able to care for tomorrow's elderly and to what extent community and government resources can or will be engaged to help.

Summary

- The *number* of elderly, as well as their *proportion* of the total U.S. population, is growing.
- Along with the impact of the baby boomers' aging and the declining proportion of children in the population, longer life expectancy has contributed to the aging of our population.
- Due mainly to differences in life expectancy, older men are much more likely to be living with their spouse than are older women.

- Among older Americans without partners, living arrangements depend on one's health, the availability of others with whom to reside, social norms regarding obligations of other family members toward their elderly, personal preferences for privacy and independence, and economics.
- Growth in Social Security benefits has resulted in dramatic reductions in U.S. poverty rates for the elderly over the last several decades, although 10 percent of older adults are living in poverty.
- Due to differences in work patterns, wage differentials, and Social Security regulations, older men are considerably better off financially than are older women.
- Most older married couples place intimacy as central to their lives, describe their unions as happy, and continue to be interested in sex into old age.
- Even when it is not an abrupt event, retirement represents a great change for individuals and couples.
- Adjustment to widowhood or widowerhood is an important family transition that married couples often must face in later life. Bereavement manifests itself in physical, emotional, and intellectual symptoms.
- Daughters are more likely than sons to have close relationships with their parents, especially with their mothers. However, a parent's divorce, separation, or repartnering often weakens the bond with adult children.
- Partly due to longer life expectancy, grandparenting became increasingly important to families throughout the twentieth century. In the twenty-first century, we see an increasing number of great-grandparents.
- As members of our families age, elder care is becoming an important feature in family life, with women providing the bulk of it.
- After spouses, adult children (usually daughters) are a preferred choice of an older family member as elder care providers.
- Elder care in families typically follows a caregiving trajectory as the care receiver ages, and it involves not only benefits but also stresses for the caregiver(s).
- Elder maltreatment—that is, abuse and neglect—by family caregivers exists in a small percentage of aging families and is addressed in public policy by either the caregiver model or the domestic violence model.
- Changes in the American family lead some policy analysts to be concerned that families will have greater difficulty providing elder care in the future. However, others point out that family relationships, though latent for long periods, can be activated when needed.
- Better elder care in the future will necessitate involving more men as caregivers and developing public policy that adequately supports expanded community services to assist families in providing elder care.

Questions for Review and Reflection

1. Discuss ways that society's age structure today affects American families.
2. Describe the living arrangements of older Americans today, and give some reasons for these arrangements.
3. Discuss some ways that the growing diversity of family forms among the older population can be expected to affect caregiving.
4. Apply the exchange, interactionist, structure-functionalist, or ecological perspective (see Chapter 2) to the process of providing elder care.
5. **Policy Question.** Describe two suggestions for policy changes that would make family elder care less difficult.

Key Terms

acculturation 438
active life expectancy 420
baby boom 419
bereavement 427
caregiver model of elder abuse and neglect 440
caregiving 432
custodial grandparent 431

domestic violence model of elder abuse and neglect 440
elder abuse 439
elder care 432
elder neglect 439
fictive kin 438
filial responsibility 434
gerontologists 432
hierarchical compensatory model of caregiving 433
informal caregiving 432
latent kin matrix 442
multigenerational family 418
multigenerational safety net function 442
noncustodial grandparent 431
private face of family 442
public face of family 442
sandwich generation 436
surrogate grandparent 430

Glossary

ABC-X model A model of family crisis in which A (the stressor event) interacts with B (the family's resources for meeting a crisis) and with C (the definition the family formulates of the event) to produce X (the crisis).

abortion See **induced abortion**.

abstinence The standard that maintains that nonmarital intercourse is wrong or inadvisable for both women and men regardless of the circumstances. Many religions espouse abstinence as a moral imperative, while some individuals are abstinent as a temporary or permanent personal choice.

accomplishment of natural growth parenting model Educational model in which children's abilities are allowed to develop naturally; this includes working-class children spending more time watching television and playing video games than children of highly educated parents.

acculturation Process whereby immigrant groups adopt the beliefs, values, and norms of their new culture.

acquaintance rape Forced or unwanted sexual contact between people who know each other, sometimes—although not necessarily—taking place on a date. See also **date rape**.

active life expectancy The period of life free of disability in activities of daily living, after which may follow a period of being at least somewhat disabled.

affectional orientation Newer term used in conjunction with **sexual identity** and **sexual orientation**; includes emotional and physical attractions beyond sexual attraction.

agentic (instrumental) character traits Traits such as confidence, assertiveness, and ambition that enable a person to accomplish difficult tasks or goals.

AIDS See **HIV/AIDS**.

arranged marriage Unions in which parents choose their children's marriage partners.

asexual, asexuality A person who is asexual does not experience sexual desire. This is different from abstinence or celibacy, which is a choice to not engage in sexual activity despite feelings of sexual desire. Asexuality may be considered a sexual orientation/identity.

assisted reproductive technology (ART) Advanced reproductive technology, such as artificial insemination, in vitro fertilization, or embryo transplantation, that enables infertile couples or individuals, including gay and lesbian couples, to have children.

assortative mating Social psychological filtering process in which individuals gradually narrow down their pool of eligible individuals for long-term committed relationships, removing those who would not make the best spouse or partner.

attachment disorder An emotional disorder in which a person defensively shuts off the willingness or ability to make emotional attachments to anyone.

attachment theory A psychological theory that holds that, during infancy and childhood, a young person develops a general style of attaching to others; once an individual's attachment style is established, she or he unconsciously applies that style to later, adult relationships. The three basic styles are **secure, insecure/anxious**, and **avoidant**.

authoritarian parenting style All decision making is in parents' hands, and the emphasis is on compliance with rules and directives. Parents are more punitive than supportive, and use of physical punishment is likely.

authoritative parenting style Parents accept the child's personality and talents and are emotionally supportive. At the same time, they consciously set and enforce rules and limits, whose rationale is usually explained to the child. Parents provide guidance and direction and state expectations for the child's behavior. Parents are in charge, but the child is given responsibility and must take the initiative in completing schoolwork and other tasks and in solving child-level problems.

baby boom The unusually large cohort of U.S. children born after the end of World War II, between 1946 and 1964.

battered woman syndrome When low self-esteem interacts with fear, depression, confusion, anxiety, feelings of self-blame , and loss of a sense of personal control to keep a physically or emotionally abused women from finding a way out of her situation.

belligerence A negative communication/relationship behavior that challenges the partner's power and authority.

bereavement A period of mourning after the death of a loved one.

bifurcated consciousness Divided perception in which someone is aware of and often troubled by two conflicting messages.

binational family An immigrant family in which some members are citizens or legal residents of the country they migrate to, while others are **undocumented**—that is, they are not legal residents.

binuclear family One family in two household units. A term created to describe a postdivorce family in which both parents remain involved and children have a home in both households.

biosocial perspective Theoretical perspective based on concepts linking psychosocial factors to anatomy, physiology, genetics, and/or hormones as shaped by evolution.

bisexuals People who are sexually attracted to both males and females.

borderwork Interaction rituals that are based on and reaffirm boundaries and differences between girls and boys.

boundary ambiguity When applied to a family, a situation in which it is unclear who is in and who is out of the family.

caregiver model of elder abuse and neglect A view of elder abuse or neglect that highlights stress on the caregiver as important to the understanding of abusive behavior.

caregiving "Assistance provided to persons who cannot, for whatever reason, perform the basic activities or instrumental activities of daily living for themselves" (Cherlin 1996, p. 762).

causal feedback loop When the workplace impacts within-family interactions and decision making—and within-family interactions and decision-making affect the workplace.

cesarean section Surgical procedure in which incisions are made in a woman's abdomen and uterus to deliver her baby.

child abuse Overt acts of aggression against a child, such as beating or inflicting physical injury or excessive verbal derogation. Sexual abuse is a form of physical child abuse. See also **emotional child abuse or neglect**.

child care The care and education of children by people other than their parents. Child care may include before- and after-school care for older children and overnight care when employed parents must travel, as well as day care for preschool children.

child maltreatment Defined by the 1974 Child Abuse Prevention and Treatment Act as the "physical or mental injury, sexual abuse, or negligent treatment of a child under the age of 18 by a person who is responsible for the child's welfare under circumstances that indicate that the child's health or welfare is harmed or threatened."

child neglect Failure to provide adequate physical or emotional care for a child. See also **emotional child abuse or neglect**.

child support Money paid by the noncustodial parent to the custodial parent to financially support children of a former marital, cohabiting, or sexual relationship.

child-to-parent violence A form of family violence involving a child's (especially an adolescent's) physical and emotional abuse of a parent.

civil union Legislation that allows any two single adults—including same-sex partners or blood relatives, such as siblings or a parent and adult child—to have access to virtually all marriage rights and benefits on the state level, but none on the federal level. Designed to give same-sex couples many of the legal benefits of marriage while denying them the right to legally marry.

coercive power One of the six power bases, or sources of power. This power is based on the dominant person's ability and willingness to punish the partner either with psychological–emotional or physical abuse or with more subtle methods of withholding affection.

cohabitation Living together in an intimate, sexual relationship without traditional, legal marriage. Sometimes referred to as *living together* or *marriage without marriage*, cohabitation can be a courtship process or an alternative to legal marriage, depending on how partners view it.

collectivist society A society in which people identify with and conform to the expectations of their relatives or clan, who look after their interests in return for their loyalty. The group has priority over the individual. A synonym is *communal society*.

commitment (to intimacy) The determination to develop relationships in which experiences cover many areas of personality, problems are worked through, conflict is expected and seen as a normal part of the growth process, and there is an expectation that the relationship is basically viable and worthwhile.

common pot system When economic resources for a family are pooled and distributed according to need regardless of biological relatedness.

communal society See **collectivist society**.

communes Groups of adults and perhaps children who live together, sharing aspects of their lives. Some communes are group marriages, in which members share sex; others are communal families, with several monogamous couples, who share everything except sexual relations and their children.

community-based resources Characteristics, competencies, and means of people, groups, and institutions outside the family that the family may call upon, access, and use to meet their demands.

commuter marriage A marriage in which the two partners live in different locations and commute to spend time together.

companionate marriage The single-earner, breadwinner–homemaker marriage that flourished in the 1950s. Although husbands and wives in the companionate marriage usually adhered to a sharp division of labor, they were supposed to be each other's companion—friends, lovers—in a realization of trends beginning in the 1920s.

concerted cultivation parenting style The parenting model, or style, according to which parents often praise and converse with their children, engage them in extracurricular activities, take them on outings, and so on, with the goal of cultivating their child's talents and abilities.

conflict perspective Theoretical perspective that emphasizes social conflict in a society and within families. Power and dominance are important themes.

Conflict Tactics Scale A scale developed by sociologist Murray Straus to assess how couples handle conflict. Includes detailed items on various forms of physical violence.

Confucian training doctrine Concept used to describe Asian and Asian American parenting philosophy that emphasizes blending parental love, concern, involvement, and physical closeness with strict and firm control.

conjugal power The ability to exercise one's will or autonomy in a marital relationship.

consensual marriages Heterosexual, conjugal unions that have not gone through a legal marriage ceremony.

consummate love A complete love, in terms of Sternberg's triangular theory of love, in which the components of passion, intimacy, and commitment come together.

contemporary sexism Modern form of sexism that denies gender discrimination persists; replaces traditional sexism.

contempt One of the **Four Horsemen of the Apocalypse** (which see), in which a partner feels that his or her spouse is inferior or undesirable.

co-parenting, co-parents Shared decision making and parental supervision in such areas as discipline and schoolwork or shared holidays and recreation. Can refer to parents working together in a marriage or other ongoing relationship or after divorce or separation.

courtesy stigma Refers to a situation in which not only the initially stigmatized individual but also her or his intimates are stigmatized by association.

covenant marriage A type of legal marriage in which the bride and groom agree to be bound by a marriage contract that will not let them get divorced as easily as is allowed under no-fault divorce laws.

criminal justice response The punitive approach for perpetrators of child maltreatment; advocates believe that one or both parents should be held legally responsible for child abuse.

criticism One of the **Four Horsemen of the Apocalypse** (which see) that involves making disapproving judgments or evaluations of one's partner.

cross-national marriages Marriages in which spouses are from different countries.

crude divorce rate The number of divorces per 1,000 population. See also **refined divorce rate**.

cultural script Set of socially prescribed and understood guidelines for relating to others or for defining role responsibilities and obligations.

custodial grandparent When a grandparent has custody of a grandchild.

cyberadultery Infidelity or adultery that occurs on the Internet.

cycle of violence Three consecutive phases in which there is the violent episode itself (Phase 1), a calm period (Phase 2), and tension buildup during which a victim feels increasingly disappointed and intimidated while the abuser's behavior is unpredictable and threatening (Phase 3).

data collection techniques Ways that data are gathered when doing research; these include interviews and questionnaires, naturalistic observation, focus groups, experiments and laboratory observation, and case studies, among others.

date rape Forced or unwanted sexual contact between people who are on a date. See also **acquaintance rape**.

dating script Highly gendered expectations that govern behavior in the getting-to-know-you stage of dating relationships, with men and women having far different expectations about what happens during and after a date.

deciding versus sliding Two ways of dealing with choices.

Defense of Marriage Act (DOMA) Federal statute declaring marriage to be a "legal union of one man and one woman," denying gay couples many of the civil advantages of marriage and relieving states of the obligation to grant reciprocity, or "full faith and credit," to marriages performed in another state.

defensiveness One of the **Four Horsemen of the Apocalypse** (which see) that means preparing to defend oneself against what one presumes is an upcoming attack.

deinstitutionalization of marriage A situation in which time-honored family definitions are changing and family-related social norms are weakening so that they "count for far less" than in the past.

displacement A passive-aggressive behavior in which a person expresses anger with another by being angry at or damaging people or things the other cherishes. See also **passive-aggression**.

divorce divide The gap in divorce rates between college-educated and less-educated men and women. The divorce rate has declined substantially among those who are college educated but not among the less educated.

divorce fallout Ruptures of relationships and changes in social networks that come about as a result of divorce.

divorce mediation A nonadversarial means of dispute resolution by which the couple, with the assistance of a mediator or mediators (frequently a lawyer–therapist team), negotiate the terms of their settlement of custody, support, property, and visitation issues.

domestic partners Partners in an unmarried couple who have registered their partnership with a civil authority and then enjoy some (although not necessarily all) rights, benefits, and entitlements that have traditionally been reserved for marrieds.

domestic violence model of elder abuse and neglect A model that conceptualizes elder abuse as a form of family violence.

double standard The standard according to which nonmarital sex or multiple partners are more acceptable for males than for females.

dramaturgy A theoretical subcategory within the interactionist-constructionist perspective, sees individuals as enacting culturally constituted scripts and socially prescribed roles in front of others (everyday-life audiences).

dripolator effect The "bottom up" operation of a stepfamily (from children to parents); see **percolator effect**.

economic hardship perspective One of the theoretical perspectives concerning the negative outcomes among children of divorced parents. From this perspective, it is the economic hardship brought about by marital dissolution that is primarily responsible for problems faced by children.

egalitarian norm (of marital power) The norm (cultural rule) that husband and wife should have equal power in a marriage.

egalitarian relationship Equal relationship in which resources and power are relatively gender-free.

elder abuse Overt acts of aggression toward the elderly, in which the victim may be physically assaulted, emotionally humiliated, purposefully isolated, or materially exploited.

elder care Care provided to older generations.

elder neglect Acts of omission in the care and treatment of the elderly.

emotional abuse Verbal threats and routine comments that damage a partner's self esteem.

emotional intelligence (1) Awareness of what we're feeling so that we can express our feelings more authentically; (2) ability and willingness to repair our moods, not unnecessarily nursing our hurt feelings; (3) healthy balance between controlling rash impulses and being candid and spontaneous; and (4) sensitivity to the feelings and needs of others.

emotional neglect When a parent or guardian is overly harsh and critical with a child, failing to provide guidance or being uninterested in the child's needs.

endogamy Marrying within one's own social group. See also **exogamy**.

equality Power or resources divided between partners so that each has the same amount.

equity A standard for distribution of power or resources of partners according to the contribution each person has made to the unit. Another way of characterizing an equitable result is that it is "fair."

ethnicity A group's identity based on a sense of a common culture and language.

evolutionary heritage In the biosocial perspective, human behavior is encoded in genetic or other biological features that come to us as members of a species.

exchange balance Balance of rewards and costs in a relationship.

exchange theory Theoretical perspective that sees relationships as determined by the exchange of resources and the reward–cost balance of that exchange. This theory predicts that people tend to marry others whose social class, education, physical attractiveness, and even self-esteem are similar to their own.

exogamy Marrying a partner from outside one's own social group. See also **endogamy**.

expectations of permanence One component of the marriage premise, according to which individuals enter marriage expecting that mutual affection and commitment will be lasting.

expectations of sexual exclusivity The cultural ideal according to which spouses promise to have sexual relations with only each other.

experience hypothesis The idea that the independent variable in a hypothesis is responsible for changes to a dependent variable. With regard to marriage, the experience hypothesis holds that something about the experience of being married itself causes certain results for spouses. See also the antonym, **selection hypothesis**.

experiment One tool of scientific investigation, in which behaviors are carefully monitored or measured under controlled conditions. Participants are randomly assigned to treatment or control groups.

expert power One of the six power bases, or sources of power. This power stems from the dominant person's superior judgment, knowledge, or ability.

expressive character traits Relationship-oriented characteristics of warmth, sensitivity, the ability to express tender feelings, and placing concern about others' welfare above self-interest.

expressive sexuality The view of human sexuality in which sexuality is basic to the humanness of both women and men, all individuals are free to express their sexual selves, and there is no one-sided sense of ownership.

extended family Family including relatives besides parents and children, such as aunts or uncles. See also **nuclear family**.

familistic (communal) values Values that focus on the family group as a whole and on maintaining family identity and cohesiveness.

family Any sexually expressive or parent-child or other kin relationship in which people live together with a commitment in an intimate interpersonal relationship. Family members see their identity as importantly attached to the group, which has an identity of its own. Families today take several forms: single-parent, remarried, dual-career, communal, homosexual, traditional, and so forth. See also **extended family, nuclear family**.

family change perspective See **family decline/family change perspectives**.

family cohesion That intangible emotional quality that holds groups together and gives members a sense of common identity.

family crisis A situation (resulting from a stressor) in which the family's usual behavior patterns are ineffective and new ones are called for.

family decline/family change perspectives Some family scholars and policy makers characterize late-twentieth-century developments in the family as "decline," while others describe "change." Those who take the "family decline" perspective view such changes as increases in the age at first marriage, divorce, cohabitation, and nonmarital births and the decline in fertility as disastrous for the family as a major social institution. "Family change" scholars and policy makers consider that the family has varied over time. They argue that the family can adapt to recent changes and continue to play a strong role in society.

family ecology perspective Theoretical perspective that explores how a family influences and is influenced by the environments that surround it. A family is interdependent first with its neighborhood, then with its social–cultural environment, and ultimately with the human-built and physical–biological environments. All parts of the model are interrelated and influence one another.

family fluidity Frequency and rate of changes in family-related experiences and outcomes; measure of children's adjustment to family changes.

family-friendly workplace policies Workplace policies that are supportive of employee efforts to combine family and work commitments.

family identity Ideas and feelings about the uniqueness and value of one's family unit.

family impact lens Way of looking at how a policy in question impacts families.

family instability perspective The thesis that a negative impact of divorce on children is primarily caused by the number of changes in family structure, not by any particular family form. A stable single-parent family may be less harmful to children than a divorce followed by a single-parent family followed by cohabitation, then remarriage, and perhaps a redivorce.

family leave A leave of absence from work granted to family members to care for new infants, newly adopted children, ill children, or aging parents, or to meet similar family needs or emergencies.

family life course development framework Theoretical perspective that follows families through fairly typical stages in the life course, such as through marriage, childbirth, stages of raising children, adult children's leaving home, retirement, and possible widowhood.

family of orientation The family in which an individual grows up. Also called *family of origin*.

family of procreation The family that is formed when an individual marries and has children.

family policy All the actions, procedures, regulations, attitudes, and goals of government that affect families.

family preservation A program of support for families in which children have been abused. The support is intended to enable the child to remain in the home safely rather than being placed in foster care.

family stress State of tension that arises when demands tax a family's resources.

family structure The form a family takes, such as nuclear family, extended family, single-parent family, stepfamily, and the like.

family systems theory An umbrella term for a wide range of specific theories. This theoretical perspective examines the family as a whole. It looks to the patterns of behavior and relationships within the family, in which each member is affected by the behavior of others. Systems tend toward equilibrium and will react to change in one part by seeking equilibrium either by restoring the old system or by creating a new one.

family transitions Expected or predictable changes in the course of family life that often precipitate family stress and can result in a family crisis.

female-demand/male-withdraw interaction pattern A cycle of negative verbal expression by one partner, followed by the other partner's withdrawal in the face of the other's demands.

femininities Culturally defined ways of being a woman. The plural conveys the idea that there are varied models of appropriate behavior.

feminist theory Feminist theories are conflict theories. The primary focus of the feminist perspective is male dominance in families and society as oppressive to women. The mission of this perspective is to end this oppression of women (or related pattern of subordination based on social class, race/ethnicity, age, or sexual orientation) by developing knowledge and action that confront this disparity. See also **conflict perspective**.

fertility Births to a woman or category of women (actual births, not reproductive capacity).

fictive kin Family-like relationships that are not based on blood or marriage but on close friendship ties.

filial responsibility A child's obligation to a parent.

flexible scheduling A type of employment scheduling that includes scheduling options such as **job sharing** or **flextime**.

flextime A policy that permits an employee some flexibility to adjust working hours to suit family needs or personal preference.

formal kinship care Out-of-home placement with biological relatives of children who are in the custody of the state.

foster care Care provided to children by other than their parents as a result of state intervention.

Four Horsemen of the Apocalypse Contempt, criticism, defensiveness, and stonewalling—marital communication behaviors delineated by John Gottman that often indicate a couple's future divorce.

free-choice culture Culture or society in which individuals choose their own marriage partners, a choice usually based at least somewhat on partners' love for one another.

friends with benefits Sexual activity between friends or acquaintances with no expectation of romance or emotional attachment; typically practiced by unattached people who want to have a sexual outlet without "complications."

gay A person whose sexual attraction is to people of the same sex. Used especially for males, but may include both sexes. This term is usually used rather than *homosexual*.

gender Attitudes and behavior associated with and expected of the two sexes. The term sex denotes biology, while *gender* refers to social role.

gender bending Behavior in which people challenge a gender mandate on purpose.

gender differentiation Cultural expectations about how men and women should behave. Humans are to some extent differentiated, or thought of as separate and different, according to gender.

gender identity the degree to which an individual sees her- or himself as feminine or masculine.

gender-modified egalitarian model Family model in which absolute equality is diminished by the symbolic importance of maintaining fairly traditional, comfortable, and familiar gender roles.

gender role Prescription for masculine or feminine behavior. The masculine gender role demands instrumental character traits and behavior, whereas the feminine gender role specifies expressive character traits and behavior.

gender schema theory of gender socialization A framework of knowledge and beliefs about differences or similarities between males and females. Gender schema shape socialization into gender roles.

gender strategy How a couple allocates paid and unpaid family labor and justifies that allocation; way of working through everyday situations that takes into account an individual's beliefs and deep feelings about gender roles, as well as her or his employment commitments.

gender structure Way in which gender roles are influenced by a society's sociocultural environment.

gender variance Expansive concept that includes people dressing and behaving more like the "other" gender.

geographic availability Traditionally known in the marriage and family literature as propinquity or proximity and referring to the fact that people tend to meet potential mates who are present in their regional environment.

gerontologists Social scientists who study aging and the elderly.

good provider role A specialized masculine role that emerged in this country around the 1830s and that emphasized the husband as the only or primary economic provider for his family. The good provider role had disappeared as an expected masculine role by the 1970s. See also **provider role**.

grandfamilies Families headed by grandparents.

habituation The decreased interest in sex over time that results from the increased accessibility of a sexual partner and the predictability of sexual behavior with that partner.

habituation hypothesis Hypothesis that the decline in sexual frequency over a marriage results from habituation.

heterogamy Marriage between partners who differ in race, age, education, religious background, or social class. Compare with **homogamy**.

heterosexism The taken-for-granted system of beliefs, values, and customs that places superior value on heterosexual behavior (as opposed to homosexual) and denies or stigmatizes nonheterosexual relations. This tendency also sees the heterosexual family as standard.

heterosexuals People who prefer sexual partners of the opposite sex.

hierarchical compensatory model of caregiving The idea that elderly people prefer their caregivers in ranked order as follows: an available spouse, adult children, siblings, grandchildren, nieces and nephews, friends, neighbors, and, finally, a formal service provider.

hierarchical parenting Concept used to describe a Hispanic parenting philosophy that blends warm emotional support for children with demand for significant respect for parents and other authority figures, including older extended-family members.

HIV/AIDS HIV is *human immunodeficiency virus*, the virus that causes AIDS, or *acquired immune deficiency syndrome.* AIDS is a sexually transmitted disease involving breakdown of the immune system defense against viruses, bacteria, fungi, and other diseases.

homogamy Marriage between partners of similar race, age, education, religious background, and social class. See also **heterogamy**.

homophobia Fear, dread, aversion to, and often hatred of homosexuals.

homosexuals People who are sexually attracted to people of the same sex. Preferred terms are *gay* or *gay man* for men and *lesbian* for women. See also **gay** and **lesbian**.

hooking up A sexual encounter between men and women with the understanding that there is no obligation to see each other again or to endow the sexual activity with emotional meaning. Usually there is a group or network context for hooking up; that is, the individuals meet at a social event or have common acquaintances. On some college campuses and elsewhere, hooking up has replaced dating, which is courtship-oriented socializing and sexual activity.

hormonal processes Chemical processes within the body regulated by such hormones as testosterone (a "male" hormone) and estrogen (a "female" hormone). Hormonal processes are thought to shape behavior, as well as physical development and reproductive functions, although experts disagree as to their impact on behavior.

hormones Chemical substances secreted into the bloodstream by the endocrine glands.

household As a Census Bureau category, a household is any group of people residing together.

hyperparenting The situation in which parents are excessively involved in their children's lives.

incest Sexual relations between closely related individuals.

income-to-needs ratio An assessment of income as to the degree it meets the needs of the individual, family, or household.

incomplete institution Cherlin's description of a remarried family due to cultural ambiguity.

independence effect Occurs when an increase in a married woman's income leads to marital dissolution because she is better able to afford to live separately.

individualism The cultural milieu that emerged in Europe with industrialization and that values personal self-actualization and happiness along with individual freedom.

individualistic society Society in which the main concern is with one's own interests (which may or may not include those of one's immediate family).

individualistic (self-fulfillment) values Values that encourage self-fulfillment, personal growth, autonomy, and independence over commitment to family or other communal needs.

individualized marriage Concept associated with the argument that contemporary marriage in the United States and other fully industrialized Western societies is no longer institutionalized. Four interrelated characteristics distinguish individualized marriage: (1) it is optional; (2) spouses' roles are flexible—negotiable and renegotiable; (3) its expected rewards involve love, communication, and emotional intimacy; and (4) it exists in conjunction with a vast diversity of family forms.

informal adoption Children are taken into a home and considered to be children of the parents, although the "adoption" is not legally formalized.

informal caregiving Unpaid caregiving, provided personally by a family member.

informational power One of the six power bases, or sources of power. This power is based on the persuasive content of what the dominant person tells another individual.

institutional marriage Marriage as a social institution based on dutiful adherence to the time-honored **marriage premise** (which see), particularly the norm of permanence. "Once ensconced in societal mandates for permanence and monogamous sexual exclusivity, the institutionalized marriage in the United States was centered on economic production, kinship network, community connections, the father's authority, and marriage as a functional partnership rather than a romantic relationship. . . . Family tradition, loyalty, and solidarity were more important than individual goals and romantic interest" (Doherty 1992, p. 33). Also referred to as *institutionalized marriage.*

instrumental traits See **agentic (instrumental) character traits**.

interaction–constructionist perspective Theoretical perspective that focuses on internal family dynamics; the ongoing action among and response to one another of family members.

interactionist perspective on human sexuality A perspective, derived from symbolic interaction theory, which holds that sexual activities and relationships are shaped by the sexual scripts available in a culture.

interethnic marriages Marriages between spouses who are not defined as of different races but do belong to different ethnic groups.

interfaith marriage Marriage in which partners are of different religions.

intermittent cohabitation Relationships in which parenting couples move in together, then out, then back in.

interparental conflict perspective One of the theoretical perspectives concerning the negative outcomes among children of divorced parents. From the interparental conflict perspective, the conflict between parents before, during, and after the divorce is responsible for the lowered well-being of the children of divorce.

interpersonal exchange model of sexual satisfaction A view of sexual relations, derived from exchange theory, that sees sexual satisfaction as shaped by the costs, rewards, and expectations of a relationship and the alternatives to it.

interracial marriages Marriages of a partner of one (socially defined) race to someone of a different race.

intersexed A person whose genitalia, secondary sex characteristics, hormones, or other physiological features are not unambiguously male or female.

intimate outsider role Stepparent role when stepfamilies reach maturation; end result is a stepparent who can be a confidante, trusted adviser, and mentor for a child on subjects and areas too threatening to share with biological parents such as sex and drug use.

intimate partner violence (IPV) Violence against current or former spouses, cohabitants, or sexual or relationship partners.

intimate terrorism Abuse that is almost entirely male and that is oriented to controlling the partner through fear and intimidation.

involved fathers Men who work fewer hours than do childless men in order to spend more time with their children.

involuntary infertility Situation of a couple or individual who would like to have a baby but cannot. Involuntary infertility is medically diagnosed when a woman has tried for twelve months to become pregnant without success.

job sharing Two people sharing one job.

joint custody A situation in which both divorced parents continue to take equal responsibility for important decisions regarding their child's general upbringing. Joint custody can include joint physical custody, which involves the children living at least part of the time with each parent.

kin Parents and other relatives, such as in-laws, grandparents, aunts and uncles, and cousins. See also **extended family**.

kin keeping Maintaining contact with family members and remembering anniversaries and birthdays, sending cards, shopping for gifts, and organizing family activities; more frequently done by women than by men.

labor force A social invention that arose with the industrialization of the nineteenth century, when people characteristically became wage earners, hiring out their labor to someone else.

latent kin matrix "A web of continually shifting linkages that provide the potential for activating and intensifying close kin relationships" (Riley 1983, p. 441).

legal custody Court determination of who has the right to make decisions with respect to a child's upbringing.

legitimate power One of the six power bases, or sources of power. Legitimate power stems from the more dominant individual's ability to claim authority, or the right to request compliance.

lesbian A woman who is sexually attracted to other women. This term is usually used rather than *homosexual.*

Levinger's model of divorce decisions This model, derived from exchange theory, presents a decision to divorce as involving a calculus of the *barriers* to divorce (e.g., concerns about children and finances; religious prohibitions), the *rewards* of the marriage, and *alternatives* to the marriage (e.g., can the divorced person anticipate a new relationship, career development, or a single life that will be more rewarding and less stressful than the marriage?).

LGBT An acronym for lesbian, gay, bisexual, or transgendered; a term commonly used when discussing sexual minorities.

life chances The opportunities that exist for a social group or an individual to pursue education and economic advancement, to secure medical care and preserve health, to marry and have children, to have material goods and housing of desired quality, and so forth.

life stress perspective One of the theoretical perspectives concerning the negative outcomes among children of divorced parents. From the life stress perspective, divorce involves the same stress for children as for adults. Divorce is not one single event but a process of stressful events—moving, changing schools, and so on.

living apart together (LAT) Emerging lifestyle choice in which a couple is committed to a long-term relationship but each partner maintains a separate dwelling.

male dominance The cultural idea of masculine superiority; the idea that men should and do exercise the most control and influence over society's members.

manipulating Seeking to control the feelings, attitudes, and behavior of one's partner or partners in underhanded ways rather than by assertively stating one's case.

marital rape A husband's forcing a wife to submit to sexual contact that she does not want or that she finds offensive.

marital sanctification A religious belief system that encourages spouses to see their marriages as ordained by God and hence having divine significance.

marriage market The sociological concept that potential mates take stock of their personal and social characteristics and then comparison shop or bargain for the best buy (mate) they can get.

marriage premise By getting married, partners accept the responsibility to keep each other primary in their lives and to work hard to ensure that their relationship continues.

martyring Doing all one can for others while ignoring one's own legitimate needs. Martyrs often punish the person to whom they are martyring by letting the person know "just how much I put up with."

menarche A girl's first menstrual cycle; age at which occurs dropped during the twentieth century for all races and ethnicities.

minority/minority group Group that is distinguishable and in some way disadvantaged within a society, regardless of size.

misogynistic Behavior that exhibits hatred, dislike, mistrust, mistreatment, or general disregard for women.

mixed message Two simultaneous messages that contradict each other; also called a *double message*. For example, society gives us mixed messages regarding family values and individualistic values and about premarital sex. People, too, can send mixed messages, as when a partner says, "Of course I always like to talk with you" while turning up the TV.

money allocation system Arrangements couples make for handling their income, wealth, and expenditures. They may involve pooling partners' resources or keeping them separate. Who controls pooled resources is another dimension of an allocation system.

motherhood penalty Negative lifetime impact on earnings for women who raise children.

multigenerational family Family that includes several generations.

multigenerational safety net function Different generations in a family providing each other with a cushion, or safety net, against potential or actual economic/health adversities.

multiple partner fertility Having children with more than one biological partner.

mutual dependency Stage of a relationship in which two people desire to spend more time together and thereby develop interdependence.

nadir of family disorganization Low point of family disorganization when a family is going through a family crisis.

National Family Violence Surveys Early and continuing research of Murray Straus, Richard Gelles, and their colleagues in shaping scientific research on family violence among the general population.

naturalistic observation A technique of scientific investigation in which a researcher lives with a family or social group or spends extensive time with them, carefully recording their activities, conversations, gestures, and other aspects of everyday life.

need fulfillment Stage of relationship development in which two people find they satisfy a majority of each other's emotional needs with the result that rapport increases and leads to deeper self-revelation, more mutually dependent habits, and still greater need satisfaction.

negative affect Showing emotion(s) defined as negative, such as anger, sadness, whining, disgust, tension, fear, and/or belligerence.

neotraditional families Families that value traditional gender roles and organize their family life in these terms as far as practicable. Formal male dominance is softened by an egalitarian spirit.

noncustodial grandparent A parent of a divorced, noncustodial parent.

nuclear family A family group comprising only the wife, the husband, and their children. See also **extended family**.

nuclear-family model monopoly The cultural assumption that the first-marriage family is the "real" model of family living, with all other family forms viewed as deficient.

occupational segregation The distribution of men and women into substantially different occupations. Women are overrepresented in clerical and service work, for example, whereas men dominate the higher professions and the upper levels of management.

opportunity costs (of children) The economic opportunities for wage earning and investments that parents forgo when raising children.

pansexual Identity of someone who has the potential to be sexually attracted to various gender expressions, including those outside the gender-conforming binary.

parallel parents Post-divorce parental relationship in which former partners parent alongside each other with minimal contact, communication, or conflict; associated with worse child outcomes.

parental adjustment perspective One of the theoretical perspectives concerning the negative outcomes among children of divorced parents. From the parental adjustment perspective, the parent's child-raising skills are impaired as a result of the divorce, with probable negative consequences for the children.

parental loss perspective One of the theoretical perspectives concerning the negative outcomes among children of divorced parents. From the parental loss perspective, divorce involves the absence of a parent from the household, which deprives children of the optimal environment for their emotional, practical, and social support.

parentification When children of divorce or those from a single parent home are forced to take on adult responsibilities before they are developmentally mature enough to handle them; linked to worse child outcomes.

passive-aggression Expressing anger at some person or situation indirectly, through nagging, nitpicking, or sarcasm, for example, rather than directly and openly. See also **displacement, sabotage**.

paternal claiming Extent to which stepfathers see their stepchildren as their own biological children.

patriarchal norm (of marital power) The norm (cultural rule) that the man should be dominant in a marital relationship.

patriarchal sexuality The view of human sexuality in which men own everything in the society, including women and women's sexuality. Males' sexual needs are emphasized while females' needs are minimized.

patriarchy A social system in which males are dominant.

percolator effect The "bottom up" operation of a stepfamily (from children to parents); see **dripolator effect**.

period of family disorganization That period in a family crisis, after the stressor event has occurred, during which family morale and organization slump and habitual roles and routines become nebulous.

permissive parenting style One of three parenting styles in this schema, permissive parenting gives children little parental guidance.

physical custody Legal determination of where a child will live and with which parent.

polyamory A marriage system in which one or both spouses retain the option to sexually love others in addition to their spouses.

polygamy A marriage system in which a person takes more than one spouse.

pool of eligibles A group of individuals who, by virtue of background or social status, are most likely to be considered eligible to make culturally compatible marriage partners.

pornification Broad social phenomenon in which sending, receiving, and viewing sexually explicit material is seen as normal activity.

pornography Sexually explicit images and descriptions as found in books, magazines, film, and cyberspace.

positive affect The expression, either verbal or nonverbal, of one's feelings of affection toward another.

postmodern family Term used to describe the situation in which (1) families today exhibit multiple forms, and (2) new or altered family forms continue to emerge or develop.

postmodern theory Theoretical perspective that largely analyzes social interaction (discourse or narrative) in order to demonstrate that a phenomenon is socially constructed.

power The ability to exercise one's will. *Personal power*, or *autonomy*, is power exercised over oneself. *Social power* is the ability to exercise one's will over others.

power politics Power struggles between spouses in which each seeks to gain a power advantage over the other; the opposite of a **no-power** relationship.

private face of family The aspect of the family that provides individuals with intimacy, emotional support, and love.

private safety net Social support from family and friends, rather than from public sources.

private sphere World inside someone's home.

pronatalist bias A cultural attitude that takes having children for granted.

provider role A term for the family role involving wage work to support the family. May be carried out by one spouse or partner only or by both.

psychological control Control over others by use of manipulative strategies, such as inducing guilt or withdrawing signs of affection.

psychological parent The parent, usually but not necessarily the mother, who assumes principal responsibility for raising the child.

public face of family The aspect of the family that produces public goods and services.

public sphere World outside someone's home.

race A group or category thought of as representing a distinct biological heritage. In reality, there is only one human race. "Racial" categories are social constructs; the so-called races do not differ significantly in terms of basic biological makeup. But "racial" designations nevertheless have social and economic effects and cultural meanings.

race socialization The socialization process that involves developing a child's pride in his or her cultural heritage while warning and preparing him or her about the possibilities of encountering discrimination.

rape myths Beliefs about rape that function to blame the victim and exonerate the rapist.

rapport Feelings of mutual trust and respect; often established by similarity of values, interests, and background.

rapport talk In Deborah Tannen's terms, this is conversation engaged in by women aimed primarily at gaining or reinforcing rapport or intimacy. See also **report talk**.

referent power One of the six power bases, or sources of power. In a marriage or relationship, this form of power is based on one partner's emotional identification with the other and his or her willingness to agree to the other's decisions or preferences.

refined divorce rate Number of divorces per 1,000 married women over age fifteen. See also crude **divorce rate**.

reframing In the family stress and coping literature, redefining stressful events to make them more manageable.

relationship-focused coping Process in relationships in which partners help or hinder one another in dealing with stresses and strain.

relationship ideologies Expectations for closeness and/or distance as well as ideas about how partners should play their roles.

replacement level (of fertility) The average number of births per woman (a total fertility rate of 2.1) necessary to replace the population.

report talk In Deborah Tannen's terms, this is conversation engaged in by men aimed primarily at conveying information. See also **rapport talk**.

reproductive coercion Behavior related to reproductive health that is used to maintain power and control in a relationship.

resilience The ability to recover from challenging situations.

resilient families Families that emphasize mutual acceptance, respect, and shared values; members rely on one another for emotional support.

resource hypothesis Hypothesis (originated by Robert Blood and Donald Wolfe) that the relative power between wives and husbands results from their relative resources as individuals.

resources in cultural context The effect of resources on marital power depends on the cultural context. In a traditional society, norms of patriarchal authority may override personal resources. In a fully egalitarian society, a norm of intimate partner and marital equality may override personal resources. It is in a transitional society that the resource hypothesis is most likely to shape marital power relations.

reward power One of the six power bases, or sources of power. With regard to marriage or partner relationships, this power is based on an individual's ability to give material or nonmaterial gifts and favors to the partner.

role ambiguity The situation in which there are few clear guidelines regarding what responsibilities, behaviors, and emotions family members are expected to exhibit.

role conflict When individuals with roles in two institutions experience conflict when they attempt to meet demands of one institution that conflict with different demands from another institution.

role-making Improvising a course of action as a way of enacting a role. In role-making, we may use our acts to alter the traditional expectations and obligations associated with a role. This concept emphasizes the variability in the ways different individuals enact a particular role.

sabotage A passive-aggressive action in which a person tries to spoil or undermine some activity another has planned. Sabotage is not always consciously planned. See also **passive-aggression**.

sandwich generation Middle-aged (or older) individuals, usually women, who are sandwiched between the simultaneous responsibilities of caring for their dependent children (sometimes young adults) and aging parents.

science "A logical system that bases knowledge on . . . systematic observation, empirical evidence, facts we verify with our senses" (Macionis 2006, p. 15).

second shift Sociologist Arlie Hochschild's term for the domestic work that employed women must perform after coming home from a day on the job.

selection effect When individuals "select" themselves into a category being investigated.

selection hypothesis The idea that many of the changes found in a dependent variable, which might be assumed to be associated with the independent variable, are really due to sample selection. For instance, the selection hypothesis posits that many of the benefits associated with marriage—for example, higher income and wealth, along with better health—are not necessarily due to the fact of being married but, rather, to the personal characteristics of those who choose—or are selected into—marriage. Similarly, the selection hypothesis posits that many of the characteristics associated with cohabitation result not from the practice of cohabiting itself but from the personal characteristics of those who choose to cohabit. See also the antonym, **experience hypothesis**.

selection perspective Posits that individuals have pre-existing traits, called *selection factors*, that sort them into groups with higher divorce risks as well as negative outcomes after divorce.

self-care An approach to child care for working parents in which the child is at home or out without an adult caretaker. Parents may be in touch by phone.

self-concept The basic feelings people have about themselves, their characteristics and abilities, and their worth; how people think of or view themselves.

self-identification theory A theory of gender socialization, developed by psychologist Lawrence Kohlberg, that begins with a child's categorization of self as male or female. The child goes on to identify sex-appropriate behaviors in the family, media, and elsewhere and to adopt those behaviors.

self-revelation Gradually sharing intimate information about oneself. Also see **self-disclosure**.

sex Refers to biological characteristics—that is, male or female anatomy or physiology. The term **gender** refers to the social roles, attitudes, and behavior associated with males or females.

sex ratio The number of men per 100 women in a society. If the sex ratio is above 100, there are more men than women; if it is below 100, there are more women than men.

sexting Using cell phones to send sexually explicit images or messages to others.

sexual abuse A form of child abuse that involves forced, tricked, or coerced sexual behavior—exposure, unwanted kissing, fondling of sexual organs, intercourse, rape, and incest—between a minor and an older person.

sexual assault Defined by U.S. legal code as any type of sexual contact or behavior that occurs without the explicit consent of the recipient and includes fondling, groping, digital penetration (with fingers), forced oral and anal sex (sodomy), forced sexual intercourse (rape), and attempted rape.

sexual coercion Use of verbal or emotional pressure (threatening violence, tricking, lying, or using guilt), one's position of power (being a boss, teacher, coach, or other adult), or other means to manipulate a victim into sexual activity. Common behavior among perpetrators of acquaintance rape.

sexual identity Whether one is attracted to one's own gender or a different gender; often described as *sexual orientation* or *sexual orientation identity*.

sexual reassignment surgery Surgical process by which transgendered men and women make full physical transformations to their preferred sex.

sexual scripts *Scripts* are culturally written patterns or "plots" for human behavior. Sexual scripts offer reasons for having sex and designate who should take the sexual initiative, how long an encounter should last, what positions are acceptable, and so forth.

shelter Physically protected and safe space that provides a woman (and often her children) with temporary housing, food, and clothing to alleviate the problems of economic dependency and physical safety.

shift work As defined by the Bureau of Labor Statistics, any work schedule in which more than half of an employee's hours are before 8 a.m. or after 4 p.m.

sibling violence Family violence that takes place between siblings (brothers and sisters).

silver divorce Phenomenon seen among older couples following long-term marriages; particularly true for baby boomers born between 1946 and 1964.

single mothers by choice Women who intentionally become mothers, although they are not married or with a partner. They are typically older, with economic and educational resources that enable them to be self-supporting.

situational couple violence Mutual violence between partners that often occurs in conjunction with a specific argument. It involves fewer instances, is not likely to escalate, and tends to be less severe in terms of injuries.

social capital perspective (on parenthood) Motivation for parenthood in anticipation of the links parenthood provides to social networks and their resources.

social class Position in the social hierarchy, such as *upper class, middle class, working class*, or *lower class*. Can be viewed in terms of such indicators as education, occupation, and income or analyzed in terms of status, respect, and lifestyle.

social fathers Males who are not a biological father but are performing the role of father, such as a stepfather.

social institution A system of patterned and predictable ways of thinking and behaving—beliefs, values, attitudes, and norms—concerning important aspects of people's lives in society. Examples of major social institutions are the family, religion, government, the economy, and education.

social learning theory (of gender socialization) According to this theory, children learn gender roles as they are taught or modeled by parents, schools, and the media.

socialization The process by which society influences members to internalize attitudes, beliefs, values, and expectations.

socioeconomic status (SES) One's position in society, measured by educational achievement, occupation, and/or income.

sociological imagination Placing an individual's or family's private troubles within a society-wide context.

spillover How pleasures or stresses associated with work affect interaction within the family and vice versa.

starter marriage First marriage that ends in divorce within the first few years, typically before a couple has children.

status exchange hypothesis Regarding interracial/interethnic marriage, the argument that an individual might trade his or her socially defined superior racial/ethnic status for the economically or educationally superior status of a partner in a less-privileged racial/ethnic group.

stay-at-home dad Man who stays at home to care for the house and family while his wife works.

stepfamily cycle Process by which new family members move from being strangers to forming nourishing and reliable relationships.

stepmother trap The conflict between two views: Society sentimentalizes the stepmother's role and expects her to be unnaturally loving toward her stepchildren but at the same time views her as a wicked witch.

Sternberg's triangular theory of love Robert Sternberg's theory that consummate love involves three components: intimacy, passion, and commitment.

stigmatization When people are subjected to negative labels, stereotyping, and cultural myths that portray them as deviant and harmful simply because they have certain social characteristics.

stonewalling One of the **Four Horsemen of the Apocalypse** (which see) that involves refusing to listen to a partner's complaints.

stress State of tension that results from the need to respond to change.

stress model of parental effectiveness The idea that stress experienced by parents causes parental frustration, anger, and depression, increasing the likelihood of household conflict and leading to poorer parenting practices.

stressors Precipitating events that cause a crisis; they are often situations for which the family has had little or no preparation. See also **ABC-X model**.

stressor overload (pileup) A situation in which an unrelenting series of small crises adds up to a major crisis.

stress-related growth Personal growth and maturity attained in the context of a stressful life experience such as divorce.

structural antinatalism The structural, or societal, conditions in which bearing and raising children is discouraged either overtly or—as may be the case in the United States—covertly through inadequate support for parenting.

structural constraints Economic and social forces that limit options and, hence, personal choices.

structure–functional perspective Theoretical perspective that looks to the functions that institutions perform for society and the structural form of the institution.

surrogate grandparent A form of fictive kin, an older individual not biologically related to the "grandchildren" who, attached to a non-biologically related family, plays a role ordinarily designated for biological grandparents.

swinging A marriage agreement in which couples exchange partners to engage in purely recreational sex.

theoretical perspective A way of viewing reality, or a lens through which analysts organize and interpret what they observe. Researchers on the family identify those aspects of families that are of interest to them, based on their own theoretical perspective.

total fertility rate (TFR) For a given year, the number of births that women would have over their reproductive lifetimes if all women at each age had babies at the rate for each age group that year; can be calculated for social or age categories as well as for nations as a whole.

traditional sexism Beliefs that men and women are essentially different and should occupy different social roles, that women are not as fit as men to perform certain tasks and occupations, and that differential treatment of men and women is acceptable.

trailing partner The spouse or partner who relocates to accommodate the needs of the other's career.

transgendered A person who has adopted a gender identity that differs from sex/gender as recorded at birth; a person who declines to identify as either male or female.

transition to parenthood The circumstances involved in assuming the parent role.

transnational family A family of immigrants or immigrant stock that maintains close ties with the sending country. Identity and behavior connect the immigrant family to the new country and the old, and their social networks cross national boundaries.

transsexuals Individuals who switch physical sexes through surgery, hormone therapy, electrolysis (hair removal), and other treatments.

transvestites People who dress as the opposite gender because it feels erotic, empowering, or rebellious or for other reasons.

two-career relationship Relationship now seen as an available and workable option; for two-career couples with children, however, family life can be hectic as partners juggle schedules, chores, and child care.

two-earner union Now majority relationship, with both partners working and displaying considerable flexibility in designing their schedules and life choices.

two-pot system When economic resources for a family economic resources are divided and distributed along biological lines and only secondarily distributed according to need.

unilateral divorce A divorce can be obtained under the no-fault system by one partner even if the other partner objects. The term *unilateral divorce* emphasizes this feature of current divorce law.

unpaid family work The necessary tasks of attending to both the emotional needs of all family members and the practical needs of dependent members, such as children or elderly parents, and maintaining the family domicile.

value of children perspective (on parenthood) Motivation for parenthood because of the rewards, including symbolic rewards, that children bring to parents.

Violence Against Women Act 1996 federal law that has resulted in greater perpetrator accountability for rape and stalking, as well as increasing rates of prosecution, conviction, and sentencing of offenders; 2010 expansion of the act includes same-gender couples, and 2013 expansion includes additional categories such as transgender individuals.

voluntarily childfree The deliberate choice not to become a parent; often used now instead of the more negative-sounding term *childless*.

vulnerable families Families that have a low sense of common purpose, feel in little control over what happens to them, and tend to cope with problems by showing diminished respect and/or understanding for each other.

wage gap The persistent difference in earnings between men and women.

War on Poverty Series of federal programs and initiatives put forth by President Lyndon Johnson in the 1960s; included the Job Corps or the Neighborhood Youth Corps, Head Start, and Adult Basic Education. Although most measures have ended, Head Start and the Job Corps continue to exist.

wheel of love An idea developed by Ira Reiss in which love is seen as developing through a four-stage, circular process, including rapport, self-revelation, mutual dependence, and personality need fulfillment.

References

AAUW Educational Foundation. 2006. *Drawing the Line: Sexual Harassment on Campus Study.* Washington, DC: American Association of University Women. January 26. Retrieved February 5, 2007 (www.aauw.org).

Aasen, Eric. 2009. "More Men Becoming Stay-at-home Dads." *Dallas Morning News,* July 29.

Abbey, L. 2009. "Elder Abuse and Neglect: When Home Is Not Safe." *Clinics in Geriatric Medicine* 25(1):47–61.

Aber, J. Lawrence. 2007. *Child Development and Social Policy.* Washington, DC: American Psychological Association.

Abma, Joyce C. and Gladys M. Martinez. 2006. "Childlessness among Older Women in the United States: Trends and Profiles." *Journal of Marriage and Family* 68(4):1045–1056.

Abraham, Margaret. 1995. "Ethnicity, Gender, and Marital Violence: South Asian Women's Organizations in the U.S." *Gender and Society* 9:450–468.

———. 2000. *Speaking the Unspeakable: Marital Violence among South Asian Immigrants in the United States.* New Brunswick, NJ: Rutgers University Press. "The Abstinence-Only Delusion" (editorial). 2007. *The New York Times,* April 28.

Achen, Alexandra C. and Frank P. Stafford. 2005. *Data Quality of Housework Hours in the Panel Study of Income Dynamics: Who Really Does the Dishes?* Ann Arbor, MI: Institute for Social Research, University of Michigan.

Ackerman, Brian P., Kristen Schoff D'Eramo, Lina Umylny, David Schultz, and Carroll E. Izard. 2001. "Family Structure and the Externalizing Behavior of Children from Economically Disadvantaged Families." *Journal of Family Psychology* 15(2):288–301.

Adam, Barry D. 2007. "Relationship Innovation in Male Couples." Pp. 122–140 in *The Sexual Self: The Construction of Sexual Scripts,* edited by Michael S. Kimmel. Nashville: Vanderbilt University Press.

Adamo, Shelley A. 2013. "Attrition of Women in the Biological Sciences: Workload, Motherhood, and Other Explanations Revisited." *BioScience* 63(1):43–48.

Adams, Lynn. 2010. *Parenting on the Autism Spectrum: A Survival Guide.* San Diego: Plural Publishers.

Adamsons, Kari and Kay Pasley. 2006. "Coparenting Following Divorce and Relationship Dissolution." Pp. 241–261 in *Handbook of Divorce and Relationship Dissolution,* edited by Mark A. Fine and John H. Harvey. Mahwah, NJ: Erlbaum.

Addo, Fenaba R. and Daniel T. Lichter. 2013. "Marriage, Marital History, and Black-White Wealth Differentials among Older Women." *Journal of Marriage and Family* 75(2):342–362.

Addo, Fenaba R. and S. Sassler. 2010. "Financial Arrangements and Relationship Quality in Low Income Couples." *Family Relations,* 59(4):408–423.

Adelman, Rebecca A. 2009. "Sold(i)ering Masculinity: Photographing the Coalition's Male Soldiers." *Men & Masculinities* 11(3):259–285.

Adimora, Adaora A., Victor J. Schoenbach, and Irene A. Doherty. 2007. "Concurrent Sexual Partnerships among Men in the United States." *American Journal of Public Health* 97(12):2230–2237.

Adler, Jerry. 1993. "Sex in the Snoring '90s." *Newsweek,* April 26, pp. 55–57.

Adler-Baeder, Francesca, Christiana Russell, Jennifer Kerpelman, Joe Pittman, Scott Ketring, Thomas Smith, Mallory Lucier-Green, Angela Bradford, and Kate Stringer. 2010. "Thriving in Stepfamilies: Exploring Competence and Well-Being among African American Youth." *Journal of Adolescent Health* 46:396–398.

Afifi, Tamara D. 2008. "Communication in Stepfamilies: Stressors and Resilience." Pp. 299–317 in *The International Handbook of Stepfamilies,* edited by J.Pryor. Hoboken, NJ: John Wiley & Sons.

Afifi, Tracie, Natalie Mota, Patricia Dasiewicz, Harriet MacMillan, and Jitender Sareen. 2012. "Physical Punishment and Mental Disorders: Results from a Nationally Representative U.S. Sample." *Pediatrics* 130(2):184–192.

African Wedding Guide. n. d. Retrieved October 15, 2006 (www.africanweddingguide.com).

Agee, Mark D., Scott E. Atkinson, and Thomas D. Crocker. 2008. "Multiple-Output Child Health Production Functions: The Impact of Time-Varying and Time-Invariant Inputs." *Southern Economic Journal* 75(2):410–428.

Agerbo, Esben, Presben Bo Mortensen, and Trine Munk-Olsen. 2012. "Childlessness, Parental Mortality and Psychiatric Illness: A Natural Experiment Based on In vitro Fertility Treatment and Adoption." *Journal of Epidemiology & Community Health.* December 5:201–245.

Ahn, Annie, Bryan Kim, and Park Yong. 2008. "Asian Cultural Values Gap, Cognitive Flexibility, Coping Strategies, and Parent-Child Conflicts among Korean Americans." *Cultural Diversity and Ethnic Minority Psychology* 14(4):353–363.

Ahrold, Tierney K., Melissa Farmer, Paul D. Trapnell, and Cindy M. Meston (2011). "The Relationship among Sexual Attitudes, Sexual Fantasy, and Religiosity." *Archives of Sexual Behavior* 40:619–630.

Ahrold, Tierney K., and Cindy M. Meston. 2011. "Ethnic Differences is Sexual Attitudes of U.S. College Students: Gender, Acculturation, and Religiosity Factors." *Archives of Sexual Behavior* 39:190–202.

Ahrold, Tierney K., Melissa Farmer, Paul D. Trapnell, and Cindy M. Meston (2011). "The Relationship Among Sexual Attitudes, Sexual Fantasy, and Religiosity." *Archives of Sexual Behavior* 40:619–630.

Ahrons, Constance. 1994. *The Good Divorce: Raising Your Family Together When Your Marriage Comes Apart.* New York: HarperCollins.

———. 2004. *We're Still Family: What Grown Children Have to Say about Their Parents' Divorce.* New York: HarperCollins.

———. 2007. "Family Ties After Divorce: Long-Term Implications for Children." Family Process 46(1):53–65.

Ahtone, Tristan. 2011. "Native American Intermarriage Puts Benefits at Risk." Retrieved May 31, 2013, from http://www.mixcloud.com/NPR/native-american-intermarriage-puts-benefits-at-risk/.

Ainsworth, Mary D. S., M. C. Blehar, E. Waters, and S. Wall. 1978. *Patterns of Attachment: A Psychological Study of the Strange Situation.* Hillsdale, NJ: Erlbaum.

Aizer, Anna. 2004. "Home Alone: Supervision After School and Child Behavior." *Journal of Public Economics* 88:1835–1848.

Albrecht, Chris and Jay D. Teachman. 2003. "Childhood Living Arrangements and the Risk of Premarital Intercourse." *Journal of Family Issues* 24(7):867–894.

Aldous, Joan. 1978. *Family Careers: Developmental Change in Families.* New York: Wiley.

———. 1996. *Family Careers: Rethinking the Developmental Perspective.* Thousand Oaks, CA: Sage Publications.

Alexander, Brian. 2010. "Lovesick: Hooking Up Over a Shared Disease." *MSNBC,* February 12. www.msnbc.msn.com.

Ali, Lorraine and Raina Kelley. 2008. "The Curious Lives of Surrogates." *Newsweek.* April 7: 45–51.

Ali, Lorraine and Julie Scelfo. 2002. "Choosing Virginity." *Newsweek,* December 9, pp. 61–71.

Allan, Graham, Graham Crow, and Sheila Hawker. 2011. *Stepfamilies.* New York: Palgrave Macmillan.

Allen, Anne Wallace. 2009. "Recession Uproots Families, Takes Toll on Children." Associated Press, May 27.

Allen, Elizabeth S., and David C. Atkins. 2012. "The Association of Divorce and Extramarital Sex in a Representative U.S. Sample." *Journal of Family Issues* 33(11):1477–1493.

Allen, Jane E. 2012. "Early Burdens: Eldercare Falls on Young Shoulders." *ABC News.* May 4. Retrieved May 7, 2013 (http://www.abcnews.go.com).

Allen, Katherine R. 1997. "Lesbian and Gay Families." Pp. 196–218 in *Contemporary Parenting: Challenges and Issues,* edited by Terry Arendell. Thousand Oaks, CA: Sage Publications.

———. 2007. "Ambiguous Loss After Lesbian Couples with Children Break Up: A Case for Same-Gender Divorce." *Family Relations* 56(April):175–183.

Allen, Kathleen R., R. Blieszner, and K. A. Roberto. 2011. "Perspectives on Extended Family and Fictive Kin in the Later Years Strategies and Meanings of Kin Reinterpretation." *Journal of Family Issues* 32(9):1156–1177.

Allen, Katherine R., Rosemary Bleiszner, Karen A. Roberto, Elizabeth B. Farnsworth, and Karen L. Wilcox. 1999. "Older Adults and Their Children: Family Patterns of Structural Diversity." *Family Relations* 48(2):151–157.

Allen, Katherine R., Christine E. Kaestle, and Abbie E. Goldberg. 2011. "More than Just a Punctuation Mark: How Boys and Young Men Learn about Menstruation." *Journal of Family Issues* 32(2):129–156.

Allen, Mike. 2002. "Law Extends Benefits to Same-Sex Couples." *Washington Post*, June 26.

Allen, Mike and Alan Cooperman. 2004. "Bush Plans to Back Marriage Amendment." *Washington Post.*

Alternatives to Marriage Project 2012 (http://www.unmarried.org/).

Altman, Irwin and Dalmas A. Taylor. 1973. *Social Penetration: The Development of Interpersonal Relations.* New York: Holt, Rinehart & Winston.

Altman, Lawrence K. 2004. "Study Finds That Teenage Virginity Pledges Are Rarely Kept." *The New York Times*, March 10.

Alvarez, Lizette. 2009. "G.I. Jane Stealthily Breaks the Combat Barrier." *The New York Times*, August 16.

Alvarez, Michelle. 2009. "Exclusive AARP Bulletin Poll Reveals New Trends in Multigenerational Housing." March 3. AARP Press Center. Retrieved November 21, 2009 (www.aarp.org).

Amato, Paul R. 1993. "Children's Adjustment to Divorce: Theories, Hypotheses, and Empirical Support." *Journal of Marriage and Family* 55(1):23–28.

———. 2000. "The Consequences of Divorce for Adults and Children." *Journal of Marriage and Family* 62:1269–1287.

———. 2003. "Reconciling Divergent Perspectives: Judith Wallerstein, Quantitative Research, and Children of Divorce." *Family Relations* 52:332–330.

———. 2004. "Tension between Institutional and Individual Views of Marriage." *Journal of Marriage and Family* 66(4):959–965.

———. 2005. "The Impact of Family Formation Change on the Cognitive, Social and Emotional Well-Being of the Next Generation." *The Future of Children* 15(2):75–96.

———. 2007. "Transformative Processes in Marriage: Some Thoughts from a Sociologist." *Journal of Marriage and Family* 69(2):305–309.

———. 2010. "Research on Divorce: Continuing Trends and New Developments." *Journal of Marriage and Family* 72(3):650–666.

———. 2012. "The Well-Being of Children with Gay and Lesbian Parents." *Social Science Research* 41: 771–774.

Amato, Paul R. and Alan Booth. 1997. *A Generation at Risk: Growing Up in an Era of Family Upheaval.* Cambridge, MA: Harvard University Press.

Amato, Paul R., Alan Booth, David R. Johnson, and S. J. Rogers. 2007. *Alone Together: How Marriage in America is Changing.* Cambridge, MA: Harvard University Press.

Amato, Paul R. and Jacob E. Cheadle. 2008. "Parental Divorce, Marital Conflict and Children's Behavior Problems: A Comparison of Adopted and Biological Children." *Social Forces* 86(3):1139–1161.

Amato, Paul R. and Bryndl Hohmann-Marriott. 2007. "A Comparison of High- and Low-Distress Marriages That End in Divorce." *Journal of Marriage and Family* 69:621–638.

Amato, Paul R. and Shelley Irving. 2006. "Historical Trends in Divorce in the United States." Pp. 41–57 in *Handbook of Divorce and Relationship Dissolution*, edited by Mark A. Fine and John H. Harvey. Mahwah, NJ: Erlbaum.

Amato, Paul R., David R. Johnson, Alan Booth, and Stacy J. Rogers. 2003. "Continuity and Change in Marital Quality between 1980 and 2000." *Journal of Marriage and Family* 65(1):1–22.

Amato, Paul R. and Jennifer B. Kane. 2011. "Life Course Pathways and the Psychosocial Adjustment of Young Adult Women." *Journal of Marriage and Family* 73(1):279–295.

Amato, Paul R., Jennifer B. Kane, and Spencer James. 2011. "Reconsidering the 'Good Divorce.'" *Family Relations* 60:511–524.

Amato, Paul R., Nancy S. Landale, Tara C. Havasevich-Brooks, and Alan Booth. 2008. "Precursors of Young Women's Family Formation Pathways." *Journal of Marriage and Family* 70:1271–1286.

Amato, Paul R., Catherine E. Meyers, and Robert E. Emery. 2009. "Changes in Nonresident Father-Child Contact from 1976 to 2002." *Family Relations* 58(1):41–53.

Ambert, A. M. 1989. *Ex-Spouses and New Spouses: A Study of Relationships.* Greenwich, CT: Jai Press.

American Academy of Pediatrics. 2013. "American Academy of Pediatrics Support Same Gender Civil Marriage." Press Release. March 21. Retrieved March 21, 2013 (http://www.aap.org/en-us/about-the-aap/aap-press-room/Pages/American-Academy-of-Pediatrics-Supports-Same-Gender-Civil-Marriage.aspx).

American College of Obstetricians and Gynecologists. 2012. "Committee Opinion: Intimate Partner Violence." February. Number 518.

American College of Pediatricians. 2007. "Corporal Punishment: A Scientific Review of Its Use in Discipline." Retrieved from http://www.acpeds.org.

———. 2013. "Guidelines for Parental Use of Disciplinary Spanking." Retrieved from http://www.BestforChildren.org.

American Moslem Society. 2010. www.masjiddearborn.org.

American Psychological Association 2005. "The Impact of Abortion on Women: What Does the Psychological Research Say?" APA Briefing Paper. Washington, DC: American Psychological Association. January 31. Retrieved November 10, 2006 (www.apa.org).

———. 2007. "Answers to Your Questions about Sexual Orientation and Homosexuality." Washington, DC: American Psychological Association. Retrieved June 12, 2007 (www.apa.org).

———. 2009. "APS Survey Raises Concern about Parent Perceptions of Children's Stress." November 3. Retrieved January 21, 2010 (http://www.apa.org).

———. 2010. "Sexual Orientation and Homosexuality." APA Online (http://www.apa.org/helpcenter/sexual-orientation.aspx).

———. 2013. *Stress in America.* February 22. Washington, DC: American Psychological Association.

American Psychological Association Task Force on Appropriate Therapeutic Responses to Sexual Orientation. 2005. "The Impact of Abortion on Women: What Does the Psychological Research Say?" APA Briefing Paper. Washington, DC: American Psychological Association. January 31. Retrieved November 10, 2006 (www.apa.org).

———. 2009. *Appropriate Therapeutic Responses to Sexual Orientation.* Retrieved on June 24, 2013 (http://www.apa.org/pi/lgbt/resources/therapeutic-response.pdf).

Ames, Barbara D., Whitney A. Brosi, and Karla M. Damiano-Teixeira. 2006. "'I'm Just Glad My Three Jobs Could Be during the Day': Women and Work in a Rural Community." *Family Relations* 55(1):119–131.

Amiraian, Dana E., and Jeffery Sobal. 2009. "Dating and Eating. Beliefs about Foods among University Students." *Appetite* 53:226–232.

Anderlini-D'Onofrio, Serina. 2004. "Introduction to Plural Loves: Bi and Poly Utopias for a New Millennium." *Journal of Bisexuality* 4(3/4):2–6.

Andersen, Julie Donner. 2002. "His Kids: Becoming a W.O.W. Stepmother." SelfGrowth.com (www.selfgrowth.com/articles/Andersen3.html).

Andersen, Margaret L. and Patricia Hill Collins. 2007. "Why Race, Class, and Gender Still Matter." Pp. 1–16 in *Race, Class and Gender: An Anthology*, 6th ed., edited by Margaret L. Andersen and Patricia Hill Collins. Belmont, CA: Wadsworth.

Andersen, Margaret L., and Howard F. Taylor. 2002. *Sociology: Understanding A Diverse Society*, 2nd ed. Belmont, CA: Wadsworth.

Anderson, Elijah. 1999. *The Code of the Street.* New York: Norton.

Anderson, Keith A., Noelle L. Fields, and Lynn A. Dobb. 2013. Caregiving and Early Life

Trauma: Exploring the Experiences of Family Caregivers to Aging Holocaust Survivors." *Family Relations* 62(2):366–377.

Anderson, Kristin L. 2010. "Conflict, Power, and Violence in Families." *Journal of Marriage and Family* 72(3):726–742.

———. 2013. "Why Do We Fail to Ask 'Why' about Gender and Intimate Partner Violence?" *Journal of Marriage and Family* 75(2):314–318.

Anderson, Sarah E. and Aviva Must. 2005. "Interpreting the Continued Decline in the Average Age at Menarche: Results from Two Nationally Representative Surveys of U.S. Girls Studied 10 Years Apart." *Journal of Pediatrics* 147(6):753–760.

Anderson, Michael A., Paulette Marie Gillig, Marilyn Sitaker, Kathy McCloskey, Katherine Malloy, and Nancy Grigsby. 2003. "'Why Doesn't She Just Leave?' A Descriptive Study of Victim Reported Impediments to Her Safety." *Journal of Family Violence* 18:151–155.

Anderson, Stephen A. and Ronald M. Sabatelli. 2007. *Family Interaction: A Multigenerational Developmental Perspective*, 4th edition. Boston and New York: Pearson.

Andersson, G. and D. Philipov. 2002. Life-Table Representations of Family Dynamics in Sweden, Hungary, and 14 Other FFS Countries: A Project of Descriptions of Demographic Behavior.*Demographic Research* [online] 7, 4. Available at http://www.demographicresearch.org/volumes/vol7/4/7-4.pdf.

Anetzberger, Georgia, Jill Korbin, and Craig Austin. 1994. "Alcoholism and Elder Abuse." *Journal of Interpersonal Violence* 9(2):184–193.

Angelo, Megan. 2010. "Fortune 500 Women CEOs." *Fortune*, April 23.

Anyiam, Thony. 2002. "Who Should Jump the Broom?" Retrieved October 2, 2006 (www.anyiams.com/jumping_the_broom).

Appleby, Julie. 2012. "Many Businesses Offer Health Benefits to Same-Sex Couples Ahead of Laws." *The Rundown: A Blog of News and Insight.* PBS News. May 14. Retrieved September 3, 2012 (http://www.pbs.org/newshour).

Arditti, Joyce A. 2003. "Incarceration Is a Major Source of Family Stress." *Family Focus* (June):F15–F17. Minneapolis: National Council on Family Relations.

Arditti, Joyce A. and Timothy Z. Keith. 1993. "Visitation Frequency, Child Support Payment, and the Father–Child Relationship Postdivorce." *Journal of Marriage and Family* 55(3):699–712.

Arditti, Joyce A., Jennifer Lambert-Shute, and Karen Joest. 2003. "Saturday Morning at the Jail: Implications of Incarceration for Families and Children." *Family Relations* 52(3):195–204.

Arendell, Terry. 1997. "Divorce and Remarriage." Pp. 154–195 in *Contemporary Parenting: Challenges and Issues*, edited by Terry Arendell. Thousand Oaks, CA: Sage Publications.

———. 2000. "Conceiving and Investigating Motherhood: The Decade's Scholarship." *Journal of Marriage and Family* 62(4):1192–1207.

Ariès, Phillipe. 1962. *Centuries of Childhood: A Social History of Family Life.* New York: Knopf.

Armstrong, Larry. 2003. "Your Mouse Knows Where Your Car Is." *Business Week* 16.

Armstrong, Elizabeth A., Paula England, and Alison C. K. Fogarty. 2012. "Accounting for Women's Orgasm and Sexual Enjoyment in College Hookups and Relationships." *American Sociological Review* 72(3):435–462.

Arnett, Jeffrey Jensen. 2000. "Emerging Adulthood: A Theory of Development from the Late Teens through the Twenties." *American Psychologist* 55(5):469–480. Retrieved March 28, 2003 (PsycARTICLES 0003-006X).

———. 2004. *Emerging Adulthood: The Winding Road from the Late Teens through the Twenties.* London: Oxford University Press.

Arnott, Teresa and Julie Matthaei. 2007. "Race, Class, and Gender and Women's Works." Pp. 283–292 in *Race, Class, and Gender*, 6th ed., edited by Margaret Andersen and Patricia Hill Collins. Belmont, CA: Wadsworth.

Aronson, Pamela. 2003. "Feminists or 'Postfeminists'? Young Women's Attitudes toward Feminism and Gender Relations." *Gender and Society* 17:903–922.

Artis, Julie E. 2004. "Judging the Best Interests of the Child: Judges' Accounts of the Tender Years Doctrine." *Law & Society Review* 38(4):769–806.

Artis, Julie E. and Eliza K. Pavalko. 2003. "Explaining the Decline in Women's Household Labor: Individual Change and Cohort Differences." *Journal of Marriage and Family* 65:746–761.

Asencio, Marysol. 2009. "Migrant Puerto Rican Lesbians Negotiating Gender, Spituality, and Ethnonationality." *NWSA Journal* 21(3):2–23.

Asexual Visibility and Education Network. 2012. Retrieved June 25, 2013 (http://www.asexuality.org/home/overview.html).

Associated Press. 2012. "Lawsuit Filed Against California Ban on Change-Gays Therapy." *Washington Times.* October 1. Retrieved October 10, 2012. (http://www.washingtontimes.com/news/2012/oct/1/lawsuit-filed-vs-calif-ban-on-change-gays-therapy/).

Association for Psychological Science. 2008. "Is Empty Nest Best? Changes in Marital Satisfaction in Late Middle Age." Retrieved April 7, 2010 (www.physorg.com).

Athenstaedt, Ursula, Gerold Mikula, and Cornelia Bredt. 2009. "Gender Role Self-Concept and Leisure Activities of Adolescents." *Sex Roles* 60(5/6):399–409.

Atkin, Bill. 2008. "Legal Structures and Re-Formed Families: The New Zealand Example." Pp. 522–544 in *The International Handbook of Stepfamilies*, edited by J.Pryor. Hoboken, NJ: John Wiley & Sons.

Atkins David. C. and Furrow, James. 2009. "Infidelity Is on the Rise: But for Whom and Why?" Paper presented at the annual meeting of the Association for Behavioral and Cognitive Therapies, Orlando, FL.

Augustine, Jennifer March and R. Kelly Raley. 2012. "Multigenerational Households and the School Readiness of Children Born to Unmarried Mothers." *Journal of Family Issues* 34(4):431–459.

Austin, Jennifer L. and Mariana K. Falconier. 2012. "Spirituality and Common Dyadic Coping: Protective Factors from Psychological Aggression in Latino Immigrant Couples." *Journal of Family Issues* 34(3):323–346.

Avellar, Sarah and Pamela J. Smock. 2003. "Has the Price of Motherhood Declined Over Time? A Cross-Cohort Comparison of the Motherhood Wage Penalty." *Journal of Marriage and Family* 65:597–607.

———. 2005. "The Economic Consequences of the Dissolution of Cohabiting Unions." *Journal of Marriage and Family* 67(2):315–327.

Aydt, Hilary and William A. Corsaro. 2003. "Differences in Children's Construction of Gender Across Culture." *American Behavioral Scientist* 46(10):1306–325.

Babbie, Earl. 2007. *The Practice of Social Research.* 11th ed. Belmont, CA: Wadsworth/Cengage.

———. 2009. *The Practice of Social Research.* 12th ed. Belmont, CA: Wadsworth/Cengage.

Babbitt, Charles E. and Harold J. Burbach. 1990. "A Comparison of Self-Orientation among College Students across the 1960s, 1970s and 1980s." *Youth & Society* 21(4):472–482.

Baca Zinn, Maxine, Pierette Hondagneu-Sotelo, and Michael A. Messner. 2004. "Gender through the Prism of Difference." Pp. 166–174 in *Race, Class, and Gender*, 5th ed., edited by Margaret L. Andersen and Patricia Hill Collins. Belmont, CA: Wadsworth.

———. 2007. "Sex and Gender through the Prism of Difference." Pp. 147–155 in *Race, Class, and Gender: An Anthology*, 6th ed, edited by Margaret L. Andersen and Patricia Hill Collins. Belmont, CA: Wadsworth.

Baca Zinn, Maxine and Barbara Wells. 2007. "Diversity within Latino Families: New Lessons for Family Social Science." Pp. 422–447 in *Family in Transition*, 14th ed., edited by Arlene S. Skolnick and Jerome H. Skolnick. Boston: Allyn & Bacon.

Bachrach, Christine, Patricia F. Adams, Soledad Sambrano, and Kathryn A. London. 1990. "Adoption in the 1980s." *Advance Data*, No. 181. Hyattsville, MD: U.S. National Center for Health Statistics, January 5.

Backstrom, Laura, Elizabeth A. Armstrong, and Jennifer Puentes. 2012. "Women's Negotiation of Cunnilingus in College Hookups and Relationships." *Journal of Sex Research* 49(1):1–12.

Badgett, M. V. Lee and Jody L. Herman. 2011. *Patterns of Relationship Recognition by Same-Sex Couples in the United States.* Retrieved April 14, 2013 (http://williamsinstitute.law.ucla.edu/wp-content/uploads/Badgett-Herman-Marriage-Dissolution-Nov-2011.pdf).

Baer, Judith C. and Mark F. Schmitz. 2007. "Ethnic Differences in Trajectories of Family Cohesion for Mexican American and Non-Hispanic White Adolescents." *Journal of Youth and Adolescence* 36:583–592.

Baham, Melinda E., Amy A. Weimer, Sanford L. Braver, and William V. Fabricius. 2008. "Sibling Relationships in Blended Families." Pp. 175–207 in *The International Handbook of Stepfamilies*, edited by J.Pryor. Hoboken, NJ: John Wiley & Sons.

Bailey, Jo Daugherty and Dawn McCarty. 2009. "Assessing Empowerment in Divorce Mediation." *Negotiation Journal* 25(3):327–336.

Bailey, Sandra J., Bethany L. Letiecq, and Mara Vannatta. 2011. "So Much for the Empty Nest: The Transitions of Grandparents Rearing Grandchildren." *Family Focus* FF49 (Summer): F17–F18.

Bainbridge, Daryl, Paul Krueger, Lynne Lohfeld, and Kevin Brazil. 2009. "Stress Processes in Caring for an End-of-Life Family Member: Application of a Theoretical Model." *Aging and Mental Health* 13(4):537–545.

Baitar, Rachid, Ann Buysse, Ruben Brondeel, Jan De Mol, and Peter Rober. 2012. "Post-Divorce Well-Being in Flanders: Facilitative Professionals and Quality of Arrangements Matter." *Journal of Family Studies* 18(1):62–75.

Bakalar, Nicholas. 2007. "Optional Caesareans Carry Higher Risks, Study Finds." *The New York Times*. March 27. Retrieved June 26, 2013 (http://www/nytimes.com)

———. 2010. "Premature Birth Rate Drops for 2nd Year." *The New York Times*, May 25.

Baker, Carrie N. 2008. *The Women's Movement Against Sexual Harassment.* New York: Cambridge University Press.

Baker, Katie. 2008. "Seeking a Samaritan." *Newsweek*. March 16.

Ball, Derek and Peter Kivisto. 2006. "Couples Facing Divorce." Pp. 145–161 in *Couples, Kids, and Family Life*, edited by Jaber F. Gubriumand James A. Holstein. New York: Oxford University Press.

Balsam, Kimberly F., Theodore P. Beauchaine, Esther D. Rothblum, and Sondra E. Solomon. 2008. "Three-Year Follow-Up of Same-Sex Couples Who Had Civil Unions in Vermont, Same-Sex Couples Not in Civil Unions, and Heterosexual Married Couples." *Developmental Psychology* 44(1):102–116.

Balsam, Kimberly F., Yamile Molina, Blair Beadnell, Jane Simoni, and Karina Walters. 2011. "Measuring Multiple Minority Stress: The LGBT People of Color Microaggressions Scale." *Cultural Diversity & Ethnic Minority Psychology* 17(2):163–174.

Baltzly, Vaughn Bryan. 2012. "Same-Sex Marriage, Polygamy, and Disestablishment." *Social Theory and Practice* 38(2): 333–354.

Banks, Ralph Richard. 2011. *Is Marriage for White People? How the African American Marriage Decline Affects Everyone.* New York: Penguin.

Banks, Stephen P. 2009. "Intergenerational Ties Across Borders: Grandparenting Narratives by Expatriate Retirees in Mexico." *Journal of Aging Studies* 23:178–187.

Banse, Rainer. 2004. "Adult Attachment and Marital Satisfaction: Evidence for Dyadic Configuration Effects." *Journal of Social and Personal Relationships* 21(2):273–282.

Banyard, Victoria L., Valerie J. Edwards, and Kathleen Kendall-Tackett, eds. 2009. *Trauma and Physical Health: Understanding the Effects of Extreme Stress and of Psychological Harm.* New York: Routledge.

Barajas, Manuel and Elvia Ramirez. 2007. "Beyond Home-Host Dichotomies: A Comparative Examination of Gender Relations in a Transnational Mexican Community." *Sociological Perspectives* 50(3):367–392.

Barash, Susan Shapiro. 2000. *Second Wives: The Pitfalls and Rewards of Marrying Widowers and Divorced Men.* Far Hills, NJ: New Horizon.

Barber, Bonnie L. and David H. Demo. 2006. "The Kids Are Alright (at Least Most of Them): Links between Divorce and Dissolution and Child Well-Being." Pp. 289–311 in *Handbook of Divorce and Relationship Dissolution*, edited by Mark A. Fine and John H. Harvey. Mahwah, NJ: Erlbaum.

Barbosa, Peter, Hector Torres, Marc Anthony Silva, and Nosh Khan.2010. "Agapé Christian Reconciliation Conversations: Exploring the Intersections of Culture, Religiousness, and Homosexual Identity in Latino and European Americans." *Journal of Homosexuality* 57(1):98–116.

Barker, Susan E. n. d. "'Cuddle Hormone': Research Links Oxytocin and Sociosexual Behaviors." Retrieved August 30, 2006 (www.oxytocin.org/cuddle-hormone/).

Barlinska, Julia, Anna Szuster, and Mikolaj Winiewski. 2012. "Cyberbullying among Adolescent Bystanders: Role of the Communication Medium, Form of Violence, and Empathy." *Journal of Community & Applied Social Psychology* 23:37–51.

Barlow, Daniel. 2005. "Vermont's — and Nation's — First Civil Union Breaking Up." *The Barre Montpelier Times Argus*, December 15. Retrieved from www.timesargus.com/apps/pbcs.dll/article?AID=/20051215/NEWS/512150369/1002.

Barnes, Medora. 2013. "Having a First Versus a Second Child: Comparing Women's Maternity Leave Choices and Concerns." *Journal of Family Issues* 34(1): 85–112.

Barnes, Norine R. 2006. "Children Need Guidance. . . . Developmentally Appropriate Interaction and Discipline." Retrieved December 30, 2009 (http://www.parentingweb.com).

Barnes, Richard E. 2009. *Estate Planning for Blended Families: Providing for Your Spouse and Children in a Second Marriage.* Berkeley, CA: Nolo Press.

Barnett, Amanda E. 2013. "Pathways of Adult Children Providing Care to Older Parents." *Journal of Marriage and Family* 75(1):178–190.

Barnett, Melissa A. 2008. "Mother and Grandmother Parenting in Low-Income Three-Generation Rural Households." *Journal of Marriage and Family* 70(5):1241–1257.

Barnett, Rosalind Chait, Karen C. Gareis, Laura Sabattini, and Nancy M. Carter. 2009. "Parental Concerns about After-School Time: Antecedents and Correlates among Dual-Earner Parents." *Journal of Family Issues* 31(5):606–625.

Barr, Ashley B., Ronald L. Simons, and Eric A. Stewart 2013. "The Code of the Street and Romantic Relationships." *Personal Relationships* 20:84–106.

Barrionuevo, Alexei. 2011. "Upwardly Mobile Nannies Move into the Brazilian Middle Class." *The New York Times.* May 20:A7.

Barron, David J. 2010. "Whether the Criminal Provisions of the Violence Against Women Act Apply to Otherwise Covered Conduct when the Offender and Victim are the Same Sex." *Memorandum Opinion for the Acting Deputy Attorney General.* Washington, DC: United States Department of Justice Office of Legal Counsel, April 27.

Barrow, Karen. 2010. "Difference Is the Norm on These Dating Sites." Retrieved July 14, 2013 (http://www.nytimes.com/2010/12/28/health/28dating.html).

Barth, Richard P. and Marianne Berry. 1988. *Adoption and Disruption: Rates, Risks, and Responses.* New York: Aldine.

Bartkowski, John P. 2001. *Remaking the Godly Family: Gender Negotiation in Evangelical Families.* Piscataway, NJ: Rutgers University Press.

Bartkowski, John P. and Jen'nan Ghazal Read. 2003. "Veiled Submission: Gender, Power, and Identity among Evangelical and Muslim Women in the United States." *Qualitative Sociology* 26(1):71–92.

Bartlett, Thad Q. 2009. *The Gibbons of Khao Yai: Seasonal Variation In Behavior and Ecology.* Upper Saddle River, NJ: Pearson.

Bartoli, Angela M. and M. Diane Clark. 2006. "The Dating Game: Similarities and Differences in Dating Scripts Among College Students." *Sexuality & Culture* 10:54–80.

Basow, Susan H. 1992. *Gender: Stereotypes and Roles.* 3rd ed. Pacific Grove, CA: Brooks/Cole.

Bass, Brenda L., Adam B. Butler, Joseph G. Grzywacz, and Kirsten D. Linney. 2009. "Do Job Demands Undermine Parenting? A Daily Analysis of Spillover and Crossover Effects." *Family Relations* 58(April):201–215.

Bastaits, Kim, Koen Ponnet, and Dimitri Mortelmans. 2012. "Parenting of Divorced Fathers and the Association with Children's Self-Esteem." *Journal of Youth and Adolescence* 41:1643–1656.

Batson, Christie D., Zhenchao Qian, and Daniel T. Lichter. 2006. "Interracial and Intraracial Patterns of Mate Selection among America's Diverse Black Populations." *Journal of Marriage and Family* 68(3):658–672.

Baucom, Donald H.,Gordon, Kristina C.,Snyder, Douglas K.,Atkins, David C., and Christensen, Andrew 2006. "Treating Affair Couples: Clinical Considerations and Initial Findings." *Journal of Cognitive Psychotherapy* 20:375–392.

Baucom, Donald H., Douglas K. Snyder, and Kristina Coop Gordon. 2009. *Helping Couples Get Past the Affair: A Clinician's Guide.* New York: The Guilford Press.

Baum, Angela C., Sedahlia Jasper Crase, and Kirsten Lee Crase. 2001. "Influences on the Decision to Become or Not Become a Foster Parent." *Families in Society* 82(2):202–221.

Baumrind, Diana. 1978. "Parental Disciplinary Patterns and Social Competence in Children." *Youth and Society* 9:239–276.

Baumrind, Diana, Robert E. Larzelere, and Philip A. Cowan. 2002. "Ordinary Physical Punishment: Is it Harmful? Comment on Gershoff." *Psychological Bulletin* 128(4):580–589.

Bausch, Robert S. 2006. "Predicting Willingness to Adopt a Child: A Consideration of Demographic and Attitudinal Factors." *Sociological Perspectives* 49(1):47–65.

Bauserman, Robert. 2002. "Child Adjustment in Joint-Custody versus Sole Custody Arrangements: A Meta-Analytic Review." *Journal of Family Psychology* 16:91–102.

———. 2012. "A Meta-Analysis of Parental Satisfaction, Adjustment, and Conflict in Joint Custody and Sole Custody Following Divorce." *Journal of Divorce & Remarriage* 53(6):464–488.

Baxter, Christine C. 1989. "Investigating Stigma as Stress in Social Interactions of Parents." *Journal of Intellectual Disability Research* 33(6):455–466.

Baxter, Janeen, Belinda Hewitt, and Michele Haynes. 2008. "Life Course Transitions and Housework: Marriage, Parenthood, and Time on Housework." *Journal of Marriage and Family* 70(2):259–269.

Baxter, Leslie A., Dawn O. Braithwaite, and Leah E. Bryant. 2006. "Types of Communication Triads Perceived by Young-Adult Stepchildren in Established Stepfamilies." *Communication Studies* 57(4):381–400.

Baxter, Leslie A., Dawn O. Braithwaite, Jody K. Kellas, Cassandra LeClaire-Underberg, Emily Lamb Normand, Tracy Routsong, and Matthew Thatcher. 2009. "Empty Ritual: Young-Adult Stepchildren's Perceptions of the Remarriage Ceremony." *Journal of Social and Personal Relationships* 26(4):467–487.

Bay-Cheng, Laina Y. and Nicole M. Fava. 2011. "Young Women's Experiences and Perceptions of Cunnilingus During Adolescence." *Journal of Sex Research* 48(6):531–542.

Bazelton, Emily. 2009. "Unwashed Coffee Mugs: How the Recession Is Affecting Family Relationships." *Slate*, February 20.

Beaman, Lori G. 2001. "Molly Mormons, Mormon Feminists and Moderates: Religious Diversity and the Latter Day Saints Church." *Sociology of Religion* 62:65–86.

Bean, Roy A. and Jason C. Northrup. 2009. "Parental Psychological Control, Psychological Autonomy, and Acceptance as Predictors of Self-Esteem in Latino Adolescents." *Journal of Family Issues* 30(11):1486–1504.

Bearman, Peter. 2008. "Exploring Genetics and Social Structure." *American Journal of Sociology* 114(Supplement):v–x.

Bearman, Peter S. and Hannah Brückner. 2001. "Promising the Future: Virginity Pledges and First Intercourse." *American Journal of Sociology* 106:859–912.

Beck, Audrey N., Carey E. Cooper, Sara McLanahan, Jeanne Brooks-Gunn. 2010. "Partnership Transitions and Maternal Parenting." *Journal of Marriage and Family* 72(2):219–233.

Becker, Arránz O. 2012. "Effects of Similarities of Life Goals, Values, and Personality on Relationship Satisfaction and Stability: Findings from a Two-Wave Panel Study." *Personal Relationships*. Retrieved May 26, 2013, from http://onlinelibrary.wiley.com/doi/10.1111/j.1475-6811.2012.01417.x/abstract.

Becker, Gay. 2000. *The Elusive Embryo: How Women and Men Approach New Reproductive Technologies*. Berkeley and Los Angeles: University of California Press.

Beer, William R. 1989. *Strangers in the House: The World of Stepsiblings and Half Siblings*. New Brunswick, NJ: Transaction Books.

Begley, Ann and Susan Piggott. 2012. "Exploring Moral Distress in Potential Sibling Stem Cell Donors." *Nursing Ethics* 20(2):178–188.

Begley, Sharon. 2007a. "Just Say No to Bad Science." *Newsweek*, May 7, pp. 57–58.

Begley, Sharon 2007b. "Of Sex and 'Soulmates.'" Retrieved July 15, 2013 (http://www.thedailybeast.com/newsweek/2010/01/27/of-sex-and-soulmates.html).

Behnke, Andrew O., Shelley M. MacDermid, Scott L. Coltrane, Ross D. Parke, Sharon Duffy, and Keith F. Widaman. 2008. "Family Cohesion in the Lives of Mexican American and European American Parents." *Journal of Marriage and Family* 70(4):1045–1059.

Beidler, Jeannie Jennings. 2012. "We Are Family: When Elder Abuse, Neglect, and Financial Exploitation Hit Home." *Generations: Journal of the American Society on Aging* 36(3):21–25.

Beins, Bernard. 2008. *Research Methods: A Tool for Life*. Boston: Allyn & Bacon.

Beitin, Ben, Katherine Allen, and Maureen Bekheet. 2010. "A Critical Analysis of Western Perspectives on Families of Arab Descent." *Journal of Family Issues* 31(2):211–233.

Belch, Michael A., Kathleen A. Krentler, and Laura A. Willis-Flurry. 2005. "Teen Internet Mavens: Influence in Family Decision-Making." *Journal of Business Research* 58(5):569–575.

Belkin, Lisa. 2010. "The Marrying Kind." *The New York Times*, March 22. Retrieved April 23, 2010 (www.nytimes.com).

Belknap, Joanne. 2012. "Types of Intimate Partner Homicides Committed by Women: Self-Defense, Proxy/Retaliation, and Sexual Proprietariness." *Homicide Studies* 16(4):359–374.

Bell, David C. 2012. "Next Steps in Attachment Theory." *Journal of Family Theory & Review* 4(4): 275–281.

Bell, Maya. 2003. "More Gays and Lesbians than Ever Are Becoming Parents." Knight Ridder /Tribune News Service, October 1.

Bellafante, Ginia. 2004. "Two Fathers, with One Happy to Stay at Home." *The New York Times*, January 12.

Bellah, Robert N., Richard Madsen, William M. Sullivan, Ann Swidler, and Steven M. Tipton. 1985. *Habits of the Heart: Individualism and Commitment in American Life*. Berkeley and Los Angeles: University of California Press.

Belluck, Pam. 2009. "In Turnabout, Children Take Caregiver Role." *The New York Times*, February 23. Retrieved March 16, 2010 (www.nytimes.com).

———. 2011. "Scientific Advances on Contraceptive for Men." *The New York Times*. July 23. Retrieved September 13, 2011 (http://www.nytimes.com).

Belsky, Jay. 2002. "Quantity Counts: Amount of Child Care and Children's Socioemotional Development." *Developmental and Behavioral Pediatrics* 23:167–170.

———. 2008. "Quality, Quantity and Type of Child Care: Effects on Child Development in the USA." In *Substitute Parenting: Alloparenting in Human Societies*, edited by G.Bentleyand R.Mace. London: Berghahn Books.

Belsky, Jay, Deborah Lowe Vandell, Margaret Burchinal, Alison Clarke-Stewart, Kathleen McCartney, Margaret Tresch Owen, and the NICHD Early Child Care Research Network. 2007. "Are There Long-Term Effects of Early Child Care?" *Child Development* 78(2):681–701.

Bem, Sandra Lipsitz. 1981. "Gender Schema Theory: A Cognitive Account of Sex Typing." *Psychological Review* 88:354–364.

Bendix, Jeffrey. 2009. "Elder Exploitation." *RN* 72(3):42–45.

Bengston, Vern L. 2001. "Beyond the Nuclear Family: The Increasing Importance of Multigenerational Bonds." *Journal of Marriage and Family* 63(1):1–16.

Bengston, Vern L., Timothy J. Biblarz, and Robert E. L. Roberts. 2002. *How Families Still Matter: A Longitudinal Study of Youth in Two Generations*. Cambridge, UK: Cambridge University Press.

Benkel, I., H. Wijk, and U. Molander. 2009. "Family and Friends Provide Most Social Support for the Bereaved." *Palliative Medicine* 23(2):141–149.

Benner, Aprile D. and Su Yeong Kim. 2009. "Intergenerational Experiences of Discrimination in Chinese American Families: Influences of Socialization and Stress." *Journal of Marriage and Family* 71(4):862–877.

———. 2010. "Understanding Chinese American Adolescents' Developmental Outcomes: Insights from the Family Stress Model." *Journal of Research on Adolescence* 20(1):1–12.

Bennett, Linda A., Steven J. Wolin, and David Reiss. 1988. "Deliberate *Family Process*: A Strategy for Protecting Children of Alcoholics." *British Journal of Addiction* 83:821–829.

Bentley, Evie. 2007. *Adulthood*. New York: Routledge.

Berenbaum, Sheri A., Judith E. Owen Blakemore, and Adriene M. Belz. 2011. "A Role for Biology in Gender-Related Behavior." *Sex Roles* 64: 804–825.

Berg, E. C. 2003. "The Effects of Perceived Closeness to Custodial Parents, Stepparents, and Nonresident Parents on Adolescent Self-Esteem." *Journal of Divorce & Remarriage*, 40:69–86.

Berge, Jerica M. and Kristen E. Holm. 2007. "Boundary Ambiguity in Parents with Chronically Ill Children: Integrating Theory and Research." *Family Relations* 56(2):123–134.

Bergen, Raquel Kennedy. 2006. "Marital Rape: New Research and Directions." The National Online Resource Center on Violence Against Women. Retrieved June 9, 2010 (new.vawnet.org).

Berger, Lawrence M., Theresa Heintze, Wendy Naidich, and Marcia Meyers. 2008. "Subsidized Housing and Household Hardship among Low-Income Single-Mother Households." *Journal of Marriage and Family* 70(4):934–949.

Berger, Peter L. and Hansfried Kellner. 1970. "Marriage and the Construction of Reality." Pp. 49–72 in *Recent Sociology No. 2*, edited by Hans Peter Dreitzel. New York: Macmillan.

Berger, Peter L. and Thomas Luckman. 1966. *The Social Construction of Reality: A Treatise in the Sociology of Knowledge*. Garden City, NY: Anchor Books.

Berger, Roni. 1998. *Stepfamilies: A Multi-dimensional Perspective*. New York: Haworth.

———. 2000. "Gay Stepfamilies: A Triple-Stigmatized Group." *Families in Society* 81(5):504–516.

Bergman, Mike. 2006. "Dramatic Changes in U.S. Aging: Highlighted in New Census, NIH Report." Press Release CB06-36, March 9. Washington, DC: U.S. Census Bureau.

Bergmann, Barbara R. 2011. "Sex Segregation in the Blue-Collar Occupations: Women's Choices or Unremedied Discrimination?" *Gender & Society* 25(1):88–93.

Berkowitz, Dana and William Marsiglio. 2007. "Gay Men: Negotiating Procreative, Father, and Family Identities." *Journal of Marriage & Family* 69(2):366–381.

Berlin, Gordon. 2007. "Rewarding the Work of Individuals: A Counterintuitive Approach to Reducing Poverty and Strengthening Families." *The Future of Children* 17(2):17–42.

Berlin, Lisa, Jean Ispa, Mark Fine, Patrick Malone, Jeanne Brooks-Gunn, Christy Brady-Smith, Catherine Ayoub, and Yu Bai. 2009. "Correlates and Consequences of Spanking and Verbal Punishment for Low-Income White, African American, and Mexican American Toddlers." *Child Development* 80(5):1403–1420.

Bernard, Jessie. 1986. "The Good-Provider Role: Its Rise and Fall." Pp. 125–144 in *Family in Transition: Rethinking Marriage, Sexuality, Child Rearing, and Family Organization*, 5th ed., edited by Arlene S. Skolnick and Jerome H. Skolnick. Boston: Little, Brown.

Bernard, Tara Siegel. 2012a. "A Family with Two Moms, Except in the eyes of the Law." *The New York Times*. July 20. Retrieved August 28, 2012 (http://www.nytimes.com).

———. 2012b. "Marriage Maintenance When Money Is Tight." *The New York Times*. March 30. Retrieved August 28, 2012 (http://www.nytimes.com).

Bernstein, Anne C. 1997. "Stepfamilies from Siblings' Perspectives." Pp. 153–175 in *Stepfamilies: History, Research, and Policy*, edited by Irene Levin and Marvin B. Sussman. New York: Haworth.

———. 1999. "Reconstructing the Brothers Grimm: New Tales for Stepfamily Life." *Family Process* 38(4):415–430.

———. 2007. "Re-visioning, Restructuring, and Reconciliation: Clinical Practice With Complex Postdivorce Families." Family Process 46(1):67–78.

Bernstein, Fred A. 2004. "On Campus, Rethinking Biology 101." *The New York Times*, March 7.

Bernstein, Jeffrey and Susan Magee. 2004. *Why Can't You Read My Mind? Overcoming the 9 Toxic Thought Patterns That Get in the Way of a Loving Relationship*. New York: Marlowe.

Berridge, K. and T. Robinson. 1998. "What Is the Role of Dopamine in Reward: Hedonic Impact, Reward Learning, or Incentive Salience?" *Brain Research Review* 28(3):309–369.

Berry, Brent. 2008. "Financial Transfers from Living Parents to Young Adult Children: Who Is Helped and Why?" *American Journal of Economics and Sociology* 67(2):207–239.

Bertram, Rosalyn M. and Jennifer L. Dartt. 2009. "Post Traumatic Stress Disorder: A Diagnosis for Youth from Violent, Impoverished Communities." *Journal of Child and Family Studies*, 294–302.

Bhalla, Vibha. 2008. "Couch Potatoes and Super-Women: Gender, Migration, and the Emerging Discourse on Housework among Asian Indian Immigrants." *Journal of American Ethnic History* 27(4):71–99.

Bianchi, Suzanne M. 2000. "Maternal Employment and *Time* with Children: Dramatic Change or Surprising Continuity?" *Demography* 37:401–614.

Bianchi, Suzanne M., and Lynne M. Casper. 2000. "American Families." *Population Bulletin* 55(4). Washington, DC: Population Reference Bureau.

Bianchi, Suzanne M., John P. Robinson, and Melissa A. Milkie. 2006. *Changing Rhythms of American Family Life*. New York: Russell Sage.

Biblarz, Timothy J. and Evren Savci. 2010. "Lesbian, Gay, Bisexual, and Transgender Families." *Journal of Marriage and Family* 72(3): 480–497.

Bickman, Keonard and Debra J. Rog. 2009. *The SAGE Handbook of Applied Social Research Methods*. 2nd ed. Los Angeles: Sage.

Bierman, Alex, Elena M. Fazio, and Melissa A. Milkie. 2006. "A Multifaceted Approach to the Mental Health Advantage of the Married." *Journal of Family Issues* 27(4):554–582.

Bigda, Carolyn. 2013. "Young Adult Caregivers." Retrieved March 2013 (http://www.seniorlivingeexperts.com).

Bilefsky, Dan. 2012. "Hard Times in Spain Force Feuding Couples to Delay Divorce." Retrieved April 12, 2013 (http://www.nytimes.com/2012/12/18/world/europe/hard-times-in-spain-force-feuding-couples-to-delay-divorce.html?_r=0&pagewanted=print).

Billingsley, Andrew. 1968. *Black Families in White America*. Englewood Cliffs, NJ: Prentice Hall.

Bindley, K. 2011, "The Mommy Wars Continue: Relations Between Nannies, Working Moms, and Stay-at-Home Moms." *Huffington Post Parents*. October 13. Retrieved February 20, 2012 (http://www.huffingtonpost.com).

"Biology of Social Bonds." 1999. *Science News*, August 7.

Bird, Gloria W., Rick Peterson, and Stephanie Hotta Miller. 2002. "Factors Associated with Distress among Support-Seeking Adoptive Parents." *Family Relations* 51(3):215–220.

"Birds of a Feather—More than a Place to Live, a Way to Live." 2010. Retrieved May 17, 2010 (flock2it.com).

Biro, Frank M., Maida P. Galvez, Louise C. Greenspan, Paul A. Succop, Nita Vangeepuram, Susan M. Pinney, Susan Teitelbaum, Lawrence H. Kushi, and Mary S. Wolff. 2010. "Pubertal Assessment Method and Baseline Characteristics in a Mixed Longitudinal Study of Girls." *Pediatrics* 126(3):e583–e590.

Biskupic, Joan. 2003. "Same-Sex Couples Are Redefining Family Law in USA." *USA Today*, February 18.

Bjelde, Kristine E. and Gregory F. Sanders. 2009. "Snowbird Intergenerational Family Relationships." *Activities, Adaptation, and Aging* 33:81–95.

Blackman, Lorraine, Obie Clayton, Norval Glenn, Linda Malone-Colon, and Alex Roberts. 2006. *The Consequences of Marriage for African Americans: A Comprehensive Literature Review*. New York: Institute for American Values.

Blakemore, Judith and Craig Hill. 2008. "The Child Gender Socialization Scale: A Measure to Compare Traditional and Feminist Parents." *Sex Roles* 58(3/4):192–207.

Blanchard, Victoria L., Alan J. Hawkins, Scott A. Baldwin, and Elizabeth B. Fawcett. 2009. "Investigating the Effects of Marriage and Relationship Education on Couples' Communication Skills: A Meta-analytic Study." *Journal of Family Psychology* 23(2):203–214.

Blankenhorn, David. 1995. *Fatherless America: Confronting Our Most Urgent Social Problem*. New York: Basic Books.

———. 2007. *The Future of Marriage*. New York: Encounter Books.

Bledsoe, Caroline and Papa Sow. 2011. "Back to Africa: Second Chances for the Children of West African Immigrants." *Journal of Marriage and Family* 73(4): 747–762.

Blieszner, Rosemary, Karen A. Roberto, Karen L. Wilcox, Elizabeth J. Barham, and Brianne L. Winston. 2007. "Dimensions of Ambiguous Loss in Couples Coping with Mild Cognitive Impairment." *Family Relations* 56(2):196–209.

Block, Joel D. and Susan S. Bartell 2001. *Stepliving for Teens: Getting Along with Step-Parents, Parents, and Siblings*. New York: Penguin Young Readers Group.

Blood, Robert O., Jr. and Donald M. Wolfe. 1960. *Husbands and Wives: The Dynamics of Married Living*. New York: The Free Press.

Blow, Adrian J. and Kelley Hartnett. 2005. "Infidelity in Committed Relationships II: A Substantive Review." *Journal of Marital and Family Therapy* 31(2):217–233.

Blum, Deborah. 1997. *Sex on the Brain: The Biological Differences between Men and Women.* New York: Penguin.

Blumberg, Rae Lesser and Marion Tolbert Coleman. 1989. "A Theoretical Look at the Gender Balance of Power in the American Couple." *Journal of Family Issues* 10:225–250.

Blumstein, Philip and Pepper Schwartz. 1983. *American Couples: Money, Work, Sex.* New York: Morrow.

Bobbitt-Zeher, Donna. 2011. "Connecting Gender Stereotypes, Institutional Policies, and Gender Composition of Workplace." *Gender & Society* 25(6): 764–786.

Bock, Jane D. 2000. "'Doing the Right Thing?' Single Mothers by Choice and the Struggle for Legitimacy." *Gender and Society* 14:62–86.

Bodenmann, Guy, Thomas Ledermann, and Thomas N. Bradbury. 2007. "Stress, Sex, and Satisfaction in Marriage." *Personal Relationships* 14(4):551–569.

Bodrug-Lungu, Valentina and Erin Kostina-Ritchey. 2013. "Impact of Parents' Migration on Moldavian Youths' Perception of Family." *Family Focus* FF56: F13–F15.

"Body Image." 2008. Children, Youth, and Women's Health Service: Kids' Health. Retrieved May 6, 2008 (www.cyh.com).

Boerner, Kathrin, Deborah Carr, and Sara Moorman. 2013. "Family Relationships and Advance Care Planning: Do Supportive and Critical Relations Encourage or Hinder Planning?" *Journals of Gerontology, Series B: Psychological and Social Sciences* 68(2):246–256.

Bogenschneider, Karen, Olivia M. Little, Theodora Ooms, Sara Benning, Karen Cadigan, and Thomas Corbett. 2012. "The Family Impact Lens: A Family-Focused, Evidence-Informed Approach to Policy and Practice." *Family Relations* 61(3): 514–531.

Bogle, Kathleen A. 2004. "From Dating to Hooking Up: The Emergence of a New Sexual Script." Unpublished PhD dissertation, Department of Sociology, University of Delaware. Newark, DE.

———. 2008. *Hooking Up: Sex, Dating, and Relationships on Campus.* New York: New York University Press.

Bolick, Kate. 2011a. "Let's Hear It for Aunthood." *The New York Times.* September 16. Retrieved September 23, 2011 (http://www.nytimes.com).

Bolick, Kate. 2011b. "What, Me Marry? All the Single Ladies." *The Atlantic.* November: 116–136.

Bolzendahl, Catherine I. and Daniel J. Myers. 2004. "Feminist Attitudes and Support for Gender Equality: Opinion Change in Women and Men, 1974–1998." *Social Forces* 83(2):759–790.

Bonach, Kathryn. 2007. "Forgiveness Intervention Model: Application to Coparenting Post-Divorce." *Journal of Divorce and Remarriage* 48(1/2):105–123.

"Boomerang Kids Contract." n. d. Retrieved December 21, 2009 (boomerangkidshelp.com).

Boon, Susan D., Vicki L. Deveau, and Alishia M. Allbhal. 2009. "Payback: The Parameters of Revenge in Romantic Relationships." *Journal of Social and Personal Relationships* 26(6–7):747–768.

Boonstra, Heather D., Rachel Benson Gold, Cory L. Richards, and Lawrence B. Finer. 2006. *Abortion in Women's Lives.* New York: Guttmacher Institute.

Booth, Alan, Karen Carver, and Douglas A. Granger. 2000. "Biosocial Perspectives on the Family." *Journal of Marriage and Family* 62(4):1018–1034.

Booth, Alanand Ann C. Crouter, eds. 2002. *Just Living Together: Implications of Cohabitation for Children, Families, and Social Policy.* Mahwah, NJ: Erlbaum.

Booth, Alan, Ann C. Crouter, and Mari Clements, eds. 2001. *Couples in Conflict.* Mahwah, NJ: Erlbaum.

Booth, Alan, Douglas Granger, Allan Mazur, and Katie Kivlighan. 2006. "Testosterone and Social Behavior." *Social Forces* 85(1):167–191.

Booth, Alan, David R. Johnson, and Douglas A. Granger. 2005. "Testosterone, Marital Quality, and Role Overload." *Journal of Marriage and Family* 67(2):483–498.

Booth, Alan, Elisa Rustenbach, and Susan McHale. 2008. "Early Family Transitions and Depressive Symptom Changes from Adolescence to Early Adulthood." *Journal of Marriage and Family* 70(1):3–14.

Booth, Alan, Mindy E. Scott, and Valarie King. 2010. "Father Residence and Adolescent Problem Behavior: Are Youth Always Better Off in Two-Parent Families?" *Journal of Family Issues* 31(5):585–605.

Booth, Cathy. 1977. "Wife-Beating Crosses Economic Boundaries." *Rocky Mountain News,* June 17.

Boss, Pauline. 1980. "Normative Family Stress: Family Boundary Changes across the Lifespan." *Family Relations* 29:445–452.

———. 1987. "Family Stress." Pp. 695–723 in *Handbook of Marriage and Family,* edited by M. B.Sussman and Suzanne K. Steinmets. New York: Plenum.

———. 1997. "Ambiguity: A Factor in Family Stress Management." St. Paul, MN: University of Minnesota Extension Service (http://www.extension.umn.edu).

———. 2002. *Family Stress Management.* 2nd ed. Newbury Park, CA: Sage Publications.

———. 2004. "Ambiguous Loss Research, Theory, and Practice: Reflections after 9/11." *Journal of Marriage and Family* 66(3):551–566.

———. 2007. "Ambiguous Loss Theory: Challenges for Scholars and Practitioners." *Family Relations* 56(2):105–111.

Bosse, Irina. 1999. "Oxytocin: A Hormone for Love." *Futureframe: International Webzine for Science and Culture.* Retrieved August 30, 2006 (www.morgenwelt.de/futureframe/9908-oxytocin.htm).

Boston Women's Health Book Collective. 2005. *Our Bodies, Ourselves: A New Edition for a New Era.* New York: Simon and Schuster.

Bottke, Allison. 2010. *Setting Boundaries with Your Aging Parents.* Eugene, OR: Harvest House Publishers.

Bouchard, Emily. n. d. "Navigating Parenting Differences." SelfGrowth.com. Retrieved September 21, 2004. (www.selfgrowth.com/articles/Bouchard2.html).

Bouchard, Genevieve, and Danielle Doucet. 2011. "Parental Divorce and Couples' Adjustment During the Transition to Parenthood: The Role of Parent-Adult Child Relationships." *Journal of Family Issues* 32(4):507–527.

Bourdieu, Pierre. 1977 [1972]. *Outline of a Theory of Practice.* New York: Cambridge.

Bouton, Katherine. 1987. "Fertility and Family." *Ms.,* April, p. 92.

Bowen, Gary L., Roderick A. Rose, Joelle D. Powers, and Elizabeth J. Glennie. 2008. "The Joint Effects of Neighborhoods, Schools, Peers, and Families on Changes in the School Success of Middle School Children." *Family Relations* 57(4):504–516.

Bowers v. Hardwick. 1986. 478 U.S. 186, 92 L. Ed. 2d 140, 106 S. Ct. 2841.

Bowlby, John. 1988. A Secure Base. London: Routledge.

Bowleg, Lisa, Jennifer Huang, Kelly Brooks, Amy Black, and Gary Burkholder. 2003. "Triple Jeopardy and Beyond: Multiple Minority Stress and Resilience among Black Lesbians." *Journal of Lesbian Studies* 7(4):87–108.

Bowleg, Lisa, Michelle Teti, Jenné S. Massie, Aditi Patel, David J. Malebranche, and Jeanne M. Tschann. 2011. "'What Does It Take to Be a Man? What Is a Real Man?': Ideologies of Masculinity and HIV Sexual Risk among Black Heterosexual Men." *Culture, Health, & Sexuality* 13(5):545–559.

Boylstein, Craig and Jeanne Hayes. 2012. "Reconstructing Marital Closeness While Caring for a Spouse with Alzheimer's." *Journal of Family Issues* 33(5):584–612.

"Boys' Academic Slide Calls for Accelerated Attention." 2003. *USA Today,* December 22.

Bracke, Piet, Wendy Christiaens, and Naomi Wauterickx. 2008. "The Pivotal Role of Women in Informal Care." *Journal of Family Issues* 29(10):1348–1378.

Bradbury, Thomas N., Frank D. Fincham, and Steven R. H. Beach. 2000. "Research on the Nature and Determinants of Marital Satisfaction: A Decade in Review." *Journal of Marriage and Family* 62(4):964–980.

Bradbury, Thomas N. and Benjamin R. Karney. 2004. "Understanding and Altering the Longitudinal Course of Marriage." *Journal of Marriage and Family* 66(6):862–879.

Bradshaw, Carolyn, Arnold S. Kahn, and Bryan K. Saville. 2010. "To Hook Up or to Date: Which Gender Benefits?" *Sex Roles* 62:661–669.

Braithwaite, Dawn O. and Leslie A. Baxter. 2006. *Engaging Theories in Family Communication: Multiple Perspectives.* Thousand Oaks, California: Sage.

Bramlett, Matthew D. and William D. Mosher. 2002. *Cohabitation, Marriage, Divorce, and Remarriage in the United States.* National Center for Health Statistics. *Vital Health Stat* 23(22).

Branden, Nathaniel. 1988. "A Vision of Romantic Love." Pp. 218–231 in *The Psychology of Love,* edited by Robert J. Sternberg and Michael L. Barnes. New Haven, CT: Yale University Press.

Brandl, Bonnie. 2007. *Elder Abuse Detection and Intervention: A Collaborative Approach.* New York: Springer.

Brandon, Emily. 2008. "Baby Boomers Moving in with Adult Children." *U.S. News & World Report,* November 20. Retrieved February 6, 2009 (www .usnews.com).

Bratter, Jenifer L. and Karl Eschbach. 2006. "'What about the Couple?' Interracial Marriage and Psychological Distress." *Social Science Research* 35(4):1025–1047.

Bratter, Jenifer L. and Rosalind B. King. 2008. "'But Will It Last?': Marital Instability among Interracial and Same-Race Couples." *Family Relations* 57(April):160–171.

Braver, Sanford, Jennesa R. Shapiro, and Matthew R. Goodman. 2006. "Consequences of Divorce for Parents." Pp. 313–337 in *Handbook of Divorce and Relationship Dissolution,* edited by Mark A. Fine and John H. Harvey. Mahwah, NJ: Erlbaum.

Braver, Sanford L., Pamela J. Fitzpatrick, and R. Curtis Bay. 1991. "Noncustodial Parent's Report of Child Support Payments." *Family Relations* 40(2):180–185.

Bray, James H. 1999. "From Marriage to Remarriage and Beyond." Pp. 253–271 in *Coping with Divorce, Single Parenting and Remarriage,* edited by E. Mavis Hetherington. Mahwah, NJ: Erlbaum.

Bray, J. H., S. H. Berger and C. L. Boethel. 1994. "Role Integration and Marital Adjustment in Stepfather Families." Pp. 69–86 in *Stepparenting: Issues in Theory, Research, and Practice,* edited by K.Pasley and M.Ihinger-Tallman. Westport, CT: Greenwood Press.

Bray, J. H. and I. Easling. 2005. Pp. 267–294 in *Family Psychology: The Art of the Science,* edited by W. M.Pinsof and J. L. Lebow. New York: Oxford Press.

Brazelton, T. Berry, and Stanley Greenspan. 2000. "Our Window to the Future." *Newsweek* Special Issue, Fall/Winter, pp. 34–36.

Breitenbecher, Kimberly Hanson. 2006. "The Relationships among Self-blame, Psychological Distress, and Sexual Victimization." *Journal of Interpersonal Violence* 21(5):597–611.

Brennan, Bridget. 2003. "No Time. No Sex. No Money." *First Years and Forever: A Monthly Online Newsletter for Marriages in the Early Years.* Chicago: Archdiocese of Chicago, Family Ministries. Retrieved September 8, 2006 (www.familyministries .org).

Brewis, Alexandra, and Mary Meyer. 2005. "Marital Coitus across the Life Course." *Journal of Biosocial Science* 37:499–518.

Brewster, Karin L. and Irene Padavic. 2002. "No More Kin Care? Change in Black Mothers' Reliance on Relatives for Child Care, 1977–94." *Gender and Society* 16:546–563.

Bricker, Jesse, Brian Bucks, Arthur Kennickell, Traci Mach, and Kevin Moore. 2012. "The Financial Crisis from the Family's Perspective." *Journal of Consumer Affairs* 46(3):537–555.

Brimhall, Andrew, Karen Wampler, and Thomas Kimball. 2008. "Learning from the Past, Altering the Future: A Tentative Theory of the Effect of Past Relationships on Couples Who Remarry." *Family Process* 47(3):373–387.

Brink, Susan. 2008. "Modern Puberty." *Los Angeles Times,* January 21.

Brock, Rebecca L. and Erika Lawrence. 2009. "Too Much of a Good Thing: Underprovision Versus Overprovision of Partner Support." *Journal of Family Psychology* 23(2):181–192.

Broderick, Carlfred B. 1979a. *Couples: How to Confront Problems and Maintain Loving Relationships.* New York: Simon and Schuster.

———. 1979b. *Marriage and the Family.* Englewood Cliffs, NJ: Prentice Hall.

Brody, Gene, Yi Fu Chen, Steven Kogan, Velma McBride Murray, Patricia Logan, and Zupei Luo. 2008. "Linking Perceived Discrimination to Longitudinal Changes in African American Mothers' Parenting Practices." *Journal of Marriage and Family* 70(2):319–31.

Brody, Jane E. 2004. "Abstinence-Only: Does It Work?" *The New York Times,* June 3.

———. 2008. "When Families Take Care of Their Own." *The New York Times,* November 11. Retrieved March 16, 2010 (www.nytimes.com).

———. "Caregiving as a Roller-Coaster Ride from Hell." *The New York Times.* April 10. Retrieved May 21, 2013 (http://nytimes.com).

Broman, Clifford L., Li Xin, and Mark Reckase. 2008. "Family Structure and Mediators of Adolescent Drug Use." *Journal of Family Issues* 29(12):1625–1649.

Bronfenbrenner, Urie. 1979. *The Ecology of Human Development: Experiments by Nature and Design.* Cambridge, MA: Harvard University Press.

Bronte-Tinkew, Jacinta, Jennifer Carrano, Allison Horowitz, and Akemi Kinukawa. 2008. "Involvement among Resident Fathers and Links to Infant Cognitive Outcomes." *Journal of Family Issues* 29(9):1211–1244.

Bronte-Tinkew, Jacinta and Allison Horowitz. 2010. "Factors Associated with Unmarried, Nonresident Fathers' Perceptions of Their Coparenting." *Journal of Family Issues* 31(1):31–65.

Bronte-Tinkew, Jacinta, Allison Horowitz, and Mindy E. Scott. 2009. "Fathering with Multiple Partners: Links to Children's Well-Being in Early Childhood." *Journal of Marriage and Family* 71(3):608–631.

Brooke, Jill. 2004. "Close Encounters with a Home Barely Known." *The New York Times,* July 22.

———. 2006. "Home Alone Together." *The New York Times,* May 4. Retrieved May 7, 2006 (www. nytimes.com).

Brooks, Robert and Sam Goldstein. 2001. *Raising Resilient Children: Fostering Strength, Hope, and Optimism in Your Child.* New York: Contemporary Books.

Brossoie, Nancy, Karen Roberto, and Katie Barrow. 2012. "Making Sense of Intimate Partner Violence in Later Life: Comments from Online News Readers." *The Gerontologist* 52(6):792–801.

Brotherson, Sean. 2003. "Time, Sex, and Money: Challenges in Early Marriage." *The Meridian.* Retrieved September 8, 2006 (www.meridian-magazine.com).

Brougham, Ruby R. and David A. Walsh. 2009. "Early and Late Retirement Exits." *International Journal of Aging and Human Development* 69(4):267–286.

Broverman, Neal. 2012. "Lieberman, Collins Domestic Partnership Bill Moves Forward." *The Advocate.* May 16. Retrieved September 3, 2012 (http://www.advocate.com).

Browder, Laura. 2010. "Impact of War." *The Washington Post.* May 24. Retrieved February 21, 2013 (http://www.voices.washingtonpost.com).

Brown, Alyssa and Jeffrey M. Jones. 2012. "Separation, divorce Linked to Sharply Lower Wellbeing: Married Americans Have Highest Wellbeing." Gallup Poll. April 20. Retrieved December 4, 2012 (http://www.gallup.com).

Brown, Carrie. 2008. "Gender-Role Implications on Same-Sex Intimate Partner Abuse." *Journal of Family Violence* 23:475–462.

Brown, Chris, Heather B. Trangsrud, and Rachel M. Linnemeyer. 2009. "Battered Women's Process of Leaving." *Journal of Career Assessment* 17(4): 439–456.

Brown, Dave and Phil Waugh. 2004. *Covenant vs. Contract.* New York: Franklin, Son Publishers.

Brown, Edna, Terri L. Orbuch, and Artie Maharaj. 2010. "Social Networks and Marital Stability among Black American and White American Couples." Pp. 318–334 in *Support Processes in Intimate Relationships,* edited by Kieran T. Sullivan and Joanne Davila. New York: Oxford University Press.

Brown, Melissa, Kerstin Aumann, Marcie Pitt-Catsouphes, Ellen Galinsky, and James T. Bond. 2010. "Working in Retirement: A 21st Century Phenomenon." *Sloan Center News.* October 6. Retrieved May 7, 2013 (http://www.bc.edu/research/aging).

Brown, Patricia Leigh. 2004. "For Children of Gays, Marriage Brings Joy." *The New York Times,* March 19.

———. 2006. "Supporting Boys or Girls When the Line Isn't Clear." *The New York Times,* December 4.

Brown, Susan L. 2004. "Family Structure and Child Well-Being: The Significance of Parental Cohabitation." *Journal of Marriage and Family* 66(2):351–367.

———. "A 'Gray Divorce' Boom." *Los Angeles Times.* Retrieved April 15, 2013 (http://www.latimes.com/news/opinion/commentary/la-oe-brown-gray-divorce-20130331,0,7982785.story).

Brown, Susan L., Jennifer Roebuck Bulanda, and Gary R. Lee. 2012. "Transitions Into and

Out of Cohabitation in Later Life." *Journal of Marriage and Family* 74(4) (August): 774–793.

Brown, Susan L., and I-Fen Lin. 2012. "The Gray Divorce Revolution: Rising Divorce Among Middle-Aged and Older Adults." *Journal of Gerontology Series B: Psychological Sciences and Social Sciences* 67(6):731–741.

Brown, Susan L. and Wendy D. Manning. 2009. "Family Boundary Ambiguity and the Measurement of Family Structure: The Significance of Cohabitation." *Demography* 46(1):85–101.

Brown, Tiffany, Miriam Linver, Melanie Evans, and Donna DeGennaro. 2009. "African American Parents' Racial and Ethnic Socialization and Adolescent Academic Grades: Teasing Out the Role of Gender." *Journal of Youth & Adolescence* 38(2):214–227.

Brown, Tony, Emily Tanner-Smith, Chase Lesane-Brown, and Michael Ezell. 2007. "Child, Parent, and Situational Correlates of Familiar Ethnic/Race Socialization." *Journal of Marriage and Family* 69(1):14–25.

Brownridge, Douglas, Tamara Tallieu, Kimberly Tyler, Agnes Tiwari, Ko Ling Chan, and Susy Santos. 2011. "Pregnancy and Intimate Partner Violence: Risk Factors, Severity, and Health Effects." *Violence Against Women* 17(7):858–881.

Brumbaugh, Stacey M., Laura A. Sanchez, Steven L. Nock, and James D. Wright. 2008. "Attitudes Toward Gay Marriage in States Undergoing Marriage Law Transformation." *Journal of Marriage and Family* 70(2):345–359.

Bruni, Frank. 2012. "Maine and Maryland Say 'We Do.'" *The New York Times.* Retrieved December 5, 2012 (http://www.nytimes.com).

Brush, Lisa D. 2008. "Book Review." *Gender and Society* 22(1):126–142.

Bryan, Willie V. 2010. *Sociopolitical Aspects of Disabilities: The Social Perspectives and Political History of Disabilities and Rehabilitation in the United States.* Springfield, IL: Charles C. Thomas.

Bubolz, Margaret M. and M. Suzanne Sontag. 1993. "Human Ecology Theory." Pp. 419–448 in *Sourcebook of Family Theories and Methods: A Contextual Approach,* edited by Pauline G. Boss, William J. Doherty, Ralph LaRossa, Walter R. Schumm, and Suzanne K. Steinmetz. New York: Plenum.

Buchanan, Christy M., Eleanor E. Maccoby, and Sanford M. Dornbusch. 1996. *Adolescents After Divorce.* Cambridge, MA: Harvard University Press.

Bucior, Carolyn. 2012. "It's Not Just the Mouse that Has to Click." *The New York Times.* Retrieved May 29, 2013 (http://www.nytimes.com/2012/10/07/fashion/when-dating-takes-more-than-the-click-of-a-mouse.html?pagewanted=all).

Bucx, Freek, Fritz Van Well, and Trudie Knijn. 2012. "Life Course Status and Exchanges of Support between Young Adults and Parents." *Journal of Marriage and Family* 74(1):101–115.

Buehler, Cheryl, Ambika Krishnakumar, Gaye Stone, Christine Anthony, Sharon Pemberton, Jean Gerard, and Brian K. Barber. 1998. "Interpersonal Conflict Styles and Youth Problem Behaviors." *Journal of Marriage and Family* 60(1):119–132.

Buehler, Cheryl and Marion O'Brien. 2011. "Mothers' Part-Time Employment: Associations with Mother and Family Well-Being." *Journal of Family Psychology* 25(6):895–906.

Bufkin 2012. "Domestic Workers Bill Killed in California by Jerry Brown Veto." *The Huffington Post Politics.* October 1. Retrieved February 20, 2013 (http://www.huffingtonpost.com).

Bukhari, Zahid Hussain. 2004. *Muslims' Place in the American Public Square: Hope, Fears, and Aspirations.* Walnut Creek, CA: AltaMira.

Bulanda, Jennifer Roebuck. 2011. "Doing Family, Doing Gender, Doing Religion: Structured Ambivalence and the Religion-Family Connection." *Journal of Family Theory & Review* 3(3) (September): 179–197.

Bulanda, Jennifer R. and S. L. Brown. 2007. "Race-Ethnic Differences in Marital Quality and Divorce." *Social Science Research,* 36:945–967.

Bulcroft, Kris, Richard Bulcroft, Linda Smeins, and Helen Cranage. 1997. "The Social Construction of the North American Honeymoon, 1800–1995." *Journal of Family History* 22(4):462–491.

Bumiller, Elisabeth. 2011. "Obama Ends, Don't Ask, Don't Tell Policy." *The New York Times.* Retrieved November 6, 2012 (http://www.nytimes.com/2011/07/23/us/23military.html).

Bumpass, Larry L. and K. Raley. 2007. Measuring Separation and Divorce. Pp. 125–144 in *Handbook of Measurement Issues in Family Research,* edited by S. L. Hofferth and L. M.Casper. Mahwah, NJ: Erlbaum.

Bumpass, Larry L., James A. Sweet, and Andrew Cherlin. 1991. "The Role of Cohabitation in Declining Rates of Marriage." *Journal of Marriage and Family* 53(4):913–927.

Burchinal, Margaret, Nathan Vandergrift, Robert Pianta, and Andrew Mashburn. 2010. "Threshold Analysis of Association Between Child Care Quality and Child Outcomes for Low-Income Children in Pre-Kindergarten Programs." *Early Childhood Research Quarterly* 25(2):166–176.

Bures, Regina M. 2009. "Living Arrangements Over the Life Course." *Journal of Family Issues* 30(5):579–585.

Bures, Regina M., Tanya Koropeckyj-Cox, and Michael Loree. 2009. "Childlessness, Parenthood, and Depressive Symptoms among Middle-Aged and Older Adults." *Journal of Family Issues* 30(5):670–687.

Burgess, Ernest and Harvey Locke. 1953 [1945]. *The Family: From Institution to Companionship.* New York: American.

Burgess, Wes. 2008. *The Bipolar Handbook for Children, Teens, and Families: Real-Life Questions with Up-to-Date Answers.* New York: Avery.

Burke, Tod W. and Stephen S. Owen. 2006. "Same-sex Domestic Violence: Is Anyone Listening?" *Gay and Lesbian Review Worldwide* 13(1):6–7.

Burns, George W. 2010. *Happiness, Healing, Enhancement: Your Casebook Collection for Applying Positive Psychology in Therapy.* Hoboken, NJ: Wiley.

Burpee, Leslie C. and Ellen J. Langer. 2005. "Mindfulness and Marital Satisfaction." *Journal of Adult Development* 12(1):1281–1287.

Burr, Jeffrey A. and Jan E. Mutchler. 1999. "Race and Ethnic Variation in Norms of Filial Responsibility among Older Persons." *Journal of Marriage and Family* 61(3):674–687.

Burr, Wesley R., Shirley Klein, and Marilyn McCubbin. 1995. "Reexamining Family Stress: New Theory and Research." *Journal of Marriage and Family* 57(3):835–846.

Burrell, Ginger L. and Mark W. Roosa. 2009. "Mothers' Economic Hardship and Behavior Problems in Their Early Adolescents." *Journal of Family Issues* 30(4):511–531.

Burton, Linda M., Andrew Cherlin, Donna-Marie Winn, Angela Estacion, and Clara Holder-Taylor. 2009. "The Role of Trust in Low-Income Mothers' Intimate Unions." *Journal of Marriage and Family* 71(December):1107–1124.

Burton, Linda M. and M. Belinda Tucker. 2009. "Romantic Unions in an Era of Uncertainty: A Post-Moynihan Perspective on African American Women and Marriage." *The Annals of the American Academy of Political and Social Science* 621(1):132–148.

Busby, Dean M. and Thomas B. Holman. 2009. "Perceived Match or Mismatch on the Gottman Conflict Styles: Associations with Relationship Outcome Variables." *Family Process* 48(4):531–545.

Buss, David M. 2009. "Darwin and the Emergence of Evolutionary Psychology." *American Psychologist* 64(2):140–148.

Buss, David M. and Todd K. Shackelford. 2008. "Attractive Women Want It All: Good Genes, Economic Investment, Parenting Proclivities, and Emotional Commitment." *Evolutionary Psychology* 6:134–146.

Buss, D. M., Todd K. Shackelford, Lee A. Kirkpatrick, and Randy J. Larsen. 2001. "A Half Century of Mate Preferences: The Cultural Evolution of Values." *Journal of Marriage and Family* 63(2):491–503.

Butler, J. (1988). Performance Acts and Gender Constitution: An Essay in Phenomenology and Feminist Theory. *Theatre Journal,* 40(4), 519–531.

Byers, E. Sandra. 2005. "Relationship Satisfaction and Sexual Satisfaction: A Longitudinal Study of Individuals in Long-Term Relationships." *Journal of Sex Research* 42(2):113–118.

Byers, E. Sandra and Heather A. Sears. 2012. "Mothers Who Do and Do Not Intend to Discuss Sexual Health with Their Young Adolescents." *Family Relations* 61(5):851–863.

Byrd, Stephanie Ellen. 2009. "The Social Construction of Marital Commitment." *Journal of Marriage and Family* 71(2):318–336.

Byrne, Kerry, Joseph Orange, and Catherine Ward-Griffin. 2011. "Care Transition Experiences of Spousal Caregivers: From a Geriatric Rehabilitation Unit to Home." *Qualitative Health Research* 21(10):1371–1387.

Byrnes, Hilary and Brenda Miller. 2012. "The Relationship Between Neighborhood Characteristics and Effective Parenting Behaviors: The Role of Social Support." *Journal of Family Issues* 33(12): 1658–1687.

Bzostek, Sharon H. 2008. "Social Fathers and Child Well-Being." *Journal of Marriage and Family* 70(4):950–961.

Bzostek, Sharon, M. J. Carlson, and Sara McLanahan. 2007. "Repartnering after a Nonmarital Birth: Does Mother Know Best?" Working Paper No. 2006-27-FF, Center for Research on Child Wellbeing. Princeton, NJ: Princeton University.

Bzostek, Sharon H., Sara S. McLanahan, and Marcia J. Carlson. 2012. "Mothers' Repartnering after a Nonmarital Birth." *Social Forces* 90(3):817–841.

Cabanes, Jason V. and Kristel A. Acedera. 2012. "Of Mobile Phones and Mother-Fathers: Calls, Text Messages, and Conjugal Power Relations in Mothers-away Filipino Families." *New Media & Society* 14(6):916–930.

Cabrera, Natasha J., Jay Fagan, and Danielle Farrie. 2008a. "Explaining the Long Reach of Fathers' Prenatal Involvement on Later Paternal Engagement" *Journal of Marriage and Family* 70(5):1094–1107.

Caldwell, John. 1982. *Theory of Fertility Decline.* London: Academic Press.

California Coalition Against Violence. n. d. "The Cycle of Violence: Why *Do* Women Stay in Violent Relationships?" Retrieved March 14, 2013 (http://www.coalitionagainstviolence.ca).

California Secretary of State's Business Programs Division. 2011. *Terminating a California Registered Domestic Partnership.* Sacramento, CA: State Government Offices, Notary Public & Special Filings Section. Retrieved December 5, 2012 (http://www.sos.ca.gov/dpregistry/forms/sf-dp2.pdf).

Call, Vaughn, Susan Sprecher, and Pepper Schwartz. 1995. "The Incidence and Frequency of Marital Sex in a National Sample." *Journal of Marriage and Family* 57(3):639–652.

Callahan, Daniel. 1985. "What Do Children Owe Elderly Parents? Toward a Policy that Promotes, not Corrupts, Family Bonds." *Hastings Center Report* (April):32–37.

Calms, Jackie. 2013. "Obama Signs Expanded Anti-Violence Law." *The New York Times.* March 7. Retrieved March 14, 2013 (http://thecaucus.blogs.nytimes.com).

Calzo, Jerel P. and L. Monique Ward. 2009. "Contributions of Parents, Peers, and Media to Attitudes Toward Homosexuality: Investigating Sex and Ethnic Differences." *Journal of Homosexuality* 56(8):1101–1116.

Campbell, Bernadette, E. Glenn Schellenberg, and Charlene Y. Senn. 1997. "Evaluating Measures of Contemporary Sexism." *Psychology of Women Quarterly* 1(1):89–102.

Campbell, Susan. 2004. *Truth in Dating: Finding Love by Getting Real.* Tiburon, CA: H. J. Kramer/New World Library.

Canary, D. J. and K. Dindia. 1998. *Sex Differences and Similarities in Communication.* Mahwah, NJ: Erlbaum.

Cancian, Francesca M. 1985. "Gender Politics: Love and Power in the Private and Public Spheres." Pp. 253–264 in *Gender and the Life Course,* edited by Alice S. Rossi. New York: Aldine.

———. 1987. *Love in America: Gender and Self-Development.* New York: Cambridge University Press.

Cancian, Francesca M. and Stacey J. Oliker. 2000. *Caring and Gender.* Walnut Creek, CA: AltaMira.

Cancian, M. and D. Meyer. 1998. "Who Gets Custody? Custody of Children in Broken Families." *Demography,* 35:147–158.

Cancian, Maria, Daniel R. Meyer, and Eunhee Han. 2012. "Full-time Father or 'Deadbeat Dad'? Does the Growth in Father Custody Explain the Declining Share of Single Parents with a Child Support Order?" Institute for Research on Poverty, University of Wisconsin. Retrieved April 26, 2013 (http://paa2012.princeton.edu/papers/120291).

Cantor, M. H. 1979. "Neighbors and Friends: An Overlooked Resource in the Informal Support System." *Research on Aging* 1:434–463.

Capizzano, Jeffrey, Gina Adams, and Jason Ost. 2006. *Caring for Children of Color: The Child Care Patterns of White, Black and Hispanic Children.* Washington, DC: The Urban Institute.

Cardiff, Ashley. 2013. *Don't Be That Guy.* Retrieved May 22, 2013 (http://www.thegloss.com/2012/12/04/sex-and-dating/canada-date-rape-campaign- 991/#ixzz2U32E1Qk4).

Careerbuilder.com.2007. "Thirty-Seven Percent of Working Dads Would Leave Their Jobs if Their Family Could Afford It." Careerbuilder.com, June 11.

Caregiver Support Blog. 2008 (October 13). Retrieved March 13, 2010 (caregiversupport.wordpress.com).

Carey v. Population Services International. 1977. 431 U.S. 678, 52 L. Ed. 2d 675, 97 S. Ct. 2010.

Carey, Benedict. 2012. "Debate on a Study Examining Gay Parents." *The New York Times.* June 11. Retrieved August 28, 2012 (http://www.nytimes.com).

Carlson, Bonnie E., Katherine Maciol, and Joanne Schneider. 2006. "Sibling Incest: Reports from Forty-One Survivors." *Journal of Child Sexual Abuse* 15(4):19–34.

Carlson, Daniel L. and Chris Knoester. 2011. "Family Structure and the Intergenerational Transmission of Gender Ideology." *Journal of Family Issues* 32(6):709–734.

Carlson, Elwood. 2009. "20th-Century U.S. Generations." *Population Bulletin* 64(1). Retrieved June 10, 2009 (www.prb.com).

Carlson, Marcia J. 2006. "Family Structure, Father Involvement, and Adolescent Behavior Outcomes." *Journal of Marriage and Family* 68(1):137–154.

———. 2011. "Adults Who Have Biological Children with More than One Partner: Patterns and Implications for U.S. Families." *Population Reference Bureau Policy Seminar.* April 15.

Carlson, Marcia J. and Frank F. Furstenberg, Jr. 2006. "The Prevalence and Correlates of Multipartnered Fertility among Urban U.S. Parents." *Journal of Marriage and Family* 68(3):718–732.

Carlson, Marcia J., Natasha V. Pilkauskas, Sara S. McLanahan, and Jeanne Brooks-Gunn. 2011. "Couples as Partners and Parents Over Children's Early Years." *Journal of Marriage and Family* 73(2):317–334.

Carlson, Ryan G., Andrew P. Daire, Matthew D. Munyon, and Mark E. Young. 2012. "A Comparison of Cohabiting and Noncohabiting Couples Who Participated in Premarital Counseling Using the PREPARE Model." *The Family Journal* 20(2):123–130.

Carlton, Erik., Jason Whiting, Kay Bradford, Patricia Hyjer Dyk, and Ann Vail. 2009. "Defining Factors of Successful University-Community Collaborations: An Exploration of One Healthy Marriage Project." *Family Relations* 58(1):28–40.

Carmalt, Julie H., John Cawley, Kara Joyner, and Jeffery Sobal. 2008. "Body Weight and Matching With a Physically Attractive Romantic Partner." 2008. *Journal of Marriage and the Family* 70:1287–1296.

Carney, Michelle, Fred Buttell, and Don Dutton. 2007. "Women Who Perpetrate Intimate Partner Violence: A Review of the Literature with Recommendations for Treatment." *Aggression and Violent Behavior* 12(1):108–115.

Carr, Anneand Mary Stewart Van Leeuwen, eds. 1996. *Religion, Feminism, and the Family.* Louisville, KY: Westminster John Knox Press.

Carr, Deborah. 2004. "The Desire to Date and Remarry among Older Widows and Widowers." *Journal of Marriage and Family* 66(4):1051–1068.

Carré, Justin M. and Cheryl M. McCormick. 2008. "Aggressive Behavior and Change in Salivary Testosterone Concentrations Predict Willingness to Engage in a Competitive Task." *Hormones and Behavior* 54(3):403–409.

Carrigan, T., B. Connell, and J. Lee. 1985. "Toward a New Sociology of Masculinity." *Theory and Society* 14(5):551–604.

Carrigan, William D. and Clive Webb. 2009. "Repression and Resistance: The Lynching of Persons of Mexican Origin in the United States, 1848–1928." Pp. 69-86 in *How the United States Racializes Latinos: White Hegemony & Its Consequences,* edited by Jose A. Cobas, Jorge Duany, and Joe R. Feagin. Boulder, CO: Paradigm Publishers.

Carroll, Jason S., Sarah Badger, and Chongming Yang. 2006. "The Ability to Negotiate or the Ability to Love? Evaluating the Developmental Domains of Marital Competence." *Journal of Family Issues* 27(7):1001–1032.

Carroll, Jason S., Chad D. Olson, and Nicolle Buckmiller. 2007. "Family Boundary Ambiguity:

A 30-Year Review of Theory, Research, and Measurement." *Family Relations* 56(2):210–230.

Carroll, Joseph. 2006. "One in Four Americans Think Most Mormons Endorse Polygamy." The Gallup Poll. September 7. Retrieved September 8, 2006 (www.galluppoll.com).

———. 2007a. "Most Americans Approve of Interracial Marriages." Gallup News Service, August 16. Retrieved February 19, 2009 (www.gallup.com/poll).

———. 2007b. "Public: 'Family Values' Important to Presidential Vote." December 26. Retrieved May 20, 2010 (www.gallup.com/poll).

———. 2007c. "Stress More Common among Younger Americans, Parents, Workers." Gallup News Service, January 24. Retrieved February 19, 2009 (www.gallup.com/poll).

———. 2008. "Time Pressures, Stress Common for Americans." *Gallup Poll*, January 2. www.gallup.com.

Carter, Betty and Monica McGoldrick. 1988. *The Changing Family Life Cycle: A Framework for Family Therapy.* 2nd ed. New York: Gardner.

Carter, Gerard A. 2009. "Book Review: Unmarried Couples with Children." *Journal of Marriage and Family* 71(2):432–434.

Cartwright, Claire. 2008. "Resident Parent-Child Relationships in Stepfamilies." Pp. 208–230 in *The International Handbook of Stepfamilies,* edited by J. Pryor. Hoboken, NJ: John Wiley & Sons.

Case, A. and C. Paxson. 2001. "Mothers and Others: Who Invests in Children's Health." *Journal of Health Economics,* 20:301–328.

Case, Anne, I-Fen Lin, and Sara McLanahan. 2000. "How Hungry Is the Selfish Gene?" *The Economic Journal* 110(October):781–804.

Casper, Lynne M., and Suzanne M. Bianchi. 2002. *Continuity and Change in the American Family.* Thousand Oaks, CA: Sage Publications.

Cass, Julia. 2007. *Katrina's Children:* Still *Waiting.* Washington, DC: Children's Defense Fund.

Cassano, Michael C. and Janice L. Zeman. 2010. "Parental Socialization of Sadness Regulation in Middle Childhood: The Role of Socialization and Gender." *Developmental Psychology* 46(5): 1214–1226.

Catalano, Shannan. 2007. *Intimate Partner Violence in the United States.* NCJ 210675 Washington, DC: U.S. Bureau of Justice Statistics, December 19. Retrieved June 9, 2010. (www.ojp.usdog.gov/bjs).

———. 2012. *Intimate Partner Violence, 1993–2010.* NCJ 239203 Washington, DC: U.S. Bureau of Justice Statistics, November. Retrieved March 13, 2013 (www.ojp.usdoj.gov/bjs).

Cataldi, Emily Forrest, Jennifer Laird, and Angelina KewalRamani. 2009. "High School Dropout and Completion Rates in the United States: 2007." NCES 2009-064. Washington, DC: National Center for Education Statistics, U.S. Department of Education.

Caughy, Margaret. 2011. "Profiles of Racial Socialization among African American Parents: Correlates, Context, and Outcome." *Journal of Child and Family Studies* 20(4):491–502.

Cavanagh, Shannon E. 2008. "Family Structure History and Adolescent Adjustment." *Journal of Family Issues* 29(7):944–980.

———. 2011. "Early Pubertal Timing and the Union Formation. Behaviors of Young Women" *Social Forces* 89(4):1217–1238.

Cave, Damien. 2012. "American Children, Now Struggling to Adjust to Life in Mexico." *The New York Times.* June 18. Retrieved June 20, 2012 (http://www.nytimes.com).

Ceballo, Rosario, Jennifer E. Lansford, Antonia Abbey, and Abigail J. Stewart. 2004. "Gaining a Child: Comparing the Experiences of Biological Parents, Adoptive Parents, and Stepparents." *Family Relations* 53(1):38–48.

Celock, John. 2013. "Kansas Sperm Donor for Lesbian Couple Faces Child Support Suit from State." *The Huffington Post.* January 2. Retrieved January 15, 2013 (http://www.huffingtonpost.com).

Censky, Annalyn. 2010. "Women in Top-Paying Jobs Still Make Less than Men." *CNN Money.com,* April 20. Atlanta, GA: Cable News Network. money.cnn.com.

Center for the Advancement of Women. 2003. "Progress and Perils: New Agenda for Women." (www.advancewomen.org).

Center for American Women and Politics. 2013. "Facts on Women in Congress 2013." Rutgers: Eagleton Institute of Politics. Retrieved March 11, 2013 (http://www.cawp.rutgers.edu).

Center for the Improvement of Child Caring. n. d. "Systematic Training for Effective Parenting Programs." Retrieved February 12, 2007 (www.ciccparenting.org).

Chabot, Jennifer M. and Barbara D. Ames. 2004. "'It Wasn't "Let's Get Pregnant and Go Do It"': Decision Making in Lesbian Couples Planning Motherhood Via Donor Insemination." *Family Relations* 53(4):348–356.

Chacko, Anil, Brian Wymbs, Lizette Flammer-Rivera, William Pelham, Kathryn Walker, Fran Arnold, Hema Visweswaraiah, Michelle Swanger-Gagne, Erin Cirio, Lauma Pirvics, and Laura Herbst. 2008. "A Pilot Study of the Feasibility and Efficacy of the Strategies to Enhance Positive Parenting (STEPP) Program for Single Mothers of Children with ADHD." *Journal of Attention Disorders* 12(3):270–280.

Chadiha, Letha A., Jane Rafferty, and Joseph Pickard. 2003. "The Influence of Caregiving Stressors, Social Support, and Caregiving Appraisal on Marital Functioning among African American Wife Caregivers." *Journal of Marital and Family Therapy* 29(4):479–490.

Chafetz, Janet Saltzman. 1989. "Marital Intimacy and Conflict: The Irony of Spousal Equality." Pp. 149–156 in *Women: A Feminist Perspective,* 4th ed., edited by Jo Freeman. Mountain View, CA: Mayfield.

Chambers, Anthony L. and Aliza Kravitz. 2011. "Understanding the Disproportionately Low Marriage Rate among African Americans: An Amalgam of Sociological and Psychological Constraints." *Family Relations* 60(5): 648–660.

Champion, Jennifer E., Sarah S. Jaser, Dristen L. Reeslund, Lauren Simmons, Jennifer E. Potts, Angela R. Shears, and Bruce E. Compas. 2009. "Caretaking Behaviors by Adolescent Children of Mothers with and without a History of Depression." *Journal of Family Psychology* 23(2):156–166.

Chandra, Anjani, Gladys M. Martinez, William D. Mosher, Joyce C. Abma, and Jo Jones. 2005. "Fertility, Family Planning, and Reproductive Health of U.S. Women: Data from the 2002 National Survey of Family Growth." *Vital and Health Statistics* 23(25). Hyattsville, MD: U.S. National Center for Health Statistics. December.

Chandra, Anjani, William D. Mosher, Casey Copen, and Catlainn Sionean. 2011. *Sexual Behavior, Sexual Attraction, and Sexual Identity in the United States: Data from the 2006–2008 National Survey of Family Growth.* National Health Statistics Reports No. 36. Hyattsville, MD: National Center for Health Statistics. 2011.

Chandra, Anjani and Elizabeth Hervey Stephen. 2010. "Infertility Service Use among U.S. Women: 1995 and 2002." *Fertility and Sterility* 93(3):725–736.

Chaney, Cassandra. 2009. "The Commitment Continuum: Cohabitation and Commitment among African America Couples." *Family Focus* (Summer):F4–F6. Minneapolis: National Council on Family Relations.

Chang, Juju and Dan Lieberman. 2012. "Swingers: Inside the Secret World of Provocative Parties and Couples Who 'Swap.'" *ABC Nightline.* May 21. Retrieved December 3, 2012 (http://www.abcnews.go.com).

Chao, R. K. 1994. "Beyond Parental Control and Authoritarian Parenting Style: Understanding Chinese Parenting through the Cultural Notion of Training." *Child Development* 65(4):1111–1119.

Chaplin, Tara M., Pamela M. Cole, and Carolyn Zahn-Waxler. 2005. "Parental Socialization of Emotion Expression: Gender Differences and Relations to Child Adjustment." *Emotion* 5(1):80–88.

Chapman, Gary. 2007. *The Five Love Languages.* Chicago: Northfield Publishing.

Chappell, Crystal Lee Hyun Joo. 1996. "Korean-American Adoptees Organize for Support." *Minneapolis Star Tribune,* December 29, p. E7.

Charles, Laurie L., Dina Thomas, and Matthew L. Thornton. 2005. "Overcoming Bias toward Same-Sex Couples." *Journal of Marital and Family Therapy* 31(3):239–249.

"Chart: State Marriage License and Blood Test Requirements." 2006. Nolo. Retrieved September 7, 2006 (www.nolo.com).

Chatzky, Jeann. 2006. "Just When You Thought It Was Safe to Retire. . . ." CNN Money.com, September 21. Retrieved October 29, 2006 (money.cnn.com).

Chaves, Mark, Shawna Anderson, and Jason Byassee. 2009. "American Congregations at the Beginning of the 21st Century." *National Congregations Study.*

Cheadle, Jacob E., Paul R. Amato, and Valerie King. 2010. "Patterns of Nonresident Father Contact." *Demography* 47(1):205–225.

Chen, Yingyu and Sarah E. Ullman. 2010. "Women's Reporting of Sexual and Physical Assaults to Police in the National Violence Against Women Survey." *Violence Against Women* 16(3):262–279.

Cherlin, Andrew J. 1978. "Remarriage as Incomplete Institution." *American Journal of Sociology* 84:634–650.

———. 1996. *Public and Private Families.* New York: McGraw-Hill.

———. 1999. "Going to Extremes: Family Structure, Children's Well-Being, and Social Science." *Demography* 36:421–428.

———. 2003. "Should the Government Promote Marriage?" *Contexts* 2(4):22–29.

———. 2004. "The Deinstitutionalization of American Marriage." *Journal of Marriage and Family* 66(4):848–861.

———. 2005. "American Marriage in the Early Twenty-First Century." *The Future of Children* 15(2):33–55.

———. 2008. "Public Display: The Picture-Perfect American Family? These Days, It Doesn't Exist." *Washington Post,* September 7. Retrieved October 28, 2008 (www.washington-post.com).

———. 2009a. *The Marriage-Go-Round: The State of Marriage and the Family in America Today.* New York: Alfred A. Knopf.

———. 2009b. "Married with Bankruptcy." *The New York Times,* May 29. Retrieved June 8, 2009 (www.nytimes.com).

———. 2010. "Demographic Trends in the United States: A Review of Research in the 2000s." *Journal of Marriage and Family* 72(3): 403–419.

Cherlin, Andrew J. and Frank F. Furstenberg, Jr. 1986. *The New American Grandparent: A Place in the Family, a Life Apart.* New York: BasicBooks.

———. 1994. "Stepfamilies in the United States: A Reconsideration." *Annual Review of Sociology,* 20:359–381.

Chesler, Phyllis. 2005 [1972]. *Women and Madness.* New York: Palgrave/Macmillan.

Chesley, Noelle. 2011. "Stay-at-Home Fathers and Breadwinning Mothers." *Gender and Society* 25(5):642–664.

Chevan, A. 1996. "As Cheaply as One: Cohabitation in the Older Population." *Journal of Marriage and the Family,* 58:656–668.

Child Abuse Prevention Council of Sacramento. n. d. "About Child Abuse: Cultural Customs." Retrieved May 5, 2007 (www.capcsac.org).

Children in the Middle. 2013. "Rules for Co-parenting." Retrieved April 13, 2013. (http://www.childreninthemiddle.com/rulescoparent.htm).

Children's Defense Fund. 2012. *The State of America's Children Handbook 2012.* Washington, DC: Children's Defense Fund.

Childs, Erica Chito. 2008. "Listening to the Interracial Canary: Contemporary Views on Interracial Relationships among Blacks and Whites." *Fordham Law Review* 76(6):2772–2786.

Child Trends. 2011. *Oral Sex Behaviors among Teens.* Retrieved November 30, 2012 (http://www.childtrendsdatabank.org/?q=node/143).

———. 2012a. *Sexually Active Teens.* Retrieved October 24, 2012 (http://www.childtrendsdatabank.org/?q=node/120).

———. 2012b. *Sexually Experienced Teens.* Retrieved October 24, 2012 (http://www.childtrendsdatabank.org/sites/default/files/24_Sexually_Experienced_Teens.pdf).

———. 2012c. *Births to Unmarried Women.* Retrieved December 4, 2012 (http://www.childtrendsdatabank.org).

———. 2013. "Adolescents Who Have Ever Been Raped." Retrieved July 14, 2013 (http://www.childtrends.org/?indicators=adolescents-who-have-ever-been-raped).

Child Welfare Information Gateway. (2011). *How many children were adopted in 2007 and 2008?* Washington, DC: U.S. Department of Health and Human Services, Children's Bureau.

Cho, Rosa M. 2011. "Understanding the Mechanism behind Maternal Imprisonment and Adolescent School Dropout." *Family Relations* (603):272–289.

"Choosing Senior Housing." 2007. Retrieved May 15, 2010 (helpguide.org).

Christakis, Nicholas A. and Paul D. Allison. 2006. "Mortality After the Hospitalization of a Spouse." *New England Journal of Medicine* 354(7):719–730.

Christensen, Andrew and Neil Jacobson. 1999. *Reconcilable Differences.* London: Guilford.

Christianity Today. 2001. "The Leadership Survey on Pastors and Internet Pornography." Retrieved August 27, 2012 (http://www.ctlibrary.com/le/2001/winter/12.89.html).

Christopher, F. Scott and Susan Sprecher. 2000. "Sexuality in Marriage, Dating, and Other Relationships." *Journal of Marriage and Family* 62:999–1017.

Christopherson, Brian. 2006. "Some Buck Trends, Marry Before Finishing College." *Lincoln Journal Star,* October 3. Retrieved October 4, 2006 (www.journalstar.com).

Christie-Mizell, C. Andre, Erin M. Pryor, and Elizabeth Grossman. 2008. "Child Depressive Symptoms, Spanking, and Emotional Support: Differences Between African American and European American Youth." *Family Relations* 57(June):335–350.

Chudacoff, Howard P. 2007. *Children at Play: An American History.* New York: NYU Press.

Ciaramigoli, Arthur P. and Katherine Ketcham. 2000. *The Power of Empathy: A Practical Guide to Creating Intimacy, Self-Understanding, and Lasting Love in Your Life.* New York: Dutton.

Cieraad, Irene. 2006. At Home: *An Anthropology of Domestic Space.* Syracuse, NY: Syracuse University Press.

Clark, Jane Bennett. 2010. "Finance Basics for Partners." *Kiplinger's Personal Finance* 64(8): 50–52.

Clark, Michele C. and Pamela M. Diamond. 2010. "Depression in Family Caregivers of Elders: A Theoretical Model of Caregiver Burden, Sociotropy, and Autonomy." *Research in Nursing and Health* 33:20–34.

Clark, Vicki, Catherine Huddleston-Casas, Susan Churchill, Denise O'Neil Green, and Amanda Garrett. 2008. "Mixed Methods Approaches in Family Science Research." *Journal of Family Issues* 29(11):1543–1566.

Clarke, L. 2005. "Remarriage in Later Life: Older Women's Negotiation of Power, Resources, and Domestic Labor." *Journal of Women and Aging* 17(4):21–41.

Claxton, Shannon E., and Manfred H. M. van Dulmen. 2013. "Casual Sexual Relationships and Experiences in Emerging Adulthood." *Emerging Adulthood* 1(2):138–150.

Claxton-Oldfield, Stephen. 2008. "Stereotypes of Stepfamilies and Stepfamily Members." Pp. 30–52 in *The International Handbook of Stepfamilies,* edited by J.Pryor. Hoboken, NJ: John Wiley & Sons.

Claxton-Oldfield, S. and Butler, B. (1998). "Portrayal of Stepparents in Movie Plot Summaries." *Psychological Reports,* 82:879–882.

Claxton-Oldfield, Stephen, Carla Goodyear, Tina Parsons, and Jane Claxton-Oldfield. 2002. "Some Possible Implications of Negative Stepfather Stereotypes." *Journal of Divorce and Remarriage,* Spring-Summer:77–89.

Clayton, Any and Maureen Perry-Jenkins. 2008. "No Fun Anymore: Leisure and Marital Quality Across the Transition to Parenthood." *Journal of Marriage and Family* 70(1):28–43.

Clayton, Obie and Joan Moore. 2003. "The Effects of Crime and Imprisonment on Family Formation." Pp. 84–102 in *Black Fathers in Contemporary American Society: Strengths, Weaknesses, and Strategies for Change,* edited by Obie Clayton, Ronald B. Mincy, and David Blankenhorn. New York: Russell Sage.

Cleek, Elizabeth, Matt Wofsy, Nancy Boyd-Franklin, Brian Mundy, and Ramika Lowell. 2012. "The Family Empowerment Program." *Family Process* 51(2):207–217.

Clements, Mari L., Scott M. Stanley, and Howard J. Markman. 2004. "Before They Said 'I Do': Discriminating among Marital Outcomes over 13 Years." *Journal of Marriage and Family* 66(3):613–626.

Clemetson, Lynette. 2006. "Adopted in China: Seeking Identity in America." *The New York Times,* March 23.

———. 2007. "Working on Overhaul, Russia Halts Adoption Applications." *The New York Times,* April 12.

Clemetson, Lynette and Ron Nixon. 2006. "Overcoming Adoption's Racial Barriers." *The New York Times,* August 17.

Clifford, Denis, Frederick Hertz, and Emily Doskow. 2007. *A Legal Guide for Lesbian & Gay Couples.* Berkeley: Nolo Press.

Clinton, Timothy and Mark Laaser. 2010. *The Quick-Reference Guide to Sexuality and Relationship Counseling.* Grand Rapids, MI: Baker Books.

Clinton, Timothy E., and John Trent. 2009. *The Quick-Reference Guide to Marriage and Family Counseling.* Grand Rapids, MI: Baker Books.

Cloud, Henry and John Sims Townsend. 2005. *Rescue Your Love Life: Changing Those Dumb Attitudes and Behaviors That Will Sink Your Marriage.* Nashville, TN: Integrity Publishers.

Clunis, D. Merilee and G. Dorsey Green. 2005. *Lesbian Couples: A Guide to Creating Healthy Relationships.* Emeryville, CA: Seal Press.

Coalition for Marriage, Family, and Couples Education. 2009. "Smart Marriages." Retrieved October 11, 2009 (www.smartmarriages.com).

Coan, James A. and John M. Gottman. 2007. "Sampling, Experimental Control, and Generalizability in the Study of Marital Process Models." *Journal of Marriage and Family* 69(1):73–80.

Cockerham, William C. 2007. "A Note on the Fate of Postmodern Theory and Its Failure to Meet the Basic Requirements for Success in Medical Sociology." *Social Theory & Health* 5(4):285–296.

Coffman, Ginger and Carol Markstrom-Adams. 1995. "A Model for Parent Education among Incarcerated Adults." Presented at the annual meeting of the National Council on Family Relations, November 15–19, Portland, OR.

Cogan, Rosemary and Bud C. Ballinger III. 2006. "Alcohol Problems and the Differentiation of Partner, Stranger, and General Violence." *Journal of Interpersonal Violence* 21(7):924–935.

Cohen, Alex and A. Martinez. 2012. "How Parents Unwittingly Fall into 'The Gender Trap' When Raising Children." *Take Two.* October 25. Retrieved November 8, 2012 (http://www.scpr.org).

Cohen, Leonard. 2008. *Runaway Youth and Multisystemic Therapy (MST): A Program Model.* West Hartford, CT: University of Hartford.

Cohen, Neil A., Thanh Tran, and Siyon Rhee. 2007. *Multicultural Approaches in Caring for Children, Youth, and Their Families.* Boston: Pearson, Allyn, and Bacon Publishers.

Cohen, Patricia. 2007a. "As Ethics Panels Expand Grip, No Research Field Is Off Limits." *The New York Times,* February 28.

———. 2007b. "Signs of Détente in the Battle between Venus and Mars." *The New York Times,* May 31.

Cohen, Robin A. and Barbara Bloom. 2005. "Trends in Health Insurance and Access to Medical Care for Children under Age 19 Years: United States, 1998–2003." *Advance Data from Vital and Health Statistics,* No. 355. Hyattsville, MD: U.S. National Center for Health Statistics.

Cohen, Ronnie. 2012. "U.S. Immigration to Treat Same-Sex Partners As Relatives." September 29. Reuters New Service, Thomson Reuters. Retrieved December 11, 2012 (http://www.reutersreprints.com).

Cohen, S. et al. 2007. "Sexual Impairment in Psychiatric Inpatients: Focus on Depression." *Pharmacopsychiatry* 40(2):58–63.

Cohn, D'Vera. 2012. "How Much Did the Foreign-Born Population Grow?" *Pew Social & Demographic Trends.* Pew Research Center. January 9. Retrieved February 16, 2012 (http://www.pewsocialtrends.org).

Cohn, D'Vera, Jeffrey Passel, Wendy Wang, and Gretchen Livingston. 2011. "Barely Half of U.S. Adults Are Married—A Record Low: Marriages Down 5 Percent from 2009 to 2010." *Pew Social & Demographic Trends.* Pew Research Center. December 14. Retrieved December 10, 2012 (http://www.pewsocialtrends.org).

Cohn, Lisa, Debbie Glasser, and Steve Mark. 2008. *The Step-Tween Survival Guide: How to Deal with Life in a Stepfamily.* Minneapolis: Free Spirit Publishers.

"Cohousing in Today's Real Estate Market." 2006. *Cohousing Magazine.* The Cohousing Association of the United States. Retrieved October 3, 2006 (www.cohousing.org).

Colapinto, John. 2000. *As Nature Made Him: The Boy Who Was Raised As a Girl.* New York: HarperCollins.

Cole, Charles Lee. 2011. "The Search for Emotional Connection in Marriage and Marriage-Like Relationships." *Family Focus* FF48 (spring): F6–F7.

Cole, Harriette. 1993. *Jumping the Broom: The African-American Wedding Planner.* New York: Henry Holt.

Cole, Thomas. 1983. "The 'Enlightened' View of Aging." *Hastings Center Report* 13:34–40.

Coleman, Marilyn, Lawrence H. Ganong, and Susan M. Cable. 1997. "Beliefs about Women's Intergenerational Family Obligations to Provide Support Before and After Divorce and Remarriage." *Journal of Marriage and Family* 59(1):165–176.

Coleman, Marilyn, Lawrence H. Ganong, and Mark Fine. 2000. "Reinvestigating Remarriage: Another Decade of Progress." *Journal of Marriage and Family* 62:1288–1307.

Coleman, Marilyn and Lynette Nickleberry. 2009. "An Evaluation of the Remarriage and Stepfamily Self-Help Literature." *Family Relations* 58(December):549–561.

Coles, M. E., L. M. Cook, and T. R. Blake. 2007. "Assessing Obsessive Compulsive Symptoms and Cognitions on the Internet: Evidence for the Comparability of Paper and Internet Administration." *Behaviour Research and Therapy* 45:2232–2240.

Coles, Roberta L. 2009. "Just Doing What They Gotta Do: Single Black Custodial Fathers Coping with the Stressors and Reaping the Rewards of Parenting." *Journal of Family Issues* 30(10):1311–1338.

Coley, Rebekah, L. Ribar, and Elizabeth Votruba-Drzal. 2011. "Do Children's Behavior Problems Limit Poor Women's Labor Market Success?" *Journal of Marriage and Family* 73(1):1–13.

Collaborative Group on Hormonal Factors in Breast Cancer. 2004. "Breast Cancer and Abortion: Collaborative Reanalysis of Data from 53 Epidemiological Studies, Including 83,000 Women with Breast Cancer from 16 Countries." *Lancet* 363(9414):1007–1016.

Collins, Patricia Hill. 2000. "Gender, Black Feminism, and Black Political Economy." *The Annals of the American Academy of Political and Social Science* 568: 41–53.

Collins, Randall and Scott Coltrane. 1995. *Sociology of Marriage and the Family: Gender, Love, and Property.* 4th ed. Chicago: Nelson-Hall.

Collymore, Yvette. 2002. "Risk of Homicide Is High for U.S. Infants." *Population Today,* May /June, p. 10–10.

Coltrane, Scott. 1998. "Gender, Power, and Emotional Expression: Social and Historical Contexts for a Process Model of Men in Marriages and Families." Pp. 193–211 in *Men in Families: When Do They Get Involved? What Difference Does It Make?* edited by Alan Booth and Ann C. Crouter. Mahwah, NJ: Erlbaum.

———. 2000. "Research on Household Labor: Modeling and Measuring the Social Embeddedness of Routine Family Work." *Journal of Marriage and Family* 62:1208–1233.

Coltrane, Scott and Randall Collins. 2001. *Marriage and the Family: Gender, Love, and Property.* Belmont, CA: Wadsworth/Cengage.

Coltrane, Scott, Erika Gutierrez, and Ross D. Parke. 2008. "Stepfathers in Cultural Context: Mexican American Families in the United States." Pp. 100–121 in *The International Handbook of Stepfamilies,* edited by J.Pryor. Hoboken, NJ: John Wiley & Sons.

Colvin, Jan, Lillian Chenoweth, Mary Bold, and Cheryl Harding. 2004. "Caregivers of Older Adults: Advantages and Disadvantages of Internet-Based Social Support." *Family Relations* 53(1):49–57.

Comerford, Lynn. 2006. "The Child Custody Mediation Policy Debate." *Family Focus* (March):F7–F12. National Council on Family Relations.

Comfort, Megan. 2008. *Doing Time Together: Love and Family in the Shadow of Prison.* Chicago: University of Chicago Press.

Concise Columbia Encyclopedia. 1994. New York: Columbia University Press.

"Conclusions Are Reported on Teaching of Abstinence." 2007. *The New York Times,* April 15.

Condon, Stephanie. 2010. "Reid: Unemployment Leads to Domestic Violence." *CBS News,* February 23 (www.cbsnews.com).

Conference for American Indian Women of Proud Nations. 2012. Retrieved October 31, 2012 (http://www.aiwpn.org/).

Conger, Rand D., Katherine J. Conger, and Monica J. Martin. 2010. "Socioeconomic Status, Family Processes, and Individual Development." *Journal of Marriage and Family* 72(3):685–704.

Congregation for the Doctrine of the Faith. 1988. "Instruction on Respect for Human Life in Its Origin and on the Dignity of Procreation." Pp. 325–331 in *Moral Issues and Christian Response,* 4th ed., edited by Paul Jersild and Dale A. Johnson. New York: Holt, Rinehart & Winston.

Conlin, Michelle. 2003. "The New Gender Gap." *Business Week,* May 26, pp. 75–82.

Connell, R. W. 2005. *Masculinities.* 2nd ed. Berkeley: University of California Press.

Conner, Karen A. 2000. *Continuing to Care: Older Americans and Their Families.* New York: Falmer.

Connidis, Ingrid Arnet. 2007. "Negotiating Inequality among Adult Siblings: Two Case Studies." *Journal of Marriage and Family* 69(2):482–499.

———. 2009. *Family Ties and Aging.* Thousand Oaks, CA: Pine Forge Press.

———. 2010. *Family Ties and Aging,* 2nd ed. Los Angeles: Pine Forge Press.

———. 2011. "Reflections on Intergenerational Relations." *Family Focus* FF50 (Fall)): F3–F5.

Connidis, Ingrid Arnet and Candace L. Kemp. 2008. "Negotiating Actual and Anticipated Parental Support: Multiple Sibling Voices in Three-Generation Families." *Journal of Aging Studies* 22:229–238.

Connor, James. 2007. *The Sociology of Loyalty.* New York: Springer-Verlag.

Conrad, Kate, Travis Dixon, and Yuanyuan Zhang. 2009. "Controversial Rap Themes, Gender Portrayals, and Skin Tone Distortion: A Content Analysis of Rap Music Videos." *Journal of Broadcasting and Electronic Media* 53(1): 134–156.

Conway, Lynn. 2012. "Transsexual Women's Successes: Links and Photos." Retrieved June 25, 2013 (http://ai.eecs.umich.edu/people/conway/TSsuccesses/TSsuccesses.html).

Conway, Tiffany and Rutledge Q. Hutson. 2007. "Is Kinship Care Good for Kids?" Center for Law and Social Policy, March 2. Retrieved September 14, 2009 (www.clasp.org).

Cook, Judith, Lynne Mock, Jessica Jonikas, Jane Burke-Miller, Tina Carter, Amanda Taylor, Carol Petersen, Dennis Grey, and David Gruenfelder. 2009. "Prevalence of Psychiatric and Substance Use Disorders among Single Mothers Nearing Lifetime Welfare Eligibility Limits." *Archives of General Psychiatry* 66(3):249–260.

Cooke, Lynn Prince and Janeen Baxter. 2010. "'Families' in International Context: Comparing Institutional Effects Across Western Societies." *Journal of Marriage and Family* 72(3) (June): 516–536.

Cooksey, E. C. and M. M. Fondell. 1996. "Spending Time with His Kids: Effects of Family Structure on Fathers' and Children's Lives." *Journal of Marriage and the Family,* 58:693–707.

Cooley, Charles Horton. 1902. *Human Nature and the Social Order.* New York: Scribner's.

———. 1909. *Social Organization.* New York: Scribner's.

Coombs-Orme, Terri and Daphne S. Cain. 2008. "Predictors of Mothers' Use of Spanking with Their Infants." *Child Abuse and Neglect* 32(6): 649–657.

Cooney, Rosemary, Lloyd H. Rogler, Rose Marie Hurrel, and Vilma Ortiz. 1982. "Decision Making in Intergenerational Puerto Rican Families." *Journal of Marriage and Family* 44:621–631.

Coontz, Stephanie. 1992. *The Way We Never Were: American Families and the Nostalgia Trap.* New York: Basic Books.

———. 1997. "Divorcing Reality." *The Nation,* November 17, pp. 21–24.

———. 2005a. "The Heterosexual Revolution." *The New York Times,* July 5.

———. 2005b. *Marriage, a History: From Obedience to Intimacy, or How Love Conquered Marriage.* New York: Viking.

———. 2005c. "The New Fragility of Marriage, for Better or for Worse." *Chronicle of Higher Education* 51(35):B7–B10. Retrieved September 29, 2006 (chronicle.com).

———. 2007. "The Romantic Life of Brainiacs." *Boston Globe,* February 18. www.boston.com.

———. 2010a. "Divorce, No-Fault Style." *The New York Times,* June 17, A29.

———. 2010b. "Why Gore Breakup Touched a Nerve." *CNN International Edition,* June 4. Retrieved June 20, 2010 (edition.cnn.com).

———. 2012a. "Cohabitation Doesn't Cause Bad Parenting." *The New York Times.* July 13. Retrieved December 11, 2012 (http://www.nytimes.com).

———. 2012b. "The Myth of Male Decline." *The New York Times.* September 29. Retrieved October 9, 2012 (http://www.nytimes.com).

———. 2013. "Why Gender Equality Stalled." *The New York Times.* February 16. Retrieved February 18, 2013 (http://www.nytimes.com).

Cooper, Carey E., Robert Crosnoe, Marie-Anne Suizzo, and Keenan A. Pituch. 2009. "Poverty, Race, and Parental Involvement During the Transition to Elementary School." *Journal of Family Issues* (October).

Cooper, Carey E., Sara McLanahan, Sarah Meadows, and Jeanne Brooks-Gunn. 2009. "Family Structure Transitions and Maternal Parenting Stress." *Journal of Marriage and Family* 71(3):558–574.

Cooper, Shauna and Vonnie McLoyd. 2011. "Racial Barrier Socialization and the Well-Being of African American Adolescents." *Journal of Research on Adolescence* 21(4): 895–903.

Coopersmith, Jared. 2009. *Characteristics of Public, Private, and Bureau of Indian Education Elementary and Secondary School Teachers in the United States: Results from the 2007–08 Schools and Staffing Survey.* NCES 2009-324. Washington, DC: National Center for Education Statistics, Institute of Education Sciences, U.S. Department of Education.

Cooter, Roger and Claudia Stein. 2010. "Positioning the Image of AIDS." *Endeavor* 34(1):12–15.

Copen, Casey E., Kimberly Daniels, Jonathan Vespa, and William Mosher. 2012. *First Marriages in the United States: Data from the 2006–2010 Survey of Family Growth.* National Health Statistics Reports no. 49. Hyattsville, MD: National Center for Health Statistics. March 22.

Corbett, Christianne, Catherine Hill, and Andresse St. Rose. 2008. *Where the Girls Are: The Facts about Gender Equity in Education.* Washington, DC: American Association of University Women.

Cordes, Henry J. 2009. "'He-Cession' Reshuffles Roles." *Omaha World-Herald,* November 29. www.omaha.com. Retrieved May 2, 2009.

Corliss, Richard and Sonja Steptoe. 2004. "The Marriage Savers." *Time,* January 19.

Cornish, Audie. 2012. "Michelle Obama Bests Ann Romney in Cookie Contest." *All Things Considered.* National Public Radio. October 4. Retrieved October 15, 2012 (http://www.npr.org.)

Cornwell, Benjamin and Edward O. Laumann. 2011. "Network Position and Sexual Dysfunction: Implications of Partner Betweenness for Men." *American Journal of Sociology* 117(1):172–208.

Cornwell, Erin York and Linda J. Waite. 2009. "Social Disconnectedness, Perceived Isolation, and Health among Older Adults." *Journal of Health and Social Behavior* 50(March):31–48.

Corra, Mamadi, Shannon K. Carter, J. Scott Carter, and David Knox. 2009. "Trends in Marital Happiness by Gender and Race, 1973 to 2006." *Journal of Family Issues* 30(10):1379–1404.

Cott, Nancy F. 2000. *Public Vows: A History of Marriage and the Nation.* Cambridge, MA: Harvard University Press.

Cotton, Sheila R., Russell Burton, and Beth Rushing. 2003. "The Mediating Effects of Attachment to Social Structure and Psychosocial Resources on the Relationship between Marital Quality and Psychological Distress." *Journal of Family Issues* 24(4):547–577.

Cottrell, Barbara and Peter Monk. 2004. "Adolescent-to-Parent Abuse: A Qualitative Overview of Common Themes." *Journal of Family Issues* 25(8):1072–1095.

Coughlin, Patrick and Jay Wade. 2012. "Masculinity Ideology, Income Disparity, and Romantic Relationship Quality among Men with Higher Earning Female Partners." *Sex Roles* 67: 311–322.

Council on Contemporary Families. 2011. "Fact Sheet: Sandwich Generation Month." Retrieved July 14, 2011, from http://www.sandwichgenerationmonth.com.

"Court Treats Same-Sex Breakup as Divorce." 2002. *Seattle Times,* November 3.

"Covenant Marriages Ministry." 1998. www.covenantmarriages.com.

Covert, Juanita J. and Travis L. Dixon. 2008. "A Changing View: Representation and Effects of the Portrayal of Women of Color in Mainstream Women's Magazines." *Communication Research* 35(2):232–256.

Cowan, Gloria. 2000. "Beliefs about the Causes of Four Types of Rape." *Sex Roles* 42(9/10):807–823.

Cowan, Philip and Carolyn Cowan. 2009. "News You Can Use: Are Babies Bad for Marriage?" *Press Release,* January 9. Chicago: Council on Contemporary Families. www.contemporaryfamilies.org.

Cowan, Philip A., Carolyn Pape Cowan, Marsha Kline Pruett, Kyle Pruett, and Jessie J. Wong. 2009. "Promoting Fathers' Engagement with Children: Preventive Interventions for Low-Income Families." *Journal of Marriage and Family* 71(4):663–679.

Cowan, Ruth Schwartz. 1983. *More Work for Mother: The Ironies of Household Technology from the Open Hearth to the Microwave.* New York: Basic Books.

Cowdery, Randi S., Norma Scarborough, Carmen Knudson-Martin, Gita Seshadri, Monique E. Lewis, and Anne Rankin Mahoney. 2009.

"Gendered Power in Cultural Contexts: Part II. Middle Class African American Heterosexual Couples with Young Children." *Family Process* 48(1): 25–39.

Cox, Erin. 2008. *Intimate Partner Violence among Pregnant and Parenting Women: Local Health Department Strategies for Assessment, Intervention, and Prevention.* Washington, DC: National Association of County & City Health Officials.

Coy, Peter, Michelle Conlin, and Moira Herbst. 2010. "The Disposable Worker." *Bloomberg Businessweek.* January 18:33–39.

Coyle, James, Thomas Nochajski, Eugene Maguin, Andrew Safyer, David DeWit, and Scott Macdonald. 2009. "An Exploratory Study of the Nature of Family Resilience in Families Affected by Parental Alcohol Abuse." *Journal of Family Issues* 30(12):1606–1623.

Coyne, Sarah M., Laura Stockdale, Dean Busby, Bethany Iverson, and David M. Grant. 2011. "'I luv u:)': A Descriptive Study of the Media Use of Individuals in Romantic Relationships." *Family Relations* 60:150–162.

Crary, David. 2007a. "More Couples Seeking Kinder, Gentler Divorces." *MSNBC,* December 18. Retrieved June 21, 2010 (www.msnbc.com).

———. 2007b. "U.S. Divorce Rate Lowest Since 1979." Associated Press, May 10. Retrieved May 10, 2007 (www.breitbart.com).

———. 2008. "Housework Gets You Laid." *The Huffington Post,* March 6. Retrieved May 20, 2010 (www.huffingtonpost.com).

Crawford, D., D. Feng, and J. Fischer. 2003. "The Influence of Love, Equity, and Alternatives on Commitment in Romantic Relationships." *Family and Consumer Sciences Research Journal* 31(3):253–271.

Crawford, Mary, and Danielle Popp. 2003. "Sexual Double Standards: A Review and Methodological Critique of Two Decades of Research." *Journal of Sex Research* 40(1):13–26.

Crawford, Lizabeth A. and Katherine B. Novak. 2008. "Parent-Child Relations and Peer Associations as Mediators of the Family Structure-Substance Use Relationship." *Journal of Family Issues* 29(2):155–184.

Crawford, Susan P. 2011. "The New Digital Divide." *The New York Times.* December 3. Retrieved September 15, 2012 (http://www.nytimes.com).

Crespo, Carla. 2012. "Families as Contexts for Attachment: Reflections on Theory, Research, and the Role of Family Rituals." *Journal of Family Theory & Review* 4(4): 290–298.

Crile, Susan. 2011. "Obama Eliminates Abstinence-Only Funding in Budget." Retrieved January 25, 2013 (http://www.huffingtonpost.com/2009/05/07/obama-eliminates-abstinen_n_199205.html?view=print&comm_ref=false).

Crittenden, Ann. 2001. *The Price of Motherhood: Why the Most Important Job in the World Is Still the Least Valued.* New York: Metropolitan.

Crook, Tylon, Chippewa M. Thomas, and Debra C. Cobia. 2009. "Masculinity and Sexuality: Impact on Intimate Relationships of African American Men." *Family Journal* 17(4):360–366.

Crooks, Robert and Karla Baur. 2005. *Our Sexuality.* 9th ed. Belmont, CA: Wadsworth.

Crosbie-Burnett, M. 1984. "The Centrality of the Step Relationship: A Challenge to Family Theory and Practice." *Family Relations,* 459–463.

Crosbie-Burnett, Margaret and Edith Lewis. 1999. "Use of African-American Family Structure and Functioning to Address the Challenges of European-American Post-Divorce Families." Pp. 455–468 in *American Families: A Multicultural Reader,* edited by Stephanie Coontz. New York: Routledge.

Crosbie-Burnett, Margaret, Edith A. Lewis, Summer Sullivan, Jessica Podolsky, Rosane Mantilla de Souza, and Victoria Mitrani. 2005. "Advancing Theory through Research: The Case of Extrusion in Stepfamilies." Pp. 213–230 in *Sourcebook of Family Theory and Research,* edited by Vern L. Bengston, Alan C. Acock, Katherine R. Allen, Peggye Dilworth-Anderson, and David M. Klein. Thousand Oaks, CA: Sage Publications.

Crosby, John F. 1991. *Illusion and Disillusion: The Self in Love and Marriage.* 4th ed. Belmont, CA: Wadsworth.

Crosnoe, Robert and Shannon E. Cavanagh. 2010. "Families with Children and Adolescents: A Review, Critique, and Future Agenda." *Journal of Marriage and Family* 72(3) (June): 594–611.

Cross-Barnet, Caitlin, Andrew Cherlin, and Linda Burton. 2011. "Bound By Children: Intermittent Cohabitation and Living Together Apart." *Family Relations* 60(December): 633–647.

Crosse, Marcia. 2008. "Abstinence Education: Assessing the Accuracy and Effectiveness of Federally Funded Programs." *Testimony Before the Committee on Oversight and Government Reform, House of Representatives* April 23. Washington, DC: United States Government Accountability Office.

Crouter, Anne C., Megan E. Baril, and Kelly D. Davis, and Susan M. McHale. 2008. "Processes Linking Social Class and Racial Socialization in African American Dual-Earner Families." *Journal of Marriage and Family* 70(5):1311–1325.

Crouter, Ann C. and Alan Booth, eds. 2003. *Children's Influence on Family Dynamics: The Neglected Side of Family Relationships.* Mahwah, NJ: Erlbaum.

Crouter, Ann C., S. D. Whiteman, S. M. McHale, and D. W. Osgood. 2007. "Development of Gender Attitude Traditionality Across Middle Childhood and Adolescence." *Child Development* 78:911–926.

Crowder, Kyle D. and Stewart E. Tolnay. 2000. "A New Marriage Squeeze for Black Women: The Role of Racial Intermarriage by Black Men." *Journal of Marriage and Family* 62(3):792–807.

Crowley, Jocelyn E. and Stephanie Curenton. 2011. "Organizational Social Support and Parenting challenges among Mothers of Color: The Case of Mocha Moms." *Family Relations* 60(1): 1–14.

Crowley, Martha, Daniel T. Lichter, and Zhenchao Qian. 2006. "Beyond Gateway Cities: Economic Restructuring and Poverty among Mexican Immigrant Families and Children." *Family Relations* 55(3):345–360.

Crowley, Michael. 2011. "The New Generation Gap." *Time.* November 14: 36–40.

Cruz, J. 2012. *Remarriage Rate in the U.S., 2010* (FP-12-14). National Center for Family & Marriage Research. Retrieved September 7, 2012 (http://ncfmr.bgsu.edu.pdf/family_profiles/file114853.pdf).

Cuber, John and Peggy Harroff. 1965. *The Significant Americans.* New York: Random House. (Published also as *Sex and the Significant Americans.* Baltimore: Penguin, 1965.)

Cui, Ming and M. Brent Donnellan. 2009. "Trajectories of Conflict over Raising Adolescent Children and Marital Satisfaction." *Journal of Marriage and Family* 71(3):478–494.

Cui, Ming, Frederick O. Lorenz, Rand D. Conger, Janet N. Melby, and Chalandra M. Bryant. 2005. "Observer, Self-, and Partner Reports of Hostile Behaviors in Romantic Relationships." *Journal of Marriage and Family* 67(5):1169–1181.

Cui, Ming, Koji Ueno, Mellissa Gordon, and Frank D. Fincham. 2013. "The Continuation of Intimate Partner Violence from Adolescence to Young Adulthood." *Journal of Marriage and Family* 75(2):300–313.

Cullen, Jennifer C., Leslie B. Hammer, Margaret B. Neal, and Robert R. Sinclair. 2009. "Development of a Typology of Dual-Earner Couples Caring for Children and Aging Parents." *Journal of Family Issues* 30(4):458–483.

Cullen, Lisa T. 2007. "Till Work Do Us Part." *Time Magazine,* September 27. Retrieved May 27, 2010 (www.time.com).

Cunningham, Mick. 2008. "Changing Attitudes toward the Male Breadwinner, Female Homemaker Family Model: Influences of Women's Employment and Education over the Lifecourse." *Social Forces* 87(1):299–323.

Cunningham-Burley, Sarah. 2001. "The Experience of Grandfatherhood." Pp. 92–96 in *Families in Later Life: Connections and Transitions,* edited by Alexis J. Walker, Margaret Manoogian-O'Dell, Lori A. McGraw, and Diana L. G. White. Thousand Oaks, CA: Pine Forge Press.

Currie, Janet, and Cathy Spatz Widom. 2010. "Long-Term Consequences of Child Abuse and Neglect on Adult Economic Well-Being." *Child Maltreatment* 15(2):111–120.

Curtis, Kristen Taylor and Christopher G. Ellison. 2002. "Religious Heterogamy and Marital Conflict." *Journal of Family Issues* 23(4):551–576.

Cyr, Mireille, Pierre McDuff, and John Wright. 2006. "Prevalence and Predictors of Dating Violence among Adolescent Female Victims of Child Sexual Abuse." *Journal of Interpersonal Violence* 21(8):1000–1017.

Dahms, Alan M. 1976. "Intimacy Hierarchy." Pp. 85–104 in *Process in Relationship: Marriage and Family,* 2nd ed., edited by Edward A. Powers and Mary W. Lees. New York: West.

Dailard, Cynthia. 2003. "Understanding 'Abstinence': Implications for Individuals, Programs and Policies." *The Guttmacher Report* 6(5). December (www.agi-usa.org).

Dailey, Rene M. 2009. "Confirmation from Family Members: Parent and Sibling Contributions to Adolescent Psychosocial Adjustment." *Western Journal of Communication* 73(3):273–299.

Dalla, Rochelle, Susan Jacobs-Hagen, Betsy Jareske, and Julie Sukup. 2009. "Examining the Lives of Navajo Native American Teenage Mothers in Context: A 12- to 15-Year Follow-Up." *Family Relations* 58(April):148–161.

Daly, Martin and Margo I. Wilson. 1994. "Some Differential Attributes of Lethal Assaults on Small Children by Stepfathers versus Genetic Fathers." *Ethology and Sociobiology* 15:207–217.

Daneback, K., A. Cooper, and S. Mansson. 2005. "An Internet Study of Cybersex Participants." *Archives of Sexual Behavior* 34:321–328.

Dang, Alain and Samjen Frazer. 2004. *Black Same-Sex Households in the United States: A Report from the 2000 Census.* New York: National Gay and Lesbian Task Force Policy Institute and National Black Justice Coalition.

Daniel, Brigid, Julie Taylor, and Jane Scott. 2010. "Recognition of Neglect and Early Response: Overview of a Systematic Review of the Literature." *Child & Family Social Work* 15(2):248–257.

Daniel, Diane. 2011. "My Husband Is Now My Wife." *The New York Times.* August 18. Retrieved September 13, 2011 (http://www.nytimes.com).

Dao, James. 2005. "Grandparents Given Rights by Ohio Court." *The New York Times,* October 11.

Davey, Adam, Jvoti Savla, Megan Janke, and Shayne Anderson. 2009. "Grandparent-Grandchild Relationships: From Families in Context to Families as Contexts." *Aging and Human Development* 69(4):311–325.

Davey, Monica. 2006. "As Tribal Leaders, Women Still Fight Old Views." *The New York Times,* February 4.

David, Deborah and Robert Brannon. 1979. *The Forty-Nine Percent Majority: The Male Sex Role.* Addison-Wesley.

Davies, Lorraine, Marilyn Ford-Gilboe, Joanne Hammerton. 2009. "Gender Inequality and Patterns of Abuse Post Leaving." *Journal of Family Violence* 24(1):27–39.

Davis, Kelly D., W. Benjamin Goodman, Amy E. Pirretti, and David M. Almeida. 2008. "Nonstandard Work Schedules, Perceived Family Well-Being, and Daily Stressors." *Journal of Marriage and Family* 70(4):991–1003.

Davis, Rebecca L. 2010. *More Perfect Unions: The American Search for Marital Bliss.* Cambridge, MA: Harvard University Press.

Davis, Shannon, Theodore Greenstein, and Jennifer Gertelsen Marks. 2007. "Effects of Union Type on Division of Household Labor." *Journal of Family Issues* 28(5):1260–1271.

Davis-Sowers, Regina. 2012. "'It Just Kind of Like Falls in Your Hands': Factors that Influence Black Aunts' Decisions to Parent Their Nieces and Nephews." *Journal of Black Studies* 43(3):231–250.

Dawkins, Richard. 1976. *The Selfish Gene.* New York: Oxford University Press.

———. 2006. *The Selfish Gene: The 30th Anniversary Edition.* New York: Oxford University Press.

Dawn, Laura. 2006. *It Takes a Nation: How Strangers Became Family in the Wake of Hurricane Katrina.* San Rafael, CA: Earth Aware Editions.

Day, Randal D. and Alan Acock. 2013. "Marital Well-Being and Religiousness as Mediated by Relational Virtue and Equality." *Journal of Marriage and Family* 75(1): 164–177.

Declercq, Eugene, Fay Menacker, and Marian MacDorman. 2004. "Rise in 'No Indicated Risk' Primary Caesareans in the United States, 1991–2001: Cross-Sectional Analysis." *British Medical Journal* (doi:10.1136/bmj.38279.705336OB). On-line First BMJ.com.

DeFrain, John. 2002. *Creating a Strong Family: American Family Strengths Inventory.* Nebraska Cooperative Extension NF01-498 (ianrpubs.unl .edu/family/nf498.htm).

DeGenova, M. K. and F. P. Rice. 2005. *Intimate Relationships, Marriages and Families,* 6th ed. New York: McGraw Hill.

DeLacey, Martha. 2013. "When WILL he say 'I love you'? Men take 88 days to say those three words—but girls make their man wait a lot longer . . ." Retrieved May 25, 2013, from http:// www.dailymail.co.uk/femail/article-2289562 /I-love-Men-88-days-say-girlfriend-women-134 -days-say-boyfriend.html#ixzz2UKq2W8Tz.

De La Lama, Luisa Batthyany, Luis De La Lama, and Ariana Wittgenstein. 2012. "The Soul Mates Model: A Seven-Stage Model for Couples Long-Term Relationship Development and Flourishing." *The Family Journal.* doi (10.1177/1066480712449797).

DeLamater, John and William N. Friedrich. 2002. "Human Sexual Development." *Journal of Sex Research* 39:10–14.

DeLamater, John D. and Morgan Sill. 2005. "Sexual Desire in Later Life." *Journal of Sex Research* 42(2):138–149.

DeLeire, Thomas and Ariel Kalil. 2005. "How Do Cohabiting Couples with Children Spend Their Money?" *Journal of Marriage and Family* 67(2):286–295.

della Cava, Marco R. 2009. "Women Step Up As Men Lose Jobs." *USA Today,* March 19, 1D.

Dell'Antonia, K.J. 2011. "Spacing Children Farther Apart Benefits Older Siblings." *The New York Times, Motherlode: Adventures in Parenting.* November 21. Retrieved December 14, 2012 (http://parenting.blogs.nytimes.com).

Dellmann-Jenkins, Mary, Maureen Blankemeyer, and Odessa Pinkard. 2000. "Young Adult Children and Grandchildren in Primary Caregiver Roles to Older Relatives and Their Service Needs." *Family Relations* 49(2):177–186.

Del Vecchio, Tamara and Susan G. O'Leary. 2006. "Antecedents of Toddler Aggression: Dysfunctional Parenting in Mother-Toddler Dyads." *Journal of Clinical Child and Adolescent Psychology* 35(2):194–202.

DeMaris, Alfred. 2001. "The Influence of Intimate Violence on Transitions out of Cohabitation." *Journal of Marriage and Family* 63(1):235–246.

———. 2009. "Distal and Proximal Influences on the Risk of Extramarital Sex: A Prospective Study of Longer Duration Marriages." *Journal of Sex Research* 46(6):597–607.

DeMaris, Alfred, Annette Mahoney, and Kenneth Pargament. 2011. "Doing the Scut Work of Infant Care: Does Religiousness Encourage Father Involvement?" *Journal of Marriage and Family* 73(2): 354–368.

D'Emilio, John and Estelle B. Freedman. 1988. *Intimate Matters: A History of Sexuality in America.* New York: HarperCollins.

Demo, David H., William S. Aquilino, and Mark A. Fine. 2005. "Family Composition and Family Transitions." Pp. 119–134 in *Sourcebook of Family Theory and Research,* edited by Vern L. Bengston, Alan C. Acock, Katherine R. Allen, Peggye Dilworth-Anderson, and David M. Klein. Thousand Oaks, CA: Sage Publications.

Demo, David H. and Mark A. Fine. 2010. *Beyond the Average Divorce.* Thousand Oaks, CA: Sage Publications.

"The Demographics of Aging." n. d. Transgenerational Design Matters. Retrieved May 5, 2010 (www.transgenerational.org/aging /demographics.htm).

Demos, J. (1966). *A Little Commonwealth: Family Life in Plymouth Colony.* New York: Oxford University Press.

DeNavas-Walt, Carmen, Bernadette D. Proctor, and Jessica C. Smith. 2011. *Income, Poverty, and Health Insurance Coverage in the United States: 2010.* U.S. Census Bureau, Current Population Reports, P60–243. Washington, DC: U.S. Government Printing Office. September.

———. 2012. *Income, Poverty, and Health Insurance Coverage in the United States: 2011.* U.S. Census Bureau, Current Population Reports, P60–243. Washington, DC: U.S. Government Printing Office. September.

Denizet-Lewis, Benoit. 2003."Double Lives on the Down Low." *The New York Times Magazine,* August 3.

———. 2004. "Friends, Friends with Benefits, and the Benefits of the Local Mall." *The New York Times,* May 30.

DeParle, Jason. 2009. "The 'W' Word, Re-Engaged." *The New York Times,* February 8.

DePaulo, Bella. 2006. *Singled Out: How Singles Are Stereotyped, Stigmatized, and Ignored, and Still Live Happily Ever After.* New York: St. Martin's Press.

———. 2012. "A New American Experiment." *The New York Times.* February 13. Retrieved November 6, 2012 (http://www .nytimes.com).

DeRose, Laura M., Mariya P. Shiyko, Holly Foster, and Jeanne Brooks-Gunn. 2011. "Associations Between Menarcheal Timing and Behavioral Developmental Trajectories for Girls from Age 6 to Age 15." *Journal of Youth and Adolescence* 40(10):1329–1342.

Desmond-Harris, Jenee. 2010. "My Race-Based Valentine." *Time Magazine.* February 22:99.

Deveny, Kathleen. . 2008. "Why Only-Children Rule." *Newsweek,* June 2. www.newsweek.com.

DeVisser, Richard and Dee McDonald. 2007. "Swings and Roundabouts: Management of Jealousy in Heterosexual 'Swinging' Couples." *British Journal of Social Psychology* 46(2):459–476.

Devitt, Kerry and Debi Roker. 2009. "The Role of Mobile Phones in Family Communication." *Children & Society* 23:189–202.

Dew, Jeffrey. 2008. "Debt Change and Marital Satisfaction Change in Recently Married Couples." *Family Relations* 57(1):60–71.

———. 2011. "Financial Issues and Relationship Outcomes among Cohabiting Individuals." *Family Relations* 60(2): 178–190.

Dew, Jeffrey, Sonya Britt, and Sandra Huston. 2012. "Examining the Relationship Between Financial Issues and Divorce." *Family Relations* 61:615–628.

Dew, Jeffrey and Jeremy Yorgason. 2010. "Economic Pressure and Marital Conflict in Retirement-Aged Couples." *Journal of Family Issues* 31(2):164–188.

Dewan, Shaila and Robert Gebeloff. 2012. "More Men Enter Fields Dominated by Women." *The New York Times*, May 21: A1, A3.

Dewilde, Caroline, and Wilfred Uunk. 2008. "Remarriage As a Way to Overcome the Financial Consequences of Divorce—A Test of the Economic Need Hypothesis for European Women." *European Sociological Review* 24(3):393–407.

Dickinson, Amy. 2002. "An Extra-Special Relation." *Time*, November 18, pp.A1+.

Dinkes, R. J.,Kemp, and K. Baum. 2009. Indicators of School Crime and Safety: 2009 (NCES 2010-012/NCJ 228478). Washington, DC: U.S. Department of Education, and U.S. Department of Justice.

Dinkmeyer, Don Jr. 2007. "A Systematic Approach to Marriage Education." *Journal of Individual Psychology* 63(3):315–321.

Dion, Karen K. 1995. "Delayed Parenthood and Women's Expectations about the Transition to Parenthood." *International Journal of Behavioral Development* 18(2):315–333.

Dion, Karen K. and Kenneth L. Dion. 1991. "Psychological Individualism and Romantic Love." *Journal of Social Behavior and Personality* 6:17–33.

Dion, M. Robin. 2005. "Healthy Marriage Programs: Learning What Works." *The Future of Children* 15(2):139–156.

DiStefano, Joseph. 2001. "Jumping the Broom." Retrieved October 2, 2006 (www.randomhouse.com).

Ditzen, Beate, Marcel Schaer, Barbara Gabriel, Guy Bodenmann, Ulrike Ehlert, and Markus Heinrichs. 2009. "Intranasal Oxytocin Increases Positive Communication and Reduces Cortisol Levels During Couple Conflict." *Biological Psychiatry* 65(9):728–732.

"Divorce Rate Drops to Lowest Since 1970." 2007. *USA Today*, May 11.

Dixon, Lee J.,GordonK. C., N. Frousakis, and J. A. Schumm. 2012. "A Study of Expectations and the Marital Quality of Participants of a Marital Enrichment Seminar." *Family Relations* 61(1): 75–89.

Dixon, Nicholas. 2007. "Romantic Love, Appraisal, and Commitment." *The Philosophical Forum* 38(4):373–386.

Dixon, Patricia. 2009. "Marriage among African Americans: What Does the Research Reveal?" *Journal of African American Studies* 13(1):29–46.

Doble, Richard deGaris. 2006. AbusiveLove.com. Retrieved August 16, 2006 (www.abusivelove.com).

Dodge, Brian, and Michael Reece, Debby Herbenick, Vanessa Schick, Stephanie A. Sanders, and J. Dennis Fortenberry. 2010. "Sexual Health among U.S. Black and Hispanic Men and Women: A Nationally Representative Study." *Journal of Sexual Medicine* 7(suppl 5):330–345.

Dodson, Jualynne E. 2007. "Conceptualization and Research of African American Family Life in the United States: Some Thoughts." Pp. 51–68 in *Black Families*, 4th ed., edited by Harriette Pipes McAdoo. Thousand Oaks, CA: Sage Publications.

Doherty, William J. 2008. "Public Policy and Couple Relationships: A Commentary on Cabrera et al. 2009." *Journal of Marriage and Family* 70(December):1114–1117.

Doherty, William, Jenet Jacob, and Beth Cutting. 2009. "Community Engaged Parent Education: Strengthening Civic Engagement among Parents and Parent Educators." *Family Relations* 58(3):303–315.

Dolan, Elizabeth M., Bonnie Braun, and Jessica C. Murphy. 2003. "A Dollar Short: Financial Challenges of Working-poor Rural Families." *Family Focus* (June):F13–F15. National Council on Family Relations.

Dolan, Frances Elizabeth. 2008. *Marriage and Violence: The Early Modern Legacy*. Philadelphia : University of Pennsylvania Press.

Dolan, Maura. 2012. "Foes of Gay Marriage Appeal Prop. 8 Ruling to U.S. Supreme Court." *Los Angeles Times*. July 31. Retrieved December 3, 2012 (http://www.latimes.com).

Dolbin-MacNab, Megan L. 2006. "Just Like Raising Your Own? Grandmothers' Perceptions of Parenting a Second Time Around." *Family Relations* 55(5):564–575.

Dolbin-MacNab, Megan I. and Margaret K. Keiley. 2009. "Navigating Interdependence: How Adolescents Raised Solely by Grandparents Experience Their Family Relationships." *Family Relations* 58(April):162–175.

"DOMA Appeal." 2012. *Huffington Post*. July 3. Retrieved November 26, 2012 (http://www.huffingtonpost.com).

Domestic Sex Trafficking of Minors. n. d. Washington, DC: U.S. Department of Justice Child Exploitation and Obscenity Section.

Dominguez, Silvia and Amy Lubitow. 2008. "Transnational Ties, Poverty, and Identity: Latin American Immigrant Women in Public Housing." *Family Relations* 57(October): 419–430.

Domitrz, Michael J. 2003. *May I Kiss You? A Candid Look at Dating, Communication, Respect, and Sexual Assault Awareness*. Greenfield, WI: Awareness Publications.

Donaldson, Doug. 2012. "The New American Super-Family." *Saturday Evening Post*. July-August. Retrieved April 10 (http://www.saturdayeveningpost.com).

D'Onofrio, Brian M. and Benjamin B. Lahey. 2010. "Biosocial Influences on the Family: A Decade Review." *Journal of Marriage and Family* 72(3)(June): 762–782.

Doohan, Eve-Anne M., Sybil Carrere, Chelsea Siler, and Cheryl Beardslee. 2009. "The Link Between the Marital Bond and Future Triadic Family Interactions." *Journal of Marriage and Family* 71(4):892–904.

Dore, Margaret K. 2004. "The 'Friendly Parent' Concept: A Flawed Factor for Child Custody." *Loyola Journal of Public Interest Law* 6:41–56.

Dorius, Cassandra R. 2012. "New Approaches to Measuring Multipartnered Fertility Over the Life Course." *Research Report* 12–769. August. University of Michigan: Population Studies Center.

Dorius, Cassandra R., Alan Booth, Jacob Hibel, Douglas Granger, and David Johnson. 2011. "Parents' Testosterone and Children's Perception of Parent-Child Relationship Quality." *Hormones and Behavior* 60: 512–519.

Dorman, Clive. 2006. "The Social Toddler: Promoting Positive Behaviour." *Infant Observation* 9(1):95–97.

Doss, Brian D., Galena K. Rhoads, Scott M. Stanley, and Howard J. Markman. 2009. "The Effect of the Transition to Parenthood on Relationship Quality: An 8-Year Prospective Study." *Journal of Personality and Social Psychology* 96(3):601–619.

Dotinga, Randy. 2009. "'Macho' Men Visit Doctor Even Less." *USA Today*, August 12. Retrieved January 16, 2010 (www.usatoday.com).

Douglas, Edward and Sharon Douglas. 2000. *The Blended Family: Achieving Peace and Harmony in the Christian Home*. Franklin, TN: Providence House.

Douglas, Susan J. 2009. "Where Have You Gone, Roseanne Barr?" Pp. 281–389 in *The Shriver Report: A Woman's Nation Changes Everything*, edited by Heather Bousheyand Ann O'Leary. Washington, DC: Maria Shriver and the Center for American Progress.

Douthat, Ross. 2009. "The Way We Love Now." *The New York Times*. Retrieved May 25, 2012 (http://www.nytimes.com/2009/06/29/opinion/29douthat.html?_r=0).

Dowd, Maureen. 2012. "Don't Tread on Us." *The New York Times*. March 13. Retrieved October 26, 2012 (http://www.nytimes.com).

Downey, Douglas B. and Dennis J. Condron. 2004. "Playing Well with Others in Kindergarten: The Benefit of Siblings at Home." *Journal of Marriage and Family* 66(2):333–350.

Downing-Matibag, Teresa, and Brandi Geisinger. 2009. "Hooking Up and Sexual Risk Taking among College Students: A Health Belief Model Perspective." *Qualitative Health Research* 19(9):1196–1209.

Downs, Barbara. 2003. *Fertility of American Women: June 2002*. Current Population Reports P20–548. Washington, DC: U.S. Census Bureau. October.

Dreger, Alice D. and April Herndon. 2009. "Progress and Politics in the Intersex Rights Movement: Feminist Theory in Action." *GLQ: A Journal of Lesbian and Gay Studies* 15(2):199–224.

Drexler, Peggy. 2012. "The New Face of Infidelity." Retrieved July 15, 2013 (http://online.wsj.com/article/SB10000872396390443684104578062754288906608.html).

Driscoll, Anne K., Stephen R. Russell, and Lisa J. Crockett. 2008. "Parenting Styles and Youth Well-Being across Immigrant Generations." *Journal of Family Issues* 29(2):185–209.

Driver, Janice L. and John M. Gottman. 2004. "Daily Marital Interactions and Positive Affect During Marital Conflict among Newlywed Couples." *Family Process* 43(3):301–314.

Drummet, Amy R., Marilyn Coleman, and Susan Cable. 2003. "Military Families Under Stress: Implications for Family Life Education." *Family Relations* 52(3):279–287.

Duba, Jill, A. Hughey, Tracy Lara, and M. Burke. 2012. "Areas of Marital dissatisfaction among Long-Term Couples." *Adultspan Journal* 11(1): 39–54.

Duba, Jill D. and Richard E. Watts. 2009. *Journal of Clinical Psychology* 65(2):210–223.

Dubbs, Shelli L., Abraham P. Buunk, and Jessica Li. 2012. "Parental Monitoring, Sensitivity Toward Parents, and a Child's Mate preferences." *Personal Relationships* 19:712–722.

Duck, Steve W. and Julia T. Wood. 2006. "What Goes Up May Come Down: Sex and Gendered Patterns in Relationship Dissolution." Pp. 169–199 in *Handbook of Divorce and Relationship Dissolution,* edited by Mark A. Fineand John H. Harvey. Mahwah, NJ: Erlbaum.

Dudley, Kathryn M. 2007. "The Social Economy of Single Motherhood: Book Review." *Social Forces* 86(1):360–361.

Duenwald, Mary. 2005. "For Them, Just Saying No Is Easy." *The New York Times,* June 9.

Dugan, Laura, Daniel S. Nagin, and Richard Rosenfeld. 2004. "Do Domestic Violence Services Save Lives?" *NIJ Journal* 250:20–25. Washington, DC: National Institute of Justice, U.S. Department of Justice.

Dugger, Celia W. 1998. "In India, an Arranged Marriage of Two Worlds." *The New York Times,* July 20, pp. A1, A10.

Duhigg, Charles. 2007. "Shielding Money Clashes with Elders' Free Will." *The New York Times,* December 24. Retrieved December 25, 2007 (www.nytimes.com).

Duncan, Simon and Miranda Phillips. 2011. "People Who Live Apart Together (LATs): New Family Form or Just a Stage?" *International Review of Sociology* 21(3):513–532.

Duncan, Stephen F., Thomas B. Holman, and Chongming Yang. 2007. "Factors Associated with Involvement in Marriage Preparation Programs." *Family Relations* 56(3):270–278.

Dunifon, Rachel and Ashish Bajracharya. 2012. "The Role of Grandparents in the Lives of Youth." *Journal of Family Issues* 33(9):1168–1194.

Dunifon, Rachel and Lori Kowaleski-Jones. 2007. "The Influence of Grandparents in Single-Mother Families." *Journal of Marriage and Family* 69(2):456–481.

Dunnewind, Stephanie. 2003. "Book Helps Impart Coping Skills, Self-Esteem to Multiracial Children." Knight Ridder/Tribune News Service, August 5.

Dunphy v. Gregor. 1994. 136 :N.J. 99.

Dupre, M. E. and S. O. Meadows. 2007. "Disaggregating the Effects of Marital Trajectories on Health." *Journal of Family Issues,* 28:623–652.

Dupuis, Sara. 2007. "Examining Remarriage: A Look at Issues Affecting Remarried Couples and the Implications Towards Therapeutic Techniques." *Journal of Divorce and Remarriage* 48(1-2):91–104.

Duquaine-Watson, Jillian M. 2007. "'Pretty Darned Cold': Single Mother Students and the Community College Climate in Post-Welfare Reform America." *Equity and Excellence in Education* 40:229–240.

Duran-Aydintug, C. 1993. "Relationships with Former In-Laws: Normative Guidelines and Actual Behavior." *Journal of Divorce & Remarriage,* 19:69–81.

Durham, Ricky. 2010. Prescription 4 Love. http://www.prescription4love.com.

Durodoye, Beth A. and Angela D. Coker. 2008. "Crossing Culture in Marriage: Implications for Counseling African American/African Couples." *International Journal of Counseling* 30:25–37.

Durr, Marlese and Shirley Ann Hill. 2006. "The Family-Work Interface in African American Households." Pp. 73–85 in *Race, Work, and Family in the Lives of African Americans,* edited by Marlese Durrand Shirley Ann Hill. Lanham, MD: Rowman & Littlefield.

Dush, Claire M. Kamp. 2011. "Relationship-Specific Investments, Family Chaos, and Cohabitation dissolution Following a Nonmarital Birth." *Family Relations* 60(5): 586–601.

Dush, Claire M. Kamp, Catherine L. Cohan, and Paul R. Amato. 2003. "The Relationship between Cohabitation and Marital Quality and Stability: Change Across Cohorts?" *Journal of Marriage and Family* 65(3):539–549.

Dush, Claire M. Kamp, Miles G. Taylor, and Rhiannon A. Kroeger. 2008. "Marital Happiness and Psychological Well-Being across the Life Cycle." *Family Relations* 57(April):211–226.

Dutton, Donald G. and Tanya L. Nicholls. 2005. "The Gender Paradigm in Domestic Violence Research and Theory: The Conflict of Theory and Data." *Aggression and Violent Behavior* 10:680–714.

Duvall, E. M. 1954. *In-Laws: Pro and Con—An Original Study of Inter-Personal relations.* New York: Association Press.

Dworkin, Shari L. and Michael A. Messner. 1999. "Just Do . . . What? Sports, Bodies, and Gender." Pp. 341–361 in *Revisioning Gender,* edited by Myra Marx Ferree, Judith Lorber, and Beth. B. Hess. Thousand Oaks, CA: Sage Publications.

Dye, Jane Lawler. 2008. *Fertility of American Women:* 2006. U.S. Census Bureau Current Population Reports P20–558.

Dyer, Carmel Bitondo, Sabrina Pickens, and Jason Burnett. 2007. "Vulnerable Elders: When It Is No Longer Safe to Live Alone." *Journal of the American Medical Association* 298(12):1448–1450.

Dyer, W. Justin, Joseph H. Pleck, and Brent McBride. 2012. "Imprisoned Fathers and Their Family Relationships: A 40-Year Review from a Multi-Theory View." *Journal of Family Theory & Review* 4(1):20–47.

Dyess, Drucilla. 2009. "Pregnancies and Sexually Transmitted Diseases on the Rise among Teens." *HealthNews,* July 21. www.healthnews.com.

Dykstra, Pearl A. and Gunhild O. Hagestad. 2007. "Roads Less Taken: Developing a Nuanced View of Older Adults Without Children." *Journal of Family Issues* 28(10):1275–1310.

Dziech,Billie Wright. 2003. "Sexual Harassment on College Campuses." Pp. 147–171 in *Academic and Workplace Sexual Harassment: A Handbook of Cultural, Social Science, Management, and Legal Perspectives,* edited by Michele Paludi and Carmen A. Paludi, Jr. Westport, CT: Praeger.

Eaklor, Vicki L. 2008. *Queer America: A GLBT History of the 20th Century.* Westport, CT: Greenwood Press.

Early, Theresa J., Thomas K. Gregoire, and Thomas P. McDonald. 2002. "Child Functioning and Caregiver Well-Being in Families of Children with Emotional Disorders." *Journal of Family Issues* 23(3):374–391.

East, Patricia L., Nina C. Chien, and Jennifer S. Barber. 2012. "Adolescents' Pregnancy Intentions, Wantedness, and Regret: Cross-Legged Relations with Mental Health and Harsh Parenting." *Journal of Marriage and Family* 74(1): 167–185.

Easterlin, Richard. 1987. *Birth and Fortune: The Impact of Numbers on Personal Welfare.* 2nd rev. ed. Chicago: University of Chicago Press.

Eaton, Danice K., Laura Kann, Steve Kinchen, Shari Shanklin, James Ross, Joseph Hawkins, William A. Harris, Richard Lowry, Tim McManus, David Chyen, Connie Lim, Nancy D. Brener, and Howell Wechsler. 2008. "Youth Risk Behavior Surveillance—United States, 2007." Surveillance Summaries. *Morbidity and Mortality Weekly Report* 57, No. SS-04, June 6.

Ebaugh, Helen Rose and Mary Curry. 2000. "Fictive Kin as Social Capital in New Immigrant Communities." *Sociological Perspectives* 43(2):189–209.

Ebeling, Ashlea. 2007. "The Second Match." *Forbes,* November 12. Retrieved April 15, 2010 (www.forbes.com).

Eccles, Jacquelynne. 2011. "Gendered Educational and Occupational Choices: Applying the Eccles et al. Model of Achievement-Related Choices." *International Journal of Behavioral Development* 35(3):195–201.

Eckel, Sara. 2010. "Role Reversal." *Working Mother* 33(2):26–33.

Eckholm, Eric. 2008. "Special Session Called on Nebraska Safe-Haven Law." *The New York Times,*

October 30. Retrieved December 12, 2008 (www.nytimes.com).

———. "'Ex-Gay' Men Fight Back Against View That Homosexuality Can't Be Changed." Retrieved June 25, 2012 (http://www.nytimes.com/2012/11/01/us/ex-gay-men-fight-view-that-homosexuality-cant-be-changed.html?_r=0)

———. 2007. "Unmarried with Children." Pp. 505–511 in *Family in Transition,* 14th ed., edited by Arlene S. Skolnickand Jerome H. Skolnick.Boston: Allyn & Bacon.

Edin, Kathryn and Rebecca Joyce Kissane. 2010. "Poverty and the American Family: A Decade in Review." *Journal of Marriage and Family* 72(3) (June): 460–479.

Edin, Kathryn and Joanna M. Reed. 2005. "Why Don't They Just Get Married? Barriers to Marriage among the Disadvantaged." *The Future of Children* 15(2):117–138.

Edwards, Margie L. K. 2004. "We're Decent People: Constructing and Managing Family Identity in Rural Working-Class Communities." *Journal of Marriage and Family* 66(2):515–529.

Edwards, Tom 2009. "As Baby Boomers Age, Fewer Families Have Children Under 18 at Home." *U.S. Census Bureau News,* February 25. Retrieved June 20, 2009 (www.census.gov/Press-Release/www/releases/).

Egelko, Bob. 2008. "Churches on Both Sides of Marriage Law Debate." *San Francisco Chronicle,* February 18: A1, A10.

———. 2009. "The Role of Religion in Adolescence for Family Formation in Young Adulthood." *Journal of Marriage and Family* 71(1):108–121.

———. 2013. "Supreme Court Sets Prop 8, DOMA Hearings." *San Francisco Chronicle.* January 7. Retrieved January 9, 2012 (http://www.sfgate.com).

Eggebeen, David. J. 2012. "What Can We Learn from Studies of Children Raised by Gay or Lesbian Parents?" *Social Science Research* 41: 775–778.

Eggebeen, David and Jeffrey Dew. 2009. "The Role of Religion in Adolescence for Family Formation in Young Adulthood." *Journal of Marriage and Family* 71(1):108–121.

Ehrenfeld, Temma. 2002. "Infertility: A Guy Thing." *Newsweek,* March 25, pp. 60–61.

Ehrenreich, Barbara. 2001. *Nickel and Dimed: On (Not) Getting By in America.* New York: Henry Holt.

Ehrensaft, Diane. 2005. *Mommies, Daddies, Donors, Surrogates: Answering Tough Questions and Building Strong Families.* New York: Guilford.

Eicher-Catt, Deborah. 2004. "Noncustodial Mothers and Mental Health: When Absence Makes the Heart Break." *Family Focus* (March): F7–F8. National Council on Family Relations.

Eisenstadt v. Baird. 1972. 405 U.S. 398.

Elder, Glen H., Jr. 1974. *Children of the Great Depression: Social Change in Life Experience.* Chicago: University of Chicago Press.

Elias, Marilyn. 2004. "Marriage Taken to Heart." *USA Today,* March 4.

———. 2005. "Orphans at Developmental 'Risk.'" *USA Today,* October 11.

———. 2009. "Mental Stress Spirals with Economy." *USA Today,* March 12. Retrieved March 8, 2010 (www.usatoday.com).

Eligon, John. 2011. "Suits Dispute City's Rule on Recording Sex Changes." Retrieved June 25, 2013 (http://www.nytimes.com/2011/03/23/nyregion/23gender.html).

Eliot, Lise. 2009. *Pink Brain, Blue Brain: How Small Differences Grow into Troublesome Gaps—and What We Can Do about It.* New York: Houghton Mifflin Harcourt.

Elizabeth, Vivienne, Nicola Gavey, and Julia Tolmie. 2012. "'. . . He's Just Swapped His Fists for the System' The Governance of Gender Through Custody Law." *Gender & Society* 26(2):239–260.

Elkind, David. 1988. *The Hurried Child: Growing Up Too Fast Too Soon.* Reading, MA: Addison-Wesley.

———. 2007a. *The Hurried Child: Growing Up Too Fast Too Soon.* 25th anniversary ed. Cambridge, MA: Da Capo Lifelong.

———. 2007b. *The Power of Play: How Spontaneous, Imaginative Activities Lead to Happier, Healthier Children.* Cambridge, MA: Da Capo Lifelong.

———. 2010. "Playtime Is Over." *The New York Times.* March 26. Retrieved July 15, 2013 (http://nytimes.com).

Ellick, Adam B. 2011. "Speed Dating, Muslim Style." Retrieved May 28, 2013, from http://www.nytimes.com/2y011/02/13/nyregion/13dating.html?pagewanted=all.

Ellin, Abby. 2009. "The Recession. Isn't It Romantic?" *The New York Times,* February 12. Retrieved June 8, 2009 (www.nytimes.com).

Elliott, Diana B. and Tavia Simmons. 2011. *Marital Events of Americans: 2009.* American Community Survey Reports, ACS-13. U.S. Census Bureau, Washington, DC.

Elliott, Sinikka and Debra Umberson. 2008. "The Performance of Desire: Gender and Sexual Negotiation in Long-Term Marriages." *Journal of Marriage and Family* 70(2):391–406.

Ellis, D. 2006. "Male Abuse of a Married or Cohabiting Female Partner: The Application of Sociological Theory to Research Findings." *Violence and Victims* 4:235–255.

Ellison, Christopher G. and Matt Bradshaw. 2009. "Religious Beliefs, Sociopolitical Ideology, and Attitudes Toward Corporal Punishment." *Journal of Family Issues* 30(3):330–340.

Ellison, Christopher G., Andrea K. Henderson, Norval D. Glenn, and Kristine E. Harkrider. 2011. "Sanctification, Stress, and Marital Quality." *Family Relations* 60(4): 404–420.

Ellison, Christopher G., Marc A. Musick, and George W. Holden. 2011. "Does Conservative Protestantism Moderate the Association Between Corporal Punishment and Child Outcomes?" *Journal of Marriage and Family* 73(5) (October): 946–961.

Ellison, Christopher G., Jenny A. Trinitapoli, Kristin L. Anderson, and Byron R. Johnson. 2007. "Race/Ethnicity, Religious Involvement, and Domestic Violence." *Violence Against Women* 13(11):1094–1112.

Ellison, Christopher G., Nicholas H. Wolfinger, and Aida I. Ramos-Wada. 2012. "Attitudes Toward Marriage, Divorce, Cohabitation, and Casual Sex among Working-Age Latinos: Does Religion Matter?" *Journal of Family Issues* 34(3):295–322.

Ellison, Peter T.and Peter B. Gray, eds. 2009. *Endocrinology of Social Relationships* .Cambridge, MA: Harvard University Press.

Elsby, Michael, Bart Hobijn, and Aysegul Sahin. 2010. "The Labor Market in the Great Recession." *Brookings Panel on Economic Activity,* March 18-19. Washington, DC: Brookings Institute.

Else-Quest, Nicole M., Janet Shibley Hyde, and Marcia C. Linn. 2010. "Cross-National Patterns of Gender Differences in Mathematics: A Meta-Analysis." *Psychological Bulletin* 136(1):103–127.

El-Sheikh, Mona and Elizabeth Flanagan. 2001. "Parental Problem Drinking and Children's Adjustment: Family Conflict and Parental Depression as Mediators and Moderators of Risk." *Journal of Abnormal Child Psychology* 29(5):417–435.

Emmers-Sommer, Tara M., Jenny Farrell, Ashlyn Gentry, Shannon Stevens, Justin Eckstein, Joseph Battocletti, and Carley Gardener. 2010. "First Date Sexual Expectations: The Effects of Who Asked, Who Paid, Date Location, and Gender." *Communication Studies* 61(3):339–355.

Employee Benefit Research Institute. 2009. "Income Statistics of the Population Aged 55 and Over." Retrieved May 8, 2010 (www.ebri.org).

Enander, Viveka and Carin Holmberg. 2008. "Why Does She Leave? The Leaving Process(es) of Battered Women." *Health Care for Women International* 29(3):200–226.

Engel, Marjorie. 2000. "The Financial (In)Security of Women in Remarriages." *Research Findings.* Stepfamily Association of America(www.saafamilies.org).

Engels, Friedrich. 1942 [1884]. *The Origin of the Family, Private Property, and the State.* New York: International.

England, Diane. 2009. *The Post Traumatic Stress Disorder Relationship: How to Support Your Partner and Keep Your Relationship Healthy.* Avon, Mass.: Adams Media.

England, Paula. 2006. "Toward Gender Equality: Progress and Bottlenecks." Pp. 245–264 in *The Declining Significance of Gender?,* edited by Francine D. Blau, Mary C. Brinton, and David B. Grusky. New York: Russell Sage.

England, Paula, Carmen Garcia-Beaulieu, and Mary Ross. 2004. "Women's Employment among Blacks, Whites, and Three Groups of Latinas: Do More Privileged Women Have Higher Employment?" *Gender and Society* 18(4):494–509.

England, Paula and Kathryn Edin(Eds.). 2007. *Unmarried Couples with Children.* New York: Russell Sage Foundation.

England, Paula, and Elizabeth A. McClintock. 2009. "The Gendered Double Standard of Aging in US Marriage Markets." *Population and Development Review* 35(4):797–816.

England,Paula and Reuben J. Thomas. 2007. "The Decline of the Date and the Rise of the College Hook Up." Pp. 151–171 in *Family in Transition,* 14th ed., edited by Arlene S. Skolnickand Jerome H. Skolnick. Boston: Allyn & Bacon.

Enos, Sandra. 2001. *Mothering from the Inside: Parenting in a Women'sPrison.* Albany: SUNY Press.

Epstein, Ann S. 2007. *The Intentional Teacher: Choosing the Best Strategies for Young Children's Learning.* Washington, DC: National Association for the Education of Young Children.

Epstein, Cynthia Fuchs. 1988. *Deceptive Distinctions: Sex, Gender, and the Social Order.* New Haven, CT: Yale University Press.

Epstein, Marina. 2011. "Exploring Parent-Adolescent Communication about Gender: Results from Adolescent and Emerging Adult Samples." *Sex Roles* 65(1/2):108–118.

Epstein, Marina, Jerel P. Calzo, Andrew P. Smiler, and L. Monique Ward. 2009. "Anything from Making Out to Having Sex: Men's Negotiations of Hooking Up and Friends With Benefits Scripts." *Journal of Sex Research* 46(5)414–442.

Erickson, Nancy S. 1991. "Battered Mothers of Battered Children: Using Our Knowledge of Battered Women to Defend Them against Charges of Failure to Act." *Current Perspectives in Psychological, Legal, and Ethical Issues,* Vol. 1A, *Children and Families: Abuse and Endangerment,* pp. 197–218.

Eriksen, Shelley and Naomi Gerstel. 2002. "A Labor of Love or Labor Itself: Care Work among Brothers and Sisters." *Journal of Family Issues* 23(7):836–856.

Essex, Elizabeth L. and Junkuk Hong. 2005. "Older Caregiving Parents: Division of Household Labor, Marital Satisfaction, and Caregiver Burden." *Family Relations* 54(3):448–460.

Etcheverry, Paul E. and Benjamin Le. 2005. "Thinking about Commitment: Accessibility of Commitment and Prediction of Relationship Persistence, Accommodation, and Willingness to Sacrifice." *Personal Relationships* 23(1):103–123.

Evans, Gary W. and Theodore D. Wachs. 2010. *Chaos and Its Influence on Children's Development: An Ecological Perspective.* Washington, DC: American Psychological Association.

Even, William E. and David A. Macpherson. 2004. "When Will the Gender Gap in Retirement Income Narrow?" *Southern Economic Journal* 71(1):182–201.

Ezzell, Matthew B. 2012. "'I'm in Control': Compensatory Manhood in a Therapeutic Community." *Gender & Society* 26(2):190–215.

Faber, Adele, and Elaine Mazlish. 2006. *How to Talk So Kids Will Listen and Listen So Kids Will Talk.* New York: Collins.

"Facts on American Teens' Sexual and Reproductive Health." 2010. In Brief, January. New York: The Guttmacher Institute.

"Factsheet: The Violence Against Women Act." n. d. Retrieved March 14, 2013 (http://www.whitehouse.gov/sites/default/files/docs/vawa_factsheet.pdf).

Fagan, Jay. 2009. "Relationship Quality and Changes in Depressive Symptoms among Urban, Married African Americans, Hispanics, and Whites." *Family Relations* 58 (July):259–274.

———. 2011. "Effect on Preschoolers' Literacy When Never-Married Mothers Get Married." *Journal of Marriage and Family* 73(5):1001–1014.

Fagan, Jay and Marina Barnett. 2003. "The Relationship between Maternal Gatekeeping, Paternal Competence, Mothers' Attitudes about the Father Role, and Father Involvement." *Journal of Family Issues* 24(8):1020–1043.

Fairchild, Emily. 2006. "'I'm Excited to Be Married, But . . .': Romance and Realism in Marriage." Pp. 1–19 in *Couples, Kids, and Family Life,* edited by Jaber F. Gubrium and James A. Holstein. New York: Oxford University Press.

Falconer, Mary Kay, Mary E. Haskett, Linda McDaniels, Thelma Dirkes, and Edward C. Siegel. 2008. "Evaluation of Support Groups for Child Abuse Prevention: Outcomes of Four State Evaluations." *Social Work With Groups* 31(2):165–182.

Falconier, Mariana K. and Norman B. Epstein. 2011. "Couples Experiencing Financial Strain: What We Know and What We Can Do." *Family Relations* 60(3): 303–317.

Families and Work Institute. 2004. *Gender and Generation in the Workplace.* Boston: American Business Collaboration. October 5.

Families with Children from China. 1999. "Mongolian Spots." Retrieved April 26, 2006 (www.fwcc.org).

Fanshel, David. 1972. *Far from the Reservation: The Transracial Adoption of American Indian Children.* Metuchen, NJ: Scarecrow.

Farr, Diane. 2011. "Bringing home the wrong race." Retrieved May 31, 2013, from http://www.nytimes.com/2011/06/05/fashion/modern-love-breaking-our-parents-rules-for-love.html?pagewanted=all.

Farrell, Betty, Alicia VandeVusse, and Abigail Ocobock. 2012. "Family Change and the State of Family Sociology." *Current Sociology* 60(3): 283–301.

Farrell, Warren. 1974. *The Liberated Man: Beyond Masculinity; Freeing Men and Their Relationships with Women.* University of Berkeley: Berkeley Books.

Farver, JoAnn M,Yiyuan Xu, Bakhtawar R. Bhadha, Sonia Narang, and Eli Lieber. 2007. "Ethnic Identity, Acculturation, Parenting Beliefs, and Adolescent Adjustment: A Comparison of Asian Indian and European American Families." *Merrill-Palmer Quarterly* 53(2):184–215.

Fass, Paula S. 2003. Review of *Anxious Parents* (www.amazon.com).

Faulkner, R. A., M. Davey, and A. Davey. 2005. "Gender-Related Predictors of Change in Marital Satisfaction and Marital Conflict." *American Journal of Family Therapy* 33:61–83.

Fausto-Sterling, Anne. 2000. "The Five Sexes Revisited." *Sciences,* July/August, pp.19–23.

Federal Bureau of Investigations.2012. "Sexting: Risky Actions and Overreactions." Retrieved September 1, 2012 (http://www.fbi.gov/stats-services/publications/law-enforcement-bulletin/july-2010/sexting).

"Federal Court Dismisses Lawsuit from Lesbian Banned from Dying Partner's Bedside." 2009. *Southern Voice,* September 29. Retrieved November 21, 2009 (www.sovo.com).

Federal Interagency Forum on Child and Family Statistics.2012. *America's Children in Brief: Key National Indicators of Well-Being, 2012.* Washington, DC: U.S. Government Printing Office.

"Federal Marriage Benefits Denied to Same-Sex Couples." n. d. http://www.nolo.com.

Feigelman, Susan, Howard Dubowitz, Wendy Lane, LesliePrescott, Walter Meyer, Kathleen Tracy, and Jeongeun Kim. 2009. "Screening for Harsh Punishment in a Pediatric Primary Care Clinic." *Journal of Child Abuse & Neglect* 33(5):269–277.

Feigelman, W. 2000. "Adjustments of Transracially and Inracially Adopted Young Adults." *Child and Adolescent Social Work Journal* 17:165–183.

Fein, Esther B. 1997. "Failing to Discuss Dying Adds to Pain of Patient and Family." *The New York Times,* March 5, pp. A1–A14.

Feinberg, Mark E., Marni L. Kan, and E. Mavis Hetherington. 2007. "The Longitudinal Influence of Coparenting Conflict on Parental Negativity and Adolescent Maladjustment." *Journal of Marriage and Family* 69(3):687–702.

Felker, J. A., D. K. Fromme, G. L. Arnaut, and B. M. Stoll. 2002. "A Qualitative Analysis of Stepfamilies: The Stepparent." *Journal of Divorce & Remarriage,* 38(1-2):125–142.

Fergusson, David M., Joseph M. Boden, and L. John Horwood. 2007. "Abortion among Young Women and Subsequent Life Outcomes." *Perspectives on Sexual and Reproductive Health* 39(1):6–12.

Ferrari, J. R. and R. A. Emmons. 1994. "Procrastination as Revenge: Do People Report Using Delays as a Strategy for Vengeance?" *Personality and Individual Differences* 17(4):539–542.

Ferree, Myra Marx. 2010. "Filling the Glass: Gender Perspectives on Families." *Journal of Marriage and Family* 72(3): 420–439.

Ferrer-Chancy, Millie. 2009. "Stepping Stones for Stepfamilies: For Step-Grandparents." University of Florida Extension. Retrieved May 5, 2010 (www.edis.ifas.ufl.edu).

Ferro, Christine, Jill Cermele, and Ann Saltzman. 2008. "Current Perceptions of Marital Rape: Some Good and Not-So-Good News." *Journal of Interpersonal Violence* 23(6):764–779.

Festinger, Trudy B. 2005. "Adoption and Disruption." Pp. 452–468 in *Child Welfare for the 21st Century: A Handbook of Practices, Policies, and Programs,* edited by G.Mallonand P.Hess. New York: Columbia University Press.

Fetsch, Robert J., Raymond K. Yang, and Matthew J. Pettit. 2008. "The RETHINK Parenting and Anger Management Program." *Family Relations* 57(5):543–552.

Few, April L. and Karen H. Rosen. 2005. "Victims of Chronic Dating Violence: How Women's Vulnerabilities Link to Their Decisions to Stay." *Family Relations* 54(2):265–279.

Fidas, Deena and Samir Luther. 2010. *Corporate Equality Index 2010: A Report Card on Lesbian, Gay, Bisexual and Transgender Equality in Corporate America.* Washington, DC: Human Rights Campaign Foundation.

Fiebert, Martin S. 2012. "References Examining Assaults by Women on Their Spouses or Male Partners: An Annotated Bibliography." Retrieved March 14, 2013 (http://www.csulb.edu).

Fields, Jason. 2004. *America's Families and Living Arrangements*: 2003. Current Population Reports P20–553. Washington, DC: U.S. Census Bureau. November.

Fields, Julianna. 2010. *Families Living with Mental and Physical Challenges.* Broomall, PA: Mason Crest Publishers.

Filinson, R. 1986. "Relationship in Stepfamilies: An Examination of Alliances." *Journal of Comparative Family Studies,* 17:43–61.

Finch, J. and J. Mason. 1990. "Divorce, Remarriage, and Family Obligations." *Sociological Review,* 38:219–246.

Fincham, Frank D. and Steven R. H. Beach. 2010. "Marriage in the New Millennium: A Decade in Review." *Journal of Marriage and Family* 72(3):630–649.

Fincham, Frank D., Ming Cui, Mellissa Gordon, and Koji Uemo. 2013. "What Comes Before Why: Specifying the Phenomenon of Intimate Partner Violence." *Journal of Marriage and Family* 75(2):319–324.

Fincham, Frank D., Julie Hall, and Steven R. H. Beach. 2006. "Forgiveness in Marriage: Current Status and Future Directions." *Family Relations* 55(4):415–427.

Fincham, Frank D., Scott M. Stanley, and Steven R. H. Beach. 2007. "Transformative Processes in Marriage: An Analysis of Emerging Trends." *Journal of Marriage and Family* 69(2):275–292.

Finer, Lawrence B. 2007. "Trends in Premarital Sex in the United States, 1954–2003." *Public Health Reports* 122(January/February):73–78. Retrieved December 20, 2006 (www.publichealthreports.org).

Finer, Lawrence B., Lori Frohwirth, Lindsay A. Dauhiphinee, Sushella Singh, and Ann M. Moore. 2005. "Reasons U.S. Women Have Abortions: Quantitative and Qualitative Perspectives." *Perspectives on Sexual and Reproductive Health* 37(3):110–118.

Fingerman, Karen L., Yen-Pi Cheng, Eric D. Wesselmann, Steven Zarit, Frank Furstenberg, and Kira S. Birditt. 2012. "Helicopter Parents and landing Pad Kids: Intense Parental Support of Grown Children." *Journal of Marriage and Family* 74(4): 880–896.

Fingerman, Karen L. and Frank F. Furstenberg. 2012. "You Can Go Home Again." *The New York Times.* May 30. Retrieved August 28, 2012 (http://www.nytimes.com).

Fingerman, Karen, Laura Miller, Kira Birditt, and Steven Zarit. 2009. "Giving to the Good and the Needy: Parental Support of Grown Children." *Journal of Marriage and Family* 71(5):1220–1233.

Finkelhor, David, Richard Ormrod, Heather Turner, and Sherry Hamby. 2005. "The Victimization of Children and Youth: A Comprehensive National Survey." *Child Maltreatment* 10:5–25.

Finley, Gordon E. 2000. "Adoptive Families: Dramatic Changes across Generations." *Family Focus* (June): F10, F12. National Council on Family Relations.

Firmin, Michael and Annie Phillips. 2009. "A Qualitative Study of Families and Children Possessing Diagnoses of ADHD." *Journal of Family Issues* 30(9):1155–1174.

Fisher, William A., Raymond C. Rosen, Ian Eardley, Michael Sand, and Irwin Goldstein 2005. "Sexual Experience of Female Partners of Men with Erectile Dysfunction: The Female Experience of Men's Attitudes to Life Events and Sexuality (FEMALES) Study." *Journal of Sexual Medicine* 2(5): 675–684.

Fisher, Linda L. 2010. *Sex, Romance, and Relationships: AARP Survey of Midlife and Older Adults.* Washington, DC: American Association of Retired Persons. Retrieved May 10, 2010 (www.aarp.org).

Fisher, Terri D., Clive M. Davis, William L. Yarber, and Sandra L. Davis. 2010. *Handbook of Sexuality-Related Measures.* Thousand Oaks, CA: Sage.

Fishman, B. 1983. "The Economic Behavior of Stepfamilies." *Family Relations,* 32:359–366.

Fishman, Margie. 2013. "Divorce Parties Mark Milestones in Uncoupling. Retrieved April 22, 2013 (http://www.usatoday.com/story/news/nation/2013/01/31/divorce-party/1881623/).

Fitzpatrick, Jacki, Elizabeth Sharp, and Alan Reifman. 2009. "Midlife Singles' Willingness to Date Partners with Heterogeneous Characteristics." *Family Relations* 58(1):121–133.

Fitzpatrick, Mary Anne. 1995. *Explaining Family Interactions.* Thousand Oaks, CA: Sage Publications.

Fivush, Robyn, Kelly Marin, Kelly McWilliams, and Jennifer Bohanek. 2009. "Family Reminiscing Style: Parent Gender and Emotional Focus in Relation to Child Well-Being." *Journal of Cognition and Development* 10(3):210–235.

Fleeson, W. and E. Noftle. (2008). "The End of the Person-Situation Debate: An Emerging Synthesis in the Answer to the Consistency Question." *Social and Personality Psychology Compass* 2:1667–1684.

Fleming, C. J. Eubanks and James V. Cordorva. 2012. "Predicting Relationship Help Seeking Prior to a Marriage Checkup." *Family Relations* 61:90–100.

Fletcher, Garth. 2002. *The New Science of Intimate Relationships.* Malden, MA: Blackwell.

Flora, Carlin. 2007. "Can Grown-Up siblings Learn to Get Along?" *Psychology Today.* March /April:48–49.

Fomby, Paula and Andrew J. Cherlin. 2007. "Family Instability and Child Well-Being." *American Sociological Review* 72(2):181–204.

Fomby, Paula and Angela Estacion. 2011. "Cohabitation and Children's Externalizing Behaviors in Low-Income Latino Families." *Journal of Marriage and Family* 73(1):46–66.

Fontaine, Romeo, Agnes Gramain, and Jerome Wittwer. 2009. "Providing Care for an Elderly Parent: Interactions among Siblings." *Health Economics* 18:1011–1029.

Foran, Heather M. and K. Daniel O'Leary. 2008. "Problem Drinking, Jealousy, and Anger Control: Variables Predicting Physical Aggression Against a Partner." *Journal of Family Violence* 23:141–148.

Ford, Melissa. 2009. *Navigating the Land of If: Understanding Infertility and Exploring Your Options.* Berkeley, CA: Seal Press.

Formoso, Diana, Nancy A. Gonzales, Manuel Barrera, Jr., and Larry E. Dumka. 2007. "Interparental Relations, Maternal Employment, and Fathering in Mexican American Families." *Journal of Marriage and Family* 69(1):26–39.

Foroohar, Rana. 2012. "Companies Are the New Countries." *Time.* February 13: 21.

Fortuny, Karina, Randy Capps, Margaret Simms, and Ajay Chaudry. 2009. *Children of Immigrants: National and State Characteristics.* Washington, DC: The Urban Institute.

Fosco, Gregory M. and John H. Grych. 2012. "Capturing the Family Context of Emotion Regulation: A Family Systems Model Comparison Approach." *Journal of Family Issues* 34(4):557–578.

Foster, Brooke Lea. 2006. "The Way They Were." *AARP Magazine,* September (www.aarp.org).

Foster, E. Michael, Damon Jones, and Saul D. Hoffman. 1998. "The Economic Impact of Nonmarital Childbearing: How Are Older, Single Mothers Faring?" *Journal of Marriage and Family* 60(1):163–174.

Foster, Holly and Jeanne Brooks-Gunn. 2009. "Toward a Stress Process Model of Children's Exposure to Physical Family and Community Violence." *Clinical Child and Family Psychology Review* 12:71–94.

Fountain, Kim, Maryse Mitchell-Brody, Stephanie A. Jones, and Kaitlin Nichols. 2009. *Lesbian, Gay, Bisexual, Transgender and Queer Domestic Violence in the United States in 2008.* New York: The National Coalition of Anti-Violence Programs. Available at www.avp.org/documents/2008NCAVPLGBTQDVReportFINAL.pdf.

Fox, Bonnie. 2009. *When Couples Become Parents: The Creation of Gender in the Transition to*

Parenthood. Buffalo, NY: University of Toronto Press.

Fox, Greer Litton, Michael L. Benson, Alfred A. DeMaris, and Judy Van Wyk. 2002. "Economic Distress and Intimate Violence: Testing Family Stress and Resources Theory." *Journal of Marriage and Family* 64:793–807.

Fox, Maggie. 2010. "Study Shows Consistent Benefit of Early Daycare." *Reuters,* May 14. Chicago: Thompson Reuters.

Fracher, Jeffrey and Michael S. Kimmel. 1992. "Hard Issues and Soft Spots: Counseling Men about Sexuality." Pp. 438–450 in *Men's Lives,* 2nd ed., edited by Michael S. Kimmel and Michael A. Messner. New York: Macmillan.

Frank, Robert H. 2011. "Gauging the Pain of the Middle Class." *The New York Times.* April 2. Retrieved September 13, 2011 (http://www.nytimes.com).

Franke-Ruta, Garance. 2013. "Obama's Minimum-Wage Gamble." *The Atlantic.* Retrieved February 20, 2013 (http://www.theatlantic.com).

Frech, Adrianne and Rachel Tolbert Kimbro. 2011. "Maternal Mental Health, Neighborhood Characteristics, and Time Investments in Children." *Journal of Marriage and Family* 73(June): 605–620.

Freedom to Marry 2012. 2012. http://www.freedomtomarry.org.

Freeman, David W. 2011. "Same-Sex Affairs: Men More Forgiving than Women." *CBS News.* Retrieved November 20, 2012 (http://www.cbsnews.com/8301-504763_162-20030042-10391704.html).

Freeman, Marsha B. 2008. "Love Means Always Having to Say You're Sorry: Applying the Realities of Therapeutic Jurisprudence to Family Law." *UCLA Women's Law Journal* 17:215–241.

Freeman, Melissa and Sandra Mathison. 2009. *Researching Children's Experiences.* New York: Guilford Press.

French, J. R. P. and Bertram Raven. 1959. "The Basis of Power." In *Studies in Social Power,* edited by D.Cartwright. Ann Arbor: University of Michigan Press.

Friedan, Betty. 1963. *The Feminine Mystique.* New York: Dell.

Friedman, Jaclyn and Jessica Valenti. 2008. *Yes Means Yes! Visions of Female Sexual Power and a World Without Rape.* Berkeley, CA: Seal Press.

Friedrich, William N., Jennifer Fisher, Daniel Broughton, Margaret Houston, and Constance R. Shafran. 1998. "Normative Sexual Behavior in Children: A Contemporary Sample." *Pediatrics* 104(April):E9 (www.pediatrics.org).

Frisby, Brandi and Melanie Booth-Butterfield. 2012. "The 'How' and 'Why' of Flirtatious Communication Between Marital Partners." *Communication Quarterly* 60(4):465–480.

Frisco, Michelle L., Chandra Muller, and Kenneth Frank. 2007. "Parents' Union Dissolution and Adolescents' School Performance: Comparing Methodological Approaches." *Journal of Marriage and Family* 69(3):721–741.

Frith, Lucy, Eric Blyth, Marilyn Paul, and Roni Berger. 2011. "Conditional Embryo Relinquishment: Choosing to Relinquish Embryos for Family-Building though a Christian Embryo 'Adoption' Programme." *Human Reproduction* 26(12):3327–3338.

Fritsch, Jane. 2001. "A Rise in Single Dads." *The New York Times,* May 20.

Fromm, Erich. 1956. *The Art of Loving.* New York: Harper & Row.

Frontline. 2012. "The Allure of Adult Content Users." American Porn. Public Broadcasting System. Retrieved July 24, 2012 (http://www.pbs.org/wgbh/pages/frontline/shows/porn/business/haveallure.html).

Frosch, Dan. "Dispute on Transgender Rights Unfolds at a Colorado School." Retrieved June 25, 2013 (http://www.nytimes.com/2013/03/18/us/in-colorado-a-legal-dispute-over-transgender-rights.html?ref=danfrosch&gwh=3886D9DE6212CE5856D99C7BF57F03BF).

Froyen, Laura, Lori Skibbe, Ryan Bowles, Adrian Blow, and Hope Gerde. 2013. "Marital Satisfaction, Family Emotional Expressiveness, Home Learning Environments, and Children's Emergent Literacy." *Journal of Marriage and Family* 75(1):42–55.

Frum, David. 2012. "Get Off the Road." *Newsweek.* July 2 & 9:46–49.

Fruhauf, Christine A., Shannon E. Jarrott, and Katherine R. Allen. 2006. "Grandchildren's Perceptions of Caring for Grandparents." *Journal of Family Issues* 27(7):887–911.

Fry, Richard, 2013. "Young Adults after the Recession: Fewer Homes, Fewer Cars, Less Debt." *Pew Research Social and Demographic Trends.* February 21. Retrieved March 20, 2013 (http://www.pewsocialtrends.org).

Fry, Richard and D'Vera Cohn. 2010. *Women, Men and the New Economics of Marriage.* Washington, DC: Pew Research Center.

———. 2011. "Living Together: The Economics of Cohabitation." *Pew Research Center Publications.* June 27. Retrieved September 24, 2011 (http://pewresearch.org).

Fryer, Roland G. 2007. "Guess who's Been Coming to Dinner: Trends in Interracial Marriage over the Twentieth Century." *Journal of Economic Perspectives* 21(2):71–90.

Fu, Xuanning and Tim B. Heaton. 2000. "Status Exchange in Intermarriage among Hawaiians, Japanese, Filipinos and Caucasians in Hawaii: 1983–1994." *Journal of Comparative Family Studies* 31(1):45–64.

———. 2008. "Racial and Educational Homogamy: 1980 to 2000." *Sociological Perspectives* 51(4):735–758.

Fujiura, Glenn T. 2010. "Aging Families and the Demographics of Family Financial Support of Adults with Disabilities." *Journal of Disability Policy Studies* 20(4):241–250.

Furdyna, Holly E., M. Belinda Tucker, and Angela D. James. 2008. "Relative Spousal Earnings and Marital Happiness among African American and White Women." *Journal of Marriage and Family* 70(2):332–344.

Furman, Wyndol, and Laura Shaffer. 2011. "Romantic Partners, Friends, Friends with Benefits, and Casual Acquaintances as Sexual Partners." *Journal of Sex Research* 48(6):554–564.

Furstenberg, Frank F., Jr. 2000. "The Sociology of Adolescence and Youth in the 1990s: A Critical Commentary." *Journal of Marriage and Family* 62(4):896–910.

———. 2003. "The Future of Marriage." Pp. 171–177 in *Family in Transition,* 12th ed., edited by Arlene S. Skolnick and Jerome H. Skolnick. Boston: Allyn & Bacon.

———. 2005. "Banking on Families: How Families Generate and Distribute Social Capital." *Journal of Marriage and Family* 67(4):809–821.

———. 2006. "Diverging Development: The Not-So-Invisible Hand of Social Class in the United States." Network on Transitions to Adulthood Research Network Working Paper. Presented at the biennial meeting of the Society for Research on Adolescence, March 23–26, 2006, San Francisco, CA.

———. 2008. "The Changing Landscape of Early Adulthood in the U.S." *Family Focus,* March:F2-F3, F18.

Furstenberg, Frank F., Jr., J. Brooks-Gunn, and S. Philip Morgan. 1987. *Adolescent Mothers in Later Life.* New York: Cambridge University Press.

Furstenberg, Frank F., Jr., Sheela Kennedy, Vonnie C. McLoyd, Rubén G. Rumbaut, and Richard A. Settersten, Jr. 2004. "Growing Up Is Harder to Do." *Contexts* 3(3):33–41.

Furstenberg, Frank F., Jr. and Kathleen E. Kiernan. 2001. "Delayed Parental Divorce: How Much Do Children Benefit?" *Journal of Marriage and Family* 63(2):446–457.

Furstenberg, Frank F., Jr, Christine Winquist Nord, James L,Peterson, and Nicholas Zill. 1983. "The Life Course of Children of Divorce." *American Sociological Review*48(5):656–668.

Furukawa, Ryoko and Martha Driessnack. 2013. "Video-Mediated Communication to Support Distant Family Connectedness." *Clinical Nursing Research* 22(1):82–94.

Gable, Shelly L., Harry T. Reis, Emily A. Impett, and Evan R. Asher. 2004. "Interpersonal Relations and Group Processes—What Do You Do When Things Go Right? The Intrapersonal and Interpersonal Benefits of Sharing Positive Events." *Journal of Personality and Social Psychology* 87(2):228–245.

Gaertner, Bridget M., Tracy I. Spinrad, Nancy Eiserberg, and Karissa A. Greving. 2007. "Parental Childrearing Attitudes as Correlates of Father Involvement During Infancy." *Journal of Marriage and Family* 69(4):962–976.

Gager, C. T., T. M. Cooney, and K. T. Call. 1999. "The Effects of Family Characteristics and Time Use on Teenagers' Household Labor." *Journal of Marriage and the Family,* 61:982–994.

Gager, Constance T. and Laura Sanchez. 2003. "Two as One? Couples' Perceptions of *Time* Spent Together, Marital Quality, and the Risk of Divorce." *Journal of Family Issues* 24(1):21–50.

Gager, Constance T. and Scott T. Yabiku. 2010. "Who Has the Time? The Relationship Between

Household Labor Time and Sexual Frequency." *Journal of Family Issues* 31(2):135–163.

Gagnon, John H. and William Simon. 2005. *Sexual Conduct: The Social Sources of Human Sexuality (Social Problems and Social Issues).* 2nd ed. New Brunswick, NJ: Transaction Books.

Galambos, Nancy I. and Harvey J. Krahn. 2008. "Depression and Anger Trajectories During the Transition to Adulthood." *Journal of Marriage and Family* 70(1):15–27.

Galinsky, Ellen, Kerstin Aumann, and James T. Bond. 2009. "Times are Changing: Gender and Generation at Work and at Home." *2008 National Study of the Changing Workforce.* New York: Families and Work Institute.

Gallagher, Charles A. 2006. "Interracial Dating and Marriage: Fact, Fantasy and the Problem of Survey Data." Pp. 141–153 in *African Americans and Whites: Changing Relationships on College Campuses,* edited by Robert M. MooreIII. New York: University Press of America.

Gallagher, Sally K. 2004. "Where Are the Antifeminist Evangelicals? Evangelical Identity, Subcultural Location, and Attitudes toward Feminism." *Gender and Society* 18(4):451–472.

Gallup Poll. 2012a. "Immigration." Retrieved August 29, 2012 (http://www.gallup.com).

———. 2012b. "Marriage." Retrieved September 10, 2012 (http://www.gallup.com).

———. 2012c. "Abortion." Retrieved January 7, 2013 (http://www.gallup.com).

Gamache, Susan J. 1997. "Confronting Nuclear Family Bias in Stepfamily Research." *Marriage and Family Review* 26(1–2):41–50.

Games-Evans, Tina. 2009. "Finding Your Personal Identity as a Mom." Babyzone, June 1. www.babyzone.com.

Ganong, Lawrence H. and Marilyn Coleman. 1993. "An Exploratory Study of Stepsibling Subsystems." *Journal of Divorce & Remarriage, 19,* 125–142.

———. 1997. "How Society Views Stepfamilies." Pp. 85–106 in *Stepfamilies: History, Research, and Policy,* edited by Irene Levinand Marvin B. Sussman. New York: Haworth.

———. 2000. "Remarried Families." Pp. 155–168 in *Close Relationships: A Sourcebook,* edited by Clyde Hendrickand Susan S. Hendrick. Thousand Oaks, CA: Sage Publications.

———.2004. *Stepfamily Relationships: Development, Dynamics, and Interventions.* New York: Kluwer Academic/Plenum.

———. 2006a. "Obligations to Stepparents Acquired in Later Life: Relationship Quality and Acuity of Needs." *Journal of Gerontology:Social Sciences,* 61B:S80–S88.

———. 2006b. "Patterns of Exchange and Intergenerational Responsibilities after Divorce and Remarriage." *Journal of Aging Studies,* 20:265–278.

Ganong, Lawrence H., Marilyn Coleman, Richard Feistman, Tyler Jamison, and Melinda Stafford Markham. 2012. "Communication Technology and Postdivorce Coparenting." *Family Relations* 61:397–409.

Ganong, Lawrence H., Marilyn Coleman, and Tyler Jamison. 2011. "Patterns of Stepchild-Stepparent Relationship Development." *Journal of Marriage and Family* 73:396–413.

Ganong, Lawrence H., Marilyn Coleman, and Tanja Rothrauff. 2009. "Patterns of Assistance between Adult Children and Their Older Parents: Resources, Responsibilities, and Remarriage." *Journal of Social and Personal Relationships* 26(2–3):161–178.

Gans, Daphne, Merril Silverstein, and Anela Lowenstein. 2009. "Do Religious Children Care More and Provide More Care for Older Parents? A Study of Filial Norms and Behaviors across Five Nations." *Journal of Comparative Family Studies* 40(2):187–201.

Gans, Herbert J. 1982 [1962]. *The Urban Villagers: Group and Class in the Life of Italian-Americans,* updated and expanded edition. New York: The Free Press.

Garasky, Steven, Craig Gundersen, Susan D. Stewart, and Brenda J. Lohman. 2010. "Toward a Fuller Understanding of Nonresident Father Involvement: A Joint Examination of Child Support, In-Kind Support, and Visitation." *Population Research and Policy Review* 29:363–393,

Garasky, Steven, Elizabeth Peters, Laura Argys, Steven Cook, Lenna Nepomnyaschy, and Elaine Sorensen. 2007. "Measuring Support to Children by Nonresident Fathers." Pp. 399–428 in *Handbook of Measurement Issues in Family Research,* edited by Sandra L. Hofferth and Lynne M. Casper. Mahwah, NJ: Lawrence Erlbaum Associates.

Garcia, Michelle. 2012. "HRC: Romney's Adoption Remark Trivializes Gay Inequalities." *Advocate.* May 15. Retrieved November 24, 2012 (http://www.advocate.com).

Gardner, Jonathan and Andrew Oswald. 2006. "Do Divorcing Couples Become Happier by Breaking Up?" *Journal of the Royal Statistical Society,* Series A 169(2):319–336.

Gardner, Margo, Christopher Browning, and Jeanne Brooks-Gunn. 2012. "Can Organized Youth Activities Protect Against Internalizing Problems among Adolescents Living in Violent Homes?" *Journal of Research on Adolescence* 22(4):662–677.

Garey, Anita I.and Karen V. Hansen(eds.) 2011. *At the Heart of Work and Family.* New Brunswick, NJ: Rutgers University Press.

Garfield, Robert. 2010. "Male Emotional Intimacy: How Therapeutic Men's Groups Can Enhance Couples Therapy." *Family Process* 49(1):109–122.

Garson, David G. n. d. "Economic Opportunity Act of 1964." Retrieved October 9, 2006 (wps.prenhall.com).

Gartrell, Nanette, Henny Bos, Heidi Peyser, Amalia Deck, and Carla Rodas. 2011. "Family Characteristics, Custody Arrangements, and Adolescent Psychological Well-Being After Lesbian Mothers Break Up." *Family Relations* 60(5): 572–585.

Gassman-Pines, Anna. 2011. "Low-Income Mothers' Nighttime and Weekend Work: Daily Associations with Child Behavior, Mother-Child Interactions, and Mood." *Family Relations* 60(1): 15–29.

Gates, Gary J. 2009. *Same-Sex Spouses and Unmarried Partners in the American Community Survey, 2008.* Los Angeles: The Williams Institute. Retrieved November 20, 2009 (www.law.ucla.edu/williamsinstitute).

———. 2011. "Family Formation and Raising Children among Same-Sex Couples." *Family Focus* FF51(Winter): F2–F4.

Gates, Gary J. and Frank Newport. 2012. "Special Report: 3.4% of U.S. Adults Identify as LGBT." Retrieved June 25, 2013 (http://www.odec.umd.edu/CD/LGBT/Special%20Report%203.4%25%20of%20U.S.%20Adults%20Identify%20as%20LGBT.pdf).

Gauchat, Gordon, Maura Kelly, and Michael Wallace. 2012. "Occupational Gender Segregation, Globalization, and Gender Earnings Inequality in U.S. Metropolitan Areas." *Gender & Society* 26(5):718–747.

Gaughan, Monica. 2002. "The Substitution Hypothesis: The Impact of Premarital Liaisons and Human Capital on Marital Timing." *Journal of Marriage and Family* 64(2):407–419.

Gault-Sherman, Martha. 2012. "What Will the Neighbors Think? The Effect of Moral Communities on Cohabitation." *Review of Religious Research* 54(1):45–67.

Gaunt, Ruth. 2006. "Couple Similarity and Marital Satisfaction: Are Similar Spouses Happier?" *Journal of Personality* 74(5):1401–1420.

———. 2013. "Breadwinning Moms, Caregiving Dads: Double Standard in Social Judgments of Gender Norm Violators." *Journal of Family Issues* 34(1):3–24.

Gavin, Lorrie, Andrea P. MacKay, Kathryn Brown, Sara Harrier, Stephanie J. Ventura, Laura Kann, Maria Rangel, Stuart Berman, Patricia Dittus, Nicole Liddon, Lauri Markowitz, Maya Sternberg, Hillard Weinstock, Corinne David-Ferdon, George Ryan. 2009. "Sexual and Reproductive Health of Persons Aged 10–24 Years—United States, 2002–2007." *Morbidity and Mortality Weekly Report Surveillance Summaries* 58(SS06):1–58. Atlanta: Centers for Disease Control and Prevention.

"Gay and Lesbian Adolescents." 2006. *Facts for Families.* Academy of Child and Adolescent Psychiatry # 63. Retrieved July 1, 2010 (www.aacap.org).

"Gay and Lesbian Rights." 2009. *Gallup Poll,* May 7–10. www.gallup.com.

Gayles, Jochebed G., J. Douglas Coatsworth, Hilda M. Pantin, and Jose Szapocznik. 2009. "Parenting and Neighborhood Predictors of Youth Problem Behaviors within Hispanic Families: The Moderating Role of Family Structure." *Hispanic Journal of Behavioral Sciences* 31(3):277–296.

"Gays Want the Right, but Not Necessarily the Marriage." 2004. *The Christian Science Monitor,* February 12.

Geewax, Marilyn. 2012a. "Paying for College: More Tough Decisions." National Public Radio. May 15. Retrieved May 15, 2012 (http://www.npr.org).

———. 2012b. "Preparing for a Future That Includes Aging Parents." National Public Radio. April 24. Retrieved May 15, 2012 (http://www.npr.org).

———. 2012c. "Sandwich Generation Must Make Tough Choices." National Public Radio. April 24. Retrieved May 8, 2012 (http://www.npr.org).

Geist, Claudia and Philip N. Cohen. 2011. "Headed Toward Equality? Housework Change in Comparative Perspective." *Journal of Marriage and Family* 73(4):832–844

Gelatt, Vicky A., Francesca Adler-Baeder, and John R. Seeley. 2010. "An Interactive Web-Based Program for Stepfamilies: Development and Evaluation of Efficacy." *Family Relations* 59:572–586.

Gelles, Richard J. 1974. *The Violent Home: A Study of Physical Aggression between Husbands and Wives.* Beverly Hills, CA: Sage Publications.

———. 1996. *The Book of David: How Preserving Families Can Cost Children's Lives.* New York: Basic Books.

———. 2005. "Protecting Children Is More Important than Preserving Families." Pp. 329–340 in *Current Controversies on Family Violence,* 2nd ed., edited by Donileen R. Loseke, Richard J. Gelles, and Mary M. Cavanaugh. Thousand Oaks, CA: Sage Publications.

Gelles, Richard J. and Mary M. Cavanaugh. 2005. "Violence, Abuse, and Neglect in Families and Intimate Relationships." Pp. 129–154 in *Families and Change: Coping with Stressful Events and Transitions,* 3rd ed., edited by Patrick C. McKenryand Sharon J. Price. Thousand Oaks, CA: Sage Publications.

Gelles, Richard J. and Murray A. Straus. 1988. *Intimate Violence: The Definitive Study of the Causes and Consequences of Abuse in the American Family.* New York: Simon and Schuster.

Gemelli, Marcella. 2008. "Understanding the Complexity of Attitudes of Low-Income Single Mothers Toward Work and Family in the Age of Welfare Reform." *Gender Issues* 25:101–113.

General Social Survey. 2012. Retrieved November 23, 2012 (http://www3.norc.org/gss+website/).

"Genetically, Race Doesn't Exist." 2003. *Washington University Magazine,* Fall, p. 4.

Gerace, Alyssa. 2012. "How Will Senior Housing Development Adapt to Multigenerational Trends?" National Public Radio. Senior Living News Wire. Retrieved May 15, 2012 (http://www.seniorlivingnewswire.com).

Gerbrandt, Roxanne. 2007. *Exposing the Unmentionable Class Barriers in Graduate Education.* Unpublished Doctoral Dissertation, University of Oregon.

Gerson, Kathleen. 2010. *The Unfinished Revolution: How a New Generation is Reshaping Family, Work, and Gender in America.* New York: Oxford University Press.

Gewertz, Catherine. 2009. "Report Probes Educational Challenges Facing Latinas." *Education Week* 29(2):12.

Gibson, Jennifer E. 2012. "Interviews and Focus Groups with Children: Methods That Match Children's Developing Competencies." *Journal of Family Theory & Review* 4(2) (June):148–159.

Gibson-Davis, Christina M. 2008. "Family Structure Effects on Maternal and Paternal Parenting in Low-Income Families." *Journal of Marriage and Family* 70(2):452–465.

———. 2009. "Money, Marriage, and Children: Testing the Financial Expectations and Family Formation Theory." *Journal of Marriage and Family* 71(1):146–160.

———. 2011. "Mothers But Not Wives: The Increasing Lag Between Nonmarital Births and Marriage." *Journal of Marriage and Family* 73(1) (February):264–278.

Giddens, Anthony. 2007. "The Global Revolution in Family and Personal Life." Pp. 26–31 in *Family in Transition,* edited by Arlene S. Skolnickand Jerome H. Skolnick. Boston: Allyn & Bacon.

Giele, Janet Z. 2007. "Decline of the Family: Conservative, Liberal, and Feminist Views." Pp. 76–91 in *Family in Transition,* edited by Arlene S. Skolnickand Jerome H. Skolnick. Boston: Allyn & Bacon.

Gilbert, Gizelle L., M. Diane Clark, and Melissa L. Anderson. 2012. "Do Deaf Individuals' Dating Scripts Follow the Traditional Dating Script?" *Sexuality & Culture* 16:90–99.

Gilbert, Neil. 2008. *A Mother's Work: How Feminism, the Market, and Policy Shape Family Life.* New Haven, CT: Yale University Press.

Gilbert, Susan. 1997. "Two Spanking Studies Indicate Parents Should Be Cautious." *The New York Times,* August 20.

Gilkey, So'Nia L., JoAnne Carey, and Shari L. Wade. 2009. "Families in Crisis: Considerations for the Use of Web-Based Treatment Models in Family Therapy." *Families in Society: Journal of Contemporary Human Services* 90(1):37–46.

Gillespie, Dair. 1971. "Who Has the Power? The Marital Struggle." *Journal of Marriage and Family* 33:445–458.

Gillmore, Mary Rogers, Jungeun Lee, Diane M. Morrison, and Taryn Lindhorst. 2008. "Marriage Following Adolescent Parenthood: Relationship to Adult Well-Being." *Journal of Marriage and Family* 70(5):1136–1144.

Gilmartin, B. 1977. "Swinging: Who Gets Involved and How." Pp. 161–185 in *Marriage and Alternatives,* edited by R. W. Libbyand R. N. Whitehurst. Glenview, IL: Scott, Foresman.

Gilrane-McGarry, Ursula and Tom O'Grady. 2012. "Forgotten Grievers: An Exploration of the Grief Experiences of Bereaved Grandparents (Part 2)." *International Journal of Palliative Nursing* 18(4):179–187.

Ginott, Haim G., Alice Ginott, and Wallace Goddard. 2003. *Between Parent and Child: The Bestselling Classic That Revolutionized Parent-Child Communication.* New York: Three Rivers Press.

Giordano, Peggy C., Monica A. Longmore, and Wendy D. Manning. 2006. "Gender and the Meanings of Adolescent Romantic Relationships: A Focus on Boys." *American Sociological Review* 71(2):260–287.

Gitner, Jess. 2012. "'All about Family': Listeners' Stories on Living in Multigenerational Households." National Public Radio. May 15. Retrieved May 15, 2012 (http://www.npr.org).

Glass, Jennifer and Leda E. Nath. 2006. "Religious Conservatism and Women's Market Behavior Following Marriage and Childbirth." *Journal of Marriage and Family* 68(3):611–629.

Glass, Shirley. 1998. "Shattered Vows." *Psychology Today* 31(4):34–52.

Glassner, Barry. 1999. *The Culture of Fear: Why Parents Are Afraid of the Wrong Things.* New York: Basic Books.

Glauber, Rebecca and Kristi L. Gozjolko. 2011. "Do Traditional Fathers Always Work More? Gender, Ideology, Race, and Parenthood." *Journal of Marriage and Family* 73(5): 1133–1148.

Glenn, Norval. 1998. "The Course of Marital Success and Failure in Five American 10-Year Marriage Cohorts." *Journal of Marriage and Family* 60(3):569–576.

Glenn, Norval and Elizabeth Marquardt. 2001. *Hooking Up, Hanging Out, and Hoping for Mr. Right: College Women on Dating and Mating Today.* New York: Institute for American Values.

Glenn, Norval D., Jeremy Uecker, and Robert Love. 2009. "Later First Marriage and Marital Success." Paper presented at the 2009 Annual Meetings of the American Sociological Association, San Francisco, August 10.

Glick, Jennifer E., Frank D. Bean, and Jennifer Van Hook. 1997. "Immigration and Changing Patterns of Extended Family Household Structure in the United States: 1970–1990." *Journal of Marriage and Family* 59:177–191.

Glick, Jennifer E. and Jennifer Van Hook. 2002. "Parents' Coresidence with Adult Children: Can Immigration Explain Racial and Ethnic Variation?" *Journal of Marriage and Family* 64(1):240–253.

Glick, Paul C. and Sung-Ling Lin. 1986. "More Young Adults Are Living with Their Parents: Who Are They?" *Journal of Marriage and Family* 48:107–112.

Glick, Paul C. and Arthur J. Norton. 1979. "Marrying, Divorcing, and Living Together in the U.S. Today." *Population Bulletin* 32(5). Washington, DC: Population Reference Bureau.

"Glossary of Senior Housing Terms." 2010. Retrieved May 17, 2010 (http://senioroutlook.com/glossary.asp).

Goble, Priscilla, Carol L. Martin, Laura D. Hanish, and Richard A. Fabes. 2012. "Children's Gender-Typed Activity Choices Across Preschool Social Contexts." *Sex Roles* 67(7–8):435–451.

Godley, Amanda. 2006. "Gendered Borderwork in a High School English Class." *English Teaching: Practice and Critique* 5(3) (Dec):4–29.

Goeke-Morey, Marcie, E. Mark Cummings, and Lauren M. Papp. 2007. "Children and Marital Conflict Resolution: Implications for Emotional Security and Adjustment." *Journal of Family Psychology* 21(4):744–753.

Goffman, Erving. 1959. *The Presentation of Self in Everyday Life.* Garden City, NY: Doubleday.

———. 1963. *Stigma.* Englewood Cliffs, N.J.: Prentice-Hall.

———. 1979. *Gender Advertisements* (1st Harper colophon ed.). New York: Harper & Row.

Golash-Boza, Tanya. 2012. *Due Process Denied: Detentions and Deportations in the United States.* New York: Routledge, Taylor and Francis.

Gold, Joshua M. 2009. "Negotiating the Financial Concerns of Stepfamilies: Directions for Family Counselors." *The Family Journal: Counseling and Therapy for Couples and Families* 17(2):185–188.

———. 2012. "Typologies of Cohabitation: Implications for Clinical Practice and Research." *The Family Journal: Counseling and Therapy for Couples and Families* 20(3): 315–321.

Gold, Steven J. 1993. "Migration and Family Adjustment: Continuity and Change among Vietnamese in the United States." Pp. 300–314 in *Family Ethnicity: Strength in Diversity,* edited by Harriette Pipes McAdoo. Newbury Park, CA: Sage Publications.

Gold, Steven J. and Mehdi Bozorgmehr. 2007. "Middle East and North Africa." Pp. 518–533 in *The New Americans: A Guide to Immigration Since 1965,* edited by Mary C. Waters, Reed Ueda, and Helen B. Marrow. Cambridge, MA: Harvard University Press.

Goldberg, Abbie E. 2007. "Talking about Family." *Journal of Family Issues* 28(1):100–131.

———. 2010. *Lesbian and Gay Parents and Their Children: Research on the Family Life Cycle.* Washington DC: American Psychological Association.

Goldberg, Abbie E., Jordan B. Downing, and April M. Moyer. 2012. "Why Parenthood, and Why Now? Gay Men's Motivations for Pursuing Parenthood." *Family Relations* 61(1): 157–174.

Goldberg, Abbie E., Deborah A. Kasby, and JuliAnna Z. Smith. 2012. "Gender-Typed Play Behavior in Early Childhood: Adopted Children with Lesbian, Gay, and Heterosexual Parents." *Sex Roles* 67: 503–515.

Goldberg, Abbie E., Lori A. Kinkler, Hannah B. Richardson, and Jordan B. Downing. 2011. "Lesbian, Gay, and Heterosexual Couples in Open Adoption Arrangements: A Qualitative Study." *Journal of Marriage and Family* 73(2): 502–518.

Goldberg, Abbie E. and Katherine A. Kuvalanka. 2012. "Marriage (In)equality: The Perspectives of Adolescents and Emerging Adults with Lesbian, Gay, and Bisexual Parents." *Journal of Marriage and Family* 74(1): 34–52.

Goldberg, Abbie E. and Aline Sayer. 2006. "Lesbian Couples' Relationship Quality across the Transition to Parenthood." *Journal of Marriage and Family* 68(1):87–100.

Goldberg, Abbie E. and JuliAnna Z. Smith. 2008. "Social Support and Psychological Well-Being in Lesbian and Heterosexual Preadoptive Couples." *Family Relations* 57(July):281–294.

Goldberg, Carey. 1998. "After Girls Get the Attention, Focus Shifts to Boys' Woes." *The New York Times,* April 23.

Goldin, Claudia. 2006. "Working It Out." *The New York Times,* March 15.

Goldscheider, Francis, Sandra Hofferth, Carrie Spearin, and Sally Curtin. 2009. "Fatherhood Across Two Generations: Factors Affecting Early Family Roles." *Journal of Family Issues* 30(5):586–604.

Goldscheider, Frances and Gayle Kaufman. (2006). "Willingness to Stepparent: Attitudes toward Partners Who Already Have Children." *Journal of Family Issues* 27(10):1415–1436.

Goldscheider, Frances, Gayle Kaufman, and Sharon Sassler. 2009. "Navigating the 'New' Marriage Market." *Journal of Family Issues* 30(6):719–737.

Goldscheider, Frances and Sharon Sassler. 2006. "Creating Stepfamilies: Integrating Children into the Study of Union Formation." *Journal of Marriage and Family* 68(2):275–291.

Goldsmith, Daena J., Jennifer J. Bute, and Kristin A. Lindholm. 2012. "Patient and Partner Strategies for Talking about Lifestyle Change Following a Cardiac Event." *Journal of Applied Communication Research* 40(1):65–86.

Goldstein, Arnold P., Harold Keller, and Diane Erne. 1985. *Changing the Abusive Parent.* Champaign, IL: Research Press.

Goleman, Daniel. 1985. "Patterns of Love Charted in Studies." *The New York Times,* September 10.

———. 1992. "Family Rituals May Promote Better Emotional Adjustment." *The New York Times,* March 11.

Gomes, Peter J. 2004. "For Massachusetts, a Chance and a Choice." *Boston Globe,* February 8. Retrieved October 6, 2006 (www.boston.com /news/globe).

Gonzales, Jaymes. 2009. "Prefamily Counseling: Working with Blended Families." *Journal of Divorce & Remarriage* 50(2):148–157.

Gonzalez, Cindy. 2006a. "Latino Leader Urges End to Concept of 'Minorities.'" *Omaha World-Herald,* November 17.

———. 2006b. "Short Supply of Visas Adds to Illegal Immigration." *Omaha World-Herald,* May 16.

Gonzalez, Cindy and Michael O'Connor. 2002. "Dialogue Key in Blending Cultures." *Omaha World-Herald,* May 4.

Good, Maria and Teena Willoughby. 2006. "The Role of Spirituality Versus Religiosity in Adolescent Psychosocial Adjustment." *Journal of Youth and Adolescence* 35:41–55.

———. 2012. "Fewer Children Are Found Exposed to Violent Crime." *The New York Times.* September 20.

Goode, William J. 1971. "Force and Violence in the Family." *Journal of Marriage and Family* 33:624–636.

———. 1982. "Why Men Resist." Pp. 131–150 in *Rethinking the Family: Some Feminist Questions,* edited by Barrie Thorneand Marilyn Yalom. New York: Longman.

———. 2007 [1982]. "*Th*eoretical Importance of the Family." Pp. 14–25 in *Family in Transition,* 14th ed., edited by Arlene S. Skolnickand Jerome H. Skolnick. Boston: Allyn & Bacon.

Goode-Cross, and David T. Tager. 2011. "Negotiating Multiple Identities: How African-American Gay and Bisexual Men Persist at a Predominantly White Institution." *Journal of Homosexuality* 58(90):1235–1254.

Goodman, W. Benjamin, Ann C. Crouter, Stephanie T. Lanza, and Martha J. Cox. 2008. "Paternal Work Characteristics and Father-Infant Interactions in Low-Income, Rural Families." *Journal of Marriage and Family* 70(3):640–653.

Goodman, W. Benjamin, Ann C. Crouter, Stephanie Lanza, Martha Cox, Lynne Vernon-Feagans, and the Family Life Project Key Investigators. 2011. "Parental Work Stress and Latent Profiles of Father-Infant Parenting Quality." *Journal of Marriage and Family* 73(3): 588–604.

Goodwin Paula, Brittany McGill, and Anjani Chandra. 2009. *Who Marries and When? Age at First Marriage in the United States, 2002.* NCHS data brief, no. 19. Hyattsville, MD: National Center for Health Statistics.

Goosby, Bridget J. 2007. "Poverty Duration, Maternal Psychological Resources, and Adolescent Socioemotional Outcomes." *Journal of Family Issues* 28(8):1113–1134.

Gootman, Elissa. 2012. "So Eager for Grandchildren, They're Paying the Egg-Freezing Clinic." *The New York Times.* May 13. Retrieved August 28, 2012 (http://www.nytimes.com).

Gordon, K. C., D. H. Baucom, and D. K. Snyder, D. K. (2004). "An Integrative Intervention for Promoting Recovery from Extramarital Affairs." *Journal of Marital and Family Therapy,* 30(2):213–231.

Gordon, Rachel A., Margaret L. Usdansky, Xue Wang, and Anna Gluzman. 2011. "Child Care and Mothers' Mental Health: Is High-Quality Care Associated with Fewer Depressive Symptoms?" *Family Relations* 60(4): 446–460.

Gordon, Thomas. 2000. *Parent Effectiveness Training: The Proven Program for Raising Responsible Children.* New York: Three Rivers Press.

Gormley, Barbara and Frederick G. Lopez. 2010. "Psychological Abuse Perpetration in College Dating Relationships: Contributions of Gender, Stress, and Adult Attachment Orientations." *Journal of Interpersonal Violence* 25(2):204–218.

Gornick, Janet C., Harriet B. Presser and Caroline Batzdorf. 2009. "Outside the 9-to-5." *The American Prospect,* June 9. Retrieved May 26, 2010 (www.prospect.org).

Gott, Natalie. 2010. "Clergy Women Make Connections." *Faith & Leadership,* February 16.

Gottesdiener, Laura. 2012. "Feminism's Next Big Step." *Alternet News and Politics.* August 23. Retrieved November 8, 2012 (http://www.salon.com).

Gottlieb, Alison Stokes. 1997. "Single Mothers of Children with Developmental Disabilities: The Impact of Multiple Roles." *Family Relations* 46(1):5–12.

Gottlieb, Laurie N., Ariella Lang, and Rhonda Amsel. 1996. "The Long-term Effects of Grief on

Marital Intimacy Following an Infant's Death." *Omega* 33(1):1–9.

Gottlieb, Lori. 2006. "How Do I Love Thee?" *Atlantic Monthly*, March, pp. 58–70.

Gotta, Gabrielle, Robert-Jay Green, Esther Rothblum, Sondra Solomon, Kimberly Balsam, and Pepper Schwartz. 2011. "Heterosexual, Lesbian, and Gay Male Relationships: A Comparison of Couples in 1975 and 2000." *Family Process* 50(3): 353–376.

Gottman, John. 2011. "John Gottman on Trust and Betrayal." Retrieved January 29, 2013 (http://greatergood.berkeley.edu).

Gottman, John M. 1979. *Marital Interaction: Experimental Investigations*. New York: Academic.

———. 1994. *Why Marriages Succeed or Fail*. New York: Simon and Schuster.

———. 1996. *What Predicts Divorce? The Measures*. Hillsdale, NJ: Erlbaum.

Gottman, John M., James Coan, Sybil Carrere, and Catherine Swanson. 1998. "Predicting Marital Happiness and Stability from Newlywed Interactions." *Journal of Marriage and Family* 60(1):5–22.

Gottman, John M. and Joan DeClaire. 2001. *The Relationship Cure: A Five-step Guide for Building Better Connections with Family, Friends, and Lovers*. New York: Crown.

Gottman, John and Julie Gottman. 2011. "How to Keep Love Going Strong: 7 Principles on the Road to Happily Ever After." Retrieved January 29, 2013 (http://www.yesmagazine.org).

Gottman, John, Julie Gottman, and Joan DeClaire. 2006. *10 Lessons to Transform Your Marriage*. New York: Random House, Three Rivers Press.

Gottman, John M. and L. J. Krotkoff. 1989. "Marital Interaction and Satisfaction: A Longitudinal View." *Journal of Consulting and Clinical Psychology* 57:47–52.

Gottman, John M. and Robert W. Levenson. 2000. "The Timing of Divorce: Predicting When a Couple Will Divorce Over a 14-Year Period." *Journal of Marriage and Family* 62(3):737–745.

———. 2002. "A Two-factor Model for Predicting When a Couple Will Divorce: Exploratory Analyses Using 14-Year Longitudinal Data." *Family Process* 41(1):83–96.

Gottman, John M., Robert W. Levenson, James Gross, Barbara Frederickson, Leah Rosenthal, Anna Ruef, and Dan Yoshimoto. 2003. "Correlates of Gay and Lesbian Couples' Relationship Satisfaction and Relationship Dissolution." *Journal of Homosexuality* 45(1):23–45.

Gottman, John M. and Clifford I. Notarius. 2000. "Decade Review: Observing Marital Interaction." *Journal of Marriage and Family* 62(4):927–947.

———. 2003. "Marital Research in the 20th Century and a Research Agenda for the 21st Century." *Trends in Marriage, Family, and Society* 25(2):283–297.

Gottman, John M. and Nan Silver. 1999. *The Seven Principles for Making Marriage Work*. New York: Crown.

Gough, Brendan, Nicky Weyman, Julie Alderson, Gary Butler, and Mandy Stoner. 2008. "'They Did Not Have a Word': The Parental Quest to Locate a 'True Sex' For Their Intersex Children." *Psychology & Health* 23(4):493–507.

Gove, Walter R., Carolyn Briggs Style, and Michael Hughes. 1990. "The Effect of Marriage on the Well-Being of Adults." *Journal of Family Issues* 11(1):4–35.

Gowan, Annie. 2009. "Immigrants' Children Look Closer for Love: More Young Adults Are Seeking Partners of the Same Ethnicity." *Washington Post*. March 8:A01.

Grace, Gerald. 2002. *Catholic Schools: Mission, Markets and Morality*. London: Routledge Farmer.

Grady, Bill. 2009. "A Marriage Blending Family and Race." Pp. 90–93 in *Social Issues First Hand: Blended Families*, edited by Stefan Kiesbye. New York: Greenhaven Press/ Cengage Learning.

Grady, Denise. 2007. "Girl or Boy? As Fertility Technology Advances, So Does an Ethical Debate." *The New York Times*, February 6.

Graefe, Deborah R. and Daniel T. Lichter. 1999. "Life Course Transitions of American Children: Parental Cohabitation, Marriage, and Single Motherhood." *Demography* 36(2):205–217.

———. 2007. "When Unwed Mothers Marry." *Journal of Family Issues* 28(5):595–622.

Graham, Elspeth and Lucy P. Jordan. 2011. "Migrant Parents and the Psychological Well-Being of Left-Behind Children in Southeast Asia." *Journal of Marriage and Family* 73(4): 763–787.

Graham, Kathy T. 2008. "Same-Sex Couples: Their Rights as Parents, and Their Children's Rights as Children." *Santa Clara Law Review* 48:999–1037.

Grall, Timothy S. 2011. *Custodial Mothers and Fathers and Their Child Support*. Washington, DC: U.S. Department of Commerce.

Grange, Christina M., Sarah Jane Brubaker, and Maya A. Corneille. 2011. "Direct and Indirect Messages African American Women Receive from Their Family Networks about Intimate Relationships and Sex: The Intersecting Influence of Race, Gender, and Class." *Journal of Family Issues* 32(5):605–628.

Grant, Jaime M. 2010. *Outing Age 2010*. Washington, DC: The National Gay and Lesbian Task Force Policy Institute. Retrieved May 15, 2010 (www.thetaskforce.org.).

Gray, Peter B., Peter T. Ellison, and Benjamin C. Campbell. 2007. "Testosterone and Marriage among Ariaal Men of Northern Kenya." *Current Anthropology* 48(5):750–755.

Greeley, Andrew. 1991. *Faithful Attraction: Discovering Intimacy, Love, and Fidelity in American Marriage*. New York: Doherty.

Green Jonathan and Michael Addis. 2012. "Individual Differences in Masculine Gender Socialization as Predictive of Men's Psychophysioogical Responses to Negative Affect." *International Journal of Men's Health* 11(1):63–82.

Green, Penelope. 2010. "Blending Like the Brady Bunch? Let's Not Go Too Far." *The New York Times*. Retrieved January 9, 2011 (http://www.nytimes.com/2010/11/18/garden/18unblended.html?pagewanted=all&_r=0).

Green, R. J. 2009. "From Outlaws to In-laws: Gay and Lesbian Couples in Contemporary Society." Pp. 197–213, 488–489, 527–530 in *Families as They Really Are*, edited by B. Risman. New York: Norton.

Greenberg, Jerrold S., Clint E. Bruess, and Debra W. Haffner. 2002. *Exploring the Dimensions of Human Sexuality*. Sudbury, MA: Jones and Bartlett.

Greenblatt, Cathy Stein. 1983. "The Salience of Sexuality in the Early Years of Marriage." *Journal of Marriage and Family* 45:289–299.

Greenhouse, Steven. 2012. "Equal Opportunity Panel Updates Hiring Policy." *The New York Times*. April 25. Retrieved August 28, 2012 (http://www.nytimes.com).

Greenstein, Theodore N. 2009. "National Context, Family Satisfaction, and Fairness in the Division of Household Labor." *Journal of Marriage and Family* 71(4):1039–1051.

Greenwald, John. 1999. "Elder Care: Making the Right Choice." *Time*, August 30, pp. 52–56.

Griswold v. Connecticut. 1965. 381 U.S. 479, 14 L. Ed.2d 510, 85 S. Ct. 1678.

Grogan-Kaylor, Andrew and Melanie D. Otis. 2007. "The Predictors of Parental Use of Corporal Punishment." *Family Relations* 56(1):80–91.

Gromoske, Andrea and Kathryn Maguire-Jack. 2012. "Transactional and Cascading Relations Between Early Spanking and Children's Social-Emotional Development." *Journal of Marriage and Family* 74(5): 1054–1068.

Gross, Jane. 2002. "U.S. Fund for Tower Victims Will Aid Some Gay Partners." *The New York Times*, May 22.

———. 2004. "Alzheimer's in the Living Room: How One Family Rallies to Cope." *The New York Times*, September 16.

———. 2006a. "As Parents Age, Baby Boomers and Business Struggle to Cope." *The New York Times*, March 25.

———. 2006b. "Forensic Skills Seek to Uncover Hidden Patterns of Elder Abuse." *The New York Times*, September 27.

———. 2006c. "Seeking Doctor's Advice in Adoptions from Afar." *The New York Times*, January 3.

———. 2007. "A Taste of Family Life in U.S., but Adoption Is in Limbo." *The New York Times*, January 13.

Gross, Michael. 2006. "Bad Advice: How Not to Have Sex in an Epidemic." *American Journal of Public Health* 96(6):964–966.

Grossman, Arnold H., and Anthony R. D'Augelli. 2006. "Transgender Youth: Invisible and Vulnerable." *Journal of Homosexuality* 51(1):111–128.

Grossman, Cathy L. 2010. "Prop 8 Ruling Drives Strong Religious Reactions: Outrage to Joy."

USA Today. August 5. Retrieved December 5, 2012 (http://content.usatoday.com).

Groves, Robert M. and Daniel L. Cork (Eds). 2008. *Surveying Victims: Options for Conducting the National Crime Victimization Survey*.Committee on National Statistics and Committee on Law and Justice, Division of Behavioral and Social Sciences and Education.Washington, DC: National Academies Press. Retrieved June 9, 2010 (www.nap.edu).

Grzywacz, Joseph G., Stephanie S. Daniel, Jenna Tucker, Jill Walls, and Esther Leerkes. 2011. "Nonstandard Work Schedules and Developmentally Generative Parenting Practices." *Family Relations* 60(1): 45–59.

Guberman, Nancy, Eric Gagnon, Denyse Cote, Claude Gilbert, Nicole Thivierge, and Marielle Tremblay. 2005. "How the Trivialization of the Demands of High-Tech Care in the Home Is Turning Family Members into Para-Medical Personnel." *Journal of Family Issues* 26(2):247–272.

———. 2009. *Analyzing Narrative Reality*. Thousand Oaks: Sage.

Gudelunas, David. 2006. "Who's Hooking Up On-Line?" *Gay and Lesbian Review Worldwide* 13(1):23–27.

Gudmunson, Clinton G., Sharon M. Danes, James D. Werbel, and Johnben Teik-Cheok Loy. 2009. "Spousal Support and Work—Family Balance in Launching a Family Business." *Journal of Family Issues* 30(8):1098–1121.

Guilamo-Ramos, Vincent, James Jaccard, Patricia Dittus, and Alida M. Bouris. 2006. "Parental Experience, Trustworthiness, and Accessibility: Parent-Adolescent Communication and Adolescent Risk Behavior." *Journal of Marriage and Family* 68(5):1229–1246.

Gültekin, Laura. 2012. "Family Homelessness, Housing Insecurity, and Health: Understanding and Acting on What We Know." *Family Focus* FF54(Fall): F4–F6.

Gupta, Sanjiv and Michael Ash. 2008. "Whose Money, Whose Time? A NonParametric Approach to Modeling Time Spent on Housework in the United States." *Feminist Economics* 14(1):93–120.

Gutis, Philip S. 1989. "New York Court Defines Family to Include Homosexual Couples." *The New York Times*, July 7.

Guttmacher Institute. 2006. "Facts on Induced Abortion in the United States." May. New York: Guttmacher Institute.

———. 2012. "Fact Sheet: Contraceptive Use in the United States." July. Retrieved April 14, 2013 (http://www.guttmacher.org).

Guynn, Jessica. 2013. "Yahoo CEO Marissa Mayer Causes Uproar with Telecommuting Ban." *Los Angeles Times*. February 26. Retrieved February 26, 2013 (http://www.latimes.com).

Guzzo, Karen Benjamin. 2009a. "Marital Intentions and the Stability of First Cohabitations." *Journal of Family Issues* 30(2):179–205.

———. 2009b. "Maternal Relationships and Nonresidential Father visitation of Children Born Outside of Marriage." *Journal of Marriage and Family* 71(3):632–649.

Guzzo, Karen Benjamin and Frank F. Furstenberg, Jr. 2007. "Multipartnered Fertility among Young Women with a Nonmarital First Birth: Prevalence and Risk Factors." *Perspectives on Sexual and Reproductive Health* 39(1):29–38.

Guzzo, Karen B. and Sarah R. Hayford. 2012. "Unintended Fertility and the Stability of Coresidential Relationships." *Social Science Research* 41(5): 1138–1151.

Guzzo, Karen Benjamin and Helen Lee. 2008. "Couple Relationship Status and Patterns in Early Parenting Practices." *Journal of Marriage and Family* 70(1):44–61.

Ha, K. Oanh. 2006. "Filipinos Work Overseas, Help Kin." *San Jose Mercury News*, April 4.

Haas, Stephen M., and Laura Stafford. 2005. "Maintenance Behaviors in Same-Sex and Marital Relationships: A Matched Sample Comparison." *Journal of Family Communication* 5(1):43–60.

Hackstaff, Karla B. 2007. "Divorce Culture: A Quest for Relational Equality in Marriage." Pp. 188–222 in *Family in Transition*, 4th ed., edited by Arlene S. Skolnickand Jerome H. Skolnick. Boston: Pearson.

Haddad, Yvonne Y. and Jane I. Smith. 1996. "Islamic Values among American Muslims." Pp. 19–40 in *Family and Gender among American Muslims: Issues Facing Middle Eastern Immigrants and Their Descendants*, edited by Barbara C. Aswadand Barbara Bilgé.Philadelphia, PA: Temple University Press.

Hagan, Frank E. 2010. *Crime Types and Criminals*. Thousand Oaks, CA: Sage.

Hagedoorn, Mariet, Nico W. Van Yperen, James C. Coyne, Cornelia van Jaarsveld, Adelita Ranchor, Eric van Sonderen, and Robbert Sanderman. 2006. "Does Marriage Protect Older People from Distress? The Role of Equity and Recency of Bereavement." *Psychology and Aging* 21(3):611–620.

Hagestad, G. 1986. "The Family: Women and Grandparents as Kin Keepers." Pp. 141–160 in *Our Aging Society*, edited by A. Piferand L. Bronte. New York: Norton.

———. 1996. "On-*Time*, Off-*Time*, Out of *Time*? Reflections on Continuity and Discontinuity from an Illness Process." Pp. 204–222 in *Adulthood and Aging*, edited by V. L. Bengston. New York: Springer.

Hagewen, Kellie J. and S. Philip Morgan. 2005. "Intended and Ideal Family Size in the United States, 1970-2002." *Population and Development Review* 31(3):507–527.

"The Hague Convention on Intercountry Adoption." 2010. *Child Welfare Information Gateway*,U.S. Department of Health and Human Services. www.childwelfare.gov.

Halford, Kim, Jan Nicholson, and Matthew Sanders. 2007. "Couple Communication in Stepfamilies." *Family Process* 46(4):471–483.

Hall, Edie Jo and E. Mark Cummings. 1997. "The Effects of Marital and Parent-Child Conflicts on Other Family Members: Grandmothers and Grown Children." *Family Relations* 46(2):135–143.

Hall, Elaine J. and Marnie Salupo Rodriguez. 2003. "The Myth of Postfeminism." *Gender and Society* 17:878–902.

Hall, Heather and J. Carolyn Graff. 2012. "Maladaptive Behaviors of Children with Autism: Parent Support, Stress, and Coping." *Issues in Comprehensive Pediatric Nursing* 35(3–4):194–214.

Hall, Scott S. and Shelley M. MacDermid. 2009. "A Typology of Dual Earner Marriages Based on Work and Family Arrangements." *Journal of Family and Economic Issues* 30(3):215–225.

Hall, Sharon K. 2008. *Raising Kids in the 21st Century*. West Sussex, UK: Wiley-Blackwell.

Halpern, Susan P. 2009. *Finding the Words: Candid Conversations with Loved Ones*. Berkeley, CA: North Atlantic Books.

Halpern-Felsher, Bonnie L., Jodi L. Cornell, Rhonda Y. Kropp, and Jeanne M. Tschann. 2005. "Oral versus Vaginal Sex among Adolescents: Perceptions, Attitudes, and Behavior." *Pediatrics* 115(4):845–851.

Halsall, Paul. 2001. *Internet Medieval Sourcebook*. Retrieved October 6, 2006 (www.fordham.edu).

Hamamci, Zeynep. 2005. "Dysfunction Relationship Beliefs in Marital Conflict." *Journal of Relational-Emotive and Cognitive-Behavior Therapy* 23(3):245–261.

Hamel, John C. and Tanya L. Nicholls. 2006. *Family Interventions in Domestic Violence: A Handbook of Gender-Inclusive Theory and Treatment*. New York: Springer.

Hamilton, Brady E., Joyce A. Martin, and Stephanie J. Ventura. 2007. *Births: Preliminary Data for 2005*. NCHS E-Stats. Hyattsville, MD: National Center for Health Statistics. Retrieved March 18, 2007 (www.cdc.gov/nchs).

———. 2009. *Births: Preliminary data for 2007*.National Vital Statistics Reports 57(12). Hyattsville, MD: National Center for Health Statistics. March 18.

———. 2010. "Births: Preliminary Data for 2008." *National Vital Statistics Reports* 58(16). Hyattsville, MD: National Center for Health Statistics. April 20.

———. 2012. "Births: Preliminary Data for 2011." *National Vital Statistics Reports* 61(5). October 3. Hyattsville MD: National Center for Health Statistics.

Hamilton, Brady E. and Stephanie J. Ventura. 2012. "Birth Rates for U.S. Teenagers Reach Historic Lows for All Age and Ethnic Groups." *NCHS Data Brief Number 89*. Hyattsville, MD: National Center for Health Statistics.

Hammel, Paul. 2012. "Court: No Inheritance for Child Conceived by Artificial Insemination." *World-Herald Bureau*. November 16. Retrieved March 31, 2013 (http://omaha.com).

Hammen, Constance, Patricia A. Brennan, and Josephine H. Shih. 2004. "Family Discord and Stress Predictors of Depression and Other Disorders in Adolescent Children of Depressed and Nondepressed Women." *Journal of the*

American Academy of Child and Adolescent Psychiatry 43(8):994–1003.

Hamon, Raeann R. and Bron B. Ingoldsby. 2003. *Mate Selection across Cultures*. Thousand Oaks, CA: Sage Publications.

Hamplova, Dana and Celine LeBourdais. 2009. "One Pot or Two Pot Strategies? Income Pooling in Married and Unmarried Households in Comparative Perspective." *Journal of Comparative Family Studies* 40(3): 355–385.

Han, Chong-suk. 2008. "A Qualitative Exploration of the Relationship Between Racism and Unsafe Sex among Asian Pacific Islander Gay Men." *Archives of Sexual Behavior* 37(5):827–837.

Han, Wen-Jui and Liana E. Fox. 2011. "Parental Work Schedules and Children's Cognitive Trajectories." *Journal of Marriage and Family* 73(5): 962–980.

Hancock, A.N. 2012. "'It's a Macho thing, Innit?' Exploring the Effects of Masculinity on Career Choice and Development." *Gender, Work, and Organization* 19(4): 392–415.

Hanna, Sharon L., Rose Suggett, and Doug Radtke. 2008. *Person to Person: Positive Relationships Don't Just Happen*. 5th ed. Upper Saddle River, NJ: Pearson/Prentice Hall.

Hans, Jason D. 2002. "Stepparenting after Divorce: Stepparents' Legal Position Regarding Custody, Access, and Support." *Family Relations* 51(4):301–307.

———. 2009. "Beliefs about Child Support Modification Following Remarriage and Subsequent Childbirth." *Family Relations* 58(February):65–78.

Hans, Jason D. and Marilyn Coleman. 2009. "The Experiences of Remarried Stepfathers Who Pay Child Support." *Personal Relationships* 16:597–618.

Hans, Jason D., Martie Gillen, and Katrina Akande. 2010. "Sex Redefined: The Reclassification of Oral-Genital Contact." *Perspectives on Sexual and Reproductive Health* 42(2):74–78.

Hans, Jason D., Lawrence H. Ganong, and Marilyn Coleman. 2009. "Financial Responsibilities Toward Older Parents and Stepparents Following Divorce and Remarriage." *Journal of Family Economic Issues* 30:55–66.

Hansen, Donald A. and Reuben Hill. 1964. "Families Under Stress." Pp. 782–819 in *The Handbook of Marriage and the Family*, edited by Harold Christensen. Chicago: Rand McNally.

Hansen, Matthew. 2012. "The Talk: A North Omaha Rite of Passage." *The Omaha World Herald*. March 25.

Hansen, Matthew and Karyn Spencer. 2009. "Safe Haven Meant Kids Finally Got Right Help." *Omaha World Herald*, February 1. Retrieved February 3, 2009 (www.omaha.com).

Hansen, Thorn, Torbjorn Moum, and Adam Shapiro. 2007. "Relational and Individual Well-Being among Cohabitors and Married Individuals in Midlife." *Journal of Family Issues* 28(7):910–933.

Haraway, Donna. 1989. *Primate Visions: Gender, Race, and Nature in the World of Modern Science*. New York: Routledge, Chapman and Hall.

Hardesty, Jennifer L. and Grace H. Chung. 2006. "Intimate Partner Violence, Parental Divorce, and Child Custody: Directions for Intervention and Future Research." *Family Relations* 55:200–216.

Harding, Rosie. 2007. "Sir Mark Potter and the Protection of the Traditional Family: Why Same Sex Marriage Is (Still) a Feminist Issue." *Feminist Legal Studies* 15(2):223–234.

Hare, Jan and Denise Skinner. 2008. "'Whose Child Is This?' Determining Legal Status for Lesbian Parents Who Used Assisted Reproductive Technology." *Family Relations* 57(3):365–375.

Hardin, M. and Jennifer Greer. 2009. "The Influence of Gender-Role Socialization, Media Use, and Sports Participation on Perceptions of Gender-Appropriate Sports." *Journal of Sport and Behavior* 32(2):1–20.

Harknett, Kristen. 2006. "The Relationship between Private Safety Nets and Economic Outcomes among Single Mothers." *Journal of Marriage and Family* 69(1):172–191.

Harknett. Kristen S. and Caroline Sten Hartnett. 2011. "Who Lacks Support and Why? An Examination of Mothers' Personal Safety Nets." *Journal of Marriage and Family* 73(4): 861–875.

Harknett, Kristen and Jean Knab. 2007. "More Kin, Less Support: Multipartnered Fertility and Perceived Support among Mothers." *Journal of Marriage and Family* 69(1):237–253.

Harmanci, Reyhan. 2006. "The Neighbordaters." *San Francisco Chronicle Magazine*, February 12, pp. 13–14.

Harmon, Amy. 2005a. "Ask Them (All 8 of Them) about the Grandkids." *The New York Times*, March 20.

———. 2005b. "Hello, I'm Your Sister, Our Father Is Donor 150." *The New York Times*, November 20.

———. 2007a. "Prenatal Test Puts Down Syndrome in Hard Focus." *The New York Times*, May 9.

———. 2007b. "Sperm Donor Father Ends His Anonymity." *The New York Times*, February 14.

Harpel, Tammy S. and Jodie Hertzog. 2010. "'I Thought My Heart Would Burst': The Role of Ultrasound Technology on Expectant Grandmotherhood." *Journal of Family Issues* 31(2):257–274.

Harper, Nevin J. 2009. "Family Crisis and the Enrollment of Children in Wilderness Treatment." *Journal of Experimental Education* 31(3):447–450.

Harris, Christine R. 2003. "A Review of Sex Differences in Sexual Jealousy, Including Self-Report Data: Psychophysiological Responses, Interpersonal Violence, and Morbid Jealousy." *Personality and Social Psychology Review* 7(2):102–108.

Harris, Deborah and Domenico Parisi. 2008. "Looking for 'Mr. Right': The Viability of Marriage Initiatives for African American Women in Rural Settings." *Sociological Spectrum* 28(4):338–356.

Harris, Marian S. and Ada Skyles. 2008. "Kinship Care for African American Children: Disproportionate and Disadvantageous." *Journal of Family Issues* 29(8):1013–1030.

Harris, Paul. 2009. "Revealed: The Shocking Rise of 'New Slavery' in US Midwest." *The Observer* November 22, p. 37.

Harrison, Kathleen McDavid, Ruiguang Song, and Xinjian Zhang. 2010. "Life Expectancy After HIV Diagnosis Based on National HIV Surveillance Data from 25 States, United States." *Journal of Acquired Immune Deficiency Syndromes* 53(1):124–130.

Harrist, Amanda W. and Ricardo C. Ainslie. 1998. "Marital Discord and Child Behavior Problems." *Journal of Family Issues* 19(2):140–163.

Hart, Joshua, Jacqueline A. Hung, Peter Glick, and Rachel E. Dinero. 2012. "He Loves Her, He Loves Her Not: Attachment Style As a Personality Antecedent to Men's Ambivalent Sexism." *Personality and Social Psychology Bulletin* 38(11):1495–1505.

Harter, James and Raksha Arora. 2008. "Social Time Crucial to Daily Emotional Well-Being in U.S." June 5. Retrieved November 24, 2009 (www.gallup.com/poll).

Hartman, Karen. 2011. "Bound in a Gay Union by a State Denying It." *The New York Times*. July 15. Retrieved September 13, 2011 (http://www.nytimes.com).

Hartman, Margaret. 2011. "Cohabitation Is Illegal in Florida, and Conservatives Want to Keep It That Way." August 31. Retrieved November 24, 2012 (http://jezebel.com/5836431/cohabitation-is-illegal-in-florida-and-conservatives-want-to-keep-it-that-way).

Hartmann, Heidi, Ashley English, and Jeffrey Hayes. 2010. "Women and Men's Employment and Unemployment in the Great Recession." *Institute for Women's Research Briefing Paper C373*. Washington, DC: Institute for Women's Research.

Hartmann, Heidi, Stephen J. Rose, and Vicky Lovell. 2009. "How Much Progress in Closing the Long-term Earnings Gap?" Pp. 125–155 in *The Declining Significance of Gender?* edited by Francine D. Blau, Mary C. Brinton, and David B. Grusky. New York: Russell Sage.

Hartocollis, Anemona. 2006. "Meaning of 'Normal' Is at Heart of Gay Marriage Ruling." *The New York Times*, July 8. Retrieved July 11, 2006 (www.nytimes.com).

Hartnett, Carolyn Sten and Emilio Parrado. 2012. "Hispanic Familism Reconsidered: Ethnic Differences in the Perceived Value of Children and Fertility Decisions." *Sociological Quarterly* 53(4):636–653.

Hartog, Henrik. 2000. *Man and Wife in America: A History*. Cambridge, MA: Harvard University Press.

Hartsoe, Steve. 2005. "Shacking Up: N.C. Anti-cohabitation Law Under Legal Attack." *Raleigh News and Observer*, May 9. Retrieved May 11, 2005 (www.newsobserver.com).

Hasan, Tabinda and Mahmood Fauzi. 2012. "If 'Women are from Venus and Men are from Mars,' Does an answer lie with Neuroanatomy?" *International Journal of Collaborative Research on Internal Medicine and Public Health* 4(5): 566–577.

Haselschwerdt, Megan. 2012. "Who Cares about the Rich Folk? An Argument for More Research on Affluent Families and Communities." *Family Focus* FF54(Fall): F14–F16.

Hassrick, Elizabeth McGhee and Barbara Schneider. 2009. "Parent Surveillance in Schools: A Question of Social Class." *American Journal of Education* 115(February):195–211.

Hatch, Laurie R. and Kris Bulcroft. 2004. "Does Long-Term Marriage Bring Less Frequent Disagreements?" *Journal of Family Issues* 25(4):465–495.

Hatfield, Elaine. 2013. Passionate Love Scale. Retrieved May 26, 2013, from http://www.elaine-hatfield.com/Passionate%20Love%20Scale.pdf.

Hatfield, Elaine and Sprecher, Susan. 2010. The passionate love scale. Pp 466–468 in *Handbook of Sexuality-Related Measures: A Compendium*, 3rd ed., edited by T. D. Fisher, C. M. Davis, W. L. Yaberand S. L. Davis (Eds.). Thousand Oaks, CA: Taylor & Francis.

Haub, Carl. 2011. "The Continuing U.S. Recession and the Birth Rate." *A Post-Recession Update on U.S. Social and Economic Trends.* Washington, DC: Population Reference Bureau. Retrieved August 29, 2012 (http://www.prb.org).

Haughney, Kathleen. 2011. "Unmarried? Living Together? You're Breaking the Law in Florida." *SunSentinel.* August 31. Retrieved November 24, 2012 (http://articles.sun-sentinel.com).

Hawkins, Alan J., Kimberly R. Lovejoy, Erin K. Holmes, Victoria L. Blanchard, and Elizabeth Fawcett. 2008. "Increasing Fathers' Involvement in child Care with a couple-Focused Intervention during the Transition to Parenthood." *Family Relations* 57(1):49–59.

Hawkins, Daniel N. and Alan Booth. 2005. "Unhappily Ever After: Effects of Long-Term, Low-Quality Marriages on Well-Being." *Social Forces* 84(1):451–471.

Hawkins, Stacy, Gabriel Schlomer, Leslie Bosch, Deborah Casper, Christine Wiggs, Noel Card, and Lynne Borden. 2012. "A Review of the Impact of U.S. Military Deployments during Conflicts in Afghanistan and Iraq on Children's Functioning." *Family Science* 3(2):99–108.

Hay, Elizabeth L., Karen L. Fingerman, and Eva S. Lefkowitz. 2007. "The Experience of Worry in Parent-Adult Child Relationships." *Personal Relationships* 14(4):605–622.

Hayden, Dolores. 1981. *The Grand Domestic Revolution: A History of Feminist Designs for American Homes, Neighborhoods, and Cities.* Cambridge: MIT Press.

Hayghe, Howard. 1982. "Dual Earner Families: Their Economic and Demographic Characteristics." Pp. 27–40 in *Two Paychecks,* edited by Joan Aldous. Newbury Park, CA: Sage Publications.

Healthy Marriage Initiative. 2009. Retrieved October 9, 2009 (www.acf.hhs.gov/healthymarriage).

Heard, Holly E. 2007. "The Family Structure Trajectory and Adolescent School Performance: Differential Effects by Race and Ethnicity." *Journal of Family Issues* 28(3):319–354.

Heaton, Tim B. 2002. "Factors Contributing to Increasing Marital Stability in the United States." *Journal of Family Issues* 23(3):392–409.

Hebert, Laura A. 2007. "Taking 'Difference' Seriously: Feminisms and the 'Man Question'." *Journal of Gender Studies* 16(1):31–45.

Heene, Els, Ann Buysse, and Paulette Van Oost. 2007. "An Interpersonal Perspective on Depression: The Role of Marital Adjustment, Conflict Communication, Attributions, and Attachment within a Clinical Sample." *Family Process* 46(4):499–514.

Heffner, Christopher L. 2003. "Counseling the Gay and Lesbian Client." *AllPsych Journal.* August 12. Retrieved March 3, 2010 (allpsych.com/journal).

Hegewisch, Ariane and Hannah Liepmann. 2010. "The Gender Wage Gap by Occupation." *Fact Sheet IWPR #C350a.* Washington, DC: Institute for Women's Policy Research.

Heilmann, Ann. 2011. "Gender and Essentialism: Feminist Debates in the Twenty-First Century." *Critical Quarterly* 53(4): 78–89.

Helfich, Christine A. and Emily K. Simpson. 2005. "Lesbian Survivors of Intimate Partner Violence: Provider Perspectives on Barriers to Accessing Service." *Journal of Gay and Lesbian Social Services* 18(2):39–59.

Helms, Heather M., Andrew J. Supple, and Christine M. Proulx. 2011. "Mexican-Origin Couples in the Early Years of Parenthood: Marital Well-Being in Ecological Context." *Journal of Family Theory & Review* 3(June): 67–95.

Hench, David. 2004. "Is Anger Management a Remedy for Batterers?" *Portland Press Herald* (Maine), October 10.

Henderson, Craig E., Bert Hayslip Jr., Leah M. Sanders, and Linda Louden. 2009. "Grandmother-Grandchild Relationship Quality Predicts Psychological Adjustment among Youth from Divorced Families." *Journal of Family Issues* 30(9):1245–1264.

Henderson, Tammy L. 2005a. "Grandparent Visitation Rights: Justices' Interpretation of the Best Interests of the Child Standard." *Journal of Family Issues* 26(5):638–664.

———. 2005b. "Grandparent Visitation Rights: Successful Acquisition of Court-ordered Visitation." *Journal of Family Issues* 26(1):107–137.

———. 2008. "Transforming the Discussion about Diversity, Policies, and Law." *Journal of Family Issues* 29(8):983–994.

Henderson, Tammy L. and Patricia B. Moran. 2001. "Grandparent Visitation Rights." *Journal of Family Issues* 22(5):619–638.

Hendrick, Clyde, Susan S. Hendrick, and Amy Dicke. 1998. "The Love Attitudes Scale: Short Form." *Journal of Social and Personal Relationships* 15(2):147–159.

Hendy, Helen, Mary Burns, Hakan Can, and Cory Scherer. 2012. "Adult Violence with the Mother and Sibling as Predictors of Partner Violence." *Journal of Interpersonal Violence* 27(11): 2276–2297.

Henly, Julia R., Sandra K. Danziger, and Shira Offer. 2005. "The Contribution of Social Support to the Material Well-Being of Low-Income Families." *Journal of Marriage and Family* 67(1):122–140.

Henry, C. S., C. P. Ceglian, and D. W. Matthews. 1992. "The Role Behaviors, Role Meanings, and Grandmothering Styles of Grandmothers and Stepgrandmothers: Perceptions of the Middle Generation." *Journal of Divorce and Remarriage,* 17:1–22.

Henry, Carolyn, Michael Merton, Scott Plunkett, and Tovah Sands. 2008. "Neighborhood, Parenting, and Adolescent Factors and Academic Achievement in Latino Adolescents from Immigrant Families." *Family Relations* 57(December):579–590.

Henry, C. S., C. P. Ceglian, and D. W. Matthews. 1992. "The Role Behaviors, Role Meanings, and Grandmothering Styles of Grandmothers and Stepgrandmothers: Perceptions of the Middle Generation." *Journal of Divorce and Remarriage,* 17:1–22.

Henry, Pamela J. and James McCue. 2009. "The Experience of Nonresidential Stepmothers." *Journal of Divorce and Remarriage* 50:185–205.

Henshaw, Stanley K. and Kathryn Kost. 2008. *Trends in the Characteristics of Women Obtaining Abortions, 1974 to 2004.* New York: Guttmacher Institute.

Hequembourg, Amy L. 2007. "Becoming Lesbian Mothers." *Journal of Homosexuality* 53(3):153–180.

Hequembourg, Amy L. and Sara Brallier. 2005. "Gendered Stories of Parental Caregiving among Siblings." *Journal of Aging Studies* 19:53–71.

Herbenick, Debby, Michael Reece, Vanessa Schick, Stephanie A. Sanders, Brian Dodge, and J. Dennis Fortenberry. 2010a. "Sexual Behavior in the United States: Results from a National Probability Sample of Men and Women Ages 14–94." *Journal of Sexual Medicine* 7(Suppl. 5): 255–265.

Herbenick, Debby, Michael Reece, Vanessa Schick, Stephanie A. Sanders, Brian Dodge, and J. Dennis Fortenberry. 2010b. "An Event-Level Analysis of the Sexual Characteristics and Composition among Adults Ages 18–59: Results from a National Probability Sample in the United States." *Journal of Sexual Medicine* 7(Suppl. 5):346–361.

Herd, Pamela. 2009. "Women, Public Pensions, and Poverty: What Can the United States Learn from Other Countries?" *Journal of Women, Politics, and Policy* 30(2–3):301–334.

Herek, Gregory M. 2009a. "Marriage Equality Attitudes: Simply Knowing Gay People Helps, But Isn't Enough." *Beyond Homophobia,* June 16. www.beyondhomophobia.com.

———. 2009b. "Sexual Stigma and Sexual Prejudice in the United States: A Conceptual Framework." Pp. 65–111 in *Contemporary Perspectives On Lesbian, Gay and Bisexual Identities: The 54th Nebraska Symposium on Motivation,* edited by D. A. Hope. New York: Springer.

Herman-Giddens, Marcia E. 2007. " Comment on: 'The decline in the age of menarche in the United States: should we be concerned?'" *Journal of Adolescent Health* 40(3):201–203.

Hernandez, Raymond. 2001. "Children's Sexual Exploitation Underestimated, Study Finds." *The New York Times,* September 10.

Herring, Jeff. 2005. "Affairs Don't Have to Be Physical." Knight/Ridder Newspapers, November 6.

Herszenhorn, David M. 2013. "In Russia, Ban on U.S. Adoptions Creates Rancor and Confusion." *The New York Times.* January 15. Retrieved January 16, 2013 (http://www.nytimes.com).

Herszenhorn, David M. and Andrew E. Kramer. 2012. "Russian Adoption Ban Brings Uncertainty and Outrage." *The New York Times.* December 28. Retrieved January 8, 2012 (http://www.nytimes.com).

Hertlein, Katherine M. 2012. "Digital Dwelling: Technology in Couple and Family Relationships." *Family Relations* 61(3) (July): 374–387.

Hertlein, K. M., and F. P. Piercy. 2006. Internet Infidelity: A Critical Review of the Literature." *Family Journal: Counseling and Therapy for Couples and Families* 14:366–371.

Hertz, Rosanna. 1997. "A Typology of Approaches to Child Care." *Journal of Family Issues* 18(4):355–385.

———. 2006. *Single by Chance, Mothers by Choice: How Women Are Choosing Parenthood without Marriage and Creating the New American Family.* New York: Oxford University Press.

Hesse-Biber, Sharlene Nagy. 2007. *Handbook of Feminist Research: Theory and Praxis.* Thousand Oaks, CA: Sage.

Hetherington, E. Mavis. 1987. "Family Relations Six Years after Divorce." Pp. 185–205 in *Remarriage and Stepparenting,* edited by K. Pasley and M. Ihinger-Tallman. New York: Guilford Press.

———. 2003. "Intimate Pathways: Changing Patterns in Close Personal Relationships across Time." *Family Relations* 52(4):318–331.

———. 2005. "The Adjustment of Children in Divorced and Remarried Families." Pp. 137–139 in *Sourcebook of Family Theory and Research,* edited by Vern L. Bengston, Alan C. Acock, Katherine R. Allen, Peggye Dilworth-Anderson, and David M. Klein. Thousand Oaks, CA: Sage Publications.

Hetherington, E. Mavis and K. M. Jodl. 1994. "Stepfamilies as Settings for Child Development." Pp. 55–80 in *Stepfamilies: Who Benefits? Who Does Not?* edited by Alan Booth and J. Dunn. Hillsdale, NJ: Erlbaum.

Hetherington, E. Mavis and John Kelly. 2002. *For Better or for Worse: Divorce Reconsidered.* New York: Norton.

Hetherington, E. Mavis and M. M. Stanley-Hagan. 1999. "Stepfamilies." Pp. 137–159 in *Parenting and Child Development in "Nontraditional" Families,* edited by M. E. Lamb. Mahwah, NJ: Lawrence Erlbaum Associates.

Hetter, Katia. 2011. "What's Fueling Bible Belt Divorces?" CNN. Retrieved April 17, 2013 (http://www.cnn.com/2011/LIVING/08/25/divorce.bible.belt/index.html).

Hewitt, Belinda and Gavin Turrell. 2011. "Short-Term Functional Health and Well-Being after Marital Separation: Does Initiator Status Make a Difference?" *American Journal of Epidemiology* 173(11):1308–1318.

Hewitt, Sally. 2009. *My Stepfamily.* Mankato, MN: Smart Apple Media.

Hewlett, Sylvia Ann. 2002. *Creating a Life: Professional Women and the Quest for Children.* New York: Hyperion.

Heyman, Richard, Ashley Hunt-Martorano, Jill Malik, and Amy M. Smith. 2009. "Desired Change in Couples: Gender Differences and Effects on Communication." *Journal of Family Psychology* 23(4):474–484.

Heyman, Richard A. and Amy M. Smith Slep. 2002. "Do Child Abuse and Interparental Violence Lead to Adulthood Family Violence?" *Journal of Marriage and Family* 64:864–870.

Heywood, Leslie and Jennifer Drake. 1997. "Introduction." Pp. 1–20 in *Third Wave Agenda: Being Feminist, Doing Feminism,* edited by Leslie Heywoodand Jennifer Drake. Minneapolis: University of Minnesota Press.

Higginbotham, Brian J., Julie J. Miller, and Sylvia Niehuis. 2009. "Remarriage Preparation: Usage, Perceived Helpfulness, and Dyadic Adjustment." *Family Relations* 58(July):316–329.

Higginbotham, Brian J., Sarah Tulane, and Linda Skogrand. 2012. "Stepfamily Education and Changes in Financial Practices." *Journal of Family Issues* 33(1):1398–1420.

Higgins, Jenny A., James Trussell, Nelwyn B. Moore, and J. Kenneth Davidson. 2010. "Virginity Lost, Satisfaction Gained? Physiological and Psychological Sexual Satisfaction at Heterosexual Debut." *Journal of Sex Research* 47(4):384–394.

Hildreth, Carolyn J., Alison Burke, and Richard Glass. 2009. "Elder Abuse." *Journal of the American Medical Association* 302(5):588.

Hill, Reuben. 1958. "Generic Features of Families Under Stress." *Social Casework* 49:139–150.

Hill, Robert B. 2003 [1972]. *The Strengths of Black Families.* 2nd ed. with "Epilogue: Thirty Years Later." Lanham, MD: University Press of America.

Hill, Shirley A. 2004. *Black Intimacies: A Gender Perspective on Families and Relationships.* Lanham, MD: Rowman and Littlefield.

Hill, Twyla. 2011. "Spousal Caregiving in Later Life: Predictors and Consequences." *Family Focus* FF48:F15–F16.

Hilliard, L. J. and L.S. Liben. 2010. "Differing Levels of Gender Salience in Preschool Classrooms: Effects of Children's Gender Attitudes and Intergroup Bias." *Child Development* 81(6) (Nov–Dec):1787–1798.

Himanek, Celeste Eckman. 2011. "Transitioning to Parenthood: The Experience of Infertility." *Family Focus* FF49:F9–F11.

Himes, Christine L. 2001. "Social Demography of Contemporary Families and Aging." Pp. 47–50 in *Families in Later Life: Connections and Transitions,* edited by Alexis J. Walker, Margaret Manoogian-O'Dell, Lori A. McGraw, and Diana L. G. White. Thousand Oaks, CA: Pine Forge.

Hinders, Dana. 2012. "Average Cost of Surrogacy." *Love to Know: Advice Women Can Trust.* Retrieved January 14, 2013 (http://www.pregnancy.lovetoknow.com).

Hines, Denise A. and Emily M. Douglas. 2009. "Women's Use of Intimate Partner Violence against Men: Prevalence, Implications, and Consequences." *Journal of Aggression, Maltreatment & Trauma* 18(6):572–586.

Hines, Melissa. 2011. "Gender Development and the Human Brain." *Annual Review of Neuroscience* 34(1): 69–88.

Hirsch, Jennifer S. 2003. *A Courtship after Marriage: Sexuality and Love in Mexican Transnational Families.* Berkeley: University of California Press.

———. 2008. "Catholics Using Contraceptives: Religion, Family Planning, and Interpretive Agency in Rural Mexico." Studies in Family Planning 39(2):93–104.

Hirsch, Jennifer S., Miguel Muñoz-Laboy, Christina M. Nyhus, Kathryn M. Yount, and José A. Bauermeister. 2009. "'They Miss More than Anything Their Normal Life Back Home': Masculinity and Extramarital Sex among Mexican Migrants in Atlanta." *Perspectives on Sexual & Reproductive Health* 41(1):23–32.

"HIV/AIDS Epidemic in the United States" 2013. Retrieved July 15, 2013 (http://kff.org/hivaids/fact-sheet/the-hivaids-epidemic-in-the-united-states/).

Hirji, Aliya. 2012. "Islam's Role in Family (and Life) Education." *Family Focus.* Issue FF53. Summer: F11–F12.

Hochschild, Arlie. 1989. *The Second Shift: Working Parents and the Revolution at Home.* New York: Viking/Penguin.

———. 1997. *Time Bind: When Work Becomes Home and Home Becomes Work.* New York: Henry Holt.

Hoelter, Lynette F., William G. Axinn, and Dirgha J. Ghimire. 2004. "Social Change, Premarital Nonfamily Experiences, and Marital Dynamics." *Journal of Marriage and Family* 66(5):1131–1151.

Hofferth, S. L. 2006. "Residential Father Family Type and Child Well-Being: Investment versus Selection." *Demography,* 43:53–77.

Hoffman, Jan. 2009. "Teenage Girls Stand by Their Man." *The New York Times.* March 19. Retrieved May 12, 2009 (www.nytimes.com).

———. 2011. "Boys Will Be Boys? Not in These Families." *The New York Times* June 10. Retrieved September 13, 2011 (http://www.nytimes.com).

Hoffman, Kristin and John N. Edwards. 2004. "An Integrated Theoretical Model of Sibling Violence and Abuse." *Journal of Family Violence* 19(3):185–200.

Hoffman, Lois W. and Jean B. Manis. 1979. "The Value of Children in the United States: A New

Approach to the Study of Fertility." *Journal of Marriage and Family* 41: 583–596.

Hogan, Dennis P. and Nan M. Astone. 1986. "The Transition to Adulthood." *Annual Review of Sociology* 12:109–130.

Hoge, Charles W. 2010. *Once a Warrior, Always a Warrior: Navigating the Transition from Combat to Home—Including Combat Stress, PTSD, and MTBI.* Guilford, CT: GPP Life.

Hohmann-Marriott, Bryndl. 2009. "Father Involvement Ideals and the Union Transitions of Unmarried Parents." *Journal of Family Issues* 30(7):898–920.

Hohmann-Marriott, Bryndl. 2011. "Coparenting and Father Involvement in Married and Unmarried Coresident Couples." *Journal of Marriage and Family* 73(1): 296–309.

Hohmann-Marriott, Bryndl and Paul Amato. 2008. "Relationship Quality in Interethnic Marriages and Cohabitations." *Social Forces* 87(2):825–855.

Holland, Rochelle. 2009. "Perceptions of Mate Selection for Marriage among African American, College-Educated, Single Mothers." *Journal of Counseling & Development* 87(Spring): 170–178.

Holloway, Angela Ann. 2008. "Pre-Marital Counseling." April 28. Retrieved October 12, 2009 (marriage.suite101.com).

Holmberg, Diane, Karen Blair, and Maggie Phillips. 2010. "Women's Sexual Satisfaction as a Predictor of Well-Being in Same-Sex Versus Mixed-Sex Relationships." *Journal of Sex Research* 47(1):1–11.

Holmes, Steven A. 1990. "Day Care Bill Marks a Turn toward Help for the Poor." *The New York Times*, January 25.

Holstein, James A. and Jaber F. Gubrium. 2008. *Handbook of Constructionist Research.* New York: Guilford Press.

Holt, Amanda. 2013. *Adolescent-to-Parent Abuse.* New York: Policy Press.

Holtzman, Mellisa. 2011. "Nonmarital Unions, Family Definitions, and Custody Decision Making." *Family Relations* 60:617–632.

Holtzworth-Munroe, Amy, Amy G. Applegate, and Brian D'Onofrio. 2009. "For the Sake of the Children: Collaborations between Law and Social Science to Advance the Field of Family Dispute Resolution: Family Dispute Resolution: Charting a Course for the Future." *Family Court Review* (Special Issue) 47(3):493–505.

Hondagneu-Sotelo, Pierrette and Michael A. Messner. 1994. "Gender Displays and Men's Power: The 'New Man' and the Mexican Immigrant." Pp. 200–218 in *Theorizing Masculinity*, edited by Harry Brodand Michael Kaufman. Newbury Park, CA: Sage Publications.

Hopper, Joseph. 1993. "The Rhetoric of Motives in Divorce." *Journal of Marriage and Family* 55:801–813.

Hornik, Donna. 2001. "Can the Church Get in Step with Stepfamilies?" *U.S. Catholic* 66(7):30–41.

Horowitz, A. 1985. "Sons and Daughters as Caregivers to Older Parents: Differences in Role Performance and Consequences." *The Gerontologist* 25:612–617.

Horwitz, Briana N. and Jenae M. Neiderhiser. 2011. "Gene-Environment Interplay, Family Relationships, and Child Adjustment." *Journal of Marriage and Family* 73(4) (August): 804–816.

House, Anthony. 2002. *A Problematic Solution: Responses to the Marriage Reform Act of 1753.* Retrieved October 5, 2006 (users.ox.ac.uk).

Houts, Carrie R. and Sharon G. Horne. 2008. "The Role of Relationship Attributions in Relationship Satisfaction among Cohabiting Gay Men." *The Family Journal: Counseling and Therapy for Couples and Families* 16(3):240–248.

Houts, Leslie A. 2005. "But Was It Wanted? Young Women's First Voluntary Sexual Intercourse." *Journal of Family Issues* 26(8):1082–1102.

Howard, Hilary. 2012. "A Confederacy of Bachelors." *The New York Times.* August 3. Retrieved November 6, 2012 (http://www.nytimes.com).

Hoyert Donna L., Xu, Jiaquan. 2012. *Deaths: Preliminary Data for 2011. National Vital Statistics Reports.* Vol. 61, no 6. Hyattsville, MD: National Center for Health Statistics.

Hsia, Annie. 2002. "Considering Grandparents' Rights and Parents' Wishes: 'Special Circumstances' Litigation Alternatives." *The Legal Intelligencer* 227(83):7.

Huang, Chien-Chung. 2009. "Mothers' Reports of Nonresident Fathers' Involvement with their Children: Revisiting the Relationship Between Child Support Payment and Visitation." *Family Relations* 58:54–64.

Huang, Chien-Chung, Ronald B. Mincy, and Irwin Garfinkel. 2005. "Child Support Obligations and Low-Income Fathers." *Journal of Marriage and Family* 67(5):1213–1225.

Huber, Joan. 1980. "Will U.S. Fertility Decline toward Zero?" *Sociological Quarterly* 21:481–492.

Huebner, Angela J., Jay A. Mancini, Ryan M. Wilcox, Saralyn R. Grass, and Gabriel A. Grass. 2007. "Parental Deployment and Youth in Military Families: Exploring Uncertainty and Ambiguous Loss." *Family Relations* 56(2):112–122.

Hughes, Mary E. and Linda J. Waite. 2009. "Marital Biography and Health at Mid-Life." *Journal of Health and Social Behavior* 50:344–358.

Hughes, Michael and Walter R. Gove. 1989. "Explaining the Negative Relationship between Social Integration and Mental Health: The Case of Living Alone." Presented at the annual meeting of the American Sociological Association, August, San Francisco, CA.

Hughes, Patrick C. and Fran C. Dickson. 2005. "Communication, Marital Satisfaction, and Religious Orientation in Interfaith Marriages." *Journal of Family Communication* 5(1):25–41.

Hughes, Robert, Jr., Jill Bowers, Elissa T. Mitchell, Sarah Curtiss, and Aaron Ebata. 2012. "Developing Online Family Life Prevention and Education Programs." *Family Relations* 61(5): 711–727.

Hull, K. G. 2008. "Broken Trust: Pursuing Remedies for Victims of Elder Financial Abuse by Agents under Power-of-Attorney Agreements." *Clearinghouse Review* 42(5/6):223–231.

Human Rights Campaign. 2007. "Which States Permit Same-Sex Parents to Be Listed on a Birth Certificate?" Washington, DC: Human Rights Campaign. Retrieved March 30, 2007 (www.hrc.org).

Human Rights Watch. 2011. *Failing Its Families: Lack of Paid leave and Work-Family Supports in the US.* NY, NY and Washington, DC: Human Rights Watch73(4) (August): 804–816. Organization. Retrieved September 10, 2012 (http://www.hrw.org).

Humble, Aine M. 2009. "The Second Time 'Round: Gender Construction in Remarried Couples' Wedding Planning." *Journal of Divorce and Remarriage* 50:260–281.

Hunt, Gail, Carol Levine, and Linda Naiditch. 2005. *Young Caregivers in the U.S: Report of Findings September 2005.* National Alliance for Caregiving.

Hunt, Janet G. and Larry L. Hunt. 1977. "Dilemmas and Contradictions of Status: The Case of the Dual-Career Family." *Social Problems* 24:407–416.

Huston, Ted L. and Heidi Melz. 2004. "The Case for (Promoting) Marriage: The Devil Is in the Details." *Journal of Marriage and Family* 66(4):943–958.

Hutchinson, Katherine M. and Julie A. Cederbaum. 2011. "Talking to Daddy's Little Girl about Sex: Daughters' Reports of Sexual Communication and Support from Fathers." *Journal of Family Issues* 32(4):550–572.

Hutson, Matthew. 2009. "With God As My Wingman." *Psychology Today* 42(1):26.

Hyde, Janet Shibley. 2005. "The Gender Similarities Hypothesis." *American Psychologist* 60(6):581–592.

———. 2007. "New Directions in the Study of Gender Similarities and Differences." *Current Directions in Psychological Science* 16(5):259–263.

Hyman, Mark. 2010. "Sports Training Has Begun for Babies and Toddlers." *The New York Times.* November 30. Retrieved January 9, 2011 (http://www.nytimes.com).

Hymowitz, Kay. 2006. *Marriage and Caste in America: Separate and Unequal Families in a Post-Marital Age.* Chicago: Ivan R. Dee.

Iacuone, David. 2005. "'Real Men Are Tough Guys': Hegemonic Masculinity and Safety in the Construction Industry." *Journal of Men'sStudies* 13(2):247–267.

Iceland, John and Kyle Anne Nelson. 2008. "Hispanic Segregation in Metropolitan America: Exploring the Multiple Forms of Spatial Assimilation." *American Sociological Review* 73(5):741–765.

"If Your Child Is Gay or Lesbian." 2008. Retrieved March 13, 2010 (www.rollercoaster.ie.).

Ihinger-Tallman, Marilyn and Kay Pasley. 1997. "Stepfamilies in 1984 and Today—A Scholarly Perspective." *Marriage and Family Review* 26(1–2):19–41.

"Improvements in Teen Sexual Risk Behavior Flatline." 2006. Press Release. Washington, DC: Advocates for Youth.

Ingersoll-Dayton, Berit, Margaret B. Neal, Jung-Hwa Ha, and Leslie B. Hammer. 2003. "Redressing Inequity in Parent Care among Siblings." *Journal of Marriage and Family* 65(1):201–212.

Ingoldsby, Bron B. 2006a. "Family Origin and Universality." Pp. 67–78 in *Families in Global and Multicultural Perspective,* 2nd ed., edited by Bron B. Ingoldsby and Suzanna D. Smith. New York: Guilford.

———. 2006b. "Mate Selection and Marriage." Pp. 133–146 in *Families in Global and Multicultural Perspective,* 2nd ed., edited by Bron B. Ingoldsby and Suzanna D. Smith. New York: Guilford.

Ingoldsby, Bron B. and Suzanna D. Smith, eds. 2005. *Families in Global and Multicultural Perspective.* 2nd ed. Thousand Oaks, CA: Sage Publications.

Ingraham, Chrys. 2008. *White Weddings: Romancing Heterosexuality in Popular Culture,* 2nd ed. New York: Routledge.

Institute for Women's Policy Research. 2009. *The Gender Wage Gap: 2008* (C350). Washington, DC: Institute for Women's Policy Research.

Interactive Autism Network. 2010. *IAN Research Report—April 2010: Grandparents of Children with ASD, Parts 1 and 2.* Retrieved May 15, 2010 (www.iancommunity.org).

Isaksen, Lise W. 2002. "Toward a Sociology of (Gendered) Disgust." *Journal of Family Issues* 23(7):791–811.

Ishii-Kuntz, Masako. 2000. "Diversity within Asian American Families." Pp. 274–292 in *Handbook of Family Diversity,* edited by David H. Demo, Katherine R. Allen, and Mark Fine. New York: Oxford University Press.

Ishii-Kuntz, Masako, Jessica N. Gomei, Barbara J. Tinsley, and Rose D. Parks. 2009. "Economic Hardship and Adaptation among Asian American Families." *Journal of Family Issues* 31(3):407–420.

Italie, Leanne. 2008. "Parents of Teens Watch Nebraska Safe Law." Associated Press, November 11. Retrieved December 12, 2008 (ap.google.com).

Jackson, Aurora P., Jeong-Kyun Choi, and Peter Bentler. 2009. "Parenting Efficacy and the Early School Adjustment of Poor and Near-Poor Black Children." *Journal of Family Issues* 30(10):1339–1355.

Jackson, Aurora P., Jeong-Kyun Choi, and T. M. Franke. 2009. "Poor Single Mothers with Young Children: Mastery, Relations with Nonresident Fathers, and Child Outcomes." *Social Work Research* 33(2):95–106.

Jackson, Pamela Braboy. 2004. "Role Sequencing: Does Order Matter for Mental Health?" *Journal of Health and Social Behavior* 45:132–154.

Jackson, Pamela Braboy, Sibyl Kleiner, Claudia Geist, and Kara Gebulko. 2011. "Conventions of Courtship: Gender and Race Differences in the Significance of Dating Rituals." *Journal of Family Issues* 32: 629–652.

Jackson, Robert Max. 2006. "Opposing Forces: How, Why, and When Will Gender Inequality Disappear?" Pp. 215–244 in *The Declining Significance of Gender?* edited by Francine D. Blau, Mary C. Brinton, and David B. Grusky. New York: Russell Sage.

Jackson, Robert M. 2007. "Destined for Equality." Pp. 109–116 in *Family in Transition,* 14th ed., edited by Arlene S. Skolnick and Jerome H. Skolnick. Boston, MA and New York: Pearson.

Jackson-Newsom, Julia, Christy M. Buchanan, and Richard M. McDonald. 2008. "Parenting and Perceived Warmth in European American and African American Adolescents." *Journal of Marriage and Family* 70(1):62–75.

Jacobs, Andres. 2006. "Extreme Makeover, Commune Edition." *The New York Times,* June 11. Retrieved June 11, 2006 (www.nytimes.com).

Jacobs, Jerry A. and Kathleen Gerson. 2004. *The Time Divide: Work, Family, and Gender Inequality.* Cambridge, MA: Harvard University Press.

Jacobsen, Linda A., Mary Kent, Marlene Lee, and Mark Mather. 2011. "America's Aging Population." *Population Bulletin* 66(1): February. Washington, DC: Population Reference Bureau.

Jacobsen, Linda A. and Mark Mather. 2010. "U.S. Economic and Social Trends Since 2000." *Population Bulletin* 65(1): February. Washington, DC: Population Reference Bureau.

———. 2011. "A Post-Recession Update on U.S. Economic and Social Trends." *Population Bulletin Update.* December. Washington, DC: Population Reference Bureau. Retrieved August 29, 2012 (http://www.prb.org).

Jacobsen, Linda A., Mark Mather, and Genevieve Dupuis. 2012. "Household Change in the United States." *Population Bulletin* 67 (1): February. Washington, DC: Population Reference Bureau.

Jacobson, Cardell K. and Bryan R. Johnson. 2006. "Interracial Friendship and African American Attitudes about Interracial Marriage." *Journal of Black Studies* 36(4):570–584.

Jacoby, Susan. 1999. "Great Sex: What's Age Got to Do with It?" *Modern Maturity* (September-October), pp. 41–47.

———. 2005. "Sex in America." *AARP The Magazine,* July/August, pp. 57–62, 114.

Jaksch, Mary. 2002. *Learn to Love: A Practical Guide to Fulfilling Relationships.* San Francisco: Chronicle Books.

Jamison, Tyler B. and Lawrence Ganong. 2011. " 'We're Not Living Together': Stayover Relationships among College-Educated Emerging Adults." 2011. *Journal of Social and Personal Relationships* 28(4): 536–557.

Jang, Soo Jung, Allison Zippay, and Rhokeun Park. 2012. "Family Roles as Moderators of the Relationship Between Schedule Flexibility and Stress." *Journal of Marriage and Family* 74(4): 897–912.

Janofsky, Michael. 2001. "Conviction of a Polygamist Raises Fears among Others." *The New York Times,* May 24.

Jansen, Mieke, Dimitri Mortelmans, and Laurent Snoeckx. 2009. "Repartnering and (Re)employment: Strategies to Cope with the Economic Consequences of Partnership Dissolution." *Journal of Marriage and Family* 71:1271–1293.

Jarrell, Anne. 2007. "The Daddy Track." *Boston Globe,* July 8. Retrieved January 18, 2009 (www.boston.com).

Jaschik, Scott. 2009. "Probe of Extra Help for Men." *Inside Higher Ed,* November 2 (www.insidehighered.com).

Jayakody, R. and A. Kalil. 2002. "Social Fathering in Low-income, African American Families with Preschool Children." *Journal of Marriage and Family* 64(2):504–516.

Jayson, Sharon. 2005. "Yep, Life'll Burst That Self-esteem Bubble." *USA Today,* February 16.

———. 2009. "Gay Couples: A Close Look At This Modern Family, Parenting." USA Today, November 5. Retrieved November 20, 2009 (usatoday.com).

———. 2010. "Delaying Kids May Prevent Financial 'Motherhood Penalty'." *USA Today,* April 16.

———. 2011. "White Women More Likely to Be Childless, Census Says." *USA Today.* May 10. Retrieved February 2, 2012 (http://yourlife.usatoday.com).

Jenkins, Nate. 2008. "Nebraska Parents Rush to Leave Kids Before Law Changes." Associated Press, November 13. Retrieved December 12, 2008 (news.yahoo.com).

Jenkins-Guarnieri, Michael A., Stephen L. Wright, and Lynette M. Hudiburgh. 2012. "The Relationships among Attachment Style, Personality Traits, Interpersonal Competency, and Facebook Use." *Journal of Applied Developmental Psychology,* 33(6):294–301.

Jensen, Alexander, Shawn Whiteman, Karen Fingerman, and Kira Birditt. 2013. "'Life Still Isn't Fair': Parental Differential Treatment of Young Adult Siblings." *Journal of Marriage and Family* 75(2):438–452.

Jeong, Jae Y. and Sharon G. Horne. 2009. "Relationship Characteristics of Women in Interracial Same-Sex Relationships." *Journal of Homosexuality* 56(4):443–456.

Jepsen, Lisa K. and Christopher A. Jepsen. 2002. "An Empirical Analysis of the Matching Patterns of Same-Sex and Opposite-Sex Couples?" *Demography* 39(3):435–453.

Jhally, Sut. 2007. *Dreamworlds 3: Desire, Sex & Power in Music Video.* Northampton, MA: Media Education Foundation.

Jin, Lei and Nicholasa Chrisatakis. 2009. "Investigating the Mechanism of Marital Mortality Reduction: The Transition to Widowhood and Quality of Health Care." *Demography* 46(3):605–625.

Jio, Sarah. 2008. "Career Couples Fight Over Who's the 'Trailing Spouse.' " *CNN,* June 26. Retrieved May 27, 2010 (www.cnn.com).

Jo, Moon H. 2002. "Coping with Gender Role Strains in Korean American Families." Pp. 78–83 in *Contemporary Ethnic Families in the United States,* edited by Nijole V. Benekraitis. Upper Saddle River, NJ: Prentice Hall.

John, Robert. 1998. "Native American Families." Pp. 382–421 in *Ethnic Families in America: Patterns and Variations,* edited by Charles H. Mindel, Robert W. Haberstein, and Roosevelt Wright, Jr. Upper Saddle River, NJ: Prentice Hall.

Johnson, Dirk. 1991. "Polygamists Emerge from Secrecy, Seeking Not Just Peace but Respect." *The New York Times,* April 9.

Johnson, Jason B. 2000. "Something Akin to Family: Struggling Parents, Kids, Move in with Their Mentors." *San Francisco Chronicle,* November 10.

Johnson, Leanor Boulin and Robert Staples. 2005. *Black Families at the Crossroads: Challenges and Prospects, 2nd ed.* San Francisco: Jossey-Bass.

Johnson, Matthew D., Joanne Davila, Ronald D. Rogge, Kieran T. Sullivan, Catherine L. Cohan, Erika Lawrence, Benjamin R. Karney, and Thomas N. Bradbury. 2005. "Problem-Solving Skills and Affective Expressions as Predictors of Change in Marital Satisfaction." *Journal of Consulting and Clinical Psychology* 73(1):15–27.

Johnson, Michael P. 2008. *A Typology of Domestic Violence: Intimate Terrorism, Violent Resistance, and Situational Couple Violence.* Lebanon, NH: Northeastern University Press.

———. 2010. "Langhinrichsen-Rolling's Confirmation of the Feminist Analysis of Intimate Partner Violence: Comment on 'Controversies Involving Gender and Intimate Partner Violence in the United States.'" *Sex Roles* 32(3–4):212–219.

Johnson, Michael P. and Kathleen J. Ferraro. 2000. "Research on Domestic Violence in the 1990s: Making Distinctions." *Journal of Marriage and Family* 62(4):948–963.

Johnson, Michael P. and Janel M. Leone. 2005. "The Differential Effects of Intimate Terrorism and Situational Couple Violence: Findings from the National Violence Against Women Survey." *Journal of Family Issues* 26(3):322–349.

Johnson, Phyllis J. 1998. "Performance of Household Tasks by Vietnamese and Laotian Refugees." *Journal of Family Issues* 10(3):245–273.

Johnson, Phyllis J. and Kathrin Stoll. 2008. "Remittance Patterns of Southern Sudanese Refugee Men: Enacting the Global Breadwinner Role." Family Relations 57(4):431–443.

Johnson, Richard W. and Joshua M. Wiener. 2006. "A Profile of Frail Older Americans and Their Caregivers." Urban Institute. March 1. Retrieved May 18, 2007 (www.urban.org/publications).

Johnson, Suzanne M. and Elizabeth O'Connor. 2002. *The Gay Baby Boom: The Psychology of Gay Parenthood.*New York: New York University Press.

Jones, A. J. and M. Galinsky. 2003. "Restructuring the Stepfamily: Old Myths, New Stories." *Social Work* 48(2):228–237.

Jones, A. R., R. M. Hyland, K. N. Parkinson, and A. J. Adamson. 2009. "Developing a Focus Group Approach for Exploring Parents' Perspectives on Childhood Overweight." Nutrition Bulletin 34(2):214–214.

Jones, Antwan. 2010. "Stability of Men's Interracial First Unions: A Test of Educational Differentials and Cohabitation History." *Journal of Family and Economic Issues* 31:241–256.

Jones, Del. 2006. "One of USA's Exports: Love, American Style." *USA Today.* February 14.

Jones, Jeffrey M. 2005. "Most Americans Approve of Interracial Dating." The Gallup Poll. Retrieved November 4, 2006, from http://www.galluppoll.com.

———. 2008. "Most Americans Not Willing to Forgive Unfaithful Spouse." Washington, DC: Gallup Poll News Service. Retrieved December 2, 2008 (www.gallup.com/poll).

———. 2011."Record-High 86% Approve of Black-White Marriages." Gallup Poll. September 12. Retrieved August 29, 2012 (http://www.gallup.com).

Jones, Nicholas A and J. Bullock. 2012. *The Two or More Races Population 2010.* U.S. Census Brief # C2010BR-13. September. Washington DC: U.S. Census Bureau.

Jones, Rachel K., Jacqueline E. Darroch, and Stanley K. Henshaw. 2002. "Patterns in the Socioeconomic Characteristics of Women Obtaining Abortions in 2000–2001." *Perspectives on Sexual and Reproductive Health* 34:226–235.

Jones, Rachel K., Lawrence B. Finer, and Susheela Singh. 2010. "Characteristics of U.S. Abortion Patients, 2008." New York: Guttmacher Institute. Retrieved January 15, 2013 (http://www.guttmacher.org).

Jong-Fast, Molly. 2003. "Out of Step and Having a Baby." *The New York Times,* October 5.

Jordan-Young, Rebecca M. 2012. "Hormones, Context, and 'Brain Gender': A Review of Evidence from Congenital Adrenal Hyperplasia." *Social Science & Medicine*74(11): 1738–1744.

Joseph, Elizabeth. 1991. "My Husband's Nine Wives." *The New York Times,* May 23.

Joshi, Pamela and Karen Bogen. 2007. "Nonstandard Schedules and Young Children's Behavioral Outcomes among Working Low-Income Families." *Journal of Marriage and Family* 60(1):139–156.

Joshi, Pamela, James M. Quane, and Andrew J. Cherlin. 2009. "Contemporary Work and Family Issues Affecting Marriage and Cohabitation among Low-Income Single Mothers." *Family Relations* 58(5):647–661.

Joyce, Amy. 2006. "Kid-Friendly Policies Don't Help Singles." *Washington Post,* September 16.

———. 2007a. "Caring for Dear Old Dad Becomes a Little Easier." *Washington Post,* March 4.

———. 2007b. "Developing Boomerang Mothers." *Washington Post,* March 11.

Joyner, Kara and Grace Kao. 2005. "Interracial Relationships and the Transition to Adulthood." *American Sociological Review* 70(4):563–581.

Joyner, Kara, Wendy Manning, and Ryan Bogle. 2013. *The Stability and Qualities of Same-Sex and Different-Sex Couples in Young Adulthood.* The Center for Family and Demographic Research Working Paper Series, Bowling Green State University. Retrieved April 14, 2013 (http://www.bgsu.edu/organizations/cfdr/file127059.pdf).

Juby, Heather, Jean-Michel Billette, BenoÎt Laplante, and Céline Le Bourdais. 2007. "Nonresident Fathers and Children: Parents' New Unions and Frequency of Contact." *Journal of Family Issues* 28(9):1220–1245.

Judge Vito Titone in Braschi v. Stahl Associates Company 1989.

Juffer, Femmie and Marinus H. van Uzendoorn. 2005. "Behavior Problems and Mental Health Referrals of International Adoptees: A Meta-Analysis." *Journal of the American Medical Association* 293(20):2501–2515.

Junn, Ellen Nan and Chris J. Boyatzis. 2005. *Child Growth and Development 05/06.*12th ed. Guilford, CT: McGraw-Hill/Dushkin.

Kader, Samuel. 1999. *Openly Gay, Openly Christian: How the Bible Really Is Gay Friendly.* Leyland Publications.

Kagan, Marilyn and Neil Einbund. 2008. *Defenders of the Heart: Managing the Habits and Attitudes That Block You from a Richer, More Satisfying Life.*Carlsbad, CA: Hay House.

Kahn, Joan R., Brittany S. McGill, and Suzanne M. Bianchi. 2011. "Help to Family and Friends: Are There Gender Differences at Older Ages?" *Journal of Marriage and Family* 73(1):77–92.

Kallivayalil, Diya. 2004. "Gender and Cultural Socialization in Indian Immigrant Families in the United States." *Feminism and Psychology* 14(4):535–559.

Kalmijn, Matthijs. 1998. "Differentiation and Stratification—Intermarriage and Homogamy: Causes, Patterns, and Trends." *Annual Review of Sociology* 24:395–427.

———. 2012. "The Educational Gradient in Intermarriage: A Comparative Analysis of Immigrant Groups." *Social Forces* 91(2):453–476.

Kalmijn, Matthijs and Frank Van Tubergen. 2010. "A Comparative Perspective on Intermarriage: Explaining Differences among National-Origin Groups in the United States." *Demography* 47(2):459–479.

Kamo, Yoshinori and Min Zhou. 1994. "Living Arrangements of Elderly Chinese and Japanese in the United States." *Journal of Marriage and Family* 56(3):544–558.

Kamp Dush, Claire M. 2011. "Relationship-Specific Investments, Family Chaos, and Cohabitation Dissolution Following a Nonmarital Birth." *Family Relations* 60:586–601.

Kan, Marni L., Susan M. McHale, and Ann C. Crouter. 2008. "Interparental Incongruence in Differential Treatment of Adolescent Siblings: Links with Marital Quality." *Journal of Marriage and Family* 70(May):466–479.

Kane, Emily W. 2012a. *The Gender Trap: Pa[illegible] and the Pitfalls of Raising Boys and Gi[illegible] New York University Press.*

———. 2012b. "Parental F[illegible] and the Gender Trap." Fro[illegible] NYU Press Blog. Retrieved N[illegible] (http://www.fromthesquare.[illegible]

Kantrowitz, Barbara. 2000. "Busy around the Clock." *Newsweek*, July 17, pp. 49–50.

Kantrowitz, Barbara and Karen Springen. 2005. "A Peaceful Adolescence." *Newsweek*, April 25, pp. 58–61.

Kantrowitz, Barbara and Peg Tyre. 2006. "The Fine Art of Letting Go." *Newsweek*, May 22, pp. 49–61.

Karney, Benjamin. 2011. "Stress Is Bad for Couples, Right?" *Family Focus* FF48: F2–F4.

Karney, Benjamin R., David S. Loughran, and Michael S. Pollard. 2012. "Comparing Marital Status and Divorce Status in Civilian and Military Populations." *Journal of Family Issues* 33(12):1572–1594.

Karraker, Amelia and John DeLamater. 2013. "Past-Year Sexual Inactivity among Older Married Persons and Their Partners." *Journal of Marriage and Family* 75:142–163.

Karasik, Rona J. and Raeann R. Hamon. 2007. "Cultural Diversity and Aging Families." Pp. 136–153 in *Cultural Diversity and Families*, edited by Bahira Sherif Traskand Raeann R. Hamon. Thousand Oaks, CA: Sage Publications.

Karim, Jamillah. 2009. *American Muslim Women Negotiating Race, Class, and Gender within the Ummah*. New York: NYU Press.

Karimzadeh, Mohammad Ali and Sedigheh Ghandi. 2008. "Early Marriage: a Policy for Infertility Prevention." *Journal of Family and Reproductive Health* 2(2):61–64.

Karre, Jennifer K. and Nina S. Mounts. 2012. "Nonresidents' Fathers' Parenting Style and the Adjustment of Late Adolescent Boys." *Journal of Family Issues* 33(12): 1642–1657.

Katz, Jackson. 2006. *The Macho Paradox: Why Some Men Hurt Women and and How All Men Can Help*. Naperville, IL: Sourcebooks, Inc.

Katz, Jennifer, Vanessa Tirone, and Erika van der Kloet. 2012. "Moving In and Hooking Up: Women's and Men's Casual Sexual Experiences During the First Two Months of College." *Electronic Journal of Human Sexuality* 15. Retrieved October 19, 2012 (http://www.ejhs.org/volume15/Hookingup.html).

Katz, Jonathan Ned. 2007. *The Invention of Heterosexuality*. Chicago: University of Chicago Press.

Kaufman, Gayle and Eva Bernhardt. 2012. "His and Her Job: What Matters Most for Fertility Plans and Actual Childbearing?" *Family Relations* 61(4): 686–697.

Kaufman, Gayle and Hiromi Taniguchi. 2006. "Gender and Marital Happiness in Later Life." *Journal of Family Issues* 27(6):735–757.

Kaufman, Gayle and Peter Uhlenberg. 1998. "Effects of Life Course Transitions on the Quality of Relationships between Adult Children and Their Parents." *Journal of Marriage and Family* 60(4):924–938.

Kaufman, Leslie. 2008. "Veterans' Families ʾeek Aid for Caregiver Role." *The New York ʾes.* November 12. Retrieved March 16, 2010 ʾ://www.nytimes.com).

Kaukinen, Catherine. 2004. "Status Compatibility, Physical Violence, and Emotional Abuse in Intimate Relationships." *Journal of Marriage and Family* 66:452–471.

Kaukinen, Catherine E., Silke Meyer, and Caroline Akers. 2013. "Status Compatibility and Help-Seeking Behaviors among Female Intimate Partner Violence Victims." *Journal of Interpersonal Violence* 28(3):577–601.

Kaye, Sarah. 2005. "Substance Abuse Treatment and Child Welfare: Systematic Change Is Needed." *Family Focus on . . . Substance Abuse across the Life Span*: FF25: F15–F16. Minneapolis: National Council of Family Relations.

Kaye, Kelleen,Suellentrop,Katherine, and-Sloup, Corinna. 2009. *The Fog Zone: How Misperceptions, Magical Thinking, and Ambivalence Put Young Adults at Risk for Unplanned Pregnancy*. Washington, DC: The National Campaign to Prevent Teen and Unplanned Pregnancy.

Kazdin, Alan E. and Corina Benjet. 2003. "Spanking Children: Evidence and Issues." *Current Directions in Psychological Science*(March): 99–103.

Kearney, Christopher A. 2010. *Casebook in Child Behavior Disorders*. 4th ed. Belmont, CA: Wadsworth/Cengage Learning.

Keaten, James and Lynne Kelly. 2008. "Emotional Intelligence as a Mediator of Family Communication Patterns and Reticence." *Communication Reports* 21(2):104–116.

Kellas, Jody Koenig, Cassandra LeClair-Underberg, and Emily Lamb Normand. 2008. "Stepfamily Address Terms: 'Sometimes They Mean Something and Sometimes They Don't.'" *Journal of Family Communication* 8:238–263.

Kelleher, Elizabeth. 2007. "In Dual-Earner Couples, Family Roles Are Changing in U.S." *USINFO* March 21. Washington, DC: Bureau of International Information Programs, U.S. Department of State. www.america.gov.

Kelly, Joan B. 2007. "Children's Living Arrangements Following Separation and Divorce: Insights from Empirical and Clinical Research." *Family Process* 46(1):35–52.

Kelly, Joan B. and Michael E. Lamb. 2003. "Developmental Issues in Relocation Cases Involving Young Children: When, Whether, and How?" *Journal of Family Psychology* 17:193–205.

Kelly, Katherine Patterson and Lawrence Ganong. 2011a. "Moving to Place: Childhood Cancer Treatment Decision Making in Single-Parent and Repartnered Family Structures." *Qualitative Health Research* 21(3):349–364.

———. 2011b. "Shifting Family Boundaries' After the Diagnosis of Childhood Cancer in Stepfamilies." *Journal of Family Nursing* 17(1):105–132.

Kelly, Maura. 2009. "Women's Voluntary Childlessness: A Radical Rejection of Motherhood?" *WSQ: Women's Studies Quarterly* 37(3/4):157–172.

Kemp, Candace L. 2007. "Grandparent-Grandchild Ties: Reflections on Continuity and Change across Three Generations." *Journal of Family Issues* 28(7):855–881.

Kempe, C. Henry, Frederic N. Silverman, Brandt F. Steele, William Droegemuller, and Henry K. Silver. 1962. "The Battered Child Syndrome." *Journal of the American Medical Association* 181:17–24.

Kenney, Catherine. 2006. "The Power of the Purse: Allocative Systems and Inequality in Couple Households." *Gender and Society* 20(3):354–381.

Kenney, Catherine and Sara S. McLanahan. 2006. "Why Are Cohabiting Relationships More Violent than Marriages?" *Demography* 43(1):127–140.

Kennedy, G. E. and C. E. Kennedy. 1993. "Grandparents: A Special Resource for Children in Stepfamilies." *Journal of Divorce & Remarriage*, 19:45–68.

Kent, Mary Mederios. 2007. "Immigration and America's Black Population." *Population Bulletin* 62(4). Washington, DC: Population Reference Bureau. Retrieved June 10, 2009 (www.prb.com).

Kephart, William. 1971. "Oneida: An Early American Commune." Pp. 481–492 in *Family in Transition: Rethinking Marriage, Sexuality, Child Rearing, and Family Organization*,edited by Arlene S. Skolnickand Jerome H. Skolnick. Boston: Little, Brown.

Kern, Louis J. 1981. *An Ordered Love: Sex Roles and Sexuality in Victorian Utopias—The Shakers, the Mormons, and the Oneida Community*. Chapel Hill: University of North Carolina Press.

Kerr, David C., Deborah Capaldi, Lee Owen, Margit Wiesner, and Katherine Pears. 2011. "Changes in At-Risk American Men's Crime and Substance Use Trajectories Following Fatherhood." *Journal of Marriage and Family* 73(5): 1101–1116.

Kershaw, Sarah. 2003. "Many Immigrants Decide to Embrace Homes for Elderly." *The New York Times*, October 20, pp. A1–A10.

———. 2009a. "Mr. Moms (by Way of Fortune 500)." *The New York Times*, April 23. Retrieved December 23, 2009 (http://www.nytimes.com).

———. 2009b. "Rethinking the Older Woman-Younger Man Relationship." *The New York Times*, October 15.

Kheshgi-Genovese, Zareena and Thomas A. Genovese. 1997. "Developing the Spousal Relationship within Stepfamilies." *Families in Society* 78(3):255–264.

Khrais, Reema. 2012. "Phone Home: Tech Draws Parents, College Kids Closer." Capitol Public Radio. September 25.

Kibria, Nazli. 2000. "Race, Ethnic Options, and Ethnic Binds: Identity Negotiations of Second-Generation Chinese and Korean Americans." *Sociological Perspectives* 43(1):77–95.

"KidsPoll." 2008. www.kidshealth.org.

Kiecolt, K. Jill, Rosemary Blieszner, and Jyoti Savla. 2011. "Long-Term Influences of Intergenerational Ambivalence on Midlife Parents' Psychological Well-Being." *Journal of Marriage and Family* 73(2): 369–382.

Kiernan, K. 1992. "The Impact of Family Disruption in Childhood on Transitions Made in Young Adult Life." *Population Studies,* 46:213–234.

Kiernan, Kathleen. 2002. "Cohabitation in Western Europe: Trends, Issues, and Implications." Pp. 3–31 in *Just Living Together: Implication of Cohabitation on Families, Children, and Social Policy,*edited by Alan Boothand A. C. Crouter. Mahwah, NJ:Erlbaum.

Kiesbye, Stefan. 2009. *Blended Families.* Detroit: Greenhaven Press.

Kilen, Mike. 2011. "'Kardashian Marriages' Rare but Aren't Unheard of in Iowa." *Des Moines Register.* Retrieved April 19, 2013 (http://www.public.iastate.edu/~nscentral/mr/11/1111/kardashian.html).

Killewald, Alexandra. 2013. "A Reconsideration of the Fatherhood Premium: Marriage, Coresidence, Biology, and Fathers' Wages." *American Sociological Review* 78:96–116.

Killian, Timothy and Lawrence H. Ganong. 2002. "Ideology, Context, and Obligations to Assist Older Persons." *Journal of Marriage and Family* 64(4):1080–1088.

Killoren, Sarah E., Shawna M. Thayer, and Kimberly A. Updegraff. 2008. "Conflict Resolution Between Mexican Origin Adolescent Siblings." *Journal of marriage and Family* 70 (December):1200–1212.

Killoren, Sarah, Kimberly Updegraff, F. Scott Christopher, and Adriana J. Umana-Taylor. 2011. "Mothers, Fathers, Peers, and Mexican-Origin Adolescents' Sexual Intentions." *Journal of Marriage and Family* 73(1):209–220.

Kilmer, Beau, Nancy Nicosia, Paul Heaton, and Greg Midgette. 2013. "Efficacy of Frequent Monitoring." *American Journal of Public Health* 103(1):E37–E43.

Kim, Haejeong and Sharon A. DeVaney. 2005. "The Selection of Partial or Full Retirement by Older Workers." *Journal of Family and Economic Issues* 26(3):371–394.

Kim, Hyoun K., Deborah M. Capaldi, and Lynn Crosby. 2007. "Generalizability of Gottman and Colleagues' Affective Process Models of Couples' Relationship Outcomes." *Journal of Marriage and Family* 69(1):55–72.

Kim, Hyoun K., and Patrick C. McKenry. 2002. "The Relationship between Marriage and Psychological Well-Being." *Journal of Family Issues* 23(8):885–911.

Kim, Hyun Sik. 2011. "Consequences of Divorce for Child Development." *American Sociological Review* 76(3):487–511.

Kim, Ji-Yeon, Susan M. McHale, Ann C. Crouter, and D. Wayne Osgood. 2007. "Longitudinal Linkages Between Sibling Relationships and Adjustment from Middle Childhood through Adolescence." *Developmental Psychology* 43(4): 960–973.

Kim, Jeounghee. 2012. "Educational Differences in Marital Dissolution: Comparison of White and African American Women." *Family Relations* 61:811–824.

Kim, Joongbaeck and Hyeyoung Woo. 2011. "The Complex Relationship Between Parental Divorce and Sense of Control." *Journal of Family Issues* 32(8):1050–1072.

Kimbro, Rachel Tolbert. 2008. "Together Forever? Romantic Relationship Characteristics and Prenatal Health Behaviors." *Journal of Marriage and Family* 70(3):756–757.

Kimbro, Rachel Tolbert and Ariela Schachter. 2011. "Neighborhood Poverty and Maternal Fears of Children's Outdoor Play." *Family Relations* 60(4) (October): 461–475.

Kimerling, Rachel, Jennifer Alvarez, Joanne Pavao, Katelyn P. Mack, Mark W. Smith, and Nikki Baumrind. 2009. "Unemployment among Women: Examining the Relationship of Physical and Psychological Intimate Partner Violence and Posttraumatic Stress Disorder." *Journal of Interpersonal Violence* 24(3):450–463.

Kimmel, Michael S. 1995. "Misogynists, Masculinist Mentors, and Male Supporters: Men's Responses to Feminism." Pp. 561–572 in *Women: A Feminist Perspective,* 5th ed., edited by Jo Freeman. Mountain View, CA: Mayfield.

———. 2000. *The Gendered Society.* New York: Oxford University Press.

———. 2001. "Manhood and Violence: The Deadliest Equation." *Newsday,* March 8.

———. 2002."'Gender Symmetry' in Domestic Violence." *Violence Against Women* 8:1332–1363.

Kimmel, Michael S. and Michael A. Messner. 1998. *Men's Lives.*4th ed. Boston: Allyn & Bacon.

Kindlon, Dan and Michael Thompson. 1999. *Raising Cain: Protecting the Emotional Life of Boys.* New York: Ballantine.

King, Anthony E. and Terrence T. Allen. 2009. "Personal Characteristics of the Ideal African American Marriage Partner." *Journal of Black Studies* 39(4):570–588.

King, Valarie. 2007. "When Children Have Two Mothers: Relationships with Nonresident Mothers, Stepmothers, and Fathers." *Journal of Marriage and Family,* 69(5):1178–1193.

———. 2009. "Stepfamily Formation: Implications for Adolescent Ties to Mothers, Nonresident Fathers, and Stepfathers." *Journal of Marriage and Family* 71(November):954–968.

King, Valarie and Mindy E. Scott. 2005. "A Comparison of Cohabiting Relationships among Older and Younger Adults." *Journal of Marriage and Family* 67(2):271–285.

King, Valarie, Katherine Stamps Mitchell, and Daniel N. Hawkins. 2010. "Adolescents with Two Nonresident Biological Parents: Living Arrangements, Parental Involvement, and Well-Being." *Journal of Family Issues* 31(1):3–30.

King, Valarie and Juliana M. Sobolewski. 2006. "Nonresident Fathers' Contributions to Adolescent Well-Being." *Journal of Marriage and Family* 68(3):537–557.

Kingston, Anne. 2009. "No Kids, No Grief." *Maclean's* 122(29/30):38–41.

Kinkler, Lori A. and Abbie E. Goldberg. 2011. "Working with What We've Got: Perceptions of Barriers and Supports among Small-Metropolitan-Area Same-Sex Adopting Couples." *Family Relations* 60(4): 387–403.

Kinsey, Alfred, Wardell B. Pomeroy, and Clyde E. Martin. 1948. *Sexual Behavior in the Human Male.* Philadelphia, PA: Saunders.

———. 1953. *Sexual Behavior in the Human Female.* Philadelphia, PA: Saunders.

Kirby, Douglas, B. A. Laris, and Lori Rolleri. 2006. *Sex and HIV Education Programs for Youth: Their Impact and Important Characteristics.* Research Triangle Park, NC: Family Health International. May 13.

Kirmeyer, Sharon E. and Brady E. Hamilton. 2011. "Childbearing Differences among Three Generations of U.S. Women." *NCHS Data Brief, No. 68.* August.U.S. Department of health and Human Services, Centers for Disease Control and Prevention, National Center for health Statistics.

Kiselica, Mark S. and Matt Englar-Carlson. 2010. "Identifying, Affirming, and Building upon Male Strengths: The Positive/Psychology Positive/Masculinity Model of Psychotherapy with Boys and Men." *Psychotherapy: Theory, Research, Practice, and Training* 47(3): 276– 287.

Kisler, Tiffani S. and F. Scott Christopher. 2008. "Sexual Exchanges and Relationship Satisfaction: Testing the Role of Sexual Satisfaction as a Mediator and Gender as a Moderator." *Journal of Social and Personal Relationships* 25(4):587–602.

Kitano, Harry and Roger Daniels. 1995. *Asian Americans: Emerging Minorities,* 2nd ed. Englewood Cliffs, NJ:Prentice Hall.

Kiviat, Barbara. 2011. "Below the Line." *Time.* November 28: 35–40.

Kjobli, John and Kristine Amlund Hagen. 2009. "A Mediation Model of Interparental Collaboration, Parenting Practices, and Child Externalizing Behavior in a Clinical Sample." *Family Relations* 58(July):275–288.

Kleber, Rolf J., Charles R. Figley, P. R. Barthold, and John P. Wilson. 1997. "Beyond Trauma: Cultural and Societal Dynamics." *Contemporary Psychology* 42(6):516–527.

Kleinfield, N. R. 2003. "Around Tree, Smiles Even for Wives No. 2 and 3." *The New York Times,* December 24.

———. 2011. "Baby Makes Four, and Complications." *The New York Times.* June 19. Retrieved September 13, 2011 (http://www.nytimes.com).

Kleinplatz, P. J., A. D. Me'nard, N. Paradis, M. Campbell, T. Dalgleish, A. Segovia, and K. Davis. 2009. "From Closet to Reality: Optimal Sexuality among the Elderly." *The Irish Psychiatrist* 10: 15–18.

Kliff, Sarah. 2010. "Outing Abortion, from Town Halls to Twitter." *Newsweek,* March 3. www.newsweek.com.

Klinenberg, Eric, Stacy Torres, and Elena Portacolone. 2012. "Aging Alone in America." A Briefing Paper Prepared for the Council on Contemporary Families for Older American[illegible] Month May 2012. Retrieved May 21, [illegible] (http://www.contemporaryfa[illegible] /aging.html).

Klohnen, E. C. and S. J. Bera[illegible] and Experiential Patterns of [illegible]

Securely Attached Women Across Adulthood: A 30-Year Longitudinal Perspective." *Journal of Personality and Social Psychology* 74:211–223.

Klohnen, Eva C. and Gerald A. Mendelsohn. 1998. "Partner Selection for Personality Characteristics: A Couple-Centered Approach." *Personality & Social Psychology Bulletin* 24(3): 268–277.

Kluger, Jeffrey. 2004. "The Power of Love." *Time,* January 19.

———. 2006. "The New Science of Siblings." *Time,* July 10, pp. 47–56.

———. 2010. "Be Careful What You Wish For." *Time,* February 22, pp. 68–72.

Kluwer, Esther S. 2011. "Psychological Perspectives on Gender Deviance Neutralization." *Journal of Family Theory and Review* 3(1): 14–17.

Kluwer, Esther S., Jose A. M. Heesink, and Evert Van de Vliert. 2002. "The Division of Labor across the Transition to Parenthood: A Justice Perspective." *Journal of Marriage and Family* 64(4):930–943.

Knapp, Mark L. and Anita L. Vangelisti. 2009. *Interpersonal Communication and Human Relationships.* Boston: Pearson Allyn & Bacon.

Knight, George, Cady Berkel, Adriana Umana-Taylor, Nancy Gonzales, Idean Ettekal, Maryanne Jaconis, and Brenna Boyd. 2011. "The Familial Socialization of Culturally Related Values in Mexican American Families." *Journal of marriage and Family* 73(5): 913–925.

Knobloch, Leanne and Lynne Knobloch-Fedders. 2010. "The Role of Relational Uncertainty in Depressive Symptoms and Relationship Quality: An Actor-Partner Interdependence Model." *Journal of Social and Personal Relationships* 27(1):137–159.

Knobloch, Leanne K., Denise H. Solomon, and Jennifer A. Theiss. 2006. "The Role of Intimacy in the Production and Perception of Relationship Talk within Courtship." *Communication Research* 33(4):211–241.

———. 2013. "Experiences of U.S. Military Couples During the Post-Deployment Transitions: Applying the Relational Turbulence Model." *Journal of Social and Personal Relationships* 29(4): 423–450.

Knoester, Chris and Alan Booth. 2000. "Barriers to Divorce: When Are They Effective? When Are They Not?" *Journal of Family Issues* 21:78–99.

Knoester, Chris, Dana L. Haynie, and Crystal M. Stephens. 2006. "Parenting Practices and Adolescents' Friendship Networks." *Journal of Marriage and Family* 68(5):1247–1260.

Knoester, Chris, Richard J. Petts, and David J. Eggebeen. 2007. "Commitments to Fathering and the Well-Being and Social Participation of New, Disadvantaged Fathers." *Journal of Marriage and Family* 69(4):991–1004.

Knoll, Melissa, Christopher Tamborini, and Kevin Whitman. 2012. "I Do . . . Want to Save: Marriage and Retirement Savings in Young Households." *Journal of Marriage and Family* 74(1):86–100.

Knox, David and Marty E. Zusman. 2009. "Sexuality in Black and White: Data from 783 Undergraduates." *Electronic Journal of Human Sexuality* 12, June 26. www.ejhs.org.

Knowlton, Brian. 2010. "Muslim Women Gain Higher Profile in U.S." *The New York Times.* December 27. Retrieved December 31, 2010 (http://www.nytimes.com).

Knudson-Martin, Carmen. 2012. "Attachment in Adult Relationships: A Feminist Perspective." *Journal of Family Theory & Review*4(4): 299– 305.

Knudson-Martin, Carmen and Martha J. Laughlin. 2005. "Gender and Sexual Orientation in Family Therapy: Toward a Postgender Approach." *Family Relations* 54(1): 101–115.

Knudson-Martin, Carmen and Anne Rankin Mahoney. 2009. *Couples, Gender, and Power: Creating Change in Intimate Relationships.* New York:Springer.

Kobliner, Beth. 2010. "Guiding a Child to Financial Independence." *The New York Times.* November 4. Retrieved January 9, 2011 (http://www.nytimes.com).

Koch, Wendy. 2009. "Savings Plan Benefits Teens Leaving Foster Care." *USA Today,* June 15. Retrieved February 10, 2010 (www.usatoday.com).

Koesten, Joy, Paul Schrodt, and Debra Ford. 2009. "Cognitive Flexibility as a Mediator of Family Communication Environments and Young Adults' Well-Being." *Health Communication* 24:82–94.

Kohl, Gillian Ferris. 2012. "Keeping Kids Connected with Their Jailed Parents." National Public Radio. Retrieved July 18, 2012 (http://www.npr.org).

Kohlberg, Lawrence. 1966. "A Cognitive–Developmental Analysis of Children's Sex-role Concepts and Attitudes." Pp. 82– 173 in *The Development of Sex Differences,* edited by Eleanor E. Maccoby. Palo Alto, CA: Stanford University Press.

Kolata, Gina. 2009. "Picture Emerging on Genetic Risks of IVF." *The New York Times,* February 17, D1.

Komsi, Niina, Katri Raikkonen, Anu-Katriina Pesonen, Kati Heinonen, Pertti Keskivaara, Anna-liisa Japvenpaa, and Timo E. Strandberg. 2006. "Continuity of Temperament from Infancy to Middle Childhood." *Infant Behavior and Development* 29(4):494–508.

Konigsberg, Ruth Davis. 2011. "Chore Wars." *Time.*August 8. Retrieved February 4, 2012 (http://www.time.com).

Kools, S. 2008. "From Heritage to Postmodern Grounded Theorizing: Forty Years of Grounded Theory." Studies in Symbolic Interaction 32: 73–86.

Kornicha, Sabino, Julie Brines, and Katrina Leupp. 2013. "Egalitarianism, Housework, and Sexual Frequency in Marriage." *American Sociological Review* 78(1):26–50.

Koro-Ljungberg, Mirka, and Regina Bussing. 2009. "The Management of Courtesy Stigma in the Lives of Families with Teenagers with ADHD." *Journal of Family Issues* 30(9):1175–1200.

Koropeckyj-Cox, Tanya and Gretchen Pendell. 2007. "The Gender Gap in Attitudes about Childlessness in the United States." *Journal of Marriage & Family* 69(4):899–915.

Koslow, Sally. 2012. *Slouching Toward Adulthood: Observations from the Not-so-empty Nest.* New York: Penguin Books.

Kossinets, Gueorgi and Duncan J. Watts. 2009. "Origins of Homophily in an Evolving Social Network." *American Journal of Sociology* 115(2): 405–450.

Kouros, Chrystyna, Christine Merrilees, and E. Mark Cummings. 2008. "Marital Conflict and Children's Emotional Security in the Context of Parental Depression." *Journal of Marriage and Family* 70(3):684–697.

Kramer, Elena. 2011. "Where the Women Are—And Aren't." *Harvard Magazine.* January–February. Retrieved February 11, 2013 (http://www.harvardmagazine.com).

Kramer, Sarah. 2011. "Three Generations Under One Roof." *The New York Times.* September 23. Retrieved September 27, 2011 (http://www.nytimes.com).

Krampe, Edythe M. 2009. "When Is the Father Really There? A Conceptual Reformulation of Father Presence." *Journal of Family Issues* 30(7):875–897.

Krampe, Edythe M., and Rae R. Newton. 2012. "Reflecting on the Father: Childhood Family Structure and Women's Paternal Relationships." *Journal of Family Issues* 33(6):773–800.

Kravets, David. 2005. "Custody, Support Applied to Gays in Calif." *USA Today,* August 23.

Kreider, Rose M. 2003. *Adopted Children and Stepchildren: 2000.* Census 2000 Special Reports CENSR-6. August.Washington, DC: U.S. Census Bureau.

———. 2006. "Remarriage in the United States." Retrieved February 20, 2013 (http://www.census.gov/hhes/socdemo/marriage/data/sipp/us-remarriage-poster.pdf).

———. 2010. "Increase in Opposite-sex Cohabiting Couples from 2009 to 2010 in the Annual Social and Economic Supplement (ASEC) to the Current Population survey (CPS)." *Housing and Household Economic Statistics Division Working Paper.* Washington, DC: U.S. Bureau of the Census, Housing and Household Economics Statistics Division.

Kreider, Rose M., and Dianna B. Elliott. 2009a. *America's Families and Living Arrangements: 2007.* Current Population Reports, P20-561. U.S. Census Bureau, Washington, DC.

———. 2009b. "The Complex Living Arrangements of Children and Their Unmarried Parents." Population Association of America 2009 Poster Presentation, Detroit, May 2.

Kreider, Rose M. and Renee Ellis. 2011a. *Living Arrangements of Children.*Washington DE: U.S. Census Bureau, Department of Commerce. June.

———. 2011b. *Number, Timing, and Duration of Marriages and Divorces: 2009.* Washington DE: U.S. Census Bureau, Department of Commerce. May.

Kreider, Rose M. and Jason M. Fields. 2005. *Living Arrangements of Children: 2001.*Current Population Reports P70-104.Washington, DC: U.S. Census Bureau.

Krugman, Paul. 2006. "The Great Wealth Transfer." *Rolling Stone,* December 14, pp. 44–48.

Krumrei, Elizabeth J., Annette Mahoney, and Kenneth I. Pargament. 2011. "Demonization of Divorce: Prevalence Rates and Links to Postdivorce Adjustment." *Family Relations* 60:90–103.

Kulkin, Heidi S., June Williams, Heath F. Borne, Dana de la Bretonne, Judy Laurendine. 2008. "A Review of Research on Violence in Same-Gender Couples." *Journal of Homosexuality* 53(4):71–87.

Kumpfer, Karol and Connie Tait. 2000. *Family Skills Training for Parents and Children.* Juvenile Justice Bulletin. April.

Kuperberg, Arielle. 2012. "Reassessing Differences in Work and Income in Cohabitation and Marriage." *Journal of Marriage and Family* 74(4): 688–707.

Kuperminc, Gabriel P., Gregory J. Jurkovic, and Sean Casey. 2009. "Relation of Filial Responsibility to the Personal and Social Adjustment of Latino Adolescents from Immigrant Families." *Journal of Family Psychology* 23(1):14–22.

Kurdek, Lawrence A. 1991. "The Relations between Reported Well-Being and Divorce History, Availability of a Proximate Adult, and Gender." *Journal of Marriage and Family* 53(1):71–78.

———. 1994. "Areas of Conflict for Gay, Lesbian, and Heterosexual Couples: What Couples Argue about Influences Relationship Satisfaction." *Journal of Marriage and Family* 56(4):923–934.

———. 2005. "Gender and Marital Satisfaction Early in Marriage: A Growth Curve Approach." *Journal of Marriage and Family* 67(1):68–84.

———. 2006. "Differences between Partners from Heterosexual, Gay, and Lesbian Cohabiting Couples." *Journal of Marriage and Family* 68(2):509–528.

———. 2007. "The Allocation of Household Labor by Partners in Gay and Lesbian Couples." *Journal of Family Issues* 28(1):132–148.

Kurdek, Lawrence A. and Mark A. Fine. 1991. "Cognitive Correlates of Satisfaction for Mothers and Stepfathers in Stepfather Families." *Journal of Marriage and Family* 53(3):565–572.

Kurtz, Stanley. 2003. "Beyond Gay Marriage: The Road to Polygamy." *The Weekly Standard,* August 11. Retrieved November 2, 2009 (www.weeklystandard.com).

———. 2006. "Big Love, from the Set." *National Review,* March 13.

Kurz, Demie. 1993. "Physical Assaults by Husbands: A Major Social Problem." Pp. 88–103 in *Current Controversies on Family Violence,* edited by Richard J. Gelles and Donileen R. Loseke. Newbury Park, CA: Sage Publications.

Kutner, Lawrence. 1988. "Parent and Child: Working at Home; or, The Midday Career Change." *The New York Times,* December 8.

Lacey, Rachel Saul, Alan Reifman, Jean Pearson Scott, Steven M. Harris, and Jacki Fitzpatrick. 2004. "Sexual-Moral Attitudes, Love Styles, and Mate Selection." *Journal of Sex Research* 41(2):121–129.

Lacy, Karyn R. 2007. *Blue-Chip Black: Race, Class, and Status in the New Black Middle Class.* Berkeley: University of California Press.

Lahaie, Claudia, Alison Earle, and Jody Heymann. 2012. "An Uneven Burden: Social Disparities in Adult Caregiving Responsibilities, Working Conditions, and Caregiver Outcomes." *Research on Aging* 35(3):243–274.

Lam, Brian Trung. 2005. "Self-Esteem among Vietnamese American Adolescents: The Role of Self-Construal, Family Cohesion, and Social Support." *Journal of Ethnic and Cultural Diversity in Social Work* 14(3/4):21–34.

Lam, Ching Man and Wai Man Kwong. 2012. "The 'Paradox of Empowerment' in Parent Education: A Reflexive Examination of Parents' Pedagogical Expectations." *Family Relations* 61(1):65–74.

Lam, Chun, Susan McHale, and Ann Crouter. 2012. "The Division of Household Labor: longitudinal Changes and Within-Couple Variation." *Journal of Marriage and Family* 74(5):944–952.

Lam, Chun, Susan McHale, and Kimberly Updergraff. 2012. "Gender Dynamics in Mexican American Families: Connecting Mothers', Fathers', and Youths' Experiences." *Sex Roles* 67(1/2):17–28.

Lamanna, Mary Ann. 1977. "The Value of Children to Natural and Adoptive Parents." PhD dissertation, Department of Sociology, University of Notre Dame, Notre Dame, IN.

Lambert, Nathaniel M. and David C. Dollahite. 2006. "How Religiosity Helps Couples Prevent, Resolve, and Overcome Marital Conflict." *Family Relations* 55(4):439–449.

Lan, Pei-Chia. 2002. "Subcontracting Filial Piety: Elder Care in Ethnic Chinese Immigrant Families in California." *Journal of Family Issues* 23(7):812–835.

Landale, Nancy S., Robert Schoen, and Kimberly Daniels. 2009. "Early Family Formation among White, Black, and Mexican American Women." *Journal of Family Issues* 31(4):445–474.

Landau, Iddo. 2008. "Problems with Feminist Standpoint Theory in Science Education." *Science & Education* 17(10):1081–1088.

Lane, Marian, Laurel Hourani, Robert Bray, and Jason Williams. 2012. "Prevalence of Perceived Stress and Mental Health Indicators among Reserve-Component and Active-Duty Military Personnel." *American Journal of Public Health* 102(6):1213–1220.

Lane, Wendy G., David M. Rubin, Ragin Monteith, and Cindy Christian. 2002. "Racial Differences in the Evaluation of Pediatric Fractures for Physical Abuse." *Journal of the American Medical Association* 288(13) (www.jama.org).

Lane-Steele, Laura. 2011. "Studs and Protest-Hypermasculinity: The Tomboyism with Black Lesbian Female Masculinity." *Journal of Lesbian Studies* 15(4):480–492.

Langeslag, Sandara J. E., Peter Muris, and Ingmar H. A. Franken. 2012. *Journal of Sex Research.* Retrieved May 25, 2013 (http://www.tandfonline.com/doi/abs/10.1080/00224499.2012.714011?journalCode=hjsr20#preview).

Langeslag, Sandra J. E., Frederik M van der Veen, and Durk Fekkes. 2012. "Blood Levels of Serotonin Are Differentially Affected by Romantic Love in Men and Women." *Federation of European Psychophysiology Societies* 26(2):92–98.

Langhinrichsen-Rohling, Jennifer, Russel E. Palarea, Jennifer Cohen, and Martin L.Rohling. 2000. "Breaking Up Is Hard to Do: Unwanted Pursuit Behaviors Following the Dissolution of a Romantic Relationship." *Violence and Victims* 15(1):73–90.

Lannutti, Pamela J. 2007. "The Influence of Same-Sex Marriage on the Understanding of Same-Sex Relationships." *Journal of Homosexuality* 53(3):135–151.

———. 2013. "Same-Sex Marriage and Privacy Management: Examining Couples' Communication with Family Members." *Journal of Family Communication* 13(1): 60–75.

LaPierre, Tracey A. 2009. "Marital Status and Depressive Symptoms Over Time: Age and Gender Variations." *Family Relations* 58(4):404–416.

———. 2011. "Parenting a Second Time Around: Grandparents Raising Their Grandchildren." *Family Focus* FF49(Summer): F15–16.

Lara, Cristina. 2013. "A Letter to College Women: On (Not) Finding Your Husband." Retrieved May 30, 2013, from http://www.huffingtonpost.com/cristina-lara/a-letter-to-college-women_b_3034014.html.

Lara, Teena and Jill Duba Onedera. 2008. "Inter-Religion Marriages." Pp. 213–228 in *The Role of Religion and Marriage and Family Counseling,* edited by Jill Duba Onedera. New York: Taylor and Francis.

Lareau, Annette. 2003a. "The Long-Lost Cousins of the Middle Class." *The New York Times,* December 20.

———. 2003b. *Unequal Childhoods: Class, Race, and Family Life.* Berkeley: University of California Press.

———. 2006. "Unequal Childhoods: Class, Race, and Family Life." Pp. 537–548 in *The Inequality Reader: Contemporary and Foundational Readings in Class, Race, and Gender,* edited by David B. Grusky and Szonja Szelenyi. Boulder, CO: Westview.

———. 2011. *Unequal Childhoods: Class, Race, and Family Life,* 2nd edition. Berkeley: University of California Press.

———. 2012. "Using the Terms Hypothesis and Variable for Qualitative Work: A Critical Reflection." *Journal of Marriage and Family* 74(4): 671–677.

LaRossa, Ralph. 2009. "Single-Parent Fam[illegible] Discourse in Popular Magazines and [illegible] Science Journals." *Journal of Mar[illegible] Family* 71(2):235–239.

Larrivée, Marie-Claude,Louise Hamelin Brabant, and Genevieve Lessard. 2012. "Knowledge Translation in the Field of Violence against Women and Children." *Children and Youth Services Review* 34:2381–2391.

Larson, Jeffry H. and Rachel Hickman. 2004. "Are College Marriage Textbooks Teaching Students the Premarital Predictors of Marital Quality?" *Family Relations* 53(4):385–392.

Larzelere, Robert E. 2008. "Disciplinary Spanking: The Scientific Evidence." *Journal of Developmental and Behavioral Pediatrics* 29(4): 334–335.

Lasch, Christopher. 1977. *Haven in a Heartless World: The Family Besieged.* New York: Basic Books.

———. 1980. *The Culture of Narcissism.* New York: Warner Books.

Lasen, Amparo and Elena Casado. 2012. "Mobile Telephony and the Remediation of Couple Intimacy." *Feminist Media Studies* 12(4): 550–559.

Latham, Melanie. 2008. "The Shape of Things to Come: Feminism, Regulation and Cosmetic Surgery." *Medical Law Review* 16:437–457.

Lau, Anna S., David T. Takeuchi, and Margarita Alegria. 2006. "Parent-to-Child Aggression among Asian American Parents: Culture, Context, and Vulnerability." *Journal of Marriage and Family* 68(5):1261–1275.

Lauby, Mary R. and Sue Else. 2008. "Recession Can Be Deadly for Domestic Abuse Victims." *Boston Globe,* December 25. Retrieved June 8, 2009 (bostonglobe.com).

Lauer, Sean R. and Carrie Yodanis. 2011. "Individualized Marriage and the Integration of Resources." *Journal of Marriage and Family* 73(June):669–683.

Laughlin, Lynda. 2011. "Maternity Leave and Employment Patterns of First-Time Mothers: 1961–2008." *Household Economic Studies.*Washington, DC: U.S. Census Bureau and U.S. Department of Commerce. October.

Laughlin, Lynda and Joseph Rukus. 2009. "Who's Minding the Kids in the Summer? Child Care Arrangements for Summer 2006." Presented at the Annual Meeting of the *Population Association of America,* April 30–May 2.

Laumann, Edward, Aniruddha Das, and Linda Waite. 2008. "Sexual Dysfunction among Older Adults: Prevalence and Risk Factors from a Nationally Representative U.S. Probability Sample of Men and Women 57–85 Years of Age." *Journal of Sexual Medicine* 5(10):2300–2311.

Laumann, Edward, John H. Gagnon, Robert T. Michael, and Stuart Michaels. 1994. *The Social Organization of Sexuality: Sexual Practices in the United States.*Chicago: University of Chicago Press.

Laumann, Edward, Sara Leitsch, and Linda Waite. 2008. "Elder Mistreatment in the United States: Prevalence Estimates from a Nationally Representative Study." *Journals of Gerontology Series B: Social Sciences* 63(4):5248–5254.

Laumann-Billings, L. and R. E. Emery. 2000. "Distress among Young Adults from Divorced Families." *Journal of Family Psychology* 14(4):671.

Lavee, Yoav and Ruth Katz. 2002. "Division of Labor, Perceived Fairness, and Marital Quality: The Effect of Gender Ideology." *Journal of Marriage and Family* 64:27–39.

Lavery, Diana. 2012. "More Mothers of Young Children in Labor Force." Population Reference Bureau. Retrieved November 13, 2012 (http://www.prb.org).

Lavin, Judy. 2003. "Smoothing the Step-Parenting Transition." SelfGrowth.com. Retrieved April 26, 2007 (www.selfgrowth.com).

Lavy, Shiri, Mario Mikulincer, Phillip R. Shaver, and Omri Gillath. 2009. "Intrusiveness in Romantic Relationships: A Cross-cultural Perspective on Imbalances Between Proximity and Autonomy." *Journal of Social and Personal Relationships* 26(6–7):989–1008.

Lawrence, K. and E. S. Byers. 1995. "Sexual Satisfaction in Long-Term Heterosexual Relationships: The Interpersonal Exchange Model of Sexual Satisfaction." *Personal Relationships* 2:267–285.

Lawrence et al. v. Texas. 2003. 539 U.S. 558.

Lawson, Willow. 2004a. "Encouraging Signs: How Your Partner Responds to Your Good News Speaks Volumes." *Psychology Today* (January/February):22.

———. 2004b. "The Glee Club: Positive Psychologists Want to Teach You to Be Happier." *Psychology Today* (January/February):34–40.

LeBlanc, Allen J. and Richard G. Wright. 2000. "Reciprocity and Depression in AIDS Caregiving." *Sociological Perspectives* 43(4): 631–649.

Le, Thao N. 2005. "Narcissism and Immature Love As Mediators of Vertical Individualism and Ludic Love Style." *Journal of Social and Personal Relationships* 22(4):543–560.

Ledbetter, Andrew M. 2009. "Family Communication Patterns and Relational Maintenance Behavior: Direct and Mediated Associations with Friendship Closeness." *Human Communication Research* 36:130–147.

Ledermann, Thomas, Guy Bodenmann, Myriam Rudaz, and Thomas N. Bradbury. 2010. "Stress, Communication, and Marital Quality in Couples." *Family Relations* 59(2):195–206.

Lee, Arlene F., Philip M. Genty, and Mimi Laver. 2005. *The Impact of the Adoption and Safe Families Act on Children of Incarcerated Parents.*Washington, DC: Child Welfare League of America.

Lee, Cameron and Judith Iverson-Gilbert. 2003. "Demand, Support, and Perception in Family-related Stress among Protestant Clergy." *Family Relations* 52(3):249–257.

Lee, Carol E. 2006. "Sibling Seeks Same to Share Apartment." *The New York Times,* January 29.

Lee, Chu-Yuan, Jared Anderson, Jason Horowitz, and Gerald August. 2009. "Family Income and Parenting: The Role of Parental Depression and Social Support." *Family Relations* 58(October): 417–430.

Lee, Elizabeth A. Ewing and Wendy Troop-Gordon.2011. "Peer Socialization of Masculinity and Femininity: Differential Effects of Overt and Relational Forms of Peer Victimization." *British Journal of Developmental Psychology* 29(2): 197–213.

Lee, Eunju, Glenna Spitze, and John R. Logan. 2003. "Social Support to Parents-in-Law: The Interplay of Gender and Kin Hierarchies." *Journal of Marriage and Family* 65(2):396–403.

Lee, F. R. 2001. "Trying to Soothe the Fears Hiding Behind the Veil." *The New York Times,* September 23.

Lee, Gary R., Chuck W. Peek, and Raymond T. Coward. 1998. "Race Differences in Filial Responsibility Expectations among Older Parents." *Journal of Marriage and Family* 60(2):404–412.

Lee, Jacrim, Mary Jo Katras, and Jean W. Bauer. 2009. "Children's Birthday Celebrations from the Experiences of Low-Income Rural Mothers." *Journal of Family Issues* 30(4):532–553.

Lee, Jennifer. 2007. "The Incredible Flying Granny Nanny." *The New York Times,*May 10. Retrieved May 11, 2007 (www.nytimes.com).

Lee, John Alan. 1973. *The Colours of Love.*Toronto: New Press.

———. 1981. "Forbidden Colors of Love: Patterns of Gay Love." Pp. 128–139 in *Single Life: Unmarried Adults in Social Context,* edited by Peter J. Stein. New York: St. Martin's.

Lee, Kristen Schultz and Hiroshi Ono. 2012. "Marriage, Cohabitation, and Happiness: A Cross-National Analysis of 27 Countries." *Journal of Marriage and Family* 74(5):953–972.

Lee, Michael. 2013. "Comparing Supports for LGBT Aging in Rural versus Urban Areas." *Journal of Gerontological Social Work* 56(2):112–124.

Lee, Mo-Yee. 2002. "A Model of Children's Postdivorce Behavioral Adjustment in Maternal- and Dual-residence Arrangements." *Journal of Family Issues* 23(5):672–697.

Lee, Rosalyn, Mikel Walters, Jeffrey Hall, and Kathleen Basile. 2012. "Behavioral and Attitudinal Factors Differentiating Male Intimate Partner Violence Perpetrators with and without a History of childhood Family Violence." *Journal of Family Violence* 28:85–94.

Lee, Sharon M. 1998. "Asian Americans: Diverse and Growing." *Population Bulletin* 53(2). Washington, DC: Population Reference Bureau.

Lee, Shawna J., Jennifer L. Bellamy, and Neil B. Guterman. 2009. "Fathers, Physical Child Abuse, and Neglect." *Child Maltreatment* 14(3):227–231.

Lee, Wai-Yung, Man-Lun Ng, Ben Cheung, and Joyce Wayung. 2010. "Capturing Children's Response to Parental Conflict and Making Use of It." *Family Process* 49(1):43–58.

Lee, Youjung and Laura Smith. 2012. "Qualitative Research on Korean American Dementia Caregivers' Perception of Caregiving: Heterogeneity between Spouse Caregivers and Child Caregivers." *Journal of Human Behavior in the Social Environment* 22(2):115–129.

Leeker, Olivia and Al Carolozzi.2012. "Effects of Sex, Sexual Orientation, Infidelity Expectations, and Love on Distress Related to Emotional and Sexual Infidelity." *Journal of Marital and Family Therapy* (DOI:10.1111/j.1752-0606.2012.00331.x).

Lees, Janet and Jan Horwath. 2009. "'Religious Parents . . . Just Want the Best for Their Kids': Young People's Perspectives on the Influence of Religious Beliefs on Parenting." *Children & Society* 23(3):162–175.

Leff, Lisa. 2006. "State Lawyer Faces Tough Questions Arguing Against Gay Marriage." Associated Press, July 11. Retrieved July 11, 2006 (www.mercurynews.com).

———. 2009. "Some Gays Seek Renewed Focus on Civil Unions." The Washington Post, November 28. Retrieved December 1, 2009 (www.washingtonpost.com).

"Legislation Introduced to Repeal Discriminatory Defense of Marriage Act." 2009. American Civil Liberties Union, September 15. Retrieved September 29, 2009 (www.aclu.org).

Lehmann-Haupt, Rachel. 2009. "Why I Froze My Eggs." Newsweek, May 18, pp. 50–52.

Leigh, Suzanne. 2004. "Fertility Patients Deserve to Know the Odds—and Risks." *USA Today,* July 7.

Leisey, Monica, Paul Kupstas, and Aly Cooper. 2009. "Domestic Violence in the Second Half of Life." *Journal of Elder Abuse and Neglect* 21(2):141–155.

Leite, Randall. 2007. "An Exploration of Aspects of Boundary Ambiguity among Young, Unmarried Fathers during the Prenatal Period." *Family Relations* 56(2):162–174.

Lento, Jennifer. 2006. "Relational and Physical Victimization by Peers and Romantic Partners in College Students." *Journal of Social and Personal Relationships* 23(3):331–348.

Leon, Kim. 2009. "Covenant Marriage: What Is It and Does It Work?" *Relationships.* Retrieved October 12, 2009 (missourifamilies.org).

Leopold, Thomas and Thorsten Schneider. 2011. "Family Events and the Timing of Intergenerational Transfers." *Social Forces* 90(2):595–616.

Lepore, Jill. 2011. "Birthright: What's Next for Planned Parenthood?" *The New Yorker.* November 4:44– 52.

Lepper, John M. 2009. *When Crisis Comes Home.* Macon, GA: Smyth & Helwys Publishers.

Lerner, Harriet.2001. *The Dance of Connection: How to Talk to Someone When You'reMad, Hurt, Scared, Frustrated, Insulted, Betrayed, or Desperate.* New York: HarperCollins.

Lessane, Patricia Williams. 2007. "Women of Color Facing Feminism—Creating Our Space at Liberation's Table: A Report on the Chicago Foundation for Women's "F" Series." *Journal of Pan African Studies* 1(7):3–10.

Letarte, Marie-Josée, Sylvie Normandeau, and Julie Allard. 2010. "Effectiveness of a Parent Training Program 'Incredible Years' in a Child Protection Service." *Child Abuse & Neglect* 34(4): 253–261.

Letiecq, Bethany L., Sandra J. Bailey, and Marcia A. Kurtz. 2008. "Depression among Rural Native American and European American Grandparents Rearing Their Grandchildren." *Journal of Family Issues* 29(3):334–356.

Letiecq, Bethany L., Sandra J. Bailey, and Fonda Porterfield. 2008."'We Have No Rights, We Get No Help': The Legal and Policy Dilemmas Facing Grandparent Caregivers." *Journal of Family Issues* 29(8):995–1012.

Levaro, Liz Bayler. 2009. "Living Together or Living Apart Together: New Choices for Old Lovers." *Family Focus* (Summer): F9–F10. Minneapolis: National Council on Family Relations.

Levin, Irene. 1997. "Stepfamily as Project." Pp. 123–133 in *Stepfamilies: History, Research, and Policy,* edited by Irene Levinand Marvin B. Sussman. New York: Haworth.

Levine, Carol. 2008. "Family Caregiving." Pp. 63– 68 in *From Birth to Death and Bench to Clinic: The Hastings Center Bioethics Briefing Book for Journalists, Policymakers, and Campaigns,*edited by Mary Crowley.Garrison, NY: The Hastings Center. Retrieved May 17, 2010 (www .thehastingscenter.org).

Levine, Judith A., Clifton R. Emery, and Harold Pollack. 2007. "The Well-Being of Children Born to Teen Mothers." *Journal of Marriage and Family* 69(1):105–122.

Levine, Kathryn. 2009. "Against All Odds: Resilience in Single Mothers of Children with Disabilities." *Social Work in Health Care* 48(4):402–419.

Levine, Robert, Suguru Sato, Tsukasa Hashimoto, and Jyoti Verma. 1995. "Love and Marriage in Eleven Cultures." *Journal of Cross-Cultural Psychology* 26(5):554–571.

Levinger, George. 1965. "Marital Cohesiveness and Dissolution: An Integrative Review." *Journal of Marriage and Family* 27:19–28.

———. 1976. "A Social Psychological Perspective on Marital Discord." *Journal of Social Issues* 32: 21–47.

Levy, Donald P. 2005. "Hegemonic Complicity, Friendship, and Comradeship: Validation and Causal Processes among White, Middle-class, Middle-aged Men." *Journal of Men's Studies* 13(2):199–225.

Levy, Pema. 2011. "How Kansas Banned Abortion." *Prospect.* July 1. Retrieved September 13, 2011 (http://www.prospect.com).

Lewin, Ellen. 2009. *Gay Fatherhood: Narratives of Family and Citizenship in America.* Chicago: University of Chicago Press.

Lewin, Tamar. 2001. "Study Says Little Has Changed." *The New York Times,* September 10.

———. 2005. "A Marriage of Unequals: When Richer Weds Poorer, Money Isn't the Only Difference." *The New York Times,* May 19.

———. 2011. "Study Undercuts View of College as a Place of Same-Sex Experimentation." *The New York Times,* March 17. Retrieved October 10, 2012 (http://www.nytimes .com/2011/03/18/education/18sex .html?_r=0).

Lewis, Thomas, M. D., Fari Amini,M. D., and Richard Lannon, M. D. 2000. *A General Theory of Love.* New York:*Random House.*

Lewontin, Richard and Richard Levins. 2007. *Biology Under the Influence: Dialectical Essays on Ecology, Agriculture, and Health.* New York: Monthly Review Press.

LGBTSS, Iowa State University.2013. "Sexual Orientation." Retrieved June 25, 2013 (http:// www.dso.iastate.edu/lgbtss/library /sexual-orientation).

Li, Jui-Chung Allen, and Lawrence L. Wu. 2008. "No Trend in the Intergenerational Transmission of Divorce." *Demography* 45(4): 875–883.

Lichter, Daniel T., J. Brian Brown, Zhenchao Qian, and Julie H. Carmalt. 2007. "Marital Assimilation among Hispanics: Evidence of Declining Cultural and Economic Incorporation?" *Social Science Quarterly* 88(3): 745–765.

Lichter, Daniel T. and Julie H. Carmalt. 2009. "Cohabitation and the Rise in Out-of-Wedlock Childbearing." *Family Focus*(Summer):F11–F13. Minneapolis: National Council on Family Relations.

Lichter, Daniel T. and Zhenchao Qian. 2008. "Serial Cohabitation and the Marital Life Course." *Journal of Marriage and Family* 70(4):861–878.

Lichter, Daniel T., Zhenchao Qian, and Leanna M. Mellott. 2006. "Marriage or Dissolution? Union Transitions among Poor Cohabiting Women." *Demography* 43(2):223–241.

Liefbroer, Aart C. and Edith Dourleijn. 2006. "Unmarried Cohabitation and Union Stability: Testing the Role of Diffusion Using Data from 16 European Countries." *Demography* 43(2): 203–221.

Limbaugh, Rush. 1992. *The Way Things Ought to Be.* New York: Pocket Books.

Lin, Chien and William T. Liu. 1993. "Relationships among Chinese Immigrant Families." Pp. 271–286 in *Family Ethnicity: Strength in Diversity,* edited by Harriette Pipes McAdoo. Newbury Park, CA: Sage Publications.

Lin, I-Fen and Susan L. Brown. 2013. "Unmarried Boomers Confront Old Age: A National Portrait." Working Paper Series 2012–2013. Bowling Green, OH: Bowling Green State University Center for Family and Demographic Research.

Lincoln, Karen,Robert Taylor, and James Jackson. 2008. "Romantic Relationships among Unmarried African Americans and Caribbean Blacks: Findings from the National Survey of American Life." *Family Relations* 57(April):254–266.

Lindau, Stacy Tessler and Natalia Gavrilova. 2010. "Sex, Health, and Years of Sexually Active Life Gained Due to Good Health: Evidence from Two U.S. Population Based Cross-Sectional Surveys of Ageing." *British Medical Journal* 340(7746):580.

Lind, Ranveig, Geir Lorem, Per Nortvedt, and Olav Hevroy. 2012. "Intensive Care Nurses' Involvement in the End-of-Life Process: Perspectives of Relatives." *Nursing Ethics*19(5):666–676.

Lindau, Stacy Tessler, L. Philip Schumm, Edward O. Laumann, Wendy Levinson, Colm A. O'Muircheartaigh, and Linda J. Waite. 2007. "A Study of Sexuality and Health among Older Adults in the United States." *New England Journal of Medicine* 357(8):762–674.

Lindberg, Laura Duberstein, Rachel Jones, and John S. Santelli. 2008. "Non-Coital Sexual Activities among Adolescents." *Journal of Adolescent Health* 43(3):231–238.

Lindner Gunnoe, Marjorie, E. Mavis Hetherington, and David Reiss. 2006. *Journal of Family Psychology* 20(4):589–596.

Lindsey, Elizabeth W. 1998. "The Impact of Homelessness and Shelter Life on Family Relationships." *Family Relations* 47(3):243–252.

Lindsey, Eric W., Yvonne M. Caldera, and Laura Tankersley. 2009. "Marital Conflict and the Quality of Young Children's Peer Play Behavior: The Mediating and moderating Role of Parent-Child Emotional Reciprocity and Attachment Security." *Journal of Family Psychology* 23(2):130–145.

Lindsey, Eric W., Jessico Campbell Chambers, James Frabutt, and Carol Mackinnon-Lewis. 2009. "Marital Conflict and Adolescents' Peer Aggression: The Mediating and Moderating Role of Mother-Child Emotional Reciprocity." *Family Relations* 58(December):593–606.

Lino, Mark. 2012. *Expenditures on Children by Families, 2011.*U.S. Department of Agriculture, Center for Nutrition Policy and Promotion. Miscellaneous Publication No. 1528-2011.

Linton, Sally, 1971."Woman the Gatherer: Male Bias in Anthropology." In Sue-Ellen Jacobs, Ed. *Women in Cross-Cultural Perspective: A Preliminary Sourcebook.*

Lips, Hilary M. 2004. *Sex and Gender: An Introduction.* New York: McGraw-Hill.

Liptak, Adam. 2002. "Judging a Mother for Someone Else's Crime." *The New York Times,* November 27.

———. 2013. "Supreme Court Bolsters Same-Sex Marriage with Two Major Rulings." *The New York Times.* June 26. Retrieved June 26, 2013 (http://www.nytimes.com).

Little, Betsi and Cheryl Terrance. 2010. "Perceptions of Domestic Violence in Lesbian Relationships: Stereotypes and Gender Role Expectations." *Journal of Homosexuality* 57(3):429–440.

Littleton, Heather, Holly Tabernik, Erika J. Canales, and Tamika Backstrom. 2009. "Risky Situations or Harmless Fun? A Qualitative Examination of College Women's Bad Hook-Up and Rape Scripts." *Sex Roles* 60:793–804.

Liu, Chien. 2000. "A Theory of Marital Sexual Life." *Journal of Marriage and Family* 62(2): 363–74.

Liu, Hesheng, Steven M. Stufflebeam, Jorge Sepulcre, Trey Hedden, and Randy L. Buckner. 2009. "Evidence from Intrinsic Activity That Asymmetry of the Human Brain Is Controlled By Multiple Factors." *Proceedings of the National Academy of Sciences of the United States of America* 106(48):20499–20503.

Liu, Hui. 2009. "Till Death Do Us Part: Marital Status and U.S. Mortality Trends, 1986–2000." *Journal of Marriage and Family* 71(December):1158–1173.

Liu, Hui and Corinne Reczek. 2012. "Cohabitation and U.S. Adult Mortality: An Examination by Gender and Race." *Journal of Marriage and Family* 74(4):794– 811.

Liu, Siwei and Kathryn Hynes. 2012. "Are Difficulties Balancing Work and Family Associated with Subsequent Fertility?" *Family Relations* 61(1): 16–30.

Livingston, Gretchen and D'Vera Cohn. 2012. "The New Demography of American Motherhood." *Pew Research Center Publications.* August 19. Retrieved December 13, 2012 (http://pewresearch.org).

Livingston, Gretchen and Kim Parker. 2011. "A Tale of Two Fathers." Pew Research Center. Retrieved September 13, 2011 (http://www.pewresearch.org).

Lleras, Christy. 2008. "Employment, Work Conditions, and the Home Environment in Single-Mother Families." *Journal of Family Issues* 29(10):1268–1297.

Lloyd, Kim M. 2006. "Latinas' Transition to First Marriage: An Examination of Four Theoretical Perspectives." *Journal of Marriage and Family* 68(4): 993–1014.

Lloyd, Sally A., April L. Few, and Katherine R. Allen. 2007. "Feminist Theory, Methods, and Praxis in Family Studies: An Introduction to the Special Issue." *Journal of Family Issues* 28(4): 447–451.

Lloyd-Thomas, Matthew and Amy Wang. 2012. "Women Underrepresented in Faculty, Report Finds." *Yale Daily News.*September 11. Retrieved October 27, 2012 (http://www.yaledailynews.com).

LoBiondo-Wood, Geri, Laurel Williams, and Charles McGhee. 2004. "Liver Transplantation in Children: Maternal and Family Stress, Coping, and Adaptation." *Journal of the Society of Pediatric Nurses* 9(2):59–67.

Lodge, Amy C., and Debra Umberson. 2012. "All Shook Up: Sexuality of Mid- to Later Life Married Couples." *Journal of Marriage and Family* 74:428–443.

Lofquist, Daphne. 2011. "Same-Sex Couple Households: American Community Survey Briefs." U.S. Census Bureau ACSBR 10-03. September.

———. 2012. "Same-Sex Couples' Consistency in Reports of Marital Status." Paper presented at the annual meeting of the Population Association of America, San Francisco: May 3–5.

Lofquist, Daphne, Terry Lugaila, Martin O'Connell, and Sarah Feliz. 2012. *Households and Families: 2010.* U.S. Census Briefs Publication Number C2010BR–14. U.S. Census Bureau, U.S. Department of Commerce. April.

Loftin, Colin, David McDowall, and Matthew Fetzer. 2008. "The Accuracy of Supplementary Homicide Report Data for Large U.S. Cities." Paper presented at the Annual Meeting of the American Society of Criminology, November 12. St. Louis Adam's Mark, St. Louis, Missouri.

Loftus, Jeni. 2001. "America's Liberalization in Attitudes toward Homosexuality, 1973 to 1998." *American Sociological Review* 66:762–782.

Logan, T. K., Robert Walker, and William Hoyt. 2012. "The Economic Costs of Partner Violence and Cost-Benefit of Civil Protective Orders." *Journal of Interpersonal Violence* 27(6):1137–1154.

Lois, Jennifer. 2010. "Gender and Emotion Management in the Stages of Edgework." Pp. 333–344 in *The Kaleidoscope of Gender: Prisms, Patterns, and Possibilities,* 3rd ed., edited by Joan Z. Spadeand Catherine G. Valentine. Newbury Park, CA: Pine Forge Press.

London, Andrew S., Elizabeth Allen, and Janet M. Wilmoth. 2012. "Veteran Status, Extramarital Sex, and Divorce: Findings from the 1992 National Health and Social Life Survey." *Journal of Family Issues* (forthcoming).

Long, George. 1875. "Patria Potestas." Pp. 873–875 in *A Dictionary of Greek and Roman Antiquities,* edited by Sir William Smith, William Wayte, and G. E. Marindin. London:J. Murray. Retrieved October 6, 2006 (penelope.uchicago.edu).

Longmore, Monica A., Abbey L. Eng, Peggy C. Giordano, and Wendy D. Manning. 2009. "Parenting and Adolescents' Sexual Initiation." *Journal of Marriage and Family* 71(4):969–982.

Lopez, Mark Hugo, AnaGonzalez-Barrera, and Seth Motel. 2011. "As Deportations Rise to Record Levels, Most Latinos Oppose Obama's Policy." December 28. Pew Research Center/Pew Hispanic Center. Retrieved February 16, 2012 (http://www.pewhispanic.org).

Loseke, Donileen R. and Demie Kurz. 2005. "Men's Violence toward Women Is the Serious Social Problem." Pp. 79–95 in *Current Controversies on Family Violence,* edited by Donileen R. Loseke, Richard J. Gelles, and Mary M. Cavanaugh. Thousand Oaks, CA: Sage Publications.

Loser, Rachel, Shirley Klein, E. Hill, and David Dollahite. 2008. "Religion and the Daily Lives of LSD Families: An Ecological Perspective." *Family and Consumer Sciences Research Journal* 37(1):52–70.

Love, Patricia. 2001. *The Truth about Love.* New York: Simon and Schuster.

Love, Patricia, and Steven Stosny. 2007. *How to Improve Your Marriage Without Talking about It: Finding Love Beyond Words.* New York: Broadway Books.

Lovell, Vicky., Elizabeth O'Neill, and Skylar Olsen. 2007. "Maternity Leave in the United States." Washington, DC: Institute for Women's Policy Research.

Lovett, Ian. 2012. "Measure Opens Door to Three Parents, or Four." *The New York Times.* July 13. Retrieved August 28, 2012 (http://www.nytimes.com).

"Loving Your Partner as a Package Deal." 2000. *Newsweek,* March 20, p. 78.

Lowan, J. M. and E. M. Dolan. 1994. "Remarried Families' Economic Behavior: Fishman's Model Revisited." *Journal of Divorce and Remarriage,* 22:103–119.

Lowe, Chelsea and Bruce M. Cohen. 2010. *Living with Someone Who's Living with Bipolar Disorder: A Practical Guide for Family, Friends, and Coworkers.* San Francisco: Jossey-Bass.

Lowe, Kendra, Katharine Adams, Blaine Browne, and Kerry Hinkle. 2012. "Impact of

Military Deployment on Family Relationships." *Journal of Family Studies* 18(1):17–27.

Lowe, Pauline and Karen McBride-Henry. 2012. "What Factors Impact upon the Quality of Life of Elderly Women with Chronic Illnesses." *Contemporary Nurse* 41(1):18–27.

Lubrano, Alfred. 2003. *Blue-Collar Roots, White-Collar Dreams.* New York: Wiley.

Lucas, Demitria L. 2013. "Princeton Mom to Female Students: Find Husband in College." Retrieved May 30, 2013, from http://www.clutchmagonline.com/2013/04/princeton-mom-to-female-students-find-husband-in-college/.

Lucas, Kristen. 2007. "Anticipatory Socialization in Blue-Collar Families: The Social Mobility-Reproduction Dialectic." Paper presented at the annual meeting of the International Communication Association, San Francisco CA, May 23. Retrieved January 19, 2010 (www.allacademic.com).

Lucier-Greer, Mallory and Francesca Adler-Baeder. 2012. "Does Couple and Relationship Education Work for Individuals in Stepfamilies? A Meta-Analytic Study." *Family Relations* 61(5):756–769.

Lugo Steidel, Angel G. and Josefina M. Contreras. 2003. "A New Familism Scale for Use with Latino Populations." *Hispanic Journal of Behavioral Sciences* 25(3):312–330.

Lumpkin, James R. 2008. "Grandparents in a Parental or Near-Parental Role: Sources of Stress and Coping Mechanisms." *Journal of Family Issues* 29(3):357–372.

Lundberg, Michael. 2009. "Our Parents' Keepers: The Current Status of American Filial Responsibility Laws." *Utah Law Review* 11(2): 534–582.

Lundquist, Jennifer Hickes. 2004. "When Race Makes No Difference: Marriage and the Military." *Social Forces* 83(2):731–757.

Lundquist, Jennifer Hickes, Michelle J. Budig, and Anna Curtis. 2009. "Race and Childlessness in America, 1988–2002." *Journal of Marriage & Family* 71(3):741–755.

Luscombe, Belinda. 2009. "Facebook and Divorce." *Time,* June 22.

———. 2011. "Latchkey Parents." Retrieved May 13, 2013 (http://www.time.com/time/magazine/article/0,9171,2093312,00.html).

———. 2013. "Confidence Woman: Facebook's Sheryl Sandberg Is on a Mission to Change the Balance of Power." *Time.* March 18:36–42.

Lustig, Daniel C. 1999. "Family Caregiving of Adults with Mental Retardation: Key Issues for Rehabilitation Counselors." *Journal of Rehabilitation* 65(2):26–45.

Luthra, Rohini and Christine A. Gidycz. 2006. "Dating Violence among College Men and Women: Evolution of a Theoretical Model." *Journal of Interpersonal Violence* 21(6):717–731.

Lynch, Jean M. (2000). "Considerations of Family Structure and Gender Composition: The Lesbian and Gay Stepfamily." *Journal of Homosexuality,* 40:81–94.

Lyons, Heidi, Peggy C. Giordano, Wendy D. Manning, and Monica A. Longmore. 2011. "Identity, Peer Relationships, and Adolescent Girls' Sexual Behavior: An Exploration of the Contemporary Double Standard." *Journal of Sex Research* 48(5):437–449.

Maccoby, Eleanor E. 1998. *The Two Sexes: Growing Up Apart; Coming Together.* Cambridge, MA: Belknap/Harvard University Press.

———. 2002. "Gender and Group Process: A Developmental Perspective." *Current Directions in Psychological Science* 11(2):54–58.

Maccoby, Eleanor E. and Carol Nagy Jacklin. 1974. *The Psychology of Sex Differences.* Stanford, CA: Stanford University Press.

Maccoby, Eleanor E. and Catherine C. Lewis. 2003. "Less Day Care or a Different Day Care?" *Child Development* 74:1069–1075.

Maccoby, E. E., and J. A. Martin. 1983. "Socialization in the Context of the Family: Parent-Child Interaction." Pp. 1–101 in *Handbook of Child Psychology,* edited by P. Mussen. New York: Wiley.

Maccoby, Eleanor E. and Robert Mnookin. 1992. *Dividing the Child: Social and Legal Dilemmas of Custody.* Cambridge, MA: Harvard University Press.

Macdonald, Cameron L. 2011. *Shadow Mothers: Nannies, Au Pairs, and the Micropolitics of Mothering.* Berkeley CA: University of California Press.

MacDonald, William L. and Alfred DeMaris. 1995. "Remarriage, Stepchildren, and Marital Conflict: Challenges to the Incomplete Institutionalization Hypothesis." *Journal of Marriage and Family* 57(2):387–398.

———. 2002. "Stepfather-Stepchild Relationship Quality: The Stepfather's Demand for Conformity and the Biological Father's Involvement." *Journal of Family Issues* 23(1):121–137.

MacDorman, M. F. and S. Kirmeyer. 2009. "The Challenge of Fetal Mortality." *NCHS Data Brief.* April 16:1–8.

MacFarquhar, Neil. 2006. "It's Muslim Boy Meets Girl, but Don't Call It Dating." *The New York Times,* September 19. Retrieved September 19, 2006 (www.nytimes.com).

Machir, John. 2003. "The Impact of Spousal Caregiving on the Quality of Marital Relationships in Later Life." *Family Focus* (September):F11–F13. National Council on Family Relations.

Macionis, John J. 2006. *Society: The Basics.* 6th ed. Upper Saddle River, NJ: Prentice Hall.

MacInnes, Maryhelen D. 2008. "One's Enough for Now: Children, Disability, and the Subsequent Childbearing of Mothers." *Journal of Marriage and Family* 70(3):758–771.

Mackey, Richard A., Matthew A. Diemer, and Bernard A. O'Brien. 2000. "Psychological Intimacy in the Lasting Relationships of Heterosexual and Same-gender Couples." *Sex Roles* (August):201–215.

Macklin, Eleanor D. 1987. "Nontraditional Family Forms." Pp. 317–353 in *Handbook of Marriage and the Family,* edited by Marvin B. Sussmanand Suzanne K. Steinmetz. New York: Plenum.

MacNeil, Sheila and Sandra E. Byers. 2009. "Role of Sexual Self-Disclosure in the Sexual Satisfaction of Long-Term Heterosexual Couples." *Journal of Sex Research* 46(1):3–14.

Magdol, Lynn, Terrie E. Moffitt, Avshalom Caspi, and Phil A. Silva. 1998. "Hitting without a License: Testing Explanations for Differences in Partner Abuse between Young Adult Daters and Cohabitors." *Journal of Marriage and Family* 60(1):41–55.

Magnuson, Katherine and Lawrence Berger. 2009. "Family Structure States and Transitions: Associations with Children's Well-Being During Middle Childhood." *Journal of Marriage and Family* 71(3):575–591.

Mahay, Jenna and Alisa C. Lewin. 2007. "Age and the Desire to Marry." *Journal of Family Issues* 28(5):706–723.

Mahoney, Annette. 2005. "Religion and Conflict in Marital and Parent-Child Relationships." *Journal of Social Issues* 61(4):689–717.

Maier, Thomas. 1998. "Everybody's Grandfather." *U.S. News & World Report,* March 30, p. 59.

Maillard, Kevin Noble. 2008. "The Multiracial Epiphany of *Loving.*" *Fordham Law Review*76(6): 2709–2732.

Mainemer, Henry, Lorraine C. Gilman, and Elinor W. Ames. 1998. "Parenting Stress in Families Adopting Children from Romanian Orphanages." *Journal of Family Issues* 19(2): 164–180.

Major, Brenda, Mark Appelbaum, Linda Beckman, Mary Ann Dutton, Nancy Felipe Russo, and Carolyn West. 2009. "Abortion and Mental Health." *American Psychologist* 64(9):863–890.

Malakh-Pines, Ayala. 2005. *Falling in Love: Why We Choose the Lovers We Choose.*New York: Routledge.

Malebranche, D. 2007. *Black Bisexual Men and HIV: Time to Think Deeper.* Paper presented at the Center for Sexual Health Promotion Sexual Health Seminar Series, Bloomington, Indiana.

Malia, Sarah E. C. 2005. "Balancing Family Members' Interests Regarding Stepparent Rights and Obligations: A Social Policy Challenge." *Family Relations* 54(2):298–319.

———. 2008. "How Relevant Are U.S. Family and Probate Laws to Stepfamilies?" Pp. 545–572 in *The International Handbook of Stepfamilies,* edited by J. Pryor. Hoboken, NJ: John Wiley & Sons.

Malinen, Kaisa, Asko Tolvanen, and Anna Ronka. 2012. "Accentuating the Positive, Eliminating the Negative? Relationship Maintenance as a Predictor of Two-Dimensional Relationship Quality." *Family Relations* 61(5): 784–797.

Manago, Adriana M., Christia Spears Brown, and Campbell Leaper. 2009. "Feminist Identity among Latina Adolescents." *Journal of Adolescent Research* 24(6):750–776.

Mandara, Jelani, Jamie S. Johnston, Carolyn B. Murray, and Fatima Varner. 2008. "Marriage, Money, and African American Mothers' Self-Esteem." *Journal of Marriage and Family* 70(5): 1188–1199.

Mandara, Jelani, Carolyn Murray, James Telesford, Fatima Varner, and Scott Richman. 2012. "Observed Gender Differences in African American Mother-Child Relationships and Child Behavior." *Family Relations* 61(1):129–141.

Mandara, Jelani, Sheba Y. Rogers, and Richard Zinbarg. 2011. "The Effects of Family Structure on African American Adolescents' Marijuana Use." *Journal of Marriage and Family* 73(June): 557–569.

Mangla, Ismat Sarah. 2013. "The Talk: How to Tackle Your Spouse's Overspending." *Money* 42(1). Retrieved January 29, 2013 (http://web.ebscohost.com).

Manisses Communications Group. 2000. "When It Comes to Handling Your Hard-to-Handle Child, Are You an Authoritative, Authoritarian or Permissive Parent?" *The Brown University Child and Adolescent Behavior Letter* 16(3):S1–S2.

Mann, Brian. 2008. "Military Moms Face Tough Choices." *National Public Radio (NPR).* May 27. Retrieved May 27, 2008 (http://www.npr.org).

Manning, Wendy D., Peggy C. Giordano, and Monica A. Longmore. 2006. "Hooking Up the Relationship Contexts of 'Nonrelationship' Sex." *Journal of Adolescent Research* 21(5):459–483.

Manning, Margaret M., Laurel Wainwright, and Jillian Bennett. 2011. "The Double ABCX Model of Adaptation in Racially Diverse Families with a School-Age Child with Autism." *Journal of Autism Development Disorder* 41:320–331.

Manning, Wendy D. 2001. "Childbearing in Cohabiting Unions: Racial and Ethnic Differences." *Family Planning Perspectives* 33(5): 217–234.

———. 2004. "Children and the Stability of Cohabiting Couples." *Journal of Marriage and Family* 66(3):674–689.

———. 2009. "Divorce-Proofing Marriage: Young Adults' Views on the Connection Between Cohabitation and Marital Longevity." *Family Focus*(Summer):F13–F15. Minneapolis: National Council on Family Relations.

Manning, Wendy D. and Susan Brown. 2006. "Children's Economic Well-Being in Married and Cohabiting Parent Families." *Journal of Marriage and Family* 68(2):345–362.

Manning, Wendy D., Jessica A. Cohen, and Pamela J. Smock. 2011. "The Role of Romantic Partners, Family, and Peer Networks in Dating Couples' Views about Cohabitation." *Journal of Adolescent Research* 26(1): 115–149.

Manning, Wendy D., Peggy C. Giordano, and Monica A. Longmore. 2006. "Hooking Up: The Relationship Contexts of 'Nonrelationship' Sex." *Journal of Adolescent Research* 21(5):459–483.

Manning, Wendy D. and Kathleen A. Lamb. 2003. "Adolescent Well-Being in Cohabiting, Married, and Single-Parent Families." *Journal of Marriage and Family* 65(4):876–893.

Manning, Wendy D. and Nancy S. Landale. 1996. "Racial and Ethnic Differences in the Role of Cohabitation in Premarital Childbearing." *Journal of Marriage and Family* 58(1):63–77.

Manning, Wendy D. and Pamela J. Smock. 1999. "New Families and Nonresident Father-Child Visitation." *Social Forces* 78:87–116.

———. 2000. "'Swapping' Families: Serial Parenting and Economic Support for Children." *Journal of Marriage and Family* 62(1):111–122.

———. 2002. "First Comes Cohabitation and Then Comes Marriage?" *Journal of Family Issues* 23(8):1065–1087.

———. 2005. "Measuring and Modeling Cohabitation: New Perspectives from Qualitative Data." *Journal of Marriage and Family* 67(4): 989–1002.

Manning, Wendy, Pamela J. Smock, and C. Bergstrom-Lynch. 2009. "Cohabitation and Parenthood: Lessons from Focus Groups and In-Depth Interviews." In *Marriage and Family: Complexity and Perspectives*, edited by E. Peters and C. Kamp-Dush. New York: Columbia University Press.

Manning, Wendy D., Susan D. Stewart, and Pamela J. Smock. 2003. "The Complexity of Fathers' Parenting Responsibilities and Involvement with Nonresident Children." *Journal of Family Issues* 24(5):645–667.

Manning, Wendy D., Deanna Trella, Heidi Lyons, and Nola Cora DuToit. 2010. "Marriageable Women: A Focus on Participants in a Community Healthy Marriage Program." *Family Relations* 59(1):87–102.

Mannis, Valerie S. 1999. "Single Mothers by Choice." *Family Relations* 48(2):121–128.

Mansson, Daniel H., Scott A. Myers, and Lynn H. Turner. 2010. "Relational Maintenance Behaviors in the Grandchild-Grandparent Relationship." *Communication Research Reports* 27(1):68–79.

Marano, Hara Estroff. 2000. "Divorced? (Remarriage in America)." *Psychology Today* 33(2):56–60.

———. 2010. "The Expectations Trap." *Psychology Today* 43(2):63–71.

Marchione, Marilynn and Lindsey Tanner. 2006. "Many U.S. Couples Seek Embryo Screening." Associated Press, September 20. Retrieved September 21, 2006 (www.apnews.myway.com).

Marikar, Sheila. 2009. "Cher is Supporting Chasity's Sex Change, Though She Doesn't Understand It." *ABC News*, June 18. Retrieved January 12, 2010 (www.abcnews.go.com).

Mark, Kristen P. and Sarah H. Murray. 2012. "Gender Differences in Desire Discrepancy as a Predictor of Sexual and Relationship Satisfaction in a College Sample of Heterosexual Romantic Relationships." *Journal of Sex & Marital Therapy* 38(2):198–215.

Markham, Melinda Stafford, and Marilyn Coleman. 2012. "The Good, The Bad, and The Ugly: Divorced Mothers' Experiences with Coparenting." *Family Relations* 61:586–600.

Markman, Howard J., Galena K. Rhoades, Scott M. Stanley, Erica P. Ragan , and Sarah W. Whitton. 2010. "The Premarital Communication Roots of Marital Distress and Divorce: The First Five Years of Marriage." *Journal of Family Psychology* 24(3):289–298.

Markman, Howard, Scott Stanley, and Susan L. Blumberg. 2001. *Fighting for Your Marriage: Positive Steps for Preventing Divorce and Preserving a Lasting Love.* San Francisco: Jossey-Bass.

Markon, Jerry. 2010. "The Baby He's Never Met; Va. Father Fights for Child His Girlfriend Sent to Utah for Adoption." *Washington Post*, April 14, p. A1.

Marks, Jaime, Chun Lam, and Susan McHale. 2009. *Sex Roles* 61(3/4):221–234.

Marks, Loren. 2012. "Same-Sex Parenting and Children's Outcomes: A Closer Examination of the American Psychological Association's Brief on Lesbian and Gay Parenting." *Social Science Research* 41: 735–751.

Marks, Loren, Katrina Hopkins, Cassandra Chaney, Pamela Monroe, Olena Nesteruk, and Diane Sasser. 2008. "'Together We Are Strong': A Qualitative Study of Happy, Enduring, African American Marriages." *Family Relations* 57(2):172–185.

Marquardt, Elizabeth. n. d. *The Revolution in Parenthood: The Emerging Global Clash between Adult Rights and Children's Needs.* Commission on Parenthood's Future. Retrieved October 4, 2006 (www.americanvalues.org).

"Marriage." 2008. The Gallup Poll. Retrieved October 9, 2009 (www.gallup.com).

"Marriage, Domestic Partnerships, and Civil Unions: An Overview of Relationship Recognition for Same-Sex Couples in the United States." 2009. National Center for Lesbian Rights. Retrieved December 20, 2009 (www.nclrights.org).

Marsh, Kris, William A. Darity Jr., Philip N. Cohen, Lynne M. Casper, and Danielle Salters. 2007. "The Emerging Black Middle Class: Single and Living Alone." *Social Forces* 86(2):735–762.

Marshall, Barbara L. 2009. "Science, Medicine and Virility Surveillance: 'Sexy Seniors' in the Pharmaceutical Imagination." *Sociology of Health & Illness* 32(2):211–224.

Marshall, Catherine A. 2010. *Surviving Cancer as a Family and Helping Co-Survivors Thrive.* Santa Barbara, CA: Praeger.

Marshall, Nancy and Allison Tracy. 2009. "After the Baby: Work-Family Conflict and Working Mothers' Psychological Health." *Family Relations* 58(October):380–391.

Marshall, Tara C., Kathrine Bejanyan, Gaia Castro, and Ruth A. Lee. 2013. "Attachment Styles as Predictors of Facebook-Related Jealousy and Surveillance in Romantic Relationships." *Personal Relationships* 20:1–22.

Marsiglio, William. 1991. "Paternal Engagement in Activities with Minor Children." *Journal of Marriage and the Family*, 53:973–986.

———. 1992. "Stepfathers with Minor Children Living at Home." *Journal of Family Issues*, 13:195–214.

———. 2004. *Stepdads: Stories of Love, Hope, and Repair.* Boulder, CO: Rowman and Littlefield.

———. 2008. "Understanding Men's Prenatal Experience and the Father Involvement Connection: Assessing Baby Steps." *Journal of Marriage and Family* 20(December):1108–1113.

———. 2012. "Interpreting a Fatherhood Legacy." *Family Focus* FF54(Fall): F2–F3.

Martin, Brittny A., Ming Cui, Koji Ueno, and Frank D. Fincham. 2013. "Intimate Partner Violence in Interracial and Monoracial Couples." *Family Relations* 62(February):202–211.

Martin, John Levi. 2005. "Is Power Sexy?" *American Journal of Sociology* 111(2):408–447.

Martin, Joyce A., Brady E. Hamilton, Paul D. Sutton, Stephanie J. Ventura, Fay Menacker, Sharon Kirmeyer, and T. J. Mathews. 2009. "Births: Final Data for 2006." *National Vital Statistics Report* 57(7). Hyattsville, MD: National Center for Health Statistics. January 7.

Martin, Joyce A., Brady E. Hamilton, Stephanie J. Ventura, M.A. Michelle, J.K. Osterman, Elizabeth Wilson, and T.J. Mathews. 2012. "Births: Final Data for 2010." *National Vital Statistics Reports* 62(1).U.S. Department of Health and Human Services, and Centers for Disease Control and Prevention. August.

Martin, Joyce A., Michelle J. K. Osterman, and Paul D. Sutton. 2010. "Are Preterm Births on the Decline in the United States? Recent Data from the National Vital Statistics System." *National Vital Statistics Report* 57(7). January 7.

Martin, Molly A. and Adam M. Lippert. 2012. "Feeding Her Children, But Risking Her Health: The Intersection of Fender, Household Food Insecurity and Obesity." *Social Science & Medicine* 74: 1754–1764.

Martin, Philip and Elizabeth Midgley. 2006. "Immigration: Shaping and Reshaping America." *Population Bulletin* 61(4). Washington, DC: Population Reference Bureau.

Martin, Sandra L., April Harris-Britt, Yun Li, Kathryn E. Moracco, Lawrence L. Kupper, and Jacquelyn C. Campbell. 2004. "Change in Intimate Partner Violence during Pregnancy." *Journal of Family Violence* 19:243–247.

Martinez, Gladys M., Anjani Chandra, Joyce C. Abma, Jo Jones, and William D. Mosher. 2006. "Fertility, Contraception, and Fatherhood: Data on Men and Women from Cycle 6 (2002) of the National Survey of Family Growth." *Vital and Health Statistics* 23(26). May. Retrieved January 27, 2007 (www.nchs.gov).

Martinez, G., C. E. Copen, and J. C. Abma. 2011. *Teenagers in the United States: Sexual Activity, Contraceptive Use, and Childbearing, 2006–2010 National Survey of Family Growth.* National Center for Health Statistics, Vital Health Stat 23(31).

Martinez, Gladys, Kimberly Daniels, and Anjani Chandra. 2012. "Fertility of Men and women Aged 15–44 Years in the United States: National Survey of Family Growth, 2006–2010." *National Health Statistics Reports* 51 (April 12). Washington, DC: U.S. Department of Health and Human Services, Centers for Disease Control and Prevention.

Martyn, Kristy, Carol Loveland-Cherry, Antonia Villarruel, Esther Gallegos Cabriales, Yan Zhou, David Ronis, and Brenda Eakin. 2009. "Mexican Adolescents' Alcohol Use, Family Intimacy, and Parent-Adolescent Communication." *Journal of Family Nursing* 15(2):152–170.

Masanori,Ishimori, Ikuo Daibo, and Yuji Kanemasa. 2004. "Love Styles and Romantic Love Experiences in Japan." *Social Behavior and Personality* 32(3):265–281.

Mason, Ashley, E., Rita W. Law, Amanda E. B. Bryan, Robert M. Portley, and David A. Sbarra. 2012. "Facing a Breakup: Electromyographic Responses Moderate Self-Concept Recovery Following Romantic Separation." *Personal Relationships* 19:551–568.

Mason, Mary Ann, Mark A. Fine, and Sarah Carnochan. 2001. "Family Law in the New Millennium: for Whose Families?" *Journal of Family Issues* 22(7):859–881.

Mason, Mary Ann, Sydney Harrison-Jay, Gloria Messick Svare, and Nicholas H. Wolfinger. 2002. "Stepparents: De Facto Parents or Legal Strangers?" *Journal of Family Issues* 23(4):507–522.

Masters, N. Tatiana, Erin Casey, Elizabeth A. Wells, and Diane M. Morrison. 2013. "Sexual Scripts among Young Heterosexually Active Men and Women: Continuity and Change." *Journal of Sex Research* 50(5):409–420.

Masters, William H. and Virginia E. Johnson. 1966. *Human Sexual Response.* Boston: Little, Brown.

———. 1976. *The Pleasure Bond: A New Look at Sexuality and Commitment.* New York: Bantam.

Masters, William H., Virginia E. Johnson, and Robert C. Kolodny. 1994. *Heterosexuality.* New York: HarperCollins.

Mateyka, Peter J., Melanie Rapino, and Liana C. Landivar. 2012. "Home-Based Workers in the United States, 2010." U.S. Department of Commerce Current Population Reports P70-132. October.

Mather, Mark. 2009. "Children in Immigrant Families Chart New Path." Washington, DC: Population Reference Bureau.

———. 2011. "More Young Adults in U.S. Postponing Marriage, Living at Home, Disconnected from Work and School." *Population Reference Bureau.* December. Retrieved August 29, 2012 (http://www.prb.org).

———. 2012. "Fact Sheet: The Decline of U.S. Fertility." *Population Reference Bureau.* July. Retrieved December 7, 2012 (http://www.prb.org).

Mather, Mark and Kelvin Pollard. 2009. "U.S. Hispanic and Asian Population Growth Levels Off." Washington, DC: Population Reference Bureau. Retrieved June 8, 2009 (www.prb.org).

Matiasko, Jennifer, Leslie Grunden, and Jody Ernst. 2007. "Structural and Dynamic Process Family Risk Factors: Consequences for Holistic Adolescent Functioning." *Journal of Marriage and Family* 69(3):654–674.

Matsunaga, Masaki, and Tadasu Todd Imahori. 2009. "Profiling Family Communication Standards." *Communication Research* 36(1):3–31.

Matta, William J. 2006. *Relationship Sabotage: Unconscious Factors That Destroy Couples, Marriages, and Family.*Westport, CT: Praeger Publishers.

Matthews, Sarah H. 2012. "Enhancing the Qualitative-Research Culture in Famil Studies." *Journal of Marriage and Family* 74(4) (August): 666–670.

May, Rollo. 1975. "A Preface to Love." Pp. 114–119 in *The Practice of Love,* edited by Ashley Montagu. Englewood Cliffs, NJ: Prentice Hall.

Mayes, Tessa. 2001. "Career Women Rent Wombs to Beat Hassles of Pregnancy." *The Sunday Times—London.* August 7. Retrieved January 14, 2013 (http://www.sunday-times.co.uk).

Mayo Clinic Health Letter. 2012. "The Power of Connection." August. Supplement 30: 1–8.

Mayo Clinic Staff. 2005. "Marriage Counseling: Working Through Relationship Problems." Retrieved May 1, 2007 (www.mayoclinic.com/health/marriage-counseling/MH00104).

———. 2011. "Family Planning: Get the Facts about Pregnancy Spacing." Retrieved December 14, 2012 (http://www.mayoclinic.com).

———. 2012. "Morning-after Pill." Retrieved December 14, 2012 (http://www.mayoclinic.com).

Mays, Vickie M. and Susan D. Cochran. 1999. "The Black Woman's Relationship Project: A National Survey of Black Lesbians." Pp. 59–66 in *The Black Family: Essays and Studies,* 6th ed., edited by Robert Staples. Belmont, CA: Wadsworth.

Mazelis, Joan Maya and Laryssa Mykyta. 2011. "Relationship Status and Activated Kin Support: The Role of Need and Norms." *Journal of Marriage and Family* 73:430–445.

McAdoo, Harriette Pipes. 2007. *Black Families,* 4th edition. Thousand Oaks, CA: Sage Publications.

McBride-Chang, Catherine and Lei Chang. 1998. "Adolescent–Parent Relations in Hong Kong: Parenting Styles, Emotional Autonomy, and School Achievement." *Journal of Genetic Psychology*159(4):421–435.

McCabe, Janice, Emily Fairchild, Liz Grauerholz, Bernice A. Pescosolido, and Daniel Tope. 2011. "Gender in Twentieth-Century Children's Books." *Gender & Society* 25(2):197–226.

McCann, Brandy Renee. 2012. "The Persistence of Gendered Kin Work in Maintaining Family Ties: A Review Essay." *Journal of Family Theory & Review* 4(3) (September):249–254.

McClain, Lauren Rinelli. 2011. "Better Parents, More Stable Partners: Union Transitions among Cohabiting Parents." *Journal of Marriage and Family* 73(5):889–901.

McClintock, Elizabeth Aura. 2011. "Handsome Wants as Handsome Does: Physical Attractiveness and Gender Differences in Revealed Sexual Preferences." *Biodemograhy and Social Biology* 57:221–257.

McClure, Robin. 2010. "Afterschool Child Care—Number of Kids Home Alone After School Has Risen." Retrieved February 13, 2013 (http://childcare.about.com).

McCoy, Kelsey, and James Oelschager. 2013. *Sexual Coercion Awareness and Prevention.* Retrieved May 27, 2013 (http://www.fit.edu/caps/documents/SexualCoercion_000.pdf).

McCubbin, Hamilton I. and Marilyn A. McCubbin. 1991. "Family Stress Theory and Assessment: The Resiliency Model of Family Stress, Adjustment and Adaptation." Pp. 3–32 in *Family Assessment Inventories for Research and Practice,* 2nd ed., edited by Hamilton I. McCubbin and Anne I. Thompson. Madison: University of Wisconsin, School of Family Resources and Consumer Services.

———. 1994. "Families Coping with Illness: The Resiliency Model of Family Stress, Adjustment, and Adaptation." Chapter 2 in *Families, Health, and Illness.* St. Louis: Mosby.

McCubbin, Hamilton I. and Joan M. Patterson. 1983. "Family Stress and Adaptation to Crisis: A Double ABCX Model of Family Behavior." Pp. 87–106 in *Family Studies Review Yearbook,* Vol. 1, edited by David H. Olsonand Brent C. Miller. Newbury Park, CA: Sage Publications.

McCubbin, Hamilton I., Anne I. Thompson, and Marilyn A. McCubbin, eds. 1996. *Family Assessment: Resiliency, Coping and Adaptation: Inventories for Research and Practice.* Madison: University of Wisconsin Press.

McCubbin, Hamilton I., Elizabeth Thompson, Anne Thompson, Jo A. Futrell, and Suniya Luthar. 2001. "The Dynamics of Resilient Families." *Contemporary Psychology* 48(2):154–156.

McCubbin, Laurie D., Hamilton I. McCubbin, Wei Zhang, Lisa Kehl, and Ida Strom. 2013. "Relational Well-Being: An Indigenous Perspective and Measure." *Family Relations* 62(2):354–365.

McCubbin, Marilyn A. 1995. "The Typology Model of Adjustment and Adaptation: A Family Stress Model." *Guidance and Counseling* 10(4):31–39.

McCubbin, Marilyn A. and Hamilton I. McCubbin. 1989. "Theoretical Orientations to Family Stress and Coping." Pp. 3–43 in *Treating Families Under Stress,* edited by Charles Figley. New York: Brunner/Mazel.

McCue, Margi Laird. 2008. *Domestic Violence: A Reference Handbook.* 2nd ed. Santa Barbara, CA: ABC-CLIO, Inc.

McDermott, Brett and Vanessa E. Cobham. 2012. "Family Functioning in the Aftermath of a Natural Disaster." *BMC Psychiatry* 12:55.

McDonough, Tracy A. 2010. "A Policy Capturing Investigation of Battered Women's Decisions to Stay in Violent Relationships." *Violence and Victims* 25(2):165–184.

McDowell, Margaret A, Debra J. Brody, and Jeffery P. Hughes. 2007. "Has Age at Menarche Changed? Results from the National Health and Nutrition Examination Survey (NHANES) 1999–2004." *Journal of Adolescent Health* 40(3) :227–233.

McGene, Juliana and Valarie King. 2012. "Implications of New Marriages and Children for Coparenting in Nonresident Father Families." *Journal of Family Issues* 33(12): 1619–1641.

McGough, Lucy S. 2005. "Protecting Children in Divorce: Lessons from Caroline Norton." *Maine Law Review* 57:13–37.

McGraw, Lori A., Anisa M. Zvonkovic, and Alexis J. Walker. 2000. "Studying Postmodern Families: A Feminist Analysis of Ethical Tensions in Work and Family Research." *Journal of Marriage and the Family* 62(February):68–77.

McHale, James, Maureen R. Waller, and Jessica Pearson. 2012. "Coparenting Interventions for Fragile Families: What Do We Know and Where Do We Need to Go Next?" *Family Process* 51(3): 284–306.

McHale, Susan, Kimberly Updegraff, and Shawn Whiteman. 2012. "Sibling Relationships and Influences in Childhood and Adolescence." *Journal of Marriage and Family* 74(5):913–930.

McHugh, Maureen F. 2007. "The Evil Stepmother." Second Wives Café: Online Support for Second Wives and Stepmothers. Retrieved May 2, 2007 (secondwivescafe.com).

McIntosh, Peggy. 1988. "White Privilege and Male Privilege: A Personal Account of Coming to See Correspondences Through Work in Women's Studies." Working paper #189. Wellesley, MA: Wellesley College Center for Research on Women.

McIntyre, Matthew H. and Carolyn Pope Edwards. 2009. "The Early Development of Gender Differences." *Annual Review of Anthropology* 38(1):83–97.

McIntyre, Matthew, Steven W. Gangestad, Peter B. Gray, Judith Flynn Chapman, Terence C. Burnham, Mary T. O'Rourke, and Randy Thornhill. 2006. "Romantic Involvement Often Reduces Men's Testosterone Levels—But Not Always: The Moderating Role of Extrapair Sexual Interest." *Journal of Personality and Social Psychology* 91(4):642–651.

McKinley, Jesse, and John Schwartz. 2010. "Court Rejects Same-Sex Marriage Ban in California." *The New York Times.* August 4. Retrieved August 10, 2010 (www.nytimes.com).

McLanahan, Sarah (Editor in chief). 2010. "Transition to Adulthood." *The Future of Children* 20(1). Princeton-Brookings.

McLanahan, S. S., andAdams, J. (1987). "Parenthood and Psychological Well-Being." *Annual Review of Immunology,* 5:237–257.

McLanahan, Sara and Marcia J. Carlson. 2002. "Welfare Reform, Fertility, and Father Involvement." *The Future of Children* 12(1): 147–165.

McLanahan, Sara, Irwin Garfinkel, Nana E. Reichman, and Julien O. Teitler. 2001. "Unwed Parents or Fragile Families? Implications for Welfare and Child Support Policy." Pp. 202–228 in *Out of Wedlock: Causes and Consequences of Nonmarital Fertility,* edited by Lawrence L. Wuand Barbara Wolfe. New York: Russell Sage.

McLaren, Rachel M, Denise H. Solomon, and Jennifer S. Priem. 2012. "The Effect of Relationship Characteristics and Relational Communication on Experiences of Hurt from Romantic Partners." *Journal of Communication* 62: 950–971.

McLoyd, Vonnie C., Ana Mari Cauce, David Takeuchi, and Leon Wilson. 2000. "Marital Processes and Parental Socialization in Families of Color: A Decade Review of Research." *Journal of Marriage and Family* 62:1070–1093.

McMahon, Thomas, Justin Winkel, Nancy Suchman, and Bruce Rounsaville. 2007. "Drug-Abusing Fathers: Patterns of Pair Bonding, Reproduction, and Paternal Involvement." *Journal of Substance Abuse Treatment* 33:295–302.

McMahon-Howard, Jennifer, Jody Clay-Warner, and Linda Renzulli. 2009. "Criminalizing Spousal Rape: The Diffusion of Legal Reforms." *Sociological Perspectives* 52(4):505–531.

McManus, Mike. 2006. "Inside/Out Dads: Helping Prisoners Reenter Society." Retrieved November 17, 2006 (smartmarriages.com).

McManus, Patricia A. and Thomas DiPrete. 2001. "Losers and Winners: The Financial Consequences of Separation and Divorce for Men." *American Sociological Review* 66:246–268.

McMenamin, Terence M. 2007. "A Time to Work: Recent Trends in Shift Work and Flexible Schedules." *Monthly Labor Review,* December. Washington, DC: U.S. Bureau of Labor Statistics.

McNamara, Melissa. 2009. "Elder Care Benefits." *CBS Evening News.* February 11. Retrieved February 11, 2013 (http://www.cbsnews.com).

McNamara, Robert Hartmann. 2008. *Homelessness in America.* Westport, CT: Praeger.

McQuillan, Julia, Arthur Greil, Karina Shreffler, Patricia Wonch-Hill, Kari Gentzler, and John Hathcoat. 2012. "Does the Reason Matter? Variations in Childlessness Concerns among U.S. Women." *Journal of Marriage and Family* 74(5):1166–1181.

McVeigh, Rory and Maria-Elena D. Diaz. 2009. "Voting to Ban Same-Sex Marriage: Interests, Values, and Communities." *American Sociological Review* 74(6):891–915.

McWey, Lenore M., Tammy L. Henderson, and Jenny Burroughs Alexander. 2008. "Parental Rights and the Foster Care System." *Journal of Family Issues* 29(8):1031–1050.

Mead, George Herbert. 1934. *Mind, Self, and Society.* Chicago: University of Chicago Press.

Meadows, Sarah O., Sara S. McLanahan, and Jeanne Brooks-Gunn. 2007. "Parental Depression and Anxiety and Early Childhood Behavior Problems Across Family Types." *Journal of Marriage and Family* 69(5):1162–1177.

Meadows, Sarah O., Sara S. McLanahan, and Jean T. Knab. 2009. "Economic Trajectories in Non-Traditional Families with Children." *The Rand Corporation,* August 21, pp. 1–41.

Medeiros, Rose A. and Murray A. Straus. 2006. "Risk Factors for Physical Violence Between Dating Partners: Implications for Gender-Inclusive Prevention and Treatment of Family Violence." Pp. 59–87 in *Family Interventions in Domestic Violence A Handbook of Gender-Inclusive Theory and Treatment,* edited by John C.Hamel and Tanya L. Nicholls. New York: Springer.

Meezan, William, and Jonathan Rauch. 2005. "Gay Marriage, Same-Sex Parenting, and America's Children." *The Future of Children* 15(2):157–175.

Mehta, Pranjal H., Amanda C. Jones, and Robert A. Josephs. 2008. "The Social Endocrinology of Dominance: Basal Testosterone Predicts Cortisol Changes and Behavior Following Victory and

Defeat." *Journal of Personality and Social Psychology* 94(6):1078–1093.

Meier, Ann and Gina Allen. 2009. "Romantic Relationships from Adolescence to Young Adulthood." *The Sociological Quarterly* 50(2): 308–335.

Meier, Ann, Kathleen E. Hull, and Timothy A. Ortyl. 2009. "Young Adult Relationship Values at the Intersection of Gender and Sexuality." *Journal of Marriage and Family* 71(3):510–525.

Melby, Todd. 2010. "Sexuality for the Young and Old." *Contemporary Sexuality* 44(1):1, 4– 6.

Menacker, Fay and Brady E. Hamilton. 2010. "Recent Trends in Cesarean Delivery in the United States." *NCHS Data Brief.* No. 35. Hyattsville, MD: National Center for Health Statistics. March.

"Men and Women in the U.S. Congress." 2012. *This Nation: American Government and Politics Online.* Retrieved October 28, 2012 (http://www.ThisNation.com).

Mendes, Elizabeth, Lydia Saad, and Kyley McGeeney. 2012. "Stay-at-Home Moms Report More Depression, Sadness, Anger." Gallup. May 18. Retrieved July 15, 2013 (http://www.gallup.com).

Menning, Chadwick L. 2008. "'I've Kept It That Way on Purpose': Adolescents' Management of Negative Parental Relationship Traits after Divorce and Separation." *Journal of Contemporary Ethnography,* 37:586–618.

Mental Health America. 2009. "Factsheet: What Every Child Needs for Good Mental Health." Retrieved January 2, 2010 (www.mentalhealthamerica.net).

Merton, Robert K. 1973 [1942]. "The Normative Structure of Science." Chapter 3 in *The Sociology of Science: Theoretical and Empirical Investigations,* edited by Robert K. Merton. Chicago: University of Chicago Press.

Merton, Robert K. 1968 [1949]. *Social Theory and Social Structure.* New York: The Free Press.

Merz, Eva-Marie and Aart C. Liefbroer. 2012. "The Attitude Toward Voluntary Childlessness in Europe: Cultural and Institutional Explanations." *Journal of Marriage and Family* 74(3): 587–600.

Messner, Michael A. 1997. *The Politics of Masculinity: Men in Movements.*Thousand Oaks, CA: Sage Publications.

Meteyer, Karen B. and Maureen Perry-Jenkins. 2009. "Dyadic Parenting and Children's Externalizing Symptoms." *Family Relati ons* 58(July):289–302.

Meyer, Cathy. 2011. "Should Premarital Counseling Be Required by Law?" March 20. Retrieved December 7, 2012 (http://divorcesupport.about.com/b/2011/03/20/should-premarital-counseling-be-required-by-law.htm).

———. 2013. "The Difference Between Domestic Abuse and Normal Marital Conflict." Retrieved March 14, 2013 (http://divorcesupport.about.com).

Meyer, D. R. and M. Cancian. 2012. "'I'm Not Supporting His Kids': Nonresident Fathers' Contributions Given Mothers' New Fertility." *Journal of Marriage and Family,* 74(1):132–151.

Meyer, Harris. 2009. "Getting Back in the Game." *The Oregonian,* November 4.

Meyer, Jennifer. 2007. "Making Good Decisions." *Omaha World Herald,* January 22.

Meyer, Julie. 2012. *Centenarians: 2010.* Washington, DC: U.S. Census Bureau Special Reports 2010, No. C2010SR-03.

Meyer, Madonna H. and Marcia L. Bellas. 2001. "U.S. Old-age Policy and the Family." Pp. 191–201 in *Families in Later Life: Connections and Transitions,* edited by Alexis J. Walker, Margaret Manoogian-O'Dell, Lori A. McGraw, and Diana L. G. White. Thousand Oaks, CA: Pine Forge.

Meyers, Dvora. 2011. "Virtual Visitation Rights." *The New York Times,* March 18. Retrieved from http://www.nytimes.com/2011/03/20/fashion/20Facebook.html?pagewanted=all.

Michaels, Marcia L. 2006. "Stepfamily Enrichment Program: A Preventive Intervention for Remarried Couples." *Journal for Specialists in Group Work* 31(2):135–152.

Michon, Kathleen, J.D. 2012. "Federal Marriage Benefits Denied to Same-Sex Couples." *Nolo.* Retrieved December 3, 2012 (http://www.nolo.com).

Middlemiss, Wendy and William McGuigan. 2005. "Ethnicity and Adolescent Mothers' Benefit from Participation in Home-Visitation Services." *Family Relations* 54(2):212–224.

Mikulincer, Mario and Phillip R. Shaver. 2012. "Adult Attachment Orientations and Relationship Processes." *Journal of Family Theory & Review* 4(4): 239–274.

Milardo, Robert M. 2005. "Generative Uncle and Nephew Relationships." *Journal of Marriage and Family* 67(5):1226–1236.

Miles, Donna. 2010. "Federal Employees' Same-Sex Domestic Partners Garn New Benefits." American Forces Press Service. June 3. Retrieved September 3, 2012 (http://www.af.mil/news).

Miles, Leonora. 2008. "The Hidden Toll—Financial Abuse Is one of the Most Common Forms of Elder Abuse." *Adults Learning* 19(9): 28–30.

Miller, Amanda J., Sharon Sassler, and Dela Kusi-Appouh. 2011. "The Specter of Divorce: Views from Working- and Middle-Class Cohabitors." *Family Relations* 60(5): 602–616.

Miller, Courtney Waite and Michael E. Roloff. 2005. "Gender and Willingness to Confront Hurtful Messages from Romantic Partners." *Communication Quarterly* 53(3):323–338.

Miller, Dawn. 2004. "From the Author. The Stepfamily Life: A Column from Life in the Blender" (www.thestepfamilylife.com).

———. n. d. "Don't Go Nuclear—Negotiate." SelfGrowth.com. Retrieved September 21, 2004 (www.selfgrowth.com/articles/Miller).

Miller, Elizabeth. 2000. "Religion and Families over the Life Course." Pp. 173–186 in *Families Across Time: A Life Course Perspective,* edited by Sharon J. Price, Patrick C. McKenry, and Megan J. Murphy. Los Angeles: Roxbury.

Miller, Laurie C. 2005a. *Handbook of International Adoptive Medicine.* New York: Oxford University Press.

———. 2005b. "International Adoption, Behavior, and Mental Health." *Journal of the American Medical Association* 293(20):2533–2535.

Mills, C. Wright. 2000 [1959]. *The Sociological Imagination.* 40th Anniversary Edition. New York: Oxford University Press.

Mills, Terry L. and Janet M. Wilmoth. 2002. "Intergenerational Differences and Similarities in Life-Sustaining Treatment Attitudes and Decision Factors." *Family Relations* 51(1):46–54.

Milner, Joel S., Cynthia J. Thomsen, Julie L. Crouch, Mandy M. Rabenhorst, Patricia M. Martens, Christopher W. Dyslin, Jennifer M. Guimond, Valerie A. Stander, and Lex L. Merrill. 2010. "Do Trauma Symptoms Mediate the Relationship Between Childhood Physical Abuse and Adult Child Abuse Risk?" *Child Abuse & Neglect* 34(3):332–344.

Min, Pyong Gap. 2002. "Korean American Families." Pp. 193–211 in *Minority Families in the United States: A Multicultural Perspective,* 3rd ed., edited by Ronald L. Taylor. Upper Saddle River, NJ: Prentice Hall.

Minino, Arialdi M. andSherry L. Murphy. 2012. "Death in the United States, 2010." NCHS*Data Brief*99 (July). Centers for Disease Control and Prevention. Retrieved May 20, 2013 (http://www.cdc.gov/nchs).

Mintzer, Dorian and Roberta Taylor. 2011. *The Couples Retirement puzzle: 10 Must-Have Conversations for Transitioning to the Second Half of Life.* New York: Lincoln Street Press.

Mishel, Lawrence, Jared Bernstein, and Heidi Shierholz. 2009. *The State of Working America 2008/2009.* New York: Cornell University Press.

Mistry, Rashmita S., Edward D. Lowe, Aprile Benner, and Nina Chien. 2008. "Expanding the Family Economic Stress Model: Insights from a Mixed-Methods Approach." *Journal of Marriage and Family* 70(1):196–209.

Mitchell, Katherine Stamps, Alan Booth, and Valarie King. 2009. "Adolescents with Nonresident Fathers: Are Daughters More Disadvantaged than Sons?" *Journal of Marriage and Family* 71:650–662.

Mitchell, Melanthia. 2007. "Cohousing for Lesbians Planned in Bremerton." Associated Press, July 8. Retrieved May 17, 2010 (www.seattlepi.com).

Mollborn, Stefanie. 2009. "Norms about Nonmarital Pregnancy and Willingness to Provide Resources to Unwed Parents." *Journal of Marriage and Family* 71(1):122–134.

Mollen, Debra. 2008. "Guidance and Guidelines for the Theory and Practice of Feminist Therapy." Sex Roles 59(11–12):900–902.

Monahan Lang, Molly and Barbara J. Risman. 2007. "A 'Stalled' Revolution or a Still-unfolding One? The Continuing Convergence of Men's and Women's Roles." Discussion Paper, 10th anniversary conference of the Council on Contemporary Families, May 4–5, University of Chicago, Chicago, IL. Retrieved June 4, 2007 (www.contemporaryfamilies.org).

Monestero, Nancy. 1990. Personal communication.

"Monogamy: Is It for Us?" 1998. *The Advocate*, June 23, p. 29.

Monroe, Barbara, and Frances Kraus. 2010. *Brief Interventions with Bereaved Children.* 2nd ed. New York: Oxford University Press.

Monserud, Maria A. 2008. "Intergenerational Relationships and Affectual Solidarity Between Grandparents and Young Adults." *Journal of Marriage and Family* 71(1):182–195.

Montgomery, Marilyn J. and Gwendolyn T. Sorell. 1997. "Differences in Love Attitudes across Family Life Stages." *Family Relations* 46:55–61.

Montoro-Rodriguez, J. and D. Gallagher-Thompson. 2009. "The Role of Resources and Appraisals in Predicting Burden among Latina and Non-Hispanic White Female Caregivers." *Aging and Mental Health* 13(5):648–658.

Montoya, R. Matthew. 2008. "I'm Hot, So I'd Say You're Not: The Influence of Objective Physical Attractiveness on Mate Selection." *Personality and Social Psychology Bulletin* 34:1315–1331.

Moody, Harry R. 2006. *Aging: Concepts and Controversies.* 5th ed. Thousand Oaks, CA: Pine Forge Press.

Moore, Abigail Sullivan.2010. "Failure to Communicate." *The New York Times.* July 22. Retrieved September 8, 2010 (http://www.nytimes.com).

Moore, David W. 2005. "Gender Stereotypes Prevail on Working Outside the Home." The Gallup Poll, August 17. Retrieved August 17, 2005 (www.gallup.com/poll).

Moore, Elena. 2012. "Paternal Banking and Maternal Gatekeeping in Postdivorce Families." *Journal of Family Issues* 33(6):745–772.

Moore, Kathleen A.,Marita P. McCabe, and Roger B. Brink. 2001. "Are Married Couples Happier in Their Relationships than Cohabiting Couples? Intimacy and Relationship Factors." *Sexual and Relationship Therapy* 16(1):35–46.

Moore, Kristin Anderson, Zakia Redd, Mary Burkhauser, Kassim Mbwana, and Ashleigh Collins. 2009. "Research Brief: Children in Poverty: Trends, Consequences, and Policy Options." April. Washington DC: Child Trends. Retrieved December 12, 2009 (www.childtrends.org).

Moore, Mignon R. 2008. "Gendered Power Relations among Women: A Study of Household Decision Making in Black, Lesbian Stepfamilies." *American Sociological Review* 73(2):335–356.

Moorman, Elizabeth and Eva Pomerantz. 2008. "The Role of Mothers' Control in Children's Mastery Orientation: A Time Frame Analysis." *Journal of Family Psychology* 22(5):734–741.

Moorman, Sara M., Alan Booth, and Karen L. Fingerman. 2006. "Women's Romantic Relationships after Widowhood." *Journal of Family Issues* 27(9):1281–1304.

"More Binding Marriage Gets a Governor's Participation." 2004. *Omaha World-Herald,* November 14.

"More New Dads Seek Time Off." 2005. *Omaha World-Herald(Baltimore Sun)*,May 9.

Morey, Jennifer N., Amy L. Gentzier, Brian Creasy, Ann M. Oberhauser, and David Westerman. 2013. "Young Adults' Use of Communication Technology within their Romantic Relationships and Associations with Attachment Style." *Computers in Human Behavior* 29(4):1771–1778.

Morford, Mark. 2006. "My Baby Has Rainbow Hair, Gay Parents, Solo Moms, Sperm-Swappin' Friends." April 12. Retrieved April 12, 2006 (www.sfgate.com).

Morgan, Elizabeth M. 2011. "Associations Between Young Adults' Use of Sexually Explicit Materials and Their Sexual Preferences, Behaviors, and Satisfaction." *Journal of Sex Research* 48(6):520–530.

Morgan, S. Philip and R. B. King. 2001. "Why Have Children in the 21st Century? Biological Predispositions, Social Coercion, Rational Choice." *European Journal of Population*17:3–20.

Morgan, S. Philip and Heather Rackin. 2010. "The Correspondence Between Fertility Intentions and Behavior in the United States." *Population and Development Review* 36(1):91–118.

Morin, Erica A. 2012. "No Vacation for Mother: Traditional Gender Roles in Outdoor Travel Literature, 1940–1965." *Women's Studies* 41(4) (June): 436–456.

Morin, Rich. 2011. "The Public Renders a Split Verdict on Changes in Family Structure." *Social & Demographic Trends.* Pew Research Center (February 16). Retrieved February 23, 2011 (http://pewsocialtrends.org http://pewsocialtrends.org/).

Morrison, Donna R. and Amy Ritualo. 2000. "Routes to Children's Economic Recovery after Divorce: Are Cohabitation and Remarriage Equivalent?" *American Sociological Review* 65(4):560–580.

Morrow, Lance. 1992. "Family Values." *Time,* August 31, pp. 22–27.

Moses, Tally. 2010. "Exploring Parents' Self-Blame in Relation to Adolescents' Mental Disorders." *Family Relations* 59(2):103–120.

Mosher, William D., Anjani Chandra, and Jo Jones. 2005. "Sexual Behavior and Selected Health Measures: Men and Women 15–44 Years of Age, United States, 2002." *Advance Data from Vital and Health Statistics,* No. 362. September 15. Hyattsville, MD: U.S. National Center for Health Statistics.

Mossaad, Nadwa. 2010. "The Impact of the Recession on Older Americans." Population Reference Bureau. March. Retrieved May 5, 2010 (www.prb.org).

Movement Advancement Project. 2011. "All Children Matter: How Legal and Social Inequalities Hurt LGBT Families." Boston and Washington, DC: Family Equality Council and Center for American Progress. October. Retrieved September 2, 2012 (http://www.lgbtmap.org/lgbt-families).

Mroz, Jacqueline. 2011. "One Sperm Donor, 150 Offspring." *The New York Times.* September 5. Retrieved September 12, 2011 (http://www.nytimes.com).

MSNBC. 2012. "Transgender Children in America Encounter New Crossroads with Medicine." Retrieved June 25, 2013. (http://insidedateline.nbcnews.com/_news/2012/07/08/12625007-transgender-children-in-america-encounter-new-crossroads-with-medicine?).

Muehlenhard, Charlene L. and Sheena K. Shippee. 2010. "Men's and Women's Reports of Pretending Orgasm." *Journal of Sex Research* 47(6):552–567.

Mueller, Margaret M. and Glen H. Elder, Jr. 2003. "Family Contingencies across the Generations: Grandparent-Grandchild Relationships in Holistic Perspective." *Journal of Marriage and Family* 65(2):404–417.

Muller, Helen Juliette. 1998. "American Indian Women Managers: Living in Two Worlds." *Journal of Management Inquiry* 7(1):4–28.

Mulvaney, Matthew K. and Carolyn J. Mebert. 2007. "Parental Corporal Punishment Predicts Behavior Problems in early Childhood." *Journal of Family Psychology* 21(3):389–397.

Mumola, Christopher J. 2000. *Incarcerated Parents and Their Children.*Washington, DC: U.S. Bureau of Justice Statistics.

Mundy, Liza. 2012. "Women, Money, and Power." *Time.* March 26: 28–31.

Munroe, Erin A. 2009. *The Everything Guide to Stepparenting.* Avon, MA: Aadams Media.

Munroe, Robert L. and A. Kimball Romney. 2006. "Gender and Age Differences in Same-Sex Aggregation and Social Behavior." *Journal of Cross-Cultural Psychology* 37(1):3–19.

Murdock, George P. 1949. *Social Structure.* New York: The Free Press.

Murphy, Tim and Loriann Hoff Oberlin. 2006. *Overcoming Passive-Aggression: How to Stop Hidden Anger from Spoiling Your Relationships, Career, and Happiness.* New York: Marlowe & Company.

Murray, Christine E. 2004. "The Relative Influence of Client Characteristics on the Process and Outcomes of Premarital Counseling." *Contemporary Family Therapy* 26(4):447–463.

Murray, Christine E. and A. Keith Mobley. 2009. "Empirical Research about Same-Sex Intimate Partner Violence: A Methodological Review." *Journal of Homosexuality* 56(3):361–386.

Murry, Velma M., P. Adama Brown, Gene H. Brody, Carolyn E. Cutrona, and Ronald L. Simons. 2001. "Racial Discrimination as a Moderator of the Links among Stress, Maternal Psychological Functioning, and Family Relationships." *Journal of Marriage and Family* 63(4):915–926.

Musick, Kelly. 2002. "Planned and Unplanned Childbearing among Unmarried Women." *Journal of Marriage and Family* 64(4):915–929.

Musick, Kelly and Ann Meier. 2012. "Assessing Causality and Persistence in Associations Between Family Dinners and Adolescent Well-Being." *Journal of Marriage and Family* 74(3): 476–493.

"Muslim Parents Seek Cooperation from Schools." 2005. CNN.com, September 5. Retrieved September 5, 2005 (www.cnn.com).

Mustillo, Sarah, John Wilson, andScott M. Lynch. 2004. "Legacy Volunteering: A Test of Two Theories of Intergenerational Transmission." *Journal of Marriage and Family* 66(2):530–541.

Myers, David G. and Letha Dawson Scanzoni. 2006. *What God Has Joined Together? A Christian Case for Gay Marriage.* San Francisco: *HarperSan Francisco.*

Myers, Jane E., Jayamala Madathil, andLynne R. Tingle. 2005. "Marriage Satisfaction and Wellness in India and the United States: A Preliminary Comparison of Arranged Marriages and Marriages of Choice." *Journal of Counseling and Development* 83(2):183–190.

Myers, Scott A. 2011."'I Have to Love Her, Even if Sometimes I May Not Like Her': The Reasons Why Adults Maintain Their Sibling Relationships." *North American Journal of Psychology* 13(1):51–62.

Myers, Scott M. 2006. "Religious Homogamy and Marital Quality: Historical and Generational Patterns, 1980–1997." *Journal of Marriage and Family* 68(2):292–304.

Myers-Walls, Judith A.2002. "Talking to Children about Terrorism and Armed Conflict." *The Forum for Family and Consumer Issues* 7(1). Retrieved September 14, 2006 (www.ces.ncsu.edu/depts/fcs/pub/2002w/myers-wall.html).

Mykyta, Laryssa, and Suzanne Macartney. 2012. "Sharing a Household: Household Composition and Economic Well-Being: 2007–2010." *Consumer Population Report* P60–242. Washington, DC: U.S. Census Bureau and U.S. Department of Commerce. June.

Nadir, Aneesah. 2009. "Preparing Muslims for Marriage." Retrieved October 11, 2009 (www.soundvision.com).

Nakonezny, Paul A. andWayne H. Denton. 2008. "Marital Relationships: A Social Exchange Theory Perspective." *The American Journal of Family Therapy* 36:402–412.

Nanji, Azim A. 1993. "The Muslim Family in North America." Pp. 229–242 in *Family Ethnicity: Strength in Diversity,* edited by Harriette Pipes McAdoo. Newbury Park, CA: Sage Publications.

Narayan,Chetna. 2008. "Is There a Double Standard of Aging?" *Educational Gerontology* 34(9): 782–787.

National Ag Safety Database. n. d. "From Family Stress to Family Strengths: Stress, Lesson 5" (www.cec.gov/niosh/nasd).

National Alliance for Caregiving. 2009. *Caregiving in the United States: A Focused Look at Those Caring for Someone Age 50 and Older, Executive Summary.* November. Retrieved May 15, 2010 (www.aarp.org).

National Alliance to End Homelessness. 2013. "Demographics of Homeless Series: The Rising Elderly Population." April 1. Retrieved May 13, 2013 (http://www.endhomelessness.org).

National Alliance to End Sexual Violence. 2012. "2012 Rape Crisis Center Survey." Retrieved March 18, 2013 (http://www.endsexualviolence.org).

National Association for Sick Child Day Care. 2010. Birmingham, AL: National Association for Sick Child Day Care. www.nascd.com.

National Association of Child Care Resource and Referral Agencies (NACCRRA). 2006. "Is This the Right Place for My Child? A Research-based Checklist." Retrieved January 29, 2013 (http://www.naccrra.org).

National Center for Elder Abuse. 2013. "Frequently Asked Questions." Retrieved May 10, 2013 (http://www.ncea.aoa.gov).

National Clearinghouse on Marital and Date Rape. n.d. Retrieved from http://www.ncmdr.org/

National Coalition Against Domestic Violence. 2009. "Domestic Violence Facts." Retrieved March 18, 2013 (http://www.ncadv.org).

National Coalition for the Homeless. 2009a. "How Many People Experience Homelessness?" Washington, DC. Retrieved January 20, 2010 (www.nationalhomeless.org).

———. 2009b. "Who Is Homeless?" July. Washington, D.C. Retrieved January 20, 2010 (www.nationalhomeless.org).

———. 2009c. "Why are People Homeless?" Washington, DC. Retrieved January 20, 2010 (www.nationalhomeless.org).

"National Compensation Survey: Employee Benefits in Private Industry in the United States, March 2007." 2007. Summary 07–05. Washington, DC: U.S. Bureau of Labor Statistics (www.bls.gov).

National Employment Law Project (NELP). 2012. "The Low-Wage Recovery and Growing Inequality." Data Brief, August. Retrieved September 10, 2012 (http://www.nelp.org).

National Employment Law Project (NELP). 2013. "Raise the Minimum Wage: Rebuilding an Economy that Works for All of Us." Retrieved February 24, 2013 (http://raisetheminimumwage.org).

National Institute of Justice. 2007. *Commercial Sexual Exploitation of Children: What Do We Know and What Do We Do about It?* Washington, DC: U.S. Department of Justice.

National Marriage Project. 2009. "Figures Supplement to *The State of Our Unions: The Social Health of Marriage in America, 2008."* Rutgers University. Retrieved October 12, 2009 (http://marriage.rutgers.edu).

National Marriage Project and the Institute for American Values. 2012. *The State of Our Unions: Marriage in America 2012.* Charlottesville, VA.

National Public Radio. 2011. "How A Divorce Helped One Couple Grow Closer." Retrieved April 17, 2013 (http://www.npr.org/2011/03/04/134236042/how-a-divorce-helped-one-couple-grow-closer).

National Resource Center on Children and Families of the Incarcerated. 2009a. "An Overview of Statistics" Retrieved March 13, 2010 (www.fcnetwork.org).

———. 2009b. "What Happens to Children?" Retrieved March 13, 2010 (www.fcnetwork.org).

National Right to Life. 2005. "Abortion's Physical Complications." Washington, DC: National Right to Life Educational Trust Fund. July.

———. 2006. "Abortion's Psycho-Social Consequences." Washington, DC: National Right to Life Educational Trust Fund, December 6. Retrieved March 29, 2007 (www.nrlc.org).

National Women's Law Center. 2010. *Congress Must Act to Close the Wage Gap for Women.* Washington, DC: National Women's Law Center.

Nazario, Sonia L. 1990. "Identity Crisis: When White Parents Adopt Black Babies, Race Often Divides." *Wall Street Journal,* September 20.

Neal, Margaret B. and Leslie B. Hammer. 2007. *Working Couples Caring for Children and Aging Parents: Effects on Work and Well-Being.* Mahwah, NJ: Erlbaum.

Neblett, Nicole Gardner. 2007. "Patterns of Single Mothers' Work and Welfare Use." *Journal of Family Issues* 28(9):1093–1112.

Neel, Aly. 2013. "Surrogate Mother Refused Abortion." *The Washington Post.*March 6. Retrieved March 26, 2013 (http://www.washingtonpost.com).

Neely-Barnes, Susan L. and J. Carolyn Graff. 2011. "Are There Adverse Consequences to Being a Sibling of a Person with a Disability? A Propensity Score Analysis." *Family Relations* 60(3):331–341.

Neff, Kristen D. and S. Natasha Beretavas. 2013. "The Role of Self-Compassion in Romantic Relationships." *Self and Identity* 12:78–98.

Neff, Kristen D. andSusan Harter. 2003. "Relationship Styles of Self-Focused Autonomy, Other-Focused Connectedness, and Mutuality Across Multiple Relationship Contexts." *Journal of Social and Personal Relationships* 20(1):81–99.

Negy, Charles and Douglas K. Snyder. 2000. "Relationship Satisfaction of Mexican American and Non-Hispanic White American Interethnic Couples: Issues of Acculturation and Clinical Intervention." *Journal of Marital and Family Therapy* 26(3):293–305.

Neimark, Jill. 2003. "All You Need Is Love: Why It's Crucial to Your Health—and How to Get More in Your Life." *Natural Health* 33(8):109–113.

Nelson, Margaret K. 2008. "Watching Children." *Journal of Family Issues* 29(4):516–538.

Nelson, Jennifer. 2012. "Feminism Gave Rise to Superwoman in Advertising." *WeNews online.* Retrieved November 8, 2012 (http://womensenews.org).

Nelson, Margaret K. 2006. "Families in Not-So-Free Fall: A Response to Comments." *Journal of Marriage and Family* 68(4):817–823.

———. 2008. "Watching Children." *Journal of Family Issues* 29(4):516–538.

———. 2011. "Between Family and Friendship: The Right to Care for Anna." *Journal of Family Theory & Review* 3(4):241–255.

Nemko, Marty. n. d. "The New Double Standard." Retrieved October 28, 2012 (http://www.martynemko.com).

Newman, Bernie Sue. 2007. "College Students' Attitudes about Lesbians: What Difference Does 16 Years Make?" *Journal of Homosexuality* 52(3/4): 249–265.

Newman, Katherine. 2012. *The Accordion Family: Boomerang Kids, Anxious Parents, and the Private Toll of Global Competition.* Boston: Beacon Press.

Newman, Katherine and James B. Knapp. 2013. "The Accordion Family." *Family Focus* FF56:F2–F3.

Newman, Susan. 2011. *The Case for the Only Child: Your Essential Guide.* Deerfield Beach, FL: Health Communications, Inc.

Newport, Frank. 2008. "Wives Still Do Laundry, Men Do Yard Work." Gallup Poll, April 4. Retrieved December 2, 2009 (www.gallup.com /poll).

———. 2009. "Extramarital Affairs, Like Sanford's, Morally Taboo." Poll Analysis. Retrieved June 25 (www.gallup.com).

———. 2012. "Half of Americans Support Legal Gay Marriage." May 8. Retrieved November 2, 2012 (http://www.gallup.com).

Newport, Frank and Brett Pelham. 2009. "Americans Least Happy in Their 50s and Late 80s." Gallup Poll, October 5. Retrieved May 6, 2010 (www.gallup.com/poll).

Nicholson v. Scopetta. 2004. N.Y. No. 113.

Nicholson, Jan M., Matthew R. Sanders, W. Kim Halford, Maddy Phillips, and Sarah W. Whitton. 2008. "The Prevention and Treatment of Children's Adjustment Problems in Stepfamilies." Pp. 485–521 in *The International Handbook of Stepfamilies,* edited by J. Pryor. Hoboken, NJ: John Wiley & Sons.

Niehuis, Sylvia, Kyung-Hee Les, A. Reifman, A. Swenson, and S. Hunsaker. 2011. "Idealization and Disillusionment in Intimate Relationships: A Review of Theory, Method, and Research." *Journal of Family Theory & Review* 3(4):273–302.

Nielsen, Rasmus. 2009. "Adaptionism—30 Years After Gould and Lewontin." *Evolution* 63(10): 2487–2490.

Nierenberg, Gerard and Henry H. Calero. 1973. *Meta-Talk: Guide to Hidden Meanings in Conversations.* New York: Trident Press.

Nir, Sarah Maslin. 2013. "You May Kiss the Computer Screen." *The New York Times.* March 5. Retrieved March 7, 2013 (http://www.nytimes).

Nixon, Elizabeth, Sheila Greene, and Diane M. Hogan. 2012. "Negotiating Relationships in Single-Mother Households: Perspectives of Children and Mothers." *Family Relations* 61(1): 142–156.

Nock, Steven L. 2001. "The Marriages of Equally Dependent Spouses." *Journal of Family Issues* 22:755–775.

———. 2005. "Marriage as a Public Issue." *The Future of Children* 15(2):13–32.

Nock, Steven M., Laura A. Sanchez, and James D. Wright. 2008. *Covenant Marriage: The Movement to Reclaim Tradition in America.* New Brunswick, NJ: Rutgers University Press.

Noland, Virginia J., Karen D. Liller, Robert J. McDermott, Martha Coulter, and Anne E. Seraphine. 2004. "Is Adolescent Sibling Violence a Precursor to Dating Violence?" *American Journal of Health Behavior* 28(Supp. 1):813–823.

Nolem-Hoeksema, S. 2004. "Gender Differences in Risk and Protective Factors and Consequences for Alcohol Use and Problems." *Clinical Psychology Review* 24: 981–1010.

Noller, Patricia and Mary Anne Fitzpatrick. 1991. "Marital Communication in the Eighties." Pp. 42–53 in *Contemporary Families: Looking Forward, Looking Back,* edited by AlanBooth. Minneapolis: National Council on Family Relations.

Nomaguchi, Kei M. 2008. "Gender, Family Structure, and Adolescents' Primary Confidants." *Journal of Marriage and Family* 70(5): 1213–1227.

Nomaguchi, Kei M. and Susan L. Brown. 2011. "Parental Strains and Rewards among Mothers: The Role of Education." *Journal of Marriage and Family* 73(3): 621–636.

Nomaguchi, Kei M. and Melissa A. Milkie. 2003. "Costs and Rewards of Children: The Effects of Becoming a Parent on Adults' Lives." *Journal of Marriage and Family* 65:356–374.

NOMAS. 2012. National Organization for Men Against Sexism. "National Conference on Men and Masculinity." Retrieved October 28, 2012 (http://www.nomas.org).

Nordwall, Smita P. and Paul Leavitt. 2004. "Court Rules for Battered Women's Rights." *USA Today,* October 24.

Norris, Jeanette et al. 2013. "How do Alcohol and Relationship Type Affect Women's Risk Judgment of Partners with Differing Risk Histories?" *Psychology of Women Quarterly* 37(2):209–223.

Notter, Megan L., Katherine A. MacTavish, and Devora Shamah. 2008. "Pathways Toward Resilience among Women in Rural Trailer Parks." *Family Relations* 57(5):613–624.

Nozawa, Shinji. 2008. "The Social Context of Emerging Stepfamilies in Japan: Stress and Support for Parents and Stepparents." In J. Pryor (Ed.), *The International Handbook of Stepfamilies* (pp. 79–99). Hoboken, NJ: John Wiley & Sons.

Nunn, Brittany. 2012. "Statistics Suggest Bleak Futures for Children Who Grow Up in Foster Care." *Amarillo Globe.* June 24. Retrieved January 27, 2012 (http://amarillo.com).

Oakley, Ann. 1972. *Sex, Gender, and Society.* London: Maurice Temple Smith.

Oberlander, Sarah E., Avril Melissa Houston, Wendy R. Miller Agostini, and Maureen M. Black. 2010. "A Seven-Year Investigation of Marital Expectations and Marriage among Urban, Low-Income, African American Adolescent Mothers." *Journal of Family Psychology* 24(1):31–40.

O'Brien, Brendan. 2012. "Wisconsin's Baldwin Becomes First Openly Gay Senator." *Chicago Tribune.* Retrieved November 9, 2012 (http://articles.chicagotribune.com/2012-11-07/news/sns-rt-us-usa-campaign-wisconsin-senatebre-8a60by-20121106_1_governor-tommy-thompson-tammy-baldwin-expensive-senate-race).

O'Brien, Karen M. and Kathy P. Zamostny. 2003. "Understanding Adoptive Families: An Integrative Review of Empirical Research and Future Directions for Counseling Psychology." *The Counseling Psychologist* 31(6):679–710.

O'Brien, M. 2005. "Studying Individual and Family Development: Linking Theory and Research." *Journal of Marriage and the Family* 67:880–890.

O'Brien, Marion. 2007. "Ambiguous Loss in Families of Children with Autism Spectrum Disorders." *Family Relations* 56(2):135–146.

O'Brien, Sharon. n. d. "All about Grandparents Day." *Senior Living.* Retrieved May 9, 2013 (http://www.seniorliving.org).

O'Connell, Martin and Daphne Lofquist. 2009. "Counting Same-Sex Couples: Official Estimates and Unofficial Guesses." Paper presented at Annual Meeting of the Population Association of America, Detroit, Michigan, April 30–May 2.

O'Connor, Joe. 2012. "Full Comment. Joe O'Connor: Trend of Couples Not Having Children Just Plain Selfish." *The National Post.* September 19. Retrieved December 13, 2012 (http://fullcomment.nationalpost.com).

Odom, Samuel L. 2009. *Handbook of Developmental Disabilities.* New York: Guilford Press.

Offer, Shira. 2013. "Family Time Activities and Adolescents' Emotional Well-Being." *Journal of Marriage and Family* 75(1): 26–41.

Office on Violence Against Women. 2010. "The Facts about Domestic Violence." *Violence Against Women Online Resources.* Retrieved December 3, 2012 (http://www.vaw.umn.edu).

Ojeda, Norma. 2011. "Living Together Without Being Married: Perceptions of Female Adolescents in the Mexico-United States Border Region." *Journal of Comparative Family Studies* 42(4):439–454.

"Older Moms' Birth Risks Called Greater." 1999. *Omaha World-Herald,* January 2.

Oldham, J. Thomas. 2008a. "Changes in the Economic Consequences of Divorces, 1958–2008." *Family Law Quarterly* 42(3):419–447.

———. 2008b. "What If the Beckhams Move to L.A. and Divorce—Marital Property Rights of Mobile Spouses When They Divorce in the United States." *Family Law Quarterly* 42(2):263–293.

Olson, David H. and Dean M. Gorall. 2003. "Circumplex Model of Marital and Family Systems." Pp. 514–547 in *Normal Family Processes,*3rd ed., edited by F.Walsh. New York: Guilford.

Olson, M., C. S. Russell, M. Higgins-Kessler, and R. B. Miller. 2002. "Emotional Processes Following Disclosure of an Extramarital Fidelity." *Journal of Marital and Family Therapy* 28:423–434.

Ono, Hiromi. 2009. "Husbands' and Wives' Education and Divorce in the United States and Japan, 1946–2000." *Journal of Family History* 34(3):292–322.

Ontai, Lenna, Yoshie Sano, Holly Hatton, and Katherine Conger. 2008. "Low-Income Rural Mothers' Perceptions of Parent Confidence: The Role of Family Health Problems and Partner Status." *Family Relations* 57(July):324–334.

Americans." *Journal of Marriage and Family* 62(1):48–60.

Treas, Judith and Sonja Drobnič. 2010. *Dividing the Domestic: Men, Women, and Household Work in Cross-National Perspective*. Stanford, CA: Stanford University Press.

Treas, Judith and Tsui-o Tai. 2012. "How Couples Manage the Household: Work and Power in Cross-National Perspective." *Journal of Family Issues* 33(8): 1088–1116.

Trevillion, K., S. Oram, G. Feder, and L.M. Howard. 2012. "Experiences of Domestic Violence and Mental Disorders: A Systematic Review and Meta-Analysis." *Plos One* 7(12). Retrieved March 12, 2013 (http://www.plosone.org).

Trickey, Helyn. 2011. "Weddings Blend Cultures as Well as Families." Retrieved May 30, 2013, from http://www.cnn.com/2011/LIVING/07/07/wedding.different.cultures/index.html.

Trillingsgaard, Tea, K. Baucom, R. Heyman, and A Elklit. 2012. "Relationship Interventions During the Transition to Parenthood: Issues of Timing and Efficacy." *Family Relations* 61(5): 770–783.

Trimberger, E. Kay. 2005. *The New Single Woman*. New York: Beacon Press.

Troilo, Jessica and Marilyn Coleman. 2008. "College Students' Perceptions of the Content of Father Stereotypes." *Journal of Marriage and Family* 70(1):218–227.

———. 2012. "Full-Time, Part-Time Full-Time, and Part-Time Fathers: Father Identities Following Divorce." *Family Relations* 61:601–614.

Troll, Lillian E. 1985. "The Contingencies of Grandparenting." Pp. 135–150 in *Grandparenthood*, edited by Vern L. Bengstonand Joan F. Robertson. Newbury Park, CA: Sage Publications.

Troxel v. Granville. 2000. 530 U.S. 57.

Troy, Adam B., Jamie Lewis-Smith, and Jean-Phillippe Laurenceau. 2006. "Interracial and Intraracial Romantic Relationships: The Search for Differences in Satisfaction, Conflict, and Attachment Style." *Journal of Social and Personal Relationships* 23(1):65–80.

Trussell, James and L. L. Wynn. 2008. "Reducing Unintended Pregnancy in the United States." *Contraception* 77:1–5.

Tshann, Jeane M., Lauri A. Pasch, Elena Flores, Barbara VanOss Marin, E. Marco Baisch, and Charles J. Wibbelsman. 2009. "Nonviolent Aspects of Interparental Conflict and Dating Violence among Adolescents." *Journal of Family Issues* 30(3):295–319.

Tucker, Corinna J., Susan M. McHale, and Ann C. Crouter. 2003. "Conflict Resolution: Links with Adolescents' Family Relationships and Individual Well-Being." *Journal of Family Issues* 24(6):715–736.

Tucker, M. Belinda. 2000. "Marital Values and Expectations in Context: Results from a 21-City Survey." Pp. 166–187 in *The Ties That Bind: Perspectives on Cohabitation and Marriage*, edited by Linda J. Waite. New York: Aldine.

Tugend, Alina. 2011. "Family Happiness and the Overbooked Child." *The New York Times*. August 12. Retrieved September 13, 2011 (http://www.nytimes.com).

Tuller, David. 2001. "Adoption Medicine Brings New Parents Answers and Advice." *The New York Times*, September 4.

Turley, Ruth N. Lopez and Matthew Desmond. 2011. "Contributions to College Costs by Married, Divorced, and Remarried Parents." *Journal of Family Issues* 32(6):767–790.

Turnbull, A. and H. Turnbull. 1997. *Families, Professionals, and Exceptionality: A Special Partnership*. 3rd ed. Upper Saddle River, NJ: Merrill.

Turner, Lynn H. and Richard L. West. 2006. *The Family Communication Sourcebook*. Thousand Oaks, California: Sage.

Turner, Ralph H. 1976. "The Real Self: From Institution to Impulse." *American Journal of Sociology* 81:989–1016.

Turney, Kristin. 2011. "Chronic and Proximate Depression among Mothers: Implications for Child Well-Being." *Journal of Marriage and Family* 73(1): 149–163.

Twenge, Jean M. 1997. "Attitudes toward Women, 1970-1995: A Meta-Analysis." *Psychology of Women Quarterly* 21(1):35–51.

Twenge, Jean M., W. Keith Campbell, and Craig Foster. 2003. "Parenthood and Marital Satisfaction: A Meta-Analytic Review." *Journal of Marriage and Family* 65:574–583.

Twenge, Jean M., W. Keith Campbell, and Brittany Gentile. 2012. "Male and Female Pronoun Use in U.S. Books Reflects Women's Status, 1900–2008." *Sex Roles* 67(9–10): 488–493.

Tyre, Peg. 2004. "A New Generation Gap." *Newsweek*, January 19, pp. 68–71.

———. 2006. "The Trouble With Boys." *Newsweek*, January 30.

———. 2009. "Daddy's Home, and a Bit Lost." *The New York Times*, January 11. Retrieved October 5, 2009 (www.nytimes.com).

Udry, J. Richard. 1974. *The Social Context of Marriage*. 3rd ed. Philadelphia, PA: Lippincott.

———. 1994. "The Nature of Gender." *Demography* 31(4):561–573.

———. 2000. "The Biological Limits of Gender Construction." *American Sociological Review* 65:443–457.

Uecker, Jeremy E. and Christopher G. Ellison. 2012. "Parental Divorce, Parental Religious Characteristics, and Religious Outcomes in Adulthood." *Journal for the Scientific Study of Religion* 51(4):777–794.

Uecker, Jeremy E. and Charles E. Stokes. 2008. "Early Marriage in the United States." *Journal of Marriage and Family* 70(November):835–846.

"UF Study: Sibling Violence Leads to Battering in College Dating." 2004. *UF News*. Gainesville, FL: University of Florida.

Uhlenberg, Peter. 1996. "Mortality Decline in the Twentieth Century and Supply of Kin over the Life Course." *The Gerontologist* 36:681–685.

Ulker, Aydogan. 2009. "Wealth Holdings and Portfolio Allocation of the Elderly: The Role of Marital History." *Journal of Family Economic Issues* 30:90–108.

Umberson, Debra, Kristin Anderson, Jennifer Glick, and Adam Shapiro. 1998. "Domestic Violence, Personal Control, and Gender." *Journal of Marriage and Family* 60(2):442–452.

Umberson, Debra, Meichu D. Chen, James S. House, Kristine Hopkins, and Ellen Slaten. 1996. "The Effect of Social Relationships on Psychological Well-Being: Are Men and Women Really So Different?" *American Sociological Review* 61:837–857.

Umberson, Debra, Tetyana Pudrovska, and Corinne Reczek. 2010. "Parenthood, Childlessness, and Well-Being: A Life Course Perspective." *Journal of Marriage and Family* 72(3) (June): 612–629.

UNICEF (United Nations Children's Fund). 2007. *Child Poverty in Perspective: An Overview of Child Well-Being in Rich Countries*. Innocenti Research Center Report Card No. 7. Florence, Italy: UNICEF Innocenti Research Center.

Unitarian Universalist Association. n. d. *Premarital Counseling Guide for Same Gender Couples*. Boston: Unitarian Universalist Association Office of Bisexual, Gay, Lesbian, and Transgender Concerns. Retrieved March 3, 2010 (www.uua.org/obgltc).

Upchurch, D. M., L. A. Lillard, and C. W. Panis. 2001. "The Impact of Nonmarital Childbearing on Subsequent Marital Formation and Dissolution." Pp. 344–382 in *Out of Wedlock*, edited by L. Wu and B. Wolfe. New York: Sage.

Updegraff, Kimberly A., Norma J. Perez-Brena, Megan E. Baril, Susan M. McHale, and Adriana J. Umana-Taylor. 2012. "Mexican-Origin Mothers' and Fathers' Involvement in Adolescents' Peer Relationships: A Pattern-Analytic Approach." *Journal of Marriage and Family* 74(5): 1069–1083.

Uruk, Ayse, Thomas Sayger, and Pamela Cogdal. 2007. "Examining the Influence of Family Cohesion and Adaptability on Trauma Symptoms and Psychological Well-Being." *Journal of College Student Psychotherapy* 22(2):51–63.

U.S. Administration on Aging. 2010. *A Profile of Older Americans:2009*. Washington, DC: U.S. Department of Health and Human Services. Retrieved May 8, 2010 (www.aoa.gov).

———. 2013. *A Profile of Older Americans:2012*. Washington, DC: U.S. Department of Health and Human Services. Retrieved May 7, 2013 (www.aoa.gov).

U.S. Bureau of Labor Statistics. 2012a. "Economic News Release: Employment Situation Summary." Washington, DC: U.S. Bureau of Labor Statistics. Retrieved October 25, 2012 (http://www.bls.gov).

———. 2012b. "Highlights of Women's Earnings in 2011." Report 1038. Washington, DC: U.S. Bureau of Labor Statistics. Retrieved October 25, 2012 (http://www.bls.gov).

———. 2013. *Women in the Labor Force: A Databook*. Report 1040. Washington, DC: U.S. Bureau of Labor Statistics. www.bls.gov.

U.S. Census Bureau. 1989. *Statistical Abstract of the United States*. 109th ed. Washington, DC: U.S. Government Printing Office.

———. 2002. *Statistical Abstract of the United States: 2006*. Washington, DC: U.S. Census Bureau.

———. 2003. *Statistical Abstract of the United States, 2003*. Washington, DC: U.S. Census Bureau.

———. 2007a. "The American Community—Pacific Islanders: 2004." Washington, DC: U.S. Census Bureau. Retrieved November 4, 2009 (factfinder.census.gov).

———. 2007b. *Statistical Abstract of the United States: 2007*. Washington, DC: U.S. Census Bureau.

———. 2008. "Current Population Survey, 2007: Annual Social and Economic Supplement." Washington, DC: U.S. Census Bureau. Retrieved June 29, 2009 (www.census.gov/population/www/socdemo/hhfam/cps2009.html).

———. 2009. "Selected Indicators of Child Well-Being—A Child's Day: 2006." Washington, DC: U.S. Census Bureau. Retrieved December 23, 2009 (www.census.gov).

———. 2010a. "Current Population Survey, March and Annual Social and Economic Supplements, 2009 and Earlier." Washington, DC: U.S. Census Bureau. Retrieved January 21, 2010 (www.census.gov/population).

———. 2010b. "Statistical Abstract of the United States." Washington, DC: U.S. Census Bureau.

———. 2011a. "Census Bureau Releases Estimates of Same-Sex Married Couples." Washington, DC: U.S. Census Bureau Public Information Office Press Release. September 27. Retrieved September 12, 2012 (http://www.census.gov.newsroom).

———. 2011b. "2010 American Community Survey 1-Year Estimates." Washington, DC: U.S. Census Bureau. Retrieved August 31, 2012 (http://factfinder.census.gov).

———. 2011c. Current Population Survey, 2011. Retrieved September 14, 2012 (http://www.census.gov).

———. 2011d. Survey of Income and Program Participation (SIPP). Retrieved February 18, 2013 (http://www.census.gov/hhes/childcare).

———. 2012a. "America's Families and Living Arrangements: 2011." *Current Population Reports*. Washington, DC: U.S. Census Bureau. Retrieved August 31, 2012 (www.census.gov/population).

———. 2012b. "Current Population Survey, 2011 Annual Social and Economic Supplement." Washington, DC: U.S. Census Bureau. Retrieved August 31, 2012 (www.census.gov/hhes).

———. 2012c. *Statistical Abstract of the United States*. Washington, DC: U.S. Census Bureau.

———. 2012d. "U.S. Census Bureau Facts for Features: Hispanic Heritage Month 2012: Sept 15-Oct.15." Retrieved August 29, 2012 (http://news.yahoo.com).

———. 2012e. "2011 American Community Survey 1-Year Estimates." Washington, DC: U.S. Census Bureau. Retrieved November 28, 2012 (http://factfinder.census.gov).

———. 2012f. "America's Families and Living Arrangements: 2010" and unpublished data, Table 66, "Families by Type, Race, and Hispanic Origin: 2010." *The 2012 Statistical Abstract*. Washington DC: U.S. Census Bureau. Retrieved September 3, 2013 (www.census.gov/prod/2011pubs/12statab/pop.pdf).

———. 2012g. "Monthly Child Support Payments Average $430 per Month in 2010." Retrieved April 22, 2013 (http://www.census.gov/newsroom/releases/archives/children/cb12-109.html).

U.S. Centers for Disease Control and Prevention. 2008. "2006 Assisted Reproductive Technology Success Rates: National Summary and Fertility Clinic Reports." Atlanta, GA: U.S. Department of Health and Human Services, Centers for Disease Control and Prevention and American Society for Reproductive Medicine, Society for Assisted Reproductive Technology.

———. 2010. *Monitoring Selected National HIV Prevention and Care Objectives by Using HIV Surveillance Data—United States and Six U.S. Dependent Areas—2010*. HIV Surveillance Supplemental Report 2013;18(No. 2, part B). Retrieved January 25, 2013 (http://www.cdc.gov/hiv/topics/surveillance/resources/reports/#supplemental).

———. 2011. Youth Risk Behavior Surveillance System: 2011 National Overview. Retrieved June 25, 2013 (http://www.cdc.gov/HealthyYouth/yrbs/pdf/us_overview_yrbs.pdf).

———. 2012a. Monitoring Selected National HIV Prevention and Care Objectives by Using HIV Surveillance Data—United States and Six U.S. Dependent areas—2010. HIV Surveillance Supplemental Report 2012;17 (No. 3, part A). Retrieved January 25, 2013 (http://www.cdc.gov/hiv/topics/surveillance/basic.htm#hivest).

———. 2012b. Estimated HIV Incidence among Adults and Adolescents in the United States, 2007-2010. HIV Supplemental Report 2012; XX(no. X). Retrieved January 25, 2013 (http://www.cdc.gov/hiv/topics/surveillance/basic.htm#hivest).

———. 2012c. HIV Surveillance Report, 2010; vol. 22. http://www.cdc.gov/hiv/topics/surveillance/resources/reports/. Published March 2012. Retrieved January 25, 2013.

———. 2012d. "Infertility." Faststats Homepage. Atlanta, GA. Retrieved January 14, 2013 (http://www.cdc.gov).

U.S. Citizenship and Immigration Services. 2011. "Battered Spouse, Children, and Parents." Retrieved March 14, 2013 (http://www.uscis.gov).

U.S. Department of Health and Human Services. 2010. *Child Maltreatment 2008*. Washington, DC: Administration for Children and Families, Administration on Children, Youth and Families, Children's Bureau. Available from www.acf.hhs.gov/programs/cb/stats_research/index.htm#can.

———. 2012a. The AFCARS Report for 2011. No. 19. Washington, DC: U.S. Department of Health and Human Services.

———. 2012b. Child Maltreatment 2011. Retrieved April 28, 2013 (http://www.acf.hhs.gov/programs/cb/research-data-technology/statistics-research/child-maltreatment).

———. 2013. *Preventing Child Maltreatment and Promoting Well-Being: A Network for Action—2013 Resource Guide*. Retrieved April 28, 2013 (https://www.childwelfare.gov/pubs/guide2013/guide.pdf).

U.S. Department of Health, Education, and Welfare. 1975. *Child Abuse and Neglect. Vol. I, An Overview of the Problem*. Publication (OHD) 75-30073. Washington, DC: U.S. Government Printing Office.

U.S. Department of Justice. 2011. "Statement of the Attorney General on Litigation Involving the Defense of Marriage Act." February 23. Washington, DC: Department of Justice Office of Public Affairs. Retrieved February 23, 2011 (http://www.justice.gov).

———. 2013. Sexual Assault. Retrieved May 27, 2013 (http://www.ovw.usdoj.gov/sexassault.htm).

U.S. Federal Interagency Forum on Aging-Related Statistics. 2013. *Older Americans 2012: Key Indicators of Well-Being*. Washington DC: U.S. Government Printing Office.

U.S. Federal Interagency Forum on Child and Family Statistics. 2005. *America's Children: Key National Indicators of Well-Being, 2005*. Washington, DC: U.S. Federal Interagency Forum on Child and Family Statistics.

———. 2006. *America's Children in Brief: Key National Indicators of Well-Being, 2006*. Washington, DC: U.S. Federal Interagency Forum on Child and Family Statistics. Retrieved August 13, 2006 (www.childtrends.org).

———. 2012. *America's Children in brief: Key National Indicators of Well-Being, 2012*. Retrieved September 3, 2012 (http://childstats.gov/americaschildren).

U.S. Food and Drug Administration. 2006. "Mifeprex (mifepristone) Information." Washington, DC: U.S. Food and Drug Administration. April 10. Retrieved March 29, 2007 (www.fda.gov).

U.S. National Center for Health Statistics. 1990. "Advance Report of Final Divorce Statistics, 1987." *Monthly Vital Statistics Report* 38(12), Suppl. April 3.

———. 1998. "Births, Marriages, Divorces, and Deaths for 1997." *Monthly Vital Statistics Report* 46(12). July 28.

———. 2006. "Births, Marriages, Divorces, and Deaths: Provisional Data for 2005." *National Vital Statistics Reports* 54(20). Hyattsville, MD: U.S. National Center for Health Statistics.

U.S. National Institute of Child Health and Human Development. 1999. "Only Small Link Found between Hours in Child Care and Mother-Child Interaction." News Release, November 7.

"U.S. Recession Causing Increase in Child Abuse Reports." 2009. Red Orbit News, April 16. Retrieved June 8, 2009 (www.redorbit.com).

U.S. Social Security Administration. 2013. "Fact Sheet: 2013 Social Security Changes." Retrieved May 8, 2013 (www.socialsecurity.gov/).

Usdansky, Margaret L. 2009a. "Ambivalent Acceptance of Single-Parent Families: A Response to Comments." *Journal of Marriage and Family* 71(2):240–246.

———. 2009b. "A Weak Embrace: Popular and Scholarly Depictions of Single-Parent Families, 1900-1998." *Journal of Marriage and Family* 71(2):209–225.

———. 2011. "The Gender-Equality Paradox: Class and Incongruity Between Work-Family Attitudes and Behaviors." *Journal of Family Theory and Review* 3(3) (September): 163–178.

Uttal, Lynet. 1999. "Using Kin for Child Care." *Journal of Marriage and Family* 61:845–857.

Vaaler, Margaret L., Christopher G. Ellison, and Daniel A. Powers. 2009. "Religious Influences on the Risk of Marital Dissolution." *Journal of Marriage and Family* 71(4):917–934.

Valentine, Kylie. 2008. "After Antagonism: Feminist Theory and Science." *Feminist Theory* 9(3):355–365.

Valiente, Carlos, Richard A. Fabes, Nancy Eisenberg, and Tracy L. Spinrad. 2004. "The Relations of Parental Expressivity and Support to Children's Coping with Daily Stress." *Journal of Family Psychology* 18(1):97–107.

Van Anders, Sari M. and Neil V. Watson. 2006. "Relationship Status and Testosterone in North American Heterosexual and Nonheterosexual Men and Women: Cross-Sectional and Longitudinal Data." *Psychoneuroendocrinology* 31(6):715–723.

———. 2007. "Testosterone Levels in Women and Men Who Are Single, in Long-Distance Relationships, Or Same-City Relationships." *Hormones and Behaviors* 51(2):286–291.

Van Campen, Kali S., and Andrea J. Romero. 2012. "How Are Self-Efficacy and Family Involvement Associated with Less Sexual Risk Taking among Minority Adolescents?" *Family Relations* 61(4):548–558.

Vandeleur, C. L., N. Jeanpretre, M. Perrez, and D. Schoebi. 2009. "Cohesion, Satisfaction with Family Bonds, and Emotional Well-Being in Families with Adolescents." *Journal of Marriage and Family* 71(5):1205–1219.

Vandell, Deborah Lowe, Jay Belsky, Margaret Burchinal, Laurence Steinberg, and Nathan Vandergrift. 2010. "Do Effects of Early Child Care Extend to Age 15 Years? Results from the NICHD Study of Early Child Care and Youth Development." *Child Development* 81(3):737–756.

Van den Haag, Ernest. 1974. "Love or Marriage." Pp. 134–142 in *The Family: Its Structures and Functions,* 2nd ed., edited by Rose Laub Coser. New York: St. Martin's.

VanderLaan, Doug P. and Paul L. Vasey. 2009. "Patterns of Sexual Coercion in Heterosexual and Non-Heterosexual Men and Women." *Archives of Sexual Behavior* 38(6):987–999.

Van der Lippe, Tanja. 2010. "Women's Employment and Housework." In *Dividing the Domestic: Men, Women, and Household Work in Cross-National Perspective,* edited by Judith Treasand Sonja Drobnič . Stanford, CA: Stanford University Press.

Vandewater, Elizabeth A. and Jennifer E. Lansford. 2005. "A Family Process Model of Problem Behaviors in Adolescents." *Journal of Marriage and Family* 67(1):100–109.

Vandivere, Sharon, Karin Malm, and Laura Radel. 2009. *Adoption USA: A Chartbook Based on the 2007 National Survey of Adoptive Parents.* Washington, DC: The U.S. Department of Health and Human Services, Office of the Assistant Secretary for Planning and Evaluation.

VanDorn, Richard A., Gary L. Bowen, and Judith R. Blau. 2006. "The Impact of Community Diversity and Consolidated Inequality on Dropping Out of High School." *Family Relations* 55(1):105–118.

Van Eeden-Moorefield, Brad, Christopher R. Martell, Mark Williams, and Marilyn Preston. 2011. "Same-Sex Relationships and Dissolution: the Connection Between Heteronormativity and Homonormativity." *Family Relations* 60(5): 562–571.

Van Eeden-Moorefield, Brad, and Kay Pasley. 2008. "A Longitudinal Examination of Marital Processes Leading to Instability in Remarriages and Stepfamilies." Pp. 231–249 in *The International Handbook of Stepfamilies,* edited by J. Pryor. Hoboken, NJ: John Wiley & Sons.

Van Eeden-Morrefield, Brad, Kay Pasley, Elizabeth Dolan, and Margorie Engel. 2007. "Financial Management and Security Among Remarried Women." *Journal of Divorce & Remarriage* 47(3–4):21–42.

VanNatta, Michelle. 2005. "Constructing the Battered Woman." *Feminist Studies* 31(2):416–429.

"Vasectomy." 2012. Planned Parenthood. Retrieved January 7, 2012 (http://www.plannedparenthood.org).

Vaughan, Ellen, Richard Feinn, Stanley Bernard, Maria Brereton, and Joy Kaufman. 2012. "Relationships Between Child Emotional and Behavioral Symptoms and Caregiver Strain and Parenting Stress." *Journal of Family Issues* 34(4):534–556.

Ventura, Stephanie J. 2009. "Changing Patterns of Nonmarital Childbearing in the United States." *NCHS Data Brief* 18. Hyattsville, MD: U.S. National Center for Health Statistics.

Ventura, Stephanie J., T. J. Mathews, and Brady E. Hamilton. 2001. "Births to Teenagers in the United States, 1940–2000." *National Vital Statistics Reports* 49(10). Hyattsville, MD: U.S. National Center for Health Statistics. September 25.

Verderber, Kathleen S., Rudolph Verderber, and Cynthia Berryman-Fink. 2010. *Inter-Act: Interpersonal Communication Concepts, Skills, and Contexts.* New York: Oxford University Press.

Verhofstadt, Lesley L., William Ickes, and Ann Buysse. 2010. "I Know What You Need Right Now:" Empathic Accuracy and Support Provision in Marriage." Pp. 71–88 in *Support Processes in Intimate Relationships,* edited by Kieran T. Sullivanand Joanne Davila. New York: Oxford University Press.

Vespa, Jonathan. 2012. "Union Formation in Later Life: Economic Determinants of Cohabitation and Remarriage Among Older Adults." *Demography* 49:1103–1125.

Vestal, Christine. 2009. "Gay Marriage Legal in Six States." Stateline.org, June 4. Retrieved November 20, 2009 (www.stateline.org).

Villar, F. and D. J. Villamizar. 2012. "Hopes and Concerns in Couple Relationships Across Adulthood and Their Association with Relationship Satisfaction." *Aging and Human Development* 75(2):115–139.

Vinciguerra, Thomas. 2007. "He's Not My Grandpa, He's My Dad." *The New York Times,* April 12.

Vogel, Erin R., Livia Haag, Mitra-Setia Tatang, Carel P. van Schaik, and Nathaniel J. Dominy. 2009. "Foraging and Ranging Behavior During a Failback Episode: *Hylobates albibarbis and Pongo pygmaeus wurmbii Compared.*" *American Journal of Physical Anthropology* 140(4):716–726.

Vogler, Carolyn. 2005. "Cohabiting Couples: Rethinking Money in the Household at the Beginning of the Twenty-First Century." *Sociological Review* 53(1):1–29.

Vogler, Carolyn, Clare Lyonette, and Richard D. Wiggins. 2008. "Money, Power and Spending Decisions in Intimate Relationships." *Sociological Review* 56(1):117–143.

Von Drehle. 2010. "Why Crime Went Away." *Time,* February 22, pp. 32–36.

Von Rosenvinge, Kristina. n. d. "Seven Tips to Improve Couple Communication." Retrieved January 26, 2010 (ezinearticles.com).

Voorpostel, M. and Rosemary Blieszner. 2008. "Intergenerational Support and Solidarity between Adult Siblings." *Journal of Marriage and Family* 70(1):157–167.

Wade, T. Joel, Ryan Kelley, and Dominique Church. 2011. "Are There Sex Differences in Reaction to Different Types of Sexual Infidelity?" *Psychology* 3(2):161–164.

Wadler, Joyce. 2009. "Caught in the Safety Net." *The New York Times,* May 14. Retrieved November 21, 2009 (www.nytimes.com).

Wadsworth, Shelley and M. MacDermid. 2010. "Family Risk and Resilience in the Context of War and Terrorism." *Journal of Marriage and Family* 72(3):537–556.

———. 2012. "Working with the Military." *Family Focus* FF52:F2.

Wagenaar, Deborah B. 2009. "Elder Abuse Education in Residency Programs: How Well Are We Doing?" *Academic Medicine: Journal of the Association of American Medical Colleges* 84(5):611–619.

Wagner-Raphael, Lynne I., David Wyatt Seal, and Anke A. Ehrhardt. 2001. "Close Emotional Relationships with Women versus Men." *Journal of Men's Studies* 9(2):243–256.

Waite, Evelyn B, Lilly Shanahan, Susan D. Calkins, Susan P. Keane, and Marion O'Brien. 2011. "Life Events, Sibling Warmth, and Youths' Adjustment." *Journal of Marriage and Family* 73(5) (October): 902–912.

Waite, Linda J. 1995. "Does Marriage Matter?" *Demography* 32(4):483–507.

Waite, Linda J., Don Browning, William J. Doherty, Maggie Gallagher, Ye Luo, and Scott M. Stanley. 2002. *Does Divorce Make People Happy? Findings from a Study of Unhappy Marriages.* New York: Institute for American Values.

Waite, Linda J. and Maggie Gallagher. 2000. *The Case for Marriage: Why Married People Are Happier, Healthier, and Better Off Financially.* New York: Doubleday.

Waite, Linda J., Ye Luo, and Alisa C. Lewin. 2009. "Marital Happiness and Marital Stability: Consequences for Psychological Well-Being." *Social Science Research* 38(1):201–212.

Wakeland, Shannon. n. d. "How to Manage Conflict Between Your Siblings." Retrieved January 29, 2013 (http://www.ehow.com).

Waldfogel, Jane. 2006. "What Do Children Need?" *Public Policy Research* 13(1):26–34.

Walker, Alexis J. 1985. "Reconceptualizing Family Stress." *Journal of Marriage and Family* 47(4):827–837.

Walker, Alexis J., Margaret Manoogian-O'Dell, Lori A. McGraw, and Diana L. G. White, eds. 2001. *Families in Later Life: Connections and Transitions.* Thousand Oaks, CA: Pine Forge Press.

Walker, Lenore E. 2009. *The Battered Woman Syndrome.* 3rd ed. New York: Springer.

Wallace, Harvey. 2008. *Family Violence: Legal, Medical, and Social Perspectives.* Boston: Pearson/ Allyn & Bacon.

Wallace, Stephen G. 2007. "Hooking Up, Losing Out?" *Healthy Teens Camping Magazine* (March/ April):26–30.

Wallach, Ian. 2010. "Mourning Has Broken." *Oprah Magazine.* October: 190–194.

Waller, Maureen R. and H. Elizabeth Peters. 2008. "The Risk of Divorce As a Barrier to Marriage among Parents of Young Children." *Social Science Research* 37(4):1188–1199.

Waller, Willard. 1951. *The Family: A Dynamic Interpretation,* revised by Reuben Hill. New York: Dryden.

Wallerstein, Judith S. 2003. "Children of Divorce: A Society in Search of Policy." Pp. 66–96 in *All Our Families: New Policies for a New Century,* 2nd ed., edited by Mary Ann Mason, Arlene Skolnick, and Stephen D. Sugarman. New York: Oxford University Press.

Wallerstein, Judith S. and Sandra Blakeslee. 1989. *Second Chances: Men, Women, and Children a Decade After Divorce.* New York: Ticknor and Fields.

———. 1995. *The Good Marriage: How and Why Love Lasts.* Boston: Houghton Mifflin.

Wallerstein, Judith S. and Joan Kelly. 1980. *Surviving the Break-Up: How Children Actually Cope with Divorce.* New York: Basic Books.

Wallerstein, Judith S. and Julia M. Lewis. 2007. "Disparate Parenting and Step-Parenting with Siblings in the Post-Divorce Family: Report from a 10-Year Longitudinal Study." *Journal of Family Studies* 13(2):224–235.

———. 2008. "Divorced Fathers and Their Adult Offspring: Report from a Twenty-Five-Year Longitudinal Study." *Family Law Quarterly* 42(4). Available through Academic Search Elite.

Wallman, Katherine K. 2012. *America's Children in Brief: Key National Indicators of Well-Being, 2012.* Washington, DC: Federal Interagency Forum on Child and Family Statistics. Retrieved September 3, 2012 (http://childstats.gov/americaschildren).

Walsh, David Allen. 2002. "Bouncing Forward: Resilience in the Aftermath of September 11." *Family Process* 41(1):34–36.

———. 2007. *No: Why Kids—of All Ages—Need to Hear It and Ways Parents Can Say It.* New York: The Free Press.

Walter, Carolyn Ambler. 1986. *The Timing of Motherhood.* Lexington, MA: Heath.

Walvoord, Emily C. 2010. "The Timing of Puberty: Is It Changing? Does It Matter?" *Journal of Adolescent Health* 47(5):433–439.

Wang, Wendy and Rich Morin. 2009. "Home for the Holidays ... and Every Other Day: Recession Brings Many Young Adults Back to the Nest." Washington, DC: Pew Research Center. Retrieved December 21, 2009 (pewresearch .org).

Wang, Wendy, Kim Parker, and Paul Taylor. 2013. "Breadwinner Moms." *Pew Social & Demographic Trends.* Washington, DC: Pew Research Center.

Ward, Margaret. 1997. "Family Paradigms and Older-child Adoption: A Proposal for Matching Parents' Strengths to Children's Needs." *Family Relations* 46(3):257–262.

Ward, Russell A., Glenna Spitze, and Glenn Deane. 2009. "The More the Merrier? Multiple Parent-Adult Child Relations." *Journal of Marriage and Family* 71(1):161–173.

Wark, Linda and Shilpa Jobalia. 1998. "What Would It Take to Build a Bridge? An Intervention for Stepfamilies." *Journal of Family Psychotherapy* 9(3):69–77.

Warne, Garry L. and Vijayalakshmi Bhatia. 2006. "Intersex, East and West." Pp. 183–205 in *Ethics and Intersex,* edited by Sharon E. Sytsma. New York: Springer.

Warner, Judith. 2006. *Perfect Madness: Motherhood in the Age of Anxiety.* New York: Riverhead Books.

Wasikowska, Mia. 2011. "A Minimum Wage Increase." *The New York Times.* March 27. Retrieved February 20, 2013 (http://nytimes. com).

Wasserman, Jason Adam. 2009. "But Where Do We Go from Here: A Reply to Tomso on the State and Direction of Postmodern Theory." *Social Theory & Health* 7(1):78–80.

Watson, Bruce. 2010. "A Hidden Crime: Domestic Violence Against Men Is a Growing Problem." Retrieved March 14, 2013 (http:// www.dailyfinance.com).

Watson, Rita. 2011. "Low Infidelity, Shock Statistics, and the Forgiveness Factor." *Psychology Today.* Retrieved November 17, 2012 (http:// www.psychologytoday.com/blog/love-and -gratitude/201109/low-infidelity-shock-statistics-and-the-forgiveness-factor).

Watson, Stephanie. 2012. "Why It's Ok to Have Just One Child." *WebMD.* February 1. Retrieved January 8, 2013 (http://www.webmd.com /parenting/features/just-one-child).

Weaver, Shannon E. and Marilyn Coleman. 2010. "Caught in the Middle: Mothers in Stepfamilies." *Journal of Social and Personal Relationships,* 27(3):305–326.

Weaver, Shannon E., Marilyn Coleman, and Lawrence H. Ganong. 2003. "The Sibling Relationship in Young Adulthood." *Journal of Family Issues* 24(2):245–263.

Webb, Amy Pieper, Christopher G. Ellison, Michael J. McFarland, Jerry W. Lee, Kelly Morton, and James Walters. 2010. "Divorce, Religious Coping, and Depressive Symptoms in a Conservative Protestant Religious Group." *Family Relations* 59:544–557.

Webster, Murray and Lisa Rashotte. 2009. "Fixed Roles and Situated Actions." *Sex Roles* 61(5/6):325–337.

Weeks, John R. 2002. *Population: An Introduction to Concepts and Issues.* 8th ed. Belmont, CA: Wadsworth.

———. 2007. *Population: An Introduction to Concepts and Issues.* 10th ed. Belmont, CA: Wadsworth.

Weeks, Linton. 2010. "A Temporary Solution for a New American Worker." National Public Radio. Retrieved February 15, 2011 (http:// www.npr.org).

Weger, H. 2005. "Disconfirming Communication and Self-Verification in Marriage: Associations among the Demand/Withdraw Interaction Pattern, Feeling Understood, and Marital Satisfaction." *Journal of Social and Personal Relationships* 22(1):19–31.

Weibel-Orlando, J. 2001. "Grandparenting Styles: Native American Perspectives." Pp. 139–145 in *Families in Later Life: Connections and Transitions,* edited by Alexis J. Walker, Margaret Manoogian-O'Dell, Lori A. McGraw, and Diana L. G. White. Thousand Oaks, CA: Pine Forge Press.

Weigel, Daniel J. 2008. "The Concept of Family: An Analysis of Laypeople's Views of Family." *Journal of Family Issues* 29(11):1426–1447.

Weil, Elizabeth. 2006. "What If It's (Sort of) a Boy and (Sort of) a Girl?" *The New York Times Magazine,* September 24. Retrieved September 6, 2007 (www.nytimes.com).

———. 2012. "Puberty Before Age 10: A New 'Normal'?" Retrieved March 25, 2013 (http:// www.nytimes.com/2012/04/01/magazine /puberty-before-age-10-a-new-normal. html?pagewanted=all&_r=0).

Weiner Jonah. 2010. "Married? A Bit Bored? See a Shootout." *The New York Times.* Retrieved November 21, 2012 (http://www.nytimes .com/2010/04/04/movies/04date.html?_r=0).

Weininger, Elliot B. and Annette Lareau. 2009. "Paradoxical Pathways: An Ethnographic Extension of Kohn's Findings on Class and

Childrearing." *Journal of Marriage and Family* 71(3):680–695.

Weissman, M. M., J. C. Markowitz, and G. L. Klerman. 2007. *Clinician's Quick Guide to Interpersonal Psychotherapy*. New York: Oxford University Press.

Weitzman, Lenore J. 1985. *The Divorce Revolution: The Unexpected Social and Economic Consequences for Women and Children in America*. New York: The Free Press.

Wejnert, Cyprian. 2008. "Strategies for Measuring and Promoting Mothers' Social Support Networks." *Marriage & Family Review* 44(2/3):380–388.

Welborn, Vickie. 2006. "Black Students Ordered to Give Up Seats to Whites." *Shreveport Times*, August 24. Retrieved October 2, 2006 (www.shreveporttimes.com).

Welch, Liz. 2011. "The Exact Words that Could Help a Friend in an Abusive Relationship." Huffpost Healthy Living. May 9. Retrieved May 8, 2013 (http://www.huffingtonpost.com).

Wells, Brooke E. and Jean M. Twenge. 2005. "Changes in Young People's Sexual Behavior and Attitudes, 1943-1999: A Cross-Temporal Meta-Analysis." *Review of General Psychology* 9(3):249–261.

Wells, Mary S., Mark A. Widmer, and J. Kelly McCoy. 2004. "Grubs and Grasshoppers: Challenge-based Recreation and the Collective Efficacy of Families with At-Risk Youth." *Family Relations* 53(3):326–333.

Wells, Robert V. 1985. *Uncle Sam's Family: Issues in and Perspectives on American Demographic History*. Albany, NY: State University of New York Press.

Wen, Ming. 2008. "Family Structure and Children's Health and Behavior." *Journal of Family Issues* 29(11):1492–1519.

Wenger, G. Clare, Pearl Dykstra, Tuula Melkas, and Kees C.P.M. Knipscheer. 2007. "Social Embeddedness and Late-Life Parenthood." *Journal of Family Issues* 28(11):1419–1456.

Wentland, Jocelyn J., and Elke D. Reissing. 2011. "Taking Casual Sex Not Too Casually: Exploring Definitions of Casual Sex Relationships." *Canadian Journal of Human Sexuality* 20(3): 75–91.

Wentzel, Jo Ann. 2001. "Foster Kids Really Are Ours." www.fosterparents.com.

Wessel, David. 2010. "Meet the Unemployable Man." *Wall Street Journal*, May 6. www.wsj.com.

Western, Bruce and Sara McLanahan. 2000. "Fathers Behind Bars: The Impact of Incarceration on Family Formation." Pp. 309–324 in *Families, Crime, and Criminal Justice*, edited by Greer Litton Foxand Michael L. Benson. New York: Elsevier Science.

Wexler, Richard. 2005. "Family Preservation Is the Safest Way to Protect Most Children." Pp. 311–327 in *Current Controversies on Family Violence*, 2nd ed., edited by Donileen R. Loseke, Richard J. Gelles, and Mary M. Cavanaugh. Thousand Oaks, CA: Sage Publications.

Wharff, Elizabeth, Katherine Ginnis, and Abigail Ross. 2012. "Family-Based Crisis Intervention with Suicidal Adolescents in the Emergency Room." *Social Work* 57(2):133–143.

"What Is Marriage and Family Therapy?" 2005. American Association for Marriage and Family Therapy. Retrieved March 3, 2010 (www.aamft.org/faqs/index_nm.asp).

Whealin, Julia and Ilona Pivar. 2006. "Coping when a Family Member Has Been Called to War: A National Center for PTSD Fact Sheet." U.S. Department of Veterans Affairs National Center for PTSD. Retrieved August 16, 2006 (www.ncptsd.va.gov).

Whisman, Mark A., Kristina Coop Gordon, and Yael Chatav. 2007. "Predicting Sexual Infidelity in a Population-Based Sample of Married Individuals." *Journal of Family Psychology* 21(2): 320–324.

Whisman, Mark A., and Douglas K. Snyder. 2007. "Sexual Infidelity in a National Survey of American Women: Difference in Prevalence and Correlates as a Function of Method of Assessment." *Journal of Family Psychology* 21(2):147–154.

Whitaker, Daniel J., Tadesse Haileyesus, Monica Swahn, and Linda S. Saltzman. 2007. "Differences in Frequency of Violence and Reported Injury between Relationships With Reciprocal and Nonreciprocal Intimate Partner Violence." *American Journal of Public Health* 97(5):941–947.

White House Conference on Aging. 2005. *Report to the President and the Congress: The Booming Dynamics of Aging—from Awareness to Action*. Retrieved May 16, 2007 (www.whcoa.gov).

White, James M. and David M. Klein. 2008. *Family Theories: An Introduction*. 3rd ed. Thousand Oaks, CA: Sage Publications.

White, Lynn K. 1998. "Who's Counting? Quasi-Facts and Stepfamilies in Reports of Number of Siblings." *Journal of Marriage and Family* 60(August):725–733.

———. 1999. "Contagion in Family Affection: Mothers, Fathers, and Young Adult Children." *Journal of Marriage and Family* 61(2):284–294.

White, Lynn K. and Alan Booth. 1985. "The Quality and Stability of Remarriages: The Role of Stepchildren." *American Sociological Review* 50:689–698.

White, Lynn K. and Joan G. Gilbreth. 2001. "When Children Have Two Fathers: Effects of Relationships with Stepfathers and Noncustodial Fathers on Adolescent Outcomes." *Journal of Marriage and Family* 63:155–167.

White, Lynn K. and Agnes Riedmann. 1992. "When the Brady Bunch Grows Up: Step/Half- and Full-Sibling Relationships in Adulthood." *Journal of Marriage and Family* 54(1):197–208.

White, Martha C. 2012. "The Booming Business of Divorce Parties." Retrieved April 22, 2013 (http://business.time.com/2012/10/15/the-booming-business-of-divorce-parties/).

White, Rebecca M. B., Mark W. Roosa, Scott R. Weaver, and Rajni L. Nair. 2009. "Cultural and Contextual Influences on Parenting in Mexican American Families." *Journal of Marriage and Family* 71(1):61–79.

Whitehead, Barbara Dafoe and David Popenoe. 2001. "Who Wants to Marry a Soul Mate?" In *The State of Our Unions 2001: The Social Health of Marriage in America*. Piscataway, NJ: RutgersUniversity, National Marriage Project.

———. 2003. "Did a Family Turnaround Begin in the 1990s?" In *The State of Our Unions 2003: The Social Health of Marriage in America*. Piscataway, NJ: Rutgers University, National Marriage Project. Retrieved September 20, 2006 (marriage.rutgers.edu).

———. 2008. "Life Without Children: The Social Retreat from Children and How It Is Changing America." Piscataway, NJ: Rutgers University, National Marriage Project. Retrieved April 16, 2010 (marriage.rutgers.edu).

Whitehurst, Lindsay. 2012. "Polygamy." Retrieved June 26, 2013 (http://www.lindsaywhitehurst.com/polygamy.html).

Whiting, Jason, Donna Smith, and Tammy Barnett. 2007. "Overcoming the Cinderella Myth: A Mixed Methods Study of Successful Stepmothers." *Journal of Divorce and Remarriage* 47(1/2):95–109.

Whitley, Bernard E., Christopher E. Childs, and Jena B. Collins. 2011. "Differences in Black and White American College Students' Attitudes Toward Lesbians and Gay Men." *Sex Roles: A Journal of Research* 64(5–6):299–310.

Whitton, Sarah W., Jan M. Nicholson, and Howard J. Markman. 2008. "Research on Interventions for Stepfamily Couples: The State of the Field." Pp. 455–484 in *The International Handbook of Stepfamilies*, edited by J. Pryor. Hoboken, NJ: John Wiley & Sons.

Whitty, M. T. 2005. "The Realness of Cybercheating: Men's and Women's Representations of Unfaithful Internet Relationships." *Social Science Computer Review* 23:5–67.

"Why Interracial Marriages Are Increasing." 1996. *Jet*, June 3, pp. 12–15.

Whyte, Martin King. 1990. *Dating, Mating, and Marriage*. New York: Aldine.

Wickersham, Joan. 2008. *The Suicide Index: Putting My Father's Death in Order*. Orlando: Harcourt.

Wickrama, K. A. S., Catherine Walker O'Neal, and Fred Lorenz. 2013. "Marital Functioning from Middle to Later Years: A Life Course-Stress Process Framework." *Journal of Family Theory & Review* 5(1):15–34.

Widmer, Eric D., Francesco Giudici, Jean-Marie LeGoff, and Alexandre Pollien. 2009. "From Support to Control: A Configurational Perspective on Conjugal Quality." *Journal of Marriage and Family* 71(3):437–448.

Wienke, Chris and Gretchen J. Hill. 2009. "Does the 'Marriage Benefit' Extend to Partners in Gay and Lesbian Relationships?" *Journal of Family Issues* 30(2):259–289.

Wiersma, Jacquelyn D., H. Harrington Cleveland, Veronica Herrera, and Judith L. Fischer. 2010. "Intimate Partner Violence in Young Adult Dating, Cohabitating, and Married Drinking Partnerships." *Journal of Marriage & Family* 72(2):360–374.

Wight, Vanessa, Suzanne Bianchi, and Bijou Hunt. 2012. "Explaining Racial/Ethnic Variation in Partnered Women's and Men's Housework: Does One Size Fit All?" *Journal of Family Issues* 34(3):394–427.

Wight, Vanessa R. and Sara B. Raley. 2009. "When Home Becomes Work: Work and Family Time among Workers at Home." *Social Indicators Research* 93(1):197–202.

Wightman, Patrick, Robert Schoeni, and Keith Robinson. 2012. "Familial Financial Assistance to Young Adults." Paper Presented at the Annual Meeting of the Population Association of America. May 3.

Wilcox, W. Bradford. 2004. *Soft Patriarchs, New Men: How Christianity Shapes Fathers and Husbands.* Chicago: University of Chicago Press.

———. 2009. "The Evolution of Divorce." *National Affairs* 1(Fall):81–94.

———. 2011. "A Shaky Foundation for Families." *The New York Times.* August 30. Retrieved December 11, 2012 (http://www.nytimes.com).

Wilcox, W. Bradford and Elizabeth Marquardt. 2009. *The State of Our Unions 2009: Marriage in America: Money & Marriage.* Virginia: National Marriage Project and the Institute for American Values.

Wilcox, W. Bradford, Elizabeth Marquardt, David Popenoe, and Barbara Dafoe Whitehead. 2011. *The State of Our Unions—Marriage in America 2011: How Parenthood Makes Life Meaningful and How Marriage Makes Parenthood Bearable.* University of Virginia, Institute for American Values: The National Marriage Project.

Wilcox, W. Bradford, Jared Anderson, William Doherty, David Eggebeen, Christopher Ellison, William Galston, Neil Gilbert, John Gottman, Ron Haskins, Robert Lerman, Linda Malone-Colon, Loren Marks, Rob Palkovitz, David Popenoe, Mark Regnerus, Scott Stanley, Linda Waite, and Judith Wallerstein. 2011. *Why Marriage Matters, Third Edition: Thirty Conclusions from the Social Sciences.* Virginia: National Marriage Project and the Institute for American Values.

Wilcox, W. Bradford and Steven L. Nock. 2006. "What's Love Got to Do with It? Equality, Equity, Commitment and Women's Marital Quality." *Social Forces* 84(3):1321–1345.

Wildeman, Christopher. 2009. "Parental Imprisonment, the Prison Boom, and the Concentration of Childhood Disadvantage." *Demography* 46(2):265–280.

———. 2012. "Despair by Association? The Mental Health of Mothers with Children by Recently Incarcerated Fathers." *American Sociological Review* 27(2):216–243.

Wildeman, Christopher and Christine Percheski. 2009. "Associations of Childhood Religious Attendance, Family Structure, and Nonmarital Fertility Across Cohorts." *Journal of Marriage and Family* 71(5):1294–1308.

Wildman, Sarah. 2010. "Children Speak for Same-Sex Marriage." *The New York Times,* January 20. Retrieved January 22, 2010 (www.nytimes.com).

———. 2011. "A Showpiece of Communal Living in Berlin." *The New York Times.* November 9. Retrieved December 5, 2012 (http://nytimes.com).

Wiley, Angela R., Henriette B. Warren, and Dale S. Montanelli. 2002. "Shelter in a Time of Storm: Parenting in Poor Rural African American Communities." *Family Relations* 51(3):265–273.

Wilkins, Amy C. 2012. "Stigma and Status: Interracial Intimacy and Intersectional Identities among Black College Men." *Gender & Society* 26(2):165–189.

Wilkins, Victoria M., Martha L. Bruce, and Jo Anne Sirey. 2009. "Caregiving Tasks and Training Interest of Family Caregivers of Medically Ill Homebound Older Adults." *Journal of Aging and Health* 21(3):528–542.

Wilkinson, Deanna, Amanda Magora, Marie Garcia, and Atika Khurana. 2009. "Fathering at the Margins of Society: Reflections from Young, Minority, Crime-Involved Fathers." *Journal of Family Issues* 30(7):945–967.

Wilkinson, Doris. 1993. "Family Ethnicity in America." Pp. 15–59 in *Family Ethnicity: Strength in Diversity,* edited by Harriette Pipes McAdoo. Newbury Park, CA: Sage Publications.

———. 2000. "Rethinking the Concept of 'Minority': A Task for Social Scientists and Practitioners." *Journal of Sociology and Social Welfare* 27:115–132.

Wilkinson, Will. 2011. "Why You Don't Believe that Kids Don't Make You Happier." *Forbes.* March 4. Retrieved December 13, 2012 (http://www.forbes.com).

Willetts, Marion C. 2003. "An Exploratory Investigation of Heterosexual Licensed Domestic Partners." *Journal of Marriage and Family* 65(4):939–952.

———. 2006. "Union Quality Comparisons between Long-Term Heterosexual Cohabitation and Legal Marriage." *Journal of Family Issues* 27(1):110–127.

Williams, Alex. 2010. "The New Math on Campus." *The New York Times.* Retrieved December 14, 2012 (http://www.nytimes.com/2010/02/07/fashion/07campus.html?pagewanted=all&_r=0).

Williams, Joan C. and Heather Boushey. 2010. *The Three Faces of Work-Family Conflict: The Poor, the Professionals, and the Missing Middle.* Center for American Progress. Retrieved May 8, 2010 (www.americanprogress.org).

Williams, K. 2003. "Has the Future of Marriage Arrived? A Contemporary Examination of Gender, Marriage, and Psychological Well-Being." *Journal of Health and Social Behavior,* 44:470–487.

Williams, Kristi, Sharon Sassler, and Lisa M. Nicholson. 2008. "For Better or For Worse? The Consequences of Marriage and Cohabitation for Single Mothers." *Social Forces* 86(4):1481–1511.

Williams, Lee M. and Michael G. Lawler. 2003. "Marital Satisfaction and Religious Heterogamy." *Journal of Family Issues* 24(8):1070–1092.

Willoughby, Brian J. and Jason S. Carroll. 2012. "Correlates of Attitudes Toward Cohabitation: Looking at the Associations with Demographics, Relational Attitudes, and Dating Behavior." *Journal of Family Issues* 33(11):1450–1476.

Willoughby, Brian J., Jason S. Carroll, and Dean M. Busby. 2011. "The Different Effects of 'Living Together': Determining and Comparing Types of Cohabiting Couples." *Journal of Social and Personal Relationships* 29(3):397–419.

Wilson, April C., and Ted L. Huston. 2013. "Shared Reality and Grounded Feelings During Courtship: Do They Matter for Marital Success." *Journal of Marriage and Family* 75:681–696.

Wilson, Brenda. 2009. "Sex Without Intimacy: No Dating, No Relationship." National Public Radio. May 18. Retrieved June 10, 2009 (www.npr.org).

Wilson, James Q. 2001. "Against Homosexual Marriage." Pp. 123–127 in *Debating Points: Marriage and Family Issues,* edited by Henry L. Tischler. Upper Saddle River, NJ: Prentice Hall.

———. 2002. "Why We Don't Marry." *City Journal,*Winter. Retrieved October 2, 2006 (www.city-journal.org).

Wilson, William Julius. 2009. *More than Just Race: Being Black and Poor in the Inner City.* New York: W. W. Norton & Company.

Wimmer, Andreas, and Kevin Lewis. 2010. "Beyond and Below Racial Homophily: ERG Models of Friendship Network Documented on Facebook." *American Journal of Sociology* 116(2): 583–642.

Winch, Robert F. 1958. *Mate Selection: A Study of Complementary Needs.* New York: Harper & Row.

Wineberg, Howard. 1996. "The Resolutions of Separation: Are Marital Reconciliations Attempted?" *Population Research and Policy Review* 15:297–310.

Winerip, Michael. 2009. "Anything He Can Do, She Can Do." *The New York Times,* November 15.

Wingert, Pat and Barbara Kantrowitz. 2010. "The Rise of the 'Silver Divorce'." *Newsweek,* June 7. Retrieved June 19, 2010 (www.newsweek.com).

Winter, Laraine, Laura Gitlin, and Marie Dennis. 2011. "Desire to Institutionalize a Relative with Dementia: Quality of Premorbid Relationship and Caregiver Gender." *Family Relations* 60(2):221–230.

Woldt, Veronica. 2010. "Elder Care Benefits: Retention and Recruitment Tools." *Corporate Wellness Magazine,* May 7. Retrieved May 30, 2010 (www.corporatewellnessmagazine.com).

Wolf, D. A., V. Freedman, and B. J. Soldo. 1997. "The Division of Family Labor: Care for Elderly Parents." *Journal of Gerontology* 52B:102–109.

Wolf, Marsha E., Uyen Ly, Margaret A. Hobart, and Mary A. Kernic. 2003. "Barriers to Seeking Police Help for Intimate Partner Violence." *Journal of Family Violence* 18:121–129.

Wolf, Naomi. 1991. *The Beauty Myth: How Images of Beauty Are Used Against Women.* New York: W. Morrow.

———. 2012. "Naomi Wolf on Third Wave Feminism." *Big Think*. Retrieved November 8, 2012 (http://bigthink.com/ideas).

Wolf, Rosalie S. 1986. "Major Findings from Three Model Projects on Elderly Abuse." Pp. 218–238 in *Elder Abuse: Conflict in the Family*, edited by Karl A. Pillemerand Rosalie S. Wolf. Dover, MA: Auburn.

Wolfers, Justin. 2006. "Did Unilateral Divorce Raise Divorce Rates? A Reconciliation and New Results." *American Economic Review* 96(5):1802–1820.

Wolfinger, Nicholas H. and Raymond E. Wolfinger. 2008. "Family Structure and Voter Turnout." *Social Forces* 86(4):1513–1528.

"Women CEOs." 2010. *Fortune Magazine*, May 3 (www.fortune.com).

Wong, Cathy L. n. d. "Filling the Void: An Autoethnographic Study of a Transracial Adoptee." Unpublished dissertation.

Wong, Sabrina, Grace Yoo, and Anita Stewart. 2006. "The Changing Meaning of Family Support among Older Chinese and Korean Immigrants." *Journal of Gerontology: Social Sciences* 61B(1):S4–S9.

Wood, Wendy and Alice H. Eagly. 2002. "A Cross-cultural Analysis of Behavior of Women and Men: Implications for the Origins of Sex Differences." *Psychology Bulletin* 128:699–727.

Wooding, G. Scott. 2008. *Stepparenting and the Blended Family: Recognizing the Problems and Overcoming the Obstacles*. Markham, Ontario: Fitzhenry and Whiteside.

Woodward, Kenneth L. 2001. "A Mormon Moment." *Newsweek*, September 10, pp. 44–51.

"World's 1st 'Test-Tube' Baby Gives Birth." 2007. CNN.com, January 15. Retrieved January 15, 2007 (http://cnn.health.com).

Wright, Eric and Brea Perry. 2006. "Sexual Identity Distress, Social Support, and the Health of Gay, Lesbian, and Bisexual Youth." *Journal of Homosexuality* 51(1):81–110.

Wright, H. Norman. 2006. *How to Speak Your Spouse's Language: Ten Easy Steps to Great Communication from One of America's Foremost Counselors*. New York: Center Street.

Wright, Robert. 1994. *The Moral Animal*. New York: Pantheon.

Wright, Susan. 2005. "Autism: Willing the World to Listen." *Newsweek*, February 28. Retrieved March 13, 2009.

Wrigley, Julia and Joanna Dreby. 2005. "Fatalities and the Organization of Child Care in the United States, 1985–2003." *American Sociological Review* 70:729–757.

Wu, Zheng and Feng Hou. 2008. "Family Structure and Children's Psychosocial Outcomes." *Journal of Family Issues* 29(12):1600–1624.

Wurzel, Barbara J. n. d. "Extension Fact Sheet: Growing Up with Yours, Mine, and Ours in Stepfamilies." Ohio State University Family and Consumer Sciences. Retrieved April 26, 2007 (ohioline.osu.edu).

Xu, Xianohe, Clarke D. Hudspeth, and John P. Bartkowski. 2006. "The Role of Cohabitation in Remarriage." *Journal of Marriage and Family* 68(2):261–274.

Yabiku, Scott T. and Constance T. Gager. 2009. "Sexual Frequency and the Stability of Marital and Cohabiting Unions." *Journal of Marriage and Family* 71(November):983–1000.

Yakushko, Oksana and Oliva M. Espin. 2010. "The Experience of Immigrant and Refugee Women: Psychological Issues." Pp. 535–558 in *Handbook of Diversity in Feminist Psychology*, edited by Hope Landrineand Nancy Felipe Russo. New York: Springer.

Yalcin, Bektas Murat and Tevfik Fikret Karaban. 2007. "Effects of a Couple Communication Program on Marital Adjustment." *Journal of the American Board of Family Medicine* 20(1):36–44.

Yancey, G. 2007. "Experiencing Racism: Differences in the Experiences of Whites Married to Blacks and Non-Black Racial Minorities." *Journal of Comparative Family Studies* 38:197–213.

Yellowbird, Michael and C. Matthew Snipp. 2002. "American Indian Families." Pp. 227–249 in *Multicultural Families in the United States*, 3rd ed., edited by Ronald L. Taylor. Upper Saddle River, NJ: Prentice Hall.

Yeung, King-To and John Levi Martin. 2003. "The Looking Glass Self: An Empirical Test and Elaboration." *Social Forces* 81(3):843–879.

Yin, Sandra. 2007. "New Restrictions Could Limit U.S. Adoptions from Top Two Countries of Origin: China and Guatemala." Washington, DC: Population Reference Bureau, March. Retrieved March 26, 2007 (www.prb.org).

Yip, Pamela. 2012. "Older Sweethearts Should Talk Money." Retrieved May 12, 2013 (http://www.dallasnews.com).

Yodanis, Carrie and Sean Lauer. 2007. "Managing Money in Marriage: Multilevel and Cross-National Effects of the Breadwinner Role." *Journal of Marriage and Family* 20(December):1307–1325.

Yodanis, Carrie, Sean Lauer, and Risako Ota. 2012. "Interethnic Romantic Relationships: Enacting Affiliative Ethnic Identities." *Journal of Marriage and Family* 74:1021–1037.

Yoo, Jeong Ah and Chien-Chung Huang. 2012. "The Effects of Domestic Violence on Children's Behavior Problems: Assessing the Moderating Roles of Poverty and Marital Status." *Children and Youth Services Review* 34:2464–2473.

Yorburg, Betty. 2002. *Family Realities: A Global View*. Upper Saddle River, NJ: Prentice Hall.

Yoshikawa, Hirokazu and Carola Suarez-Orozco. 2012. "Deporting Parents Hurts Kids." *The New York Times*. April 20. Retrieved August 28, 2012 (http://www.nytimes.com).

Yoshikawa, Hirokazu and Jenya Kholoptseva. 2013. *Unauthorized Immigrant Parents and Their Children's Development: A Summary of the Evidence*. March. Migration Policy Institute. Retrieved March 20, 2013 (http://mpi.org).

Yoshioka, Marianne R., Louisa Gilbert, Nabila El-Bassel, and Malahat Baig-Amin. 2003. "Social Support and Disclosure of Abuse: Comparing South Asian, African American, and Hispanic Battered Women." *Journal of Family Violence* 18:171–180.

"Young Caregivers." 2009 (April 29). Retrieved March 13, 2010 (www.girlshealth.gov).

Young, Molly. n. d. "Native Women Move to the Front of Tribal Leadership." *Native Daughters*. Retrieved October 31, 2012 (http://cojmc.unl.edu/nativedaughters/leaders/native-women-move-to-the-front-of-tribal-leadership).

Young, Stacy L. 2010. "Positive Perceptions of Hurtful Communication: The Packaging Matters." *Communication Research Reports* 27(1):49–57.

"Your Baby Can Read Set." 2011. Amazon.com.

Yuan, Anastasia S. Vogt and Hayley A. Hamilton. 2006. "Stepfather Involvement and Adolescent Well-Being." *Journal of Family Issues* 27(9):1191–1213.

Yuval-Davis, Nira. 2006. "Intersectionality and Feminist Politics." *European Journal of Women's Studies* 13(3):1193–2009.

Zablocki v. Redhail. 1978. 434 U.S. 374, 54 L. Ed. 2d 618, 98 S. Ct. 673.

Zacchilli, Tammy L., Clyde Hendrick, and Susan S. Hendrick. 2010. "The Romantic Partner Conflict Scale: A New Scale to Measure Relationship Conflict." *Journal of Social and Personal Relationships* 26(8):1073–1096.

Zaman, Ahmed. 2008. "Gender Sensitive Teaching: A Reflective Approach for Early Childhood Education Teacher Training." *Education* 129(1):110–118.

Zang, Xiaowel. 2008. "Gender and Ethnic Variation in Arranged Marriages in a Chinese City." *Journal of Family Issues* 29(5):615–638.

Zeitz, Joshua M. 2003. "The Big Lie about the Little Pill." *The New York Times*, December 27.

Zernike, Kate. 2006. "The Bell Tolls for the Future Merry Widow." *The New York Times*, April 30. Retrieved April 30, 2007 (www.nytimes.com).

———. 2009. "And Baby Makes How Many?" *The New York Times*, February 9. Retrieved December 5, 2009 (www.nytimes.com).

———. 2011. "Fast-Tracking to Kindergarten?" *The New York Times*. May 15:1,10.

———. 2012. "Court's Split Decision Provides Little Clarity on Surrogacy." *The New York Times*. October 24. Retrieved November 2, 2012 (http://www.nytimes.com).

Zezima, Katie. 2006. "When Soldiers Go to War, Flat Daddies Hold Their Place at Home." *The New York Times*, September 30.

Zhang, Jing, Suzanna Smith, Marilyn Swisher, Danling Fu, and Kate Fogarty. 2011. "Gender Role Disruption and Marital Satisfaction among Wives of Chinese International Students in the United States." *Journal of Comparative Family Studies* 42(4):523–542.

Zhang, Shuangyue. 2009. "Sender-Recipient Perspectives of Honest but Hurtful Evaluative Messages in Romantic Relationships." *Communication Reports* 22(2):89–101.

Zhang, Shuangyue and Susan L. Kline. 2009. "Can I Make My Own Decision? A Cross-Cultural Study of Perceived Social Network Influence in Mate Selection." *Journal of Cross-Cultural Psychology* 40(1):3–23.

Zhang, Yuanting and Jennifer Van Hook. 2009. "Marital Dissolution among Interracial Couples." *Journal of Marriage and Family* 71(February):95–107.

Zhao, Yilu. 2002. "Immersed in 2 Worlds, New and Old." *The New York Times,* July 22.

Zhou, Yanqui Rachel. 2012. "Space, Time, and Self: Rethinking Aging in the Contexts of Immigration and Transnationalism." *Journal of Aging Studies* 26:232–242.

Zielinski, David S. 2009. "Child Maltreatment and Adult Socioeconomic Well-Being." *Child Abuse & Neglect* 33(10):666–678.

Zimmerman, Gregory M. and Steven F. Messner. 2013. "Individual, Family Background, and Contextual Explanations of racial and Ethnic Disparities in youths' Exposure to Violence." *American Journal of Public Health* 103(3):435–442.

Zimmerman, Gregory M. and Greg Pogarsky. 2011. "The Consequences of Parental Underestimation and Overestimation of Youth Exposure to Violence." *Journal of Marriage and Family* 73(1):194–208.

Zink, D. W. 2008. "The Practice of Marriage and Family Counseling and Conservative Christianity." Pp. 55–72 in *The Role of Religion and Marriage and Family Counseling,* edited by Jill Duba Onedera. New York: Taylor and Francis.

Zsembik, Barbara A. and Zobeida Bonilla. 2000. "Eldercare and the Changing Family in Puerto Rico." *Journal of Family Issues* 21(5):652–674.

Zuang, Yuanting. 2004. "Why Foreign Adoption?" Presented at the annual meeting of the American Sociological Association, August 15, San Francisco, CA.

Name Index

Note: Page numbers with a *b* indicate box material, with an *f* indicate a figure, and with a *t* indicate a table.

Subject Index

Note: Page numbers with a *b* indicate box material, with an *f* indicate a figure, and with a *t* indicate a table.